CONSTRUCTION
PROJECT MANAGEMENT

THEORY AND PRACTICE

SECOND EDITION

KUMAR NEERAJ JHA

ASSOCIATE PROFESSOR
DEPARTMENT OF CIVIL ENGINEERING,
IIT, DELHI

Sr. Editor–Acquisitions: Anita Yadav
Editor–Production: Vipin Kumar

Copyright © 2015 Pearson India Education Services Pvt. Ltd

This book is sold subject to the condition that it shall not, by way of trade or otherwise, be lent, resold, hired out, or otherwise circulated without the publisher's prior written consent in any form of binding or cover other than that in which it is published and without a similar condition including this condition being imposed on the subsequent purchaser and without limiting the rights under copyright reserved above, no part of this publication may be repro- duced, stored in or introduced into a retrieval system, or transmitted in any form or by any means (electronic, mechanical, photocopying, recording or otherwise), without the prior writ- ten permission of both the copyright owner and the publisher of this book.

Although the author and publisher have made every effort to ensure that the information in this book was correct at the time of editing and printing, the author and publisher do not assume and hereby disclaim any liability to any party for any loss or damage arising out of the use of this book caused by errors or omissions, whether such errors or omissions result from negligence, accident or any other cause. Further, names, pictures, images, characters, businesses, places, events and incidents are either the products of the author's imagination or used in a fictitious manner. Any resemblance to actual persons, living or dead or actual events is purely coincidental and do not intend to hurt sentiments of any individual, community, sect or religion.

In case of binding mistake, misprints or missing pages etc., the publisher's entire liability and your exclusive remedy is replacement of this book within reasonable time of purchase by similar edition/reprint of the book

ISBN 979-83-325-4201-3

First Impression, 2015

Published by Pearson India Education Services Pvt. Ltd, CIN: U72200TN2005PTC057128

Head Office: 15th Floor, Tower-B, World Trade Tower, Plot No. 1, Block-C, Sector 16, Noida 201 301, Uttar Pradesh, India.

Registered Office: 7th Floor, SDB2, ODC 7, 8 & 9, Survey No.01 ELCOT IT/ ITES SEZ, Sholinganallur, Chennai – 600119, Tamil Nadu, India. Website: in.pearson.com, Email: companysecretary.india@pearson.com

Compositor: D Book Media

Digitally printed in India by Trinity Academy for Corporate Training Ltd, New Delhi in the year of 2022.

Dedicated to all the construction professionals

CONTENTS

PREFACE xxi
PREFACE TO THE SECOND EDITION xxiv
FOREWORD xxv
ABOUT THE AUTHOR xxvi

Chapter 1 Introduction 1

1.1	GENERAL	1
1.2	A CONSTRUCTION PROJECT	2
1.2.1	Phases of a Construction Project	3
1.3	IMPORTANCE OF CONSTRUCTION AND CONSTRUCTION INDUSTRY	9
1.4	INDIAN CONSTRUCTION INDUSTRY	9
1.5	CONSTRUCTION PROJECT MANAGEMENT AND ITS RELEVANCE	11
1.5.1	Role of Project Management	11
1.5.2	Why Construction Project Management?	12
1.6	PARTICIPANTS/STAKEHOLDERS OF A CONSTRUCTION PROJECT	13
1.6.1	Architect	13
1.6.2	Client (Owner)	13
1.6.3	Constructor	14
1.6.4	Engineer (Consultant)	14
1.6.5	Subcontractor/Supplier/Vendor	14
1.6.6	Lawyer, Insurer, etc.	14
1.7	ORGANIZATION OF THE BOOK	15
	References	16
	Review Questions	17

Chapter 2 Project Organization 18

2.1	INTRODUCTION	18
2.2	CONSTRUCTION COMPANY	18
2.3	FORMS OF BUSINESS ORGANIZATION	19
2.3.1	Sole Proprietorship	19
2.3.2	Partnership	19
2.3.3	Corporation	20
2.3.4	Limited Liability Company (LLC)	20
2.3.5	Private Limited Company	20

	2.3.6	Public Limited Company	20
	2.3.7	Government Enterprises	21
	2.3.8	Joint Ventures	21
2.4		STRUCTURE OF CONSTRUCTION ORGANIZATION	21
	2.4.1	Centralized Functional	24
	2.4.2	Decentralized Multidivisional	25
2.5		ORGANIZING FOR PROJECT MANAGEMENT	26
	2.5.1	Classical (Functional)	26
	2.5.2	Pure Project or Product Management	27
	2.5.3	Matrix Organizations	29
2.6		MANAGEMENT LEVELS	31
	2.6.1	Director Level	31
	2.6.2	President Level	32
	2.6.3	Construction Management Level	32
	2.6.4	Project Management Level	33
	2.6.5	Functional Management Level	33
2.7		TRAITS OF A PROJECT MANAGER	34
	2.7.1	Strategies for Enhancing the Performance of a Project Manager	37
2.8		IMPORTANT TRAITS OF A PROJECT COORDINATOR	38
	2.8.1	Team Building Skill	40
	2.8.2	Contract Implementation Skill	40
	2.8.3	Project Organization Skill	41
2.9		ETHICAL CONDUCT FOR ENGINEERS	41
2.10		FACTORS BEHIND THE SUCCESS OF A CONSTRUCTION ORGANIZATION	43
		References	45
		Review Questions	47

Chapter 3 Construction Economics 49

3.1	INTRODUCTION	49
3.2	ECONOMIC DECISION-MAKING	50
3.2.1	Out-of-Pocket Commitment	50
3.2.2	Payback Period	50
3.2.3	Average Annual Rate of Return	51
3.3	TIME VALUE OF MONEY	52
3.4	CASH-FLOW DIAGRAMS	53
3.4.1	Project Cash-flow and Company Cash-flow Diagrams	56
3.4.2	Using Cash-flow Diagrams	66
3.5	USING INTEREST TABLES	74

	3.6	EVALUATING ALTERNATIVES BY EQUIVALENCE	75
	3.6.1	Present Worth Comparison	75
	3.6.2	Future Worth Comparison	77
	3.6.3	Annual Cost and Worth Comparison	78
	3.6.4	Rate of Return Method	78
	3.7	EFFECT OF TAXATION ON COMPARISON OF ALTERNATIVES	80
	3.8	EFFECT OF INFLATION ON CASH-FLOW	81
	3.9	EVALUATION OF PUBLIC PROJECTS: DISCUSSION ON BENEFIT–COST RATIO	82
	3.9.1	Benefit/Cost Criteria	82
		References	84
		Solved Examples	85
		Review Questions	100

Chapter 4 Client's Estimation of Project Cost — 104

	4.1	INTRODUCTION	104
	4.2	APPROXIMATE METHODS OF ESTIMATION	104
	4.2.1	Preliminary Estimate for Buildings	105
	4.2.2	Preliminary Estimate for Industrial Structures	107
	4.3	TYPES OF ESTIMATES	108
	4.3.1	Rough Order of Magnitude Estimates	108
	4.3.2	Client's Indicative Cost Estimate	108
	4.3.3	Client's Preliminary Cost Estimate	109
	4.3.4	Client's Detailed Estimate	109
	4.3.5	Client's Definitive Estimate	110
	4.3.6	Revised Estimate	111
	4.3.7	Supplementary Estimates	111
	4.3.8	Project Closure Cost	111
	4.4	METHODS OF STRUCTURING PROJECT COSTS	111
	4.5	ILLUSTRATIVE CASES IN PREPARATION OF ESTIMATE	111
	4.5.1	Case1: Multi-level Car Parking Facility	111
	4.5.2	Case 2: Preliminary Estimate for Construction of Sewage Treatment Plant	117
		References	121
		Solved Examples	121
		Review Questions	122

Chapter 5 Construction Contract — 125

	5.1	CONSTRUCTION CONTRACT	125
	5.2	CONTRACT DOCUMENT	126
	5.2.1	The Contract Drawings	126

5.2.2	The Specifications	126
5.2.3	The General Conditions of Contract (GCC)	127
5.2.4	The Special Conditions of Contract (SCC)	127
5.2.5	The Bill of Quantities (BOQ)	128
5.3	CLASSIFICATION OF ENGINEERING CONTRACTS	128
5.3.1	Separated Contract	129
5.3.2	Management Contract	132
5.3.3	Integrated Contract	133
5.3.4	Discretionary Contract	135
5.4	BIDDING PROCESS	136
5.4.1	Pre-qualification Process	136
5.4.2	Notice Inviting Tender	142
5.4.3	Submission of Bids	143
5.4.4	Analysis of Submitted Tenders	144
5.4.5	Basis for Evaluation and Acceptance	145
5.4.6	Letter of Intent	145
5.4.7	Work Order	145
5.4.8	Agreement	145
5.5	CPWD CONTRACT CONDITIONS	146
5.6	FIDIC FORM OF CONTRACT AGREEMENT	149
5.6.1	Need and Principles of FIDIC Contracts	151
5.6.2	Salient Features of FIDIC Form of Contract	152
5.7	SUBCONTRACTING	154
5.7.1	Classification of Subcontractors	155
5.7.2	Selection of Subcontractors	156
5.7.3	Work Order	157
5.7.4	Terms and Conditions	158
5.7.5	Subcontractor Management—Some Guidelines	158
	References	159
	Review Questions	160

Chapter 6 Construction Planning 162

6.1	INTRODUCTION	162
6.2	TYPES OF PROJECT PLANS	164
6.2.1	Time Plan	164
6.2.2	Manpower Plan	164
6.2.3	Material Plan	164
6.2.4	Construction Equipment Plan	165
6.2.5	Finance Plan	165
6.3	WORK-BREAKDOWN STRUCTURE	165
6.3.1	Methodology of WBS	166

	6.4	PLANNING TECHNIQUES—TERMINOLOGIES USED	166
	6.4.1	Event and Activity	166
	6.4.2	Dummy Activity	168
	6.4.3	Network	168
	6.4.4	Precedence	171
	6.4.5	Network Logic	171
	6.4.6	Duration of an Activity	174
	6.4.7	Forward and Backward Pass	175
	6.4.8	Float or Slack Time	178
	6.4.9	Path and Critical Path	179
	6.5	BAR CHARTS	180
	6.6	PREPARATION OF NETWORK DIAGRAM	180
	6.7	PROGRAMME EVALUATION AND REVIEW TECHNIQUE (PERT)	184
	6.8	CRITICAL PATH METHOD (CPM)	189
	6.9	LADDER NETWORK	190
	6.10	PRECEDENCE NETWORK	191
	6.11	THE LINE-OF-BALANCE (LOB)	203
	6.12	NETWORK TECHNIQUES ADVANTAGES	209
		References	210
		Solved Examples	211
		Review Questions	215

Chapter 7 Project Scheduling and Resource Levelling — 224

	7.1	INTRODUCTION	224
	7.2	RESOURCE LEVELLING	225
	7.3	RESOURCE ALLOCATION	229
	7.4	IMPORTANCE OF PROJECT SCHEDULING	235
	7.5	OTHER SCHEDULES DERIVED FROM PROJECT SCHEDULES	237
	7.5.1	Preparing Invoice Schedule	239
	7.5.2	Schedule of Milestone Events	240
	7.5.3	Schedule of Plant and Equipment	241
	7.5.4	Schedule of Project Staff	242
	7.5.5	Schedule of Labour Requirement	242
	7.5.6	Schedule of Materials Requirement	243
	7.5.7	Schedule of Specialized Agencies	244
	7.5.8	Schedule of Direct Costs	244
	7.5.9	Schedule of Overheads	245

7.5.10	Schedule of Cash Inflow	245
7.5.11	Schedule of Cash Outflow	246
7.6	NETWORK CRASHING AND COST-TIME TRADE-OFF	246
	References	252
	Review Questions	252

Chapter 8 Contractor's Estimation of Cost and Bidding Strategy — 256

8.1	CONTRACTOR'S ESTIMATION AND BIDDING PROCESS	256
8.1.1	Get Involved in Pre-qualification Process	256
8.1.2	Study the Tender Document, Drawings and Prepare Tender Summary	258
8.1.3	Decisions to Take	258
8.1.4	Arrange for Site Visit and Investigation	259
8.1.5	Consultation, Queries and Meetings, and Other Associated Works	259
8.1.6	Prepare Construction Schedule and Other Related Schedules	260
8.1.7	Collect Information	260
8.1.8	Determining Bid Price	262
8.1.9	Analysis of Rates	267
8.1.10	Fix Mark-up	272
8.1.11	Computing Bid Price	275
8.1.12	Submit Bid	277
8.1.13	Post-submission Activities	277
8.2	BIDDING MODELS	277
8.2.1	Game Theory Models	278
8.2.2	Statistical Bidding Strategy Models	278
8.2.3	Cash Flow-Based Models	286
8.3	DETERMINATION OF OPTIMUM MARK-UP LEVEL	287
8.4	BIDDING AND ESTIMATION PRACTICES IN INDIAN CONSTRUCTION INDUSTRY	287
8.4.1	Prevailing Estimation Practices	288
8.4.2	Use of Statistical/Mathematical Tools in Estimation	288
8.4.3	Breakup of Mark-up	288
8.4.4	Labour Cost Estimation	289
8.4.5	Plant and Equipment Cost Estimation	289
8.4.6	Dealing with Uncertainties	289
8.4.7	Average Range of Mark-up	289
8.4.8	Mark-up Distribution	289

8.4.9	Mark-up Range	290
8.4.10	Summary and Conclusion from the Study	290
	References	291
	Solved Examples	292
	Review Questions	318

Chapter 9 Construction Equipment Management 322

9.1	INTRODUCTION	322
9.2	CLASSIFICATION OF CONSTRUCTION EQUIPMENT	323
9.3	FACTORS BEHIND THE SELECTION OF CONSTRUCTION EQUIPMENT	323
9.3.1	Economic Considerations	324
9.3.2	Company-specific	324
9.3.3	Site-specific	325
9.3.4	Equipment-specific	325
9.3.5	Client- and Project-specific	325
9.3.6	Manufacturer-specific	325
9.3.7	Labour Consideration	326
9.4	EARTHWORK EQUIPMENT	326
9.5	CONCRETING EQUIPMENT	326
9.6	HOISTING EQUIPMENT	330
9.6.1	Hoists	330
9.6.2	Cranes	330
9.7	PLANT AND EQUIPMENT ACQUISITION	336
9.8	DEPRECIATION	337
9.9	DEPRECIATION AND TAXATION	338
9.10	METHODS OF CALCULATING DEPRECIATION	340
9.10.1	Straight-line Method	340
9.10.2	Sum of Years Digit Method	341
9.10.3	Declining Balance Method	342
9.10.4	Sinking Fund Method	343
9.10.5	Accelerated Depreciation	345
9.11	EXAMPLE OF DEPRECIATION CALCULATIONS FOR EQUIPMENT AT A SITE	346
9.12	THE EFFECT OF DEPRECIATION AND TAX ON SELECTION OF ALTERNATIVES	348
9.13	EVALUATING REPLACEMENT ALTERNATIVES	353
9.14	ADVANCED CONCEPTS IN ECONOMIC ANALYSIS	355
9.14.1	Sensitivity Analysis	355
9.14.2	Breakeven Analysis	362

	9.14.3	Incorporating Probability in Cash Flows	365
	9.14.4	Determining Expected Values, Variance, Standard Deviation, and Coefficient of Variation	365
	9.14.5	Determining Expected Values and Variance in Aggregated Cash Flows	367
	9.14.6	Selection of Alternatives Based on Expected Value and Variance of Returns	375
		Solved Examples	377
		References	384
		Review Questions	385

Chapter 10 Construction Accounts Management — 390

10.1	GENERAL	390
10.2	PRINCIPLES OF ACCOUNTING	391
10.3	ACCOUNTING PROCESS	392
10.4	CONSTRUCTION CONTRACT REVENUE RECOGNITION	394
10.4.1	Cash Method of Revenue Recognition	394
10.4.2	Straight Accrual Method of Revenue Recognition	395
10.4.3	Completed Contract Method of Revenue Recognition	395
10.4.4	Percentage of Completion Method of Revenue Recognition	395
10.5	CONSTRUCTION CONTRACT STATUS REPORT	395
10.6	LIMITATIONS OF ACCOUNTING	397
10.7	BALANCE SHEET	397
10.7.1	Liabilities	400
10.7.2	Assets	401
10.8	PROFIT AND LOSS ACCOUNT	402
10.9	WORKING CAPITAL	404
10.9.1	Need for Working Capital	405
10.9.2	Operating Cycle	405
10.9.3	Components of Working Capital	406
10.9.4	Determination of Working Capital	407
10.9.5	Financing Sources of Working Capital	408
10.10	RATIO ANALYSIS	412
10.10.1	Liquidity Ratios	413
10.10.2	Capital Structure Ratios	414
10.10.3	Profitability Ratios	415
10.10.4	Activity Ratios	418
10.10.5	Supplementary Ratios	419
10.11	FUNDS FLOW STATEMENT	419
10.11.1	Changes in Working Capital or Funds	421

	10.11.2	Determining Funds Generated/Used by Business Operations	422
	10.11.3	Application of Funds Flow Statement	422
		References	423
		Review Questions	423

Chapter 11 Construction Material Management — 427

	11.1	INTRODUCTION	427
	11.2	MATERIAL PROCUREMENT PROCESS IN CONSTRUCTION ORGANIZATION	429
	11.3	MATERIALS MANAGEMENT FUNCTIONS	431
	11.3.1	Materials Planning	431
	11.3.2	Procurement	431
	11.3.3	Custody (Receiving, Warehousing and Issuing)	431
	11.3.4	Materials Accounting	432
	11.3.5	Transportation	432
	11.3.6	Inventory Monitoring and Control	433
	11.3.7	Materials Codification	433
	11.3.8	Computerization	434
	11.3.9	Source Development (Vendor Development)	435
	11.3.10	Disposal	435
	11.4	INVENTORY MANAGEMENT	435
	11.4.1	Inventory-Related Cost	436
	11.4.2	Functions of Inventory	436
	11.4.3	Inventory Policies	437
	11.4.4	Selective Inventory Control	439
	11.4.5	Inventory Models	444
		Solved Examples	448
		References	452
		Review Questions	453

Chapter 12 Project Cost and Value Management — 454

	12.1	PROJECT COST MANAGEMENT	454
	12.1.1	Resources Planning Schedules	454
	12.1.2	Cost Planning	454
	12.1.3	Cost Budgeting	455
	12.1.4	Cost Control	455
	12.2	COLLECTION OF COST-RELATED INFORMATION	456
	12.2.1	Labour Cost	457
	12.2.2	Material Cost	458
	12.2.3	Plant and Equipment Cost	460
	12.2.4	Subcontractor Cost	460

12.2.5	Overhead Cost	460
12.3	COST CODES	461
12.4	COST STATEMENT	461
12.5	VALUE MANAGEMENT IN CONSTRUCTION	462
12.6	STEPS IN THE APPLICATION OF VALUE ENGINEERING	465
12.7	DESCRIPTION OF THE CASE	468
12.8	VALUE-ENGINEERING APPLICATION IN THE CASE PROJECT	469
12.8.1	Foundation Design	470
12.8.2	Flooring System	474
12.8.3	Precast vs in-situ Construction	477
12.8.4	Discussion of Results	478
	References	478
	Review Questions	479

Chapter 13 Construction Quality Management — 481

13.1	INTRODUCTION	481
13.2	CONSTRUCTION QUALITY	482
13.2.1	Definition of Quality	482
13.2.2	Evolution of Quality	484
13.3	INSPECTION, QUALITY CONTROL AND QUALITY ASSURANCE IN PROJECTS	485
13.3.1	Inspection	486
13.3.2	Quality Control (QC)	487
13.3.3	Quality Assurance (QA)	487
13.4	TOTAL QUALITY MANAGEMENT	488
13.5	QUALITY GURUS AND THEIR TEACHINGS	489
13.5.1	Deming	489
13.5.2	Juran	491
13.5.3	Philip Crosby	493
13.6	COST OF QUALITY	493
13.7	ISO STANDARDS	496
13.7.1	Benefits of ISO 9000	496
13.7.2	Principles of Quality Management Systems	497
13.7.3	ISO 9001–2000 Family of Standards	498
13.8	CONQUAS—CONSTRUCTION QUALITY ASSESSMENT SYSTEM	501
13.9	AUDIT	502
13.10	CONSTRUCTION PRODUCTIVITY	503

	13.10.1	Typical Causes of Low Labour Productivity	504
		References	505
		Review Questions	506

Chapter 14 Risk and Insurance in Construction — 509

14.1	INTRODUCTION	509
14.2	RISK	511
14.3	RISK IDENTIFICATION PROCESS	511
14.3.1	Preliminary Checklist	512
14.3.2	Risk Events Consequences Scenario	512
14.3.3	Risk Mapping	512
14.3.4	Risk Classification	512
14.3.5	Risk Category Summary Sheet	513
14.4	RISK ANALYSIS AND EVALUATION PROCESS	513
14.4.1	Data Collection	514
14.4.2	Modelling Uncertainty	514
14.4.3	Evaluation of Potential Impact of Risk	514
14.5	RESPONSE MANAGEMENT PROCESS (RISK TREATMENT STRATEGIES)	514
14.5.1	Risk Avoidance	515
14.5.2	Loss Reduction and Risk Prevention	515
14.5.3	Risk Retention and Assumption	515
14.5.4	Risk Transfer (Non-insurance or Contractual Transfer)	516
14.5.5	Insurance	516
14.6	INSURANCE IN CONSTRUCTION INDUSTRY	516
14.6.1	Fundamental Principles of Insurance	516
14.6.2	Insurance Policies for a Typical Construction Organization	516
14.6.3	Project Insurance	516
14.6.4	Marine-cum-Erection Insurance	519
14.6.5	Contractor's All-risk Insurance (CAR Insurance)	519
14.6.6	Marine/Transit Insurance	520
14.6.7	Fire Policy	521
14.6.8	Plant and Machinery Insurance	521
14.6.9	Liquidity Damages Insurance	521
14.6.10	Professional Indemnity Policy	522
14.7	COMMON EXAMPLES OF BUSINESS AND PROJECT RISK	522
14.8	RISKS FACED BY INDIAN CONSTRUCTION COMPANIES ASSESSING INTERNATIONAL PROJECTS	522
14.8.1	Risks in International Construction	522
	References	526
	Review Questions	527

Chapter 15 Construction Safety Management — 529

15.1	INTRODUCTION	529
15.2	EVOLUTION OF SAFETY	531
15.3	ACCIDENT CAUSATION THEORIES	532
15.4	FOUNDATION OF A MAJOR INJURY	535
15.4.1	Unsafe Conditions	535
15.4.2	Unsafe Acts	538
15.5	HEALTH AND SAFETY ACT AND REGULATIONS	541
15.5.1	Building and Other Construction Workers (Regulation of Employment and Condition of Services) Act 1996	542
15.5.2	Building and Other Construction Workers (Regulation of Employment and Conditions of Service) Central Rules, 1998	543
15.6	COST OF ACCIDENTS	544
15.7	ROLES OF SAFETY PERSONNEL	545
15.8	CAUSES OF ACCIDENTS	546
15.9	PRINCIPLES OF SAFETY	551
15.10	SAFETY AND HEALTH MANAGEMENT SYSTEM	552
15.10.1	Safety Policy and Organization	552
15.10.2	Safety Budget	553
15.10.3	Safety Organization	553
15.10.4	Education and Training	554
15.10.5	Safety Plan	555
15.10.6	Safety Manual	555
15.10.7	Safety Committee	555
15.10.8	Incentive Programmes	556
15.10.9	Accident Reporting, Investigation and Record Keeping	556
15.10.10	Incident Investigation and Analysis	557
15.10.11	Accident Statistics and Indices	560
15.10.12	Safety Inspection	562
15.10.13	Safety Audit	562
15.10.14	Workers' Health and First-aid Facilities	563
15.11	RESEARCH RESULTS IN SAFETY MANAGEMENT	563
	References	564
	Review Questions	566

Chapter 16 Project Monitoring and Control System — 570

16.1	INTRODUCTION	570
16.2	UPDATING	570
16.2.1	Updating Using Bar Chart	571
16.2.2	Updating Using PERT/CPM	572
16.2.3	Updating Using Precedence Network	574

16.3	PROJECT CONTROL	575
16.4	SCHEDULE/TIME/PROGRESS CONTROL	577
16.4.1	Monthly Progress Report	577
16.4.2	Measuring Progress at Site	577
16.4.3	Typical Reports to Aid the Progress Review	578
16.5	COST CONTROL	578
16.5.1	Profit/loss at the Completion of Contract	579
16.5.2	Stage-wise Completion of Cost	579
16.5.3	Standard Costing	579
16.5.4	S-Curve	579
16.5.5	Unit Costing	582
16.6	CONTROL OF SCHEDULE, COST AND TECHNICAL PERFORMANCE—EARNED VALUE METHOD	586
16.6.1	Terminologies of Earned Value Method	588
16.7	ILLUSTRATIONS OF COST CONTROL SYSTEM	594
16.8	MANAGEMENT INFORMATION SYSTEM	597
	References	600
	Review Questions	600

Chapter 17 Construction Claims, Disputes, and Project Closure — 603

17.1	CLAIM	603
17.1.1	Sources of Claims	603
17.1.2	Claim Management	604
17.1.3	Some Guidelines to Prepare the Claims	605
17.2	DISPUTE	605
17.2.1	Causes of Disputes	605
17.2.2	Dispute Avoidance Vs Dispute Resolution	607
17.2.3	Mechanisms of Dispute Resolution	608
17.2.4	Causes Leading to Arbitration	609
17.2.5	Advantages of ADR over Legal Proceedings in a Court	609
17.2.6	Some Do's and Don'ts to Avoid Dispute	610
17.3	CORRESPONDENCE	611
17.4	PROJECT CLOSURE	612
17.4.1	Construction Closure	612
17.4.2	Financial Closure	614
17.4.3	Contract Closure	614
17.4.4	Project Manager's Closure	615
17.4.5	Lessons Learned from the Project	615
	References	615
	Review Questions	615

Chapter 18 Computer Applications in Scheduling, Resource Levelling, Monitoring, and Reporting — 618

- 18.1 INTRODUCTION — 618
- 18.2 POPULAR PROJECT MANAGEMENT SOFTWARE — 619
  - 18.2.1 Primavera — 619
  - 18.2.2 Milestone Professional — 620
  - 18.2.3 'Candy'—Construction Project Modelling and Project Control — 620
  - 18.2.4 AMS Realtime Projects — 620
  - 18.2.5 Project KickStart — 620
  - 18.2.6 MS Project — 620
- 18.3 FUNCTIONS OF PROJECT MANAGEMENT SOFTWARE — 621
  - 18.3.1 Scheduling Function — 621
  - 18.3.2 Resource Management Function — 621
  - 18.3.3 Tracking or Monitoring Function — 622
  - 18.3.4 Reporting Function — 623
  - 18.3.5 Additional Functions — 623
- 18.4 ILLUSTRATION OF MS PROJECT — 624
  - 18.4.1 Definitions of Some Terminologies — 624
  - 18.4.2 Working with MS Project — 626
- 18.5 ILLUSTRATION OF PRIMAVERA — 635
  - 18.5.1 Adding a New Project — 635
  - 18.5.2 Preparing Schedule — 636
  - 18.5.3 Resource Levelling — 638
  - 18.5.4 Tracking the Project — 644
  - 18.5.5 Reporting — 645
  - 18.5.6 Some Additional Features — 645
  - References — 651
  - Review Questions — 651

Chapter 19 Factors Behind the Success of a Construction Project — 655

- 19.1 GENERAL — 655
- 19.2 PROJECT PERFORMANCE MEASUREMENT — 656
- 19.3 CRITERIA FOR PROJECT PERFORMANCE EVALUATION — 657
- 19.4 PROJECT PERFORMANCE ATTRIBUTES — 658
  - 19.4.1 Success Attributes/Factors — 659
  - 19.4.2 Failure Attributes — 659
- 19.5 EFFECT OF OTHER ELEMENTS ON PROJECT PERFORMANCE — 661

	19.6	THE THEORY OF 3CS AND THE IRON TRIANGLE	662
		References	673
		Review Questions	676

Chapter 20 Linear Programming in Construction Management — 677

	20.1	INTRODUCTION	677
	20.2	LINEAR PROGRAMMING	678
	20.3	FORMULATION OF LINEAR PROGRAMMING PROBLEMS	679
	20.4	GRAPHICAL SOLUTION OF LINEAR PROGRAMMING PROBLEMS	681
	20.5	SIMPLEX METHOD	689
	20.5.1	Understanding Simplex Iterations	692
	20.5.2	Illustration of Simplex Method—A maximization problem	693
	20.6	PRIMAL DUAL	696
	20.6.1	Construction of the Dual from the Primal	697
	20.6.2	Rules for Constructing the Dual Problem	698
	20.7	BIG-M METHOD	701
	20.8	TWO PHASE METHOD	710
	20.9	DUAL SIMPLEX METHOD	712
	20.10	SPECIAL CASES IN SIMPLEX	719
	20.10.1	Infeasibility	720
	20.10.2	Unboundedness	722
	20.10.3	Degeneracy	724
	20.10.4	Multiple or Alternative Optima	726
	20.11	ALTERNATIVE METHOD OF SIMPLEX TABLE COMPUTATIONS	728
	20.11	SENSITIVITY ANALYSIS	736
	20.11.1	Changes in the Constraints	736
	20.11.2	Changes in the Objective Function Coefficients	742
		References	752
		Review Questions	752

Chapter 21 Transportation, Transshipment, and Assignment Problems — 758

	21.1	INTRODUCTION	758
	21.2	TRANSPORTATION PROBLEM	758
	21.2.1	Formulation as LP problem	759
	21.2.2	The Transportation Algorithm	760

21.2.3	Phase I Obtaining an Initial Feasible Solution		760
21.2.4	Phase II—Moving toward optimality		768
21.3	TRANSSHIPMENT PROBLEM		772
21.4	ASSIGNMENT PROBLEM		780
21.5	UNBALANCED PROBLEM AND RESTRICTIONS ON ASSIGNMENT		788
21.6	SOME MORE APPLICATIONS		791
21.6.1	Maximization Problem		791
21.6.2	Crew Assignment Problem		793
21.6.3	Travelling Salesman Problem		798
	References		800
	Review Questions		801

Answers to the Objective Questions 805

Appendices 809

Index 869

PREFACE

It gives me great pleasure to present the first edition of this book. The book is the result of my twelve years of field experience with Larsen and Toubro Limited and six years of experience in teaching undergraduate and graduate students at IIT Kanpur and IIT Delhi, consulting, researching and organizing training programmes for teachers and practitioners.

At the beginning of my career, I used to feel that I have been pushed from a highly theoretical world to a highly practical world. There was no smooth transition from theory to practice. Many of the things which could have been covered in classrooms, especially in the area of construction project management, for this smooth transition were missing. It takes months, sometimes years, to understand what can be covered in a few lectures with the help of a good instructor having field experience and good reference material.

When I got into academics, I was trying all the time to look for a text/reference book which could cover the different facets of construction project management in the Indian context. At the end of different training programmes conducted for teachers and practitioners, one of the most commonly asked queries was to suggest the participants a good text/reference book. I could not name a single book which covers the different facets of construction project management and deals with both the theoretical and practical aspects.

In fact, now I firmly believe that enough progress in the discipline has not been made due to the absence of good text and reference books at undergraduate and graduate levels. In the absence of such books, academicians also feel reluctant to teach this course and this prevents even interested students from opting for this important discipline. This has led to a situation where even some respected academicians do not think of construction project management beyond mere PERT/CPM. This is one of the reasons why I chose to restrict the number of pages on PERT/CPM.

The book introduces the readers to the different facets of construction project management and is designed to meet the following objectives:

i. To present a framework for the different aspects of construction project management. The theoretical framework has been supported with practical applications and case studies from the field.
ii. To enhance the interest of students in taking up research works in construction project management. That is why reference details of several relevant books have been provided at the end of each chapter.
iii. And, of course, to silence those critiques who do not consider construction project management beyond PERT/CPM

The book is divided into 19 chapters. The first two chapters introduce the readers to the construction project and the construction organization. Thereafter, the book has been organized in a manner similar to the different stages or phases through which a typical construction project passes. In Chapter 3, the readers are introduced to the construction economics, which may be useful for the selection of projects. The concept of 'time value of money' has been explained and different methods of project evaluation have been discussed. Chapter 4 contains descriptions and illustrations of project cost estimation by client organizations. The concepts have been emphasized primarily with the help of real life cases. In Chapter 5, the different aspects of construction contracts have been discussed. In Chapters 6 and 7, the various planning and scheduling techniques are explained. Chapter 8 deals with the contractor's estimation of cost and bidding strategy. In Chapter 9 issues related to construction equipment management are discussed, besides an overview of advanced concepts in economic analysis. Chapter 10 deals with construction accounts, while in Chapter 11, the salient features of construction material management are presented. In Chapter 12, the various aspects of project cost management are discussed and the concepts of value management have been illustrated with the help of a real life case project. Construction quality management, risk and insurance, and construction safety management are the subject matters for Chapter 13, 14 and 15, respectively. In Chapter 16, project monitoring and control systems are discussed, while in Chapter 17, construction claims, disputes and project closure issues are addressed. In Chapter 18, the computer applications in project scheduling, resource leveling, monitoring and reporting have been covered. The examples considered in Chapter 6, 7 and 16 have been solved using commonly used project management softwares, such as Microsoft Project and Primavera Project Planner. Finally, at the end, the factors for the success of a construction project are discussed which have been drawn from the author's Ph.D. work. A number of appendices have been provided wherever felt necessary. References of books have been provided at the end of each chapter to assist the readers in further study. Also, each chapter has review questions–objective, short answer and long answer type. Depending on the requirements of the chapters, certain solved examples have also been provided. It is hoped that readers will find these useful.

ACKNOWLEDGEMENTS

I wish to thank all the people, without whose contribution, this book would not have come into existence.

I would like to thank the reviewers who provided valuable comments for the betterment of the book. I would like to thank my colleagues at the Department of Civil Engineering, IIT Delhi, who contributed their time on a number of occasions discussing the contents of the book. I would like to thank my colleagues from other IITs, especially IIT Kanpur, IIT Chennai, and IIT Guwahati. I would like to place on record the encouragement I received from Prof. S. N. Sinha, Prof. A. K. Jain, Prof. B. Bhattacharjee, Prof. K. C. Iyer, Prof. A. K. Mittal, Prof. G. S. Benipal, Prof. A. K. Singh, Prof. Koshy Varghese, and Prof. K. N. Satyanarayana on various occasions during the preparation of the manuscript.

I am thankful to Prof. Sudhir Misra, who helped me immensely in preparing the first draft of the Table of Contents. Besides, he also edited and contributed in the draft of Chapters 3, 5, 6, and 10. I am thankful to him for all his encouragements and efforts. The comments on the Table of Contents received from Prof. A. K. Singh of IIT Guwahati were quite useful and led to almost restructuring of the entire manuscript. The book in its present form looks much better and I am thankful to him for his efforts. I value the comments received on a few chapters from Prof. Anil Sawhney and I am thankful to him.

The comments received from my colleagues at Larsen and Toubro Limited on the practical contents provided in book are thankfully acknowledged. I am thankful to Mr. V. P. Sinha for going through some of the practical examples of the book. I would also like to acknowledge the support provided by the people at Pearson Education, namely Vipin Kumar, and Anita Yadav.

Several other people have given me much to thank for. I am thankful to my teaching assistants, Satish Kumar Reddy, Anil Kumar, Akash Karak, and Priya Chandrayan for assisting me in drawing the various figures both in Autocad and MS Excel. I wish to acknowledge the contribution of Abhishek Kumar Singh, a research scholar in the Department of Civil Engineering. He was instrumental in formulating the review questions for different chapters. Let me also acknowledge the efforts of my students of different batches who attended my courses on Construction Economics and Finance (CEL779), Project Planning and Implementation (CEL 769), Construction and Contract Management (CEL767) and Computational Methods in Construction Project Management (CEP775), delivered at IIT Delhi. The help received from Gayatri Sachin Vyas, Lecturer, College of Engineering, Pune and Dilip Patel, Assistant Professor of SVNIT Surat for proofreading the manuscript.

I cannot forget my friend Prof. Lalit Manral for inspiring me to take up the academic profession. I thank him for being the source of inspiration for me at all times.

I can not find words to describe the efforts of my parents for all they have done for my upbringing. Without their blessings this book would not have been completed. And of course, there are no words for the gratitude and love I feel for my family members, Arti (wife), Srijan (elder son), and Sajal (younger son) who, from time to time, encouraged me to complete this book even at the cost of my time to be spent with them.

Kumar Neeraj Jha

PREFACE TO THE SECOND EDITION

I am extremely pleased to present the second edition of *Construction Project Management: Theory and Practice*. This revised edition contains two new chapters—Chapter 20 and Chapter 21 and some additional topics in Chapter 9. Chapter 20 presents an introduction to linear programming in construction management while Chapter 21 deals with transportation, transshipment, and assignment problems. These two topics are very important to construction managers as they deal with scarce resources and, thus, must know how to utilize them in an optimal manner. In Chapter 9, concept of probability in cash flow analysis and selection of alternatives, based on expected values and variances of returns, have been added. In addition, the printing and calculation mistakes of the previous edition have been rectified. The book in the revised form looks much better and I am hopeful that the readers would find it more useful. Some contents have also been shifted to the website of the book. For preparing the second edition, I thankfully acknowledge the contribution of my research scholars Amit Chandra, Dilip Patel, Satish and Manish. Readers are requested to mail their feedbacks on the content and presentation at jhakneeraj@rediffmail.com.

Kumar Neeraj Jha

FOREWORD

India is becoming increasingly noticed in the international arena for its infrastructure development. The country's inherent infrastructure deficit, coupled with growth-oriented measures by the Government and increased private capex, drives the activities in the infrastructure sector many folds. With the Public–Private Partnership (PPP) model going full steam, fund is no more a constraint. Major deterrent, from my viewpoint, is the domestic project execution capabilities/project management consultancy (PMC) services.

In construction industry, the project management is traditionally learnt 'on-the-job', which is a long-drawn process and typically takes 3–4 projects before someone gets a grip of it. This was working fine till now, but going forward, with the kind of growth prospects being envisaged, this method would fall short of the demand. To meet the demand, many universities have started courses in Construction Technology/Management in engineering post-graduation. This helps the candidates to enter the construction industry with adequate knowledge about project management and, thereby, expediting the process of getting mastery over project management skills.

Most of the books in construction management used in India are written by foreign authors and, naturally, they give examples of methods and systems practised in their respective countries. Many a times, this results in a dichotomy of appreciating the system and doubting its applicability in Indian conditions. Ready application of knowledge is more important keeping in view the talent gaps mentioned above.

Kumar Neeraj Jha's attempt to bridge this gap deserves a generous appreciation. It is quite clear that this book has been written after thorough discussions with the industry players and their vital inputs add immense value to the book. As the title suggests, the book focuses on both theory and practice. A lot of practical examples throughout the book will ensure that the concepts are learnt properly and retained in one's memory. More importantly, all examples and jargons used in this book very much relate to domestic conditions and, therefore, the Indian readers will be able to appreciate the nuances of it easily.

The book covers all aspects of construction management from 'concept to commissioning'. And, therefore, it not only caters to the needs of construction companies but also to concessionaires, consultants, independent engineers, subcontractors and other stakeholders in the construction industry. This book will also be a very good resource for the candidates pursuing post-graduate programme in Construction Management.

The knowledge management in the Indian construction industry needs marked improvement and this book will go a long way in meeting that objective.

K. V. Rangaswami
Member of the Board and President (Construction)
Larsen and Toubro Ltd

ABOUT THE AUTHOR

Kumar Neeraj Jha is an Associate Professor in the Department of Civil Engineering, IIT Delhi. He started his career with Larsen and Toubro Ltd as Graduate Engineer Trainee and subsequently held a number of responsible positions in the same company. He has been involved in a number of construction projects in different capacities and specialized in project management and formwork. He also taught at IIT Kanpur for a brief period. He has authored two more books: (1) *Formwork for Concrete Structures*; and (2) *Determinants of Construction Project Success in India*. He has contributed papers to a number of international and national journals, and has presented research papers in many international conferences. He teaches undergraduate and graduate students in the area of construction technology and management. He has supervised a large number of Masters and PhD students. He has conducted a number of training programs for academicians and practitioners in the area of project management and has also been involved with a number of consultancy projects. His research interests are—the project success factors, the organization success factors, formwork, project risks, and construction safety.

1

Introduction

> General, a construction project, importance of construction and construction industry, Indian construction industry, construction project management and its relevance, participants/stakeholders of a construction project, organization of the book

1.1 GENERAL

Construction, in one form or the other, has been practised since the dawn of civilization. In the Indian context, people were familiar with construction using burnt bricks as far back as 2600 BC. The townships of Mohen-jo-daro and Harappa provided its citizens with the comfort and luxury that was unheard of in other parts of the world during those days. The Ashoka Pillars erected around 250 BC, also known as monoliths, were one of the finest examples of craftsmanship. These monoliths of height in the range of 13 m to 15 m were made of a single block of sandstone. The polish to the sandstone was such that the monoliths appeared as if they were made of metals. The top of these monoliths was crowned with figures of the lion, the elephant and the bull. The Grand Anicut situated in southern India is one of the oldest water diversion projects in the world. This is an example of the familiarity of the people of second century BC with dam construction. That the dam is still in use speaks of the quality of construction practised in those days. The strength, symmetry and aesthetics of Qutab Minar in Delhi prove beyond doubt that the Indians of thirteenth century were familiar with the nuances of construction. The Taj Mahal, completed in 1653, is an outstanding example of architecture. Under the British regime, some of the most challenging construction of roads, railways and irrigation projects were undertaken. Some of the important construction works undertaken during this period are—construction of light railway lines in 1845, construction of the first railway bridge over Thane Creek in 1854, construction of the first narrow gauge line in 1862, construction of the Ganges Canal consisting of main canals and distribution channels between 1842 and 1854, construction of a major bridge in Dehri-on-Sone in 1900, and construction of Juhu Aerodrome in Mumbai in 1928.

Chander (1989) reports that during the British period a number of specialized departments such as the Public Works, Survey of India, etc. were established. Construction on large scales for secretariats, residential complexes and other building projects were undertaken in Delhi when the national capital was shifted from Kolkata (earlier Calcutta). After the completion of this massive project, the need was felt to look after these constructed facilities. Central Public Works Department (CPWD) was, thus, established in 1930 principally to take care of these facilities. However, the pressure of the Second World War saw the need of constructing structures for military purposes as well. CPWD was entrusted with this job and an expansion of CPWD to other major cities—namely Mumbai (earlier Bombay), Kolkata and Chennai (earlier Madras)—took place.

Post Independence, a number of notable construction projects have been undertaken, namely construction of the first nuclear reactor in 1956; construction of Hirakud Dam in 1957, Bhakra-Nangal Dam in 1970 and Idukki Dam in 1976; and construction of Mumbai Pune Expressway in 2000, Tehri Dam in 2005 and Bandra Worli Sea Link Bridge in 2009. A number of organizations were created post-Independence to construct and maintain specific constructions. For example, the construction related to defence is undertaken by the Military Engineering Services; Border Roads Organisation (established in 1963) undertakes the expansion and maintenance of road networks in border areas; National Highways Authority of India (established in 1988) undertakes the development, maintenance and management of national highways; and Indian Railways undertakes the construction and maintenance of rail network in India. The works related to irrigation projects and power projects are under the supervision of Central Water Commission and Central Electricity Authority, respectively. They provide advisory services to different state governments, which in turn have separate public works departments for executing construction of different types.

Construction in India is the second largest industry next to agriculture, and it provides employment to about 33 million people. It is an integral part of the country's infrastructure and industrial development. As per the Tenth Five Year Plan (2002–07) of the Government of India, it constitutes about five to six per cent of gross domestic product (GDP) and is vital for the growth of the overall economy. The performances of construction projects in India have not been very encouraging due to time and cost overruns and disputes in various contracts. With the opening up of the Indian economy to the outside world, India has been facing increased competition with other countries. A large number of infrastructure facilities are being created and development works are undertaken to facilitate a comfortable atmosphere—and here, the construction industry has got a great role to play. It is, therefore, far more important to understand the current problems with the overall performance of the construction projects and find out appropriate remedial measures to keep pace with the required growth.

In this chapter, we define the concept of construction project and discuss different phases involved in a construction project. Further, the different tasks required to be performed in different phases are discussed. The importance of construction projects and a bird's eye view of the Indian construction industry are presented next. Construction project management and its relevance are discussed thereafter. A brief introduction to the different stakeholders involved in a construction project has also been presented. Towards the end of the chapter, the organization of the book is presented briefly.

1.2 A CONSTRUCTION PROJECT

The Guide to the Project Management Body of Knowledge (PMBOK) published by Project Management Institute (PMI) defines project as a temporary endeavor undertaken to provide a unique product or service. The product in case of a construction project is the constructed facility such as building, assembling of some infrastructure and so on. The constructed facilities are

> **Box 1.1** Unique features of a construction project
>
> ## Unique Features of a Construction Project
>
> - One-time activity—it must be performed correctly the first time every time
> - Complexity—it is multidisciplinary because it involves a set of interrelated tasks to be done by specialists
> - High cost and time for execution
> - High risk of failure
> - Difficulty in defining quality standards
> - Uniqueness of people relationship
> - Feedback mechanism
> - Lack of experience of client or owner
> - Untrained workforce

supposed to adhere to some predetermined performance objectives. Examples of service in the context of a construction project would be design, planning, execution, and so on.

Construction projects involve varying manpower and their duration can range from a few weeks to more than five years. Each one of them is 'unique' and 'temporary' in nature, and so is the management involved. Here, the term 'unique' means that every project is different in some way from other projects, and the term 'temporary' means that every project has a definite beginning and an end (PMBOK 2000). A summary of unique features of a construction project is given in Box 1.1.

A project involves a series of complex or interrelated activities and tasks that consume resources to achieve some specific objectives. It has to be completed within a set of specifications under a limited budget (Munns and Bjeirmi 1996, Pinto and Slevin 1988a). For many organizations, projects are a means to respond to those requests that cannot be addressed within the organization's normal operational limits. A project may involve a single unit of one organization or may extend across organizational boundaries, as in case of joint ventures and partnering. A project is regarded as key to accomplish the business strategy of any organization, as it is the means by which strategy is implemented (PMBOK 2000), and a project is, therefore, not an isolated event but a realization of objectives through concerted efforts of different participants in various phases of the project life cycle.

1.2.1 Phases of a Construction Project

Just as the human life cycle has different phases or stages—such as conception phase, birth phase, childhood phase, adolescence phase, and so on—a construction project also passes through different phases in its life cycle. The term 'phases', 'stages' and 'steps' are used interchangeably in project management literature and, accordingly, no distinction has been made in this book as well.

The phases can be defined as the top-level breakdown of an entity, and a construction project is distinctly characterized by a number of phases or stages during its life cycle, though there

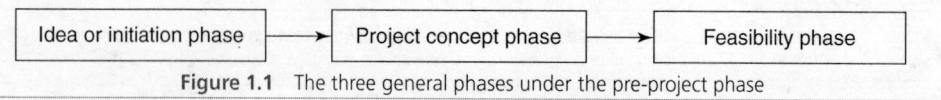

Figure 1.1 The three general phases under the pre-project phase

may be some overlaps between the characteristics of two phases. Considering the definition of construction project as unique, it is really very difficult to identify some common stages across all the construction projects, since depending on the type of project, the type of organization and other parameters, the phases as well as the tasks performed in each of the phases may vary. For example, some of the stages in a technical project such as a petroleum refinery project or a chemical project would be different from that of a building construction project. Researchers and practitioners do not adopt a common nomenclature to distinguish different phases of a project. Thus, there is no single standard nomenclature available to describe the different phases of a construction project. In the following paragraphs, we discuss a generic approach to describe the different project phases under three broad categories—the pre-project phase, the project phase and the post-project phase.

Pre-Project Phase

The three sub-phases under Pre-Project Phase are shown in Figure 1.1 and are discussed below in brief.

Initiation or idea phase The pre-project phase aims to identify all possible projects based on the examination of needs and the possible options. This stage is also sometimes referred to as *initiation phase* or *idea phase*. A possible example would be a municipal authority concerned with the growing parking problems near a prominent city milestone. The municipal authority may explore different options to address the parking problem. The options shall be evaluated against the mission and vision of the municipal authority and the limits to which funding is available with the municipal agency.

Project concepts The initiation phase aims to sort out all the mentioned information to identify some *project concepts*. As many project concepts as possible are identified, and using some selection procedure (such as the benefits for the organization that intends to employ them) in line with the objectives of the organization, several project concepts are selected. The project concept phase of a new construction project is most important, since decisions taken in this phase tend to have a significant impact on the final cost. It is also the phase at which the greatest degree of uncertainty about the future is encountered.

The selected project concepts, then, are used as the inputs for the feasibility phase.

Feasibility This phase aims to analytically appraise project concepts in the context of the organization, taking into consideration factors such as the needs of the organization, the strategic charter of the organization, and the capabilities and know-how of the organization. With this information, the decision makers should be able to decide whether or not to go ahead with the project concept proposed. The feasibility phase can be broadly characterized into the following:

Conceptual For the selected project concepts, the preliminary process diagrams and layouts are prepared. Design basis or design briefs are also formulated.

Project strategy The strategy in terms of selection of an in-house design team or the contractor's design team is deliberated upon. The resources required and their availability is discussed. Further, the number and type of contractors required for the execution of project is also formulated. Besides all these, the project strategy also contains the overall project schedule, the project scope and the overall project plan.

Estimate A preliminary estimate is prepared with reasonable accuracy by first breaking down the project into work packages/elements. The estimates may be prepared for each of the work packages using the established historical database and the resources estimated for each of the work packages.

Approval Approval consists of financial evaluation, identifying details of funding and their timing, capital/revenue, etc., besides evaluation of different options.

The feasibility phase has sub-phases such as market feasibility analysis (to confirm the viability of the project concept from a purely marketing point of view), technical feasibility analysis (to demonstrate whether the project is technically feasible and to estimate the cost of project concept), environmental analysis (to ensure that the project does not go against ecological issues and regulations), and financial feasibility analysis (to establish whether the project once materialized would generate profits for the organization). It is only after the first three sub-phases are found to be positive that a financial feasibility analysis is performed. If the feasibility analysis is positive, one can go ahead, but if it is negative, the project can be abandoned and eliminated from the 'project concept' definition.

The feasibility phase is terminated when a decision maker decides to transform the project concept into a project. At this point, in the case of a plant, the capacity is decided, the locations are chosen, the financing is agreed upon, the overall budget and schedule determined, and a preliminary organization is established.

Some of the tasks related to pre-project phase (Bubshait and Al-Musaid 1992) are summarized below:

1. Assigning a task force to conduct preliminary studies for the proposed project
2. Studying the users' requirements
3. Defining the technical specifications and conditions that determine the quality of the required work
4. Studying how to secure funds to finance the project
5. Estimation of the project cost and duration
6. Approval of the project cost
7. Studying and determining the technical specifications of the materials
8. Studying the impact of the project on the safety and health of the community and environment
9. Establishing criteria for the selection of project location
10. Advising members of the task force (consultant, engineering, etc.) on the approved funds for the project
11. Establishment of milestones for the project for review and approval

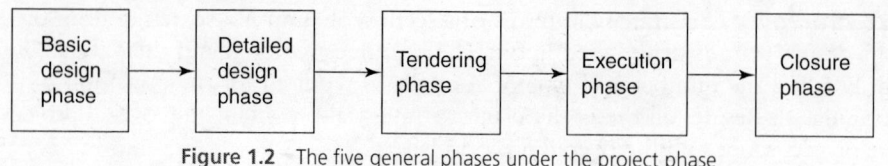
Figure 1.2 The five general phases under the project phase

12. Describing the responsibilities and authority of project parties
13. Pre-establishing a system to prepare for 'change order' procedures
14. Establishment of design criteria for structural specifications
15. Conducting a feasibility study of the proposed project

Project Phase

The project phase is also referred to as project implementation phase, project realization phase, or project materialization phase. It can be broken down into five sub-phases as shown in Figure 1.2. The sub-phases are discussed below in brief.

Basic design phase The activities in this phase are carried out by an engineering organization or an architect. During this phase, the documentation for tendering and contracting the physical construction or for procuring equipment is prepared. It involves performing basic design calculation, preparing tender drawings, preparing design and material specification, etc. The changes, if any, from the initial scope of work are recorded. Regular design and specification review meeting is one of the important features of this stage of the project.

Detailed design phase Detailed design may be carried out in-house or through contracting. In some cases, such as 'item rate' contract, it may be required to carry out the detailed design before starting the tendering process. However, in some cases, such as 'design build' contract or 'lump sum' contract, the tendering process can start immediately after the completion of basic design and specifications.

Tendering phase Tenders are issued if it is decided to execute the project through contracting. The preparation of clear and precise documents is essential to eliminate any dispute about scope of work at the contract stage. The tender preparation includes preparing the specifications and agreement conditions, preparing bill of quantities and estimating the contract value. It also includes issuing of tender document to the interested applicants, holding meetings, receiving bids and evaluating them. After the bids are evaluated, recommendations are made for the successful contractor and approvals sought to place a contract, and finally the contract is awarded.

Execution or construction phase Immediately after the contract is awarded, construction phase begins. In cases where the detailed drawings and designs were not available as part of the tender document, the contractor proceeds with the preparation of detailed design and drawings, and follows it up with the construction. In some cases, the preparation of detailed design and construction may proceed simultaneously with milestone-wise deliverables for both design and construction. After the construction work has started, the progress is closely monitored and regular meetings held with the contractor to assess cost and schedule. The variations in cost, quality and schedule are noted and corrective measures are taken to bring them to the desired level.

Closure or completion phase   In this phase, the major equipment are tested and commissioned, and the constructed facility in totality is handed over to the client for use. Client issues approval of work and a completion certificate after all the work has been checked and found to be in order.

Some of the typical tasks performed in the project phase are listed below:

Design-related tasks (Bubshait and Al-Musaid 1992)

1. Arranging the documents of the construction contract
2. Qualifying of design professionals
3. Performing technical and financial analysis of offers from competing contractors
4. Selecting the design team
5. Negotiating with the qualified design professionals
6. Providing the qualified design professionals with the needed information
7. Monitoring the design progress of the proposed project
8. Evaluating the design and making the necessary decisions
9. Updating design documents
10. Reviewing design documents
11. Conducting design peer review
12. Monitoring design quality
13. Updating drawings and specifications to reflect the requirements of location or environment
14. Using technical standards (e.g., Indian standard, American standard, British standard, etc.) to describe materials quality and construction methods

Tendering- and construction-related tasks (Bubshait and Al-Musaid 1992)

1. Pre-qualifying contractors
2. Holding a pre-bid conference and providing the necessary information
3. Negotiating contract price with qualified contractor
4. Reviewing at frequent intervals documents submitted by the contractor (e.g., work schedules, manpower qualifications, equipment)
5. Interpreting and clarifying ambiguities in the contract documents
6. Taking necessary precautions to prevent the loss of project data
7. Making necessary decisions against contractor claims during project implementation
8. Monitoring and controlling implementation methods, cost, schedule and contractor productivity

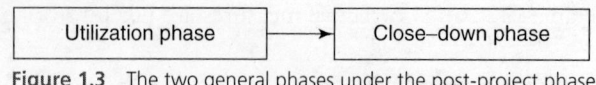

Figure 1.3 The two general phases under the post-project phase

9. Enforcing quality and safety principles during project implementation
10. Assigning personnel to supervise, monitor and control the implementation of quality (quality control programme)
11. Establishing quality assurance programme
12. Enforcing the quality assurance programme
13. Conducting regular visits to project site
14. Establishing acceptance criteria
15. Receiving record (as-built) contract drawings and specifications

Post-Project Phase

The post-project phase is also known as the turnover phase or the start-up phase. During this phase, the responsibility of the materialized deliverable is transferred from the engineers, the architects and/or the general contractors to the owners. The two general phases under Post-project phase are shown in Figure 1.3 and are discussed below in brief.

Utilization phase During this phase, the client or the end user makes use of the finished project. The performance of the constructed facility is monitored at regular intervals, and maintenance at regular intervals is performed.

Close-down phase Once the project has lived its intended life, it is dismantled and disposed of. The entire cycle explained under different phases is repeated.

It is very difficult to have a clear-cut demarcation among the different phases. This is truer in today's context when there is a tendency to cut down the project duration, thus requiring considerable overlapping in different phases of a project.

Normally, each of the mentioned phases requires different skills and, hence, should be carried out by different people with different skills. Pre-project phase requires creativity, while a project phase requires rigour. Only a few people are involved in pre-project phases over considerably long time periods, while hundreds or thousands of people can be involved in project phases over the shortest possible time periods. The phases play a dominant role in several decision-making areas, as given below:

- Phases of a construction project decide the relative importance to be assigned to various project performance attributes in its life cycle (Pinto and Slevin 1989).
- Phases of a construction project govern the importance of various project success criteria (de Wit 1988).
- Change in the phase of a project calls for different skill requirements of a project manager (Spitz 1982).

The objective of a project participant is to achieve success in all the phases to make the project an overall success, which can be assumed as the cumulative success in different phases.

1.3 IMPORTANCE OF CONSTRUCTION AND CONSTRUCTION INDUSTRY

Construction is an integral part of infrastructure such as houses, offices, townships, schools, hospitals, urban infrastructure (water supply, sanitation), highways, ports, railways, airports, power, irrigation, industrial infrastructure and so on. The importance of construction can be gauged from the simple fact that cost of construction of certain infrastructure projects may be as high as 60 per cent–80 per cent of the project cost.

Construction is recognized as the basic input for socio-economic development. It generates substantial employment. It employs not only engineers, managers and skilled workers, but also unskilled male and female workers from rural and urban areas. Since construction industry is dependent on a number of industries, its growth propels growth in other industries as well. For example, it positively contributes in the development of building materials industry and construction equipment industry. The demands for cement, steel, paints, chemicals, aluminium, glass, etc., from the construction industry provide positive impetus to the growth of these industries. While the forward movement of construction industry is an indicator of growth of the country, its backward movement creates widespread impact on employment and income, and the growth in GDP is also affected. The multiplier factor between growth rate of construction and growth rate of GDP has been in the range of 1.5 to 1.6. Construction products such as buildings, roads and power plants have a long life when compared to the products of other industries. In general, the construction industry of a country depicts the health of the economy of the country, and it is imperative that the industry is properly nurtured for the growth of the overall economy.

1.4 INDIAN CONSTRUCTION INDUSTRY

The construction industry was accorded Industrial Concern Status under the Industrial Development Bank of India (Amendment). Now, the construction industry is the second largest industry, next only to agriculture. Its contribution to the GDP at factor cost in 2006–07 was ₹196,555 crore, which is about 6.9 per cent of the country's GDP. It employed 31.46 million personnel comprising both skilled and unskilled workers, technicians, foremen, clerical staff and engineers.

With increasing thrust on developing infrastructure and attractive concessions appeasing private partnership in infrastructure projects, the Indian construction industry is already booming and is poised to see a bigger growth in near future. Some of the factors in favour of the Indian construction industry are availability of cheap labour, availability of qualified professionals, excellent opportunities at present and a large number of construction companies (nearly 28,000 organized companies and 75 large contracting companies).

Some of the factors that go against the Indian construction industry are low productivity, low ratio of skilled to unskilled workers, high cost of finance and complicated tax structure, and the presence of mostly small contractors who lack financial and technological backup, have low technology base, have negligible investment in R&D, and show little regard for systems.

A host of factors such as acute housing shortage, upturn in industrial sector, restructuring of state electricity boards (SEBs) and expansion of power grid will contribute to the growth of the Indian construction industry. There is a big upturn in commercial production, creating a correspondingly big market for commercial buildings. Factories are being put up at mass scale at some locations that have been declared as 'industrial area' by the state governments. The concept of tax holiday has given further boost to establishing factories. There is also a huge market in building residential units for both the private and public sectors. Real estate investment has shot up in recent times due to the tax benefits announced by the government. A lot of housing projects are being undertaken not only in metropolitan cities but in other major cities as well.

In roads sector, the execution of the golden quadrilateral and the north–south and east–west corridor has created a lot of opportunity. Further, work on Pradhan Mantri Gram Sadak Yojana (PMGSY) proposes an investment to the tune of ₹ 60,000 crore. This will connect to the rural area with certain habitat strengths. The operation and maintenance of the huge road network will spawn extensive opportunity to work in this sector.

The public–private partnership initiative through build–operate–transfer (BOT) and build–own–operate–transfer (BOOT) route is proving to be a blessing in disguise for the upgrade of existing airports in the country. Operation and maintenance of these airports are sure to attract bulk investment.

There are 11 major ports and 163 minor ports in the country at present. The average ship turnaround time is six days in India, and this is very high when we compare it to other better-managed ports with high-class infrastructure in the developed world. For example, the average ship turnaround time in Singapore is just six hours. Traffic is expected to increase further, which will throw up a huge potential for investment and construction. This sector is set to grow since development of small ports and inland waterways is a part of the prime minister's investment policy decision.

The water and effluent treatment sector is also bound to grow in the near future. Already, there is an increase in investments in improvement of civic facilities in rural and urban areas, in order to improve water supply and address sanitation-related issues as envisaged in the Tenth Five Year Plan. There is a plan to link the major rivers of the country, which will bring huge investments into the construction sector. There are many projects in the planning stages to bring water from rivers to cities through pipelines. Further, enforcement of stringent norms for discharge from industries under Environmental and Pollution Act will also spur growth in the effluent treatment sector.

The restructuring of state electricity boards, setting up of new substations and switch yards, implementation of accelerated power development and reforms programme (APDRP), and revamping of transmission lines to reduce transmission and distribution losses from 35 per cent–40 per cent to international norms of 8 per cent–10 per cent are sure to bring momentum to electrical projects. Privatization of transmission and distribution systems will further increase investments.

India suffers from perennial power shortage and there is a huge potential in the field of power, as new additions in power-generation capacity are required. National Hydroelectric Power Corporation (NHPC) and Nuclear Power Corporation of India Limited (NPCIL) have plans for huge capacity building in the hydel power sector and nuclear power sector, respectively. In thermal power, large private investments are in the pipeline and a number of independent power producers (IPPs) are in final stages of financial closure. Also, there are increased investments in captive power plants by large industries in cement, aluminium, sugar, etc.

India has considerable natural gas reserves. There are plans to invest in pipelines to save on transportation costs. Similarly, there are plans to lay pipelines from ports to refineries to carry crude and finished products to different destinations, apart from projects for expansion of refineries and LNG storage facilities. All these are bound to throw up opportunities in the hydrocarbon sector.

In addition to the above-mentioned opportunities, there are promising prospects in bulk materials-handling projects. While development of ports would require investment in port-handling equipment, investment in thermal power sector would call for more coal-handling plants, and increased food production would mean more opportunity for grain-handling plant construction.

Here, it must be noted that the indigenous construction industry is bound to face increasing competition from multinational companies. Already, projects including major portions of Delhi Metro, some high-value hydroelectric jobs such as Uri Project, high-tech projects such as Bangalore Infotech project and Chennai Tidel Park project, and power projects such as Enron have seen a major chunk of construction going into the hands of multinational companies. A number of MNCs such as Skanska, Hyundai, Bechtel, Kumi Gumi, Obayashi and Toyo are now operating in India. The future is going to be quite competitive and the domestic companies who can face the challenges posed by the MNCs will emerge stronger.

1.5 CONSTRUCTION PROJECT MANAGEMENT AND ITS RELEVANCE

1.5.1 Role of Project Management

Project management as defined by PMBOK 2000 is the application of knowledge, skills, tools and techniques to a broad range of activities to meet the requirements of the particular project. Munns and Bjeirmi (1996) define project management as the process of controlling the achievement of the project objectives. Utilizing the existing organizational structure and resources, it seeks to manage the project by applying a collection of tools and techniques, without adversely disturbing the routine operations of the company. The function of project management includes defining the requirement of work, establishing the extent of work, monitoring the progress of the work and adjusting deviations from the plan.

The main elements of project management according to Austen and Neale (1995) are given in Figure 1.4, and the cycle of activities to achieve the project goals is shown in Figure 1.5.

Project management aims to achieve the stated goals of the project leading to completed facility, by virtue of planning, executing and controlling time, funds and human and technical resources. The planning essentially consists of setting objectives, identifying resources and forming strategy. Executing consists of allocation of resources, guiding execution, coordinating efforts and motivating the staff. Controlling consists of measuring achievement goals, reporting, and resolving problems. The planning, executing and controlling are performed on a continuous basis till the goals of the project are realized.

Project management knowledge and practices are best described in terms of their component processes. These processes can be placed into five process groups (initiating, planning, executing, controlling and closing) and nine knowledge areas (project integration management, project scope management, project time management, project cost management, project quality management, project human resource management, project communications management, project risk management and project procurement management).

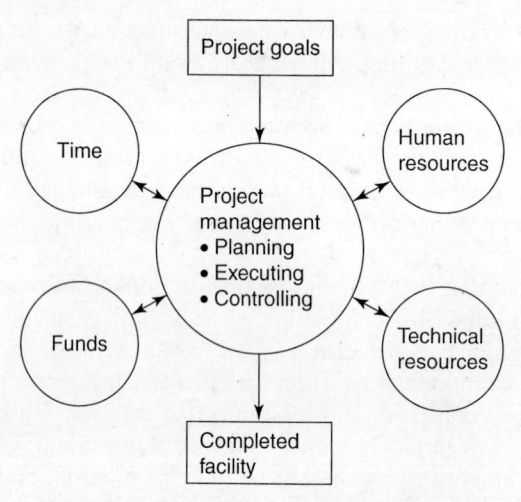

Figure 1.4 The elements of project management

Project management is essentially about managing a project from an idea through to completion. Projects today are getting increasingly complicated than they have ever been, embracing multiple disciplines and including increasingly larger sums of money. The basic ingredient of bringing together ideas and successfully executing them remains, even though the new techniques in project management are continuously being deployed.

Project management, as we know it today, first emerged in the early 1950s on large defence projects. It was adopted gradually by smaller organizations, and currently, even the smallest construction organizations are known to operate project management in some form.

1.5.2 Why Construction Project Management?

As we go through the subsequent chapters, we will find that there is probably no other discipline that is more difficult than construction project management. The statement may come as

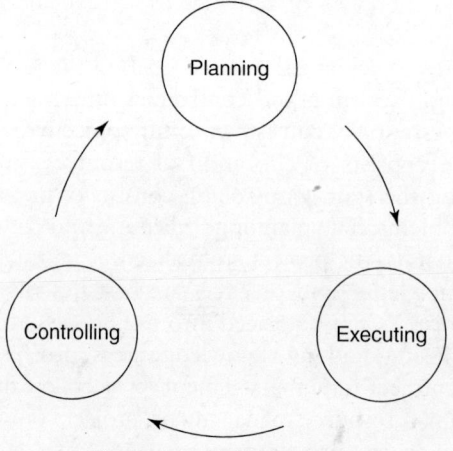

Figure 1.5 Cycle of activities

a surprise to readers; after all, the general goal of a construction project seems simple enough—building a project on time, within budget, with the stated quality standards, and in a safe environment. Looks so simple!

Yet, research shows that less than 20 per cent of most construction projects meet the four requirements mentioned above. Not surprisingly, then, the contracting companies or agencies in charge of construction are not held in high esteem. Construction project management is known for continual problems and the companies involved have a continual record of poor performance.

According to Austen and Neal (1995), managing a project is quite different from managing a 'steady state' organization. A project has a distinct beginning and an end. On the other hand, the 'steady state' organizations run continuously. Some examples of the 'steady state' organization are mass-production factories where mostly routine works are carried out.

Construction utilizes a lot of natural resources that are scarce. This calls for even more care in management of construction projects if the resources are not to be wasted.

The last two decades have seen the growth of construction project management in terms of knowledge, management skills and increased performance and quality. Today, there are many excellent contractors who perform well on time, within budget, and by adhering to quality and safety standards. So, how do they do it? This will be the question we shall be exploring throughout the text. It will be our endeavour to see that scientific principles are applied in practice, and not rely solely on experience gained at the project site.

1.6 PARTICIPANTS/STAKEHOLDERS OF A CONSTRUCTION PROJECT

The following section describes the position and role of various people involved in a construction project; we take a viewpoint based on project execution processes as categorized by internationally accepted methods. It should be noted that the roles, rights, responsibilities and principles of action of these people, even when in the same position, sometimes differ highly in each nation and region.

1.6.1 Architect

Traditionally, architects were responsible both in pre-project and project phases, and they were acting on client's behalf. In pre-project phase, they were responsible for preparation of drawings, preparation of tender document and contractor selection. In project phase, they were responsible for checking the measurement, certification of bills and overall project management functions. Over the years, however, most of their roles have been performed by construction managers. In the case of Public Sector, there may be a separate department (Engineering Department) which is responsible for the tender preparation and contractor selection.

1.6.2 Client (Owner)

The client is often the person or organization that will manage the facilities or structures upon completion of the project. The client is in a position to judge the use of funds to execute the project and they are at his discretion. Examples of clients are national and local governments, public corporations, public enterprises, the army, stock companies, cooperative societies, enterprise groups, legal entities, and individuals.

1.6.3 Constructor

The constructor is sometimes called the construction contractor. The job is an important one: completing the project on schedule according to the contract concluded with the client, and in accordance with design drawings and specifications. Constructors are of a vast variety of forms and sizes. Some are big firms that go about business by taking on large-scale projects involving many subcontractors and specialized contractors. They may actively conduct research and development. Some constructors are firms who develop by specializing in a specific construction technology, by possessing certain construction machinery, or by having skilled workers at their disposal; these would receive subcontracted work. Others are small-scale firms who develop by mobilizing workers at construction sites for labour-intensive construction work.

Constructors are generally profit-making firms, so their aim is to obtain as much contracted money as possible from the client at the earliest possible time and to pay subcontractors as little as possible for what they do at the latest possible time.

1.6.4 Engineer (Consultant)

Generally, the term 'engineer' means the consulting engineer who works with the client to conclude the contract. He provides technical services on behalf of the client (as a partner and/or agency). Design alterations and cost and schedule changes are often unavoidable as a construction project proceeds because working conditions may change unexpectedly. When they are necessary, the consultant plays an important role as an unbiased arbiter, acting for client and contractor. The position, role, responsibilities and duties of an engineer differ greatly across countries, as one may well imagine.

The consultant is also referred to as 'Construction Management Consultant', 'Construction Supervision Consultant', or 'Project Management Consultant'. The nature of services offered by these consultants varies from project to project. However, in general, they are responsible for undertaking project feasibility study, preparation of cost estimates, geo-technical investigation, reviewing and coordinating engineering drawings, helping client in bidding process, and coordinating in execution phase of the project.

1.6.5 Subcontractor/Supplier/Vendor

In large projects, very often no single contracting company has adequate expertise and/or resources to be able to undertake all the activities on their own. Under such situations, they employ small contractors for certain specialized items of work, for either execution purpose or material procurement purpose, or both. These are referred to as subcontractors or vendors, and are a very important party in any construction project.

1.6.6 Lawyer, Insurer, etc.

It may be a little surprising to find the lawyer as participant in a construction project. While in some countries lawyers play a much more minor role during construction projects, in some countries, lawyers are part and parcel of a construction project and it is normal for a lawyer to be retained. Lawyers specializing in claims settlements and disputes play an important role in domestic projects undertaken in India, as well as in international projects. Some of the construction companies in India do have a separate legal cell comprising lawyers specializing in construction disputes. Insurers also play an important role in construction projects as shall be seen subsequently in latter chapters.

1.7 ORGANIZATION OF THE BOOK

The book is intended to be used by students of construction management as well as practising professionals. Further, among practising professionals, personnel of both the client organization and the contracting organization would find the book useful.

The book is presented in 19 chapters. The first two chapters introduce the readers to the construction project and the construction organization. Thereafter, the book has been organized in a manner similar to the different stages or phases through which a typical construction project passes. In Chapter 3, readers are introduced to the construction economics, which may be useful for selection of projects. The concept of 'time value of money' has been explained and different methods of project evaluation have been discussed.

In Chapter 4, the process of preparation of estimates of the proposed project by a client or an owner organization has been discussed. These estimates serve a number of functions as outlined in this chapter. Owner/client organization may choose to get the project executed through a contracting organization and, accordingly, the various aspects of contract management have been presented in Chapter 5.

In Chapter 6, different planning techniques, which may be useful to both the client organization and the contracting organization, have been discussed. After the project plans are made, the ways to convert the plans into schedules are put forth. The project-scheduling and resource-levelling aspects have been covered in Chapter 7.

The estimation of cost and preparation of bid are important functions of any contracting organization. In Chapter 8, the estimation process from the point of view of the contracting organization has been discussed in detail. The bidding strategy and the process to determine the optimal markup have also been covered in this chapter.

Chapter 9 to Chapter 17 mostly cover topics pertinent during the execution stage of the project. In Chapter 9, the aspects of construction equipment management are covered. Here, the important factors in the selection of equipment and the issues in selection of major plant and equipment are discussed, in addition to presenting the advanced concepts in economic analysis.

Construction accounting is different from the accounting practices adopted for other industries. In Chapter 10, the salient features of construction accounting are explained. The emphasis is not on preparation of balance sheet and profit-and-loss account for a construction company, but on the interpretation of figures contained in the financial statements.

Materials constitute a significant portion of the cost of a construction project, and Chapter 11 addresses all the salient aspects of materials management. Chapter 12 covers cost management and value engineering aspects in the context of a construction project.

In a recent survey, a majority of Indian construction professionals have rated quality performance in a construction project as one of the topmost performance parameters in terms of importance. The different facets of construction quality have been covered in Chapter 13. Construction being a risky business, with incidences of a number of failures of construction companies year after year, it is pertinent to cover construction risk management. In Chapter 14, construction project risk management has been covered with special emphasis on insurance. Risk in international construction has also been discussed in this chapter.

Construction safety is of paramount importance in modern construction, more so in the light of the number of legislations being enacted presently. In Chapter 15, the major issues in construction safety management are elaborated.

Projects, if left on their own, are most likely to get completed at extra cost and beyond the stipulated time frame. It is, thus, imperative to monitor them on a regular basis so that suitable corrective measures can be taken appropriately and in good time. In Chapter 16, the monitoring and control systems pertinent to construction projects are explained.

In Chapter 17, construction claims, disputes and project closures have been taken up.

With so much of emphasis being placed on software in the area of construction management, it was felt that it would not be out of place to include a chapter on computer applications in construction management. Accordingly, Chapter 18 features computer applications in scheduling, resource levelling, monitoring and reporting. The illustrations on working of MS Project and Primavera—two of the most popular softwares used in the construction industry—have also been provided in this chapter.

In Chapter 19, a detailed discussion on the factors crucial for the success of a construction project are presented. The chapter contains recommendations for project professionals for better achievement of schedule, cost and quality performance in a construction project.

In Chapter 20, an introduction to linear programming in construction management is presented. The chapter contains linear programming problem formulation, and the solution methods, namely, graphical and simplex method. An introduction to dual problem and sensitivity analysis is also presented.

In Chapter 21, transportation, transshipment, and assignment problems are formulated and their solution methods have been illustrated. Some special problems applicable in construction have also been presented. Most of the chapters have a number of practical examples that readers may find quite useful. Also, in all the chapters, review questions have been provided which may prove to be useful in the learning process.

The text contains a number of appendices that interested readers may refer to.

REFERENCES

1. Austen, A.D. and Neale, R.H., 1995, *Managing Construction Projects: A guide to process and procedures*, International Labour Organization.
2. Bubshait, A.A. and Al-Musaid, A.A., 1992, 'Owner involvement in construction projects in Saudi Arabia', *Journal of Management in Engineering*, 8(2), pp. 176–185.
3. Chander, K., 1989, 'OR Applications in Strategic Sector Constructions with Capital Budgeting', Doctoral thesis submitted at Department of Civil Engineering, Indian Institute of Technology, Delhi.
4. Construction Industry Development Council, available online at http://www.cidc.in, retrieved on 15 September 2009 (for data pertaining to construction industry).
5. de Wit, A., 1988, 'Measurement of project success', *International Journal of Project Management*, 6(3), pp. 164–170.
6. http://cpwd.nic.in, viewed on 13 September 2009 (for history of CPWD).
7. http://en.wikipedia.org, viewed on 15 September 2009 (for details of major construction events in India).
8. http://indianarmy.nic.in, viewed on 15 September 2009 (for History of Indian Railways—major events).
9. Munns, A.K. and Bjeirmi, B.F., 1996, 'The role of project management in achieving project success', *International Journal of Project Management*, 14(2), pp. 81–87.

10. Pinto, J.K. and Slevin, D.P., 1989, 'Critical success factors in R&D projects', *Research Technology Management*, 32(1), pp. 31–35.
11. Pinto, J.K. and Slevin, D.P., 1988a, 'Project success: Definitions and measurement techniques', *Project Management Journal*, 19(1), pp. 67–72.
12. Pinto, J.K. and Slevin, D.P., 1988b, 'Critical success factors across the project life cycle', *Project Management Journal*, 19(3), pp. 67–74.
13. PMBOK, 2000, *A guide to the project management body of knowledge*, Project Management Institute, http://www.pmi.org.
14. Spitz, C.J., 1982, 'The project leader: A study of task requirements, management skills and personal style', *Doctoral Dissertation*, Case Western Reserve University, USA.

REVIEW QUESTIONS

1. State whether True or False:
    a. Construction projects are gradually becoming multidisciplinary with multiple objectives.
    b. The three phases of a construction project are—pre-project phase, project phase and post-project phase.
    c. Lawyers are not a participant in construction projects.
2. Define 'project' and 'project management'. What are the different stages in a typical construction project?
3. Explain the different phases of a project specifying the activities to be carried out in each of the phases.
4. How do you define a construction project?
5. Mention some of the features of a construction project which make it unique compared to other project types.
6. What are the different phases through which a construction project passes?
7. What typical activities are performed in pre-project phase?
8. What important tasks/activities are performed in project phase?
9. What tasks are performed by a client organization in tendering?
10. Name the two most important activities in post-project phase?
11. Why is it important to understand the different phases of a construction project?
12. Why is construction management needed?
13. Who are the major participants/stakeholders involved in a construction project?
14. Find out the contribution of construction activities in the GDP of a country.
15. 'Construction projects may have different scopes of work, but the core nature of project implementation is very similar.' Discuss this statement with respect to: (a) phases of projects and (b) stakeholders involved in the project.
16. 'Execution of engineering contracts is different from setting up of a manufacturing project in many ways.' Discuss this statement with respect to (a) phases of projects and (b) stakeholders involved in the project.

2

Project Organization

Introduction, construction company, forms of business organization, structure of construction organization, organizing for project management, management levels, traits of a project manager, important traits of a project coordinator, ethical conduct for engineers, factors behind the success of a construction organization

2.1 INTRODUCTION

According to an estimate, the Indian construction industry comprises nearly 200 firms in the corporate sector and about 120,000 class 'A' contractors registered with different government construction bodies. In addition to these, there are numerous small contractors including individuals who execute small jobs including repairing and maintenance. They are working mainly as subcontractors to the large contractors.

Besides large companies employing millions of workers, there are equally large numbers of self-employed individuals engaged in the actual construction work and allied activities, like whitewashing, painting, plumbing and fixing of mechanical or electrical fixtures.

While the corporate companies work in an organized fashion, most of the works of small companies are done in an unorganized manner. The construction industry was accorded Industrial Concern Status under the Industrial Development Bank of India (Amendment) Act very recently. Since the volume of construction works in the hands of corporate groups is minuscule when compared to the large chunk of construction being controlled by small companies or individuals, we can safely conclude that the Indian construction industry is still to assume the status of an organized industry. In this chapter, the various forms of organization and their structure, different management levels in a construction organization and other issues are discussed.

2.2 CONSTRUCTION COMPANY

According to Tenah (1986), a construction company can be defined as a group of people sharing specialized knowledge to design, estimate, bid, procure/purchase, and obtain resources to complete a construction project. Thus, the construction company, whether it involves a one-person organization or a larger firm with several departments, is primarily devoted to giving service with the aim of making profit.

In general, the common functions of a construction company are general administration, estimating, managing contracts and personnel, design, engineering, purchasing/procurement, accounting, and managing field construction. The organization of a construction company is the conceptual framework of the resources that carries out the above-mentioned functions. These functions are not limited to areas within the construction organization itself, but include the interwoven relationships with other stakeholders as well. The stakeholders could be architects, engineers, other general contractors, subcontractors, manufacturers, material suppliers or vendors, equipment distributors, labour, government agencies and the general public.

The broad aim of an organization is twofold—to divide responsibility according to the technical knowledge and to divide that responsibility by degrees of executive ability. The extent to which authority, work, or responsibility is delegated (the entrusting of authority and responsibility) is, therefore, an important feature of any organization. For the purpose of the upcoming discussion, we follow the definitions given by Kerzner (2004) for authority, responsibility and accountability.

Authority is the power granted to individuals (possibly by their positions) so that they can make final decisions for others to follow.

Responsibility is the obligation incurred by individuals in their roles in the formal organization in order to effectively perform assignments.

Accountability is the state of being totally answerable for the satisfactory completion of a specific assignment.

2.3 FORMS OF BUSINESS ORGANIZATION

A construction organization can take any of the following forms of business organization.

2.3.1 Sole Proprietorship

This kind of organization is owned by a single person. The owner is licensed with the government. This form of organization is widely used in service industries. No formal charter of operation is required in such arrangements. Also, there are few government regulations to which such organizations are subjected. Such organizations do not have to pay corporate income taxes and their earnings are subject to personal income tax. The proprietor of such organizations has unlimited personal liability for debts and they also find it difficult to obtain a large sum of money for the business. The organization lasts as long as the proprietor lasts.

2.3.2 Partnership

When two or more persons associate to conduct business, a partnership is said to exist. This ranges from informal oral understandings to a formal agreement filed with the respective ministry/body. Similar to proprietorship, in partnership there is ease and economy of formation as well as freedom from special government regulations. The profits generated out of the business are taxed as personal income tax in proportion to the partners' claims, whether they are distributed or not. In the event that a new partner joins the business, the old partnership ceases and a new one is created. In case of dissolution due to disputes, the distribution of assets can be made based on the agreement formulated while forming the partnership organization. The major disadvantage in such organization is that of impermanence, the difficulty faced in transferring ownership, and the unlimited liability.

2.3.3 Corporation

A corporation is an artificial being, invisible, intangible and existing only in contemplation of the law (Marshall 1819, available at http://www.1700s.com). Being a mere creation of the law, it possesses only those properties that the charter of its creation confers upon it either expressly or as incidental to its very existence. It exists as a separate legal entity and apart from its owners. Consequently, owner's liability is limited to his or her investment. Also, the capital required for business can be raised in the name of the corporation, without exposing owners to unlimited liability. The ownership is evidenced by shares of stocks that are transferable, and so is the ownership. Thus the organization continues to exist even if the owner dies or sells his stock. One of the major disadvantages of such organizations is the double taxation that they are subjected to. While company pays tax on the income it earns, the stockholders are also taxed when they receive income in the form of dividend.

2.3.4 Limited Liability Company (LLC)

This form of organization is the combination of corporation and partnership. It provides its owners with corporate-style limited liability and the tax treatment of a partnership. This is well suited for small- and medium-sized firms. It has a few restrictions and greater flexibility. This form of organization has unlimited life. Complete transfer of ownership and interest is usually subject to approval of at least a majority of the other LLC members.

2.3.5 Private Limited Company

It is a type of incorporated firm that (like a public firm) offers limited liability to its shareholders but which (unlike a public firm) places certain restrictions on its ownership. These restrictions are spelled out in the firm's 'articles of association' or bylaws, and are meant to prevent any hostile takeover attempt. The major restrictions are:

- Stockholders (shareholders) cannot sell or transfer their share without offering them first to the other stockholders (shareholders) for purchase.
- Stockholders cannot offer their shares or debentures to the general public over a stock exchange.
- The number of stockholders cannot exceed a fixed figure (commonly 50).

More details can be found in Company Act.

2.3.6 Public Limited Company

Public limited company is an incorporated, limited liability firm whose securities or shares are traded on a stock exchange (national, regional and international) and can be bought and sold by anyone. Public limited companies are strictly regulated and are required by law to publish their complete and true financial position so that investors can determine the true worth of its stock (shares). It is also referred to as publicly held company. For more details, readers may refer to the Companies Act, 1956, and Business Dictionary Online.

2.3.7 Government Enterprises

A government-owned corporation, a state-owned enterprise, or a government business enterprise is a legal entity created by a government to undertake commercial or business activities on behalf of an owner government. The defining characteristics are that they have a distinct legal form and they are established to operate in commercial affairs. While they may also have public-policy objectives, government companies should be differentiated from other forms of government corporation or entity established to pursue purely non-financial objectives that have no need or goal of satisfying shareholders with return on their investment through price increase or dividends. In India, PSU is a term used to refer to companies in which the government (union/state) owns a majority (51 per cent or more) of company equity.

2.3.8 Joint Ventures

According to Shreshtha (Shreshtha 1993, cited in Ogunlana et al. 2003), construction joint ventures combine certain attributes of one venture with complementary features of another, for the purpose of engaging in a specific individual or multiple construction undertakings either as a one-team project or on long-term basis. The enterprise is co-owned and co-managed by the JV partners. Joint ventures between contractors from developed and developing countries are recognized mechanisms for technology transfer and, therefore, one way of improving the skills that are lacking (Ofori 1994). In the construction industry, creation of joint ventures may mostly result from complementary needs of technology, capital management and human resource. A number of JVs operated in the construction of Delhi Metro. Another variant of joint venture is unincorporated joint venture, in which the legal means of dividing the project's equity is by shareholdings in a company.

2.4 STRUCTURE OF CONSTRUCTION ORGANIZATION

The organizational structure is about how to use one of the basic resources, people, and how to facilitate overcoming the communication barriers at organizational interfaces (Enshassi 1997). It refers to the organizational and administrative patterns. For example, the organizational structure indicates the arrangement of different departments and the division of labour. The arrangement has a bearing on the response time for delivering decisions.

It is rare for two construction companies to have an exactly similar kind of organization structure. In fact, it is often jokingly said that there are as many organization structures as there are construction companies. Even for the same construction company, the organization structure may not remain same over a long period of time. The organization structure keeps on evolving and it depends on a number of factors such as technology, complexity, resource availability, products and services, competition and decision-making requirements (Kerzner 2004).

Most of today's large construction companies had a modest beginning in terms of undertaking small construction activities. Some of them started as subcontractors or petty contractors. During the initial days, these companies were able to get relatively simple construction assignments with little or no complications, and were basically labour-intensive. Such assignments were executed with the direct involvement of owner, with the help of managers/supervisors. These managers/supervisors were directly overseeing the works carried out by their subordinates. One such arrangement is shown in Figure 2.1. The communication follows ver-

CONSTRUCTION PROJECT MANAGEMENT

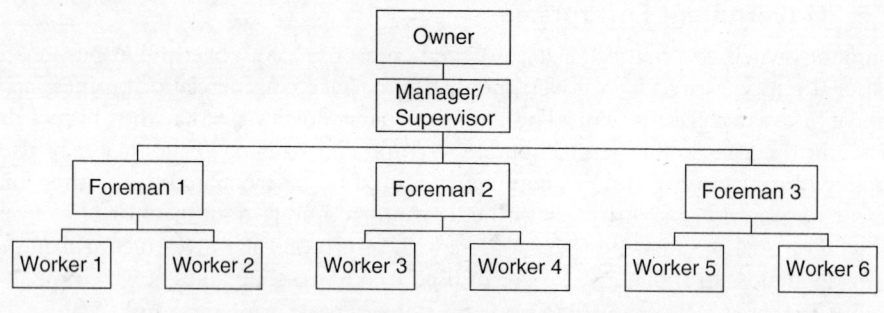

Figure 2.1 Military- or line-type organization

tically in such an arrangement. For example, owner will command manager/supervisor, who in turn will command the foremen 1, 2 and 3. The foreman, in turn, will command workers directly under them. For example, Foreman 1 will command workers 1 and 2, Foreman 2 will command workers 3 and 4, and so on. There is negligible horizontal flow of communication. For example, Foreman 1 will neither receive any instruction nor issue any command to Foreman 2. This type, in which the line of authority is direct from one level to another level of hierarchy, is similar to the military or line organization.

Such military- or line-type organizations have the following advantages and disadvantages:

Advantages:

- These are the easiest to establish.
- They are one of the simplest to explain to employees.
- In this structure, there is a unity of control.
- There is a strong sense of discipline.
- Each of the employees is assigned a fixed role and responsibility.
- Decisions can be quickly taken and the organization can adjust to changing needs in no time.

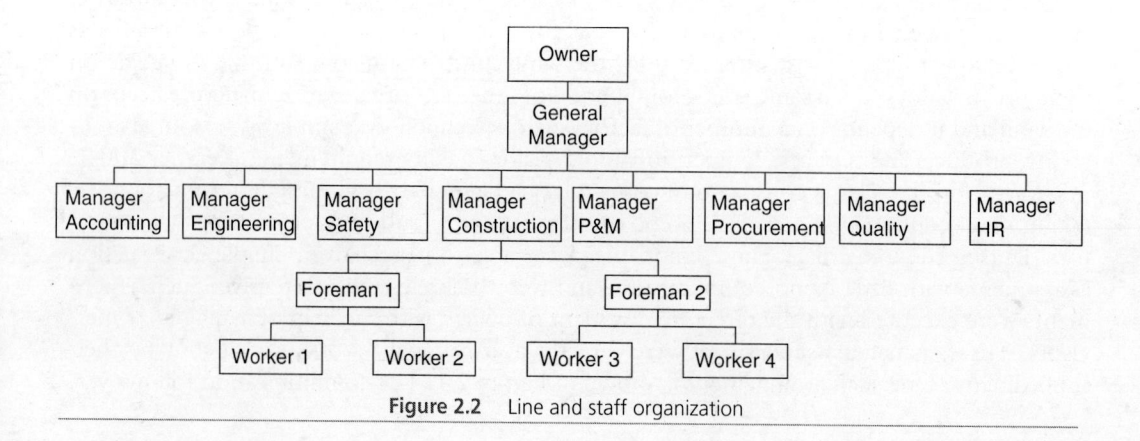

Figure 2.2 Line and staff organization

Disadvantages:

- There is a lot of expectation from the person holding the authority. The efficiency of the structure is heavily dependent on the person in authority.
- The advice of a smart employee at the bottom of the hierarchy may go unheeded, as there is no communication from an employee belonging to a lower hierarchy.
- The structure suffers from lack of specialized skill of experts. For example, Foreman 1 may not be skilled in all aspects of work; yet, he is supposed to give guidance to workers 1 and 2 under him. Also, there may be chances of favouritism creeping in such a structure.

When the construction companies started getting relatively bigger projects, the owners realized that they would not be able to handle such projects with the traditional line- or military-type organization. The owners felt the need to consult specialists on occasions. Thus, there were additions of people with staff responsibilities, and that led to the 'line and staff' type of organization, as shown in Figure 2.2. The staff specialists advise the line managers in performing their duties. These staff positions are purely advisory. The staff personnel can recommend but they do not have authority to implement those recommendations. For example, manager-safety plays the role of staff in Figure 2.2. In practice, it is very difficult to make a distinction between staff and line departments. For making such distinction, just pose this simple question: Is the person or department directly contributing to the achievement of the company's overall objectives? If the answer is 'yes', then the concerned person or the department is performing line function; else, it is performing staff function.

Some of the advantages and disadvantages of line and staff organizations are listed below:

Advantages:

- In the line and staff structure, line employees are responsible for execution while staff employees play the advisory role.
- The line and staff structure offers ample opportunity for the growth of employees. It also offers good training opportunity to the employees.
- The quality of decisions arrived at in a problem situation is high, as careful thought is given to arriving at the decision.

Disadvantages:

- There is a lack of well-defined authority structure.
- The structure is suitable mostly for large organizations where there is constant need for employing people with specialized skills.
- There is always a possibility of conflict arising out of various reasons, which may prove to be detrimental for the growth of the company.
- The distinction between line function and staff function is difficult to make.

With the increase in size and complexity of projects, the line and staff type of organization fails to deliver. This is because line and staff organization tends to load a few men at the top of the hierarchy with more duties than they can handle efficiently. This eventually gave rise to adoption of the departmental organization. One typical departmental organization is shown in Figure 2.3. The departmental or functional organization has departments with department heads. The department heads have control of the functions allocated to them and they are free

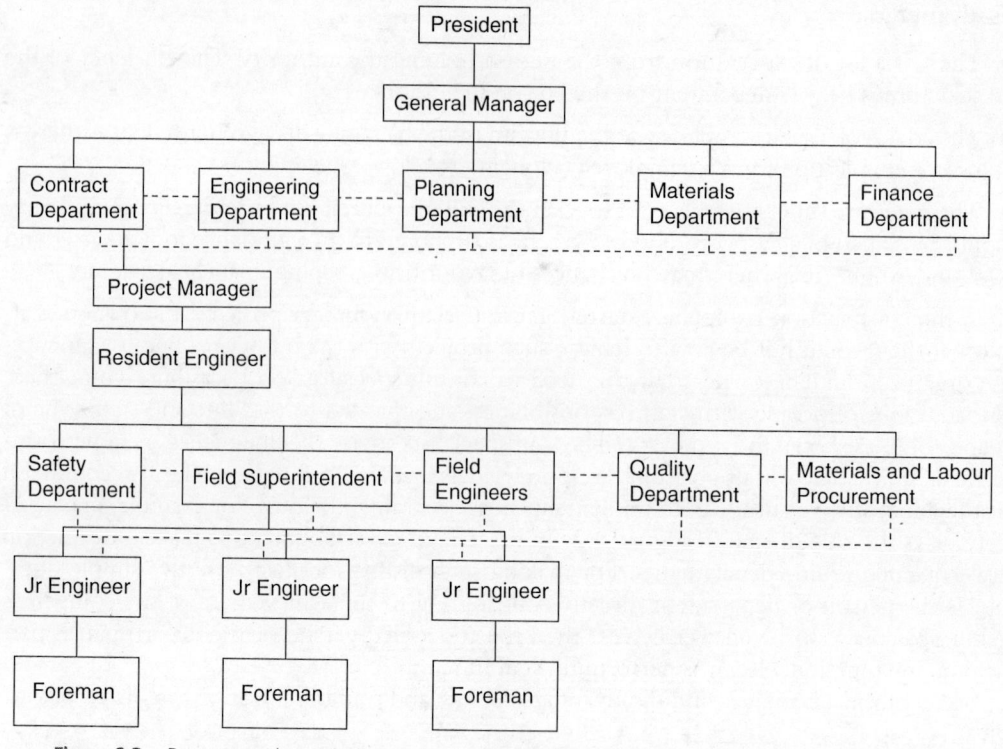

Figure 2.3 Departmental organization, authority, communication and contact (adapted from Tenah 1986)

to communicate directly with the field forces. In this organization structure, the top-hierarchy people are relieved from much of the heavy burden. The structure combines the best of the military- or line-type structure and the 'line and staff' structure. It also leaves functional managers in direct charge of the work and ensures them the support of a highly trained technical staff.

2.4.1 Centralized Functional

In this form, the centre of power is concentrated at the top of the organization. The departments are arranged by important functions, each headed by a manager who reports to a chief executive officer. The chief executive officer coordinates the activities of functional departments.

The advantages of the centralized functional structure (Thomas et al. 1983) are given below:

- It has a simple reporting mechanism and, thus, administration is easier.
- Functions are organized logically and in a cost-effective manner.
- Power and prestige of major functions are maintained, and strategic direction-setting is easier to attain.
- Duplication of efforts is minimized.
- Tight control at the top can be ensured.

The disadvantages (Thomas et al. 1983) are:

- The over-specialization and narrow vision of key personnel are problematic.
- There is a possibility of limited development for project managers.
- There may be difficulty in achieving economic growth if the firm attempts to diversify, as the functional management may not be quick to react.
- The lack of coordination between functional departments may be a problem area. This increases the burden on the chief executive officer. As the number of functional departments increases, the burden also increases.

2.4.2 Decentralized Multidivisional

In this arrangement, the departments are separated on the basis of the project market or region. Each unit is relatively self-contained in that it has the resources to operate independently of other divisions. The division manager has almost total authority to establish division strategy and to manage internal operation. It is like dividing a company into a number of smaller companies, except that each division manager is generally subject to some degree of evaluation and control by the central corporate office.

The advantages of the decentralized multidivisional structure (Thomas et al. 1983) are given below:

- The decision-making is quicker and simpler. The various divisions can take advantage of their own functional organization.
- The divisional focus is on the end result rather than the work required to produce the end result. The management can react to changing business conditions quickly and adapt to it.
- The output and responsibility can be easily identified.
- In this arrangement, each of the divisions can be operated as a profit centre.
- Corporate executives can be removed from the operating details to concentrate on overall corporate matters and long-range planning.
- Motivation and development prospects of project managers are enhanced, and there is greater likelihood of innovation and creativity.

The disadvantages (Thomas et al. 1983) are:

- The increased number of managerial and functional people may be unnecessary and may not prove to be cost-effective.
- The divisions may have little real incentive to cooperate with other divisions and, as such, lack of coordination among different divisions may be an issue.
- Coordination of customer relations and research-and-development activities may be difficult because of divisional autonomy.

Owing to its various disadvantages, the departmental organization is now giving way to the matrix organization. In the matrix organization, as we shall see later, the normal vertical hierarchy is overlaid by some form of lateral authority and communication. This, in turn, allows maximum efficiency in the utilization of resources as opposed to the departmental organization.

One of the distinct features of the matrix organization is that a dual rather than a single chain of command is followed. For example, some of the managers under matrix structure may report to two superiors instead of a single superior.

There are two alternatives through which a construction company can create a matrix organization. In the first alternative, the company can abandon the departmental groupings and adopt a structure solely on a project basis. In the second alternative, it can overlap construction/project management on top of the existing departmental organization. Most of the construction companies follow the second alternative. The details of organizing for construction/project management are discussed in the next section.

2.5 ORGANIZING FOR PROJECT MANAGEMENT

One of the primary functions of a construction company is to execute construction projects. This function is performed by companies following different organizational structures at project level. The organization at project level and at corporate level is not the same. Construction companies follow different organizational forms at project level. It is not an easy task to integrate the project management structure into a corporate structure. The inherent challenge here is to organize people from different specializations and different departments into an effective project team. The challenge gets further compounded when a number of different organizations with different objectives are involved in the project. Depending on the extent to which the authority is delegated to a project manager and the mode in which the power is delegated, the project management structure can be divided into categories such as—(1) functional, (2) matrix, and (3) project forms of organization. While in the first form the project manager has virtually nil authority, in the third form complete authority is vested with the project manager. The matrix method is a compromise between the two extremes.

The factors that are important for choosing a particular project-management authority structure, as identified by Thomas et al. (1983), are—(1) project size and duration; (2) organizational experience; (3) resources; (4) difference; (5) importance; (6) technology uncertainty; (7) financial uncertainty; (8) number of projects; and (9) cost and schedule control. The three most frequently used organizational forms for managing construction projects are—classical or functional, pure project, and matrix.

2.5.1 Classical (Functional)

Traditionally, classical or functional organizations are marked by a vertical structure with long lines of communication and a long chain of command. A typical functional organization is shown in Figure 2.4. In this form, each employee has one clear superior. For example, the general manager is reporting to the owner, the senior project manager is reporting to the general manager, and so on.

The employees are grouped by speciality—for example, human resources development, construction, engineering, tendering, finance, and so on. Each of these speciality groups works under one executive. The groups are further subdivided into sections. For example, engineering can be subdivided into civil, electrical and mechanical sections. Sections under other groups are not shown for clarity in diagram.

Figure 2.4 Functional organization

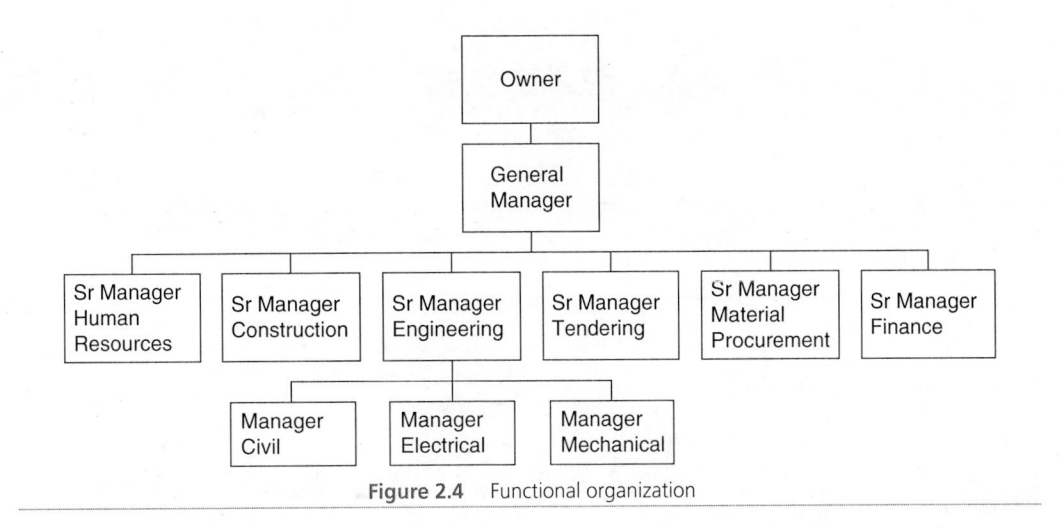

In a functional organization, project-related issues are resolved by the functional head. For example, construction-related issues would be sorted out by senior manager, construction. The functional organization structure assumes that the common bond supposed to be there between an employee and his superior would enhance the cooperation and effectiveness of the individual and the group (Enshassi 1997).

Some of the advantages of functional organizations are:

- The degree of efficiency is high since the employees have to perform a limited number of activities.
- There is a greater division of labour and, thus, the advantages of functional organizations are inherent in this structure.
- The specialized groups can enhance the possibility of mass production.

Some of the disadvantages are:

- The structure as such is unstable as it lacks disciplinary control.
- The structure is slightly complicated as it has several layers of sections.
- The responsibility for unsatisfactory results may be difficult to fix under such structure.
- There may be conflict among employees of equal rank.

In recent years, efforts have been made to reshape functional organizations having long lines of communication, by incorporating a horizontal structure with the aim of creating a shorter line of command. Such horizontally structured organizations are seen to be more flexible and able to quickly and effectively adapt to a changed environment.

2.5.2 Pure Project or Product Management

Pure project or product organizations can be formed to support a steady flow of ongoing projects. One such typical pure project organization is shown in Figure 2.5. In a project organization, employees are grouped by project. The majority of the organization's resources is

Figure 2.5 Project organization

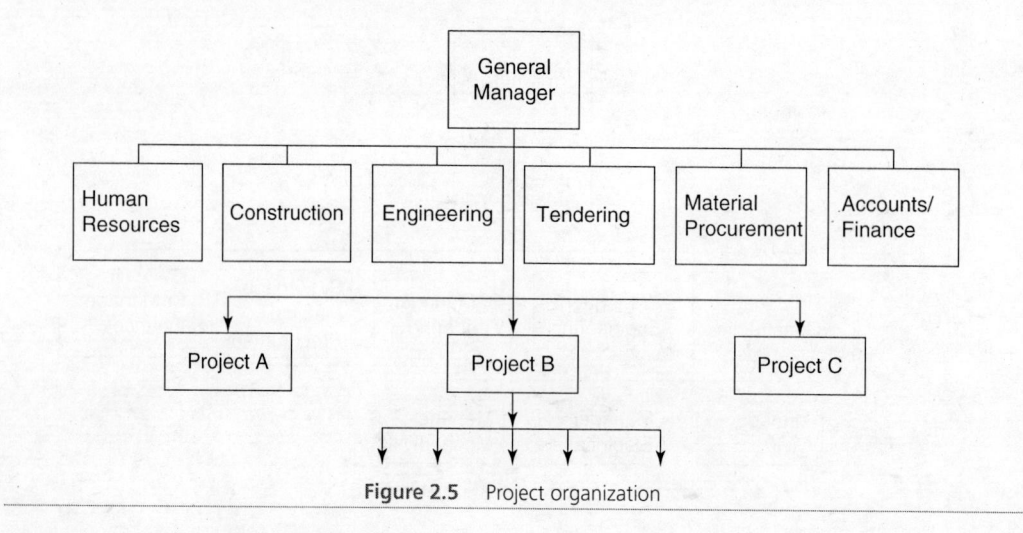

directed towards successful completion of projects. The project managers enjoy a great deal of independence and authority. In such structures, the different organizational units called departments either report directly to the project manager or provide supporting roles to the projects.

The advantages of pure project organizations are:

- The project manager maintains complete authority over the project and has maximum control over the project.
- The lines of communication are strong and open, and the system is highly flexible and capable of rapid reaction times. Thus, the structure can react quickly to the special and changing project needs.
- The project is the only real concern of the project employees. The pure project structure provides a unity of purpose in terms of effectiveness. It brings together all the administrative, technical and support personnel needed to bring a project from the early stages of development through to operational use.
- The appraisal of employees is based upon the performance of the project.
- The focus of resources is towards the achievement of organization goals rather than the provision of a particular function.

The disadvantages are:

- There could be a duplication of efforts.
- It is very difficult to find a project manager having both general management expertise and diverse functional expertise.
- The administrative duties of a project manager may be demanding and the job could be quite stressful.
- Due to the fear of impediments in career growth, some employees may not prefer to leave their departments.

The pure project or product organizations are unusual in the construction industry (Newcombe et al. 1990, cited in Uher and Toakley 1999). In order to address some of the disadvantages of the pure project structure, 'partial projectization' has been used in some cases. Here, the critical functions such as engineering and construction are assigned to the project manager, while functions such as procurement and accounting are performed within the functional departments.

The advantages of such partial projectization are:

- The project manager can spend more time directing the project-related activities of the major functional group.
- The project manager need not be an expert in all functional areas, as is the case with the pure project structure.

The disadvantages are:

- The project manager would still be under significant administrative burden despite some of the functions being carried out by the functional departments.

2.5.3 Matrix Organizations

The matrix organizations have evolved from the classical functional model. They combine the advantages of both the classical and the pure project/product structures. A matrix organization can take on a wide variety of specific forms, depending on which of the two extremes (functional or pure project) it most resembles.

In the matrix organization, the human resources are drawn from within various functional departments to form specific project teams. Every functional department has a pool of specialists. For example, the plant and machinery functional department may have the experts P_1, P_2, P_3, P_4, and so on. Similarly, the safety department may have experts S_1, S_2, S_3, S_4, and so on. When the project team for executing a project X is organized, the experts from different functional departments join together under the leadership of a project manager. For example, P_1 and S_1 may represent plant and machinery department and safety department, respectively. Similarly, depending on the requirement, personnel from different functional departments may join the project manager. The functional representatives such as P_1 and S_1 are referred to as the project engineers. On completion of projects, the project engineers return to their functional units within the vertical organization structure.

The matrix organization recognizes the dynamic nature of a project and allows for the changing requirements. Such structures can cater to the varying workload and expertise demanded by the project. For example, if at any stage it is found that the services rendered by P_1 are inadequate and needs strengthening in the form of more persons, the project manager can request the functional head of plant and machinery department to send a few more persons such as P_2, P_3, and so on. Similarly, during less workload the project manager can request for demobilization of these project engineers. The project engineers P_1, P_2, P_3, and so on usually contact their parent functional departments for getting advice on complex technical matters or when they encounter unusual problems. Otherwise, for all practical purposes the project engineers are under the control of the project manager. One of the features of the matrix organization is that the knowledge gained by the project engineers P_1, P_2, and others while executing projects can be shared vertically upwards for the benefit of future projects.

Figure 2.6 Typical matrix organization

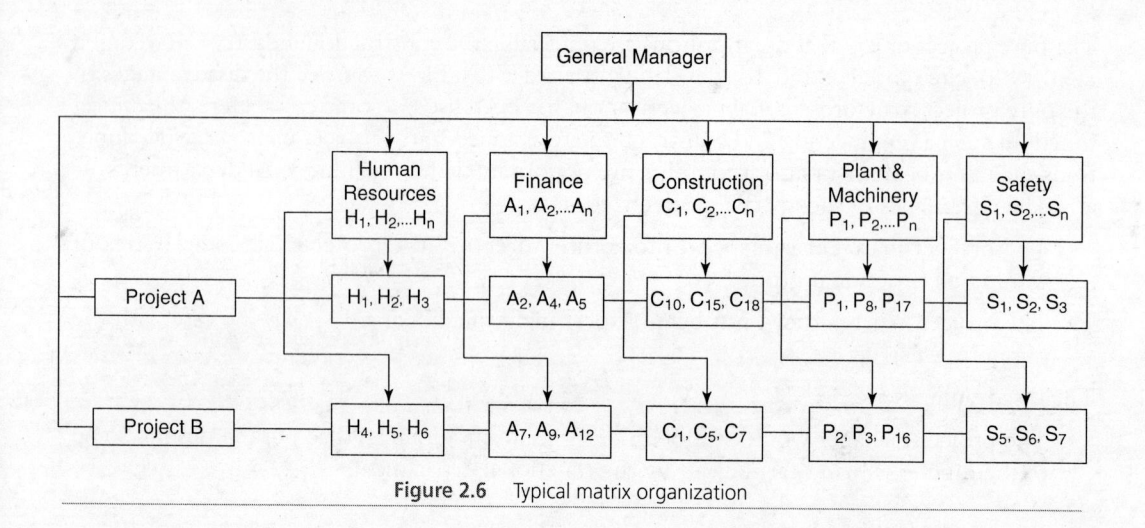

One typical matrix organization is shown in Figure 2.6. As mentioned earlier, the project manager heads a team of personnel sent in by their respective functional specialist departments. The role of project manager is that of a coordinator with considerable authority. While the main advantages are in the shorter and shared lines of command and communication, there is the potential of conflicts arising from dual lines of authority and dual reporting.

The advantages and disadvantages identified by Thomas et al. (1983) are given below. Advantages of the matrix structure are:

- The structure facilitates quick response to changes, conflicts and project needs.
- There is a flexibility of establishing independent policies and procedures for each project, provided that they do not contradict company policies and procedures.
- There is a possibility of achieving better balance between time, cost and performance than is possible with the other structures such as functional or project forms.
- The project manager has authority to commit company resources provided the schedule does not cause conflicts with other projects.
- The strong base of technical expertise is maintained.

The disadvantages are:

- Successful matrix authority application tends to take years to develop, especially if the company has never used dual authority relationships before.
- Initially, more effort and time is needed to define policies, procedures, responsibilities and authority relationships.
- The balance of power between functional and project authority must be carefully monitored.
- Functional managers may be biased according to their own set of priorities.
- Reaction times in a fast-changing project are not as fast as in the pure project authority structure.

The matrix organization can be further expanded into three different forms:

(a) **Functional matrix:** A person is formally designated to oversee the project across different functions. This person has limited authority over the functional people involved, and primarily plans and coordinates the project. The functional managers retain prime responsibility for their specific segments of the project.
(b) **Balanced matrix:** A person is assigned to oversee the project and interacts on an equal basis with functional managers. This person and the functional managers jointly plan and direct workflow segments and approve technical and operations decisions.
(c) **Project matrix:** A manager is assigned to plan, direct and oversee the project, and is responsible for the completion of the project. The functional managers' involvement is limited to assigning personnel as needed and providing advisory expertise.

A number of criticisms were levelled against matrix organizations in the 1980s. In fact, some companies reportedly dropped the use of matrix structure. Poirot (1991) recommended the following points for successful implementation of the matrix organization:

- Allowing time to define responsibilities and authorities
- Committing senior management time to explain the system
- Developing people who want the matrix to work
- Making decisions based on what is good for the client and the firm
- Promoting open communication with no secrets
- Eliminating politics and resolving conflicts at high levels
- Committing energy to evaluate and compensate on a common basis
- Using consensus management. Although it may take time and energy to do this, in the long run it is better since decisions taken by consensus require little time for implementation
- Hiring top-quality people having integrity and willingness to place the interests of the client, the firm and other people before their own interests
- Consolidating net income at corporate level and rewarding everyone in the firm

2.6 MANAGEMENT LEVELS

Management level can be defined as 'a position in management that is stratifiably (layer or level) differentiable (degree) in terms of power, authority, responsibility, and accountability over resources required to achieve defined objective(s).' Tenah (1986) identifies five management levels—(a) the board of directors' level; (b) the president's level; (c) the construction management level; (d) the project management level; and (e) the functional management level. The levels are discussed with respect to a typical construction organization.

2.6.1 Director Level

The different functions at this level are:

- setting plans,
- formulating objectives, and
- deciding among different courses of action.

CONSTRUCTION PROJECT MANAGEMENT

Figure 2.7 Positions in director level

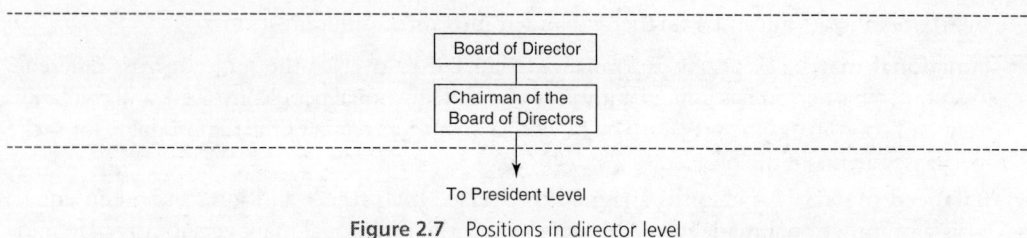

The three basic types of information required at this level to perform the above functions are:

- environmental information,
- competitive information, and
- the company's financial status and general performance data.

The various positions in Director level and their hierarchies are given in Figure 2.7.

2.6.2 President Level

The primary functions at this level are acquiring business and formulating the company's immediate objective, in line with the board's plans and strategies. This level requires a much more detailed and departmentalized format of the environmental, competitive and internal information. It also requires progress reports that summarize, for each project, its status, current and future cost and schedule performance, and problems, with management actions underway to resolve them. The various positions in President level and their hierarchies are given in Figure 2.8.

2.6.3 Construction Management Level

The major functions at this level are—obtaining and monitoring work for the company at the district or divisional level. This level requires summary formats of the three basic types of information

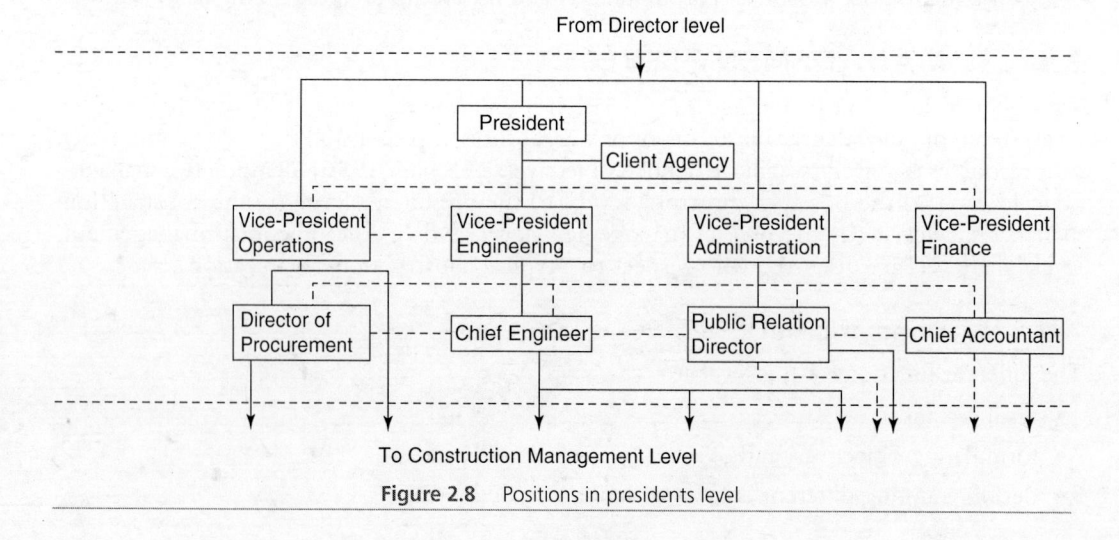

Figure 2.8 Positions in presidents level

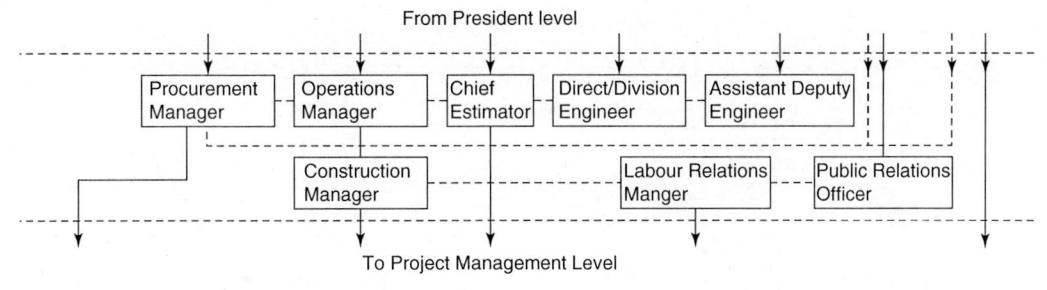

Figure 2.9 Positions in construction management level

(environmental, competitive and internal) that apply to the geographic areas. In addition, this level requires a clear and straightforward summary format of information on general progress, financial status, schedule status, procurement status and engineering status on each project under their jurisdiction. The various positions in Construction Management level and their hierarchies are given in Figure 2.9.

2.6.4 Project Management Level

The major functions of the project management level are:

- managing the day-to-day operations of all aspects of a project, and
- watching closely the development of the project as a group.

The information needs of this level are the same as those of the construction management level, backed up with the following details—(a) field costs, (b) summary and/or detail construction schedules, (c) list of critical or near-critical items in the network, (d) detailed prediction of future accomplishments, (e) current working estimates, and (f) cash-flow summaries. The various positions in Project Management level and their hierarchies are given in Figure 2.10.

2.6.5 Functional Management Level

The functional management level directly organizes, supervises and coordinates the workers, materials, equipment and services, to ensure that the project is built within the required

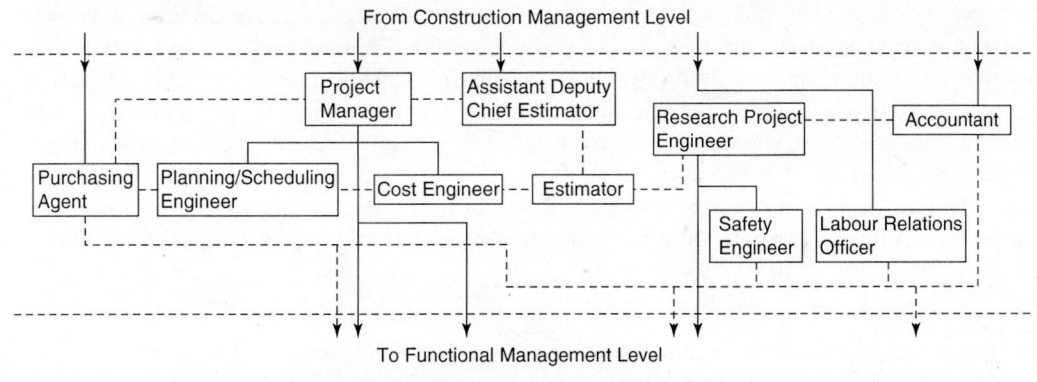

Figure 2.10 Positions in project management level

```
                    From Project Management Level
─────────┬──┬──────────────┬──────────────┬──┬──────────────────┬──────
         ▼  ▼              ▼              ▼  ▼                  ▼
      ┌─────────────────────┐          ┌──────────────┐   ┌──────────────┐
      │General Superintendent│          │Field Office  │───│Field Engineer│
      └─────────────────────┘          │Engineer      │   └──────────────┘
         ┌────────────┐  ┌─────────────┐└──────────────┘
         │Superintendent├──│Subcontractor│
         └────────────┘  └─────────────┘
            ┌─────────┐
            │General  │
            │Foreman  │
            └─────────┘
```

Figure 2.11   Positions in functional management level

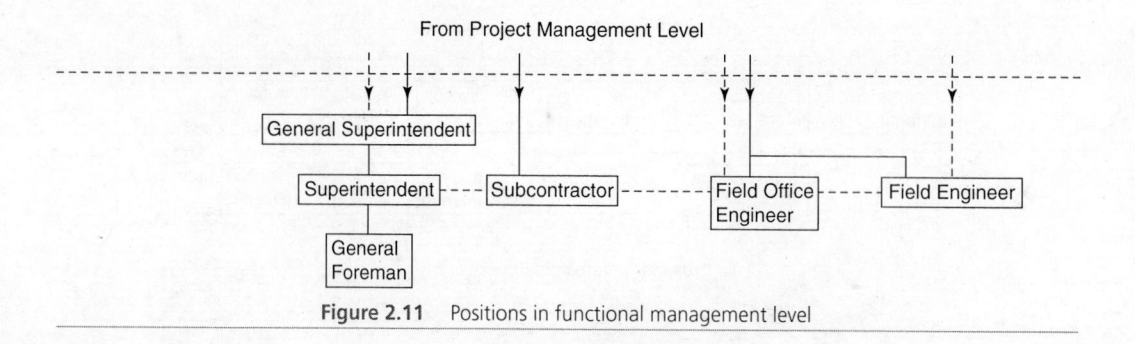

time, budget and safety and quality standards. All data are collected at these levels to be used for costing, estimating and scheduling purposes. The functional management level requires performance and productivity information on the organizational units that each manager or supervisor handles, as well as detailed analyses of the problem areas. The various positions in Functional Management level and their hierarchies are given in Figure 2.11.

2.7   TRAITS OF A PROJECT MANAGER

Project manager is a person formally appointed to manage a project with specific accountability for achieving defined project objectives with allocated resources. A project manager has access to, and a formally defined relationship with, the project leader to whom the specific project has been assigned (www.tbs-sct.gc.ca). Role of a project manager is very critical to the success of a project and recognizing this, a number of studies have been conducted to find the required traits of a project manager.

The terms 'project manager', 'project coordinator', 'construction manager', 'project administrator' and 'project controller' are used quite interchangeably, and all of them appear to have very similar kinds of role, but the intensity of their job requirement and the expectations from them vary (Kerzner 2004). For example, the typical responsibilities of a project manager and a project coordinator include—*coordinating and integrating of subsystem tasks; assisting in determining technical and manpower requirements, schedules and budgets;* and *measuring and analysing project performance regarding technical progress, schedules and budgets.* However, a project manager is supposed to play a stronger role in project planning and controlling. A project manager is also responsible for negotiating; developing bid proposal; establishing project organization and staffing; and providing overall leadership to the project team, in addition to profit generation and new business development. A project coordinator is seldom entrusted with these responsibilities. In fact, the project coordinator's role is to augment the project managers' visibility for larger projects (Forsberg et al. 1996). A project coordinator is chartered as a representative of the project manager who proactively ensures future events will occur as planned. They signal problem areas and recommend solutions. According to Forsberg et al. (1996), a project coordinator's function is to:

- know how the organization 'works';
- provide expediting help to the project and support organizations;
- provide independent assessment of project information and status to the project manager;

- ensure planning and milestones are satisfied; and
- ensure control procedures are being adhered to.

Katz and Kahn (1978) have suggested that an effective project manager should possess essentially three skills—technical skills, human relationships skills and conceptual skills. While technical skills include the ability to apply knowledge in a given field, such as engineering and finance, human relationships skills involve the ability to communicate efficiently and to maintain a harmonious working group. The ability to motivate employees also falls in the category of human relationships skills. Finally, conceptual skills include the ability to perceive the project as a system by keeping a global perspective and not thinking of only one aspect at once.

The above model has led to a number of debates on the extent to which a project manager needs technical skills. While it is understandable to have a technical expert as a project manager in case of a small project that involves knowledge of only one small specialist area, for larger projects involving multiple disciplines, searching for a technical expert may not be a wise option (Goodwin 1993). This is not to say that technical skills are not needed at all in larger projects, but the emphasis should be more on managerial skills of a project manager. Technical skills in larger projects are needed to appreciate the full implications of the project, which a project manager obtains as expert advice on 'as and when' basis. Some researchers are also of the opinion that project manager should not be a technical expert due to the apprehension that the project manager would engage himself in too much technical details and may not be able to do justice to other aspects of the project (Katz 1974, Goodwin 1993, Kerzner 2002).

El-Sabaa (2001) has analysed the relative importance of the three skill groups advocated by Katz and Kahn. Human skill with a percentile score of 85.30 has emerged to be the most essential project manager skill. Conceptual and organizational skill with a percentile score of 79.60 represents a second essential project manager skill. Technical skill with a percentile score of 50.46 has emerged as the least essential project manager skill relatively.

Odusami (2002) concludes through the analysis of a questionnaire survey conducted among clients, consultants and contractors that for a client the most important skill of an effective project leader is decision-making; for a consultant, the most important skill is leadership and motivation, and for the contractors, communication is the most important skill. Laufer et al. (1999) opine that the project manager's principal role is to manage his/her team's decision-making and not to make his own decisions.

It is advisable to appoint the project manager as early as possible, preferably in the feasibility stage itself, so that he is aware of all the aspects of the project and can take control of the project. However, the early appointment may not be possible in all situations.

The project manager should be responsible for coordinating all project activities; making project recommendations; fixing a design and preparing drawings and specifications for tender and construction; preparing all estimates; and administering all contracts and issuing certificates.

Various researchers have stressed the need for different types of skills required by a project manager in order to make the project successful. Their findings are either based on their experiences or based on empirical researches. Amongst the first category, we have the skills suggested by Stuckenbruck (1976), Kerzner (2002), Fryer (1979), and Adams and Barndt (1978). The details of these suggested skills are presented below from the works of Gaddis (1959), Katz and Kahn (1978), Stuckenbruck (1976), Adams and Barndt (1978), Fryer (1979) cited in Odusami (2002), Anderson (1992), Project Management Body of Knowledge (2001), and Kerzner (2002).

- Human skills: These include mobilization, communication, coping with situations, delegation of authority, political sensitivity, high self-esteem and enthusiasm. He should have the capability to motivate and integrate his team members to achieve goals for the projects.
- Conceptual skills: These include planning, organizing, having strong goal orientation, ability to see the project as a whole, ability to visualize the relationship of the individual project to the industry and the community, and strong problem orientation.
- Technical skills: These include specialized knowledge in the use of tools and techniques, project knowledge, understanding methods, process and procedures, understanding the technology required, and skill in the use of computer. He should have necessary knowledge of tools and techniques used in engineering and construction processes. He should be able to understand the technology trends and their evolution. He should also have skills to synchronize different technologies. He should have expertise connected with product and process used in the project.
- Attitude: This refers to an open, positive and 'can do' attitude, which encourages communication and motivation, and fosters cooperation.
- Common sense: This refers to a strong ability to spot sensible, effective, straightforward, least risky and least complex solutions—i.e., 90 per cent right on time is better than 100 per cent far too late!
- Open-mindedness: This refers to an approach where one is always open to new ideas, practices and methods, and in particular, gives equal weight to the various professional disciplines involved in the project.
- Adaptability: This refers to a propensity to be flexible where necessary and avoid rigid patterns of thinking or behaviour, and to adapt to the requirements of the project, the needs of the sponsors, its environment and people working on it. He should be able to adapt to change. He should be able to manage the changes and, in the process, recognize the opportunities.
- Inventiveness: This refers to an ability to discover innovative strategies and solutions either from within oneself or through interaction with other members of the project team, and to identify ways of working with disparate resources to achieve the project objectives.
- Prudent risk-taker: This refers to a willingness and ability to identify and understand risks but not to take a risky approach in an unwise or reckless fashion.
- Fairness: This refers to a fair and open attitude, which respects all human values.
- Commitment: This refers to a very strong overriding commitment to the project's success, user satisfaction and team coordination.
- Conflict resolution: The project manager should be able to resolve conflicts arising among his team members.
- Project manager needs solid basic experience in the relevant field. He should be multidiscipline-oriented. He should also be global-problem-oriented, i.e., he must consider the external, political, legal and environmental aspects.
- Project manager should have leadership skills. He should be a clear leader and director with sufficient authority. He should have the capability to elicit commitment from

his team members. He should be able to identify and solve the problem, and be able to balance technical, economics and human factors. He should also have the spirit of entrepreneurship.
- The project manager should master the basics of planning, budgeting, coordinating and assessing financial reports, and following up.
- A good manager and administrator, he should have the knowledge and understanding of estimating systems, cost control, scheduling control, quality and safety. He should be able to develop procedures and be able to implement them.
- He should be an effective problem solver and decision maker. He should possess good analytical abilities and should be proficient in the social skills.
- Effective communication skill is identified as one of the most important skills for a project manager. Other researchers such as Goodwin (1993) have also stressed the importance of verbal and written communication skills. The project manager should have the requisite interpersonal and negotiating skills.
- The project manager should have the right temperament, and should be able to keep his calm. He should be realistic, dedicated, generous, stable, quick-thinking, disciplined and persistent.
- The project manager should be creative in dealing with information and problems. Information-processing skill allows the manager to obtain, use and disseminate information.

Among the various literature cited above, the empirical research has been by Spitz (1982). She concluded that the priority of skills of a project manager vary depending on the phase in which the project presently exists. She has also tried to assess the skill needed in each of the phase of a project.

2.7.1 Strategies for Enhancing the Performance of a Project Manager

Tarricone (1992) believes that although some are born leaders, most managers are actually created. Some of the ways in which a future project manager can be groomed start right from undergraduate training, continuing through graduate training, employer in-house training programmes, and by seat-of-the-pants experience. Some of the strategies identified by Anderson (1992) in enhancing the performance of project manager's effectiveness are presented in the following paragraphs.

To Increase the Pre-appointment Training

At the entry level, potential project managers have only a technical background in areas such as civil engineering, mechanical engineering and electrical engineering. Most of the times, they are supposed to be learning the required managerial skills in 'on-the-job training' and 'sink or swim' mode. It is desirable to begin the induction of project managers after providing them with 'in-house' training or sending them to some specialized project management training institutes.

To Establish a Mentor System

The potential project manager should be attached to a good role model in the organization who may be from the middle to upper management. This way, he or she can learn the requisite project management skills.

Careful Selection of Potential Project Manager

Successful project managers have a good blend of technical and managerial skills. Concentrating fully on one aspect and overlooking the other can be disastrous for the project. Successful companies do have different schemes to identify potential project managers suitable for project management.

To Identify the Career Path Requirement Early

Depending on the skills possessed by an individual, the organization should identify his or her career path commensurate with the skills. Putting wrong men at wrong positions may not be good for either the organization or the individual.

Determining Which Key Skills to Foster

There are different aspects in project management skills, under the broad heads of human skills, administrative skills and leadership skills. It would not be appropriate or even necessary to impart training in all the aspects. The organization must make an effort to identify the key skills that the individual should be trained in for successful management of the project.

2.8 IMPORTANT TRAITS OF A PROJECT COORDINATOR

As discussed earlier, a number of studies have been undertaken to address the skill requirements of a project manager. However, very few studies have been undertaken to explore the skill requirements of a person coordinating a construction project. Although a project manager is also responsible for project coordination to an extent, in a large project, the project manager needs assistance from a project coordinator to take care of coordination aspects of the project. Past studies have also recognized the need for a project coordinator, and authors have tried to distinguish the roles and responsibilities of a project manager from that of a project coordinator (Forsberg et al. 1996, Kerzner 2002).

Any typical, large multidisciplinary project needs coordination among the personnel of different departments such as civil, electrical, mechanical, plant and machinery, HVAC, accounts, materials, design, construction method, quality, safety and HRD, totalling at least 12. Besides, a number of designers, subcontractors, construction managers, consultants and specialists from different disciplines are also involved in these projects, making coordination even more complex. To coordinate among the intra-organization departments with the above-mentioned 12 function lines, there will be $12 \times (12 - 1)$ possible coordination routes, i.e., 132 routes, and the difficulty any project manager would have in coordinating resources for sites can easily be imagined. The complexity of the task will be demanding and the project manager may not be able to attend to important project requirements. It is in these circumstances that the role of a project coordinator is considered vital.

In terms of hierarchy, Kerzner (2002) places project coordinator in between project administrator and technical assistants, and locates planning, coordinating, analysing and understanding of organization as the required skills to carry out his responsibility. In the subsequent sections, the skill requirement of a project coordinator has been examined.

Jha (2004) identified 24 attributes (refer to Table 2.1) of a project coordinator. These attributes were then ranked based on the responses of a questionnaire survey conducted among top Indian construction professionals. It was found that distinct differences existed between

Table 2.1 Glossary of project coordinator's traits (Katz 1974, Pettersen 1991, Goodwin 1993, Kerzner 2002, El-Sabaa 2001)

Description of Traits	Definition
Team-building skills	
1. Concern for conciliation	The act of placating and overcoming distrust and animosity
2. Concern for other's ego	Not to remain self-centred; respect other's individuality; regard for other's interest, power and happiness
3. Understanding of human psychology	Understanding the science of the human soul, specifically the systematic or scientific knowledge of the powers and function of the human soul
4. Analytical skills	Ability to look logically at a technical situation
5. Motivating skills	Ability to influence others to contribute to attaining firm's goals
6. Belief in team-playing spirit	The ability to integrate people from many disciplines into an effective team
7. Timeliness	Ability to successfully manage multiple tasks within given time constraints
8. Facilitating skills	Skill to make easy or less difficult the execution of a task
9. Interpersonal skill	Skill to mix in; being friendly; ability to encourage conversation
10. Communication skill	Ability to interact effectively with others at all levels within and outside the organization
11. Technical knowledge of the subject	The capacity to manage the technological innovation and integration of solutions for the success of the project; understanding of complex elements required to effectively complete tasks associated with a given profession
12. Resource-utilization skills	The programme manager needs to work out specific agreements with all key contributors and their superiors on the tasks to be performed and the associated budgets and schedules
Contract-implementation skills	
13. Reliance on systematic approach	Skill to do things methodically and not in a haphazard manner; a series of orderly actions at regular hours
14. Understanding of contract clauses	The power to understand and rationally approach contract clauses
15. Concern for safety, health and welfare of labour and employees	Interest in safety, health and welfare of labour and employees
16. Monitoring skills	Ability to observe something (and sometimes keeping a record of it), showing quick and keen perception
17. Maintaining records	Skill of keeping a diary and keeping notes
18. Follow-up quality	Pursuance, or skill for the continuance, of something begun with a view to its completion
19. Forecasting skills	Skill of predicting or foretelling about the future by looking at the present status
20. Planning skills	This involves the preparation of a project summary plan before the project starts, and requires communication and information processing skills

(Continued)

Table 2.1 Continued

Description of Traits	Definition
Project-organization skills	
21. Relationship with client, consultant and contractor	Skill in maintaining good human relations with client, consultant and other contractors
22. Coordination for achieving quality	Ability to manage production of goods or services within a clearly defined set of expectations
23. Liaison skill	Ability to channelize communication between groups
24. Knowledge of project finance	Ability to understand financial statements and financial ratios, and to deal with accounting firms and financial institutions

the attributes of the project coordinators in projects that were considered successful and that were considered failures. Project coordinators of successful projects were found to excel in certain important attributes such as relationship with client, consultant and contractor; timeliness; technical knowledge of the subject; belief in team-playing spirit; and coordination for achieving quality. Interestingly, the required attributes remain the same for project coordinators whether they work for the contracting organization or the owner's organization.

Jha (2004) further classified the 24 skills needed of a project coordinator in terms of the following three categories.

2.8.1 Team Building Skill

It can be observed from any construction project site that a project coordinator has to carry out his work within limited authority. Unless his team members have confidence in him, things are not likely to work for the coordinator. Traits emerging under this skill group encompass the human relationships skill suggested by Katz (1974) for the project manager. This skill involves the ability to communicate effectively, maintain a harmonious working group and the ability to motivate employees. The ability to motivate employees also falls in human relationships skill as suggested by Katz.

Team building requires a conciliatory approach rather than a confrontational approach. A coordinator needs to show concern for the other person's ego and should have a sound understanding of psychology.

Most importantly, a coordinator must believe in the team spirit; must be able to communicate properly, both through verbal and written communication; and must be proficient in interpersonal skill.

It is to be kept in mind that in the course of fulfilment of his/her duties, a coordinator has to interact with different departments that may not be under his/her direct control, and in such situations, he/she must demonstrate team-building skills and project himself or herself as a team member.

2.8.2 Contract Implementation Skill

Contract-implementation skill is one of the major requirements for a project coordinator. A coordinator is supposed to assist the project manager in fulfilling the contractual promises.

Reliance on a systematic approach and a sound understanding of contract clauses make a project coordinator understand his responsibilities towards fulfilling this duty.

A project coordinator with monitoring and forecasting skills can keep a close watch on schedule and cost of the project, and apprise the project manager of any deviation from the same. Subsequently, with follow-up skill, he can push his team members to correct the deviations to bring the project back on track as far as time and cost requirements are concerned.

Safety, health and welfare of employees are one of the important contract requirements, and the project coordinator's concern for the same cannot be underestimated. Maintaining records of all important events is also an important function and it helps in reducing disputes at a later date.

2.8.3 Project Organization Skill

The project coordinator must be able to perceive the project as a system by keeping a global perspective and not thinking about one aspect at a time. The project coordinator must be good at keeping good working relationships with client and consultants. A project coordinator has to work with many different groups or departments to perform his duties, and this requires a cordial and cooperative relationship, which can be achieved through good interpersonal and communication skills.

The project coordinator must be good at liaison and he should also ensure proper quality of workmanship. The project coordinator must be able to plan, should have requisite knowledge of project finance, and should be able to ensure timeliness.

2.9 ETHICAL CONDUCT FOR ENGINEERS

In day-to-day life, an engineer faces a number of situations in which he has to choose between the well-being of the project he is working on and the well-being of the society at large. An action taken by him may be beneficial for the project but it may endanger the ecosystem. For example, the alignment of a highway may be passing through a dense forest and construction may harm the trees, animal lives, etc. In such cases, he will be in a dilemma regarding the well-being of the project versus the well-being of the entire ecosystem. Should the engineer emphasise on project at the cost of different forms of life and the environment?

In some situations, he may be advised to expedite the construction works even at the cost of compromising on sound engineering practices. As another example, let us take the case of a transit mixer full of concrete reaching the concrete location after initial setting time of the concrete is over. Should the concrete be poured or should it be thrown out? There may be situations in which some of the construction materials have been rejected by the client. In such situations, should you try to use them in the project in a clandestine manner or should you waste the money of your employer?

Consider another situation in which you have to choose, from your project team, between a stronger colleague and a weaker colleague in terms of their positions in the hierarchy. In that situation, let us assume that the stronger colleague is at fault. Should you be aligning with the stronger colleague or with the weaker colleague?

In delicate or confusing situations, the concerned engineer may have the option to consult some experts and act on their recommendations, or he may act according to the wishes of the superior ignoring the ethical part of the problem, or he may himself weigh different arguments for and against, and take a decision.

The decision taken by an engineer has far-reaching consequences and, therefore, he should be even-handed in taking different points of view into consideration. Many temptations come in the way of discharging duties in the course of one's career. The engineer should be able to resist these temptations by controlling his desires in view of the larger public interest and the welfare of mankind.

An engineer should not only be having expertise in his/her professional field, but also be familiar with ethical rules and codes of conduct of his/her profession. Ethical rules and codes of conduct vary across business organizations. What is acceptable in one organization may not be acceptable in another. This may even vary with time and culture. Some of the keywords included in newly drafted ethical rules pertain to environment and sustainable development. The commonly preached ethical rules extracted from American Society of Civil Engineers (ASCE), Japan Society of Civil Engineers JSCE, German Engineering Association, and International Association for Bridge and Structural Engineering (IABSE) are given below:

- The engineer should contribute to the development of the nation and the promotion of human welfare through their professional knowledge. He/she should honour human life without discriminating against cast, creed, social position and religion. He/she should serve the society and should not have any intention adverse to the national development and public welfare. He/she should hold paramount the safety, health and welfare of the public in performance of their professional duties.

- The engineer should aim at development of technology and should strive to improve his/her techniques and put the results into wide practice. He/she should continue their professional development in research field and practical utilization throughout their careers, and should contribute to the engineering society through the publication of their results. He/she should evaluate the works of his/her colleagues in the same way as he/she wants to be evaluated himself/herself.

- The engineer should support the professional and technical societies of his/her disciplines. The engineer should issue public statements only in an objective and truthful manner. He/she should act according to well-balanced reasoning as regards their social, ecological and economic responsibilities. He/she should endeavour to conserve limited natural resources and reduce harmful pollution. He/she should obtain a comprehensive education, learn how to work with complex relationships, and acquire the ability to participate in interdisciplinary discussion. The engineer needs to develop and promote a sustainability ethics.

- The engineer should act in a manner to uphold and enhance the honour, integrity and dignity of the engineering profession above the economic advantages. He/she should not make excessive design or excessive cost-cutting at the cost of safety. He/she should not damage the prestige of the profession.

- The engineer should be honest and impartial, and should not behave suspiciously. He/she should avoid associating himself/herself with dubious work or lending his/her name to such work. He/she should perform services only in areas of their competence.

2.10 FACTORS BEHIND THE SUCCESS OF A CONSTRUCTION ORGANIZATION

Construction is a risky business and a large number of organizations go bankrupt every year. The construction industry is often characterized as having low entry barriers. So, every year more and more construction industries are entering the market and then, after a few years, go bankrupt because of one or the other reason.

Traditionally, construction companies completing projects in time within an established budget and meeting required quality considerations have been considered as successful companies. But if construction projects are successful, then it is not always necessary that the construction organization will also be successful. It can even fail and go bankrupt in near future. So, a shift in emphasis from project success to corporate success should be examined for construction organizations. A number of studies are needed to understand the sector completely. Although a great deal of work has been carried out in the area of construction project success factors, very few studies are reported about understanding the success parameters applicable in case of construction organizations. In this section, we present the findings of the study conducted by Garg (2007) to identify and evaluate the parameters affecting the success of a construction organization.

A total of 38 success parameters were listed out through literature survey and interviews with selected professionals from construction industries, and a set of questions was formulated. Pilot survey was then undertaken and necessary modifications in the questionnaire were carried out. A six-point scale was used in the questionnaire, intended to measure the level of importance of organizational success variables. In the scale, 0 represented 'Insignificant/Not at all required'; 1 represented 'marginal';2 represented 'significant';3 represented 'desirable'; 4 represented 'essential'; and 5 represented 'Vital/Most critical variable'.

A total of 600 questionnaires were mailed to top Indian construction industry professionals covering large- and medium-size organizations, in both private and government sectors, selected randomly from across the country. A total of 111 completed responses were collected, including 63 responses from government and 48 from private professionals. From the six-point scale used in the questionnaire, relative importance index (RII) was calculated for each of the success variables in order to know their rank based on their criticality. The RII was evaluated using the following expression:

$$\text{Relative importance index (RII)} = \frac{(\Sigma w)}{(A \times N)} \qquad (2.1)$$

where w is the weight given to each variable by the respondents and ranging from 0 to 5,
A is the highest weight (i.e., 5 in this case) and N is the total number of respondents.

Table 2.2 shows the success variables arranged in the descending order of RII values. The highest RII indicates the most critical variable with rank 1; the next indicates the next most critical variable with rank 2; and so on.

Table 2.2 Success variables for construction organization

S. No.	Organization success variables	RII	Rank
1.	Supportive top management	0.863	1
2.	Acquiring proper and adequate equipment for construction	0.863	1
3.	Acquiring new and up-to-date technology for construction	0.850	3
4.	Availability of resourceful project managers/project leaders	0.848	4
5.	Effectiveness of the project management in improving schedule, cost and quality of the construction output	0.839	5
6.	Developing team-working approach	0.832	6
7.	Implementing an effective quality assurance and control programme	0.828	7
8.	The clarity of customer/client requirements	0.821	8
9.	Achievement of goals set by the organization	0.818	9
10.	Providing training to staff to enable them to perform their roles	0.816	10
11.	Assessment of the 'value' for money and 'quality' for customers	0.812	11
12.	Competitive strategies used by the organization which keep it apart from others in a market	0.810	12
13.	Effectiveness of human resource (HR) in its functioning (such as recruitment and monitoring of performance)	0.805	13
14.	The clarity of the responsibilities towards organization's success	0.803	14
15.	Constant motivation to staff by the managers	0.803	14
16.	Regular review of the feedback from employees, customers and all stakeholders, and continuous monitoring of their satisfaction levels	0.789	16
17.	Developing an appropriate organization structure	0.789	16
18.	Regular monitoring of organizational strengths, weaknesses, opportunities and threats	0.781	18
19.	Identifying and continuously developing management skills and linking them to the organization's objectives	0.780	19
20.	Adopting latest project-execution techniques	0.778	20
21.	Regular and systematic evaluation for operational staff	0.767	21
22.	Presence of effective risk-management capabilities in the organization	0.756	22
23.	Benchmarking the procedures and construction productivity, and improving them	0.756	22
24.	Developing relevant technology	0.747	24
25.	Clear identification of the critical actions that are to be taken in business	0.747	24
26.	The extent of knowledge the employees have regarding the key purpose of the organization	0.745	26
27.	Relationship with customers, material suppliers and subcontractors	0.740	27
28.	Considering the scope for growth in organization and the need for change (if any), and balancing against their costs	0.738	28

(*Continued*)

Table 2.2 Continued

S. No.	Organization success variables	RII	Rank
29.	Award system by the organization to show appreciation for good work	0.735	29
30.	Flexibility of the organization towards implementation of latest trends	0.733	30
31.	Proper selection of projects/type of projects (keeping in mind political, environmental, technical and social issues)	0.729	31
32.	Regular and systematic evaluation for administrative staff	0.729	31
33.	Arranging and managing additional sources of finance	0.720	33
34.	Capability of organization in pulling out of unprofitable regions	0.720	33
35.	Dealing with organizational, departmental and individual conflicts openly rather than ignoring them	0.706	35
36.	Offering the option of variable compensation to employees, and the same should be decided on the basis of performance as the main criterion	0.693	35
37.	Proper selection of clients	0.691	37
38.	Flexibility of working hours	0.544	38

The columns 2 and 3 in Table 2.2 show the most important variables to be *supportive top management* with RII = 0.863, followed by *acquiring proper and adequate equipment for construction, acquiring new and up-to-date technology for construction, availability of resourceful project managers/project leaders,* and *effectiveness of the project management in improving schedule, cost and quality of the construction output*. The importance of the variable *supportive top management* can be gauged from the fact that the dynamic, project-based nature of the construction industry results in extreme fluctuations in organizations' workload and requires teams to form, develop and disband relatively quickly. Thus, the importance of efficient management of employee resourcing activities cannot be understated. Moreover, timely help from top management in getting the resources or arriving at critical decisions can have far-reaching implications on the performance of the organization as a whole.

The reason why the variables *acquiring proper and adequate equipment for construction* and *acquiring new and up-to-date technology for construction* are given importance is that most of the times they affect the schedule and cost of the projects and the quality of the constructed facility, and these, in turn, affect the performance of a construction organization. It is due to the same reason that the variable *effectiveness of project management in improving schedule, cost and quality of the construction output* is given highrank (in top 5).

The success variable, *availability of resourceful project managers/project leaders*, getting high rank (in top 5) implies its importance for an organization. Leadership seeks to lay the foundation for the transformation of the construction organization, particularly in relation to the creation of a culture for continuous improvement and customer/market focus.

REFERENCES

1. Adams, J.R. and Barndt, S.E. 1978, 'Organizational life cycle implication for major projects', *Project Management Quarterly*, 9(4), pp. 32–39.

2. Anderson, S.D., 1992, 'Project quality and project managers', *International Journal of Project Management*, 10(3), pp. 138–144.
3. El-Sabaa, S., 2001, 'The skills and career path of an effective project manager', *International Journal of Project Management*, 19(2001), pp. 1–7.
4. Enshassi, A., 1997, 'Site organization and supervision in housing projects in the Gaza Strip', *International Journal of Project Management*, 15(2), pp. 93–99.
5. Fryer, B., 1979, 'Management development in the construction industry', Building Technology and Management, May, pp. 16–18.
6. Gaddis, P.O., 1959, 'The project manager', *Harvard Business Review*, May/Jun, pp. 89–97.
7. Garg, S., 2007, 'Factors for the success of a construction organization', Masters dissertation, Indian Institute of Technology, Delhi, India.
8. Goodwin, R.S.C., 1993, 'Skills required of effective project managers', *Journal of Management in Engineering*, ASCE, 9(3), pp. 217–226.
9. Goodwin, R.S.C., 1993, 'Skills required of effective project managers', *Journal of Management in Engineering*, ASCE, 9(3), pp. 217–226.
10. http://www.en.wikipedia.org, viewed on 14 September 2009 (for information on Indian construction industry).
11. http://www.History1700s.com (for definition of corporation by Chief Justice John Marshall in 1819) viewed on 07.03.2010.
12. Jha, K.N., 2004, 'Factors for the success of a construction project: An empirical study', Doctoral dissertation, Indian Institute of Technology, Delhi, India.
13. Katz, D. and Kahn, R.L., 1978, *The Social Psychology of Organisations*, John Wiley, USA.
14. Katz, R.L., 1974, 'Skills of an effective administrator', *Harvard Business Review*, September/October, pp. 90–102.
15. Kerzner, H., 2004, *Project Management: A Systems Approach to Planning, Scheduling and Controlling*, CBS Publishers & Distributors, New Delhi, India.
16. Laufer, A. Woodward H. and Howell, G.A., 1999, 'Managing the decision-making process during project planning', *Journal of Management in Engineering*, ASCE, 15(2), pp. 79–84.
17. Newcombe, R., Langford, D. and Fellows, R., 1990, *Construction management organization systems*, Mitchell, London, p. 219.
18. Odusami, K.T., 2002, 'Perceptions of construction professionals concerning important skills of effective project leaders', *Journal of Management in Engineering*, ASCE, 18(2), pp. 61–67.
19. Ofori, G., 1994, 'Construction industry development: Role of technology transfer', *Construction Management and Economics*, 12, pp. 379–392.
20. Ogunlana, S.O., Li, H. and Sukhera, F.A., 2003, 'System dynamics approach to exploring performance enhancement in a construction organization', *Journal of Construction Engineering and Management*, 129(5), pp. 528–536.

21. Pettersen, N., 1991, 'Selecting project managers: An integrated list of predictions', *Project Management Journal*, 22(2).
22. Pettersen, N., 1991, 'What do we know about the effective project manager?', *International Journal of Project Management*, 9(2), pp. 99–104.
23. Poirot, J.W., 1991, 'Organising for quality: Matrix organization', *Journal of Management in Engineering*, ASCE, 7(2), pp. 178–186.
24. PMBOK, 2001, *A guide to the project management body of knowledge*, Project Management Institute, http//www.pmi.org.
25. Spitz, C.J., 1982, 'The project leader: A study of task requirements, management skills and personal style', Doctoral dissertation, Case Western Reserve University, USA.
26. Stuckenbruck, L.C., 1976, 'The ten attributes of the proficient project manager', Proceedings of the 8th Project Management Institute Seminar/Symposium, Montreal, Canada.
27. Tarricone, P., 1992, 'Portrait of a Manager', *Civil Engineering*, ASCE, 62(8), pp. 52–54.
28. Tenah, K.A., 1986, 'Management level as defined and applied within a construction organization by some US contractors and engineers', Butterworth & Co. (Publishers) Ltd, 4(4), pp. 195–204.
29. Thomas, R., Keating, J.M. and Bluedorn, A.C., 1983, 'Authority Structures for Construction Project Management', *Journal of Construction Engineering and Management*, 109(4), pp. 406–422.
30. Uher, T.E. and Toakley, A.R., 1999, 'Risk management in the conceptual phase of a project', *International Journal of Project Management*.

REVIEW QUESTIONS

1. State whether True or False:

   a. The amount of authority given to a project manager can be related with the organizational structure.

   b. Project manager has maximum authority in matrix organization structure.

   c. Project manager has least authority in pure project organization.

   d. Authority is the power granted to individuals so that they can make final decisions for others to follow.

   e. Responsibility is the obligation incurred by individuals in their roles in the formal organization in order to effectively perform assignments.

   f. Accountability is the state of being totally answerable for the satisfactory completion of specific assignments.

   g. The main advantage of sole proprietorship is its unlimited personal liability for debts and the ease in obtaining a large sum of money for the business.

   h. The major advantage of partnership organization is that of impermanence, the ease in transferring ownership, and its limited liability.

   i. A corporation is a real being, visible, tangible and existing as per the law.

j. Private limited company is a type of incorporated firm and offers limited liability to its shareholders.
k. Public limited company is an incorporated, limited liability company having tradable shares, etc.
l. Limited liability companies are a combination of corporation and partnership.
m. Project management can be divided into three categories—matrix, functional and project forms of organization.
n. Matrix organization has three different forms—functional matrix, balanced matrix and project matrix.
o. Project manager's effectiveness can be ensured by three essential skills—technical skills, human relationship skills and conceptual skills.
p. Ethical conduct and professionalism are not required for the success of a construction project.
q. Important traits required for project coordination are—team-building skills, contract-implementation skills and project-organization skills.

2. What do you mean by a construction company? Distinguish between authority, responsibility and accountability.
3. Discuss different forms of business organization firms.
4. Comment on the joint venture form of businesses. What are the merits and demerits of such forms?
5. Why is structuring required for a construction company? Discuss different structures of a construction company.
6. Differentiate between centralized and decentralized forms of a construction company.
7. What are the key features of the matrix organizational structure?
8. What are the important factors for choosing a particular project management authority structure?
9. Distinguish between vertically and horizontally structured organizations.
10. The common bond between individuals and their supervisors will enhance cooperation and effectiveness of an individual group. Discuss in the context of the functional organization.
11. What do you mean by steady flow of projects?
12. Why are pure project organizations least common nowadays?
13. Discuss partial projectization.
14. Matrix organizations are a blend of functional and projectized characteristics. Explain.
15. Discuss different hierarchies of management level.
16. Discuss different functions and skills required of a project coordinator.
17. Discuss strategies for enhancing the performance of a project manager.
18. Discuss the important traits of a project coordinator.
19. What are the essential ethical conducts expected from project engineers?
20. Define organization, management, levels and roles of management, and types of information required at different levels of management.

3

Construction Economics

Introduction, economic decision-making, time value of money, cash-flow diagrams, using interest tables, evaluating alternatives by equivalence, effect of taxation on comparison of alternatives, effect of inflation on cash-flow, evaluation of public projects: discussion on benefit–cost ratio

3.1 INTRODUCTION

The relationship between engineering and economics is very close, and has in fact been heightened in the present scenario, wherein engineers are expected to not only create technical alternatives but also evaluate them for economic efficiency. Highlighting the economic aspect of engineering decision-making, Wellington (Wellington 1887, cited in Riggs et al. 2004) in his book *The Economic Theory of the Location of Railways* wrote, '… it would be well if engineering were less generally thought of, and even defined, as the art of constructing. In a certain important sense it is rather the art of not constructing; or, to define it rudely but not ineptly, it is the art of doing that well with one dollar which any bungler can do with two …'.

With the growing maintenance cost, especially in the infrastructure sector, it has been realized that not only the initial cost, but also the overall life-cycle cost of a project should be taken into account when evaluating options. In fact, in the case of construction projects, economics affects decision-making in many ways—from the cost of materials to purchase and scrapping of equipment, bonus and/or penalty clauses, and so on. It is, therefore, extremely important that engineers and construction managers have a working knowledge of economic principles, terminology and methods.

The effort in this chapter is directed to provide some important tools relating to different aspects of economic decision-making. In the construction industry, decisions involved include deferred payments or receipts, payments (or receipts) in installments, etc. Thus, effort is directed here to discuss the concepts of value of money and cash-flow diagrams in some detail. Further, in order to be able to compare alternatives, it is important that they are reduced to a common platform. Equivalence is one of the methods commonly used. All these concepts have been discussed at some length in the following sections. The chapter also briefly discusses the use of interest tables, relationship between concepts such as inflation and taxation, and economic decision-making in the context of construction projects.

3.2 ECONOMIC DECISION-MAKING

There are various situations—such as (a) comparison of designs or elimination of over–design; (b) designing for economy of production/maintenance/transportation; (c) economy of selection; (d) economy of perfection; (e) economy of relative size; (f) economy and location; and (g) economy and standardization and simplification—in which an engineer has to take a decision among the competing alternatives. These are referred to as *problems of present economy*. In such situations, the decision maker may not consider the time value of money.

The three most common methods of evaluation—'out-of-pocket commitment' comparison, 'payback period', and the 'average annual rate of return' methods—in which the time value of money is not considered are discussed in the following sections.

3.2.1 Out-of-Pocket Commitment

Suppose a pre-cast concrete factory has to produce 100,000 railway sleepers per year. An economic choice has to be made between using steel formwork and wooden formwork. The life of steel formwork is estimated to be one year, while that of wooden formwork is one month. The costs of preparing one set of steel formwork and one set of wooden formwork are ₹ 400,000 and ₹ 50,000, respectively. It is further estimated that the labour costs for fixing and removing the steel and wooden formwork are ₹ 10 and ₹ 9 per sleeper, respectively. Now, this is a situation wherein we may not need to consider time value of money, and use the lowest out-of-pocket commitment criteria to choose the most economical alternative.

The out-of-pocket commitment is the total expense required for an alternative. For example, the out-of pocket commitment for steel formwork option would be the sum of total labour cost incurred for production of one lakh sleepers/year plus the cost of the steel formwork/year, that is, ₹ 100,000 × 10 + ₹ 400,000 = ₹ 1,400,000. Similarly, the out-of-pocket commitment for wooden formwork option is = ₹ 9 × 100,000 + ₹ 50,000 × 12 = ₹ 1,500,000.

Since the out-of-pocket commitment for steel formwork is lesser than that for wooden formwork, the decision would be to choose the 'steel formwork' option.

3.2.2 Payback Period

The payback period for an investment may be taken as the number of years it takes to repay the original invested capital, and serves as a very simple method for evaluation of projects and investments. Though this method does not take into account the cash flows occurring after the payback period, given the ease of computation involved, the method is widely used in practice, and it is understood that the shorter the payback period, the higher the likelihood of the project being profitable. In other words, upon comparing the two projects or alternatives, the option with the shorter payback period should be selected.

For example, let a contractor have two brands of excavators, A and B, to choose from. Both the brands are available for a down payment of four lakh rupees. Both brands can be useful for a period of four years. Brand A is estimated to give a return of ₹ 50,000 for the first year, ₹ 150,000 for the second year, and ₹ 200,000 for the third and fourth year. Brand B, on the other hand, is expected to give a return of ₹ 150,000 for all the four years. The payback period for both the brands is calculated in the following manner.

CONSTRUCTION ECONOMICS | 51 |

The payback period for Brand A = 3 years, as the initial investment of four lakh rupees is recovered in three years (50,000 + 150,000 + 200,000 = 400,000). The method does not consider the returns after the payback period. Hence, in this case the return of fourth year (₹ 200,000) is simply overlooked.

For Brand B, the return is ₹ 300,000 up to the end of second year and in the third year it equals ₹ 450,000. Thus, the investment amount of ₹ 400,000 is recovered somewhere between second year and third year, which can be found out by interpolation.

$$\text{Hence, the payback period for Brand B} = 2 + (3-2) \times \frac{(400{,}000 - 300{,}000)}{(450{,}000 - 300{,}000)}$$

$$= 2.67 \text{ Years} = 2 \text{ Years and 8 months.}$$

Here also, as in the first case we neglect the return that is expected beyond the payback period.

Hence, it is beneficial to buy Brand B excavator as it has a lesser payback period.

Although we have taken the equal initial investment and time period for the two alternatives, the payback period can also be used when these values are not same. That is, we can use payback period for problems involving unequal initial investment and service life.

3.2.3 Average Annual Rate of Return

In this method, the alternatives are evaluated on the basis of only the average rate of return as expressed in terms of a percentage (of the original capital). Since this method does not distinguish the period at which transactions take place, or the timings of cash flow, it is not possible to account for the concept of 'time value of money' in this approach. We use the previous example (used in payback period comparison method) to illustrate the computations involved.

$$\text{The average annual return from Brand A} = \frac{(50{,}000 + 150{,}000 + 200{,}000 + 200{,}000)}{4}$$

$$= \frac{600{,}000}{4} = 150{,}000$$

Here, 4 is the number of years. The above-average annual return is converted into percentage to get the average annual rate of return.

Average annual rate of return for Brand A in % = $\frac{150{,}000}{400{,}000} \times 100 = 37.5\%$.
Here, 400,000 is the original invested capital.

$$\text{The average annual return from Brand B} = \frac{(150{,}000 + 150{,}000 + 150{,}000 + 150{,}000)}{4}$$

$$= \frac{6000{,}000}{4} = 150{,}000$$

The average annual rate of return for equipment B in % = $\frac{(150{,}000)}{(400{,}000)} \times 100 = 37.5\%$.

This is also the same as Brand A. Thus, as far as average annual rate of return is concerned, both the brands are equivalent.

However, given a choice between the two brands A and B, which one would you prefer? You guessed it right! We would go for Brand B. This is because we are getting higher returns in the initial years for Brand B when compared to Brand A. The basis for taking such decisions is inherent in the 'time value of money' concept, which is discussed in the next section.

3.3 TIME VALUE OF MONEY

Although monetary units (rupee, dollar, etc.) serve as an excellent tool to compare otherwise incomparable things, e.g., a bag of cement and tons of sand, we know that the real worth of a certain amount of money is not invariant on account of factors such as inflation, dynamic interactions between demand and supply, etc. Now, time value of money is a simple concept that accounts for variations in the value of a sum of money over time.

In most decisions, the change in the value of money needs to be accounted for, and this chapter seeks to present simple tools of analysis to help a construction engineer make logical decisions based on sound economic principles. The most important principle involved is that of 'interest', which could be looked upon as the cost of using capital (Riggs et al. 2004), or the (additional) money (in any form) paid by the borrower for the use of funds provided by the lender. According to Riggs et al. (2004), the interest represents the earning power of money, and is the premium paid to compensate a lender for the administrative cost of making a loan, the risk of non-repayment, and the loss of use of the loaned money. A borrower pays interest charges for the opportunity to do something now that otherwise would have to be delayed or would never be done.

In simple terms, interest could be simple or compound, where in the former case, the interest component does not attract any interest during the repayment period, whereas in the latter case, the interest amount itself also attracts further interest, as we know from traditional arithmetic. The concept is explained briefly in the following paragraph to illustrate some other related ideas.

Let us consider the following statement by a bank: Interest on the deposit will be payable at the rate of eight per cent compounded quarterly. Here, the period of compounding represents the interval of time at which interest will be added to the principal, and the revised principal used to calculate the interest on the next time step. Thus, in a year, for quarterly compounding of interest, four periods (of three months each) should be considered. Now, if the principal amount was ₹ 100, at the end of the first three-month period, the new principal (to calculate interest on the second quarter) should be taken as = 100 + 100 × $\frac{0.08}{4}$ = ₹ 102.00, whereas the principal amount for the third quarter would be 102 + 102 × $\frac{0.08}{4}$ = ₹ 104.04, and so on. At the end of one year, the principal would become ₹ 108.24. In other words, it can be stated that the value of ₹ 100 changes to ₹ 108.24 under the given conditions of eight per cent nominal interest and quarterly compounding.

Now, in the above example the eight per cent is sometimes referred to as the nominal rate of interest, which is the annual interest for a one-year period with no compounding. From the above example, it is clear that ₹ 8.24 can be seen as the interest amount attracted by ₹ 100 in a one-year period under the given rate (eight per cent) and condition of quarterly compounding. This modified rate of interest is sometimes referred to as the effective rate of interest (i_{eff}), which in addition to the nominal rate of interest (i_{nom}) also depends on the period of compounding. One can work out that the effective rate of interest taking i_{nom} to be 5 per cent and 10 per cent, will be 5.09 per cent and 10.38 per cent, respectively, for quarterly compounding, and 5.12 per cent and 10.47 per cent for monthly compounding. In general, given the number of periods in a year to be taken for compounding (m) and the nominal rate of interest (i_{nom}), the effective rate of interest, i_{eff}, can be calculated as below:

$$i_{eff} = \left(\left(1 + \frac{i_{nom}}{m \times 100}\right)^m - 1\right) \times 100 \qquad (3.1)$$

3.4 CASH-FLOW DIAGRAMS

Any organization involved in a project receives and spends different amounts of money at different points in time, and a cash-flow diagram is a visual representation of this inflow and outflow of funds. Although in practice this inflow and outflow does not necessarily follow any pattern, it is sometimes assumed that all transactions (inwards or outwards) take place either at the beginning or end of a particular period, which may be a week, a month, a quarter, or a year, simply to simplify the analysis. In other words, if it is decided that all transactions in a month will be recorded as having occurred on the last day of that month, or the first day of the next month, either approach can be followed provided the system is consistently followed.

In a cash-flow diagram (see Figure 3.1), usually time is drawn on the horizontal (X) axis in an appropriate scale, in terms of weeks, months, years, etc., whereas the Y-axis represents the amount involved in the transaction, with the receipts and disbursements being drawn on the positive and negative side, respectively, of the Y-axis. While scale is maintained for the time axis, the representation on the Y-axis is sometimes not to scale; however, effort should be made to maintain a semblance of balance. Thus, it is a practice to actually write the amount of each transaction next to the arrow. Some of the other aspects related to drawing and interpreting a cash-flow diagram are explained by way of the following example.

Example 3.1
The details of the financial transactions during the months of April, May and June for M/s Alpha Industries are given in Table 3.1. Draw a neat cash-flow diagram for the transactions

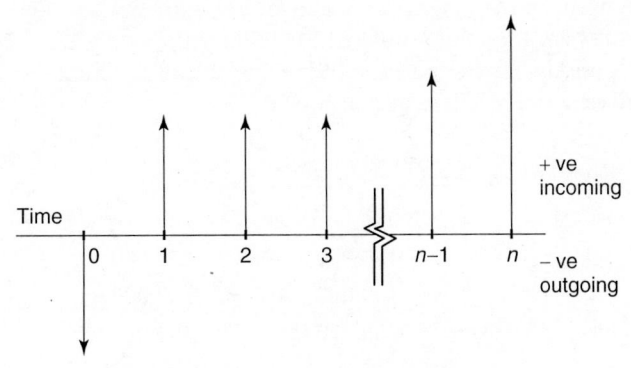

Figure 3.1 Typical cash-flow diagram

Table 3.1 Details of transactions carried out by M/s Alpha Industries for April to June

Date	Description	Amount (₹)
April 5	Receipt for running account bill # 9	150,000
April 10	Salary disbursement	80,000
April 16	Payment for supply of aggregates	20,000
April 21	Payment for supply of stationery for office use	5,000
May 7	Receipt for running account bill # 10	180,000
May 10	Salary disbursement	80,000
May 28	Payment for supply of cement	50,000
June 6	Receipt for running account bill # 11	250,000
June 10	Salary disbursement	80,000
June 16	Payment for supply of structural steel	100,000
June 28	Payment of rent of premises for July and August	100,000

using a month as a single unit, showing all transactions during a month at the end of that month.

Now, since a month is the basic unit to be used, the given transactions can be summarized as given in Table 3.2.

Now, this summary can be represented as a cash-flow diagram as shown in Figure 3.2 or Figure 3.3, depending on whether only the net transaction {algebraic sum of the receipt (taken positive) and the disbursement (taken negative)} for the month is to be shown or whether both the inflow and outflow are to be shown for each month. Obviously, no matter what convention is followed, it is clear that the total receipts and expenditures for the said three months are ₹ 580,000 and ₹ 515,000, respectively.

In construction economics, we come across two main types of problem: income expansion and cost reduction. Correspondingly we can distinguish the cash-flow diagrams also into revenue-dominated and cost-dominated cash-flow diagrams. In the former case, incomes or savings are emphasised, while the latter diagrams largely concentrate on periodic costs or expenditures. Illustrative representations of the diagrams are shown in Figure 3.4 and Figure 3.5.

Cash-flow diagrams are frequently utilized for finding the equivalence of money paid or received at different periods, for the purpose of comparison of different alternatives and for a host of other objectives, as we shall see in due course.

Table 3.2 Summary of transactions for April to June

Month	Receipts (₹)	Expenditures (₹)
April	150,000	105,000
May	180,000	130,000
June	250,000	280,000
Total	**580,000**	**515,000**

CONSTRUCTION ECONOMICS | 55

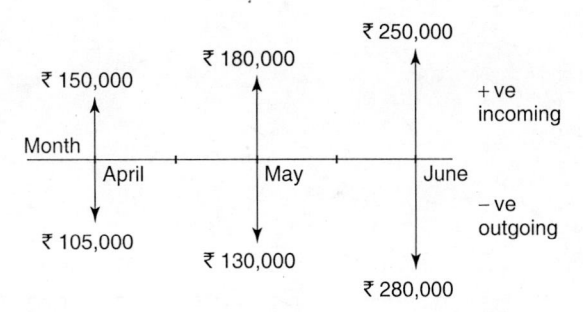

Figure 3.2 Cash-flow diagram for example problem (Alternative 1)

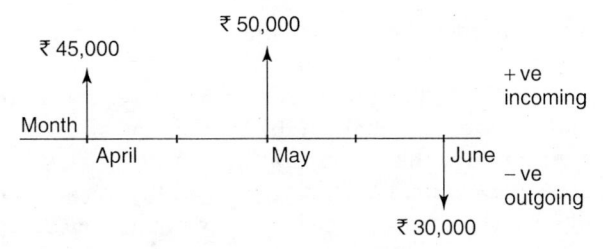

Figure 3.3 Cash-flow diagram for example problem (Alternative 2)

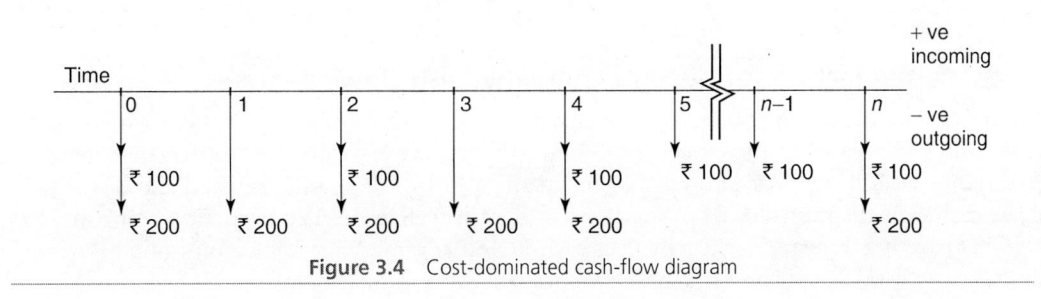

Figure 3.4 Cost-dominated cash-flow diagram

Clearly, as one party spends an amount of money, another party has to earn it, and it is inevitable that the cash-flow diagram would depend on which perspective is taken. Thus, an entry corresponding to an expenditure of ₹ 100,000 for the supply of structural steel on June 16 (refer to Example 3.1) would figure as a receipt in the cash-flow diagram drawn for the

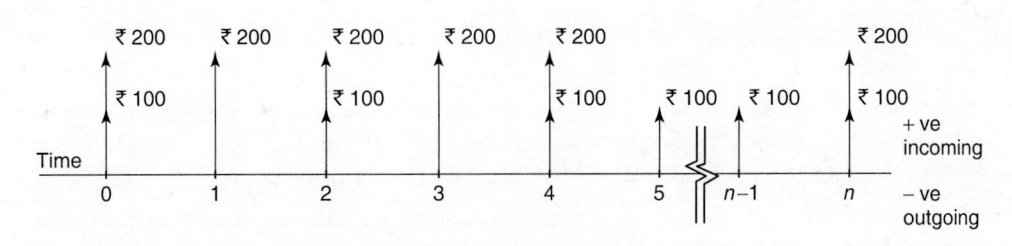

Figure 3.5 Revenue-dominated cash-flow diagram

Figure 3.6 Cash-flow diagrams from lender's as well as borrower's perspectives

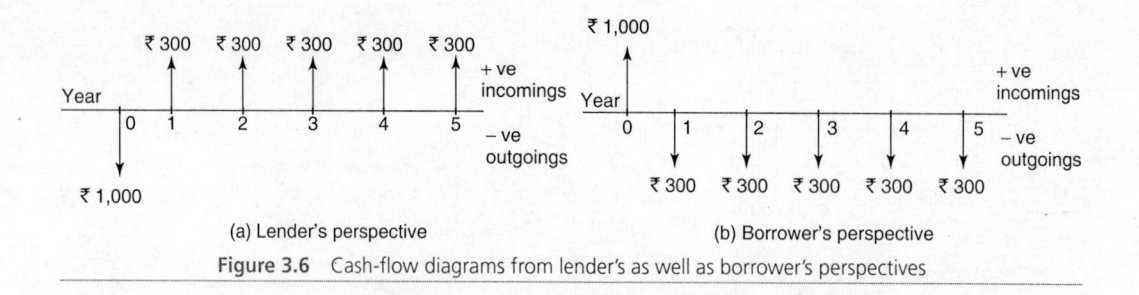

supplier of structural steel. Another illustrative example showing the difference in representation of cash-flow diagrams from different points of view is given below.

Suppose a lender has lent ₹ 1,000 to a borrower (at t = 0), and is expected to get this money back in equal instalments of ₹ 300 payable at the end of every year for a five-year period. The cash-flow diagram for this transaction from both lender's perspective and borrower's perspective is drawn in Figure 3.6. It may be noted that repayment of a ₹ 1,000 loan is amounting to ₹ 1,500, which is expected taking into account the concept of time value of money as already discussed above, and the computation of rates of interest and compounding periods under conditions such as these (payment in instalments) are discussed in subsequent sections.

3.4.1 Project Cash-flow and Company Cash-flow Diagrams

We have discussed the problems wherein we were given information on the cash transactions and the period of their occurrence. In practice, we have to derive such information from some conditions. Based on the concepts of cash-flow diagram discussed above, we can draw cash-flow diagrams for a construction project as well as for a construction company. In the first case it is referred to as a project cash-flow diagram, while the second one is known as company cash-flow diagram. These two are briefly explained below.

Project Cash-flow Diagram

The project cash flow is basically a graph (pictorial representation) of receipts and disbursements versus time. The project cash flow can be prepared from different perspectives—of contractor, owner, etc. It is usual to represent time in terms of month for project cash-flow diagram. For executing a construction project, a constructor spends and receives the money at different points of time.

In order to draw a project cash-flow diagram, the following details are required.

1. The gross bill value and its time of submission
2. The measurement period. It is usual for contractors to be paid on a monthly basis. The payment can be made fortnightly or sometimes bimonthly as well. These conditions can be found under **'terms of payment'** given in the tender document
3. The certification time taken by the owner. In normal conditions, the owner takes about three to four weeks to process the bill and release the payment to the constructor or the contractor

4. The retention money deducted by the owner and the time to release the retention money
5. The mobilization advance, the plant and equipment advance, and the material advance, and the terms of their recovery
6. The details of cost incurred by the contractor for raising a particular bill value. This break-up of costs should be in terms of labour cost, materials cost, plant and equipment cost, subcontractors cost and project overheads
7. The credit period (delay between incurring a cost and the actual time at which the cost is reimbursed) enjoyed by the contractor in meeting the costs towards labour, materials, plant and equipment, and overheads

For illustrating the project cash-flow diagram, we take up an example in which a month-wise invoice is estimated and this is produced in Table 3.3. The process of preparing the estimate of monthly invoice has been explained later in the text.

Table 3.3 Monthly invoice for the example problem (values in ₹ lakh)

Item description	M1	M2	M3	M4	M5	M6	M7	M8	M9	M10
Earthwork: All soils	0.25	0.75	1.00	1.25	1.25	0.50				
Concrete works		8.00	16.00	16.00	20.00	12.00	8.00			
Formwork		4.20	8.40	8.40	10.50	6.30	4.20			
Reinforcement works		11.25	22.50	22.50	28.125	16.875	11.25			
Brickwork			3.40	5.10	6.80	6.80	6.80	3.40	1.70	
Plastering: All types				1.50	2.250	3.00	3.00	3.00	1.50	0.75
Painting: All types				0.50	0.75	1.00	1.00	1.00	0.50	0.25
Flooring: All types				11.00	16.50	22.00	22.00	22.00	11.00	5.50
Waterproofing works							6.00	6.00		
Aluminium work					3.375	3.375	3.375	3.375		
Electrical work						8.75	8.75	8.75	5.25	3.50
Sanitary and plumbing works				1.00	2.00	3.00	4.00	3.00	3.00	2.00	2.00
Road works						8.00	8.00			
Total	0.25	24.20	52.30	68.25	92.55	92.60	85.375	50.525	21.95	12.00
Cumulative	0.25	24.45	76.75	145.00	237.55	330.15	415.525	466.05	488.00	500.00

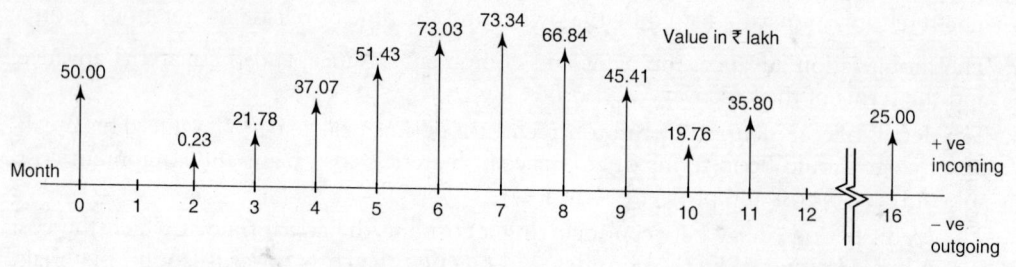

Figure 3.7 The cash-inflow diagram for the contractor for the example problem

Let us assume that the contractor has been awarded this contract on the following terms:
- Advance payment of ₹ 50 lakh, to be recovered in five equal instalments from the third running account bill onwards
- The total cost for the contractor to execute a particular item is 90 per cent of the quoted rate
- The total cost for a particular item consists of labour (20 per cent), material (55 per cent), plant and machinery (10 per cent), subcontractor cost (10 per cent) and project overheads (5 per cent)
- Assume that there is no delay in payment towards labour costs and overhead costs, but a delay of one month occurs in paying to the subcontractors, material suppliers, and plant and machinery supplier
- Retention is 10 per cent of billed amount in every bill. Fifty per cent retention amount is payable after one month of practical completion, while the remaining 50 per cent is payable six months later

For the given condition, the computations for the cash inflow and the cash outflow are given in Table 3.4. The table is self-explanatory. The resulting cash inflow and outflow diagrams are shown in Figure 3.7 and Figure 3.8, respectively. The resultant cash-flow diagram is shown in Figure 3.9.

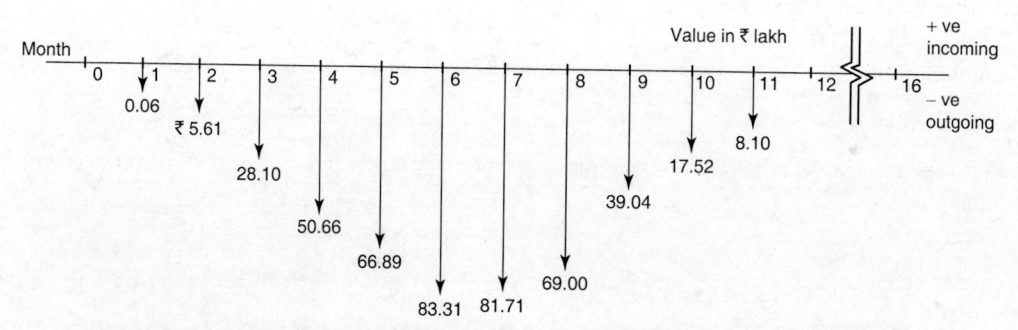

Figure 3.8 The cash-outflow diagram for the contractor for the example problem

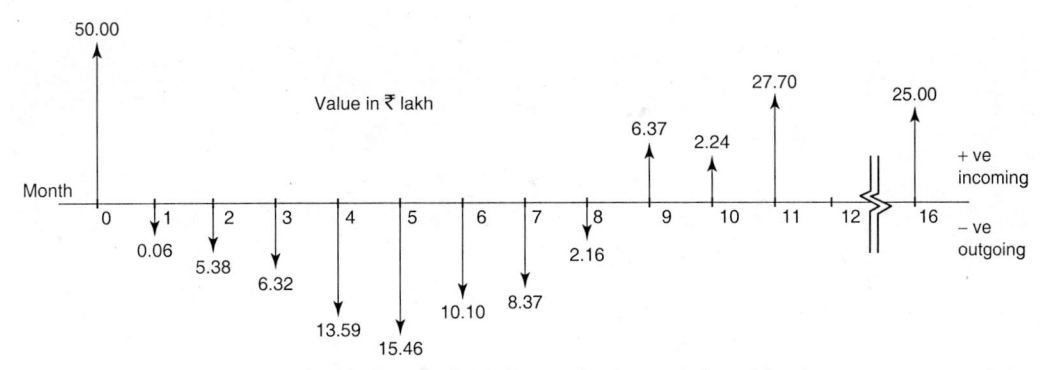

Figure 3.9 The project cash-flow (inflow–outflow) diagram for the example problem (contractor's perspective)

Assuming that the contractor is not receiving any mobilization advance, we can perform a similar analysis as shown in Table 3.5 and compute the cumulative cash flow for the total duration of the project. This is shown in column 3 of Table 3.5. In this table, column 2 summarizes the cumulative cash flow obtained for the condition wherein the contractor has received a mobilization advance of ₹ 50 lakh at time t = 0.

In the previous discussion, we have seen the impact of mobilization advance on the cash-flow position. There are many other variables that affect the cash flow of a contractor. Some of these are briefly discussed in the next section.

Factors Affecting Project Cash-flow

In addition to the mobilization advance, there are many factors that affect the project cash flow and it would be of interest to know such factors and their impact on project cash flow. These factors are—(1) the margin in a project, (2) retention, (3) extra claims, (4) distribution of margin such as front loading and back loading, (5) certification type such as over-measurement and under-measurement, (6) certification period, and (7) credit arrangement of the contractor with labour, material, and plant and equipment suppliers, and other subcontractors.

Margin The margin (profit margin or contribution) is the excess over costs. Thus, the higher the margin in a project, the better it is for the contractor's cash flow.

We illustrate this aspect with the help of a fictitious project of 10 months duration. It is also assumed that ₹ 100 (cost) is incurred every month and the margin is 10 per cent of the cost. The retention amount is five per cent of the bill value and the whole retention money is to be released six months after practical completion (i.e., 10 months). Further, it is also assumed that the owner delays the payment by one month. The cash-flow computations for this fictitious example are shown in Table 3.6.

Similar computations for cash flow can be performed assuming different margin percentages—say, five per cent and 15 per cent—keeping all the conditions same as in the case of 10 per cent margin. The results associated with different margin percentages are summarized in Table 3.7. It can be clearly seen that as the margin increases, there is betterment in the contractor's cash-flow position, keeping all other factors. same.

Retention Retention tends to reduce the margin obtained from a project. In case of very low margins, the retention can even reduce the margin to zero or less. Thus, retention affects

Table 3.4 Computation of cash inflow and cash outflow for the example problem

	Description	t = 0	M1	M2	M3	M4	M5	M6	M7	M8	M9	M10	M11	M15	M16
1	Work done by measurement		0.25	24.20	52.30	68.25	92.55	92.60	85.375	50.525	21.95	12.00			
2	Recoveries towards advance payment of ₹ 50 lakh		0.00	0.00	10.00	10.00	10.00	10.00	10.00	0.00	0.00	0.00			
3	Retention @10% of gross invoice (0.1× row 1)		0.03	2.42	5.23	6.83	9.26	9.26	8.54	5.05	2.20	1.20			
4	Payment due (1−2−3)		0.23	21.78	37.07	51.43	73.30	73.34	66.84	45.47	19.76	10.80			
5	Cash received (one-month delay)			0.23	21.78	37.07	51.43	73.30	73.34	66.84	45.47	19.76	10.80		
6	Release of retention												25.00	0.00	25.00
7	Total incomings	50.00	0.00	0.23	21.78	37.07	51.43	73.30	73.34	66.84	45.47	19.76	35.80	475.00	25.00
8	Cumulative incomings	50.00	50.00	50.23	72.01	109.08	160.50	233.80	307.14	373.97	419.45	439.20	475.00	475.00	500.00
	Outgoings														
9	Total cost (0.9 of 1)		0.23	21.78	47.07	61.43	83.30	83.34	76.84	45.47	19.76	10.80			
10	Cost towards labour @20% of total cost		0.05	4.36	9.41	12.29	16.66	16.67	15.37	9.09	3.95	2.16			
11	Cost towards material @55% of total cost		0.12	11.98	25.89	33.78	45.81	45.84	42.26	25.01	10.87	5.94			
12	Cost towards subcontractors @10% of total cost		0.02	2.18	4.71	6.14	8.33	8.33	7.68	4.55	1.98	1.08			

(Continued)

Table 3.4 Continued

	Description	t = 0	M1	M2	M3	M4	M5	M6	M7	M8	M9	M10	M11	M15	M16
13	Cost towards plant and machinery @10% of total cost		0.02	2.18	4.71	6.14	8.33	8.33	7.68	4.55	1.98	1.08			
14	Project overheads @5% of total cost		0.01	1.09	2.35	3.07	4.16	4.17	3.84	2.27	0.99	0.54			
15	Labour payment (no delay)		0.05	4.36	9.41	12.29	16.66	16.67	15.37	9.09	3.95	2.16			
16	Material suppliers' payment (one-month delay)			0.12	11.98	25.89	33.78	45.81	45.84	42.26	25.01	10.87	5.94		
17	Subcontractors' payment (one-month delay)			0.02	2.18	4.71	6.14	8.33	8.33	7.68	4.55	1.98	1.08		
18	Plant and machinery supplier's payment (one-month delay)			0.02	2.18	4.71	6.14	8.33	8.33	7.68	4.55	1.98	1.08		
19	Overhead cost payment (no delay)		0.01	1.09	2.35	3.07	4.16	4.17	3.84	2.27	0.99	0.54			
20	Outgoings total		0.06	5.61	28.10	50.66	66.89	83.31	81.71	69.00	39.04	17.52	8.10		
21	Cumulative outgoings		0.06	5.67	33.77	84.43	151.32	234.63	316.34	385.34	424.38	441.90	450.00	450.00	450.00
22	Cumulative cash flow (8–21)	50.00	49.94	44.56	38.23	24.64	9.18	−0.84	−9.21	−11.37	−4.94	−2.70	25.00	25.00	50.00

Table 3.5 Cumulative cash-flow with and without mobilization advance

End of month	Cumulative cash-flow with mobilization advance	Cumulative cash-flow without mobilization advance
0	50.00	0.00
1	49.94	−0.06
2	44.56	−5.45
3	38.23	−11.77
4	24.64	−15.36
5	9.18	−20.82
6	−0.84	−20.84
7	−9.21	−19.21
8	−11.37	−11.37
9	−4.94	−4.94
10	−2.70	−2.70
11	25.00	25.00
12	25.00	25.00
13	25.00	25.00
14	25.00	25.00
15	25.00	25.00
16	50.00	50.00

the contractor's cash flow in a negative manner. The higher the retention, the bigger is the cash-flow problem. In order to address this issue, some contractors request the owners to get away with retention amount in lieu of bank guarantee. Thus, instead of cash retention, bank guarantee is in vogue these days. The effect of different percentage of cash retention is explained with the fictitious example discussed in the previous section. Recall that the margin was 10 per cent and retention was five per cent in the original problem. The retention percentage is changed to 0 per cent and 10 per cent from 5 per cent, and the resultant cash flow is computed. These are summarized in Table 3.8.

Extra claims Extra claims in a project may result on account of many reasons such as extra work, changes in quantity and specification and so on. In general claims take a long time to settle. Sometimes it may be settled even after the completion of project. Thus in practice the extra claims tend to worsen the cash flow position of the contractor, though after the claim is settled the contractor may be even able to achieve the original intended level of profit margin.

Distribution of margin This aspect is covered in detail in later chapters. Keeping the overall margin amount same, the margin can be distributed across different items of a project, either in a uniform manner or it can be front- or back-loaded. In case of uniform loading, all the items of the project carry equal margin percentage, while in front-end-rate loading, the items to be executed early in the project carry a higher margin than the later items. By doing so, the

Table 3.6 Cash-flow computations for the fictitious example

S. No.	Description	\multicolumn{12}{c}{Months}
		1	2	3	4	5	6	7	8	9	10	11	15	16
A	Cost	100.00	100.00	100.00	100.00	100.00	100.00	100.00	100.00	100.00	100.00			
B	Margin @10% of cost (A*0.1)	10.00	10.00	10.00	10.00	10.00	10.00	10.00	10.00	10.00	10.00			
C	Value (A + B)	110.00	110.00	110.00	110.00	110.00	110.00	110.00	110.00	110.00	110.00			
D	Retention @5% (C*0.05)	5.50	5.50	5.50	5.50	5.50	5.50	5.50	5.50	5.50	5.50			
E	Monies due (C − D)	104.50	104.50	104.50	104.50	104.50	104.50	104.50	104.5	104.50	104.50	104.50		55.00
F	Money receipt delay by one month	0.00	104.50	104.50	104.50	104.50	104.50	104.50	104.50	104.50	104.50	104.50		55.00
G	Net cash flow (F − A)	−100.00	4.50	4.50	4.50	4.50	4.50	4.50	4.50	4.50	4.50	104.50		55.00

Table 3.7 Summary of net cash-flow for different margins

End of month	Variation in net cash-flow for different margin %		
	5%	10%	15%
1	−100.00	−100.00	−100.00
2	−0.25	4.50	9.25
3	−0.25	4.50	9.25
4	−0.25	4.50	9.25
5	−0.25	4.50	9.25
6	−0.25	4.50	9.25
7	−0.25	4.50	9.25
8	−0.25	4.50	9.25
9	−0.25	4.50	9.25
10	−0.25	4.50	9.25
11	99.75	104.50	109.25
12	—	—	—
13	—	—	—
14	—	—	—
15	—	—	—
16	52.50	55.00	57.50

cash-flow position of the contractor improves, even though the overall margin derived from the project would be the same as that obtained in case of uniform loading. In the back-end-rate loading, the items to be executed later in the project carry higher margin, while the items to be executed early in the project carry lower margin. This has a tendency to increase the negative cash flow and contractors resort to such means in a market where the inflation rate is higher than the interest rate. This way, the contractors hope to recover a substantial amount on account of price escalation.

Certification type
Over-measurement Over-measurement is the device whereby the amount of work certified in the early months of a contract is greater than the amount of work done. This is compensated for in later measurements. Thus, over-measurement has the same effect as front-end loading; it improves the cash in the early stages and reduces the capital lock-up.

Under-measurement Under-measurement is the situation whereby the amount of work certified in the early months of a contract is less than the amount actually done. This has the effect of increasing the negative cash flow for the contractor in a project.

Delay in receiving payment from client The time between interim measurement, issuing the certificate, and receiving payment is an important variable in the calculation of cash

Table 3.8 Summary of net cash-flow for different margins for different retention conditions

End of month	Variation in net cash-flow for different retention conditions (Margin 10%)		
	0%	5%	10%
1	−100.00	−100.00	−100.00
2	10.00	4.50	−1.00
3	10.00	4.50	−1.00
4	10.00	4.50	−1.00
5	10.00	4.50	−1.00
6	10.00	4.50	−1.00
7	10.00	4.50	−1.00
8	10.00	4.50	−1.00
9	10.00	4.50	−1.00
10	10.00	4.50	−1.00
11	110.00	104.50	99.00
12	—	—	—
13	—	—	—
14	—	—	—
15	—	—	—
16	0.00	55.00	110.00

flows. Any increase in the delay in receiving this money delays all the income for the contract, with a resulting increase in the capital lock-up.

Delay in paying labour, plant hires, materials suppliers and subcontractors The time interval between receiving goods or services and paying for these is the credit the contractor receives from his suppliers.

The Company Cash-flow Diagram

For running the projects efficiently, a company usually maintains a head office and a number of regional offices or branch offices. The expenses incurred and the revenues generated by these offices are commonly known as head office outgoings or incomings.

Some examples of head office outgoings would be rents, electricity charges, water charges, telephone and tax bills, hire charges for office equipment, payment to shareholders, taxes, etc. The examples of head office incomings would be claims made in past projects, realization of retention money not settled for past projects, etc.

A construction company executes a number of projects at any time. The company cash-flow diagram is an aggregation (sum) of all the outgoings and incomings for all the projects that the company is executing, besides the head office outgoings and incomings.

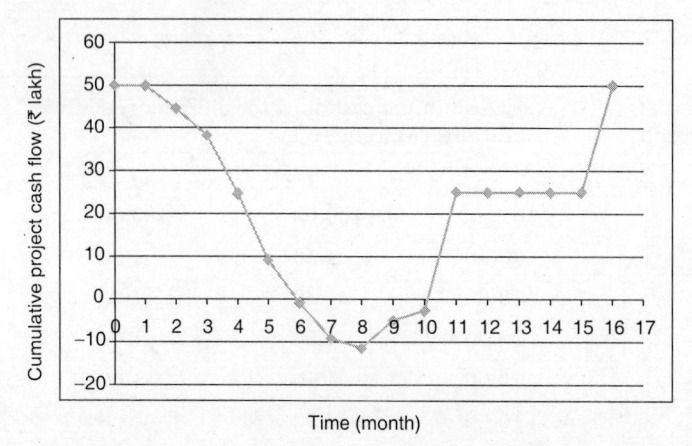

Figure 3.10 Cumulative cash-flow curve with mobilization advance

3.4.2 Using Cash-flow Diagrams

Determining Capital Lock-up

While estimating the effect of margin and retention money on contractor's cash-flow diagram, one can notice that the contracting company sometimes faces negative cash flow in the early stages of the project. This negative cash flow experienced in the early stages of projects represents locked-up capital that is either supplied from the contracting company's cash reserves or borrowed. If the company borrows the cash, it will have to pay interest charged to the project; if the company uses its own cash reserves, it is being deprived of the interest-earning capability of the cash and should, therefore, charge the project for this interest loss. A measure of the interest payable is obtained by calculating the area between the cash-out and the cash-in.

The negative cash flow indicates that the contractor has to mobilize this much fund to execute the project. The area under the 'negative cash flow' period is used to calculate the financing charges for the project by the contractor. The total area would have a unit of ₹ x months and is also known as *captim*, standing for *capital × time*. In order to calculate the interest charges for financing the project, we use the captim in the following manner:

Interest on the capital required for the project

$$= \frac{(Captim \text{ in } ₹ \text{ } Month \times interest \text{ } charges \text{ } per \text{ } annum \text{ } \%)}{12 \times 100} \quad (3.2)$$

For example, if the captim value is 10,000 ₹ month and the interest rate is 12 per cent per annum, the interest charges for financing the project would be $= \dfrac{10,000 \times 12}{12 \times 100} =$ ₹ 100.

The result of Table 3.5 is shown graphically in Figure 3.10 and Figure 3.11. It can be noticed that when the contractor has received the mobilization advance, there is no negative cash-flow situation in the early stages of the project—a very happy situation, indeed. In fact, there is negative cash-flow situation only for a brief period between month 6 and month 10.

Turning to Figure 3.11, which shows the cumulative cash-flow situation corresponding to the case when there is no mobilization advance received by the contractor, we find that the

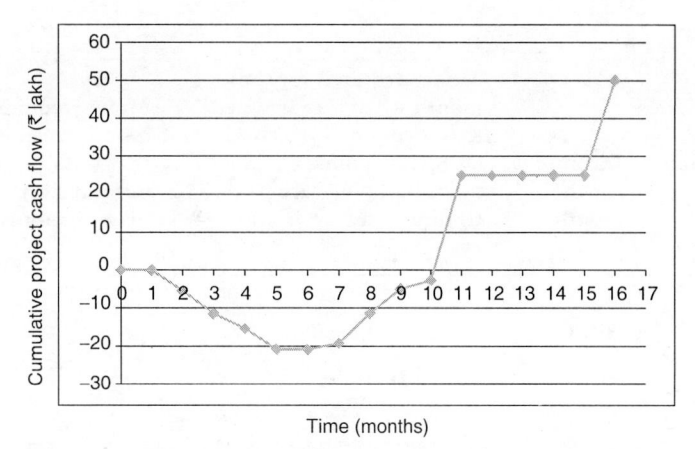

Figure 3.11 Cumulative cash-flow curve without mobilization advance

contractor experiences huge negative cash flow during most of the stages of the project. In fact, the positive cash-flow situation comes only after the completion of project, which is a very grim scenario.

It is usual for contractors to neglect the positive cash-flow portion. This is justified because the cash available with the contractor may not fetch the same interest as the interest he has to pay for the borrowed amount. However, an exact analysis would demand the subtraction of this area from the negative cash-flow area.

Determining the Cash Requirement of a Project

With the knowledge of cumulative cash outflow and cumulative cash receipt, we can determine the maximum cash required for a project. For illustration, let us take the same problem that was used for drawing a project cash-flow diagram. The computations for month-wise cash requirement are performed in Table 3.9. The cumulative outgoings and cumulative incomings data have been taken as they are for the case in which there was no mobilization advance given to the contractor.

Month-wise requirement of cash is determined by deducting the cumulative cash outgoings for that month less the cumulative cash receipt till previous month. For example, the cash requirement for month 2 (M2) is cumulative outgoings for month 2 − cumulative cash receipt of month 1 = 5.61 − 0.00 = 5.61; for month 3 the cash requirement is 28.10 − 0.23 = 27.88; and so on. We find that the maximum cash of ₹ 28.88 (50.66 − 21.78) is needed immediately before payment is received at month 6. This is based on the assumption that incomings are received at discrete points of time while outgoings are paid continuously (refer to column 4 of Table 3.9).

The month-wise cash requirement would change if we assume that incoming and outgoing both are occurring at discrete points of time. The maximum cash required in this case would be for month 3 (M3) and is equal to 6.32 (refer to column 5 of Table 3.9).

Equivalence of Alternatives

In economic comparisons, very often there are situations wherein a comparison has to be made between alternatives, involving payments (or receipts) of different amounts of money at different points in time. Drawing cash-flow diagrams makes such comparisons easy to understand,

Table 3.9 Computation of month-wise cash requirement for the project (all values in ₹ lakh)

Month (1)	Outgoings for the month (2)	Incomings for the month (3)	Money required for the month when it is assumed that incoming is received at discrete points of time while outgoings are paid continuously (4)	Money required for the month when it is assumed that both incoming and outgoing are occurring at discrete points of time (5)=(2) − (3)
M1	0.06	0.00	0.06	0.06
M2	5.61	0.23	5.61− 0.00 = 5.61	5.39
M3	28.10	21.78	28.10 − 0.23 = 27.88	6.32
M4	50.66	47.07	50.66 − 21.78 = 28.88	3.59
M5	66.89	61.43	66.89 − 47.07 = 19.82	5.47
M6	83.31	83.30	83.31 − 61.43 = 21.88	0.01
M7	81.71	83.34	81.71 − 83.30 = −1.58	−1.63
M8	69.00	76.84	69.00 − 83.34 = −14.34	−7.84
M9	39.04	45.47	39.04 − 76.84 = −37.79	−6.43
M10	17.52	19.76	17.52 − 45.47 = −27.96	−2.24
M11	8.10	35.80	8.10 − 19.76 = − 11.66	−27.70
M12	0.00	0.00	0.00 − 35.80 = −35.80	0.00
M13	0.00	0.00	0.00 − 0.00 = 0.00	0.00
M14	0.00	0.00	0.00 − 0.00 = 0.00	0.00
M15	0.00	0.00	0.00 − 0.00 = 0.00	0.00
M16	0.00	25.00	0.00 − 0.00 = 0.00	−25.00

as illustrated in the following example. The principle used is that of equivalence, which reduces different alternatives to a common baseline.

Consider making a choice between the following two alternatives of receiving payment:

1. a lump of ₹ 1,000 in a single instalment immediately, and
2. four instalments of ₹ 400 every year for four years, with payments being received at the end of years 1, 2, 3 and 4.

A cash-flow diagram representative of these options is drawn in Figure 3.12 and Figure 3.13.

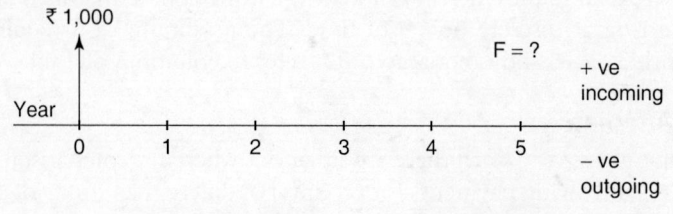

Figure 3.12 Lump of ₹ 1,000 in single instalment

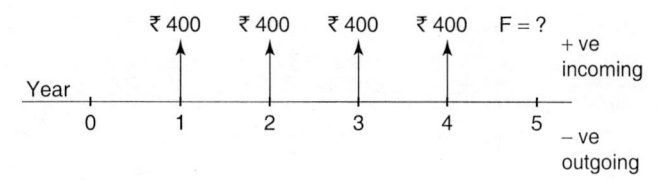

Figure 3.13 ₹ 400 every year for four years

Given the statement of the problem, it may be noted that ₹ 1,000 has been the inflow at t = 0, whereas the inflow of ₹ 400 has been drawn for four years as given (1, 2, 3, 4), and the amounts being inflows have been plotted above the X-axis. Now, the total money received in the two cases is ₹ 1,000 and ₹ 1,600 (400 × 4), respectively. Two things are said to be equivalent only when they have the same effect, and in this case, since money has time value, the two amounts cannot be simply compared, as they are received at different periods. The different ways of reducing alternatives to a common base are discussed later on, but for illustrative purposes, let us assume that we work out the total funds available at the end of, say, five years using an (nominal) interest rate of 10 per cent and six-monthly (half-yearly) compounding. In terms of the cash-flow diagram, the problem boils down to calculating A_1 and A_2, the amounts in the two cases, 1 and 2, as shown in the figures.

For case 1, the issue is simply calculating the compound interest on ₹ 1,000 for a period of five years, at the rate of 10 per cent compounded at six-monthly intervals. It can be verified that the total amount would be ₹ 1,628.89. For case 2, however, the amounts accruing for a principal value of ₹ 400 for periods of 4 years, 3 years, 2 years and 1 year (calculated under identical conditions) need to be added, and it will be seen that the overall sum is ₹ 2,054.22. Thus, it is clear that under the given conditions governing the time value of money, option 2 is preferable to option 1. However, if the instalment payment was reduced to ₹ 300, the total sum available in Case 2 at the end of five years would be only ₹ 1,540.67, making option 1 more preferable. It may be noted that simplistically speaking, even here the total amount received (1,200) in case 2 is higher than that in case 1 (1,000). As an extension to the problem discussed here, it can be worked out that if the instalment was ₹ 317.15, the two options would be equivalent. It should be borne in mind that the point at which the final comparisons are being made is also critical, and the calculations given would change if the numbers were worked out at another base line (say, at the end of 10 years).

Formulations for Interest Computation

It is clear from the above exercise that equivalence can be established or alternatives compared only when the applicable conditions for compounding and the rates of interest are known. The example also illustrates a possible approach to handling payments (or receipts) in instalments, etc. In this section, the more widely used formulations for interest calculations will be covered in greater detail. For the sake of convenience, three categories addressing eight commonly used interest formulations have been discussed in the following paragraphs.

1. Category A: With a single payment (SP)
   (a) Compound amount factor (SPCAF)
   (b) Present worth factor (SPPWF)

2. Category B: With an equal payment series (EPS) or several equal instalments. Equal payment series is also referred to as uniform series (US) in literature
   (a) Compound amount factor (EPSCAF)
   (b) Present worth factor (EPSPWF)
   (c) Sinking fund deposit factor (EPSSFDF)
   (d) Capital recovery factor (EPSCRF)
3. Category C: With unequal payment series
   (a) Arithmetic gradient factor (AGF)
   (b) Geometric gradient factor (GGF)

As is clear from the above classification, in category A only single lump-sum amounts are involved, while category B handles a series of uniform (equal) number of payments (or disbursements). On the other hand, category C addresses cases wherein there is a uniform increase or decrease in receipts (or disbursements).

In the following discussion, i, n, P, F and A abbreviate the interest rate per period, the number of interest periods, a present sum of money, a future sum of money, and a periodic instalment or payment, respectively. As will be seen in interest problems, 4 of these 5 parameters are always present and 3 of the 4 must be known and given as input data, and the object is to compute the missing parameter. In fact, it is convenient to discuss the problem in terms of factors, where the required information is obtained using other available information. Tables have since been developed to facilitate the calculations for different interest rates and periods, as will be illustrated subsequently. It should be noted that these tables are conventionally based upon transactions assumed to be occurring at the end of a period.

Single payment compound amount factor (SPCAF) This factor, denoted by SPCAF and read as 'single payment compound amount factor', is one by which a single payment (P) is multiplied to find its compound amount (F) at a specified time in future, as represented schematically on the cash-flow diagram. Another way of representing this factor is (F/P, i, n), which is read as 'F given P at an interest rate of i for a period n'. From our understanding of compound interest, effectively

$$(F/P, i, n) = (1 + i)^n \tag{3.3}$$

Single payment present worth factor (SPPWF) When a single future payment (F) is multiplied by this factor, the present worth (P) is obtained. This factor, SPPWF, is calculated as 1/

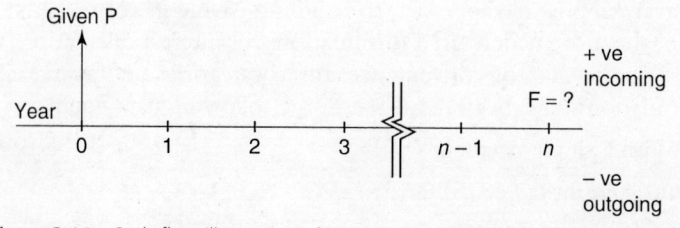

Figure 3.14 Cash-flow illustration of single payment compound amount factor

CONSTRUCTION ECONOMICS

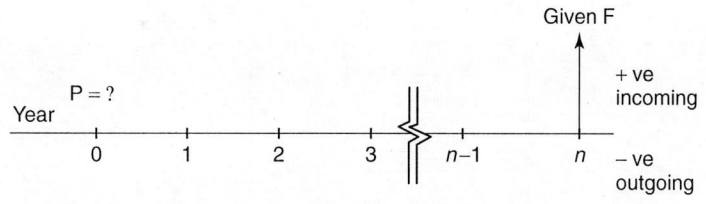

Figure 3.15 Cash-flow illustration of single payment present worth factor

$(1 + i)^n$, and also represented as $(P/F, i, n)$, which is read as 'P given F at an interest rate of i for a period n'. Hence, effectively

$$(P/F, i, n) = \frac{1}{(1+i)^n} \qquad (3.4)$$

The cash-flow diagram representation of the problem in this case is essentially the converse of the SPCAF, and is shown in Figure 3.15.

Uniform series compound amount factor (EPSCAF or USCAF) This factor converts a uniform series payment (A) to its compound amount (F), as shown in the cash-flow diagram in Figure 3.16. This factor is sometimes represented as $(F/A, i, n)$, which is read as 'F given A at an interest rate of i for a period n', and is effectively equal to

$$(F/A, i, n) = \frac{(1+i)^n - 1}{i} \qquad (3.5)$$

Uniform series present worth factor (EPSPWF or USPWF) This factor is used to determine the present worth (P) for a uniform series payment of A, as shown in Figure 3.17. This

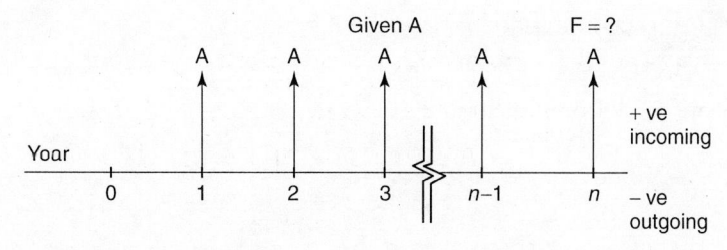

Figure 3.16 Cash-flow illustration of uniform series compound amount factor

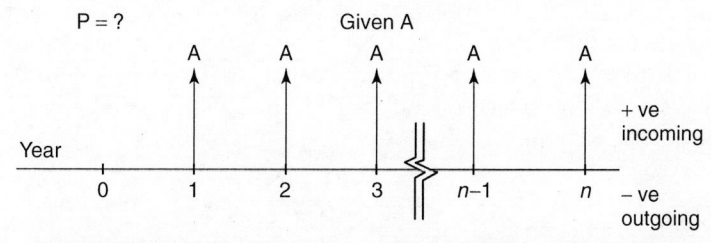

Figure 3.17 Cash-flow illustration of uniform series present worth factor

![Figure 3.18 cash-flow diagram]

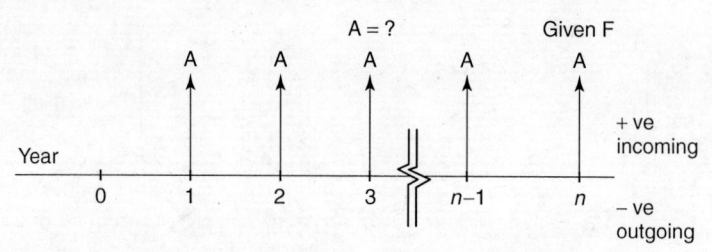

Figure 3.18 Cash-flow illustration of sinking fund deposit factor

factor is represented as $(P/A, i, n)$, which is read as 'P given A at an interest rate of i for a period n'. Effectively,

$$(P/A, i, n) = \frac{(1+i)^n - 1}{i(1+i)^n} \quad (3.6)$$

Sinking fund deposit factor (EPSSFDF) It is the factor by which a future sum (F) is multiplied to find a uniform sum (A) that should be set regularly, such that the final value of the funds set aside is F. The sinking fund deposit factor, SFDF, is also represented as $(A/F, i, n)$, which is read as 'A given F at an interest rate of i for a period n'. Mathematically, the EPSSFDF can be shown to be equal to

$$(A/F, i, n) = \frac{i}{(1+i)^n - 1} \quad (3.7)$$

The cash-flow diagrammatic representation of the factor is given in Figure 3.18. The formula is derived in the following manner:

$$F = A(1+i)^{n-1} + A(1+i)^{n-2} + \cdots + A(1+i) + A \quad (3.8)$$

$$\Rightarrow F = A[(1+i)^{n-1} + (1+i)^{n-2} + \cdots + (1+i) + 1] \quad (3.9)$$

Multiplying by $(1+i)$ on both sides of Equation 3.9, we get

$$F(1+i) = A[(1+i)^n + (1+i)^{n-1} + (1+i)^{n-2} + \cdots + (1+i)^2 + (1+i)] \quad (3.10)$$

Subtracting Equation 3.9 from Equation 3.10, we get,

$$F(1+i-1) = A[(1+i)^n - 1]$$

$$\Rightarrow F = \frac{A[(1+i)^n - 1]}{i}$$

Thus,

$$(A/F, i, n) = \frac{i}{(1+i)^n - 1} \quad (3.11)$$

Capital recovery factor (EPSCRF) The capital recovery factor can be used to find the uniform payments of A to exactly recover a present capital sum (P) with interest. This factor is also represented as $(P/A, i, n)$, which is read as 'P given A at an interest rate of i for a period n'. Mathematically, EPSCRF equals

$$(A/P, i, n) = \frac{i(1+i)^n}{(1+i)^n - 1} \quad (3.12)$$

and Figure 3.19 shows a schematic representation of the problem statement.

CONSTRUCTION ECONOMICS | 73

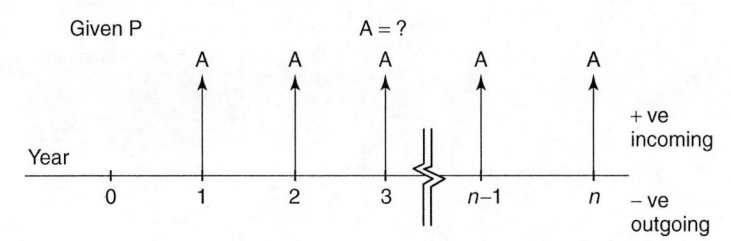

Figure 3.19 Cash-flow illustration of capital recovery factor

Arithmetic gradient factor (AGF) In this case, the increase/decrease in instalments, whether it is payments or disbursements, follows an arithmetic pattern, as shown in Figure 3.20. This factor, AGF, is sometimes denoted as $(A/G, i, n)$ and read as 'A given G at an interest rate of i for a period n'. Mathematically, the factor is equal to

$$(A/G, i, n) = \left[\frac{1}{i} - \frac{n}{(1+i)^n - 1} \right] \tag{3.13}$$

The formula basically converts the arithmetic increase/decrease amount into a uniform series.

Geometric gradient factor (GGF) In this case, the increase/decrease in instalments, whether it is payments or disbursements, follows a geometric pattern, as shown in Figure 3.21. This factor, GGF, is sometimes denoted as $(P/g, i, n)$ and read as 'P given g at an interest rate of i for a period n'. Mathematically, the factor is equal to

$$\left(\frac{P}{g}, i, n \right) = \frac{1 - \frac{(1+g)^n}{(1+i)^n}}{(i-g)} \tag{3.14}$$

where g is equal to the rate of increase/decrease and c is the initial amount.

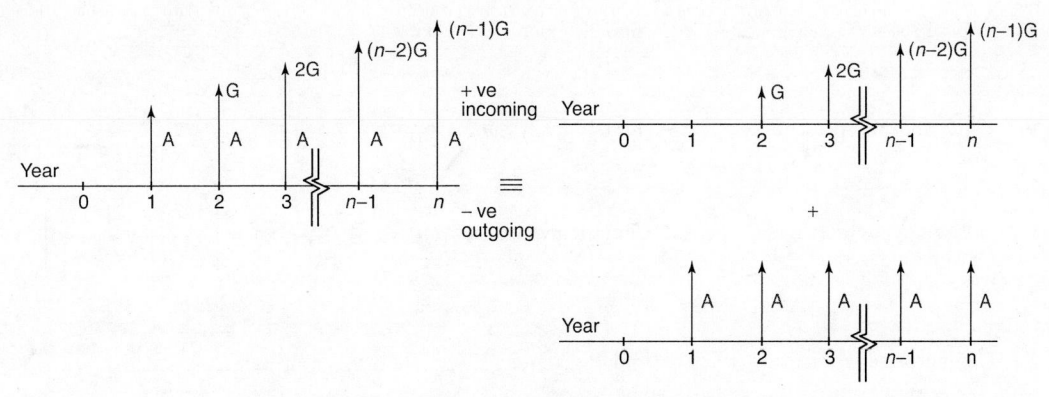

Figure 3.20 Cash-flow illustration of arithmetic gradient factor

| 74 | CONSTRUCTION PROJECT MANAGEMENT

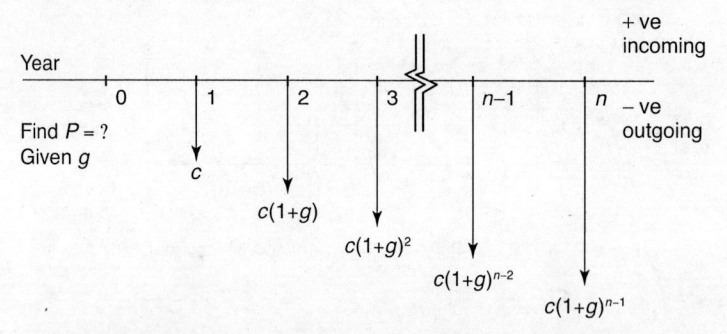

Figure 3.21 Cash-flow illustration of geometric gradient factor

3.5 USING INTEREST TABLES

A simplified summary of the discussion on interest formulations in the above section is presented in Table 3.10. As mentioned earlier, ready-to-use tables are available, and reproduced in the appendices to find the different interest factors at varying interest rates and time periods. Tables in Appendix 1 have been provided for interest rates from 0.5 per cent

Table 3.10 Summary sheet of interest formulae

1	Single payment compound amount factor	F/P (find F, given P)	$\text{SPCAF} = (F/P, i, n) = (1+i)^n$
2	Single-point present worth factor	P/F (find P, given F)	$\text{SPPWF} = (P/F, i, n) = \dfrac{1}{(1+i)^n}$
3	Uniform series compound amount factor	F/A (find F, given A)	$\text{USCAF} = (F/A, i, n) = \dfrac{(1+i)^n - 1}{i}$
4	Sinking fund deposit factor	A/F (find A, given F)	$\text{SFDF} = (A/F, i, n) = \dfrac{i}{(1+i)^n - 1}$
5	Capital recovery factor	A/P (find A, given P)	$\text{CRF} = (A/P, i, n) = \dfrac{i \times (1+i)^n}{(1+i)^n - 1}$
6	Uniform series present worth factor	P/A (find P, given A)	$\text{USPWF} = (P/A, i, n) = \dfrac{(1+i)^n - 1}{i \times (1+i)^n}$
7	Arithmetic gradient conversion factor	A/G (find A, given G)	$\text{AGF} = (A/G, i, n) = \left[\dfrac{1}{i} - \dfrac{n}{(1+i)^n - 1}\right]$
8	Geometric series factor	P/g (find P, given g)	$\text{GGF} = (P/g, i, n) = \dfrac{1 - \dfrac{(1+g)^n}{(1+i)^n}}{(i-g)}$

to 30 per cent, and period n up to 100. In order to find an interest factor, say compound amount factor $(F/P, 15, 5)$, the user needs to look for 5 in the n column corresponding to interest table of 15 per cent, and then read across to the 'compound amount factor' column to find 2.01136. Similarly, other interest factors can also be found out. Illustrative examples have been included in the 'solved examples' section to show the physical significance of the different factors discussed above.

3.6 EVALUATING ALTERNATIVES BY EQUIVALENCE

Once the concept of a cash-flow diagram is well understood, the next step is to proceed to the extension of these diagrams to compare different engineering alternatives from an economic point of view. The principle involved is that of 'equivalence', wherein the alternatives are determined to be similar. It should be recalled that the concept has been briefly discussed in Section 3.4 above, and it was pointed out that equivalence exists only at a certain given rate of return, and if this rate of return is changed, the alternatives are likely to change.

Construction management often involves cost comparisons between alternatives of different engineering efficiency, namely, one with a high initial cost and low operation and maintenance costs, compared to another with a low initial cost but high operating and maintenance costs. Using the time value of money and the cash-flow diagrams for illustrative purposes, equivalence is studied to identify the better alternative, using a common basis. Some of the frequently applied methods used for the purpose in engineering economic analysis are listed below and discussed in the following paragraphs.

1. Present worth comparison
2. Future worth comparison
3. Annual cost and worth method
4. Rate of return method

3.6.1 Present Worth Comparison

In this method, the present worth (at time zero) of the cash flow in terms of equivalent single sum is determined using an interest rate, sometimes also called the discounting rate. The method is based on the following assumptions:

(a) Cash flows are known.
(b) Cash flows do not include effect of inflation. The discussion on inflation follows later in the text.
(c) The interest rate (discounting rate) is known.
(d) Comparisons are made with before-tax cash flows. The concept of tax has been discussed in later sections.
(e) Comparisons do not include intangible considerations.
(f) Comparisons do not include consideration of the availability of funds to implement alternatives.

Figure 3.22 shows three typical types of problems that are encountered in present worth analysis. The three sets of problems are—alternatives with equal lives; alternatives with unequal lives; and alternatives with infinite lives.

```
                    ┌─────────────────┐
                    │ Present worth   │
                    │ problems        │
                    └─────────────────┘
         ┌─────────────────┼─────────────────┐
┌─────────────────┐ ┌─────────────────┐ ┌─────────────────┐
│ Type 1:         │ │ Type 2:         │ │ Type 3:         │
│ Alternatives    │ │ Alternatives    │ │ Alternatives    │
│ with equal lives│ │ with unequal    │ │ with infinite   │
│                 │ │ lives           │ │ lives           │
└─────────────────┘ └─────────────────┘ └─────────────────┘
                           │                     │
                    ┌──────┴──────┐              │
            ┌──────────────┐ ┌──────────────┐ ┌──────────────────┐
            │ Common       │ │ Study period │ │ Capitalized      │
            │ multiple     │ │ method       │ │ equivalent       │
            │ method       │ │              │ │ method           │
            └──────────────┘ └──────────────┘ └──────────────────┘
```

Figure 3.22 Types of problems in present worth analysis

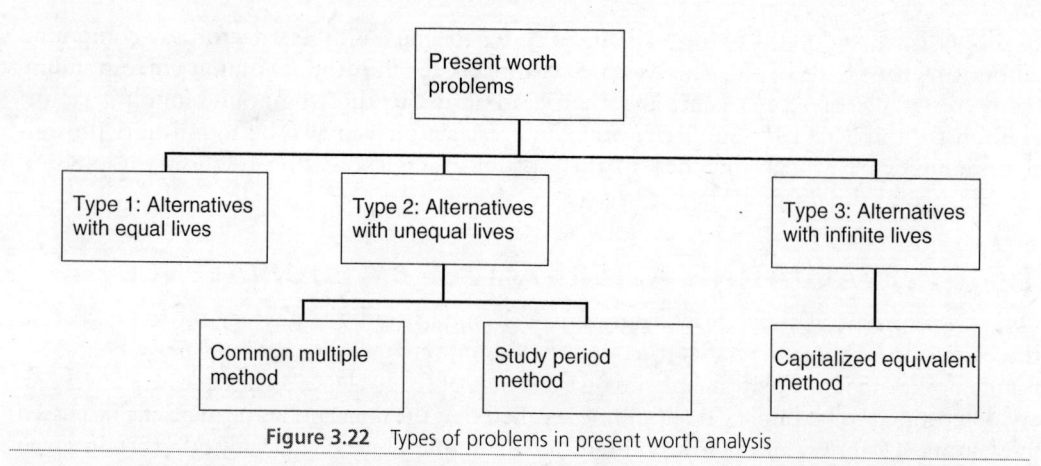

Type 1: Alternatives with Equal Lives

As the name suggests, in such problems the competing alternatives have equal lives. For evaluating the alternatives, the present worth of both the competing alternatives are found out. The alternative with the maximum present worth is the most economical alternative. For cost-dominated cash-flow diagrams, the alternative with the lowest present cost is chosen. In case of cash-flow diagrams involving both costs and revenues, the net or difference of present worth of revenues and costs are found. This is referred to as net present worth or net present value (NPV). The method of comparison of NPV is quite popular for evaluation of alternatives. NPV is also used to calculate profitability for an investment alternative. Profitability index (PI) is the ratio of NPV and capital cost (CC) for an investment. In other words,

$$PI = \frac{NPV}{CC} \qquad (3.15)$$

Type 2: Alternatives with Unequal Lives

In such problems, the alternatives do not have an equal life period of service—in other words, they are not coterminous. A relevant example would be a decision to choose between two batching plants that may have different service lives—say, 5 years and 10 years. Needless to say, a simple comparison of the two alternatives would not be accurate, as in one case there would be a need to replace the plant at the end of five years, and any cost likely to be incurred at that point in time should be appropriately accounted for in the budgeting at the outset. The common multiple method and the study period method are two approaches to discuss this class of problems.

Common multiple method In this method, a coterminous life period is chosen for the alternatives using the least common multiples of the different life periods. For example, if the alternatives have life periods of 2, 3, 4 and 6 years, they will be put in use for a period equal to the least common multiple of their life periods. In this case, it is 12 years. This means that the alternative with a two-year life period shall be replaced 6 times, the one with three years shall be replaced four times, and so on, for achieving a coterminous life period for all the alternatives. It is assumed here that the alternatives shall be replaced after their service life with same cost characteristics. This assumption stands valid if the common multiple of alternative life periods is small, and the possibility of a technically different option emerging during this time is ignored.

Study period method In this method, a study period is chosen on the basis of the length of the project or the service lives of the alternatives. An appropriate study period reflects the replacement circumstances. Thus, study period may be chosen as the shortest life period of the alternatives as a protection against technological obsolescence. In this method, we assume that all the assets will be disposed of at the end of the analysis period.

Type 3: Alternatives with Infinite Lives

Such problems are encountered in a real-life situation when the alternatives involved have long lives—for example, the appraisal of different alternatives involving construction of civil engineering structures such as dams, power projects and tunnel projects, which have a reasonably long life. A very popular approach used in such problems is the application of capitalized equivalent (CE) method, which is described in the following section.

Capitalized equivalent method This method is based on the concept of capitalized equivalent, which is the present (at time zero) worth of cash inflows and outflows. In other words, CE is a single amount determined at time zero, which at a given rate of interest will be equivalent to the net difference of receipts and disbursements if the given cash flow pattern is repeated in perpetuity (in perpetuity the period assumed is infinite). In other words, CE can be looked upon as equal to the present worth, with the added rider that the cash flow extends forever. Thus, to determine CE, the normal procedure is to first convert the actual cash flow into an equivalent cash flow of equal annual payment.

Mathematically,

$$CE = A \times (EPSWF)^i_\infty$$

$$CE = A\,(EPSWF)^i_\infty = A\left(\frac{P}{A}, i, n = \infty\right) = A\left[\frac{(1+i)^\infty - 1}{i \times (1+i)^\infty}\right]$$

$$\Rightarrow CE = \frac{A}{i} \tag{3.16}$$

CE analysis is very useful to compare long-term projects. In fact, for projects such as roads, where the cash flow may contain negative terms, only this method can be used to compare the relative merits of different types of road alignments or surfaces, etc.

3.6.2 Future Worth Comparison

In this method, the future worth of each component of cash flow is evaluated and the algebraic sum of such future worth becomes the basis for comparison. Also, there is no discounting each component of cash flow to the present, as is done in the present worth method.

Although there is no special advantage in using this method over the present worth method, it is frequently used in cases where the owner expects to sell or liquidate an investment at some future date, and wants an estimate of net worth at that point in time. In everyday life, the method is useful in situations such as planning for retirement. A comparison and evaluation using this method seems to be more meaningful as it provides some insight into future receipts also.

3.6.3 Annual Cost and Worth Comparison

This method is perhaps most widely used for comparing alternatives because it is basically simple and easy to understand and explain. Also, the computations involved are easy to carry out. In this method, all payments and disbursements are converted into an annualized cost series—annual cost is the cost pattern of each alternative converted into an equivalent uniform series of annual cost at a given interest rate. Needless to say, the alternative that yields the least cost is chosen. The method could be used to compare alternatives with equal and unequal lives. In the latter case, common multiple method and study period method could be adopted, which is similar in concept as discussed under 'present worth method' problems with unequal lives.

3.6.4 Rate of Return Method

This is another method for evaluation of different competing alternatives, especially in the area of investments. In general, rate of return may be regarded as an index of profitability, and terms such as minimum attractive rate of return (MARR), internal rate of return (IRR), incremental rate of return (IRoR) and ERR (external rate of return) are commonly encountered. The following paragraphs briefly explain some of these, through illustrative examples wherever required.

Minimum Attractive Rate of Return (MARR)

This is the minimum rate of return below which a company would not be interested in the proposed investment alternative. In other words, an investor's interest in an alternative is awakened only when the rate of return is at least equal to or greater than MARR, which itself depends on a number of factors such as conditions of the market, level of competition and cost of capital. It is interesting to note that in the context of construction projects, though the MARR values are obviously different for different companies, as they are willing to work under different rates of return, even within the same company it is always possible to have different yardsticks of MARR that could be adopted for comparison of investment alternatives. For example, in a contracting company working for construction of buildings, bridges and tunnels, the MARR may be the lowest for the building business and the largest in tunnels, on account of the level of competition faced in the different segments of the construction industry. In other words, since the competition is likely to be highest in the building sector, it will also have lower MARR, and so on.

Internal Rate of Return Method (IRR)

Internal rate of return, sometimes represented by the symbol i^* or the acronym IRR, is defined as the interest rate that reduces the present worth of given cash flow to zero. It represents the percentage or rate of interest earned on the unrecovered balance of an investment, at any point of time, and further, the earned recovered balance is reduced to zero at the end of the project.

An illustrative computation showing how a part of the cash flow at the end of a year goes towards payment of interest due on the outstanding (unrecovered) balance of investment, and the remainder liquidates the outstanding investment, is shown in Table 3.11, and the cash-flow diagram is shown in Figure 3.23. In this example, at the end of the proposal's life (four years), the entire investment has just been recovered, and the applicable rate of interest (10 per cent) is a special and unique rate called the IRR.

CONSTRUCTION ECONOMICS | 79

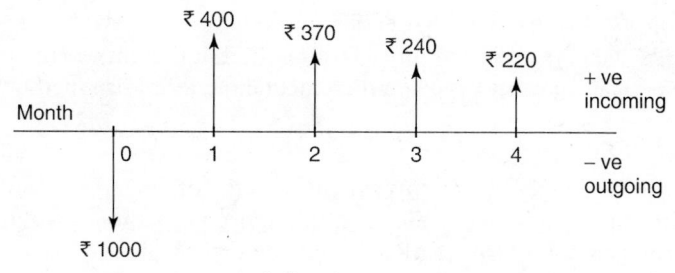

Figure 3.23 Cash-flow diagram to illustrate the IRR

In principle, *IRR* can be determined by equating the net present worth of the cash flow to zero, i.e., setting the difference of the benefits and cost of the present worth to zero, as shown below:

$$(PW)_{benefits} - (PW)_{cost} = 0 \qquad (3.17)$$

It may be noted that the above equation is too complex to solve directly, and usually the *IRR* is determined by a trial-and-error procedure as outlined below:

Step 1 Assume a trial rate of return (i^*).

Step 2 Counting the cost as negative and the income as positive, find the equivalent net worth of all costs and incomes.

Step 3 If the equivalent net worth is positive, then the income from the investment is worth more than the cost of investment and the actual percentage return is higher than the trial rate, and vice versa.

Step 4 Adjust the estimate of the trial rate of return and go to step 2 again until one value of i is found that results in a positive equivalent net worth, and another higher value of i is found with negative equivalent net worth.

Step 5 Solve for the applicable value of i^* by interpolation.

It is not possible to calculate the rate of return for the cash flows involving cost alone or revenue alone, as can be observed from Equation 3.17. Also, the *IRR* method should not be used

Table 3.11 Illustration for IRR

End of year t	Cash-flow at EOY, t	Unrecovered balance at the beginning of year t	Interest earned on the unrecovered balance during the year	Unrecovered balance at the beginning of the year (t + 1)
0	−1,000	−	−	−1,000
1	400	−1,000	−100	−700
2	370	−700	−70	−400
3	240	−400	−40	−200
4	220	−200	−20	0

10% is the IRR

for ranking of projects as it may give erroneous results, not in line with the results obtained from other methods of analysis such as annual cost method or present worth method. For such cases, we need to evaluate alternatives using incremental rate of return method.

Incremental Rate of Return (IRoR)

If an alternative requires a higher initial investment than the other and evaluation is of the rate of return on the increment of initial investment, the return yielded on this extra investment is called the incremental rate of return (IRoR). The incremental analysis is based on the principle that every rupee of investment is as good as the other.

The analysis makes the assumptions that sufficient funds are available to finance the alternatives with the highest investment, and that there are opportunities available to utilize the surplus funds at a rate higher than the MARR, should there be any excess funds after financing the alternative with the lower investment costs. Some of the steps involved in the incremental analysis are summarized below.

Step 1 List out all the alternatives in ascending order of their first cost or initial investment. It may be pointed out at this stage that in most cases, alternatives with the lowest investment are likely to turn out to be the 'do nothing' alternative.

Step 2 Compare the rate of return of all alternatives with the assumed MARR, and check if the rate of return is at least equal to the MARR. If not, the alternative is dropped and not considered in the further analysis.

Step 3 Prepare the cash-flow diagram on incremental basis between the alternatives being examined and the current alternative (to begin with, we have taken the alternative with the lowest initial investment).

Step 4 When an alternative that has just been examined is acceptable (rate of return is more than the MARR), it becomes the current best replacing the earlier one. The new best is examined with the next higher investment alternative.

Step 5 In case rate of return is less than MARR, the alternative under examination is ruled out and the current alternative remains the lucrative one. The current best is compared to the next higher investment.

Step 6 The above process is repeated till all the alternatives have been looked into and the best alternative is selected.

3.7 EFFECT OF TAXATION ON COMPARISON OF ALTERNATIVES

In the discussion so far, any effect that a taxation regime may have as far as comparison and evaluation of economic alternatives is considered has not been taken into account. It should, however, be borne in mind that prevailing tax laws impose taxes on companies on the basis of the annual income generated through conduct of business, and not only the rates of taxes vary from time to time, but also certain incentives and relief in taxes are provided on certain expenditures for various reasons. Companies often take advantage of such schemes and use available funds in a manner so as to optimize their interests. While it is not the intention here to discuss

Table 3.12 Illustration of effect of taxation on comparison of alternatives

S. No.	Item description	A	B	Remarks
1.	Gross earnings	₹ 150 lakh	₹ 150 lakh	
2.	Admissible expenses (AE)	₹ 50 lakh	₹ 25 lakh	
3.	Pre-tax profits	₹ 100 lakh	₹ 125 lakh	(1) − (2)*
4.	Tax payable	₹ 40 lakh	₹ 50 lakh	40% of (3)
5.	Net profit	₹ 60 lakh	₹ 75 lakh	(3) − (4)

*In this example, for simplicity a 100% deduction has been made for the admissible expenses. At times, only a certain percentage of these expenses is allowed to be deducted from the gross earnings for the purpose of tax calculations. For example, in case this percentage was 50%, the pre-tax profits for A and B would be 125 and 137.5, respectively, and the subsequent figures modified accordingly.

taxation rules at any length, the example below is included only to place in perspective the discussion on taxation and alternative comparison.

Example
Basic details of gross earnings, etc., for two companies A and B are given in Table 3.12. As can be seen from the table, apart from the gross earnings, there are 'admissible *expenses*', whose deduction from the former is allowed before working out the tax liability (the amount of taxes to be paid). The simple computations shown in the table assume a flat 40% tax rate (though this percentage is often actually a function of the level of gross or net earnings).

It can be seen from Table 3.12 that for the same gross earnings, the tax payable and the net profit can be quite different depending upon the 'admissible expenses', which often include expenses on equipment purchase, research and development, etc. Depending on the prevailing regulations at any point in time, this becomes an important consideration in the comparison of different alternatives. More on taxation and their implication on selection of alternatives are given in Chapter 9.

3.8 EFFECT OF INFLATION ON CASH-FLOW

Construction projects, by and large, take a number of months and during this period, the cost of labour, materials, plant and machinery may undergo an inflationary trend. Inflation in general is defined as the increase in the price level resulting in decrease in purchasing power of money.

Since general inflation results in price rise in all goods, the relative prices remain constant. Hence, it is possible to disregard escalation. In India, it is normal practice to use 12 per cent as discount rate. According to IRC:SP:61-2004, where there is a large difference between the rates of inflation and interest, the discount rate is evaluated using the following expression:

$$\text{Modified discount rate} = \left[\left(\frac{1 + \frac{\text{interest rate (\%)}}{100}}{1 + \frac{\text{inflation rate (\%)}}{100}}\right) - 1\right] \times 100\% \qquad (3.18)$$

Considering an interest rate of 12% and an inflation rate of 8%, the modified discount rate = [{(1 + 12%) ÷ (1 + 8%) − 1}] × 100% = 3.70%

3.9 EVALUATION OF PUBLIC PROJECTS: DISCUSSION ON BENEFIT–COST RATIO

In all the preceding discussions, we focused on private projects, where the objective was to maximize the profits while adhering to the contractual requirements of the projects. The objective of public projects is to provide goods/services to the public at the minimum cost. The benefit accrued from such projects should at least recover the cost of the projects. Public projects here mean those funded by government (state or centre). The government has a responsibility towards public welfare. Some examples of public projects are construction of dams and highways, and projects related to defence, education and public health.

Many government projects cannot be evaluated strictly in commercial terms. Profit, taxes and payoff periods take a backseat in public projects, unlike in private projects. Yet, the resources of the government, however large, are also limited, and it has to choose between different alternatives based on some criterion. The objectives could be as given below:

- To check the viability of the project economically
- To select the most viable project from a set of economically viable projects
- To rank or order the economically viable alternatives for a given situation

One of the most commonly used criteria to evaluate public projects is benefit-cost (B/C) ratio. B/C ratio is defined as the ratio of benefit to public and cost to the government. B/C ratio can be obtained using Present Worth analysis, Future Worth method, or Annual Worth method. Classification of benefits or costs is mostly arbitrary and causes confusion. When should particular project consequences be classified as benefits and not as a cost? Different classification of impacts as benefits and costs can result in significantly different benefit-cost ratio values and can even result into negative and zero values for projects with a net positive economic impact (Au 1988, Halvorsen and Ruby 1981, both cited in Lund 1992).

The numerator of all the B/C ratios is usually taken to mean the net benefit, which is the sum of all benefits minus all the disbenefits. Due care is required to determine the numerator and denominator, since some confusion may arise in a problem. However, if the definition is clearly understood, the ratio can be correctly obtained. When comparisons of several alternatives are to be made, a basis of incremental analysis should be made. B/C ratio of more than one indicates that benefit outweighs cost and, thus, the investment is justifiable. B/C ratio of 1.5 means that each rupee in cost yields ₹ 1.5 in benefits over the lifetime of the project. B/C ratio of less than one indicates that benefit accrued from the project is less than the cost required to be invested and, thus, the investment is not justifiable. B/C analysis per se does not select an alternative; however, it is a useful tool to simply distinguish between an acceptable alternative and an unacceptable one.

B/C criterion is not designed to rank or order the project. It sets only a minimum level of acceptability and does not pretend to identify the source of investment funds. Basically, it develops information useful for the purpose of decision–making, depending on the availability of investment funds, capital rationing and the special features of social merit and economic objectives.

3.9.1 Benefit/Cost Criteria

Some of the benefit/cost criteria for evaluation of alternatives are:

- Minimum investment
- Maximum benefit

- Aspiration level
- Maximum advantage of benefits over cost (B−C)
- Highest B/C ratio
- Largest investment that has a benefit/cost ratio greater than 1
- Maximum incremental advantage of benefit over cost (ΔB−ΔC)
- Maximum incremental benefit/cost ratio (ΔB/ΔC)
- Largest investment that has an incremental B/C ratio greater than 1.0

Some of the limitations in the application of benefit/cost analysis relate to estimating the monetary benefits, especially in the case of intangible aspects such as recreation and the saving of human lives. Further, as in other methods, the determination of proper interest rates and time horizon for discounting future benefits and costs has long been a source of controversy.

Example

A government is planning for a hydroelectric project that will also provide flood control, irrigation and recreation benefits. The established benefits and cost of three alternatives are given in Table 3.13. The interest rate to be used for the analysis is five per cent, and the life of each of the alternatives X, Y and Z is to be assumed as 50 years. Choose the best alternative.

Solution

The computation of B/C using annual cost/worth analysis is given in Table 3.14.

Table 3.13 Data for B/C ratio computation (all values in ₹ million)

Alternatives	X	Y	Z
Initial cost	250.00	350.00	500.00
Annual power sales	10.00	12.00	18.00
Annual flood control	2.50	3.50	5.00
Annual irrigation benefit	3.50	4.50	6.00
Annual recreation benefits	1.00	2.00	3.50
Annual operation and maintenance	2.00	2.50	3.50

Table 3.14 Solution for B/C ratio problem

Annual Cost, C, Computation	X	Y	Z
Annual cost equivalent = P × (A/P, 5%, 50)	250 × 0.0548 = 13.70	350 × 0.0548 = 19.18	500 × 0.0548 = 27.40
Operation and maintenance cost	2.00	2.50	3.50
Receipt on power sales	−10.00	−12.00	−18.00
Total annual cost equivalent	5.70	9.68	12.90
Annual benefit, B, computation			
Annual flood control	2.50	3.50	5.00
Annual irrigation benefit	3.50	4.50	6.00
Annual recreation benefits	1.00	2.00	3.50
Total benefit	7.00	10.00	14.50
B/C ratio	1.23	1.03	1.12

Table 3.15 Computation for incremental analysis

Annual Cost, C, Computation	Y − X	Z − X
Annual cost equivalent = P × (A/P, 5%, 50)	100 × 0.0548 = 5.48	250 × 0.0548 = 13.7
Operation and maintenance cost	0.50	1.50
Receipt on power sales	−2.00	−8.00
Total annual cost equivalent	3.98	7.20
Annual benefit, B, computations		
Annual flood control	1.00	2.50
Annual irrigation benefit	1.00	2.50
Annual recreation benefits	1.00	2.50
Total benefit	3.00	7.50
B/C ratio	3.00/3.98 = 0.75	7.50/7.20 = 1.04
Remarks	B/C < 1, Reject Y	B/C > 1, Accept Z

B/C ratio of the three alternatives is more than 1 and, thus, all the above alternatives are economically viable. In the second stage, incremental analysis is performed to identify the best alternative. The computation is shown in Table 3.15.

REFERENCES

1. Au, T., 1988, 'Profit measures and methods of economic analysis for capital project selection', *Journal of Management in Engineering*, ASCE, 4(3), pp. 217–228.
2. Blank, L. and Tarquin, A., 1989, *Engineering Economy*, 3rd ed., New York: McGraw-Hill.
3. Halvorsen, R. and Ruby, M.G., 1981, *Benefit-cost analysis of air pollution control*, Lexington: Lexington Books.
4. Harris, F. and McCaffer, R., 2005, *Modern Construction Management*, 5th ed., Blackwell Publishing, India.
5. IRC:SP: 61-2004, *An Approach Document on Whole Life Costing for Bridges in India*, Indian Road Congress, New Delhi.
6. Lund, J.R., 1992, 'Benefit-cost ratios: Failures and alternatives', *Journal of Water Resources Planning and Management*, ASCE 118(1), pp. 94–100.
7. Panneerselvam, R., 2005, *Engineering Economics*, 4th print, New Delhi: Prentice Hall.
8. Riggs, J.L., Bedworth, D.D. and Randhawa, S.U., 2004, *Engineering Economics*, 4th ed., New Delhi: Tata McGraw Hill.
9. Taylor, G.A., 1980, *Managerial and Engineering Economics*, 3rd ed., D. Van Nostrand Company.
10. Thuesen, G.J. and Fabrycky, W.J., 1989, *Engineering Economy*, 7th ed., Englewood Cliffs, NJ: Prentice-Hall.
11. Van Horne, J.C. and Wachowicz J.M., 2001, *Fundamentals of Financial Management*, 11th ed., New Delhi: Pearson Prentice Hall.

CONSTRUCTION ECONOMICS | 85 |

12. Wellington, A.M., 1887, *The Economic Theory of the Location of Railways*, New York: John Wiley and Sons.

SOLVED EXAMPLES

Example 3.1

For the following data of a project, (a) prepare the month-wise running account bill, (b) prepare the cash inflow diagram for the contractor, and (c) prepare the cash outflow diagram for the owner.

Contractor has prepared the construction schedule, which has been approved by the owner. The construction schedule is shown in Table Q3.1.1. Also shown are the estimated quantities that are likely to be executed during each month.

- Value of contract: ₹ 7,625,000 (Seventy-six lakh twenty-five thousand rupees only)
- Duration: Four months
- The owner makes an advance payment of ₹ 5 lakh, which is to be recovered in four equal instalments.
- The owner also supplies materials worth ₹ 3.2 lakh, which is also to be recovered equally from each running account (RA) bill.
- The owner will recover from the payments made to the contractor two per cent of the value of the work done as income tax deducted at source, and deposit this amount with the Reserve Bank of India (RBI).

Table Q3.1.1 Construction schedule

S. No.	Item description	Unit	Total quantity	Rate (₹)	Amount (₹)	Month 1	Month 2	Month 3	Month 4
1	Earthwork in excavation	m³	500	50	25,000	500			
2	R.C.C	m³	1,000	4,000	4,000,000	250	500	250	
3	Brickwork	m³	2,000	1,000	2,000,000	500	600	900	
4	Sanitary works	L.S	—	—	200,000			50%	50%
5	Electrical works	L.S	—	—	200,000			50%	50%
6	Woodwork	L.S	—	—	250,000			50%	50%
7	Finishing work	m²	4,750	200	950,000				4,750

Table S3.1.1 Computation for RA bill (Value in ₹)

S. No.	Item description	Months 1	2	3	4
1	Earthwork in excavation	25,000			
2	R.C.C.	1,000,000	2,000,000	1,000,000	
3	Brickwork	500,000	600,000	900,000	
4	Sanitary works			100,000	100,000
5	Electrical works			100,000	100,000
6	Woodwork			125,000	125,000
7	Finishing work				950,000
	Total work progress	1,525,000	2,600,000	2,225,000	1,275,000

Additional conditions and assumptions:

- The cost for the contractor to execute a particular item is 90 per cent of their quoted rates.
- The total cost for a particular item consists of labour (20 per cent), material (60 per cent), plant and machinery (10 per cent), and subcontractor cost (10 per cent).
- Assume that there is no delay in payment to labour, but a delay of one month occurs in paying to the subcontractors, material suppliers, and plant and machinery supplier.
- Retention is 10 per cent of billed amount in every bill. Fifty per cent retention amount is payable after one month of practical completion, while the remaining 50 per cent is payable six months later.

Solution

The R.A. bill computation is shown in Table S3.1.1. The computation for cash inflow for the contractor is shown in Table S3.1.2.

Reconciliation

Payment to be received for work done		7,625,000
Less TDS at 2%		(−) 152,500
Less advance payment	(−) 500,000	
Less materials issued by owner	(−) 320,000	
Actual payments to be received	6,652,500	
Actual payments made	1,137,000 + 2,083,000 + 1,753,000 + 917,000 + 762,500 = 6,652,500 (Hence, reconciled)	

The cash inflow diagram for the contractor is as shown in Figure S3.1.1.

A contractor is also supposed to pay to his subcontractors, suppliers for the materials, and labour for the work done during a month. These factors are taken into account while preparing the cash outflows of a contractor. The computation for cash-flow diagram is performed in Table S3.1.3 using the conditions of the problem given earlier.

Based on the above computation, the cash outflow diagram for the contractor is shown in Figure S3.1.2.

Table S3.1.2 Computation for cash inflow

	Item description	0	1	2	3	4	5	6	...	10
9	Total work progress (from above)		1,525,000	2,600,000	2,225,000	1,275,000				
	Recoveries									
a	Towards advance payment	500,000	125,000	125,000	125,000	125,000				
b	Towards material advance	320,000	80,000	80,000	80,000	80,000				
c	TDS		30,500	52,000	44,500	25,500				
d	Retention @10% of gross invoice (0.1* row 9)		152,500	260,000	222,500	127,500				
e	Total deductions (a + b + c + d)		388,000	517,000	472,000	358,000				
f	Payment due (9-e)		1,137,000	2,083,000	1,753,000	917,000				
f	Release of retention							381,250		381,250
g	Cash received (one-month delay)	820,000		1,137,000	2,083,000	1,753,000	1,298,250			381,250

The R.A. bill will look thus:

	1st R.A. bill	2nd R.A. bill	3rd R.A. bill	4th bill
Work done by measurement	1,525,000	4,125,000 (cumulative)	6,325,000 (cumulative)	7,625,000 (cumulative)
Recoveries as per statement (TDS, mobilization advance, material advance and retention	388,000	905,000 (cumulative)	1,377,000 (cumulative)	1,735,000 (cumulative)
Payment due	1,137,000	3,220,000 (cumulative)	4,973,000 (cumulative)	5,890,000 (cumulative)
Payment already made	Nil	1,137,000	3,220,000	4,973,000
Payment to be made now	1,137,000	2,083,000	1,753,000	917,000

88 | CONSTRUCTION PROJECT MANAGEMENT

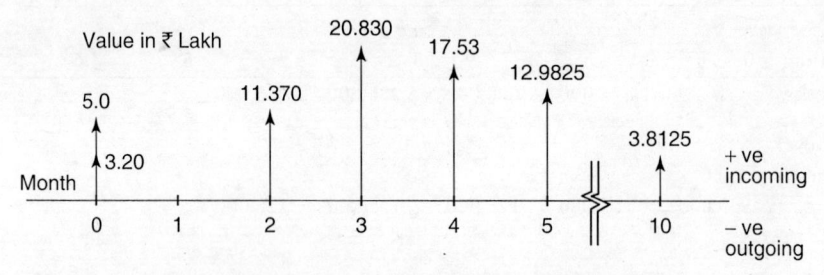

Figure S3.1.1 Cash-inflow diagram for the contractor

Table S3.1.3 Computation for cash outflows

	Item description	1	2	3	4	5	6	...	10
9	Total work progress (from above)	1,525,000	2,600,000	2,225,000	1,275,000				
10	Total cost (0.9 of above value)	1,372,500	2,340,000	2,002,500	1,147,500				
11	Cost towards labour @20% of total cost	274,500	468,000	400,500	229,500				
12	Cost towards material @60% of total cost	823,500	1,404,000	1,201,500	688,500				
13	Cost towards subcontractors @10% of total cost	137,250	234,000	200,250	114,750				
14	Cost towards plant and machinery @10% of total cost	137,250	234,000	200,250	114,750				
15	Labour payment (no delay)	274,500	468,000	400,500	229,500				
16	Material suppliers' payment (one-month delay)		823,500	1,404,000	1,201,500	688,500			
17	Subcontractors' payment (one-month delay)		137,250	234,000	200,250	114,750			
18	Plant and machinery supplier's payment (one-month delay)		137,250	234,000	200,250	114,750			
19	Outgoings total	274,500	1,566,000	2,272,500	1,831,500	918,000			

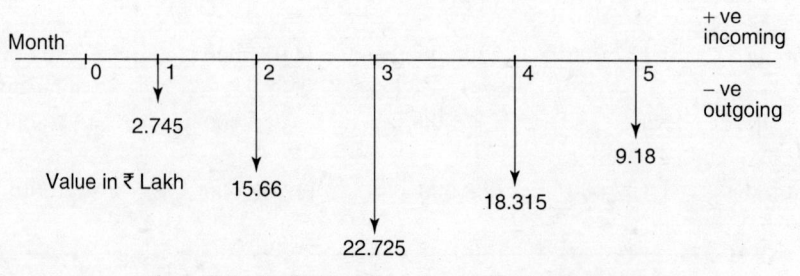

Figure S3.1.2 Cash-outflow diagram for the contractor

Example 3.2

A manufacturing company purchases materials worth ₹ 50 lakh every year. Calculate the present worth of material purchase for a five-year period, if the material price follows a geometric pattern with (a) g = −5%, (b) g = 0%, and (c) g = 5%. The interest rate can be assumed to be eight per cent.

Solution

Given c = ₹ 50 lakh. The cash-flow diagram for the given data is given in Figure S.3.2.1.

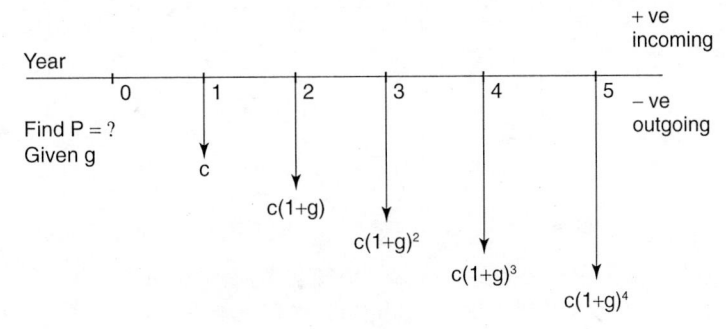

Figure S3.2.1 Cash-flow diagram

(a) The value of g = −5%

Geometric gradient factor corresponding to an interest rate of 8% = 3.641 (from Equation 3.14)
Thus, present worth of material purchase = 5,000,000 × 3.641 = ₹ 18,206,834.45

(b) The value of g = 0%

This implies there is no change in the annual cost and, thus, the cash-flow diagram would be as shown in Figure S3.2.2.
Thus, the present worth of material purchase = 5,000,000 (P/A, 8%, 5) = ₹ 19,963,500.00

(c) The value of g = 5%

Geometric gradient factor corresponding to an interest rate of 8% = 4.379 (from Equation 3.14)
Thus, present worth = 5,000,000 × 4.379 = ₹ 21,897,368.98

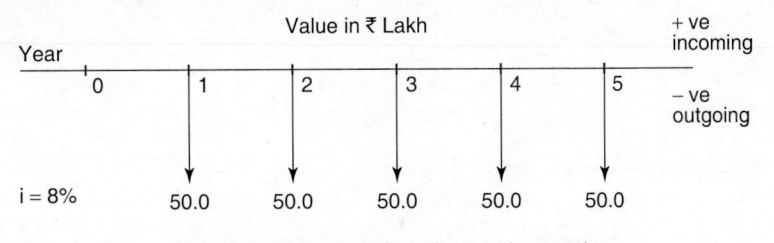

Figure S3.2.2 Cash-flow diagram (g = 0%)

Example 3.3

An alternative, A, requires an initial investment of ₹ 500,000 and an annual expense of ₹ 250,000 for the next 10 years. Alternative B, on the other hand, requires an initial investment of ₹ 750,000 and an annual expense of ₹ 200,000 for the next 10 years. Which alternative would you prefer if interest rate were 10 per cent?

Solution
The cash-flow diagram corresponding to alternative A is given in Figure S3.3.1.

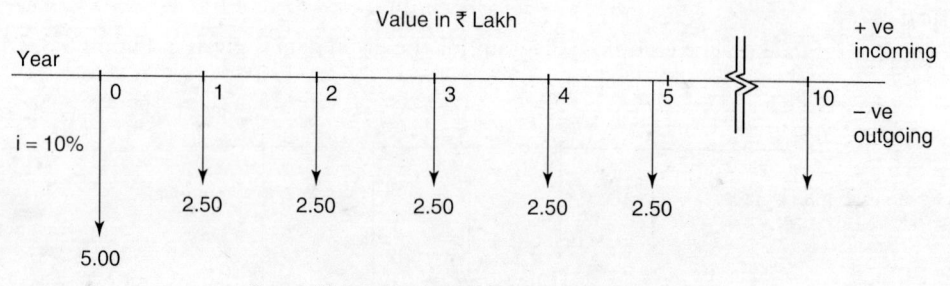

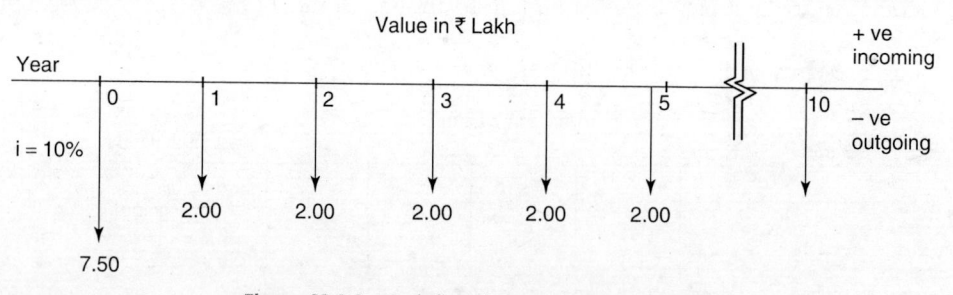

Figure S3.3.1 Cash flow diagram for Alternative A

Present cost of alternative A = 500,000 + 250,000 (P/A, 10%, 10)
                             = 500,000 + 250,000 (6.1446)
                             = ₹ 2,036,150.00

The cash-flow diagram corresponding to alternative B is given in Figure S3.3.2.
Present cost of alternative B = 750,000 + 200,000 (P/A, 10%, 10)
                             = 750,000 + 200,000 (6.1446)
                             = ₹ 1,978,920.00

It can be noticed from the two cash-flow diagrams that the cost data alone are provided. Thus, alternative with the lowest cost at present time would be the most preferable. In this case, since the present cost for alternative B is less than that of alternative A, it is preferable to choose alternative B.

Example 3.4
What is the present equivalent value of ₹ 50,000, five years from now at 14 per cent compounded semi-annually?

Figure S3.3.2 Cash flow diagram for Alternative B

Solution
The cash-flow diagram is shown in Figure S3.4.1. For an interest rate of 14 per cent compounded semi-annually, the effective interest rate can be obtained by using the formula

CONSTRUCTION ECONOMICS

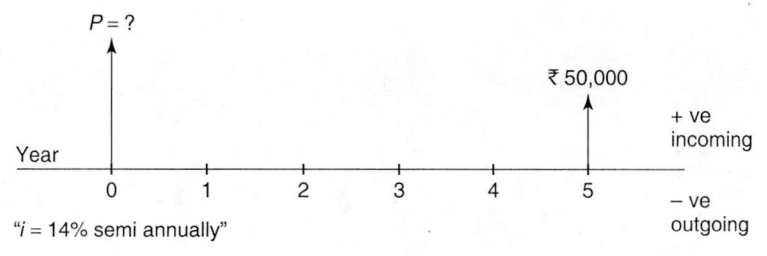

Figure S3.4.1 Cash flow diagram for Example 3.4

Here, $m = 2$, $i_{nom} = 14\%$

$$i_{eff} = \left[\left(1 + \frac{i_{mom}}{m \times 100}\right)^m - 1\right] \times 100$$

Thus,

$$i_{eff} = \left[\left(1 + \frac{14}{2 \times 100}\right)^2 - 1\right] \times 100$$

$$= 14.49\% = 0.1449$$

Thus, present equivalent $P = 50,000\ (P/F, i_{eff}, 5)$
The value of $(P/F, i_{eff}, 5)$ can be obtained by the formula $P = F/(1 + i_{eff})^5$
$$= 50,000\ (.5085)$$
$$= ₹ 25,417.46$$

In order to use the interest table, one needs to interpolate for an interest rate of i = 14.49%. For this, select (P/F, 14%, 5) and (P/F, 15%, 5), and interpolate linearly for an approximate value of (P/F, 14.49%, 5).

Example 3.5
An investor has to choose between the following two investment options:

(a) Investing in a bond that earns him a rate of return of 10 per cent on an eight-year investment

(b) Depositing ₹ 7.5 lakh for first three years and earning ₹ 5 lakh every year from the end of 4th year to the end of 8th year

Which investment option is desirable?

Solution
The cash-flow diagram for investment option 'b' is shown in Figure S3.5.1.
The net present worth of option 'b' at i = 10% (rate of return of option 'a') is found out. If it happens to be positive, it would indicate that option 'b' yields a return higher than 10%. If the net present worth were negative, it would indicate that return of option 'b' is less than 10%.

The net present worth = total benefits − total costs
The net present worth of option 'b' = + 500,000/(1 + i)⁴ + 500,000/(1 + i)⁵
+ 500,000/(1 + i)⁶
+ 500,000/(1 + i)⁷ + 500,000/(1 + i)⁸
− 750,000/(1 + i)¹ − 750,000/(1 + i)²
− 750,000/(1 + i)³
= − 440,975

Figure S3.5.1 Cash-flow diagram for option 'b'

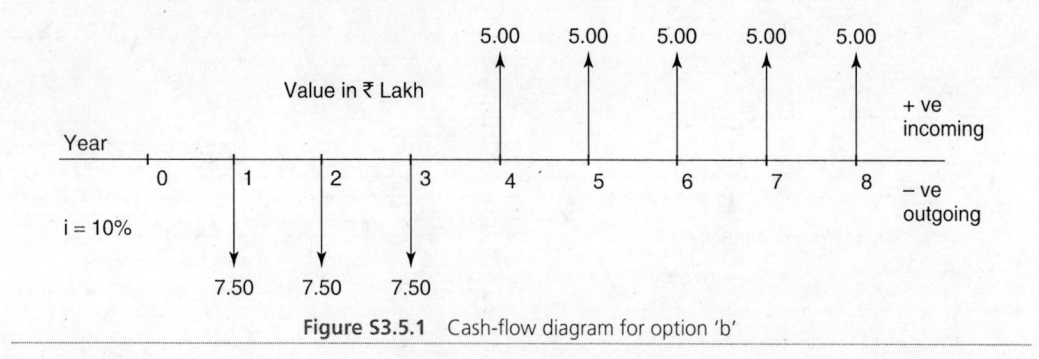

Since the net present worth of option 'b' yields negative net present worth at an interest rate of 10%, option 'a' is better than option 'b'. It may be noted that the above problem can be solved using interest tables also.

Example 3.6
Type 'A' design of a dam costs ₹ 50 crore to construct and an expense of ₹ 7.5 crore every year to operate and maintain it. Type 'B' design of the dam, on the other hand, would require ₹ 75 crore to construct and an annual expense of ₹ 5 crore to operate and maintain. Both the designs have considered 100 years as the design life of the dam. The minimum required rate of return is five per cent. Which design should be given a go-ahead?

Solution
Both the dam options can be assumed to be permanent as the design life is 100 years (significantly large).

The cash-flow diagram corresponding to type 'A' design of dam is given in Figure S3.6.1.

$$\text{The net present cost for this case} = 500{,}000{,}000 + 75{,}000{,}000/0.05$$
$$= ₹\ 2{,}000{,}000{,}000.00$$

The cash-flow diagram of type 'B' design of dam is given in Figure S3.6.2.

$$\text{The net present cost for this case} = 750{,}000{,}000 + 50{,}000{,}000/0.05$$
$$= ₹\ 1{,}750{,}000{,}000.00$$

The net present cost of type 'B' design of dam is less. Hence, it is preferable to go for type 'B' design of dam (note that all the values given pertain to cost). It may be noted that the above problem can be solved using interest tables also.

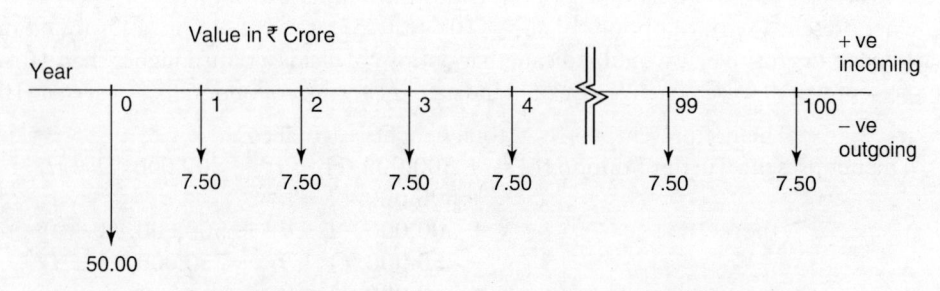

Figure S3.6.1 Cash-flow diagram for type 'A' dam design

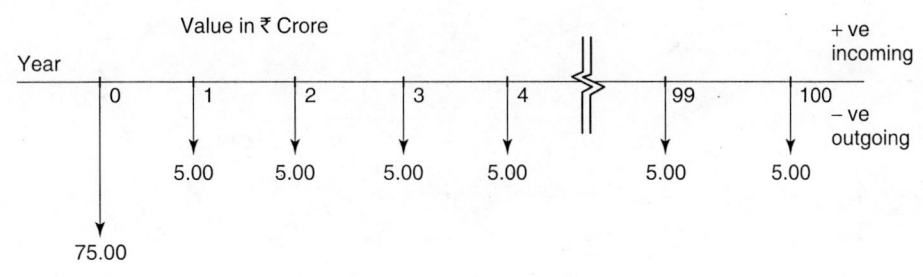

Figure S3.6.2 Cash-flow diagram for type 'B' dam design

Example 3.7

A contractor has been awarded to do a job that requires procurement of an equipment. Two brands A and B are available to perform the job. Brand A requires an investment of ₹ 450,000, while brand B requires an investment of ₹ 725,000. The annual savings generated by brands A and B are given in Table Q3.7.1. Which brand of equipment should the contractor choose if the interest rate is eight per cent?

Table Q3.7.1

Option	Saving details year-wise		
	1	2	3
Brand A	225,000	225,000	225,000
Brand B	300,000	300,000	300,000

Solution

The cash-flow diagram associated with brand 'A' is as shown in Figure S3.7.1.
The net present worth of brand 'A' is computed thus:
The net present worth = present worth of savings − present worth of cost
$$= 225{,}000 \, (P/A, 8\%, 3) - 450{,}000$$
$$= 225{,}000 \, (2.5771) - 450{,}000$$
$$= 579{,}847.50 - 450{,}000$$
$$= ₹ \, 129{,}847.50$$

The cash-flow diagram associated with brand 'B' is as shown in Figure S3.7.2. The net present worth of brand 'B' is computed thus:

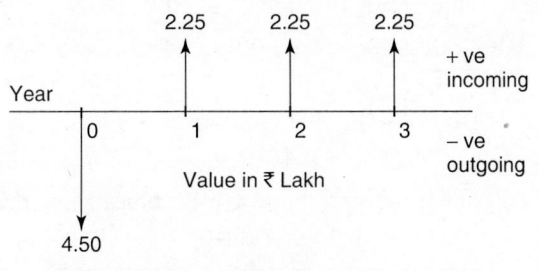

Figure S3.7.1 Cash-flow diagram for brand 'A'

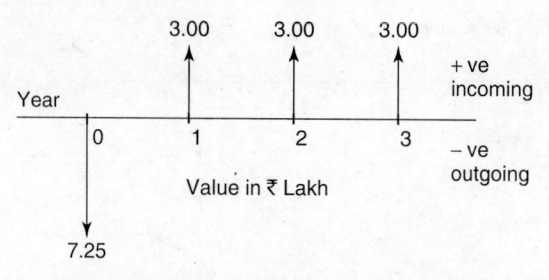

Figure S3.7.2 Cash-flow diagram for brand 'B'

The net present worth = present worth of savings − present worth of cost
= 300,000 (P/A, 8%, 3) − 725,000
= 300,000 (2.5771) − 725,000
= 773,130 − 725,0000
= ₹ 48,130.00

The net present worth associated with brand A is greater than the net present worth associated with brand B. So, select brand A.

Example 3.8
A piece of land has been purchased at ₹ 40 lakh. An investment of an additional ₹ 20 lakh has been made to construct a small shopping complex on this piece of land. It is expected to fetch an annual rental of ₹ 75,000 to the owner, while the cost towards its upkeep, tax, etc., is expected to be ₹ 30,000 annually. The owner plans to sell the entire plot with constructed facilities at an expected price of ₹ 120 lakh at the end of five years. What percent rate of return will be earned by the owner on this investment?

Solution
The cash-flow diagram for the given data is shown in Figure S3.8.1
Assuming a trial rate of return of 15%,

Net present worth = −₹ 6,000,000 − ₹ 30,000 × (P/A, 15%, 5) + ₹ 75,000 × (P/A, 15%, 5) + ₹ 12,000,000 × (P/F, 15%, 5)

Thus, net present worth = −₹ 6,000,000 − ₹ 30,000 × 3.3522 + ₹ 75,000 × 3.3522 + ₹ 12,000,000 × 0.4972 = 117,249

The positive net present worth indicates that the rate of return is higher than 15%.
Let us assume a higher value of rate of return, say 20%.

Net present worth = −₹ 600,000 − ₹ 30,000 × (P/A, 20%, 5) + ₹ 75,000 × (P/A, 20%, 5) + ₹ 12,000,000 × (P/F, 20%, 5)

Net present worth = −₹ 6,000,000 − ₹ 30,000 × 2.9906 + ₹ 75,000 × 2.9906 + ₹ 12,000,000 × 0.4019 = − 1,042,623

Thus, the rate of return at which net present worth becomes zero, that is, the value of i at which exactly the same return is met, lies somewhere between 15% and 20%, which can be found out by interpolation.

CONSTRUCTION ECONOMICS | 95

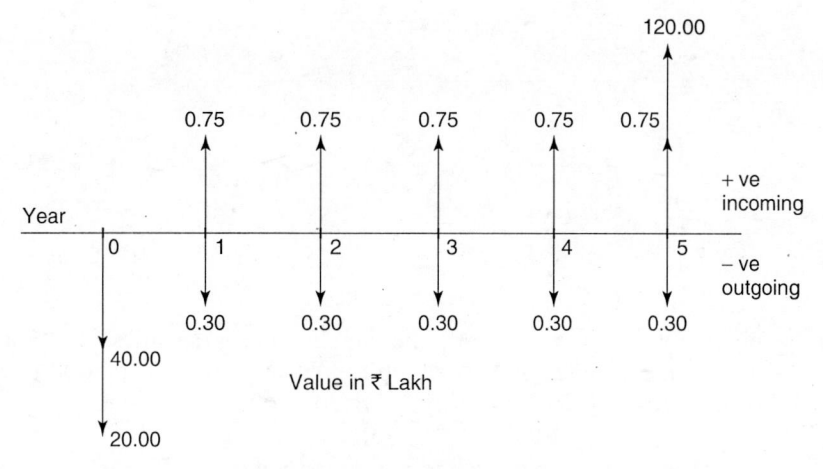

Figure S3.8.1 Cash-flow diagram

$$i = 15 + \frac{(20-15) \times 117,249}{(117,249 - (-1,042,623))} = 15.50\%$$

Example 3.9

A supplier of prefabricated railway sleepers procures each piece of sleeper for ₹ 4,000. The demand for sleepers is 350 units, and it is estimated that a similar demand would prevail for another three years. Equipment to manufacture sleepers is available for ₹ 18 lakh. The annual operating cost for producing 350 sleepers is estimated to cost ₹ 7 lakh for year 1, with 10 per cent increase every year for years 2 and 3. If the equipment has no salvage value at the end of three years, should the supplier continue to outsource it or should he buy the equipment and start producing the sleepers on his own? The minimum attractive rate of return is 15 per cent.

Solution

There are two options before the supplier: (1) option of outsourcing, and (2) buy the equipment and start manufacturing on his own.

The cash-flow diagram for option 1 is shown in Figure S3.9.1. The total cost to outsource = ₹ 4,000 × 350 = 1,400,000 for years 1, 2 and 3.

The cash-flow diagram for option 2 is shown in Figure S3.9.2.

As can be seen from both the cash-flow diagrams, only cost-related information is available. Thus, it is not possible to calculate the rate of return for individual options. In view of this, it is proposed that the incremental rate of return method be used for comparison of alternatives. The incremental cash-flow diagram is shown in Figure S3.9.3.

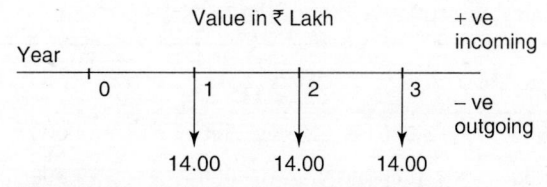

Figure S3.9.1 Cash-flow diagram for option of outsourcing

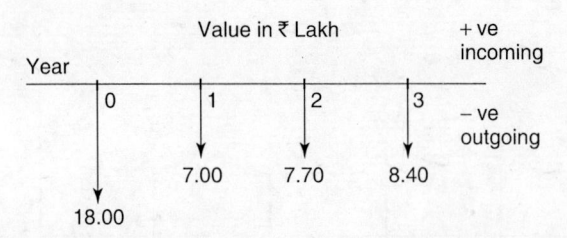

Figure S3.9.2 Cash-flow diagram for option of manufacturing on his own

It can be observed that at time zero, the supplier needs to invest ₹ 1,800,000 in equipment, which will give savings of ₹ 1,400,000 − ₹ 700,000 = ₹ 700,000 in year 1; ₹ 1,400,000 − ₹ 770,000 = ₹ 630,000 in year 2; ₹ 1,400,000 − ₹ 840,000 = ₹ 560,000 in year 3.

A trial-and-error process is adopted to find out the rate of return.
Assume a trial rate of return equal to minimum attractive rate of return, that is, $i = 15\%$
The net present worth at $i = 15\%$ = −₹ 1,800,000 + ₹ 700,000 × (P/F, 15%, 1) + ₹ 630,000 × (P/F, 15%, 2)
+ ₹ 560,000 × (P/F, 15%, 3)

Thus, net present worth $\doteq$ −₹ 346,700

Since the net present worth at $i = 15\%$ is negative, it is not desirable to invest in purchase of equipment. In other words, it is advisable to continue with the existing system of outsourcing.

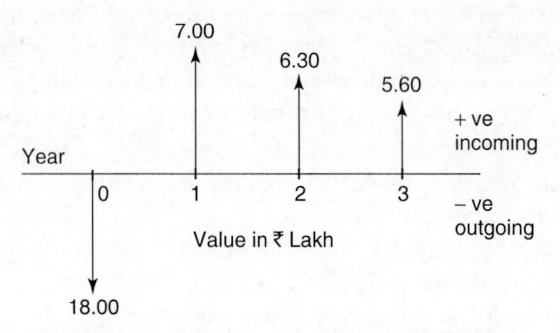

Figure S3.9.3 Cash-flow diagram on incremental basis

Example 3.10

Solve the problem using (a) present worth method, (b) annual worth method, and (c) internal rate of return method for the given cash-flow data. Check your results with incremental rate of return method. The minimum attractive rate of return is 10 per cent.

End of Year	0	1	2	3	4
Project X	−50,000	5,000	17,500	30,000	42,500
Project Y	−50,000	40,000	15,000	15,000	15,000

Solution

The cash-flow diagrams for projects X and Y are shown in Figure S3.10.1 and Figure S3.10.2, respectively.

(a) Present worth method

The net present worth for project X = $-50,000 + 5,000 \times (P/F, 10\%, 1) + 17,500 \times (P/F, 10\%, 2) + 30,000 \times (P/F, 10\%, 3) + 42,500 \times (P/F, 10\%, 4)$

NPW = $-50,000 + 5,000 \times 0.9091 + 17,500 \times 0.82645 + 30,000 \times 0.75131 + 42500 \times 0.6830$

= ₹ 20,575

The net present worth for project Y = $-50,000 + 40,000 \times (P/F, 10\%, 1) + 15,000 \times (P/F, 10\%, 2) + 15000 \times (P/F, 10\%, 3) + 15,000 \times (P/F, 10\%, 4)$

NPW = $-50,000 + 40,000 \times 0.9091 + 15,000 \times 0.8264 + 15,000 \times 0.7513 + 15,000 \times 0.6830$ = ₹ 20,274

Hence, choose X since the net present worth for project X is more.

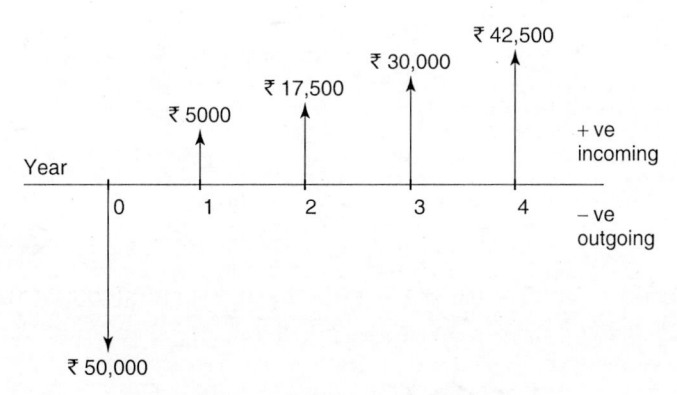

Figure S3.10.1 Cash-flow diagram for project X

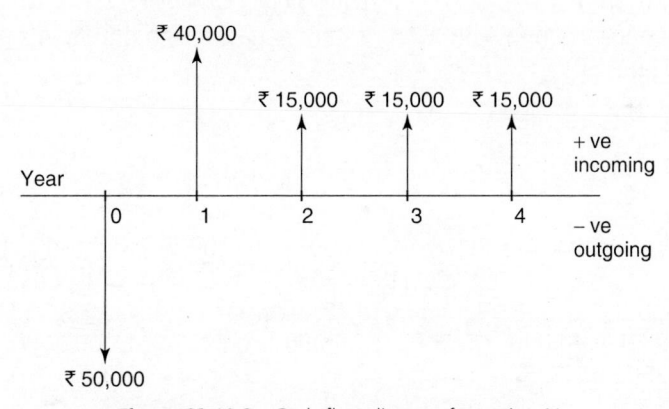

Figure S3.10.2 Cash-flow diagram for project Y

(b) Annual worth method

The equivalent annual cost for project X = $-50{,}000 \times$ (A/P, 10%, 4) + $5{,}000 \times$ (P/F, 10%, 1) (A/P, 10%, 4) + $17{,}500 \times$ (P/F, 10%, 2) (A/P, 10%, 4) + $30{,}000 \times$ (P/F, 10%, 3) (A/P, 10%, 4) + $42{,}500 \times$ (P/F, 10%, 4) (A/P, 10%, 4)

Equivalent annual cost = $-50{,}000 \times 0.3155 + 5{,}000 \times 0.9091 \times 0.3155 + 17{,}500 \times 0.8264 \times 0.3155 + 30{,}000 \times 0.7513 \times 0.3155 + 42{,}500 \times 0.6830 \times 0.3155$

Equivalent annual worth = ₹ 6,491.09

The equivalent annual cost for project Y = $-50{,}000 \times$ (A/P, 10%, 4) + $40{,}000 \times$ (P/F, 10%, 1) (A/P, 10%, 4) + $15{,}000 \times$ (P/F, 10%, 2) (A/P, 10%, 4) + $15{,}000 \times$ (P/F, 10%, 3) (A/P, 10%, 4) + $15{,}000 *$ (P/F, 10%, 4) (A/P, 10%, 4)

Equivalent annual worth = $-50{,}000 \times 0.3155 + 40{,}000 \times 0.9091 \times 0.3155 + 15{,}000 \times 0.8264 \times 0.3155 + 15000 \times 0.7513 \times 0.3155 + 15000 \times 0.6830 \times 0.3155$

Equivalent annual cost = ₹ 6,396.60

Hence, choose X since equivalent annual worth for project X is higher.

(c) Internal rate of return method

For project X,
Assume i = 20%
The net present worth at i = 20% = $-50{,}000 + 5{,}000 \times$ (P/F, 20%, 1) + $17{,}500 \times$ (P/F, 20%, 2) + $30{,}000 \times$ (P/F, 20%, 3) + $42{,}500 \times$ (P/F, 20%, 4)

NPW = $-50{,}000 + 5{,}000 \times 0.8333 + 17{,}500 \times 0.6944 + 30{,}000 \times 0.5787 + 42500 \times 0.4823 = 4177.25$

Assume i = 25%
The net present worth at i = 25% = $-50{,}000 + 5{,}000 \times$ (P/F, 25%, 1) + $17{,}500 \times$ (P/F, 25%, 2) + $30000 \times$ (P/F, 25%, 3) + $42{,}500 \times$ (P/F, 25%, 4)

NPW = $-50{,}000 + 5{,}000 \times 0.8 + 17{,}500 \times 0.64 + 30{,}000 \times 0.512 + 42{,}500 \times 0.4096 = -2032.00$

By interpolation,
i = 23.371%

For project Y,
Assume i = 30%
The net present worth at i = 30% = $-50{,}000 + 40{,}000 \times$ (P/F, 30%, 1) + $15{,}000 \times$ (P/F, 30%, 2) + $15{,}000 \times$ (P/F, 30%, 3) + $15{,}000 \times$ (P/F, 30%, 4)

NPW = $-50{,}000 + 40{,}000 \times 0.7692 + 15{,}000 \times 0.5917 + 15{,}000 \times 0.4552 + 15{,}000 \times 0.3501 = 1723$

Assume i = 35%

The net present worth = $-50,000 + 40000 \times (P/F, 35\%, 1) + 15,000 \times (P/F, 35\%, 2) + 15,000 \times (P/F, 35\%, 3) + 15,000 \times (P/F, 35\%, 4)$

NPW = $-50,000 + 40,000 \times 0.741 + 15,000 \times 0.5509 + 15,000 \times 0.4098 + 15,000 \times 0.3052 = -1371.5$

By interpolation,
i = 32.78%

Hence, choose project Y since it gives a higher rate of return.

(d) Incremental rate of return method

Let project X be preferred over project Y. Then, the cash-flow diagram on incremental basis would be as shown in Figure S3.10.3.

We perform trial-and-error to find the rate of return for the cash-flow diagram shown in Figure S3.10.3.

Assume i = 10%

The net present worth = $0 - 35,000/(P/F, 10\%, 1) + 2,500(P/F, 10\%, 2) + 15,000 (P/F, 10\%, 3)^3 + 27,500 (P/F, 10\%, 4) = 300.40$ (check)

Assume i = 15%

The net present worth = $0 - 35,000(P/F, 15\%, 1) + 2,500(P/F, 15\%, 2) + 15,000 (P/F, 15\%, 3) + 27,500(P/F, 15\%, 4) = -2958.46$ (check)

By interpolation, $i = 10 + \dfrac{(15-10) \times 300.4}{(300.4 - (-2958.46))} = 10.46\%$

Since 10.46% (exact value of *i*) > 10% (MARR), prefer X over Y.

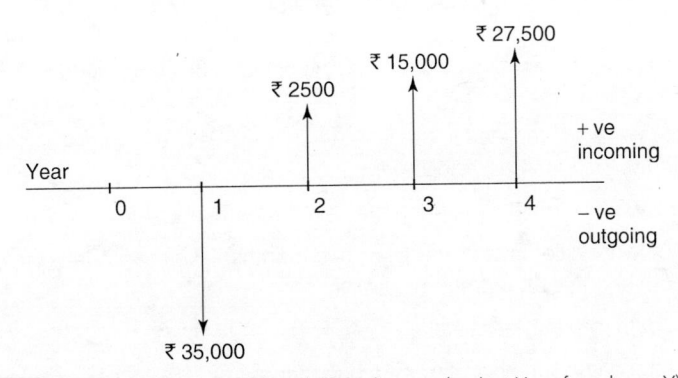

Figure S3.10.3 Incremental cash-flow diagram (project X preferred over Y)

Remarks

With the help of this example, it has been emphasised that the results could vary if internal rate of return method is applied for ranking the projects. In this example, we saw that although project Y has higher internal rate of return, it is not the best alternative as suggested by present worth method, annual worth method and incremental rate of return method. Thus, internal rate of return method should not be used for ranking, and we should always use incremental rate of return for the reasons described inside the text.

REVIEW QUESTIONS

1. Fill in the blanks.
   a. ₹ 50,000 now is equivalent to ₹ _____ after 10 years. at a rate of interest 10%.
   b. ₹ 50,000 is equivalent to an annual payment of ₹ _____ for a period of 10 years at an interest rate of 10%.
   c. The present value of a sum of ₹ 129,700 to be received 10 years hence at an interest rate of 10% is ₹ _____.
   d. ₹ 8,137.50 every year for next 10 years at a rate of interest of 10% is equivalent to a future sum of ₹ _____ at the end of 10 years.
   e. ₹ 50,000 payable at the end of 10 years is equivalent to ₹ _____ at present at an interest rate of 10%.
   f. At an interest rate of _____ %, ₹ 5 lakh invested today will be worth ₹ 10 lakh in 9 years.

2. Write short notes on the payback period method.

3. A construction company is contemplating purchase of a crawler excavator costing ₹ 6,000,000. The finance officials have identified the cost and benefits of the investments, which are given below. Company proposes to use the excavator for four years. The required rate of returns is 16%. Ascertain the payback period and discounted payback period.

Initial investment	1st year	2nd year	3rd year	4th year
₹ 6,000,000	₹ 1,500,000	₹ 2,200,000	₹ 2,800,000	₹ 3,500,000

4. A project is estimated to cost ₹ 2.4 crore and is expected to be completed in 12 months. You may assume the following:

   - 10% of the work is completed in the first two months. Take the progress to be 3% and 7% in the first and second months, respectively.
   - 80% of the work is completed during the next 8 months.
   - 10% of the work is completed in the last two months. Take the progress to be 8% and 2% in the eleventh and twelfth months, respectively.

   The following data may also be useful:

   - The contractor has to submit an earnest money deposit of 10% of the estimated cost at the time of submitting his tender in the form of demand draft. This amount is to be retained by the client as initial security deposit.
   - The contractor needs to be paid a mobilization advance of ₹ 16 lakh initially, which can be recovered in four equal instalments, beginning with the third running bill.
   - An amount equivalent to 5% of the gross amount due is to be retained in each bill towards (additional) security deposit.
   - 50% of the total security deposit is to be returned to the contractor at the end of 6 months after the project has completed, and the remaining part after another 6 months.

- 10% recovery is to be made towards income tax. Payment to the department of income tax is to be made at the end of the project.
- The contractor submits monthly bills on the first day of the following month.
- On the client's side, it takes about 20 days to process a bill, and payment is made only on the last day of the same month in which the bill is submitted.
- Client maintains his funds with the bank and is paid an interest of 5% compounded monthly.
- Rate of interest on borrowing from the bank is 15% per annum, compounded monthly.

Based on the information provided: (a) Clearly show how much cash the client needs at different points in time as the project progresses. (b) How much money does the client need to borrow at different points of time, and what will be his total liability at the completion of the project (say, at the end of month 24)? Clearly outline all the assumptions you make in your analysis.

5. A newly opened toll bridge has a life expectancy of 25 years. Considering a rise in construction and other costs, a replacement bridge at the end of that time is expected to cost ₹ 60,000,000. The bridge is expected to have an average of 100,000 toll-paying vehicles per month for the 25 years. At the end of every month, the tolls will be deposited into an account bearing annual interest of 11 per cent compounded monthly. The operating and maintenance expenses for the toll bridge are estimated at Re 0.80 per vehicle.

   a. How much toll must be collected per vehicle in order to accumulate ₹ 60,000,000 by the end of 25 years?

   b. What should be total toll per vehicle in order to include operating and maintenance cost as well as replacement cost?

   c. What annual effective interest rate is being earned?

6. A small shopping complex can be constructed for ₹ 5,000,000. The financing requires ₹ 500,000 down (you pay) and a ₹ 4,500,000 loan at 10%, with equal annual payments. The gross income for the first year is estimated at ₹ 600,000 and is expected to increase 5% per year thereafter. The operating and maintenance costs and taxes should average about 40% of the gross income and rise at the same 5% annual rate. The resale value in 30 years is estimated at ₹ 10,000,000. Find the rate of return on this investment over the 30-year life. Solve the problem assuming that the increase follows (i) arithmetic gradient and (ii) geometric gradient.

7. Due to increasing age and downtime, the productivity of a contractor's excavator is expected to decline with each passing year. Using the given data, calculate the price in terms of rupees per m³ that the contractor must charge to cover the cost of buying and selling the excavator.

   Cost new = ₹ 2,200,000

   Resale price at the end of year 6 = ₹ 900,000

   Contractor borrows at i = 10%

   Production decline at r = (−) 6%

   Production for year 1 = 160,000 m³

   Assume all funds are credited at end of year

8. A contractor has a contract to construct a 3,600-metre-long tunnel in 30 months. He is trying to decide whether to do the job with his own forces or subcontract the job. You are to calculate the equivalent monthly cost under each alternative. His i = 1.75% per month. Production under both alternatives will be 120 metres per month.

    Alternative A: Buy a tunnelling machine and work with own force

    a. Cost of tunnelling machine = ₹ 20,000,000
    b. Salvage value of machine at end of month 30 = ₹ 4,000,000
    c. Cost of labour and material = ₹ 10,000/m for the first 10 months, and increasing by 0.5% per month at EOM each month thereafter (i.e., ₹ 10,050/m at EOM11, ₹ 10,100.25 at EOM 12, etc.)

    Alternative B: Subcontract the work

    Cost is ₹ 27,000 per metre of tunnel

9. A sewage treatment plant has three possible schemes of sewage carriage and treatment systems. If the life of the scheme is 20 years, which scheme should be recommended as the most economic?

Scheme	Installation Cost (₹)	Annual Running Cost (₹)
A	18,250,000	7,250,000
B	20,200,000	4,600,000
C	24,200,000	4,000,000

    Use 15% to represent the cost of capital. If the cost of capital were 10%, would the recommendation alter?

10. A town built on a river is considering building an additional bridge across the river. Two proposals have been put forward for bridges at different sites. The costs of each proposal are summarized as follows:

Description	Bridge A	Bridge B
Initial cost of bridge	₹ 6,500,000	₹ 5,000,000
Initial cost of road works	₹ 3,500,000	₹ 3,000,000
Annual maintenance of bridge	₹ 50,000	₹ 90,000 for first 15 years and 110,000 thereafter
Annual maintenance of roads	₹ 30,000	₹ 25,000
Life of bridge	60 years	30 years
Life of roads	60 years	60 years

    With the cost of capital at 9%, which proposal should be adopted? (Assessment of the proposals to be carried out by a comparison of their present worth)

11. Your firm owns a large earthmoving machine that may be renovated to increase its production output by an extra 8 m³/hour, with no increase in operating costs. The renovation will cost ₹ 500,000. The earthmoving machine is expected to last another eight years, with zero salvage value at the end of that time. Earthmoving for this machine is currently being contracted at

₹ 14/m³. The company is making an 18 per cent return on their invested capital. You are asked to recommend whether or not this is a good investment for the firm. Assume the equipment works 1,800 hours per year. (Hint i = 18%)

12. For the following data of a project, calculate the payment received by the contractor at different time periods. Also prepare (a) the month-wise running account bill, (b) the cash inflow diagram for the contractor, and (c) the cash outflow diagram for the owner.
    - Value of contract: ₹ 10,000,000 (one crore rupees)
    - Time period: 4 months
    - Payment schedule: First month 15% of the contract value, 2nd month 40% (cumulative), 3rd month 80% (cumulative) and 4th month 100% (cumulative)
    - The owner makes an advance payment of ₹ 5 lakh, which is to be recovered in 4 equal instalments
    - The owner also supplies materials to the value of about ₹ 3.2 lakh, which value is also to be recovered equally from each running account bill
    - In addition, the owner will recover from the payments made to the contractor 2% of the value of the work done as income tax deducted at source, and deposit this amount with the Reserve Bank of India

13. Two equipment 'A' and 'B' have the capability of satisfactorily performing a required function. Equipment 'B' has an initial cost of ₹ 160,000 and expected salvage value of ₹ 20,000 at the end of its four-year service life. Equipment 'A' costs ₹ 45,000 less initially, with an economic life one year shorter than that of 'B'; but 'A' has no salvage value, and its annual operating costs exceed those of 'B' by ₹ 12,500. When the required rate of return is 15%, state which alternative is preferred when comparison is by
    a. the common multiple method
    b. A 2 year study period (assuming the assets are needed for only 2 years).

4

Client's Estimation of Project Cost

Introduction, approximate methods of estimation, types of estimates, methods of structuring project costs, illustrative cases in preparation of estimate

4.1 INTRODUCTION

An accurate estimate of the cost involved is of great importance, especially to a contractor, as this forms a basis for his bid proposals, procurement plans and control of the cost being incurred in a job. Estimating in construction projects is a very complex process due to inherent interactions and interdependence involved and the absence of standard norms. Given the highly competitive environment at present, it is important that special attention is paid to cost estimation so that a contractor can win a job and still maintain a reasonable margin of profit.

It may be noted that the client (or owner) organizations and contracting agencies draw up their own estimates from the information and data available to them. Also, for different purposes, the required precision in the estimates is different – for example, for budgetary purposes an owner may be happy with even ±10 per cent variation, but such an error for a contractor bidding for the job could be simply disastrous and unacceptable. There is, thus, a clear difference in perception towards estimation in the client and contracting organizations. An effort has been made to bring the various issues in client's estimation of project cost in this chapter. The issues inherent in project cost estimation from contractor's perspective are presented in Chapter 8.

It is obvious that the accuracy of an estimate can be improved once the nature of the project is clearly defined, and all quantities, quality of material and workmanship, logistics, etc., are well understood. Since this is often not the case, rule of thumb and approximate methods have also evolved, and reference has been made to some of the more commonly used approaches at appropriate places. For building works, for example, whereas approximate estimates are at times drawn up on the basis of plinth area, floor area, enclosed volume, length of wall, etc., detailed estimates would require specification, programme of works, available drawings, material, manpower and equipment productivity, and the indirect costs.

4.2 APPROXIMATE METHODS OF ESTIMATION

The actual cost involved in any project can be accurately and simply worked out once the exact quantities of different items and the unit cost associated with each of them is known; however,

Table 4.1 Estimation of preliminary costs of different civil engineering jobs

S. No.	Structure	Basis for preliminary estimates
1	Buildings	Depending upon the type of the building, the basis could be number of occupants (e.g., per student for schools and hostels, per bed for hospitals, or per seat for theatres). Some of the other commonly used bases are also discussed separately in subsequent paragraphs.
2	Roads and highways	The estimates are usually made on a per-kilometre basis, depending upon the nature and classification of the road, its width and the cross-section of the road. Obviously, the estimates are different for concrete and bitumen-topped roads.
3	Irrigation works	Estimates for these could be on the basis of length involved, the capacity of the channel, or also the area of land to be covered.
4	Bridges and culverts	This is done on the basis of the span of the structure, depending on the roadway, the nature and depth of foundation, the type of structure, etc. For small culverts, an approximate cost may also be available in handbooks on a 'per culvert' basis for culverts of different spans. At times, approximate costs of bridges are also worked out separately for the substructure and the superstructure.
5	Sewage and water supply	Estimates for these could be made on the basis of per head of population served or the area covered.
6	Overhead water tank	Estimates could be made on the basis of the capacity of the tank, depending on the type of structure, height of staging, etc.

while the latter is known with reasonable accuracy, the former is very difficult to determine till it is fairly late into the project. It may be borne in mind that the specifications used for the materials and construction methods are also of relevance as far as estimation of the cost is concerned. However, preliminary estimates (including estimates that may not be very accurate) of various aspects of a work are essential for planning, mobilization of resources, policy decisions and administrative approvals. Such preliminary estimates may be prepared differently for different structures, as summarized in Table 4.1.

Table 4.1 gives an idea of the diverse bases that need to be adopted for estimation of preliminary costs of civil engineering jobs, and, indeed, experienced clients/contractors have reasonably accurate thumb rules for the purpose. Some of the other methods used for building and industrial constructions are briefly discussed below.

4.2.1 Preliminary Estimate for Buildings

Plinth Area Method

For similar specifications, a bigger building costs more to build than a smaller building. This dictum forms the basis of this method and estimates are prepared on the basis of plinth area of the building. The rate is being deduced from the cost of similar buildings having similar specifications, heights and types of construction in the same locality. Plinth area estimates are made by determining the plinth area of a building and multiplying it by the plinth area rate. The former should be calculated for the covered area by taking external dimensions of the building at the floor level, excluding areas such as a courtyard or other open areas, which are not likely to count towards the construction cost.

In cases where the plan of the building is not ready or available, and the plinth area cannot be accurately determined, the estimate may be made by estimating the proposed floor area of the rooms from the functional requirements, and about 30 per cent to 40 per cent of this total added to account for walls, circulation area, etc., and an approximate cost may be obtained by multiplying the unit rate and the estimated plinth area rate.

Plinth Area Rate (PAR) method is one of the popular methods for preparing preliminary estimates. PAR is prepared by Central Public Works Department (CPWD) for preparation of preliminary estimates based on plinth area of a building. In PAR-2007, buildings are classified in residential and non residential categories such as houses of different types (e.g. Type I, Type II, Type III, Type IV, Type V, Type VI, and Servant quarters), hostels, offices, college, schools, and hospitals. In PAR-2007, typical specifications such as type of foundation (based on safe bearing capacity), type of structure (such as load-bearing or RCC-framed structure), flooring (such as mosaic, Kota stone, or marble), other finishing items and electrical fixtures are specified for a standard (nominal) building specification. The rates specified in PAR-2007 are based on prevailing market rates as on 1.10.2007, and the base index for year 2007 is assumed to be 100. At regular intervals, the indices are revised based on the current market rates. Correction factors need to be applied if PAR-2007 is to be used in places other than Delhi.

The step-by-step method of preparing preliminary estimates based on PAR-2007 is given below:

Step 1: Preparation of preliminary architectural drawings based on requirement of area, specification, function, etc.

Step 2: Fixing of architectural features, type of foundation system, type of superstructure, general specification pertaining to civil works, electrical works, sanitary, water supply, external development, etc., and special structural requirements

Step 3: Based on the specification, classification of proposed building into one of the categories available in PAR-2007. Additional amount, as the case may be, need to be applied for a richer specification than those specified in PAR-2007

Step 4: Calculation of plinth area of the building in accordance with the specified methods

Step 5: Multiplication of the plinth area and the specified rate with additional cost for richer specification, if any

Step 6: Cost-index correction is applied based on the prevailing cost index

Step 7: Addition of contingency and architect's fee (if applicable) in the total sum calculated as above

Cube Rate Estimate

The cube rate estimate based on the overall volume of enclosed space comes handy, especially in cases such as multi-storey buildings, with the rate per cubic metre being deduced on the basis of location, specifications and type of construction. The volume of the building is calculated from the external dimensions of the building (with the length and breadth being taken at the floor level), excluding the foundation, the plinth and the parapet above roof. Needless to say, this method is more accurate than the plinth area estimate. In fact, some engineers also use an approximate estimate based on the total floor area to be provided on a per-square-metre basis. This method also includes the height of the building and is fairly accurate.

Approximate Quantity Method Estimate
This method is based on the total length of the walls in a building, which is found from the drawings, and then using a per-metre rate, the total cost is estimated. In effect, for this method the building may be considered as divided into two parts—the foundation including plinth and the superstructure, and the cost per running metre is calculated using appropriate values for the two portions separately. In determining the rate for the foundation part, items such as excavation, brick work up to plinth level and the damp-proof course should be included, and the rate for the superstructure should appropriately account for items such as brickwork, woodwork, flooring and roof, finishing works, etc. Needless to say, the plan or line plan of the structure should be available for this method.

4.2.2 Preliminary Estimate for Industrial Structures

Ratio Method
Rather than a single method, these are actually a class of methods that provide only a crude estimate since they ignore several important details and are good only for very preliminary cost estimates, such as those an investor may require for studying opportunities and screening options for further study. Some of the commonly used ratio methods are briefly explained below.

Investment per annual tonne capacity Here, the cost of an upcoming plant or project is estimated on the basis of a similar plant of similar or even a different capacity elsewhere.

Turnover ratio and capital ratio The ratio between annual sales and investment expressed in monetary units (rupees, dollars, etc.) is known as turnover ratio, and its inverse (ratio of plant investment and annual sales) is known as capital ratio. Now, if the capital ratio relating to a particular process and plant size is C, then for the proposed plant, knowing the sales volume and price, the installed cost R_2 can be estimated as

$$R_2 = C \times V_1 \times P_1 \qquad (4.1)$$

where V_1 is proposed projected annual sales volume and P_1 is the price per unit of sales volume.

Six-tenth factor This is a modification of the investment per annual t capacity method, where in order to find the cost of a component, a system, or a plant, the size of the new item is determined using a reference item and a 'factor'. Mathematically, the method can be represented as follows:

$$\text{New component's cost} = \left(\frac{\textit{Attribute of new component}}{\textit{Attribute of reference component}}\right)^{\textit{factor}} \times \textit{Cost of reference component} \qquad (4.2)$$

Although any value between 0.20 and 1.37 can be taken for the factor, it is usually taken as 0.6 (and hence the name of the method). This method is a fast way to develop an estimate and sometimes good enough to be used even for preparing a definitive estimate.

Table 4.2 Types of estimates used in the construction industry

S. No.	Estimate type	Comments
1	Project proposal indicative cost estimate	Feasibility stage
2	Preliminary estimate	Budgeting costs toward the end of planning and design phase
3	Detailed estimate	Controlling costs during the execution of the project
4	Definitive estimate	To assess cost at completion
5	Final closure cost estimate	Final cost

4.3 TYPES OF ESTIMATES

Cost estimates of different accuracies are required by different agencies for different purposes in different stages of the project! The primary objective of an estimate is to enable one to know beforehand the cost of the work, though the actual cost is only known after the completion of the work from the accounts of the completed work. If the estimate is prepared carefully, there will not be much difference in the estimated cost and the actual cost. For accurate estimate, the estimator should be experienced and fully acquainted with the method of construction.

The estimate may be prepared approximately in a manner explained earlier or may be made in detail, item-wise. In general, a choice of the method of estimation to be used depends upon the nature of the project, the life-cycle phase, the purpose for which the estimate is required, the degree of accuracy desired and the estimating effort employed. The essential steps in estimation are calculation of quantity and cost. While the working of quantities can be done based on the available drawings and design data, a number of other things are needed to estimate the cost. While rough estimates are prepared based on some thumb rule, the detailed estimates are prepared based on a number of factors such as specification, programme of works, available drawings, material, manpower and equipment productivity, and indirect costs.

Table 4.2 summarizes some of the types of estimates that are used in the construction industry and their basic functions. Some of these are explained in greater detail in subsequent paragraphs.

4.3.1 Rough Order of Magnitude Estimates

These estimates are also referred to as ROM estimates and are useful for 'go/no-go' kind of decision making, which essentially refers to whether the project should or should not be pursued. Some of the methods that can be useful for such estimates are investment per annual t, capacity, turnover and capital ratio, and six-tenth factor. The accuracy levels in such estimates are usually only of the order of ± 60 per cent.

4.3.2 Client's Indicative Cost Estimate

A client on the basis of the indicative cost estimates prepared during the project feasibility stage may decide to proceed with the engineering phase of the project, which involves detailed design and plans for execution. At this stage, a more refined indicative estimate is also required, and there are several methods available for the purpose—cost per function estimate, square metre method, cubic metre method, cost index method, schedule-based estimates and S-curve forecast. The choice of a method at this stage depends on the nature of project, as the execution of the process (i.e., getting the estimate) needs little time and effort (may be just a few hours) provided the past performance data and prevailing trend/market rates are available. The range of accuracy at this stage is about 40 per cent and, therefore, further refinement is needed, which can be done once the architectural and structural drawings and other related information become available.

4.3.3 Client's Preliminary Cost Estimate

During the project planning stage, a technical team, whether from the client's side or the consultant's, develops the design, specifications and drawings, from which a bill of quantities (BOQ) can be formulated. The BOQ contains estimated quantities of different items of work and also the approximate unit cost, and thus, a reasonably refined estimate can be arrived at. The unit rates contain input from the planning and the design team based on their cost analysis and previously tendered costs. The preliminary cost estimates are also prepared based on available plinth area rates. The cost estimates at this stage usually have an accuracy range of −15 per cent to +30 per cent, and the time involved in working the estimates out depends on the nature and scope of the project, the availability of the past performance data and the method of execution. Apart from the BOQ, similar estimates can also be arrived at using parameter cost estimate method and factor cost estimate method. It may be noted that this estimate becomes the basis for making the final go-ahead decision for execution of the project, which essentially means tendering action, and of course, the acceptance of a bid implies the client's commitment for the payment of the quoted costs and the execution of the works.

4.3.4 Client's Detailed Estimate

The detailed estimate is an accurate estimate and consists of working out the quantities of each item of work and the associated costs. For working out the quantities, the dimensions—length, breadth, height or thickness—of each item are taken correctly from drawings. A typical format used for the measurement and calculations of quantities are shown in Table 4.3.

As a second step, the cost for each item of work is calculated from the quantities already computed. The rates obtained from schedule of rates published by different agencies such as CPWD, state PWDs, state irrigation departments, etc., can be used for the cost estimate of

Table 4.3 Details of measurement form

Item No.	Description or particulars	No.	Length (m)	Breadth (m)	Height or depth (m)	Quantity (m³)	Total quantity (m³)
1	Excavation long wall	2	50	1.0	0.6	60	
							90
2	Excavation	2	25	1.0	0.6	30	
3	—	—	—	—	—	—	—

Table 4.4 Abstract of estimated cost or bill of quantity

Item No.	Description or particulars	Quantity	Unit	Rate (₹/unit)	Total cost (₹)	Remarks
1	Earthwork	90	m³	100	9,000	Rate obtained from Delhi Schedule of Rates 2007
2	Curtain wall	200	m²	12,000	2,400,000	Rate derived from analysis based on market rates
3	—	—	—	—	—	—
				Total	Sum of the above, say X	

[1] During the construction of a building or a project, a client requires a certain number of work supervisors, *chowkidars*, clerks, etc. Their salaries are paid from the amount of work-charged establishment provided in the estimate. Usually, 1½ per cent to 2 per cent of the estimated cost is provided for this and is included in the estimate. The work-charged employees, or temporary staff, are employed for a specific period. Their services are terminated at the expiry of the sanction period. Fresh sanctions shall have to be taken from the competent authority in case their services are further required.

standard or scheduled items. Market rates can be obtained for non-scheduled items. The rates so obtained are filled up in the format as shown in Table 4.4. The cost is summed up for all the items. Let it be X.

A margin of 3 per cent to 5 per cent of X is added for contingencies, unforeseen expenditure, changes in design, changes in rates, etc., which may occur during the execution of work. A margin of 1½ per cent to 2 per cent is also added to meet the expenditure of work-charged establishment[1]. The grand total thus obtained is the estimated cost of work. For large projects, a margin of 1 per cent to 1½ per cent of the estimated cost is provided in the estimate for the purchase of tools and plants that will be required for the execution of work. Normally, the contractor has to arrange and use his own tools and plants.

The detailed estimate is usually supplemented with (1) report or history, (2) detailed specification, (3) drawings such as plan, elevation, sectional elevation and site plan, (4) calculation and designs, and (5) analysis of rates. Clients prepare detailed estimate for getting the sanction of the competent authority. The estimate is also required to be prepared for the bidding process. The detailed cost worked out is used to compare and evaluate the rates quoted by bidders.

4.3.5 Client's Definitive Estimate

Most contracts have an escalation clause and invariably, some additional items of work and deviations from specifications are involved, and all have financial implications. Therefore, it is important that the estimation process continue into the execution stage, when the client's cost accountant, based on information from the site, analyses actual payments made and deviations from original estimates, and predicts the final cost of a project incorporating prevailing cost trends. When the project is 70 per cent to 80 per cent complete, a reasonably definitive estimate can be prepared, including the detailed estimate of work quantities that also takes into account deviation orders and anticipated final bills of the contractors and suppliers. It should be noted that such estimates obviously exclude payments that may be made at a later date on account of disputes, claims, etc.

4.3.6 Revised Estimate

It is always possible that in spite of all precautions in the planning stages, it becomes clear during execution that the actual cost of a project will exceed the original estimates. Now, generally a certain cushion (say, 10 per cent) of the cost is available; if the exceedance is higher, fresh estimates are prepared with appropriate justifications for financial approvals from competent authority. It is prepared on the basis of estimate on which sanction was obtained, showing the existing sanction and the progress made up to date. The revised estimate should be accompanied by a comparative statement showing the original and revised rates and quantities, along with variation and the reasons of variations for each item of work.

4.3.7 Supplementary Estimates

There is always a likelihood that while executing a certain project, it may be considered worthwhile to carry out additional work, which was not foreseen in the initial stages and, therefore, not accounted for in the preliminary estimates. Execution of such works requires drawing up and approval of supplementary estimates, and the exercise is essentially similar to that of drawing up the estimates for the main work. It is naturally expected that the costs of such additional works will be much smaller than the main works. The abstract of such estimate should show the amount of original estimate (X) and the total of the sanction required (X + Y) including the supplementary estimate (Y).

4.3.8 Project Closure Cost

Upon completion of a project, the final bill prepared by the contractor is closely scrutinized by the client and the consultant/cost engineer, and this scrutinized bill along with documents in support of actual cost data is used to prepare the final estimate of project costs, excluding unresolved disputes and claims.

4.4 METHODS OF STRUCTURING PROJECT COSTS

The cost estimate, prepared by a client during the planning stage, generally follows a structured approach. The structure of this estimate takes many forms. These forms vary with the nature of the project, the type of contract, the company policy, the type of contract and the client requirements. The three typical well-known forms in structuring estimation items in the United States, besides the one used predominantly in India, are given in Table 4.5.

In particular, the bill of quantities of a detailed estimate can be suitably divided under the heads given above. In the bid estimates, the contractor-quoted price includes direct and indirect costs and profit margin suitably distributed across the various items of the bill of quantities.

4.5 ILLUSTRATIVE CASES IN PREPARATION OF ESTIMATE

4.5.1 Case1: Multi-level Car Parking Facility

A municipal body of the country is planning to construct a multilevel car-parking facility in a locality. The study constructed in the locality suggests the requirement of parking facility for about 1,000 car units. Further, there is a requirement of parking space for about 400 two-wheelers.

Table 4.5 The prevalent structured approach

	Prevalent in USA		Prevalent in India
RSMeans	Engineering news recorder	Construction specification institute	CPWD-Delhi Schedule of Rates for building work
Substructure	Site work	General requirements	Earthwork
Superstructure	Foundations	Site work	Mortar such as lime mortar, cement mortar and cement lime mortar
Exterior enclosure	Floor systems	Concrete	Concrete work
Interior construction	Interior columns	Masonry	Reinforced cement concrete, formwork, reinforcement
Conveying system	Roof systems	Metals	Brickwork
Plumbing system	Exterior wall	Wood and plastics	Stonework
HVAC system	Exterior glazed openings	Thermal moisture	Marble work
Electrical system	Interior wall systems	Doors and windows	Woodwork
Fixed equipment	Doors	Finishes	Steel work
Special foundation	Specialities	Specialities	Flooring
Site construction	Equipment	Equipment	Roofing
General contingencies	Conveying systems	Furnishings	Plastering
Related costs	Plumbing	Special construction	Repairs
	HVAC	HVAC	Dismantling and demolition
	Electrical systems	Mechanical	
	Fixed equipment	Electrical	
	Special electrical		
	Markup		
	General contingencies		
	Related costs		

Usually, such estimates are accompanied with the description of need for the project. The need for this illustrative project is stated thus:

Increase in number of vehicles and stringent parking regulations coming into force are making parking in public areas and in new upcoming projects a very important design consideration. Solutions are required to create sufficient parking areas to enable maximum cars to be parked safely, in minimum time and in minimum land area.

The municipal authority has proposed the preliminary estimate on plinth area basis, which is shown in Table 4.6. This estimate has also been used to obtain administrative and financial approval from the competent authority. The schematic sketch of proposed parking is given in Figure 4.1 and Figure 4.2.

The schematic diagram (Figure 4.1) provides only the entry and exit schemes of the proposed car park, besides providing the overall area involved in the project. The designer at this

Table 4.6 Plinth area calculation

S. No.	Description	Dimension from preliminary drawing	Area (in m²)
1	**Lower basement**		
1.1	Lower basement—car parking	214 m × 60.50 m	12,947
1.2	Ramp	58.4 m × 4.0 m	233.6
1.3	Total for lower basement		**13,180.6**
2	**Upper basement**		
2.1	Upper basement—two-wheeler parking	57.35 m × 50.0 m	2,867.5
2.2	Ramp	29.2 m × 4.0 m	116.8
2.3	Deduct floor panel	14.6 m × 12.0 m	−175.2
2.4	Total for upper basement		**2,809.1**
3	**Total plinth area**		**15,989.7**
			Say, 16,000 m²

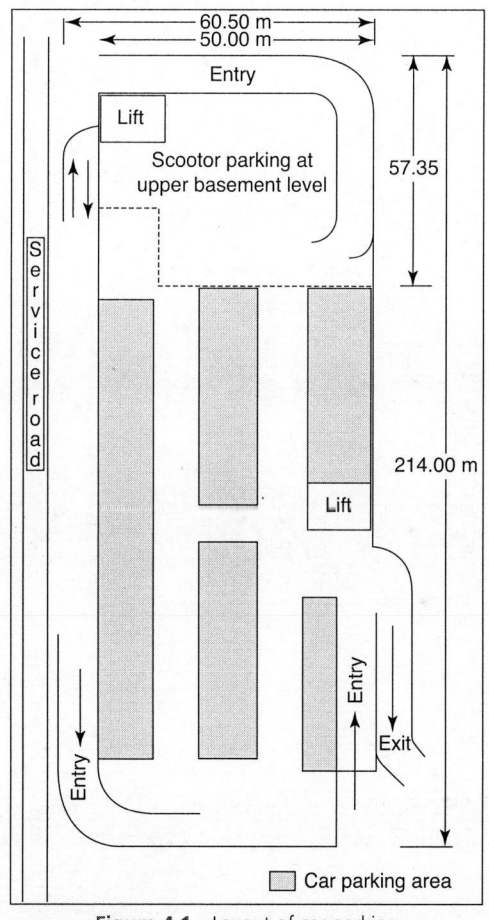

Figure 4.1 Layout of car parking

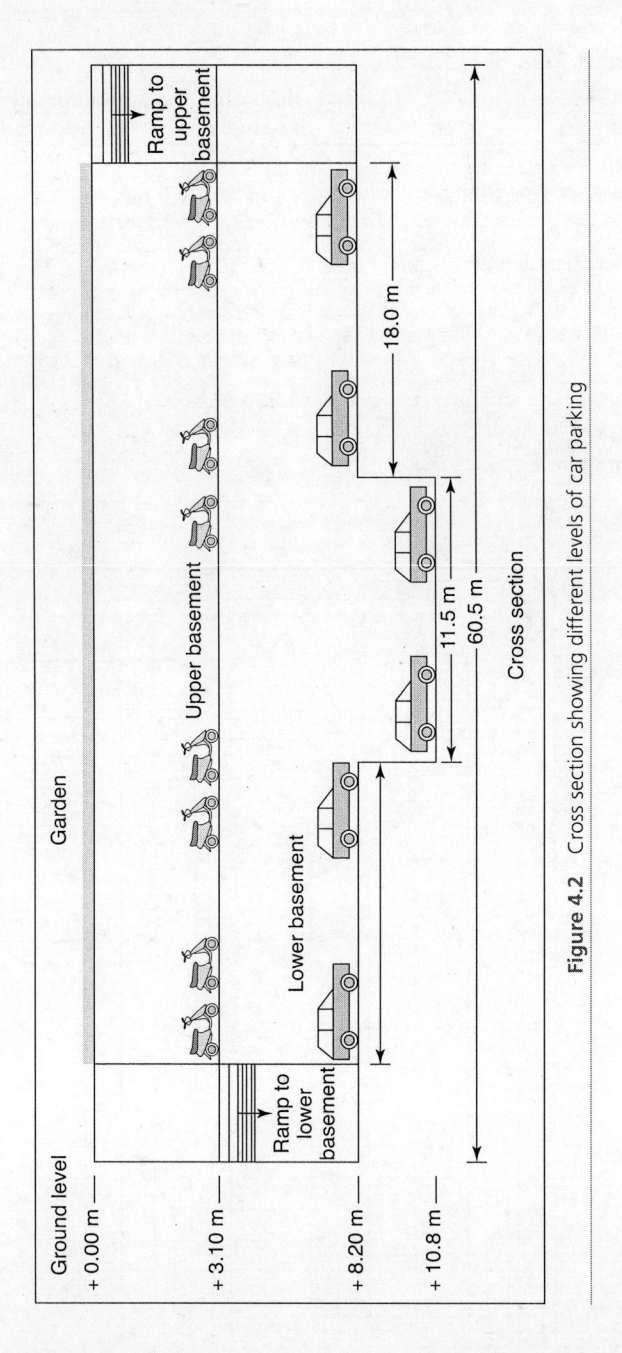

Figure 4.2 Cross section showing different levels of car parking

stage is familiar with broad specifications of the project and he would also know the preliminary scheme of car parking. For example, at this stage he is expected to know the number of levels in which cars will be stacked (Figure 4.2) and, thus, the construction height involved in the project. Based on the broad specifications and the preliminary sketch, the quantities are worked out and schedules are referred to in order to arrive at the project cost. The preliminary plan and sectional drawing shown in Figure 4.1 and Figure 4.2 form the basis of plinth area calculation, which is the basis of estimate preparation at this stage. The plinth area calculation for this example is given in Table 4.6.

Now, based on the plinth area calculated above, the broad specification for the project can be laid out, and on basis of CPWD plinth area rate as on 01.10.2007, the estimate can be prepared. The estimate is made in two parts—for schedule items and for non-schedule items. While the rates for schedule items are readily available in the schedule of rates, the non-schedule items are estimated based on the market rates prevailing at the time of preparation of estimate.

The estimate for schedule items is given in Table 4.7. The estimate contains the following major parts:

RCC Frame Structure

The total area for RCC frame structure is 16,000 m^2, and the corresponding rate given in schedule of rates is ₹ 18,035/m^2. On this base rate, the provision for mastic asphalt in the basement, earthquake resistant design and stronger structural members to carry heavy loads are to be added. Further, the schedule specifies additional cost provisions for modules over 35 m^2 (which in our case is applicable), and for additional heights above 3.35 m (this is again applicable in our case), and accordingly, these provisions have been kept as given in Table 4.7.

Fire Fighting, Automatic Fire Alarm System and Pressurized Mechanical Ventilation System

The estimate for fire fighting is based on plinth area of 16,000 m^2 and the applicable schedule of rate is ₹ 450/m^2. Similarly, automatic fire alarm system and pressurized mechanical ventilation system in the basements with supply of exhaust blowers are based on PAR-2007 and the applicable rates are ₹ 300/m^2 and ₹ 50/m^2, respectively.

Services Such as Internal Water Supply and Sanitary Installation, External Service Connection, Internal Electric Installation and Quality Assurance

The schedule of rates specifies a certain percentage of total cost obtained from the above two heads. In case there is a basement involved in the project, the percentages are reduced to 50 per cent of the given percentage. Thus, for internal water supply and sanitary installation, 50 per cent of the specified 4 per cent of the subtotal cost given at Sl. No. (C) of Table 4.7 is taken, which is equal to ₹ 15,579,977.00. Similarly, other estimates have also been derived. A provision of 1 per cent of cost mentioned at Sl. No. (C) of Table 4.7 is taken for quality assurance.

Lifts

Four passenger lifts of 20 persons capacity are envisaged for the project, and the unit rates as mentioned in the schedule of rates are taken for estimate preparation.

Water Tanks

For estimating the cost of water tanks, the volume of the tanks is computed in an approximate manner by means of approximate knowledge of the length, breadth and height of the tank. The schedule of rates provides the rate on per-litre basis (₹ 9/ℓ as per CPWD PAR-2007).

The estimate for non-schedule items is given in Table 4.8. There are only two items—automatic parking system and Qualideck PU flooring. In order to arrive at the market rates for

Table 4.7 Preparation of estimate for schedule items based on C.P.W.D. PAR–2007

S. No.	Code No.	Description	Quantity	Rate (₹/unit)	Amount (₹)
1	1	**RCC frame structure**			
2	1.3.1	Basement of floor height 3.35 m with normal water proofing treatment with bituminous felt	16,000 m^2	18,035	288,560,000.00
3	1.3.2.1	Extra for basement with mastic asphalt WPT	16,000 m^2	1,144	18,304,000.00
4	1.2.8	Extra for resisting earthquake forces	16,000 m^2	630	10,080,000
5	1.2.11	Extra for stronger structural members to take heavy load above 500 kg per m^2	16,000 m^2	850	13,600,000.00
6	1.2.12	Extra for larger modules over 35 m^2	16,000 m^2	990	15,840,000.00
7	1.3.2.2	Extra for every 0.3 m additional height (above 3.35 m) 25 heights (25 × 1274 – ₹ 31,850)	13,181 m^2	31,850	419,814,850.00
(A)		Sub total of Sl. No. 2 to Sl. No. 7			766,198,850.00
8	1.4.2	Fire fighting with sprinkler system	16,000 m^2	450	7,200,000.00
9	1.5.2	Automatic fire alarm system	16,000 m^2	300	4,800,000.00
10	1.7	Pressurized mechanical ventilation system in the basements (with supply of exhaust blowers)	16,000 m^2	50	800,000.00
(B)		Subtotal of Sl. No. 8 to Sl. No. 10			12,800,000.00
(C)		Subtotal of Sl. No. (A) and Sl. No. (B)			778,998,850.00
11	3	**Services** (rates are taken 50% due to basement)			
12	3.1	Internal water supply and sanitary installation	50% of 4.00% of (C)	2.00% of (C)	15,579,977.00
13	3.2	External service connection	50% of 5.00% of (C)	2.50% of (C)	19,474,971.25
14	3.3	Internal electric installation	50% of 12.50% of (C)	6.25% of (C)	48,687,428.13
15	3.6.7	Quality assurance		1% of (C)	7,789,988.50
16	4	**Lifts**			
17	4.1	Lift for 20 Passenger capacity	4 Nos.	2,700,000.00	10,800,000.00
18	5	**Water tanks**			
19	5.5	Under ground sump			
20	1	General	50,000 l	9	450,000.00
21	2	Fire fighting	200,000 l	9	1,800,000.00
(D)		Sub total of items under services, lifts, and water tanks			
(E)		Total of scheduled items (C)+(D)			885,381,214.90

Table 4.8 Preparation of estimate for non-scheduled items

S. No.	Description	Unit	Quantity	Rate (₹/unit)	Amount (₹)
1	Automatic parking system	Nos	952	300,000	285,600,000.00
2	Qualideck PU flooring	m²	16,000	2,000	32,000,000.00
(F)	**Total of non-scheduled items**				**317,600,000.00**

these items, the broad specification must be frozen. For example, the level of automation, the time of retrieval, the level of automation, the security system, etc., would be required for estimating the cost of an automatic car-parking system. Market rate at this point need not be based on formal quotation. One can give reference to the various discussions with manufacturers and installing agencies, as well as the actual costs of similar systems in other nearby projects. Too much emphasis should not be given on the discussion held with a single agency; one should use his own judgement to arrive at an appropriate estimate.

The abstract of cost estimate for Multi Level Car Parking Facility is presented in Table 4.9. The sum total of estimate of schedule items is taken from Table 4.7. The cost index prevailing as on the date of preparation of estimate is added in the sum total of schedule items. For this case, cost index value of 20% has been assumed. The sum total of estimate of non-schedule items is taken from Table 4.8. The total cost of schedule items, increase on account of cost index, and the total of non-schedule items have been summed up in Sl. No. 4 of Table 4.9. On this a contingency amount of 3% of total cost is added to arrive at the estimated project cost of Multi Level Car Parking Facility (See entry at Sl. No. 6 of Table 4.9).

4.5.2 Case 2: Preliminary Estimate for Construction of Sewage Treatment Plant

Let us take another example project of construction of a sewage treatment plant. The plant is to treat the sewage generated from a township, and the estimated capacity of the plant is 2 MLD. For illustration, let us prepare the preliminary estimate only for civil works associated with the sewage treatment plant. The preliminary design envisages the structures/utilities as

Table 4.9 Abstract of cost estimate for Multi Level Car Parking Facility Project

S. No.	Item of work	Amount (₹)
1	Schedule items estimated on the basis of PAR-2007 (Sl. No. (E) of Table 4.7)	885,381,214.90
2	Add prevailing cost index (say 20% over PAR-2007 estimate) as on the date of estimate preparation	177,076,242.98
3	Total of non-schedule items based on the market rate (Sl. No. (F) of Table 4.8)	317,600,000.00
4	Total (Sl. No. 1 + Sl. No. 2 + Sl. No. 3)	1,380,057,457.88
5	Add 3% contingency on Total at Sl. No. 4	41,401,723.74
6	Total estimated project cost for Multi Level Car Parking Facility (Sl. No. 4 + Sl. No. 5)	**1,421,459,181.62**

given in Table 4.10. The plan for the sewage treatment plant is given in Figure 4.3. The estimate is prepared in three major heads (see Table 4.11):

Load-bearing Construction

These items are estimated based on the total plinth area of load-bearing construction. The project envisages construction of single-storey quarters of Type I, Type II and Type

Table 4.10 Summary of civil works

S. No.	Unit	Size	Material of construction	Numbers
1a	Inlet chamber	1.0 m × 0.5 m × 0.8 m	RCC with epoxy-coated steel	1
1b	Coarse screen	4.2 m × 0.25 m × 0.8 m	Stainless steel	2
2a	Grit chamber (mech.)	2.5 m × 1.7 m × 1.4 m	RCC with epoxy-coated steel	1
2b	Grit chamber (manually)	1.4 m × 1.0 m × 2.0 m	RCC with epoxy-coated steel	1
3	Oil and grease trap	2.5 m × 1.55 m × 2.0 m	RCC with epoxy-coated steel	1
4	Equalization tank	14.5 m × 8.6 m × 4.5 m	RCC with epoxy-coated steel	1
5	Fine screen	2.8 m × 0.53 m × 0.8 m	Stainless steel	2
6	Anoxic tank	10.0 m × 5.0 m × 4.5 m	RCC with epoxy-coated steel	1
7	Aeration tank	12.0 m × 10.0 m × 4.5 m	RCC with epoxy-coated steel	1
8	Membrane bioreactor tank	5.5 m × 7.0 m × 4.5 m	RCC	2
9	Treated water tank	8.5 m × 6.0 m × 4.5 m	RCC	1
10	Constructed wet land	100.0 m × 20.0 m × 1.0 m	Brickwork in cement motor + 30 mm–40 mm media	1
11	Thicken feed sump	2.4 m × 1.6 m × 3.0 m	RCC with epoxy-coated steel	1
12	Sludge thickener	3.85 m × 3.0 m	RCC with epoxy-coated steel	1
13	Decanter feed sludge sump	2.6 m × 1.8 m × 3.0 m	RCC with epoxy-coated steel	1
14	Decanter shed	5.0 m × 5.0 m × 6.0 m	Brick work	1
15	Supernatant sump	3.0 m × 2.0 m × 2.8 m	RCC with epoxy-coated steel	1
16	Pump/blower house	7.0 m × 5.0 m × 4.0 m	RCC with epoxy-coated steel	1
17	CIP tank	2.5 m × 2.3 m × 4.0 m	RCC	1
18	Plant general/maintenance room laboratory	7.0 m × 6.0 m × 4.0 m	Brickwork in cement motor with acid alkali-resistant flooring	1
19	Plant control/SCADA room	9.0 m × 6.0 m × 4.0 m	RCC	1
20	Chemical store room	8.5 m × 6.0 m × 4.0 m	Brickwork in cement motor with acid alkali-resistant flooring	1
21	Plant toilet	4.0 m × 6.0 m × 4.0 m	Brickwork	1
22	UV unit	2.0 m × 1.0 m × 0.8 m	RCC with epoxy-coated steel	1
23	Plant roads	320 m × 5.0 m	Metallic	1
24	Watchman cabin	2.0 m × 3.0 m × 3.0 m	Brickwork	1

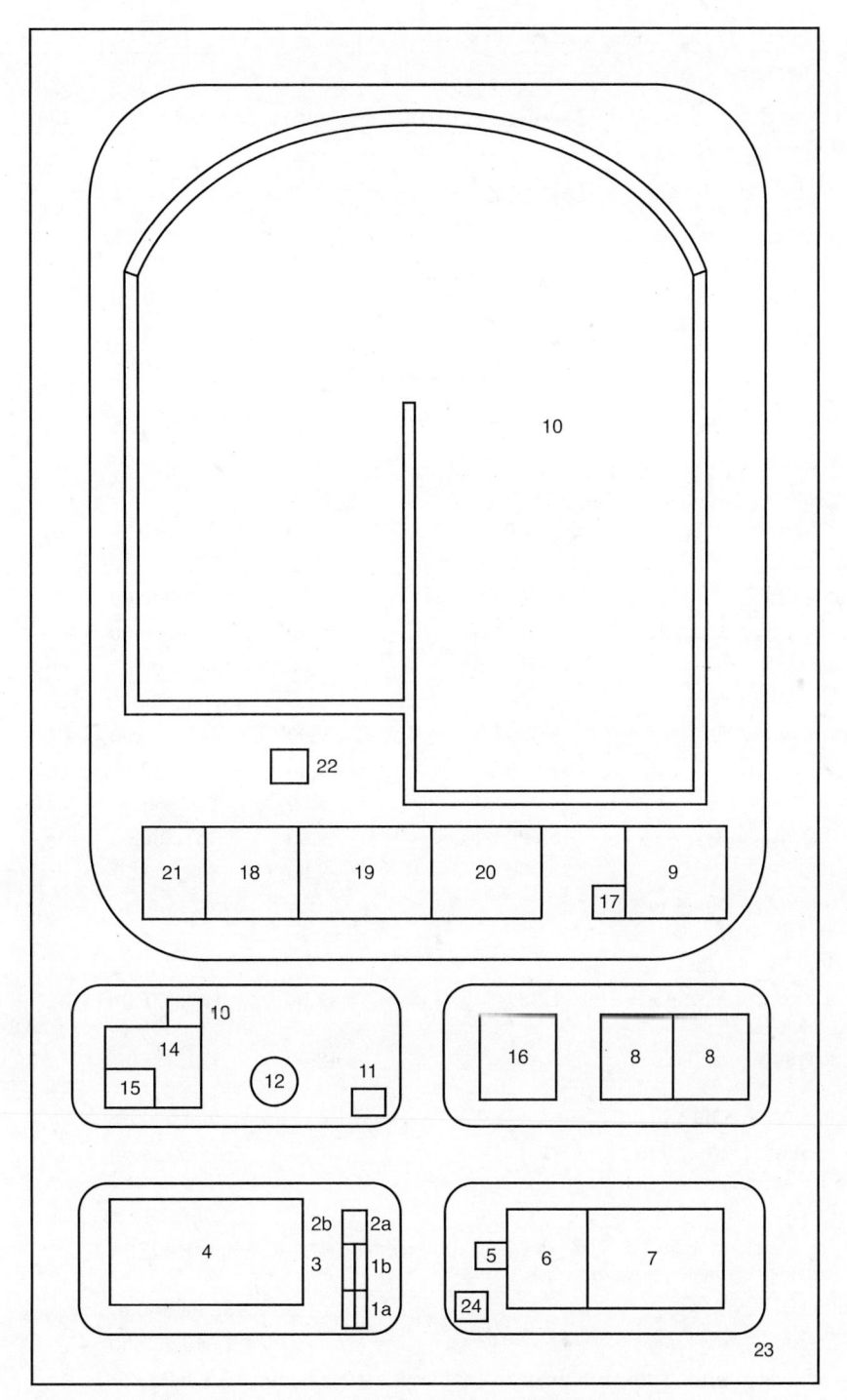

Figure 4.3 Layout of sewage treatment plant for a township

Table 4.11 Abstract of cost estimate for civil works for a sewage treatment plant

S. No.	Item of work	Quantity	Unit	Rate (₹/unit)	Amount (₹)	Ref. of PAR-2007
1	Load-bearing construction					
2	Floor Height 2.9 m (residential quarters Type I, II, III and servant quarters), single-storey	239	m²	6,390.00	1,527,210.00	Sl. No. 2.2.1, Col. 6
(A)	Sub total				1,527,210.00	
3	Water tank					
4	Inlet chamber	400.00	ℓ	9.00	3,600.00	Sl. No. 5.5 of PAR-2007
5	Grit chamber (mech.)	5,950.00	ℓ	9.00	53,550.00	-do-
6	Grit chamber (manually)	2,800.00	ℓ	9.00	25,200.00	-do-
7	Oil and grease Trap	7,750.00	ℓ	9.00	69,750.00	-do-
8	Equalization tank	561,150.00	ℓ	9.00	5,050,350.00	-do-
9	Anoxic tank	225,000.00	ℓ	9.00	2,025,000.00	-do-
10	Aeration tank	540,000.00	ℓ	9.00	4,860,000.00	-do-
11	Membrane bioreactor tank	173,250.00	ℓ	9.00	1,559,250.00	-do-
12	Treated water tank	229,500.00	ℓ	9.00	2,065,500.00	-do-
13	Constructed wet land	2,000,000.00	ℓ	9.00	18,000,000.00	-do-
14	Thicken feed sump	11,520.00	ℓ	9.00	103,680.00	-do-
15	Sludge thickener	34,906.99	ℓ	9.00	314,162.91	-do-
16	Decanter feed sludge sump	14,040.00	ℓ	9.00	126,360.00	-do-
17	Supernatant sump	16,800.00	ℓ	9.00	151,200.00	-do-
18	CIP Tank	23,000.00	ℓ	9.00	207,000.00	-do-
(B)	Sub total of Sl. No. 4 to Sl. No. 18				34,614,602.91	
19	Development at site					
20	Internal roads and paths	1,837	m²	83.00	152,471.00	Sl. No. 6.2 of PAR-2007
21	Horticulture operations	3,465	m²	47.00	162,855.00	Sl. No. 6.6 of PAR-2007
22	Area lighting in plant area		LS		200,000.00	
(C)	Sub total of Sl. No. 20 to Sl. No. 22				515,326.00	
(D)	Total = (A) + (B) + (C)				36,657,138.91	
(E)	Add prevailing cost index (say 20% over PAR-2007 estimate (D)) as on the date of estimate preparation				7,331,427.78	
(F)	Total = (D) + (E)				43,988,566.69	
(G)	Add 3% contingency on (F)				1,319,657.00	
(H)	Total estimated project cost for civil works of sewage treatment plant ((F) + (G))				45,308,223.69	

III, and servant quarters of floor height 2.90 m. The plinth area rate specified for such construction, based on PAR-2007, is ₹ 6,390/m².

Water Tank

A large number of water tanks are required in a sewage treatment plant. The sizes are fixed based on the preliminary design available at this stage. Based on the preliminary sizing, the volumes of each of the tanks are calculated. The schedule of rates provides per-litre rate (₹ 9ℓ based on PAR-2007) for such construction.

Development at Site

Under this head, internal roads, horticultural works and area lighting in plant area have been considered. While per-square-metre rates are available in the schedule of rates for internal roads and horticultural works, for area lighting a lump-sum cost has been assumed.

A cost index of 20 per cent is assumed as on the date of preparation of estimate and, accordingly, the total cost of the above three heads has been increased by 20 per cent. A contingency value equal to three per cent of the total cost is also added to get the estimated cost of the project as shown in Table 4.11.

REFERENCES

1. Chitkara, K.K., 2006, *Construction Project Management: Planning, Scheduling and Controlling*, 10 reprint, New Delhi: Tata McGraw-Hill.
2. Choudhary, S., 1988, *Project Management*, New Delhi: Tata McGraw-Hill.
3. Datta, B.N., 2008, *Estimating and Costing in Civil Engineering – Theory and Practice (Including Specifications and Valuations)*, Ubs Publishers' Distributors (p) Ltd.
4. DSR 2007, Central Public Works Department – Delhi Schedule of Rates 2007, Jain Book Agency, New Delhi.
5. PAR-2007, Central Public Works Department, Plinth Area Rate, Jain Book Agency, New Delhi.

SOLVED EXAMPLES

Example 4.1

Derive the rate per cubic metre for providing and laying in position cement concrete 1:2:4 (1 cement: 2 coarse aggregate: 4 graded stone aggregate 20 mm nominal size) for all work up to plinth level excluding the cost of centering and shuttering.

Solution

The approach adopted here for the analysis of rate is mostly adopted by government departments for small jobs or projects. The labour and material coefficients per m³ or per 10 m³ are specified along with the rates of material and labour as on the base year of the schedule of rates. The coefficient for material carriage is also specified in them. The working for this example (see Table S3.4.1) is based on per m³ basis according to the Delhi Schedule of Rates for the base year 2007.

Table 4.1.1 Working of rates for 1 m³ of cement concrete

	Unit	Quantity	Rate	Amount
Material				
Stone aggregate 20 mm	m³	0.67	700.00	469.00
Stone aggregate 10 mm	m³	0.22	700.00	154.00
Carriage of aggregate	m³	0.89	53.21	47.36
Coarse sand	m³	0.445	600.00	267.00
Carriage of coarse sand	m³	0.445	53.21	23.68
Cement	t	0.32	4,500.00	1,440.00
Carriage of cement	t	0.32	47.29	15.13
Labour				
Mason	d	0.10	146.55	14.66
Beldar	d	1.63	135.25	220.46
Bhisti	d	0.70	138.45	96.92
Mixer	d	0.07	400.00	28.00
Vibrator	d	0.07	200.00	14.00
Sundries	L.S	14.30	1.00	14.30
Total				2,804.54
Add 1% for water charges				28.05
Add 15% for contractor's profit and overheads				424.88
Cost of 1 m³				3,257.44
			Say	3,257.45

REVIEW QUESTIONS

1. State whether True or False:

    a. Estimating costs in construction projects is a very complex process due to the inherent interactions and interdependence involved and the absence of standard norms.

    b. Number of occupants is the basis for preparing preliminary estimate of a building project.

    c. The preliminary estimate of roads and highway projects is prepared based on per-kilometre basis and nature of classification of the road.

    d. For irrigation projects, the preliminary estimates are prepared based on the length involved.

    e. The preliminary estimate of bridges and culverts is prepared on the basis of the span of structure.

    f. The preliminary estimate for sewage and water supply projects is on the basis of per head of population.

    g. The preliminary estimate of overhead water-tank project is prepared on the basis of capacity of tank.

    h. Methods for preparing preliminary estimate of industrial structures are—plinth area method, cube area estimate and approximate quantity method estimate.

i. Methods for preparing preliminary estimate of buildings are—ratio method, investment per annual t capacity, turnover ratio and capital ratio, and six-tenth factor.
2. Match the following with proper options.

No.	Types of Estimate Used in Construction Industry	No.	Stages of Project
A	Project proposal indicative cost estimate	1	Final cost
B	Preliminary estimate	2	To assess cost at completion
C	Detailed estimate	3	Controlling cost during the execution of project
D	Definite Estimate	4	Budgeting costs toward the end of planning and design phase
E	Final closure cost estimate	5	Feasibility stage

3. Prepare a preliminary estimate of a school building proposed to have 10 classrooms and a total of 500 students.
4. Why client's estimate of cost differs from that of contractor's? Identify which estimate is more critical, and why?
5. Give a detailed account of different methods involved in preliminary estimate of cost involved in a building project and an industrial project.
6. Define capital ratio and turnover ratio. How is it helpful in preliminary estimate of cost of an industrial project.
7. Identify different stages involved in a construction project and give a detailed account of types of cost estimate associated with each stage.
8. Visit a construction site and get the project's drawing and specification for that project, and on the basis of this, prepare a detailed client's cost estimate? Compare your estimate with the client's prepared cost estimate.
9. Why is the revised estimate necessary?
10. What are the different factors that determine the form of structure of an estimate?
11. What are the different types of costs involved in a project, and how is the overall project cost calculated? Taking an example of a project you are working with or are familiar with, give the rough way of working out how the overall project cost was derived.
12. What is preliminary estimate? Explain its process.
13. Write short notes on detailed estimates.
14. Derive the rate per cubic metre for 1:5:10 cement concrete in foundations with brick ballast 40 mm size.
15. Derive the rate per cubic metre of 2nd class brick work in foundation with 20 cm × 10 cm × 10 cm nominal size with cement sand mortar 1:6.
16. Derive the rate of one MT of reinforcement cost including cutting, bending and tying all complete. Assume any suitable data required for deriving the rate.

17. Derive the rate of one square metre of shuttering/centring to be done for wall, column, beam and slab. Assume any suitable data required for deriving the rate.
18. Derive the quantity of materials required for producing one m³ of cement concrete 1:4:8 in foundation. Assume total dry mortar for 1 m³ of cement concrete = 1.54 m³.
19. Derive the quantity of dry materials required for producing one m³ of brick masonry in cement sand mortar. Ratio 1:5. Assume total dry mortar required for 1 m³ of brick masonry = 0.30 m³.

5
Construction Contract

> Construction contract, contract document, classification of engineering contracts, bidding process, CPWD contract conditions, FIDIC form of contract agreement, subcontracting

5.1 CONSTRUCTION CONTRACT

Contract as per the Indian Contract Act 1872 means 'agreements which are enforceable as such having been made by free consent of the parties, by persons competent to contract for a lawful consideration and lawful object and which are not expressly declared to be void by any statute.' From the definition, we can infer the criteria required for a contract to be valid. These criteria are:

- There must be mutual agreement between the two parties.
- There must be an offer made by one party called the *promisor*.
- The other party, called the *promisee*, must accept the offer.
- There must be *considerations*, which is usually payment in the form of money for doing of an act or abstinence from doing a particular act by *promisor* for *promisee*.
- The offer and acceptance should relate to something that is not prohibited by law.
- The offer and acceptance constitute an agreement that when enforceable by law becomes a contract.
- The contracting parties entering into agreement should be competent, i.e., not disqualified by either infancy or insanity to make such agreement.

Theoretically speaking, there could be an oral contract (one that is not in written form). However, it is virtually impossible to enforce it in the context of construction since keeping track of the scope of agreement reached between the parties will be difficult. Construction contracts are invariably in the written form. The purpose of construction contract essentially is to help achieve a quality construction project within stipulated time and cost, while adhering to all the safety norms.

Various types of contracts have been evolved to suit the various subject matters of contracts complying with the legal requirements. However, the discussions on contracts will be restricted to construction contracts. Important sections of the Indian Contract Act 1872 are covered in

Appendix 2. Construction contracts also have many variants and these vary from country to country. In Appendix 3, a brief note on some of the important acts applicable to establishments engaged in building and other construction works have been provided.

5.2   CONTRACT DOCUMENT

A construction contract comprises essentially the following documents:
- The contract drawings
- The specifications
- The general conditions of contract (GCC)
- The special conditions of contract (SCC)
- The agreement
- The bill of quantities (BOQ) if applicable

The turnkey tender documents may be having only the preliminary system drawings and may not have the bill of quantities.

5.2.1   The Contract Drawings

The contract drawings are the means through which the physical, quantitative and visual descriptions of the project are conveyed to the contractor. These are normally provided in the form of a two-dimensional diagram, referred to as the plan or the blueprint; however, in some cases, the drawings could be provided in the form of a softcopy consisting of 'read only' Autocad drawing files. The drawings are classified into— (a) site drawings, (b) architectural drawings giving all the details to convey to the contractor an overall picture of the total work, (c) structural drawings, (d) HVAC—heating, venting and air conditioning, and other services drawings, (e) electrical drawings, and (f) special details. Depending on the nature of work, there could be fire-fighting details, public-announcement system details, building automation details, etc.

5.2.2   The Specifications

Specifications, or technical provisions, are written instructions to carry out a work. It also contains information not possible to show on a piece of drawing. Drawings mentioned earlier together with specifications furnish the complete instructions to convert an architect's and a designer's imagination into reality. The drawings and specifications are also useful for preparing the cost estimates of work items of a project. Specifications commonly deal with the following aspects:
- The quality of materials
- The quality of workmanship
- The frequency of testing
- The approved manufacturers
- The relevant Indian standards describing the material
- The inspection and installation method

The specification could be of any type mentioned in Figure 5.1.

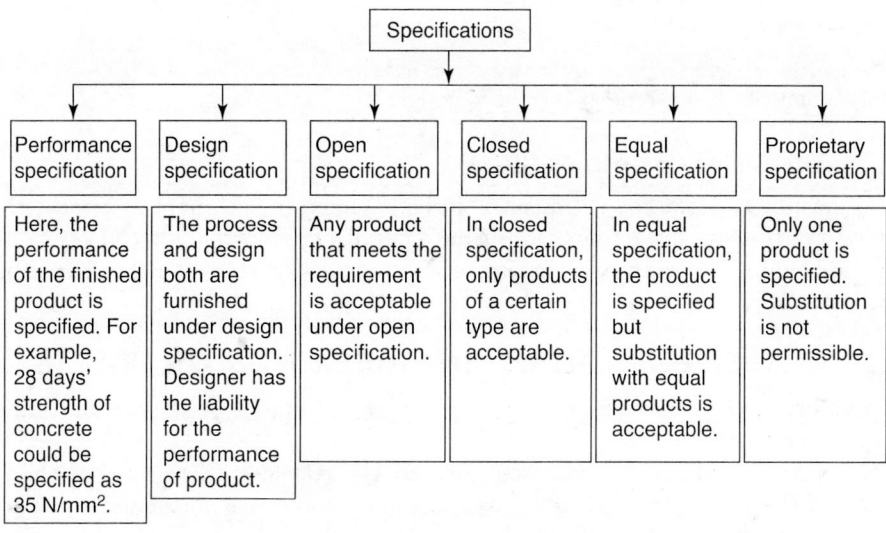

Figure 5.1 Types of specifications

Performance specification	Design specification	Open specification	Closed specification	Equal specification	Proprietary specification
Here, the performance of the finished product is specified. For example, 28 days' strength of concrete could be specified as 35 N/mm².	The process and design both are furnished under design specification. Designer has the liability for the performance of product.	Any product that meets the requirement is acceptable under open specification.	In closed specification, only products of a certain type are acceptable.	In equal specification, the product is specified but substitution with equal products is acceptable.	Only one product is specified. Substitution is not permissible.

Figure 5.1 Types of specifications

5.2.3 The General Conditions of Contract (GCC)

The general conditions of contract are an essential part of the contract. The term 'general' implies that the document is a standard one used in all the contracts entered by a party (the owner). Different owners such as CPWD, MES and IOCL have evolved standard forms of general conditions. The GCC evolved by American Institute of Architects is a popular document and many owners have formulated their GCC along these lines.

The general conditions of contract set out the responsibility and obligation of parties to the contract. It spells out the scope and performance of the contract, valuation and payment terms, arbitration and laws, labour regulations, safety code, various forms used for the tender and required deeds under the general conditions of contract. It is advisable to use standard general conditions of contract since most of these conditions have been tested in court over a period of time.

In India, for government jobs, CPWD conditions of contract are most widely used, though there is a growing trend of use of FIDIC contract conditions in large projects, especially those funded by World Bank (WB) and Asian Development Bank (ADB). The FIDIC conditions of contract are discussed elsewhere in the text.

5.2.4 The Special Conditions of Contract (SCC)

Certain amendments/additions/deletions are made in general conditions of contract in order to make it suitable for a particular project. These amendments are contained in a separate document called special conditions of contract (SCC). SCC may commonly address the following issues depending on the requirement of a project:

- Materials provided by the owner
- Site visits

- Mobilization advance
- Start date of construction
- Requirement of various reports related to progress

5.2.5 The Bill of Quantities (BOQ)

The bill of quantities shows the net quantity to be executed in each item of work. Items are classified into earthwork, anti termite treatment, waterproofing, brickwork, concreting, whitewashing and painting, flooring and finishing, doors and windows, structural steel, aluminium works, stonework, etc.

5.3 CLASSIFICATION OF ENGINEERING CONTRACTS

This section focuses on a brief description of some of the commonly used types of contracts in the construction industry.

It should be pointed out that the discussion here is particularly relevant to large and complex construction projects that are multidisciplinary in nature. It may further be borne in mind that the 'owner' organization, which finally owns and operates the facility, need not have specialized knowledge related to the very diverse engineering issues that may be involved in the design and construction of the project. For example, if a business house wants to enter the oil business and set up a refinery to refine crude oil, it may find that instead of handling all of it in-house, it may be much easier to hire the services of a consultant for the technical details such as process and instrumentation diagram, drawing up of appropriate specifications, design of equipment, identification of suitable contractors and vendors, and supervision of construction and commissioning works.

While the above example is for a refinery project, it is not difficult to see that a similar breakdown can be drawn up for other major civil engineering projects such as construction of a power plant and other industrial complexes. In fact, in the construction of a bridge or a housing complex also, there is an involvement of several agencies.

The activities in a construction project can be taken to comprise largely the following classes:

Engineering
These activities include issues related to process finalization, structural analysis and design, technical issues related to equipment design and selection, etc.

Procurement
The procurement of materials equipment, etc., comes under this category, which may also be taken to include identification of suitable vendors.

Construction
This covers the construction, installation and test run of a constructed facility before it is handed over to the owner for actual operations.

Thus, it is clear that in a construction project several agencies are involved, and the owner needs to have well-defined 'contracts' with each of them, clearly defining the scope of goods and services to be rendered, and the payment to be made to the contactor in return for these services. Apart from issues related to (partial) payments, etc., contracts also need to address aspects related to risk allocation, compliance with schedule, safety and labour norms, liabilities for delay in completion, rectification of defects, performance guarantees, arbitration, etc.

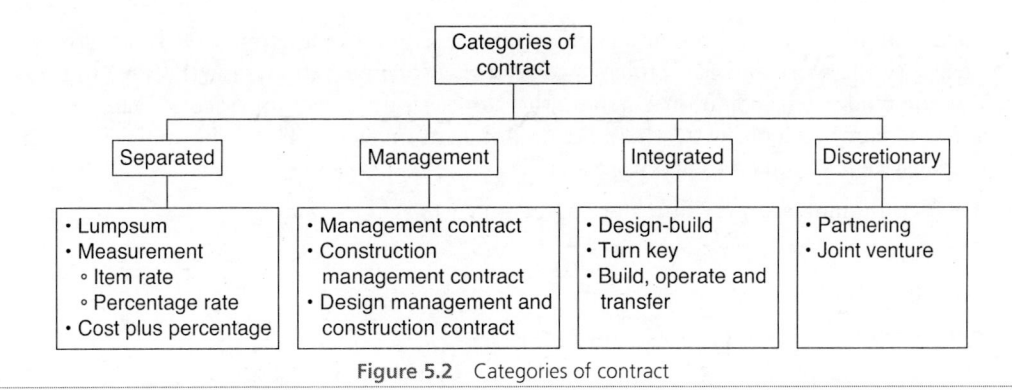

```
                        ┌─────────────────────┐
                        │ Categories of        │
                        │ contract             │
                        └─────────────────────┘
          ┌───────────────┬──────────┴──────┬──────────────┐
    ┌──────────┐   ┌──────────────┐  ┌────────────┐  ┌──────────────┐
    │ Separated│   │ Management   │  │ Integrated │  │ Discretionary│
    └──────────┘   └──────────────┘  └────────────┘  └──────────────┘
```

- Lumpsum
- Measurement
  ○ Item rate
  ○ Percentage rate
- Cost plus percentage

- Management contract
- Construction management contract
- Design management and construction contract

- Design-build
- Turn key
- Build, operate and transfer

- Partnering
- Joint venture

Figure 5.2 Categories of contract

Traditionally, contracts were made with a clear description of different measurable items, and payments made for the quantities of work actually executed. Although the contracting process has been explained separately, it should be recalled that at the outset the client carries out a preliminary estimate based on diverse factors such as past experience and the likely quantities and rates of different items involved, etc. Now, based on the method of contracting to be followed, the contractors submit a 'bid', which is evaluated by the client before the job is awarded to a contractor.

While it is not the intention here to trace the evolution of the different modes or types of contracts that are commonly used in the construction industry, it may be noted that almost all over the world, civil engineering projects were largely in the domain of varying degrees of state control till quite recently, and each society developed different contracting procedures. The ongoing privatization and globalization in the construction industry have led to a sea change in contracting procedures, and the following paragraphs only briefly describe some of the more important categories of contracts that are used in the construction sector today.

Broadly speaking, in execution of civil engineering works, the categories of contract systems used can be classified as given in Figure 5.2.

In addition to the above, sometimes other methods such as rate contracts or term contracts are also used. A brief discussion on some other forms of contract, such as BOT, is also included in the following paragraphs.

5.3.1 Separated Contract

The separated contract, which is a sequential process, has been the traditional system adopted for construction contracts. In separated contracts, there is a clear division between the design and construction responsibilities. The design phase comprising project briefing, feasibility studies, outlining proposals, scheme design and detail design is taken care of by some entity, while the construction is taken care of by some other entity. The preparation and approval of drawings, and the mistakes found in design documentation are frequent causes of delay in the design phase. In this method, sufficient time is needed (sometimes, it may need several months) for the preparation of full documentation by all consultants and for the quantity surveyors to complete a final estimate prior to calling tenders. The construction phase normally does not begin until the design is completed. As a result, the whole of the development process gets delayed.

Such type of traditional procurement requires a lengthy tendering period, to allow for complexity of the work and for tenderers to read the documentation, visit the site, and prepare for the tender. The traditional system, therefore, is often recommended for fairly simple small- to medium-sized projects, where time is not a critical factor. The major criticisms of the traditional system identified in literature are:

- Time-consuming aspects of the development process
- The effect of cost uncertainty
- The effect of buildability
- Fragmentation of organizational interfaces

Some of the variants of separated contracts are discussed briefly in the following paragraphs.

Lump-sum Contract

In this form of contracting, from the drawings and other details of the project provided by the client, the contractors quote a single lump-sum figure, which is the total contract value of the work. Obviously, the contractor arrives at this figure on the basis of his own analysis of rates and estimated quantities. This lump-sum amount refers to the total sum of money for which the contractor agrees to build the required facility, accepting all responsibility for factors relating to the supply of raw materials, uncertainties relating to construction hazards, and other difficulties. From the point of view of the client, this form of contract has one big advantage—he knows the exact amount of funds required for the completion of the structures.

For a lump-sum contract to be successful, it should be ensured that:

(a) The quantities of the different items involved are calculable at the stage of tendering itself. In other words, the design and specification must be fully developable.

(b) The nature of the work to be done must be reasonably measurable.

(c) The contractor must be given all the facilities to which he is contractually entitled.

It may be noted that the amount in a lump-sum contract is also subject to a revision under certain conditions, which may be considered outside the contractor's control. A change in the design or specifications made by the client could be one condition requiring a revision of the lump-sum amount. Also, a change in the cost of certain important items, such as statutory wages, transportation charges and customs duties, would justify a change in the lump-sum value. Normally, the stages or milestones are specified for the payment of bills to the contractor on a certain predefined percentage.

Measurement Contracts

In turnkey and lump-sum contracts, by the very nature of the agreement, there is no need to do any detailed measurement of the work carried out by the contractor. However, a large body of contracts requires that payments be made according to the actual work carried out, which should be determined on the basis of physical measurements. Such contracts are referred to as 'measurement contracts' and could be either (a) item rate or (b) percentage rate, as briefly discussed below.

Item rate contract This contract is so called because more than the total amount or the quantity of work in any item, it is the rate of the item quoted by the contractor that is held

Table 5.1 Sample illustrative extract from a BOQ

S. No.	Item description	Unit of measurement	Quantity	Rate	Amount
1 (a)					
1 (b)					
2					
3					

sacrosanct. In other words, it is held that the contractor agrees to carry out a unit quantity of a particular work (may be in units of cubic metres, square metres, numbers) for a particular sum of money. This form of contract is in contrast to a lump-sum contract, where the agreement is for delivering the entire project for a certain sum.

In the item rate contract, the tender document contains a detailed bill of quantities (BOQ), where an estimated quantity of the work for each item involved in the particular work is listed, along with a detailed description. The contractor carries out a detailed analysis to determine the rate of each individual item and writes the same in a column next to each item. The total contract value of the work is found out by multiplying the quantity of each item by the quoted rate of the contractor and adding the cost of all items. A sample of a few items in a BOQ is given in Table 5.1, along with the applicable units of measurements. As the details of the estimated quantities are also available, the contractor works out his rates for each item and fills them under the rate column.

Although the list of items in the BOQ is supposed to be exhaustive, the possibility of an item not being covered in the BOQ, but essential for the completion of a job, cannot be ruled out. In such cases, the applicable rates are determined on the basis of labour, material, equipment and overheads involved.

Naturally, in such contracts, an accurate account of the actual work (for each item) is kept, and the payment is made only for the actual work carried out. In other words, there can be a difference in the quantities of work actually executed and those foreseen in the estimate.

Percentage rate contract In this form of contract, tender documents also contain the analysed schedule of rates for each item, in addition to the detailed estimated quantities expected in the execution of the works. Thus, an estimate of the total value of the work is clearly available to the contractor. Now, the contractor works out his rates for the items in the usual manner, and arrives at his total price, which is converted to a percentage (positive or negative) by which his amount differs from the estimate given. This percentage is submitted as a quotation by the contractor—in other words, there is an overall modification in the rates of the contractor with this factor. An illustrative example showing the operation of this method is given in Table 5.2.

In this contract, for additional quantities of work done, and for items not included in the bill of quantities, payment is made on the basis of actual costs worked out on the basis of appropriate analysis of rates, which are then modified in accordance with the percentage agreed upon.

The method requires a detailed analysis of the rates to be carried out by client organizations, and usually only government departments or large organizations adopt this system. From the point of view of a client, the method results in tenders that are easier to evaluate

Table 5.2 Sample illustrative example of a percentage rate contract

S. No.	Item description	Unit of measurement	Quantity	Rate	Amount
1 (a)	abc…		100	x	100x
1 (b)	xyz…		1000	y	1000y
				Total	(100x + 1000y) say w

The contractor has worked out the total amount to be z.
If $z >$ the given amount of w, the % rate quoted would be $= \frac{z-w}{w} \times 100\%$ above w.
If $z <$ the given amount of w, the % rate quoted would be $= \frac{w-z}{w} \times 100\%$ below w.

and removes problems such as front-loading. However, it is important that the rates used are frequently updated lest there are anomalies in the escalation clause, or the percentages quoted become too high.

Cost Plus Percentage

In this kind of contract, the client agrees to pay the contractor a certain percentage of the (actual) cost incurred by the contractor while completing a job, in addition to the cost itself. Thus, the tenderer only quotes this percentage. Often, the client makes a part of the material available to the contractor, who is otherwise required to keep a detailed account of the expenses incurred in order to be able to claim them in his (subsequent) bills. This type of contracts is usually used in an emergency, when time may not be available to draw up an estimate and work out details of items involved. The contract may also be used for very small jobs where the traditional forms of contract may not be justified. Since all costs related to material, labour, etc., are borne by the client, only the final measurements are taken and the cost of materials involved is worked out on the basis of established guidelines. In certain cases, the cost of the material brought to site is directly paid for (against appropriate bills), and any material remaining after the completion of the project is retained by the owner for future consumption.

5.3.2   Management Contract

There are, in general, three variants under management contract.

Management Contract

In this type of contract, the managing contractor is appointed at the earliest possible time. This helps the client to avoid dealing with a large number of small contractors. In management contract, the client has to deal with a single (principal) contractor besides a designer. The principal contractor provides planning, management and coordination services to the client. The design services are provided by the designer, who is separately appointed by the client. Some of the responsibilities assigned to the management contractor are:

- Preparation of overall construction schedule
- Preparation of work package schedule
- Coordinating with the designer to steer through the design stage

- Assistance in subcontractor(s) selection
- Coordinating among different subcontractors

Usually, the principal contractor is barred from executing the construction work himself, though in some cases the principal contractor can contribute some resources such as formwork, cranes, etc., to the subcontractors.

Construction Management Contract

In the construction management contract, construction manager is appointed by the client at an early stage to provide planning, management and coordination. The owner also appoints the designer and contractors for different works. The role of construction manager, therefore, is mainly coordination among different contractors, besides ensuring timely completion of project within the budgeted cost according to the specifications. Some of the responsibilities assigned to the construction manager are:

- Advising the designer
- Advising on drawing suitable work package
- Assisting in procurement
- Managing the bidding process

Similar to the management contract, here too, the construction management firm is barred from executing the construction work on its own.

Design, Management and Construction Contract

In this arrangement, the client appoints a single (principal) contractor to take care of design and construction. Thus, the client has to deal with a single agency for both design and construction. The basic design concepts may be provided by the client himself or through an independent agency. After the basic design concepts are frozen, the client calls for the tender and selects the appropriate agency for providing design, management and construction services. In practice, the design, management and construction contractor sublets the design and construction work to subcontractors and suppliers, and coordinates among them.

5.3.3 Integrated Contract

Design–Build

This is a form of contract in which the contractor takes up the responsibility for both design and construction, based on basic plans drawn up by the client. In other words, design and construction are handled within a single organizational structure, and a perennial conflict between the designer and the contractor is avoided. This also facilitates application of uniform standards. In most cases, a cost-plus-fee contract or a lump-sum contract that includes both design and construction costs may be adopted. Contracts of this form are often adopted when the client has no in-house design and engineering departments, and when subcontracting (or outsourcing) only the design to a separate agency is considered inappropriate. Obviously, the contracting agency in such cases should have expertise in both design and construction. In very large projects, however, separate companies specializing in design and construction can always form a joint venture and bid for such a project, with appropriate financial and legal arrangements.

Apart from encouraging a holistic and comprehensive approach that tends to bring the costs down, the method also stimulates development of technical prowess in contractors, and reduces the number of disputes and lawsuits.

The Turnkey Contract

As the name suggests, this comprehensive contract entrusts the responsibility of all activities involved to the contracting agency, and the owner simply wants to 'turn the key' at completion to take over the facility. Thus, in such contracts all activities related to surveying, drawing up specifications, design, project planning, construction and test operation are entrusted to one large contracting organization, which may break the activities down and engage other agencies to carry out specific jobs. At times, the contract may also include operating the facility for a limited period. Such contracts have been found to be especially useful in projects involving a combination of civil, electrical, mechanical, chemical and mining engineering, and are seen typically in design and construction of industrial complexes including petrochemical plants and nuclear power stations.

The following developments have contributed to the growing popularity of this method of contract:

(a) Modern construction has become very complex and the client prefers to deal with a single organization rather than with a multiple of specialist contractors, each with his own contractual peculiarity.

(b) Large contracting firms have both the technical and managerial skills to take up such works. Several large public-sector contracting agencies like Engineers India Limited (EIL), Bechtel, Larsen & Toubro, and Hindustan Construction Company (HCC) often handle turnkey projects in India and abroad.

At the outset of such a project, the client first prepares documents stating the requirements of the facilities to be constructed, and either selects the best proposal from those submitted by multiple bidders or designates a specific contractor from the beginning and enters into a contract when negotiations begin.

From the viewpoint of a client, the system has the merit of clearly laying out responsibility. These comprehensive contracts may include not only civil engineering and building works, but also procurement and installation of equipment and systems. In fact, depending upon the scope, a contract may also include training of operators in the operation of a facility. The terms 'package deal contract' and 'general turnkey contract' are often used to describe the kind of contract described here.

Build, Operate and Transfer (BOT) Contract

Apart from the responsibilities of the turnkey contract, this throws in the responsibility of fundraising for the project in the contractor's court. In return, the contractor is allowed to 'operate' the facility for an agreed period of time to recover the cost incurred in the design and construction of the facility. This system of contracting is useful when the client does not want to invest directly in the project, and wants to encourage development projects through external funding and investment. It is also a method of attracting and involving the private sector in public projects and infrastructure development, which typically involve very heavy capital investment. For example, a power corporation may ask bidders to set up a power plant on BOT basis, wherein the bidder agrees to design and construct the plant in return for rights to operate the plant for, say, 10 years, during which the contractor can generate and sell the power. Design and construction of certain toll highways or airports can also be similarly done on a BOT basis. Very often, financial institutions are an integral part of such a contract, precisely to take care of long-term financial implications.

Since these contracts often involve long-term relationships and commitments, it is crucial that the contractor carries out his own research into not only the economic and technical feasibility but also the social and administrative aspects of the project. A judicious system for risk allocation is called for to address some of the concerns of a contracting agency. Often, the client somehow guarantees the contractor's operating income. For example, the contractor needs an assurance that the power produced will be used and paid for, at a rate to be determined in an agreed manner, with or without a base minimum. A contractor may also seek a mechanism to redress a situation when estimates go awry — for example, the traffic on a toll highway does not grow at an anticipated rate for whatever reason, and the contractor is unable to 'collect' at the estimated rate.

5.3.4 Discretionary Contract

The important variants under discretionary contract such as partnering and joint venture are discussed in the following sections.

Partnering

This is a new form of agreement or system, adopted within normal construction contracts or design–build contracts, in which the client and the contractor together form a project team based on mutual confidence and then work together to manage the project to a successful conclusion, yielding a profit for both parties.

The fundamental philosophy behind partnering is the mutual trust among parties involved in the partnering. The concerned parties meet before the start of the project to set out the project goals and then strive to achieve them. The dispute arising during the execution of project and afterwards are settled based on the agreed method of dispute resolution. The relationship between the parties is called a partnership or an alliance. One agreement is usually valid for a number of years, but agreements for shorter periods or those for a single project are also possible. The agreement typically covers planning, design, engineering, procurement and construction supervision. Payment provision is made based on a cost-plus-fee basis.

The advent of partnering was fundamentally to avoid contractual confrontations and disputes. How far this has been achieved is yet to be reported, as partnering has been in use since the early 1980s only. Besides, there are very few examples of partnering in public works.

Joint Venture

In large projects, very often no single contracting company has adequate expertise and/or resources to be able to bid alone and become the main contractor. In such cases, several contractors pool in their resources and form a joint venture, and bid for the project together. Very often, a company is formed especially for that particular project.

The companies usually sign an MOU and form such a venture. Naturally, the MOU spells out the terms and conditions of this 'mini-merger' or 'part-merger', including the individual shares of the participating companies. The company providing the project leader (resident manager) is also specified in the MOU. In a manner of speaking, the MOU defines a kind of *common minimum programme*—which is usually a one-point agenda of bidding and completing a particular project. A copy of the MOU is submitted along with the tender document, and thus, the client is also aware of the objective-specific new company.

In the case of a JV, the proportions of shareholding among partners vary. A partner with 51 per cent or more shareholding usually controls ownership of the joint venture.

It may also be pointed out that the different participating companies contribute staff to a JV, as may be clearly spelt out in the MOU. Such personnel are often treated as staff of the JV, and treated as *on deputation* from their parent organization. Of course, once the project is completed, they return to their parent cadre.

5.4 BIDDING PROCESS

For the purpose of discussion, the process that starts with the owner inviting parties to 'bid' for a project and culminates in a contract being signed between the owner and a party identified to carry out the job has been called the bidding process.

Open Bidding

Open bidding is adopted for small-value projects that involve typical nature of work. The risk involved in the project is less here. The owner specifies some minimum eligible criteria for issue of the tender document. If these criteria are satisfied by a contractor, the tender document is issued to him. Upon issue of tender document, the owner may invite separate technical and financial bids or may ask for a single-package bid.

Selective Bidding (Limited Tender)

Selective bidding is adopted for very specialized projects. In this approach, a two-tier procedure is adopted. The first step is the pre-qualification process for selection of a set of contractors. In selective bidding, the tender document is issued only to selected bidders who had qualified from the pre-qualification process. The selective bidding is normally adopted for projects having large contract value, difficult construction, etc. Indeed, the risk involved in such projects is substantial.

As is discussed in the following paragraphs, the bidding process can be looked upon as the sum total of the following—notice inviting tenders (NIT), submission of completed bids, analysis of submitted tenders, acceptance of tender, letter of intent (LOI), work order and agreement in case of an open bidding, whereas the pre-qualification of contractor also gets added to the bidding process in case of selective bidding. In case of selective bidding, after the completion of pre-qualification process the tender documents are issued to the selected contractors. The selected contractors bid for the project and the contract is awarded usually to the lowest responsive bidder. In the following sections, we discuss about the pre-qualification process.

5.4.1 Pre-qualification Process

Construction procurement is a risky proposition. An owner has a lot at stake. He tries to make every move cautiously. He realizes that a wrong move in the very beginning itself such as choosing the wrong contractor for his proposed project may not augur well for his project. The terms 'right' and 'wrong' contractor are subjective and have to be dealt with on a project-to-project basis. How to choose a set of right contractors for the project, is the essence of the pre-qualification process.

The term 'right contractor' signifies 'fitness of purpose' for the proposed project. The term 'right contractor' has nothing to do with a large or a small contractor, since it may so happen sometimes that the large contractor may not be the right contractor for a proposed project if it is of low value. Similarly, a contractor, even if he is a leader in the heavy civil construction sector, may not be the 'right contractor' for a project that involves buildings with complex architectural features. The process of selecting a pool or set of right contractors is the purpose of the pre-qualification process.

A typical pre-qualification process would take anything between 8 weeks and 10 weeks, and may involve considerable efforts on the part of the owner organization. Selection of the 'wrong contractor' has been identified as one of the causes of project failures. Hence, the gains in long terms that result from pre-qualification process are worth the time and effort spent on it.

There are other terminologies and processes that closely serve the function of the pre-qualification process. These are licensing, registration of contractors, enlistment of contractors, and rating or grading of contractors. Some organizations, instead of resorting to the pre-qualification process again and again, enlist or register some contractors for doing a particular type of work, and they also specify the limit of contract value (say, up to ₹ 5 crore, ₹ 5 crore–₹ 25 crore, more than ₹ 25 crore, and so on) for which the contractors are eligible. As and when any project of certain value is undertaken by these organizations, the tender document is issued to the contractors enlisted for the said contract value. Indeed, this process saves time and effort for the owner as well as the contracting organization. The enlistment or registration is done for a particular period. Upon the expiry of the registration period, fresh application may have to be submitted in order to be registered.

In some countries, there is a system of providing license to the contractors. Under this system, a project beyond a certain value can be executed by a licensed contractor only. Under the licensing system, the contractors are awarded license in different categories, such as common contractors, and special contractors for different types of works such as civil works, plumbing and sanitary works, and electrical works. These classifications are done based on the amount of work executed by the applicant and a number of other factors including experience of the contractor in relevant construction work; available staff strength; sales volume of completed projects; financial parameters such as ratio of current assets to current liabilities, ratio of fixed assets to capital, ratio of net profit to total liabilities and net worth; construction machinery owned by the contractor; and safety and labour relations record.

As explained earlier, the enlistment or registration system helps in saving time as every time the pre-qualification process need not be repeated. However, for any unusual or specialized kind of work, pre-qualification process is carried out afresh. Pre-qualification of contractors is done by government as well as private organizations.

The announcement of pre-qualification process is advertised in leading dailies, trade journals, etc., and sometimes also intimated individually to reputed contractors.

Notice for Pre-qualification

Upon receipt of pre-qualification document from the owner, the contractor fills up the different information required for the purpose. Although the information required for pre-qualification varies from project to project and from owner to owner, there are certain aspects that are typically desired by an owner or employer for selecting the prospective bidders for a proposed project. These are discussed in the following sections.

Typical Documents Required for Pre-qualification

Letter of transmittal Sometimes, in order to maintain uniformity, even the sample letter to be submitted along with the pre-qualification document is provided. The applicant has to write his name and address in the specified places in the letter of transmittal. The essence of the letter of transmittal is to convey to the owner that the contractor has read the documents carefully and the information provided by the contractor is true to the best of his knowledge. Letter of transmittal also gives a list of enclosures with the pre-qualification document.

Power of attorney A copy of power of attorney stating the signing authority is also required to be furnished along with the pre-qualification document.

Financial information This statement shows the financial information of the applicant organization. Information such as gross annual turnover details of the applicant in the last 3 or 5 or 7 years as well as profits earned for the same period are asked.

Further, the financial arrangement for carrying out the proposed work is also asked from the applicant. The applicant can utilize his own funds through reserves or can show proof of credit limit enjoyed by the applicant with different banks. For the latter, the applicant needs to show the banker's certificate. A typical format is shown in Figure 5.3.

Owners also ask for income tax clearance certificate (ITCC), usually for a three- or five-year period. However, some Government organizations do not ask for ITCC these days. Sometimes, a certificate from the chartered accountant indicating the value of liquid assets and a solvency certificate from nationalized banks are also asked from the applicant. In addition to the above, the owners may also prefer to review the balance sheet and the profit-and-loss

---

TO WHOMSOEVER IT MAY CONCERN

At the request of ...., the following banking reference is being given.

1. Name of the institution: State Bank of India
2. Nationality: Incorporated in India
3. Address: Express Towers, 20th Floor, Nariman Point, Mumbai
4. Date and type of operation performed: A scheduled commercial bank, constituted under State Bank of India Act, 1955
5. Maximum amount of credit eligibility of the bidder company

--, enjoy the following working capital credit facilities from consortium of banks, with our bank being the lead bank.

Fund-based:	Rs 1000 crore
Non-fund-based:	Rs 1500 crore

6. Type of collateral for the above finance: Hypothecation charge on the entire current assets of the company

Signature of competent authority

Figure 5.3 Typical certificate issued by the bank showing credit limit enjoyed by the applicant

S. no.	Name of work	Owner or sponsoring organization	Cost of work (in ₹ crore)	Date of commencement	Stipulated date of completion	Actual date of completion	Litigation/arbitration details	Name/address/telephone no. of officers	Remarks (configuration, no. of storeys)

Figure 5.4 A typical format for details of similar works done

account of the applicant. For a typical government organization, all of the above-mentioned information are given in a typical form called Form 'A'.

Details of similar works Usually, owners ask for details on the experience of the applicant in executing projects of similar nature in the last five or seven years. This is expected since owners would like to award the project to those contractors who have sufficient experience and, hence, will be in a better position to anticipate problems and sort them out. A typical format in which such details are desired is reproduced in Figure 5.4.

Concurrent commitment 'Concurrent commitment' means the projects under execution by the applicant or just awarded to the applicant. This information is required to assess the bid capacity of the applicant as well as to know the intention or willingness of the applicant to take up the proposed project. A typical format in which such details are desired is reproduced in Figure 5.5.

The applicants establish their claim of concurrent commitment by producing letter of intent or agreement copy for running projects.

Certificates for completed jobs The applicants are supposed to establish their claim of experience by producing completion certificates of the completed projects. Sometimes, the applicants also furnish photographs of the projects executed by them in order to create a better impression.

Structure and organization Through this questionnaire, information related to structure and organization is obtained. Some of the answers that an owner typically looks for are:

- Name and address of applicant, contact details such as telephone and fax numbers, email ID
- Legal status—this information is required to know the legal status of the applicant, such as whether the applicant is an individual, a proprietary firm, a partnership firm, or a limited company or corporation. In order to verify the legal status, owners ask for copies of documents such as incorporation certificate defining the legal status

S. no.	Name of work and location	Owner or sponsoring organization	Cost of work (in ₹ crore)	Date of commencement	Stipulated date of completion	Up-to-date percentage progress	Cost of balance work	Name/address/telephone no. of officers	Remarks (slow progress, if any)

Figure 5.5 A typical format for showing concurrent commitment

S. no.	Designation	Total number	Number available for this work	Name	Qualification	Professional experience and details of work carried out	How they would be involved in this work	Remarks

Figure 5.6 A typical format for showing details of technical and administrative personnel

- Particulars of registration with various government bodies such as CPWD, MES, or department of sales tax. For verifying this, attested photocopy of the enlistment certificate is submitted by the applicant
- Names and titles of directors and officers who are going to be concerned with the proposed work
- Designation of individuals authorized to act for the organization

Details of technical and administrative personnel The owners may wish to know about the proposed organization structure for the project, with details of qualification, responsibility and experience of the key members. A typical format in which such details are desired is reproduced in Figure 5.6.

Although for a smaller organization it may be easier to provide exact names, qualifications and experiences of the personnel proposed to be employed, for large contractors a tentative list of personnel proposed for the project is furnished along with the total list of staff available with the contractor.

Details of plant and equipment The owner may wish to know about the availability of required plant and equipment with the applicants. A typical format used for extracting such information is given in Figure 5.7.

Some other questions In addition to the above information, the pre-qualification document contains some typical questions as given below:

- Was the applicant ever required to suspend construction for a period of more than six months continuously, after you commenced the construction? If so, give the name of the project and the reasons for suspension of work.

S. no.	Name of equipment	No. of units of equipment required for the work	Capacity	Age	Condition	Ownership status			Current location	Remarks
						Presently owned	Leased	To be purchased		

Figure 5.7 A typical format for showing details of plant and equipment

- Has the applicant or any constituent partner ever abandoned the awarded work before its completion? If so, give name of the project and the reasons for abandonment.
- Has the applicant or any constituent partner, in case of a partnership firm, ever been debarred/blacklisted from tendering in any organization at any time? If so, give details.
- Has the applicant or any constituent partner, in case of a partnership firm, ever been convicted by a court of law? If so, give details.
- In which field of civil engineering construction do you claim specialization and interest?
- Specify the minimum and maximum values of contracts executed by you.
- Would you be prepared to work with nominated subcontractors? If yes, what type of arrangement do you propose?
- Specify the list of disciplines that can be executed in-house.
- Specify a list of subcontractors.
- Do you have any minimum value of contract that will be acceptable to you?
- Why should you be hired for construction for the proposed project?
- Any other information considered necessary but not included above
- References

Other details The following details may also be desired in some cases of pre-qualification process.

- *Quality assurance plan:* Contains questionnaire regarding quality policy, responsibility and statement of purpose
- *ISO certification (if any)*
- *Safety, health and environment management plan:* Questionnaire related to policy, responsibility and statement of purpose
- *Planning, scheduling and monitoring plan, and reporting methodology*
- *Commissioning and handing over plan*
- *Project close-out strategy*
- *Questionnaires related to penalties in case the time schedule, the agreed quality standards, and the safety, health and environment standards are not adhered to/complied with*

Construction methodology A typical construction methodology for a project is given in Appendix 4. Needless to say, construction methodology will change from project to project.

Upon receipt of the completed pre-qualification document from interested applicants, owners—either themselves or by appointing a consultant—evaluate each application in the light of the predetermined criteria. A panel of reputed contractors is drawn up for issue of the complete tender document.

The criteria are framed in such a way (see Box 5.1) that it results in neither too many applicants qualified for bidding for the project nor too less applicants. The idea is to encourage fair competition among the selected contractors. Usually, the number of applicants is restricted between five and seven. In a nutshell, the pre-qualification process provides a level-playing field for competitors and eliminates the odd one among the various applicants.

> **Box 5.1** Suggested guidelines for establishing pre-qualification criteria
>
> 1. Minimum annual financial turnover for construction works in any one year (from among last five years of operations) should usually be not less than two-and-a-half times of the estimated annual payments under the contract.
> 2. The contractor should have satisfactorily completed three works costing not less than the amount equal to 40% of the estimated cost, two similar works costing not less than the amount equal to 50% of the estimated cost, or one similar completed work costing not less than the amount equal to 80% of the estimated cost.
> 3. The contractor must show proof of carrying out concrete and excavation items usually 80 per cent of the expected peak rate for the proposed construction.
> 4. The contractor must show the availability a project manager with not less than five years experience in implementation/construction of similar work for the proposed project.
> 5. The contractor must show credit lines/letter of credit/certificates from banks for meeting the funds requirements, etc., usually the equivalent of the estimated cash flow for three months in peak construction period.
> 6. Bidders who meet the minimum qualification criteria will be qualified only if their available bid capacity is more than the total bid value. The available bid capacity is calculated as under:
>
> $$\text{Assessed available bid capacity} = A \times N \times 2 - B$$
>
> where N = Number of years prescribed for completion of the subject contract
>
> A = Maximum value of works executed in any one year during last five years (at current price level)
>
> B = Value at current price level of existing commitments and ongoing works to be completed in the next N years
>
> 7. In spite of contractor's meeting all the criteria, they can be disqualified if it is established that the contractor made misleading statements and has dubious records in terms of abandoning the works, not properly completing the contract, inordinate delays in completion, litigation history, or financial failures.

5.4.2 Notice Inviting Tender

For execution of work through contract, especially in an open bidding system, the jobs need to be given due publicity. A common practice is to publish a formal 'notice inviting tender' (NIT), with the following details:

- Name of the authority inviting the bids
- Name of the project
- Conditions for eligibility of contracting agencies to submit a bid
- Brief details of the project
- Estimated cost and time of completion of the project
- The cost of the tender documents
- Earnest money to be deposited with the completed tender

> **Box 5.2** Notice inviting tender, ABC Institute, New Delhi
>
> Job No. XXX
> Date of NIT 05.05.2003
>
> *Last date for sale of tender forms*
> *15.05.2008 up to 2 pm*
>
> *Last date for receipt of completed tender forms*
> *16.05.2008 up to 3 pm*
>
> *Date and time of opening of tender forms*
> *16.05.2008 at 3:15 pm*
>
> Sealed tenders are invited by the works department of our institute for earthwork involving cutting, filling, compacting and making level a part of the campus located at Hauz Khas, New Delhi.
>
> Tender forms along with terms and conditions and specifications can be obtained from the office of the undersigned upon payment of ₹ 2,000 as tender fee payable in cash between 10 am and 2 pm on any working day till 15.05.2008.
>
> Duly completed tender forms along with earnest money, ITCC and all other required documents should be submitted at the office of the executive engineer before 3 pm on 16.05.2008. Tenders will be opened in the presence of the officers of the institute and tenderers present at 3.15 pm on the same day. The institute reserves the right to reject any or all the tenders without assigning any reason whatsoever. Conditional tenders may be summarily rejected.
>
> Executive Engineer
> Works and Estate Division
> ABC Institute, New Delhi

- Date and time by which the bids are to be submitted and the place of submission
- The date and time of opening of the bids

The detail of a typical NIT is given in Box 5.2.

Notices inviting tenders are generally publicized through press or the Internet. In the latter case, owner organizations upload the required information on their websites. Some websites are also dedicated to hosting information related to business-related opportunities in the construction industry. It may be pointed out that an online bidding process is still not very common, though there are growing numbers of cases where tender documents, along with the drawings and other conditions of the job, can be downloaded directly from a website.

5.4.3 Submission of Bids

Once a contracting agency, through an NIT or otherwise, learns of the availability of an opportunity, and decides to make an offer, it obtains the required tender documents and other details, carries out its own analysis of the job, and determines the cost at which it is willing to carry out the project. Normally, contracting agencies carry out a survey by visiting the site to check the availability of water, labour, power, transport, etc., and study issues like the kind of construction methodology and temporary infrastructure that would be required to be set up.

Depending upon the nature and size of the project, the bids may be submitted as a single package or in two parts containing the technical and financial parts separately.

5.4.4 Analysis of Submitted Tenders

As mentioned above, contracting agencies may be required to submit their offers in a single package or break it up into technical and financial packages (with the two being submitted separately). Evaluation of offers is generally carried out by an evaluation committee usually consisting of three persons, with one person being from the finance department. The seniority of members of the committee depends upon the value of the contract. The committee scrutinizes the submitted tenders, prepares a comparative statement containing the rates of all the offers and conditions, if any, and submits a recommendation for the award of the job. The committee is authorized to negotiate with the tenderer, wherever necessary, to lower the rates and also regarding any conditions if included by the tenderer. Normally, negotiations are first done with the lowest tenderer (L_1). At times, if the rates remain high even after negotiations, the next lowest tenderer is called for negotiation, and so on. The objective is to make all possible efforts to save the resources spent in the bidding process, though the owner clearly reserves the right to reject all offers, without assigning any reason. The final recommendation is based on the rates and conditions agreed upon after negotiations, after the committee has scrutinized the financial and technical competence of the agencies and their experience. It may be noted that the committee is usually not empowered to enter into the contract, and only make the final recommendation, which itself may or may not be binding on the competent authority. In case of public works, negotiation with only L_1 is permitted under specified conditions.

The process needs to be appropriately modified in cases where the technical and financial bids are submitted separately. It should be reiterated that such separate bids are invited only in large projects, where the client is desirous of holding negotiations on technical issues related to the project, without the negotiations being at least directly influenced by the financial details. Further, this practice can be followed in both open and designated systems of bidding.

In such cases, as a first step, only the technical bids are opened, and negotiations held between the owner and the contracting agencies on the technical details. This exercise is followed by the opening and analysis of financial bids, and the issue of letter of intent (LOI) to the contractor chosen to carry out the job. A brief description of the technical and financial bids is given below.

Technical Bid

The package should contain all the information that may be required to establish the credentials of the tenderer, and exclude financial requirements and conditions relating to the particular project. Therefore, the technical package usually contains (a) earnest money deposit (EMD), (b) copy of power of attorney, (c) valid financial papers such as income and sales tax clearance certificates, (d) details of concurrent commitments and past experience, (e) proposed project schedule, (f) the proposed organization chart for the project with appropriate details, including description of personnel responsibilities and bio-data of key personnel, (g) detailed cash-flow projections, (h) details of subcontractors proposed to be used, (i) list of plant and machineries to be deployed, and (j) details of materials proposed to be used, including their source and brand names, where applicable.

Financial Bid

The financial package consists of total price of the contractor to complete the project. The total price and discount or rebate, if any, is conveyed to the owner by means of a cover letter included in the financial bid. Further, the financial bid also contains filled-up bill of quantities in which the contractors enter their rates in words and figures for all the items of the project.

5.4.5 Basis for Evaluation and Acceptance

Indeed, while the quoted cost is perhaps the most widely used basis for drawing up a comparative statement, other aspects of past performance of a contracting agency such as safety, compliance with quality standards, and dispute resolution are also being increasingly considered.

It should be noted that a system based purely on the lowest cost is highly vulnerable to

(a) issues like safety and quality being compromised by an agency due to the intense competition that the method seeks to generate

(b) formation of pre-bidding 'ring' among contracting agencies, which is the equivalent of 'cartels'. This tendency is essentially an outcome of a conviction among the contracting agencies that the owner has no alternative but to execute the works, and that all contractors benefit if they 'agree' to a certain minimum cost

(c) a tendency among the contracting agencies to first submit a low bid to get the job, and then try to obtain additional payments through dispute resolution

It is obvious that these factors can vitiate the atmosphere of goodwill between the contracting agency and the client, and are detrimental to maintaining of schedule, quality and safety. Besides, there is also the danger of the project turning out to be more costly than estimated. Therefore, acceptance of an offer needs to be done very carefully, keeping a comprehensive view of the situation in mind. It should, however, be pointed out that increasing efforts are being made to develop other criteria than cost.

5.4.6 Letter of Intent

If the competent authority approves the recommendations of the tender committee, a letter of intent is issued to the contractor requesting him to submit necessary documents like partnership deed in case of partnership firm, and income tax clearance (if not submitted earlier). At this stage, the earnest money deposit of the successful tenderer is converted to a security deposit, and the contracting agency is requested to pay any balance amount towards the security deposit. At times, there is also a provision in the conditions of the contract that such amounts could be adjusted against the initial running bills.

5.4.7 Work Order

After the contracting agency accepts the offer and submits necessary documents, a work order is issued detailing the special terms and conditions, the mode of payment, the payment of security deposit, the total value of the contract, etc. In the work order, the contractor is asked to enter into an agreement with the owner and initiate the work.

5.4.8 Agreement

At this stage, the contractor contacts the engineer-in-charge of the project, and while preparing to start work at the site, enters into an agreement with the owner. The agreement includes previous relevant documents like the letter of intent, the work order, the general conditions of contract, the special conditions of contract, and the specifications and drawings. After the documents are signed by both parties, it becomes a contract, which is legally binding.

5.5 CPWD CONTRACT CONDITIONS

As pointed out earlier, CPWD conditions of contract are widely used and, hence, our discussion keeps these as a reference point. In Appendix 4, a brief description of contract clauses from CPWD Form No. 7 and Form No. 8 is given. Some of the important contract clauses that ordinarily find place in various projects are:

- Compensation for delay and incentive for early completion (Clause 2 and 2A)
- Determination and/or rescission of contract (Clause 3)
- Time and extension for delay (Clause 5)
- Payment on intermediate certificate to be regarded as advances (Clause 7)
- Completion certificate and completion plans (Clause 8)
- Escalation clauses (Clause 10C, 10CA, 10CC)
- Deviation/Variations (Clause 12)
- Action in case work not done as per specification (Clause 16)
- Work not to be sublet and action in case of insolvency (Clause 21)
- Settlement of disputes and arbitration (Clause 25)
- Employment of technical staff and employees (Clause 36)
- Return of material and recovery for excess material issued (Clause 42)

Compensation for delay and incentive for early completion (Clauses 2 and 2A) These clauses refer to recovery of compensation from the contractor for delays and defaults on his part. This clause can be divided into three parts:

i. Observation of time allowed for completion of work
ii. Payment of compensation by contractor for non-commencement, not finishing in time, or slow progress during execution
iii. The decision of the superintending engineer regarding compensation payable by the contractor shall be final

Clause 2A provides for incentive payable to the contractor in case of early completion of work.

Determination and/or rescission of contract in the event of breach (Clause 3) This clause is very important. It empowers the owner organization to determine or rescind the contract in the event of breach of contract by the contractor. It further allows the department to complete the balance work either departmentally or through another contractor at the risk and cost of the original contractor. In the event of recourse to this clause, the security deposit of the contractor is forfeited.

Time and extension for delay (Clause 5) Time is the essence of the contract on the part of the contractor. This is, however, not the case as far as the department is concerned. Accordingly, there is a provision for granting extension of time for the delay that may be caused by the department in meeting its obligation to the contractors in terms of handing over the site, furnishing of drawings/designs, etc., in taking appropriate decisions from time to time, and in issuing departmentally materials required for the execution of works. Extension of time may also be given in the event of increase in scope of work.

Payment on intermediate certificate to be regarded as advances (Clause 7) This clause deals with the circumstances under which the intermediate payments can be made to the contractor. To claim the payment, the contractor has to submit the bill.

Completion certificate and completion plans (Clause 8) According to this clause, a completion certificate is to be given by the engineer-in-charge to the contractor on completion of the work. This completion certificate is also a prerequisite for the contractor claiming the final bill.

Action in case work not done as per specification (Clause 16) This clause is important as it casts an obligation on the contractor and the departmental staff to ensure execution of good-quality work. Under this clause, the contractor can be asked to make good the defects in work at his own expenses, or re-execute the work if it is not in accordance with the specification, design, etc.

Work not to be sublet and action in case of insolvency (Clause 21) This clause specifies the circumstances under which tender accepting authority can rescind the contract. Under Clause 21, permission to sublet or assign the contract to another party should not be given to a contractor by the divisional officer without prior reference to the authority who accepted the tender.

Settlement of disputes and arbitration (Clause 25) This clause provides for appointment of an arbitrator in case of questions and disputes arising at any stage between the parties. This clause, however, does not apply to actions taken under clauses for levy of compensation for delay, determination of contract in the event of breach, extension of time in the event of delay caused by the department, and derivation of rates for additional and altered items. The contractor cannot go to a court of law for the redressal of his grievances unless he has exhausted the channel of arbitration. The Government of India has appointed a panel of arbitrators in the ministry of works and housing, and the disputes between the Government and the contractors are referred to arbitration by one of them. The authority of an appointed arbitrator can be revoked only with the order of the court. The award of the arbitrator is final and binding on the parties to the contract, unless it is set aside by the court.

Employment of technical staff and employees (Clause 36) This clause casts an obligation on the contractor to deploy well-trained, qualified and skilled professionals at site of work to execute quality work, and spells out the consequences that would arise on his failure to do so.

Return of material and recovery for excess material issued (Clause 42) This clause imposes an obligation on the contractor to manage an effective inventory control of the expensive and essential stipulated materials, and spells out the consequences in case of non-observance of diligence in their usage by the contractor. The intention behind the clause is to ensure that the contractor shall take only the required quantity of materials, and if any such materials remain unused at the time of completion or determination of the contract, it has to be returned to the engineer-in-charge.

Escalation clauses (Clause 10C, 10CA, 10CC) According to Clause 10C, the contractor can be reimbursed due to increase — caused as a direct result of coming into force of any fresh law or statutory rule or order (but not due to any change in sales tax/VAT)—in price of material incorporated in the work and/or wages of labour, compared to prevailing rates at the time of receipt of tender for the work. Clause 10CA provides for varying the amount of contract due to increase or decrease in prices of various materials pertaining to the work. In the contract where clause 10CC is applicable, this clause shall not be operational. For materials covered under clause 10CA, price variation under clause 10C shall not be applicable.

Price escalation clause According to this clause, the contractor is entitled to compensation for such increase as per provisions detailed in the clause for all major works. In the event of escalation in the price of material, labour and petroleum, oil and lubricant (POL), the cost of work on which escalation is possible is reckoned as 85 per cent of the cost of work as per the bills, minus the amount of the value of materials supplied departmentally as per the terms of the agreement. Thereafter, the compensation is worked out as per the increase in material cost index, consumer price index, consumer price index for industrial labour, and average index number of wholesale price for fuel, etc. The components of material, labour and POL are taken as certain percentages, and are predetermined for every work and incorporated in the tender document. In the case of building works, the component of POL is negligible and the components for material and labour are taken as 75 per cent and 25 per cent, respectively.

The compensation for escalation for materials, labour and POL is worked out as per the formula given below:

1. $$VM = W \times \frac{X}{100} \times \frac{(MI - MIO)}{MIO} \qquad (5.1)$$

   where, VM = Variation in material cost, i.e., increase or decrease in the amount (in rupees) to be paid

   W = Cost of work done

   X = Component of materials expressed as percent of the total value of work

   MI and MIO = All-India wholesale index for commodities for the period under reckoning, as published by Economic Advisor to Government of India, Ministry of Industry and Commerce, for the period under consideration

2. $$VL = W \times \frac{Y}{100} \times \frac{(LI - LIO)}{LIO} \qquad (5.2)$$

   where, VL = Variation in labour cost, i.e., increase or decrease in the amount (in rupees) to be paid or recovered

   W = Cost of work done

   Y = Component of labour expressed as percent of the total value of work

   LI and LIO = Consumer price index for industrial labour (all-India) declared by Labour Bureau, Government of India, as applicable for the period under consideration and tenders, respectively

3. $$VF = W \times \frac{Z}{100} \times \frac{(FI - FIO)}{FIO} \qquad (5.3)$$

where, W = Cost of work done

Z = Component of POL expressed as percent of total value of work, as indicated under the special conditions of contract

FI and FIO = Average index number of wholesale price for group (fuel, power, light and lubricants) as published weekly by the Economic Advisor to Government of India, Ministry of Industry, for the period under reckoning and valid at the time of receipt of tenders

4. The following principles are adhered to while working out the indices mentioned above.
   - The index relevant for any month is the arithmetic average of the indices relevant to the three calendar months preceding the month in question.
   - The base index is the one relating to the month in which the tender was stipulated to be received.
   - The composition for escalation is worked out at quarterly intervals and it is with respect to the cost of work done during the previous three months. The first such payment will be made at the end of three months interval.

5. In the event the price of materials and/or the wages of labour required for execution of the work decrease/s, there is downward adjustment of the cost of work, so that such price of materials and/or wages of labour are deductible from the cost of work under the contract.

6. The escalation is normally not applicable for project duration of less than 18 months. Earlier this duration was taken as 6 months.

Addition, alternation, substitution, derivation of rates This clause empowers the department to order addition, alternation and substitution as may be required during the execution of the work. The rates for the altered, additional, or substituted items are to be derived in accordance with the priority set out in the clause—such as derivation of rates from tendered rates for similar items, scheduled rates plus enhancement, or based on market rates.

5.6 FIDIC FORM OF CONTRACT AGREEMENT

FIDIC stands for Federation Internationale des Ingenieurs Conseils (International Federation of Consulting Engineers). It was founded in the year 1913 in Europe, and now has about 70 countries as members. The secretariat is situated in Switzerland. The FIDIC form of contract has evolved over a period. The contract conditions are equally suitable for use on domestic contracts. Prior to 1999, FIDIC had three forms of building and engineering contracts—the Red Book for civil engineering construction, the Yellow Book for electrical and mechanical works, and the Orange Book for design and build contracts. In September 1999, FIDIC published four new editions of the forms of contract (see Box 5.3). They are briefly described in the following sections.

Conditions of contract for construction These include the set of conditions recommended for building or engineering works where the employer provides most of the design. However, the works may also include some contractor-designed civil, mechanical and/or electrical construction works (Red logo).

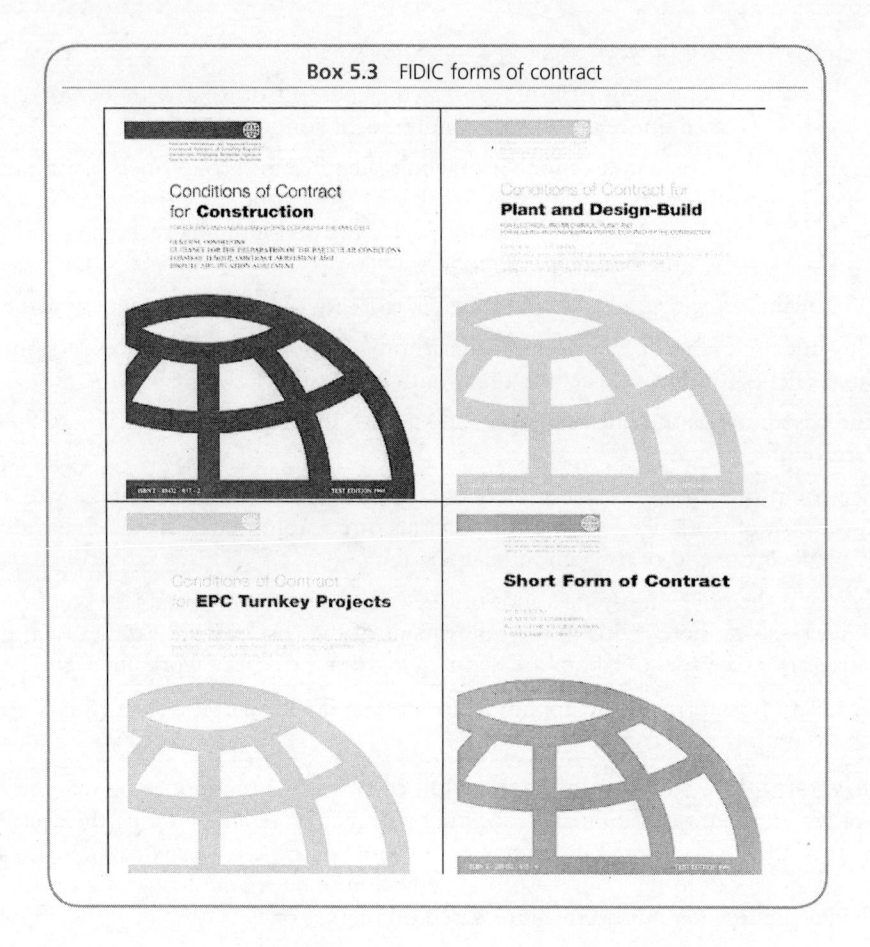

> **Box 5.3** FIDIC forms of contract
>
> Conditions of Contract for **Construction**
>
> Conditions of Contract for **Plant and Design-Build**
>
> Conditions of Contract for **EPC Turnkey Projects**
>
> **Short Form of Contract**

These days, it is very common for the contractor to design a significant portion of works on his own and, accordingly, the conditions of contract for construction contains more provisions that are applicable under such cases.

Conditions of contract for plant and design–build These include the set of conditions recommended for the provision for electrical and/or mechanical plant and for design and construction of building or engineering works (Yellow logo). Under the usual arrangements for this type of contract, the contractor designs and provides, in accordance with the employer's requirements, the plant and other works; this may include any combination of civil, mechanical, electrical and/or construction works.

The conditions are also suitable for the design and construction of building and engineering works. These contract conditions require the employer to appoint 'the engineer to administer the contract.'

Conditions of contract for EPC (engineering-procurement-construction)/turnkey projects These are suitable for use in projects of turnkey basis, such as projects for a power plant or a process factory, an infrastructure project, or any developmental works. In this type of contract, there is a higher degree of certainty on the price and time. The contractor undertakes total responsibility for the design and the execution of the project, including the guarantees for the performance (Grey logo). Under the usual arrangements for this type of contract, the entity carries out the engineering, procurement and construction, providing a fully equipped facility ready for operation (at the turn of a key). This type of contract is usually negotiated between the parties.

In such contracts, the contractor has a greater freedom to satisfy the requirements of the end user as specified in the contract. The contractor enters into such a contract with the expectation that it would be more profitable than under the traditional procurement principles and is, thus, prepared to accept a greater degree of risk.

Short form of contract This form of contract is suitable for small works (small capital value) of short duration, or for relatively simple and repetitive works. This form of contract is suitable for any discipline of engineering irrespective of who provides the engineering (Green logo). Depending on the type of work and the circumstances, this form may also be suitable for contracts of greater value.

In this form of contract, the contractor constructs the works in accordance with the design provided by the employer or by his representative (if any), but this form may also be suitable for a contract that includes, or wholly comprises, contractor-designed civil, mechanical, electrical and/or construction works.

5.6.1 Need and Principles of FIDIC Contracts

The conventional contract forms being used in various government departments in India are considered to be one-sided. Due to globalization of economy and many multinational companies contracting for various infrastructure and developmental projects in India, the conventional form of contract is not considered to be suitable and the global contract form such as that of FIDIC is in vogue. The FIDIC form of contract is considered to be a well-balanced and equitable form that clearly defines the role and responsibility of all parties to a contract. It has a fair apportioning of risks, rights and obligations between the parties. It is in wide use for international contracts and is supported and recommended by various development banks such as World Bank and Asian Development Bank. It contains a set of effective, clear and complete conditions. It has time limits specified for different actions to be taken by different parties. Besides, it also has effective provisions for adjudication.

The basic principles behind all FIDIC contracts are given below:

- To achieve optimum results by not expecting contractors to quote for risks that could not be reasonably foreseen or evaluated
- For the employer to assume responsibility for costs arising from events that may never occur, which lie outside the contractor's control or which cannot be covered by insurance at a reasonable premium (i.e., employer's risks)
- Close cooperation and teamwork between employer/contractor and engineer within the framework of the contract, with a mutual desire to produce a satisfactory end product
- To remove mistrust or lack of confidence, with all parties performing their duties under the contract responsibly and correctly
- The use of independent 'engineer', who is required to exercise his discretion with impartiality, even if he is an employee of the employer

5.6.2 Salient Features of FIDIC Form of Contract

Obligations

In FIDIC, the obligations of contractor as well as owner/employer and engineer are set out clearly. The obligations under a building contract for the contractor, employer and engineer are given below:

Employer
- To give possession of site
- To make payment
- To nominate a supervising officer (engineer/architect)
- To supply instructions to carry out the works
- To supply necessary plans, drawings and data
- Not to interfere with the progress of the works
- To nominate specialist subcontractors and suppliers
- To supply materials for use in the works (where applicable)
- To permit the contractor to carry out the whole of the works

Contractor
- To complete the works (including defects liability period)
- To ensure suitability or effectiveness of the works using the materials specified
- Design responsibility (where not specified) especially in relation to materials and workmanship (e.g., concrete mix and reinforcement)
- To warn the employer in relation to an impracticable design

Engineer Under FIDIC conditions, the engineer may be the employee of the employer department, but he is not a signatory or party to the contract between the employer and the contractor. He has to perform as an interpreter of the contract and as a judge of its fulfilment by both the parties. Decision/opinion taken by the engineer may be opened up, reviewed, or revised by the arbitrator if challenged within the stipulated time.

Delay and Extension

Procurement of all materials, plant, equipment and other things required for execution of work is the responsibility of the contractor. However, if the delay is caused due to failure or inability of the engineer to issue any drawing or instruction for which notice has been given by the contractor, or due to failure on the part of the employer to give possession of the site, then the contractor is entitled for (1) extension of time and (2) additional amount (compensation).

Performance Security

Performance security shall be refunded to the contractor within 14 days of the issue of the defect liability certificate. The performance security/guarantee shall be in the form annexed to these conditions or in such other form as may be agreed upon between the employer and the contractor. Generally, the guarantees are conditional.

Suspension of Work

The engineer can order suspension for any reason. Contractor shall *not* be entitled for any compensation if suspension is—(1) otherwise stated in the contract; (2) due to default of contractor; (3) necessary by reasons of climatic conditions on the site; or (4) necessary for proper execution of work or for safety of the works (except in the case of defaults by engineer or employer, or when involving employer's risk).

If such suspension exceeds 112 days (after 84 days of suspension, contractor gives notice for restoration within 28 days), the contractor may elect to treat the suspension, where it affects the whole of the work, as an event of default by the employer and terminate his employment. In the event of such termination, the contractor is entitled for reasonable compensation.

Deviation

The engineer can make any variation in the form, quality, or quantity of the work, or any part thereof that in his opinion is necessary. No deviation limit is specified; however, if deviations exceed 15 per cent of the effective contract price, then the contract price shall be adjusted.

Sharing of Risks

This form contains various conditions regarding obligation of both the parties towards the risk. 'Employer's risks' are generally events or circumstances over which neither party will have any control (e.g., war, hostilities and the like), or events or circumstances caused by the employer, directly or indirectly. For example, some of the employer's risks are—loss or damage due to the use or occupation by the employer, loss or damage due to defective design provided by the employer or the engineer, and so on.

The contractor is required to take full responsibility for the care of the works, materials and plant, from the commencement date until the taking-over certificate is issued for the works. If any of the contractor's equipment, plant, or temporary works on or near or in transit to the site sustain destruction or damage by special risk, then the contractor shall be entitled to payment for replacing or rectifying such equipment.

If during the execution of the works the contractor encounters physical obstructions or conditions, which an experienced contractor cannot foresee, then the contractor shall be entitled to (1) extension of time and (2) reasonable compensation.

Additional Claims

FIDIC describes the procedure for additional claims under Section 53 as given here:

- Notice of claims by the contractor
- Contractor to keep contemporary records to support claims
- Inspection of such contemporary records by the engineer
- Substantiation of claims
- Payment of claims

Payment of Bills

The contractor shall submit to the engineer after the end of each month a statement showing the amount to which the contractor considers himself to be entitled. The engineer shall, within 28 days of receiving such statement, deliver to the employer an interim payment certificate, and the payment is to be made by the employer within 28 days after receipt of the engineer's certificate.

Taking Over of Completed Work
The engineer shall issue a taking-over certificate within 21 days after receipt of the contractor's notice and undertaking, if work is substantially completed in his opinion.

Dispute Resolution
Unless the parties otherwise agree, reference to arbitration may be made before completion of work (during execution of work) or after completion of work.

Contract Price and Payment
Contract price is to be agreed to and determined under Sub-clause 12.3, and subject to adjustments under the contract. The price is inclusive of all taxes, duties and fees, and not to be adjusted except for the reasons as stated in Clause 13.7 (adjustments for changes in legislation). The quantities mentioned in the bill of quantities are estimated quantities.

Employer is to make advance payment in the form of an interest-free loan for mobilization. The engineer shall issue the first interim payment certificate after receiving the application for the same and the performance security and guarantee for advance payments, with such advance payments to be repaid through percentage deductions.

The engineer shall issue to the employer an interim certificate of payment within 28 days of receiving a statement and supporting documents, which shall stipulate an amount that the engineer fairly determines to be due, with supporting documents. Withholding of the interim payment certificate shall not be done, but in cases where anything supplied or work done by the contractor is not in accordance with the contract, the cost of rectification or replacement may be withheld till it has been completed, and if the contractor has been failing to perform any obligation under the contract and has been notified earlier, the value of work or obligation may be withheld until performed.

Measurement and Evaluation
The engineer is to proceed in accordance with Sub-clause 3.5 (determinations) to agree to or determine the contract price by evaluating each item of work, applying the measurement agreed to or determined in accordance with 12.1 and 12.2, and the appropriate price for each item.

5.7 SUBCONTRACTING

The major construction agencies are essentially civil engineering organizations. More often than not, they are required to engage a variety of vendors and subcontractors for execution of specialized works in a modern-day project. Any project involves a number of items. For example, a multi-storey building may have works similar to the ones listed below that need to be executed:

- Waterproofing work, woodwork, painting work, aluminium works, plumbing (internal and external) and sanitation, flooring and tile work, false ceiling, electrical works (internal and external), air conditioning, interior works, networking, telephone wiring, horticulture and landscaping, thermal insulation, automation, acoustic control, external development, fit-out, façade work, building management services (BMS), and mechanical, electrical, and plumbing (MEP) services etc.

As can be seen from the above list, it is very difficult to be a specialist for all the activities/works involved in a project. In such cases, a part of activities is sublet or subcontracted by the main contractor to other contractors, known as subcontractors to the main contractor for this project.

Works are subcontracted to avail the expertise of other agencies in order to do it economically, within a given time schedule and given quality standards. Subcontracting normally results in speedy mobilization. It also increases the productivity of staff of the main contractor, as with the limited staff a large quantum of invoicing or billing can be done. Sometimes, subcontracting is resorted to in order to avoid legal hurdles, to do away with directly employing a huge fleet of workforce, and to honour contractual commitments entered into by the main contractor for the project. Subcontracting also helps in risk-sharing.

For ensuring the engagement of quality vendors, it is imperative to provide a list of reputed vendors in the contract itself, wherein the contractor may be given a choice of three vendors each. Alternatively, the qualification of each vendor may be defined in the contract and the contractor may be given the freedom to employ a vendor who fulfils the eligibility criteria.

5.7.1 Classification of Subcontractors

There are different types of subcontractors, depending on the type of agreement that the main contractor has entered into with subcontractors. The subcontract may also be of different types such as — item rate contract; work with materials, labour and plant; work with material and labour; work with labour alone; and lump-sum contract. The classification of subcontractors can be understood from Figure 5.8.

Labour Only

The subcontractors undertaking such works are known as **general subcontractors** (piece-rate workers [PRW]) and sometimes also referred to as petty contractors.

Material Only

When the requirement is that of supplying materials alone, the supplying agencies are also known as vendors or suppliers, and the contract is known as supply contract.

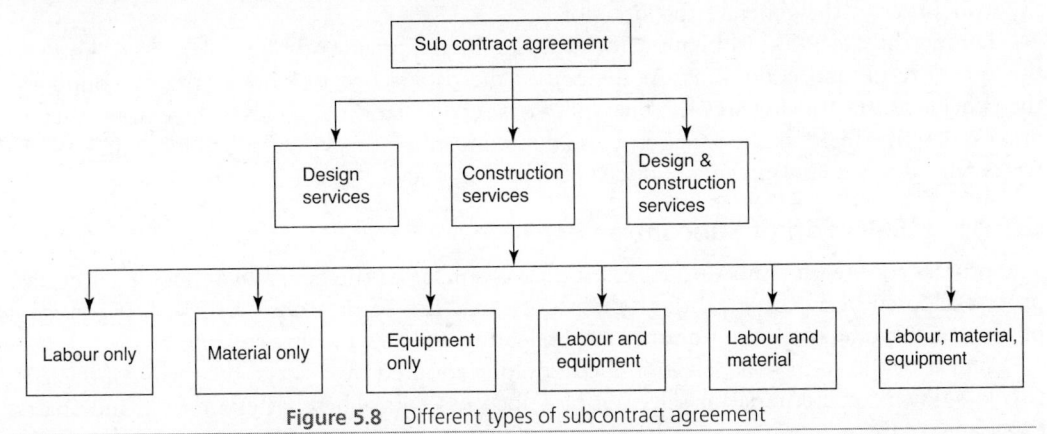

Figure 5.8 Different types of subcontract agreement

Equipment Hiring

General contractors hire equipment for a variety of reasons. The hiring rate depends on market conditions, duration of hire, payment terms and other terms and conditions, besides a host of other factors. The terms and conditions of hiring should be clearly spelt out. It should contain information such as equipment model and capacity, responsibility for providing operators and helpers for maintenance, minimum working hours, responsibility of fuel, lubricants, other consumables and spares, responsibility of mobilization cost, and period of mobilization. It is always better to check hire charges from different agencies before finalizing the hiring rates.

Back-to-Back

Sometimes, general contractors also resort to subcontracting a work package in totality. Needless to say, these subcontractors are cash-rich contractors and command respect in the industry. Such arrangements should be thoughtfully considered, though. The terms and conditions should be spelt out clearly, and details such as performance guarantee, mobilization advance and bank guarantee, retention, item description and general conditions of contract be made very clear right in the beginning. The responsibility towards preparing drawings and maintaining the desired quality should be well-defined.

Labour, Material and Equipment

They are referred to as specialized subcontractors or speciality contractors. Sometimes, these contractors are specified by the owners themselves. In such cases, they are called nominated subcontractors.

After going through the tender documents and studying the scope of work, the general contractor splits the entire bill of quantities into two broad heads — one that would be performed by the general contractor himself, and the other that would be performed by the subcontractors. The subcontracting involves selection of subcontractors, inviting quotations from them, and selecting the most suitable rates for the subcontractors. The first set of role played by the subcontractor ends here.

The next important role is played out after the award of the contract in favour of the general contractor. At this stage, some general contractors re-invite the quotations from normally those subcontractors who had participated during the tendering process. Out of these subcontractors, the most suitable subcontractor is selected and the work awarded to him by issuing the work order by the general subcontractor.

During the execution, subcontractor's measurement is made and he is paid according to the agreed terms and conditions. At the end of the total scope of work in the subcontract, the general contractor prepares the final bill, does reconciliation and closes the contract. The process described here is the usual process adopted in the industry for subcontracting. Some issues related to subcontracting are discussed below in detail.

5.7.2 Selection of Subcontractors

The process adopted for subcontractor selection is similar to the general contractor selection process adopted by an owner, though the process may not be rigorously followed. It consists of floating enquiries, collecting quotations, evaluating and short-listing subcontractors.

Similar to the contractor selection methodology adopted by owners, the main contractor also lays down some criteria for selecting his subcontractors. These criteria may include past experience of the subcontractor, technical capability to execute the work, mobilization capacity in terms of human resources and plant and equipment, financial capacity, the quality of works executed in the past, record in terms of working relationships with the main contractor, record of work safety and workers' welfare, and capability to chalk out construction methodology and work as per the same. Indeed, the rate offered by subcontractors is an important criterion as well.

The general contractor may also wish to inspect the facilities of subcontractors, if required. Finally, the rates submitted by the subcontractors are negotiated. Before negotiating the rates, the general contractor usually prepares his own estimates for the subcontracted items.

Usually, these estimates are a rough order of magnitude estimates based on past experience, existing rates in the project locality, or productivity. However, if required, the general contractor may sometimes carry out a detailed rate analysis of the subcontracted items as well. This helps him in comparing the price quoted by the subcontractors for these items. While negotiating on the subcontract after the general contractor has been awarded the contract, he also keeps the tender provision in mind.

Further, the registration of the subcontractors with the provident fund (PF) and sales tax (ST) authorities are also noted.

5.7.3 Work Order

Once the subcontractor has been finalized for a work, work order is prepared. This is a legally binding agreement between the main contractor and his subcontractor, and should be treated as a sacrosanct document in order to avoid disputes at a later date. The preparation of work order should be given considerable thought and all the terms and conditions clearly spelt out.

The work order should contain scope of work (what it includes and what it does not), clear and unambiguous specification, relevant drawings, provision of samples, mock-ups and inspection, and clear mobilization and delivery programme/schedule. The work order should mention the mode of measurement as many a time, disputes arise due to ambiguity in mode of measurement.

It should also mention all the taxes and deductions that are applicable. The taxes and deductions are made to account for income tax, workmen compensation, administrative expenses and retention money. Table 5.3 shows typical details of taxes and deductions that are usually there in a work order. The amounts shown may vary from company to company, and from work order to work order.

There can be a number of types of work order issued to a subcontractor. The work order may be a running work order; a first and final work order; a labour supply work order; a work order for individual bill of quantity items such as concrete, erecting and dismantling scaffolding, formwork and reinforcement; or even a work order for non-bill of quantities activities such as construction of temporary structures, housekeeping and miscellaneous repairs.

Table 5.3 Tax and deductions

Retention	5%
Income tax	1.1% to 2.2% for specialized subcontractor
Workmen compensation	1.0%
Administrative expenses	0.5%

Before issuing the work order to the subcontractor, it has to be approved by the competent authority within the general contractor organization. Depending on the value of the work order, different levels of management personnel such as resident engineer, construction manager, or project manager are given authorization to approve a work order within a general contractor organization. Work order is issued after getting their approval. Sometimes, it may be required that work orders are issued to a subcontractor who is not the lowest bidder; in such cases, the specific reasons for doing so should be mentioned in order to get the approval of the competent authority.

5.7.4 Terms and Conditions

The following terms and conditions are usually mentioned along with any subcontract to spell out the obligation and responsibility of each party (general contractor and subcontractor) to the subcontract. These are:

- Performance guarantee to be deposited by the subcontractor
- Commencement and completion dates
- Applicable taxes and duties on works contracts, insurance, value-added tax and income tax deduction
- Variation in rate
- Site working and access
- Method of measurement
- Variation in scope of work
- Liquidity damages
- Maintenance and defect liability period
- Payment terms
- Obligation towards labour
- Subcontractor's obligations and responsibility
- Facilities to be provided by general contractor
- Arbitration clause in case any dispute arises

5.7.5 Subcontractor Management—Some Guidelines

Subcontractors are important team members for any multidisciplinary project. It is important that mutual trust and faith exist between the general contractor and the subcontractors. Although there are instances of bitter fight among them during execution, it is essential to maintain good coordination not only among the general contractor crew and the subcontractor, but also among different subcontractors working at a project site. The process of managing the subcontractor right from their selection till the subcontract is closed out is essential for the success of a project, given the fact that in some projects more than half of the project value is subcontracted.

Screening of subcontractors is an essential first step in subcontract management. Care should be taken that run-of-the-mill and fly-by-night subcontractors are *not* selected even if the rates offered by them are relatively low. These subcontractors do not show any commitment towards the subcontract assigned to them, and they run away at the slightest instance of loss in sight during the execution of contract. It is always better to lay down a set of guidelines for subcontractor selection. Only those subcontractors should be encouraged who have registered themselves with the provident fund and income tax authorities. It is preferable to maintain a data bank of subcontractors who have left the subcontract in between and have been blacklisted in some other project sites.

Once a subcontractor has been selected, and his rates negotiated, it is a good practice to get the performance security from the subcontractor as well as get the work order signed from the competent authority, before the work is started at site by the subcontractor. These actions save a lot of harassment later during the execution and payment stage.

In order to maintain better coordination at sites, it is preferable to award the subcontract to a single subcontractor which involves interface at work. For example, a single subcontractor should be engaged for formwork, reinforcement and concreting, since it becomes very difficult to coordinate, say, three subcontractors for the purpose. The involvement of a number of subcontractors for these operations results in idle time for one or the other subcontractor, leading to different kinds of dispute at the time of payment.

During the execution of the subcontract, the representative of the general contractor should ensure that necessary fronts are released and all inputs given to the subcontractor to avoid idling of the subcontractor's crew. The work carried out by the subcontractor should be promptly measured and bills of the subcontractor promptly made. These things keep the subcontractor motivated enough to get involved with the work and show his commitment. Further, before making payments the general contractor should also ensure that the workmen of the subcontractors are paid their due. No miscellaneous measurements shall be made by site engineer for adjusting any payments, as this sets a wrong precedence and subcontractors try to take advantage of it.

General contractors do engage some labour from the subcontractor on supply basis, in which the number of labour days is counted for making the payment. As far as possible, this practice should be avoided since it is very difficult to monitor the productivity of the supply workers. Subcontractors sometimes even engage physically unfit workers as well as some child workers for these supply works, exposing them to hazards of construction activities.

Finally, all the final bills pertaining to a subcontract should be prepared before the closing of site.

REFERENCES

1. Central Public Works Department (CPWD), http://www.cpwd.gov.in.
2. Federation Internationale des Ingenieurs Conseils (International Federation of Consulting Engineers), FIDIC, http://www.fidic.org.
3. Harris, F., and McCaffer, R., 2005, Modern Construction Management, 5th ed., Blackwell Publishing.

4. Indian Contract Act (1872).
5. Lutz, J.D., and Halpin, D.W., 1992, 'Analyzing linear construction operations using simulation and line of balance'.
6. Schexnayder, C.J., and Mayo, R.E., 2004, *Construction Management Fundamentals*, Singapore: McGraw-Hill.

REVIEW QUESTIONS

1. State whether True or False:
   a. When time is the essence of contract, the completion time can be extended.
   b. When time is the essence of contract, liquidated damages clause can apply.
   c. When time is the essence of contract, penalty can be charged for delayed completion.
   d. When time is the essence of contract, full payment by employer is delayed.
   e. The agreement is voidable under Indian Contract Act when both parties are under mistake.
   f. The agreement is voidable under Indian Contract Act when awarded without consideration.
   g. The agreement is voidable under Indian Contract Act when the objective is unlawful.
   h. The agreement is voidable under Indian Contract Act when agreement is without free consent.
2. Explain the common types of engineering contracts and compare the following types of contract:
   a. Item rate contract v/s lump-sum contract
   b. Cost plus contract v/s turnkey contract
3. What is pre-qualification of contractors, and what criterion is applied for taking a decision by the client? Suggest weightage rate for merit rating.
4. Write short notes on CPWD contract conditions and special conditions of contract.
5. How is a contractor normally qualified to carry out certain types of work for government's departmental work?
6. What are the conditions of contract in a contract document? Why is it recommended to use standard forms of contract? What are the matters to be defined in the conditions of contract? Elaborate.
7. What are the contents of a tender document? Enumerate the complete tendering procedure with illustrations.
8. Write short notes on (a) price escalation clause and (b) liquidated damages and penalty.
9. Explain the necessity of involving private participation in infrastructure development. What are the structure and salient features of BOT form of public–private partnership? You may explain with the background of road construction on BOT basis.
10. What is a bid? How is it submitted? What are the various stages and types? How does the presentation of the bid and its evaluation take place?

11. What is the content of specification? What are the different types of specification?
12. What are the criteria that ensure validity of a contract?
13. Discuss in brief different contract documents.
14. Collect a few contract documents and list out the common general contract conditions.
15. Analyse some special contract conditions for a few projects being executed in and around you.
16. What do you mean by bill of quantity and how is it important for a construction project? Prepare a bill of quantity from the set of drawings for a project under construction in your locality.
17. Discuss different categories of contract in detail and differentiate them with respect to their important characteristics.
18. Differentiate between design–build, BOT and turnkey contracts.
19. What is meant by earnest money deposit?
20. Why do general contractors use subcontractors rather than performing all the work on a project?
21. What risks do general contractors incur by using subcontractors?
22. Why might a project manager require performance and payment bonds from a subcontractor?

6

Construction Planning

Introduction, types of project plans, work-breakdown structure, planning techniques—terminologies used, bar charts, preparation of network diagram, programme evaluation and review technique (PERT), critical path method (CPM), ladder network, precedence network, the line-of-balance (LOB)

6.1 INTRODUCTION

Planning can be defined as 'drawing up a method or scheme of acting, doing, proceeding, making, etc., developed in advance.' The last phrase ('in advance') is a key operational part of the definition, and requires experience and foresight. Construction projects involve using different resources—human, equipment and material, money, etc., and at all times, the task of a construction planner is to draw up plans for optimum utilization of all these resources and to ensure appropriate preparedness at all times.

As mentioned earlier, construction of a large project involves diverse agencies—government regulators, clients/owners, designers, consultants and contractors. It is obvious that each agency carries out its own planning exercise, and it is important to ensure proper coordination to ensure that the agencies do not work at cross-purposes, and that the common goal is served. It should be noted that plans are drawn up at each of the stages or phases of a project, though different terminologies are used at times depending upon the stage of the project. For example, initially when the project is at the inception stage, the plan could be referred to as a feasibility plan, while in the engineering and execution stages, terms such as preliminary plan and construction plan, respectively, are commonly used. This chapter largely focuses on the features of a construction or execution plan, though the principles outlined can be used for other stages of the project as well.

Some of the activities involved in construction planning are briefly discussed below:

Defining the scope of work Since all activities involve consumption of different resources to different extents, it is important that the scope of work involved is properly and, to the extent possible, completely defined. Any addition, deletion, or modification in the scope could have serious repercussions in terms of time of completion and cost, and even be the root of litigation, besides souring the relationships between different agencies. For example, if felling trees and getting environment clearances is added (at a later date) to the scope of a contractor who has been awarded a job for construction of roads, it would obviously cause difficulties.

Identifying activities involved This part of planning is very closely linked to defining the scope, and involves identifying activities in a particular job. Since different activities involved consume different physical resources to varying extents, it is crucial that these activities are exhaustively listed, along with the resources required. For example, though different agencies may be concerned with 'environmental impact assessment', it is important for them to identify the tools or parameters each will be using so as to plan effectively.

Establishing project duration This can be done only with a clear knowledge of the required resources, productivities and interrelationships. This information is used to prepare a network and other forms of representations outlining the schedules. It may be remembered that the duration required for any activity is related to the resources committed, and it may be possible to reduce the project duration by increasing the resource commitment, even at additional cost. Thus, a balance between time and project cost is required to arrive at an optimum level of resource commitment.

Defining procedures for controlling and assigning resources It is important that the planning document prepared is followed by others involved in the execution of the project, or in its individual phases. Thus, the procedures to be followed for procurement and control of resources for different activities—manpower, machines, material and money—are also laid down.

Developing appropriate interfaces The planner needs to devise an appropriate system for management information system (MIS) reporting. Tools such as computers and formats for reporting are widely used, and it may be noted that several software are readily available to aid the planner.

Updating and revising plans Although a construction plan needs to be continuously updated and revised during monitoring, some basic issues should be borne in mind before drawing up a full-fledged plan. For example, the planner should clearly understand the product to be produced in terms of scope and expected performance, the input required and the process involved, including the issues in quality control and tolerances at different steps. At the same time, the time and productivity aspects involved in the different activities should also be understood, besides the interdependence of activities. The planning should also identify milestones and targets for the different agencies to facilitate proper monitoring during execution. Inclusion of features identifying risks associated with a project, and the appropriate responses for mitigation enhance the quality of the project plan.

Considerable effort is needed to gather the information outlined above, as it involves identifying and determining key success indicators for the project, applicable corporate or industry standards, prevailing norms of productivity, and understanding any security or operational constraints. Although the importance of a careful study of all dimensions of a project at the outset cannot be over-emphasised, it should be borne in mind—given the uncertainties involved in almost all activities—that it is a task that is virtually impossible to achieve. For example, factors such as a possibility of change in government policies, minor changes in the user's requirement, extra features emerging during project execution (requiring an expansion in scope), or inability of contractors or suppliers to keep their commitment during the course of the project cannot be ruled out. While some aspects related to dealing with uncertainties in terms of probabilities, etc., are explained in greater detail later in this chapter, it should be reiterated that given their nature, it is virtually impossible to establish a fixed time frame or requirement of funds, especially for large projects.

Involvement of all key stakeholders of a project in the planning phase goes a long way in ensuring that frequent changes are not required during execution. This list can include not only representatives of the management, sponsors, client, customers, users and other associated key technical disciplines, but also legal and procurement departments.

6.2 TYPES OF PROJECT PLANS

Schedule, cost, quality and safety can be identified as specific items on which the success of any (construction) project is evaluated. Although there is a complex interrelationship between these, it is possible to discuss them independently—a statement such as a project being completed with very high quality but with different levels of cost and time overruns can at least be technically understood. Thus, at times it makes sense to have different plans for each of these criteria—and draw up (separately) a time plan (or schedule), cost plan, quality plan and safety plan. Of course, depending upon the nature and stage of the project, one may also need to deal with a plant and equipment plan, a maintenance plan and a staff deployment plan. It should be emphasised that all 'independent' plans should always be in line with an overall 'master plan' that lays down the overall plan of the project. A brief description of some of the commonly used plans in the construction industry is given in the following paragraphs.

6.2.1 Time Plan

Time is the essence of all construction projects, and contracts often have clauses outlining awards (bonus payments) or penalties (as liquidated damages) for completing a work ahead or later than a scheduled date. While effort is made to ensure timely completion of work, it should be noted that some of the common reasons for delays could be a sluggish approach during planning, delay in award of contract, changes during execution, alterations in scope of work, delay in payments, slow decision-making, delay in supply of drawings and materials, and labour trouble. Several reasonably well-established techniques are available and commonly used for time planning (or 'scheduling') activities—for example, critical path method (CPM), programme evaluation and review technique (PERT), precedence network analysis (PNA), line-of-balance technique (LOB), linear programme chart (LPC) and time scale network (TSN). The choice of the method to be used in a particular case depends on the intended objective, the nature of the project, the target audience, etc. Some of these methods are discussed in greater detail elsewhere.

6.2.2 Manpower Plan

This plan focuses on estimating the size of workforce, division in functional teams and scheduling the deployment of manpower. It may be noted that manpower planning also involves establishing labour productivity standards, providing suitable environment and financial incentives for optimum productivity, and grouping the manpower in suitable functional teams in order to get the optimum utilization.

6.2.3 Material Plan

The material plan involves identification of required materials, estimation of required quantities, defining specification and forecasting material requirement, besides identification of appropriate source(s), inventory control, procurement plans and monitoring the usage of materials.

6.2.4 Construction Equipment Plan

Modern construction is highly mechanized and the role of heavy equipment in ensuring timely completion of projects cannot be over-emphasised. Machines are used in modern construction for mass excavation, trenching, compacting, grading, hoisting, concreting, drilling, material handling, etc. Induction of modern equipment could improve productivity and quality, besides reducing cost. At the same time, it should be borne in mind that heavy equipment are very costly and should be optimally utilized in order to be productive. It is also important that the characteristics of equipment are kept in mind when drawing up an equipment plan.

6.2.5 Finance Plan

Given the fact that large construction projects require huge investments, and a long time to complete, it is obvious that all the money is not required at any one point in time. Contractors fund their projects from their working capital, a part of which is raised by the contractors using their own sources (e.g., bank loans secured against assets, deployment of resources from their inventory), whereas the rest comes from a combination of avenues such as mobilization advance for the project, running-account bills paid by the client, secured advances against materials brought at site, advance payments, and credits from suppliers against work done. Thus, a careful analysis needs to be carried out to determine how the requirement of funds varies with time. It is little wonder that capital inflow can be looked upon as the lifeline of any large project. Careful planning for funds and finances has achieved added significance in cases when projects are funded by the private sector or financial institutions that view the project as a financial investment and seek returns in monetary terms also.

6.3 WORK-BREAKDOWN STRUCTURE

'Work-breakdown structure' (or WBS), or simply 'work breakdown', is the name given to a technique in project management in which the project is broken down into 'manageable chunks'. WBS represents 'a task-oriented "family tree" of activities and organizes, defines, and graphically displays the total work to be accomplished in order to achieve the final objectives of the project.' This provides a central organizing concept for the project and serves as a common framework for other exercises such as planning, scheduling, cost estimating, budgeting, configuring, monitoring, reporting, directing and controlling the entire project. Thus, it should be intuitively clear that for a complex project, greater care is required in formulating a successful WBS.

A work-breakdown structure (usually triangular in shape) progresses downwards in the sense that it works from pursuing general to specific objectives—much like a family tree, it provides a framework for converting a project's objectives into specific deliverables. It should also be pointed out that any advantage of the technique will not be easily visible in simple or small projects, and even in medium-sized projects, the work-breakdown structure may not be needed as it may suffice to model the project in terms of 'activities' and organizing them into logical arrangements and sequence. However, as the project grows in complexity and several contractors and subcontractors, and possibly even consultants, get involved, the number of 'activities' and their interrelationships become simply too complex to handle using conventional methods. It is in such cases that the power and utility of the WBS method in effective management of the work is clearly demonstrated.

6.3.1 Methodology of WBS

A project is split into different levels from top to bottom. An illustrative example of such an exercise is shown schematically in Figure 6.1, which shows the 'whole to part' relation between the project, sub-projects, work packages, tasks and activities. The WBS does not go into the details of activity at the operational level. The term 'subprojects', 'work packages', and 'tasks' are used interchangeably. However, in this text, the sub-projects are assumed to be at higher level than the work packages, and tasks are assumed at lower level than the work packages. Work packages are the smallest self-contained grouping of work tasks considered necessary for the level of control needed, and may typically last a week to a month in duration.

Figure 6.1 shows a concrete example of this breakdown in the case of an information technology park project. The project consists of building three multi-storey towers, road works and area development works. These works are considered as sub-projects in the work-breakdown structure. The further breakdown is limited to only the second sub-project: Tower 2 for maintaining clarity in the figure. This subproject has further been sub-divided in work packages such as: T2.1 (civil works), T2.2 (architectural finishing), T2.3 (plumbing and sanitary), T2.4 (mechanical works), T2.5 (fire-protection works), T2.6 (electrical works), and T2.7 (LV system). These work packages have further been broken down in different tasks. For example, work package T2.1 has been broken into T2.1.1 (preliminary and general works), T2.1.2 (earthwork), T2.1.3 (piling works), T2.1.4 (CC and RCC works), and T2.1.5 (water-proofing works). Similarly, other work packages have also been broken down into different tasks associated with the corresponding work package. The tasks are broken down into activities that are the lowest level of a work-breakdown structure. For clarity in the diagram, the activities under Task T2.1.4 only have been shown in Figure 6.1.

It should be borne in mind that once this breakdown is exhaustive, operations such as development of the time schedules, resource allocation and project monitoring become simplified.

6.4 PLANNING TECHNIQUES—TERMINOLOGIES USED

Some of the terms commonly used in construction project planning techniques are discussed in the following paragraphs.

6.4.1 Event and Activity

Event is a point in time when certain conditions have been fulfilled, such as the start or completion of one or more activities. An event consumes neither time nor any other resource. Hence, it only expresses a state of system/project.

Activities take place between events. Activity is an item of work involving consumption of a finite quantity of resources and it produces quantitative results. An exception to this rule is the dummy activity as defined below. When breaking down a project into tasks, and so on, it is important to bear in mind that activities should be defined and organized in a manner that there are tangible outputs so that progress can be objectively monitored.

An example of activity could be 'laying of concrete floors' (shown in Figure 6.2). This could be represented as activity i-j. The start (node i) and the completion (node j) of this activity can be considered as events.

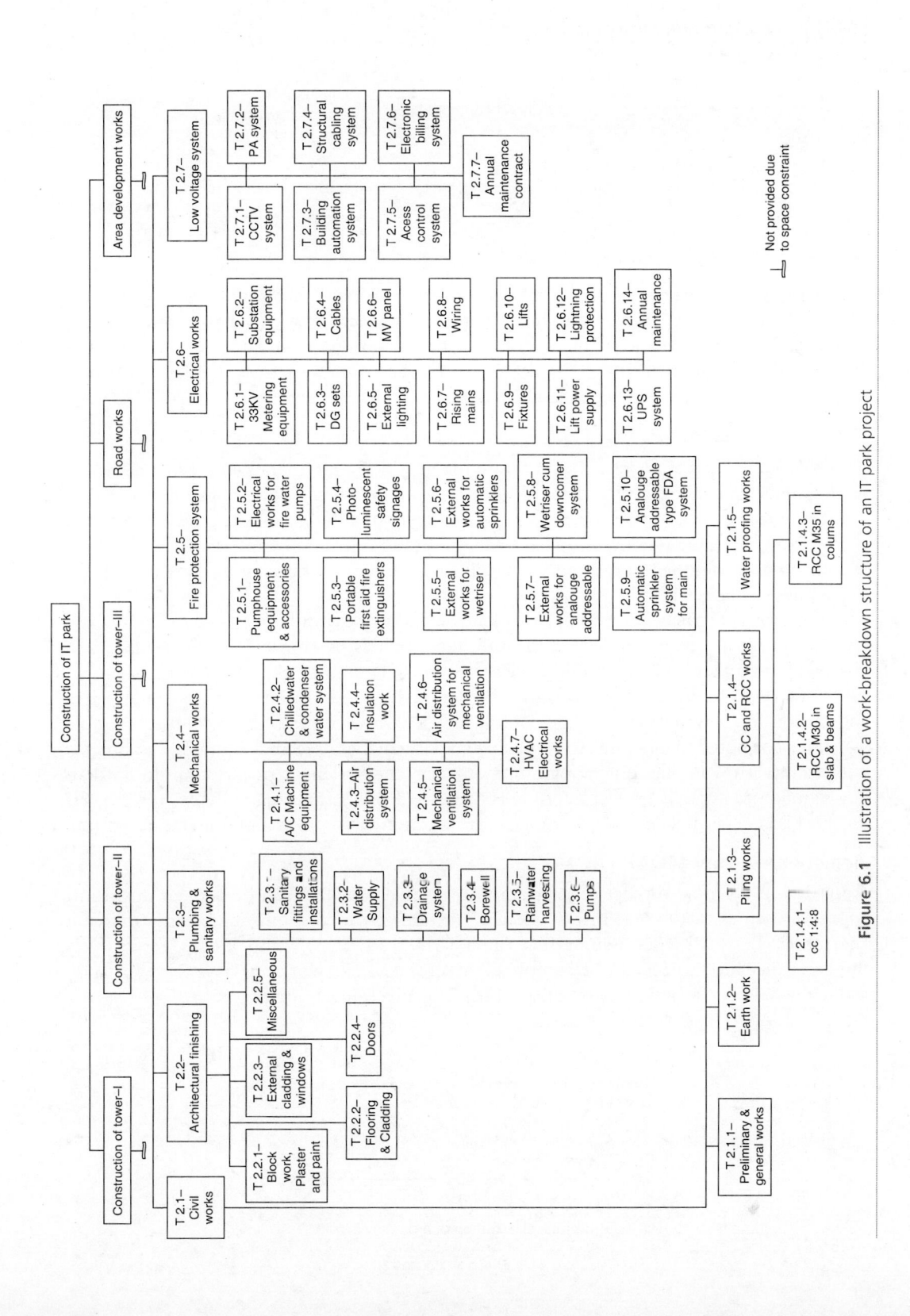

Figure 6.1 Illustration of a work-breakdown structure of an IT park project

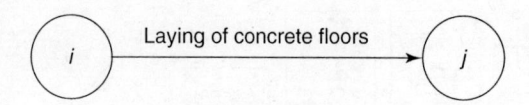

Figure 6.2 Example of events and an activity

6.4.2 Dummy Activity

This activity does not involve consumption of resources, and therefore does not need any time to be 'completed'. It is used to define interdependence between activities and included in a network for logical and mathematical reasons, as will be shown later on in this section. In Figure 6.3, a dummy activity (30, 40) is shown. The dummy activity $E(30, 40)$ has been introduced to represent the dependence of activity D on activities A and C.

6.4.3 Network

Networks consisting of nodes and arrows are the graphical representation of activities, showing logical dependence between them. While drawing a network, certain rules are followed for numbering the events or nodes. For example—same node number is not to be used twice in the network; tail node number is smaller than the head node; numbering starts from lefthand top and ends in righthand bottom. For details on numbering of nodes, the text on network techniques can be referred to.

Now, for construction planning, two kinds of networks can be used—activity-on-arrow (AOA) and activity-on-node (AON). In AOA, the activities are shown as arrows leading from one node to another node, and nodes here can be looked upon as either the starting or the end point of an activity. In AON, the activities are denoted by nodes and the immediate predecessor relationship between two activities is shown by an arrow connecting two nodes.

It may be noted that before an activity begins, all activities preceding it must be completed, and that an arrow implies only a logical precedence, and its length or direction do not have any significance as far as project duration is concerned. While drawing networks for projects, an effort is generally made to depict time as flowing from left to right and from top to bottom.

Activity-on-Arrow (AOA)

Considering a simple project (construction of a small wall), it can be broken down into activities such as earthwork, brickwork and plastering, which should necessarily follow in that order. Let us further assume that these activities take 3, 7 and 2 days, respectively, implying that the project can be completed in a total of 12 (working) days. Now, activity-on-arrow network representation of this project is given in Figure 6.4.

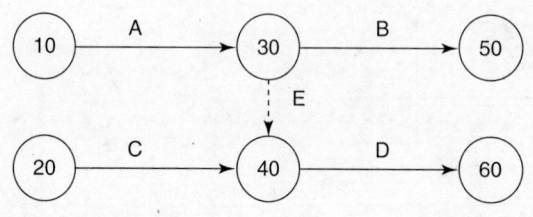

Figure 6.3 Illustration of a dummy activity

```
      EW         BW         PL
(10)─────▶(20)─────▶(30)─────▶(40)
       3          7          2
```

Figure 6.4 Example of AOA network

In this model, each activity (say, A) is defined in terms of two nodes (say, i, j), such that i and j represent the start and the end of the activity. This, however, does not mean that i or j has any significance on the time scale. The starting and end timings of any activity (i, j) need to be calculated separately. In the above example (Figure 6.4), the activities excavation, brickwork and plastering can, thus, be represented as $(10, 20)$, $(20, 30)$ and $(30, 40)$, respectively. When drawing such a network, it should be remembered that each activity has a unique (i, j) pair. Thus, a representation of the form given in Figure 6.5, where two activities have the same (i, j) pair, is not valid. However, several activities can start at the same node (i), or end at the same node (j), as shown in Figure 6.6, which are both valid representations. Further, it should be ensured that $j > i$, and that arrows point towards j.

The use of arrows as activities seems to be the most intuitive when creating a simple network manually. There is, however, a need for providing dummy activities when the project contains groups of two or more activities that have some, but not all, of their immediate predecessors in common. This has been explained with the help of Figure 6.7, which is a correct representation of Figure 6.5.

Also, if an activity has more than one predecessor, and one or more of these predecessors is/are also a predecessor for some other activity/activities, then dummy activities are generally required to make the connections. This aspect has been illustrated in Figure 6.8.

The addition of dummy activities to an AOA network is a cumbersome procedure and it adds to extra work in computational process (as will be seen, the computation takes both the

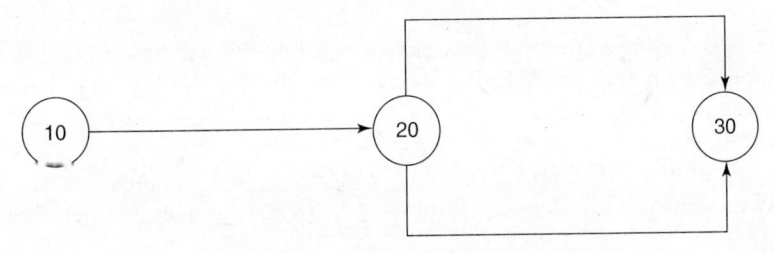

Figure 6.5 Incorrect activity representation

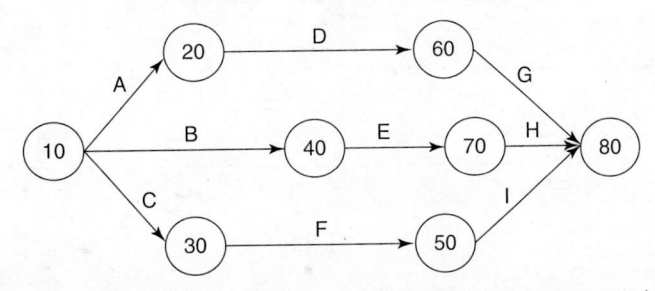

Figure 6.6 Several activities starting from Node 10 and terminating at Node 80

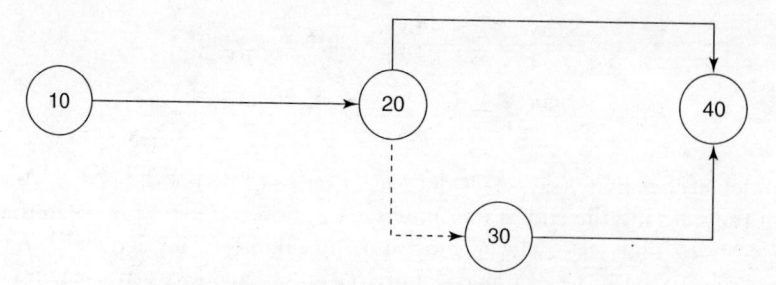

Figure 6.7 Correct representation of network of Figure 6.5

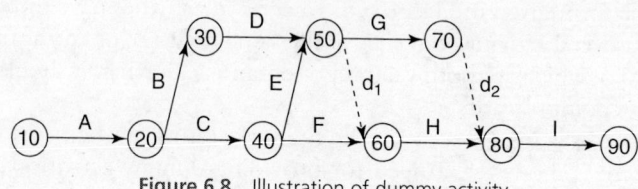

Figure 6.8 Illustration of dummy activity

actual and dummy activities into account) in network techniques, in addition to making the network look lengthy. In order to avoid these shortcomings, the AON alternative has been devised.

Activity-on-Node (AON)

In this type of network, the activities are denoted by circles or boxes called nodes, and the immediate predecessor relationship between the two activities is shown by an arrow connecting the two nodes.

The network shown in Figure 6.9 presents an example of 'activity-on-node network', for the AOA example shown in Figure 6.8. It can be noticed that for the same number of activities and similar relationships, AON has not used a single dummy, while AOA has used 2 dummy activities d_1 and d_2.

People often compare AOA and AON, and wonder which is the most convenient to use. The use of AOA seems to be the most intuitive when creating a simple network manually,

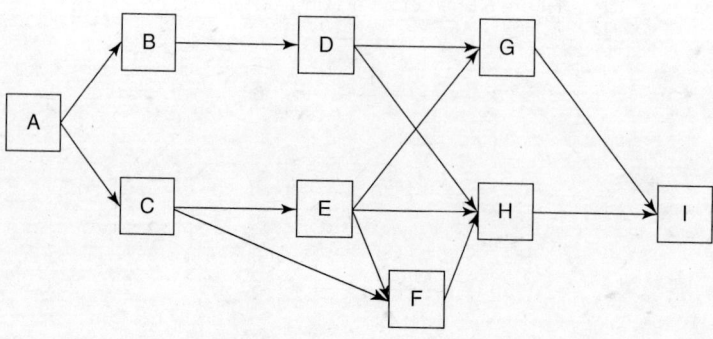

Figure 6.9 Example of an AON

while AON is more convenient if the network is large, complex and with many relationships. It is also much easier for setting up computation. AON is simple to draw and revise when compared to AOA (though computer programmes can make this also a simple affair). Further, AON is simpler to explain and can be understood even by a non-technical person. In recent times, the application of commercial scheduling software such as MS Project and Primavera in construction industry has made AON very popular. In practice, it is not very difficult to convert an AOA into an AON, and vice versa, as shown in Figure 6.9.

6.4.4 Precedence

This is the logical relationship implying that an activity needs one activity (or more activities) to be completed, before this activity can start. For example, in order to be able to start plastering, the brickwork needs to have been completed, i.e., logically, brickwork precedes plastering. It is a common practice in most construction projects to represent the precedence of activities in the form of a table, called the precedence table. For preparing the precedence table, a list of activities that should precede a given activity is given. It should also be mentioned that this concept (of precedence) is sometimes referred to as 'dependence'. Now, simply speaking, this implies that if it is identified that activities A, C and D must precede activity X (in other words, X depends on A, C and D), in the parlance of network analysis, precedence can be stated as follows—to initiate an activity (i, j), all activities having (i) as the end event should have been completed. A variation of this concept could be for an activity that can be started so long as another activity (which should logically precede) has at least started—for example, though painting a wall should, indeed, be preceded by plastering, it is not necessary that the latter be completed before the former can be taken up. Painting can be taken up even as plastering is being carried out, provided, of course, enough work front is available. These variations in the concept of precedence are discussed in greater detail subsequently in the chapter.

6.4.5 Network Logic

Some of the common logical ways useful in preparing a network are shown below:

Figure 6.10 shows an example of a 'burst' situation wherein two activities A and B are starting in parallel, while Figure 6.11 shows the example of a 'merge' situation wherein two activities C and D are getting completed together.

Figure 6.12 shows the incorrect way of showing three parallel activities A, B and C. The three activities have the same initial node number (1) and final node number (2). This has been corrected (see Figure 6.13) by adding two dummy activities, (2, 4) and (3, 4). In

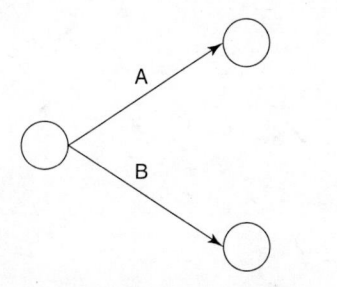

Figure 6.10 Example of a burst situation

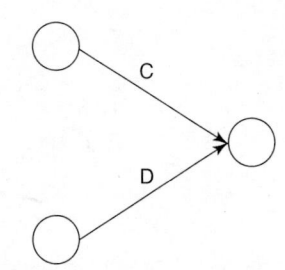

Figure 6.11 Example of a merge situation

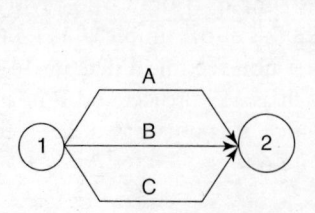

Figure 6.12 Incorrect representation

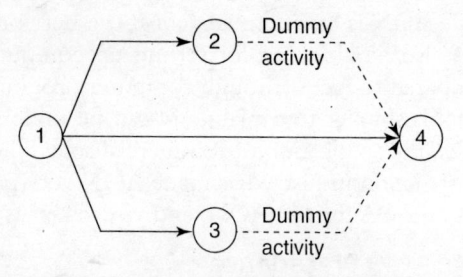

Figure 6.13 Correct representation of Figure 6.12

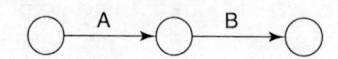

Figure 6.14, activity A precedes activity B. In other words, activity A is the predecessor of activity B.

The situation of activity C having two predecessors, A and B, is shown in Figure 6.15, while activity A having two successors, B and C, is shown in Figure 6.16. In other words, in the former, C is controlled by A and B, while in the latter case, activity A controls activities B and C.

The situation of activities C and D having two predecessors, A and B, is shown in Figure 6.17. In other words, activities C and D are controlled by activities A and B.

Some complicated situation such as activity C being controlled by activities A and B, and activity D being controlled by B alone, is represented in Figure 6.18. Some more complicated situations are shown in Figures 6.19 to 6.21.

Figure 6.14 Activity A precedes activity B

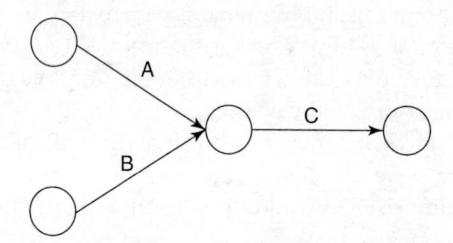

Figure 6.15 Activity C has A and B as predecessors

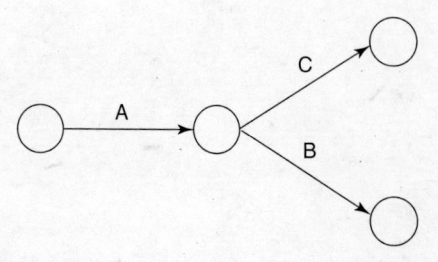

Figure 6.16 Activities B and C having predecessor A

CONSTRUCTION PLANNING

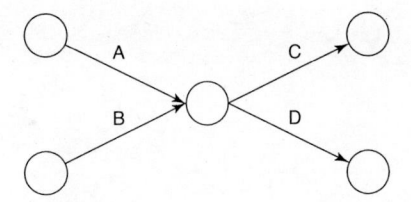

Figure 6.17 Activities C and D have two predecessor, activities A and B

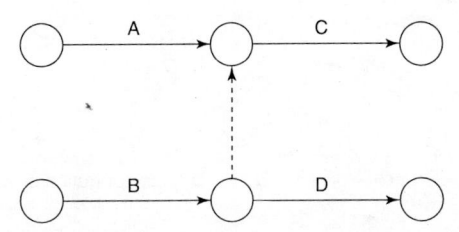

Figure 6.18 Activity C controlled by activities A and B, and activity D controlled by B alone

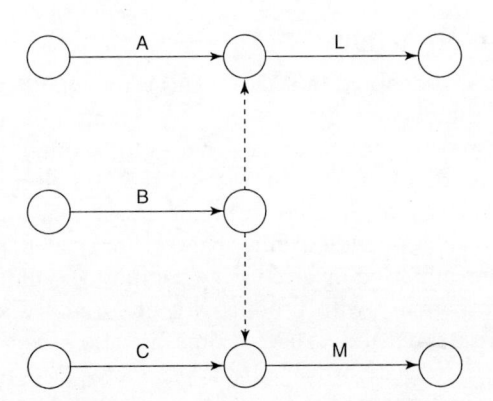

Figure 6.19 Activities A and B control L, and activities B and C control M

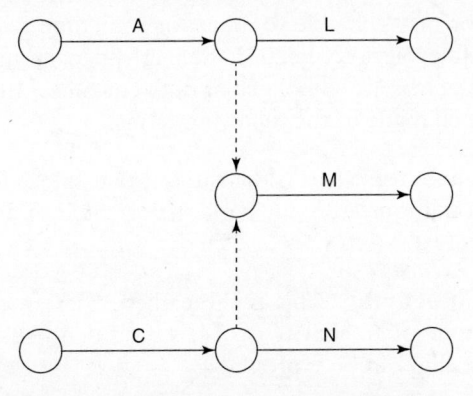

Figure 6.20 Activity A controls activities L and M, and activity C controls M and N

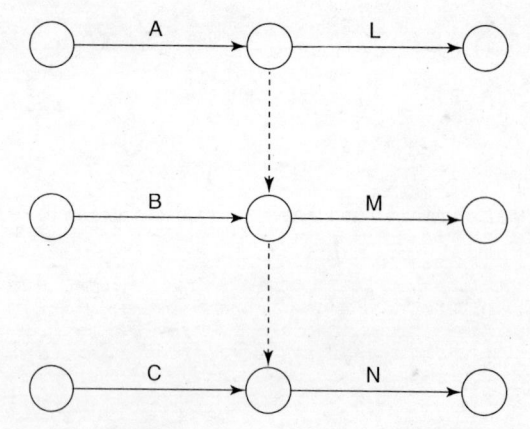

Figure 6.21 *L, M and N are controlled by A; M and N are controlled by B; N is controlled by C*

6.4.6 Duration of an Activity

Duration of an activity (i, j) is denoted by $D(i, j)$. This is the length of time required to carry out an activity (i, j) from the beginning to its end. Depending upon the activity and the level of detail, the duration may be expressed in days, weeks, or months. It should be noted that though the actual duration depends, in principle, on the quantum of work involved in the activity and the resources deployed, it is not really necessary that the relationship be exactly linear, as will be discussed subsequently in this chapter. Further, it should be borne in mind that a duration cannot be really fixed or given as a final number, and as such remains only an estimate, based on past experience with productivity, etc. Terms such as 'most likely time', 'optimistic time', 'pessimistic time' and 'expected time' are also used in the context of defining the duration of an activity. These are explained at appropriate places in the chapter.

Start and Finish Times

In principle, an activity can be started as soon as the groundwork involved has been completed, but the client or contractor may (be able to) wait for sometime before starting the activity without affecting the overall project completion. Similarly, depending upon the starting time and the duration, the activity may be completed at different times. In the context of start and finish times, reference is often made to the following terms:

Earliest start time of an activity This is the earliest, that the activity (i, j) can be started, i.e., all the necessary preconditions are met. Earliest start time of an activity (i, j) has been denoted by $EST(i, j)$ in this text.

Earliest finish time of an activity This is the earliest, that an the activity (i, j) can be completed. Earliest finish time of an activity (i, j) has been denoted by $EFT(i, j)$ in this text. Mathematically, the relationship can be expressed as

$$EFT(i, j) = EST(i, j) + D(i, j) \qquad (6.1)$$

Latest finish time of an activity This is the latest time that an activity needs to be completed in order that there is no delay in the project completion. Latest finish time of an activity (i, j) has been denoted by $LFT(i, j)$ in this text.

Latest start time of an activity This is the latest time when an activity must be started, in order that there is no delay in the project completion. Latest start time of an activity (i, j) has been denoted by $LST(i, j)$ in this text. Mathematically, the relationship can be expressed as:

$$LST(i, j) = LFT(i, j) - D(i, j) \qquad (6.2)$$

6.4.7 Forward and Backward Pass

The forward pass moves from the 'start' node towards the 'finish' node, and basically calculates the earliest occurrence times of all events. Considering that the project starts at time zero, the earliest occurrence time at each node is found by going from node to node in the order of increasing node numbers, keeping in mind the logical relationships between the nodes as shown by the connecting arrows. The earliest occurrence time for any node can be estimated from the (maximum) time taken to reach that node from the different incoming arrows.

Consider the network given in Figure 6.22, defining the relationship between activities (1, 2), (2, 3), (2, 4), (3, 5), (4, 5), and (5, 6), such that A is the first activity. Activities B and C can be carried out simultaneously, but only when activity A is completed. Activity D can start only when activity B is completed. Similarly, activity E can start only after C is completed. The last activity, F, can commence only when both the activities D and E get completed. Assuming the durations involved for the six activities to be $D(1, 2)$, $D(2, 3)$, $D(2, 4)$, $D(3, 5)$, $D(4, 5)$ and $D(5, 6)$, and the earliest occurrence times for nodes 1, 2, 3, 4, 5 and 6 to be E_1, E_2, E_3, E_4, E_5 and E_6, the earliest time for different nodes can be found out depending on whether there is a single incoming arrow for a node (for example, node 2, 3, 4 and 6) or there are more than one arrow entering into a node (for example, node 5 in Figure 6.22). The calculation of early occurrence times for nodes 1, 2, 3, 4, 5 and 6 is given below.

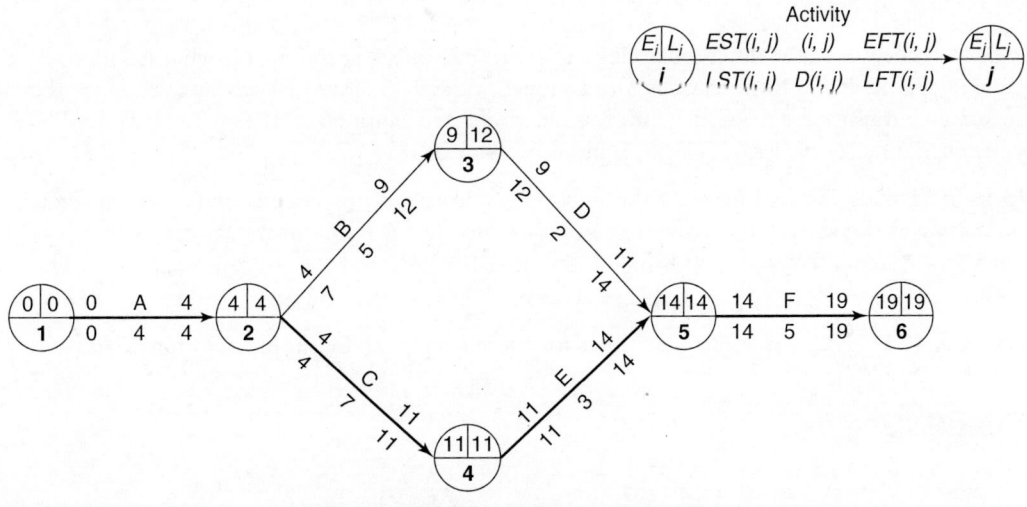

Figure 6.22 Example of a small network

Node 1 This is a starting node. Early occurrence time is conventionally taken as 0, though we can take some other value also. Assume $E_1 = 0$ in this case.

Nodes 2, 3, 4 The earliest times for reaching these nodes can be simply determined as

$$E_j = E_i + D(i, j) \tag{6.3}$$

Thus

$$E_2 = E_1 + D(1, 2) = 0 + 4 = 4$$
$$E_3 = E_2 + D(2, 3) = 4 + 5 = 9$$
$$E_4 = E_2 + D(2, 4)) = 4 + 7 = 11$$

Node 5 For node 5, there are two incoming arrows—one coming from node 3 and another from node 4. *Under such* situations the expression for computing earliest occurrence time is given as below:

$$E_j = \max_i [(E_i + D(i, j)], \text{ where the maximization is over all nodes } i \text{ that precede node } j \tag{6.4}$$

Thus, applying Equation (6.4), the earliest occurrence time for node 5 would be the maximum of early occurrence times of node 3 plus the duration activity (3, 5), and early occurrence time of node 4 plus the duration of activity (4, 5).

That is,

$$E_5 = \max[\{E_3 + D(3, 5)\}, \{E_4 + D(4, 5)\}] = \max\{(9 + 2), (11 + 3)\} = \max(11, 14) = 14$$

Node 6 Node 6 is the final (finish) node and has a single incoming arrow. Thus, as before:

$$E_6 = E_5 + D(5, 6) = 14 + 5 = 19$$

Thus, once the final (finish) node is reached, the earliest time for reaching that node also represents the minimum time it will take to complete the project.

Now, the backward pass is made in a similar manner to that of the forward pass, except that the process is carried out in reverse through the nodes, starting from the end node and finishing at the start node. Using the example illustrated in Figure 6.22, and defining L_6, L_5, L_4, L_3, L_2 and L_1 to be the late occurrence times for nodes 6, 5, 4, 3, 2 and 1, respectively, the late occurrence time for different nodes can be found out, depending on whether there is a single outgoing arrow from a node (for example, nodes 1, 3, 4 and 5) or there are more than one arrow outgoing from a node (for example, node 2 in Figure 6.22). The calculation of late occurrence time for nodes 1, 2, 3, 4, 5 and 6 is given below:

Node 6 This is the end node of the network. The late occurrence time is conventionally taken same as the early occurrence time of this node, i.e., E_6, though we can take some other value also. In real life projects, this may sometimes be dictated by the requirements of owner. Assume $L_6 = E_6 = 19$ for the example problem.

Nodes 5, 4, 3 The late occurrence times for these nodes can be simply determined as—

$$L_i = L_j - D(i, j) \tag{6.5}$$

Thus,

$$L_5 = L_6 - D(5, 6) = 19 - 5 = 14$$
$$L_4 = L_5 - D(4, 5) = 14 - 3 = 11$$
$$L_3 = L_5 - D(3, 5) = 14 - 2 = 12$$

Node 2 For node 2, there are two outgoing arrows—one going to node 3 and another to node 4. Under such situations the expression for computing late occurrence time is given as below:

$$L_i = \underset{j}{Min} [L_j - D(i, j)], \text{ where the minimization is over all nodes } j \text{ that succeed node } i \quad (6.6)$$

Thus, applying Equation (6.6), the late occurrence time for node 2 would be the minimum of late occurrence times of node 3 minus the duration of activity (3, 5), and late occurrence time of node 4 minus the duration of activity (4, 5).

That is,

$$L_2 = min[\{L_3 - D(2, 3)\}, \{L_4 - D(2, 4)\}] = min\{(12 - 5), (11 - 7)\} = min(7, 4) = 4$$

Node 1 Node 1 is the start node and has a single outgoing arrow. Thus, as before:

$$L_1 = L_2 - D(1, 2) = 4 - 4 = 0$$

The backward process gets completed at node 1.

It will be seen that for some events (nodes) in the network, the two values (E and L) will be the same if the latest project completion time is taken as the earliest project completion time. Now, these events constitute the critical events, and the continuous path through them gives the critical path.

It is possible, to compute the activity times from the event times calculated above. For example, given E_i, L_i, E_j and L_j, for events i and j, and $D(i, j)$ it is possible to determine the $EST(i, j)$, $EFT(i, j)$, $LST(i, j)$ and $LFT(i, j)$ of the activity (i, j) from the following expressions.

$$EST(i, j) = E_i \quad (6.7)$$

Once the $EST(i, j)$ is determined from Equation (6.7), the $EST(i, j)$ can be determined from Equation (6.1).

$LFT(i, j)$ is determined from the following expression:

$$LFT(i, j) = L_j \quad (6.8)$$

Once the $LFT(i, j)$ is determined from Equation (6.8), the $LST(i, j)$ can be determined from Equation (6.2).

For the example network shown in Figure 6.22, let us take activity (2, 3) for illustration of computation of various activity times.

For nodes 2 and 3, i.e. activity B, we have $E_2 = 4$, $L_2 = 4$, $E_3 = 9$, and $L_3 = 12$

Thus,

$$EST(2, 3) = E_2 = 4$$
$$EFT(2, 3) = EST(2, 3) + D(2, 3) = 4 + 5 = 9$$
$$LFT(2, 3) = L_3 = 12$$
$$LST(2, 3) = LFT(2, 3) - D(2, 3) = 12 - 5 = 7$$

The computations for event times and activity times are shown in Figure 6.22.

From the values obtained for E's and L's for various events of Example 6.22, it is clear that for some events (nodes) in the network, the two values (E's and L's) are the same, for example the event marked 1, 2, 4, 5, and 6. However for event marked 3, there is a difference between the values of E_3 and L_3. Since the events essentially represent the start (or the end) of an activity, we consequently find that for some activities such as B and D there is a difference in the EST's &

LST's and EFT's & LFT's. However for some activities such as A, C, D, E and F, there are no differences between the EST's & LST's and EFT's & LFT's. The activities where we do not find any difference between the start times and finish times are termed as critical activities. Mathematically speaking, an activity qualifies as critical activity where the following three conditions are fulfilled.

$$E_i = L_i \tag{6.9}$$

$$E_j = L_j \tag{6.10}$$

$$L_j - E_i - D(i, j) = 0 \tag{6.11}$$

The activities not fulfilling the above three conditions are referred to as non critical activities. It may be noticed that all non critical activities have certain cushion or flexibility in scheduling them. For example if we see activity B i.e., (2, 3) of Example 6.22, we notice that the EST and EFT are 4 and 9 respectively, and LST and LFT are 6 and 11 respectively. In other words even if the activity is delayed by 2 days, it will not affect the project completion time of 19 days. The term float or slack is used in the context of the extent to which a given activity can be delayed. The floats in individual activities can be calculated using the early and late occurrence times, activity times (earliest and latest), and the duration of the activities as explained in the next section.

6.4.8   Float or Slack Time

An activity need not be started as soon as it can physically be started, without adversely affecting project completion. In other words, activities have some additional time available, which can be used in different ways as illustrated in a following example. As a corollary, it stands to reason that all activities on the critical path have no float. In fact, another way of defining the critical path could be in terms of the floats available—it can be defined as the set of activities connecting the start and the end of a project and having zero float. Reference is made to variations of the 'float' defined here in literature, and some of the important ones are discussed below.

Calculation of different types of floats can easily be done using a diagram as depicted in Figure 6.23. In Figure 6.23, (i, j) represents activity A. E_i and L_i represent early and late occurrence times of event i. Similarly, E_j and L_j represent early and late occurrence times for event j.

Total float in an activity   Total float of an activity is the amount of time by which the start of the activity may be delayed without causing a delay in the completion of the project. This is calculated as

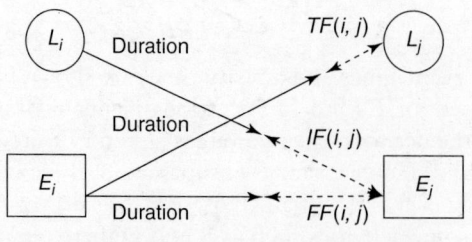

Figure 6.23   Illustration for TF, IF and FF calculation

$$TF(i, j) = LST(i, j) - EST(i, j) \tag{6.12}$$

Or

$$TF(i, j) = LFT(i, j) - EFT(i, j) \tag{6.13}$$

The values of $TF(i, j)$ calculated from Equation (6.12) and Equation (6.13) are referred to as start float and finish float respectively. In terms of event times, the $TF(i, j)$ can be defined as the late occurrence time L_j of the succeeding event minus the early occurrence time E_i of the preceding event minus the duration of the activity defined between these events.

Thus,

$$TF(i, j) = L_j - E_i - D(i, j) \tag{6.14}$$

Free float Free float is the amount of time by which the start of an activity may be delayed without delaying the start of a following activity. Free float is defined as the earliest occurrence time E_j of the following event minus the earliest occurrence time E_i of the preceding event minus the duration of the activity defined between these events. Free float for an activity (i, j) is denoted by $FF(i, j)$ and is calculated from the following expression:

$$FF(i, j) = E_j - E_i - D(i, j) \tag{6.15}$$

Independent float Independent float is the amount of time by which the start of an activity may be delayed without affecting the preceding or the following activity. Independent float is defined as the earliest occurrence time E_j of the following event minus the latest occurrence time L_i of the preceding event minus the duration of the activity defined between these events. Independent float for an activity (i, j) is denoted by $IF(i, j)$ and is calculated from the following expression:

$$IF(i, j) = E_j - L_i - D(i, j) \tag{6.16}$$

Interference float It is defined as the difference in total float and free float. In other words,

$$\textit{Interference Float} = TF(i, j) - FF(i, j) \tag{6.17}$$

It may be noticed that the term earliest occurrence time and early occurrence time are used interchangeably in our discussion. Similarly the term latest occurrence time and late occurrence time are used interchangeably in the text in the context of event times. These are also referred to as Early Event Time (EET) and Late Event Time (LET) in some texts. The terms 'Node' and 'Events' are also used interchangeably in the text.

6.4.9 Path and Critical Path

Any series of activities connecting the starting node to the finishing node can be said to define a 'path' and, indeed, in a project having several activities, several such 'paths' can be identified. Among these paths, the 'critical path' is defined as one that gives the longest time of completion (of the project), which also defines the shortest possible project completion time. For the example problem (see Figure 6.22), there are two paths: 1-2-3-5-6 and 1-2-4-5-6. The path

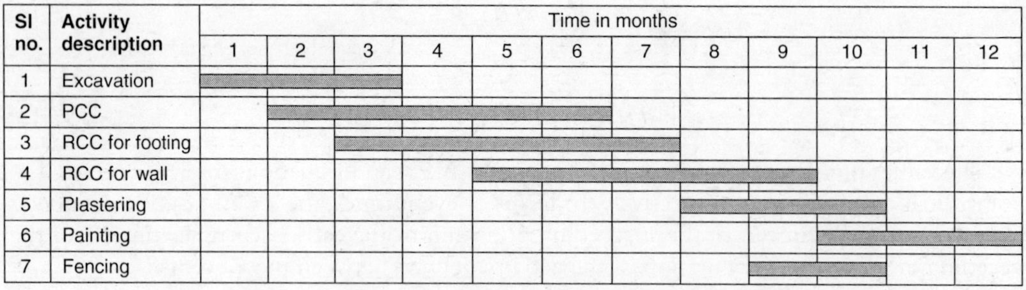

Sl no.	Activity description	Time in months											
		1	2	3	4	5	6	7	8	9	10	11	12
1	Excavation												
2	PCC												
3	RCC for footing												
4	RCC for wall												
5	Plastering												
6	Painting												
7	Fencing												

Figure 6.24 Bar chart for the construction of boundary wall

marked 1-2-4-5-6 is the critical path and the project completion time is 19 days. The critical path is marked with bold line in the text.

6.5 BAR CHARTS

The bar chart is a graphical representation of a project, and given the fact that activities are shown on a real-time scale, they are easy to understand and very useful in reviewing progress. It is one of the oldest methods and an effective technique for overall project planning. These charts were developed by Henry L. Gantt during World War I and, accordingly, these are also sometimes referred to as Gantt charts. They give an idea of duration of activities/project and, hence, can be useful in preparing strategy for working.

In a bar chart, the activities are shown as horizontal bars on a horizontal time scale, where the start and end locations of the bars coincide with the start and finish dates of the activities. A bar chart representation of the illustrative example of the construction of a boundary wall is given in Figure 6.24. It can be seen from the bar chart that excavation starts at the beginning of month 1 and is supposed to get completed at the end of month 3. Similarly, PCC is planned to start at the beginning of month 2 and is supposed to get completed at the end of month 6. As is clear, while bar charts are an excellent tool for visual representation of the plan of the project, there is no way of knowing logical interdependency of activities using these charts. Also, the criticality of an activity or any cascading effect of delay in an activity cannot be easily assessed from bar charts.

Some of the shortcomings of Bar/Gantt charts are partially overcome if linked Gantt charts are used. Further, the mechanics of generating and updating Gantt charts have also considerably improved in the last couple of decades as a result of developments in scheduling software. Some of these software packages may even let the scheduler update or adjust the schedule by making changes directly on the Gantt chart, while the system interacts with the network module built into the system and makes sure that precedence relationships between activities are not overlooked in this process.

6.6 PREPARATION OF NETWORK DIAGRAM

As outlined in the previous sections, creating a network diagram involves preparing a work-breakdown structure for the project, determining the interdependency among the activities, estimating the duration for each activity and, finally, drawing the network. Once an initial draft of the network is prepared, the analysis is carried out using a forward pass, followed by a backward pass to determine the critical path in the project and the activities lying on it. The draft network is then revised to finalize the details. Some additional details involved in these steps are discussed in the following paragraphs.

Preparing the Work Breakdown Structure and Activity Identification

As mentioned above, this involves defining the constituting activities of the project. It may be recalled that work-breakdown structure divides the work based on similarity of nature of works. For example, works requiring similar labour, similar plant and equipment, etc., may be classified in the same group. Each activity under different divisions of work-breakdown structure should be in a manageable unit of work that can be clearly identified, and its relationship with other activities suitably defined. Resources should be considered while defining the activities, and their requirements estimated.

Interdependence of Activity

Identification of (inter) dependence relationship among the activities is the next step in developing a network. Now, such dependence could arise on account of a variety of reasons—the need for earthwork to precede brickwork in foundation is an example of a requirement for physical and logical sequencing of activities. However, in a situation where a crane or a dozer at site can be used for activity A and activity B, whether the equipment is used for A or B (which leads to a 'dependence', in the sense that one can be taken up only when the equipment is free from the other one) is an example where the need to share common resources is at the root of dependence. For each activity the planner must know which activity precedes or succeeds a particular activity, and which activity can be taken up concurrently with this particular activity. The answer to these questions will furnish the dependency relationships between the activity in question and its immediately following activities. Such considerations define certain logic for the network and the plan of construction of the project. The interdependence is often represented in the form of a table for ease of referencing and understanding. The ability to define appropriate dependence requires substantial experience, intuition and judgement on part of the planner. Several difficulties and revisions in the network can be avoided through foresight in defining dependences.

Estimating Duration for an Activity

The time required to complete an activity should depend not only on the quantum of work to be executed (Q), but also on the resources allocated (R) and the (unit) productivity of the resources (P). The experience of the planner comes to the fore not only in cases when such data is not available, but also when the data is to be interpreted to account for geographical differences, age of equipment, operating conditions, etc. Simply put, the time required (T) to complete an activity can be calculated using the following relationship, provided care is taken to ensure proper units for all the quantities.

$$T \times \frac{Q}{R \times P} \qquad (6.18)$$

For example, the time taken to paint 100 m² (Q) using 2 painters (R), assuming each painter can do 5 m² in an hour (P), is simply 10 hours (T). Life in real projects is, however, not as simple, and there could be uncertainties involved in the estimation of quantities, resources and productivity of a resource. For example, in a project, in the absence of accurate ground data, it would be very difficult to estimate the quantities of activities associated with underground works. Similarly, in dewatering works the time taken to dewater up to a particular depth is difficult to estimate.

The productivity of a resource is dependent on a number of parameters discussed elsewhere in the text (refer to Chapter 13, Section 10), and it is really a very challenging task to estimate the productivity of a particular resource. Needless to say, in the absence of accurate estimate of productivity figures, there could be uncertainties involved in the estimate of resource requirement as well. Statistical data stored over a period of time for a number of similar construction projects plays an important role in the estimation of the three parameters Q, R and P, and consequently helps in estimating the duration of an activity. The duration for an activity can be estimated using several approaches, and some of the commonly used ones are listed below:

'Time study' approach   In this approach, the time,

$$T = \frac{Q}{p \times n} \tag{6.19}$$

where Q = total quantity of work, p = productivity factor, and n = normal size of crew. It can be noticed that Q, p and n are all dependent on the availability of the information or data, and at the time of estimate, information on all the three variables would be difficult to get.

Previous project data   Data from completed domestic projects or overseas projects with some necessary modifications are utilized in this approach without giving attention to the values of Q, p and n.

Guesstimating approach   Under this, experienced project professionals are asked to guess the duration. For example, if we need to estimate the duration for construction of brickwork at ground level, the experts with considerable experience in brickwork will be told to estimate the duration for the given quantity. Sometimes, information on the three time estimates—optimistic, pessimistic and most likely time—is sought, from which the planner can make an estimate of the expected time.

Range estimates   These estimates may, again, be guesstimates or even past data. No two 'time' data from past projects for any work will be the same; they can be better expressed by a range. Vendors quite often quote deliveries like 6 to 8 weeks or 7 to 9 days, indicating a range, though in reality the two figures may be a part of the number game vendors play—the short one to give hope to the buyer and the longer one to defend themselves against slippage. A single estimate for any piece of work is arrived at, either through computation or negotiation; it is otherwise alien to the whole estimating process (Choudhary 1988).

Estimates from vendors and subcontractors   Vendors and subcontractors are asked to indicate the time estimates as they are often asked to quote budgetary cost estimates. These estimates in a competitive situation are supposed to provide a realistic estimate. But this approach provides no better estimates than what has been discussed in previous paragraphs.

A vendor may quote for 18 months for a product whose manufacturing life-cycle time may be only six months. During negotiation, this duration may crash to six months or to a slightly higher duration without any extra price. The vendor quotations taken together may at best provide a range estimate and are used for the development of a preliminary schedule.

It may be noted that the various approaches discussed above, except the guesstimating approach, talk of how much time it will take to complete a particular activity. Actual activity time, true to Parkinson's Law, always expands to consume the time allowed. So, past data cannot provide a true estimate of the required time. The guesstimating approach, on the other hand, results in conservative estimates—too tight or too relaxed. If the estimator has any responsibility for implementation, he would like to provide all sorts of contingencies to cover up delays that may occur because of change in assumptions regarding scope of work, inefficiency, uncertainties, etc. The reverse happens if the estimator has no responsibility for implementation. The estimator, in that case, tends to take the job too lightly and, so, provides a very tight and unrealistic estimate.

In real-life situations, there are very few activities such as curing of concrete, etc., in a project that would require a fixed duration like an incubation period. The durations could be changed, within a limit, to meet the requirement of the project. So, what is necessary in fixing up the activity duration is to allocate the time the project would spare for that activity and ensure that this allocation is within the limit of compressibility of the activity. The allocation must, of course, be reasonable, but one need not be too concerned with accuracies.

In practice, therefore, instead of trying to accurately estimate the duration, a reasonable duration is allocated and commitment obtained from the agency that will be held responsible for its implementation. If the duration is not acceptable to any agency, then it may be changed. It has been found to be a pragmatic approach provided there is not much horse-trading between the planning and implementing agencies.

Drawing the Network

An arrow diagram drawn as a network follows the decision about dependency relationships. All activities having no preceding activities are started from a common (start) node and can be called 'starting' activities. Similarly, all (terminal) activities, which have no following (or succeeding) activities, are terminated at a common (end) node. By definition, therefore, for all other activities there must be at least one preceding activity and one succeeding (or following) activity. Mathematical ease and visual clarity require that the network use a minimum number of nodes and is an elegant representation of the activities and their dependence. Now, such networks can be drawn manually only with sufficient experience and ingenuity.

Analysis of Construction Networks

Having drawn an activities network from the information on the activities involved and their interdependence, calculations need to be carried out to identify duration of the project, activities on the critical path, floats of activities, etc. As mentioned above, the critical path is defined as the longest chain of activities from the start node to the finish node of the network. For small networks, this can be identified simply by inspection of all alternative paths and identifying the longest paths. However, for larger networks this becomes cumbersome, and the method of making a 'forward pass' followed by a 'backward pass' on the network is used, as briefly discussed in the previous sections. It may be pointed out that since the real focus in this text is not construction networks and scheduling, the topic has been only briefly dealt with here, and reference should be made to specialist literature for more details.

Fit in and Review

It should be reiterated that the durations discussed in the above paragraphs are closely related to the resources committed for the activity—and, therefore, the preliminary estimates of finishing a project may not always meet the 'desired' time of completion for a project. In such an event, the exercise needs to be reviewed keeping in mind the desired time, and once the discrepancy is removed and the activities on the critical path are identified, the network is frozen for the purpose of resource allocation and monitoring, as briefly discussed below.

Recast Network

Based on the finalized network, resources are assigned to all activities, keeping in mind especially those on the critical path. In the event that the resources required are in excess of a realistic estimate of availability, alternative methods and resources are sometimes identified, which may at times necessitate reworking of the network.

Review and Monitoring

This phase reflects the monitoring phase, when the project is already underway—and the objective is to ascertain whether the progress is as per the predetermined schedule and if any readjustment is required. At the time of review, it may at times happen that the duration of an activity may need to be revised—in such cases, the network may need to be re-analysed and this may also include carrying out the forward and backward passes to ascertain any change in the critical path. During the review and monitoring, the task of a planning engineer is made easy by presenting the information at a given date in the form of activities started but yet to be completed, activities that should have been in progress but have not started, and future activities to be taken in the next week, fortnight, month, quarter, etc., the revised list of critical activities, and the list of activities with ascending order of floats, etc. (to be used in resource levelling). It may be pointed out that as far as a contractor is concerned, these are some of the efforts that are made before it is finally concluded that an extension in the project duration should be sought.

6.7 PROGRAMME EVALUATION AND REVIEW TECHNIQUE (PERT)

PERT is more commonly used in the manufacturing industry, especially in the research and development types of programmes. It is assumed that activities and their interdependence are well-defined, though it recognizes uncertainty in the time estimate of an activity. PERT incorporates uncertainties in activity durations in its analysis, requiring three durations for each activity, which are the most probable, the optimistic (shortest) and the pessimistic (longest) duration. In practice, a competent person, such as an engineer, is asked to give two other time estimates—the pessimistic time estimate (t_p) and the optimistic time estimate (t_o), other than what is perceived to be the most likely estimate (t_m) for (completing) an activity. As the names imply, the pessimistic time is the best guess of the maximum time that may be required to complete the activity, and the optimistic time refers to the best guess of the minimum time required to complete the activity. Indeed, extremely unlikely events such as a fire or an earthquake may

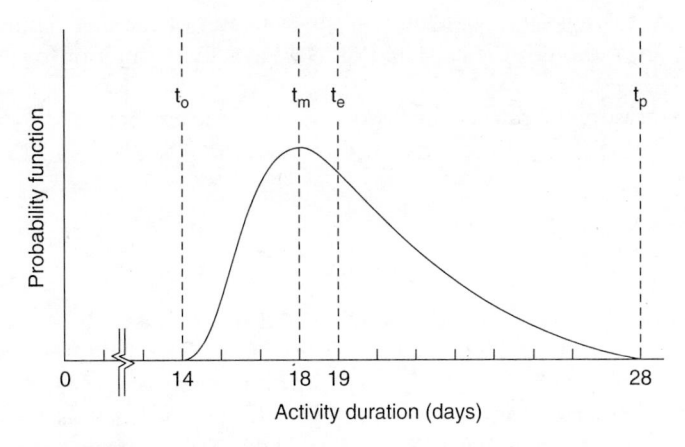

Figure 6.25 Beta distribution for the activity 'design foundation'

not be considered in giving these estimates. However, depending upon the purpose of the exercise and various other issues, it may at times be prudent to include effects of factors such as a flood or other natural calamity.

For example, let there be an activity 'design foundation', and the optimistic, the most likely and the pessimistic time estimates for completing this activity be 14 days, 18 days and 28 days, respectively. The PERT technique assumes that the three time estimates of an activity are random variables, and the frequency distribution of duration of an activity takes the shape of beta distribution (shown in Figure 6.25). The probability that the duration would be less than the optimistic duration (t_o) is about 1 per cent, and the probability that the duration would be more than the pessimistic duration (t_p) is also about 1 per cent.

The average or expected time, t_e, is calculated as a weighted average of the three estimates, assuming that the optimistic and the pessimistic times are about equally likely to occur, and that the most probable activity time is four times more likely to occur than the other two. Mathematically, the average, or expected, time is given by

$$t_e = \frac{(t_0 + 4t_m + t_p)}{6} \tag{6.20}$$

For the case of 'design foundation', t_e can be worked out to be 19 days [(14 + 4 × 18 + 28)/6]. The fact that $t_e > t_m$ in this case is a reflection of the extreme position of t_p and the asymmetry in the beta distribution (see Figure 6.25), even though computationally the weights given to t_o and t_p are the same.

There has been a lot of criticism on the approach of obtaining three 'valid' time estimates to put into the PERT formulas. It is often difficult to arrive at one activity-time estimate; three subjective definitions of such estimates do not help the matter (how optimistic and pessimistic should one be). Nevertheless, the three time estimates also provide the advantages of ascertaining the variability or uncertainty associated with a particular set of estimates. For example, suppose we have two sets of estimates provided by different estimators for the same 'design foundation' activity. In order of (t_e, t_m, t_p), let the first set of estimates be (14, 18, 28) and the other set of estimates be (17, 18, 25). Looking at these two sets of estimates, we can make a guess that there is large variability in the estimates of the first estimator compared to the second one, even though the expected or average activity duration turns out to be 19 in both the cases.

In order to measure the uncertainty associated with the estimate of duration of an activity, the standard deviation (S_t) and the variance V_t are determined, which in PERT are defined as:

$$S_t = \frac{(t_p - t_o)}{6} \quad (6.21)$$

$$V_t = (S_t)^2 \quad (6.22)$$

The formula for S_t indicates that it is one-sixth of the difference between the two extreme time estimates. Further, the greater the uncertainty in time estimates, the greater the value of $(t_p - t_o)$, and the more spread out will be the distribution curve. A high S_t represents a high degree of uncertainty regarding activity times. In other words, there is a greater chance that the actual time required to complete the activity will differ significantly from the expected time t_e.

For the two sets of estimates used in our example of 'design activity', the S_t and V_t would be 2.33 days and 5.44, respectively, for the first set of estimates, while 1.33 days and 1.77 are the corresponding values of S_t and V_t for the second set of estimates.

The expected length or duration of project T_e is calculated by summing up the expected duration t_e's of activities on the critical path. The critical path is determined following the forward pass and backward pass explained earlier. The variance associated with the critical path is the sum of variances associated with the activities on the critical path. In case there is more than one critical path in a project network, then the path with the largest variance is chosen to determine the V_T and S_T. Mathematically,

$$T_e = \Sigma t_e \quad (6.23)$$

$$V_T = \Sigma V_t \quad (6.24)$$

$$S_T = \sqrt{V_T} \quad (6.25)$$

V_T and S_T represent variability in the expected project duration. The higher the V_T and S_T values, the more likely it is that the time required to complete the project will differ from the expected project length T_e.

As was pointed out earlier, t_o, t_m and t_p are assumed to be random variables following beta distribution in the PERT technique. Now that t_e is the weighted sum of t_o, t_m and t_p, it is also treated as a random variable. Since T_e is the sum of t_e's, it is, indeed, a random variable. However, the distribution of T_e follows normal distribution according to the 'central limit' theorem of statistics.

The behaviour of normal distribution is well known and a number of inferences can be drawn which can be useful to a planner or a project manager. For example, it is possible to compute the probability (index) of whether a project (or a key stage therein) will be completed on or before their scheduled date(s). Of course, all this is possible under the assumption that the activities of the projects are statistically independent of each other.

Let us suppose that it is required to compute the probability of completing the project within a target duration of TD days. Now, given the T_e of the project, it is possible to calculate the deviation of TD from T_e in units of standard deviation. This is calculated from the

Table 6.1 Expected duration, standard deviations and variances for activities

	Duration (days)			Expected duration (days) $t_e = (t_o + 4t_m + t_p)/6$	Standard deviation $S_t = (t_p - t_o)/6$	Variance $V_t = (S_t)^2$
Activity ID	Optimistic duration t_o	Most likely duration t_m	Pessimistic duration t_p			
Col. 1	Col. 2	Col. 3	Col. 4	Col. 5	Col. 6	Col. 7
A(10, 20)	3	12	21	12	3	9
B(20, 30)	2	5	14	6	2	4
C(20, 40)	6	15	30	16	4	16
D(30, 40)	1	2	3	2	1/3	1/9
E(30, 50)	5	14	17	13	2	4
F(40, 50)	2	5	14	6	2	4
G(40, 60)	4	5	12	6	4/3	16/9
H(50, 60)	1	4	7	4	1	1

normal distribution table given in Appendix 5 of the book. To adopt the table, a ratio called the standardized deviation or, more often, the normal deviate, Z, is derived. Z is defined as the ratio of the difference in TD and T_e to S_T.

$$Z = \frac{TD - T_e}{S_T} \qquad (6.26)$$

We take up an example to illustrate the application of PERT to a small project. The example project consists of 8 activities (A to H). The t_o, t_m and t_p for each of the activities are given in columns 2, 3 and 4, respectively, in Table 6.1. The computations for calculating t_e, S_t and V_t are shown in columns 5, 6 and 7, respectively, and are self-explanatory. The activity-on-arrow network is shown in Figure 6.26 for this PERT example, although PERT problems are also represented on activity-on-node diagram. The t_e for each of the activities is shown in Figure 6.26, on the bottom of arrows connecting the two nodes. For example, for activity (10, 20), the expected or average duration t_e is 12.

Once the t_e's of all the activities are calculated, forward and backward passes are performed in a manner explained under the section on network terminologies. Using the forward pass, the early occurrence time (E_i) for all the events is determined, and using

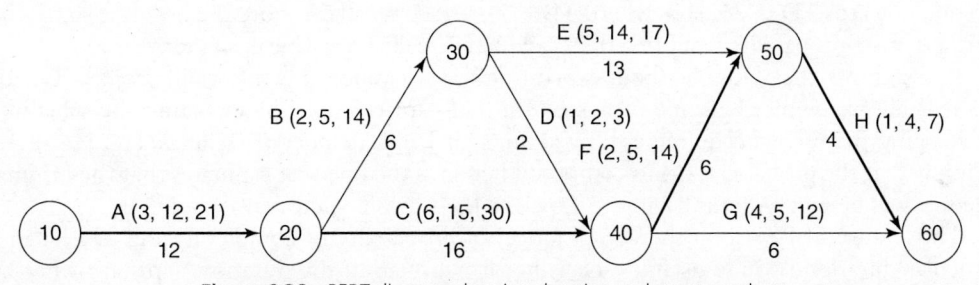

Figure 6.26 PERT diagram showing duration and event numbers

Table 6.2 Computation of early occurrence and late occurrence times for the example problem

Node	Early occurrence time	Late occurrence time	Slack
10	0	12 − 12 = 0	0
20	0 + 12 = 12	Min. of [(21 − 6) = 15 and (28 − 16) = 12] = 12	0
30	12 + 6 = 18	Min. of [(34 − 13) = 21 and (28 − 2) = 26] = 21	3
40	Max. of [(12 + 16) = 28 and (18 + 2) = 20] = 28	Min. of [(38 − 6) = 32 and (34 − 6) = 28] = 28	0
50	Max. of [(18 + 13) = 31 and (28 + 6) = 34] = 34	38 − 4 = 34	0
60	Max. of [(34 + 4) = 38 and (28 + 6) = 34] = 38	38	0

the backward pass, the late occurrence time (L_i) for all the events is determined. These are shown in Table 6.2, and the computations are self-explanatory. The slacks are determined and the critical path is thus determined. For the example problem, the critical path is 10-20-40-50-60 (verify). The expected duration of project $T_e = \Sigma t_e = t_e(10, 20) + t_e(20, 40) + t_e(40, 50) + t_e(50, 60) = 12 + 16 + 6 + 4 = 38$ days. The variance $V_T = \Sigma V_t = V_t(10, 20) + V_t(20, 40) + V_t(40, 50) + V_t(50, 60) = 9 + 16 + 4 + 1 = 30$, and thus, $S_T = \sqrt{V_T} = \sqrt{30} = 5.48$ days.

Now, the problem of computing the probability of meeting target duration (*TD*), such as 42 days is quite simple. For this case, $Z = (TD − T_e)/S_T = (42 − 38)/5.48 = 0.73$ *standard deviations*. In other words, the target duration (*TD*) is 0.73 standard deviations greater than the expected time $T_e = 38$ *days*. The equivalent probability $P(Z = 0.73)$ can be read off a normal probability distribution (Appendix 5). This corresponds to a probability of 0.7673 (76.73 per cent), which implies that there is an 76.73 per cent chance that the project will get completed within 42 days.

Assuming that time now is zero, one may expect this project to end at time 38 days (corresponding probability of achieving this target being 50 per cent, verify. Hint: TD = 38, T_e = 38); and the probability that it will end on or before the target duration of 42 days, without expediting the project, is approximately 76.73 per cent. On the other hand, if one were to schedule towards TD = 35 days, where TD < T_e, i.e., Z = −0.55 (note the negative sign), the corresponding probability would be 0.291, which is really a very bleak situation.

In the above, the phrase 'without expediting' is very important. In certain projects, schedules may always be met by some means or another—for example, by changing the schedule, by changing the project requirement, by adding further personnel or facilities, etc. However, we imply that the probability being computed herein is the one that estimates that the original schedule will be met without having to expedite the work in some way or another.

The feature in PERT on the computation of probability of completing the project in a particular duration is quite useful, especially for negotiating the duration with an owner by the executing agency. For example, while agreeing on a particular duration, the executing agency would like to judge his chances of completing the project in that duration. For being reasonably sure of a particular duration, he would like to attain a probability of more than 95 per cent. Thus, for the same example, suppose that the executing agency is asked to provide the projected duration for the project. The agency would find out the duration corresponding to Z(P = 0.95) = 1.65. The target duration for this case could be TD = T_e + 1.65 × S_T = 38 + 1.65 × 5.48 = approximately 47 days. In other words, the executing agency would be quite confident of completing the project in 47 days.

6.8 CRITICAL PATH METHOD (CPM)

NASA used the critical path method (CPM) to help determine an efficient schedule for the tasks that led to the moon-landing. CPM was the result of a joint effort to develop a procedure for scheduling maintenance shutdowns in chemical-processing plants. CPM undermines the uncertainty involved in any type of project and, hence, it ignores usage of probabilistic activity times; as a result, CPM is a 'deterministic' rather than a 'probabilistic' model. Nevertheless, CPM takes into account uncertainty or variations involved in a job at the planning stage itself.

In CPM also, the workflow can be shown schematically by means of an arrow, where the logical relationships between the various activities can clearly be seen. For CPM, activity durations are considered more deterministic. Accordingly, instead of the three time estimates for an activity, in CPM a single time estimate for an activity is adopted which is assumed to be proportional to the resources allocated for each activity. In CPM also, the similar process of forward and backward paths calculations to find out the start and finish times, the floats, the critical activities, and the length of the critical path are adopted.

CPM is mostly used for construction projects to address the following main issues:

- How long will the entire project take to complete? CPM formally identifies tasks that must be completed on time for the whole project to be completed on time
- Which activities determine total project time? It identifies which tasks can be delayed for a while if resource needs to be reallocated to catch up on missed tasks. It helps to identify the minimum length of time needed to complete a project.
- Which activity times should be shortened, if possible, or in other words, how many resources should be allocated to each activity?

Critical path method is a procedure for using network analysis to identify those tasks that are on the critical path, i.e., where any delay in the completion of these tasks will lengthen the project timescale, unless action is taken. For all tasks off the critical path, a degree of tolerance is possible (e.g., late start, late completion, early start, etc.). Earlier, CPM analysis was carried out by hand computations. Software is now available to perform CPM calculations.

The following steps are needed in CPM:

- List the activities and relationships
- Create a start node
- Draw arrows from start node to the first activity's node
- Sequentially arrange all activities from 'Start'
- Repeat process from successors for all activities
- Check if there is any relationship that is missing

Table 6.3 Example problem for CPM illustration

Activity	Duration (days)	Dependency
A	7	–
B	3	–
C	6	A
D	3	B
E	3	D, F
F	2	B
G	3	C
H	2	E, G

An illustration is provided here in which a small project has eight activities, A to H. The estimated duration and dependence for each of these activities are given in Table 6.3.

The network corresponding to the data given in Table 6.3 is shown in Figure 6.27.

The computations of early start time (EST), early finish time (EFT), late start time (LST), late finish time (LFT) and total float (TF) are given in Table 6.4 and are self-explanatory.

CPM does have some limitations when applied to detailed engineering design work during the early stages of a project because it requires an extensive description of the interrelationships of activities (Oberlender 1993).

6.9 LADDER NETWORK

A ladder network is more or less an extension of the arrow network, and is useful in cases of repetitive works, such as piping jobs, railways electrification and construction of storm-water drains, or roads. In such projects, the number of activities is small, but each activity is repeated several times. A number of dummy activities are used to complete the network, and the com-

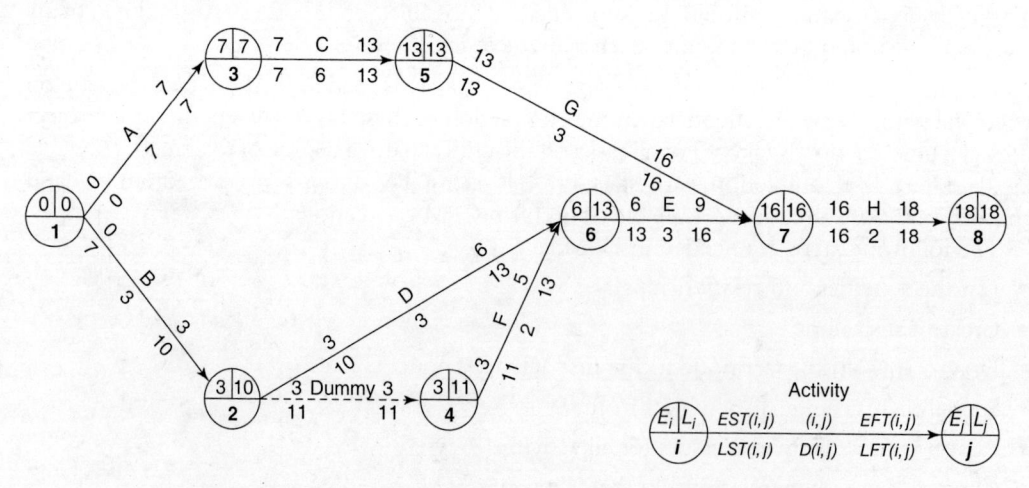

Figure 6.27 Network for example problem based on data of Table 6.3

Table 6.4 Computations for the CPM example problem

Activity	$D(i,j)$ (days)	$EST(i,j) = E_i$	$EFT(i,j) = EST(i,j) + D(i,j)$	$LST(i,j) = LFT(i,j) - D(i,j)$	$LFT(i,j) = L_j$	$TF(i,j) = LST(i,j) - EST(i,j) = LFT(i,j) - EFT(i,j)$
A(1, 3)	7	0	7	0	7	0
B(1, 2)	3	0	3	7	10	7
C(3, 5)	6	7	13	7	13	0
D(2, 6)	3	3	6	10	13	7
E(6, 7)	3	6	9	13	16	7
F(4, 6)	2	3	5	11	13	8
G(5, 7)	3	13	16	13	16	0
H(7, 8)	2	16	18	16	18	0

pleted network looks almost like a ladder, and hence the name of this network. For illustrating a ladder network, let us take an example of a repetitive work of construction of retaining wall having these activities—earthwork (EW), cement concrete (CC), concrete raft (CR), concrete wall (CW) and fencing (FE). Let the length of the retaining wall be 500 m and it is desired to divide it in a stretch of 100 m, giving a total of five stretches.

Let the activities be named as EW_1, EW_2, EW_3, EW_4 and EW_5 for earthwork activity in stretch 1, 2, 3, 4 and 5, respectively. Similarly, let the activities CC_1 to CC_5, CR_1 to CR_5, CW_1 to CW_5, and FE_1 to FE_5 represent cement concrete in the base of wall, concrete raft, concrete wall and fencing activities in stretch 1 to 5, respectively. Further, assume that each stretch of wall takes 2 days for earthwork, 1 day for cement concrete, 7 days for concrete raft, 10 days for concrete wall and 2 days for fencing.

The network for the data presented above is shown in Figure 6.28 and the corresponding ladder network is given in Figure 6.29. The ladder network is drawn in terms of lead and lag of a given activity with its predecessor activity. For example, the lead for activity 'cement concrete' with its predecessor 'earthwork' is 2 days and lag for activity 'cement concrete' with its predecessor 'earthwork' is 1 day.

As can be observed, the ladder network is simple to draw and understand, especially for projects of repetitive types. Further, it is possible to separate out the activities belonging to one type of work. One major limitation with the ladder network, however, is the difficulty in identifying activities on the critical path.

6.10 PRECEDENCE NETWORK

The most common type of precedence relationship used in PERT, CPM and ladder network is the *finish-to-start* (FS) relationship. In real life, we come across many situations where the two activities are related in relationships other than the *finish-to-start* relationship. These are *start-to-start* (SS), *finish-to-finish* (FF) and *start-to-finish* (SF) relationships. Precedence networking incorporates all the four types of relationships—FS, SS, FF and SF.

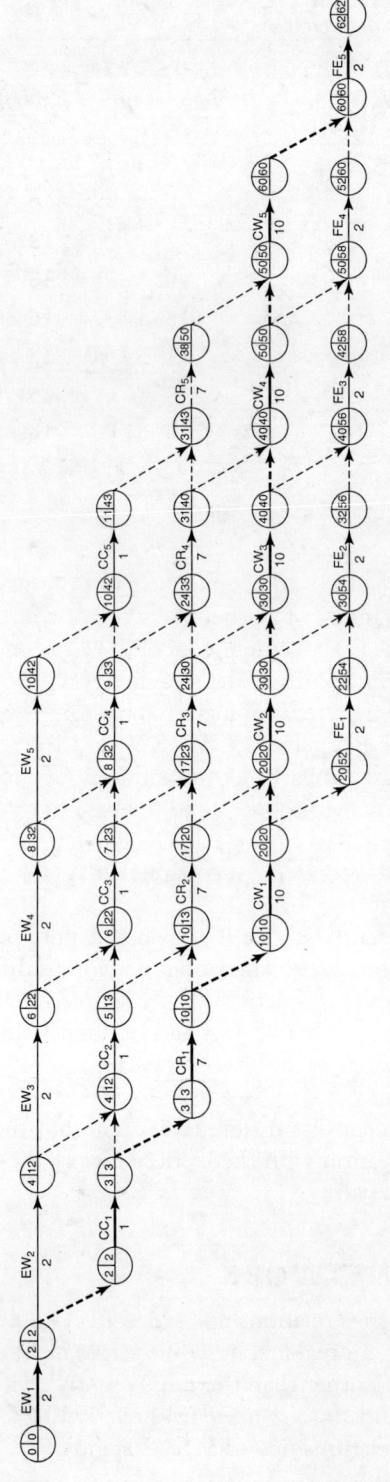

Figure 6.28 Arrow network for repetitive works

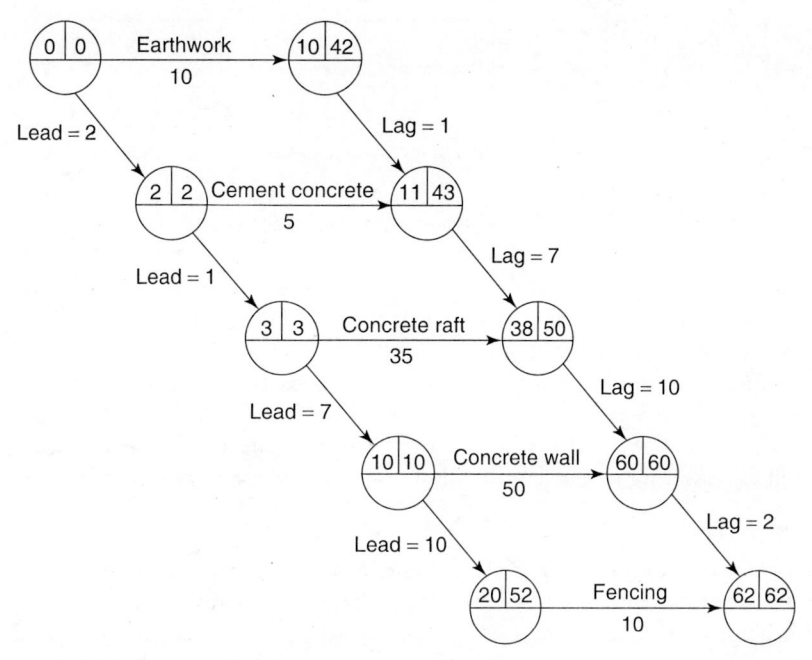

Figure 6.29 Ladder network corresponding to the network of Figure 6.28

Finish-to-start (FS)

Task B cannot start unless Task A is completed. For example, suppose in a project we have two tasks— 'laying bricks' and 'plaster'. In this case, 'plaster' cannot start until 'laying bricks' finishes (see Figure 6.30 a). This is the most common type of dependency. If it is desired to commence the activity 'plaster' seven days after the completion of 'laying bricks', then its representation in the precedence network is as given in Figure 6.30 b. Similarly, if the 'plaster' is to commence five days before the completion of the activity 'laying bricks', its representation in the precedence network is as given in Figure 6.30 c. The number displayed in Figure 6.30 a to c adjacent to equality sign on the line joining the two boxes represents

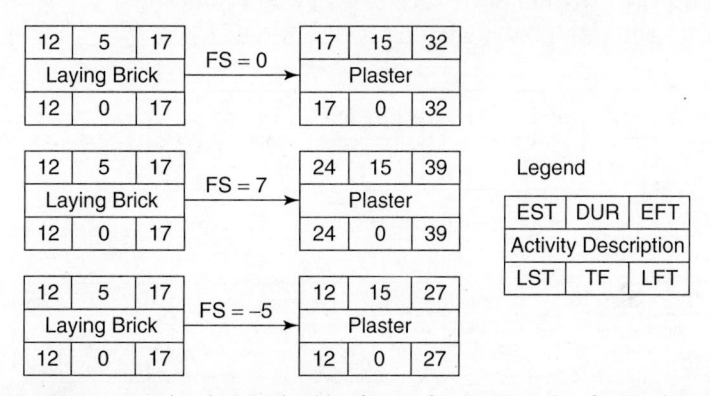

Figure 6.30 A, B and C showing lead-lag factor of 0, 7 and –5 days for FS relationship

12	5	17		17	7	24		24	15	39
Laying brick			→	Pseudo activity			→	Plaster		
12	0	17		17	0	24		24	0	39

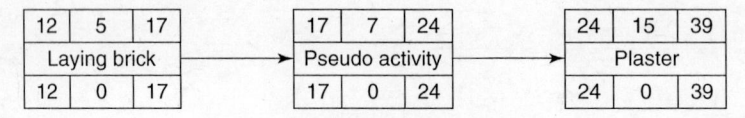

Figure 6.31 Introduction of pseudo method to represent lead-lag factor of 7 days in FS relationship

lead-lag factor. In Figure 6.30 a, there is no lead, while there is a lead-lag factor of 7 and −5 days for Figure 6.30 b and Figure 6.30 c, respectively.

In precedence network, there are two ways in which *finish-to-start* relationship with lead-lag factor can be portrayed. In the first method, the lead-lag factors are shown on the arrow joining the two boxes, as shown in Figure 6.30 a to c.

In the second method, we can introduce a 'pseudo activity' in between the two activities— 'laying brick' and 'plaster'—as shown in Figure 6.31. This method is similar to the approach adopted in PERT and CPM, with similar disadvantages of having to deal with a number of 'pseudo activities' resulting into a lengthy network and increase in computational efforts.

Finish-to-finish (FF)

Task B cannot finish until task A finishes, if A and B are associated in a FF manner. Here also, the lead-lag factor can be introduced. For example, if it is assumed that 'inspect electrical work' cannot finish before 7 days of completion of 'add wiring', then they can be represented as in Figure 6.32.

The given relationship can be represented using a pseudo activity, as in Figure 6.33.

Start-to-start (SS)

Task B cannot start until task A starts if they are associated in a SS relationship. For example, if we have two tasks 'pour foundation' and 'level concrete', then 'level concrete' cannot begin until 'pour foundation' begins.

A start-to-start relationship can also be associated with the lead-lag factor, as shown in Figure 6.34. There are two ways in which a start-to-start relationship with lead and lag factor can be represented in precedence network. One way is to split the preceding activity in two components, A_1 and A_2, and proceed as shown in Figure 6.35. The disadvantage would be the additional list of activities and a lengthy network. This is similar to the approach used in PERT/CPM in order to accommodate such type of relationships.

The same can again be represented using two activities.

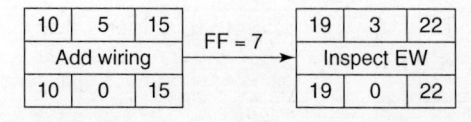

10	5	15		19	3	22
Add wiring			FF = 7 →	Inspect EW		
10	0	15		19	0	22

Figure 6.32 FF relationship with lead-lag factor of 7 days

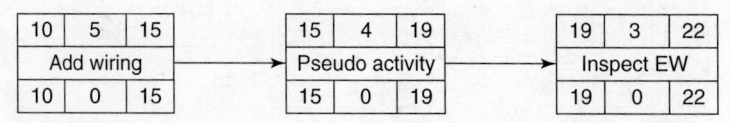

10	5	15		15	4	19		19	3	22
Add wiring			→	Pseudo activity			→	Inspect EW		
10	0	15		15	0	19		19	0	22

Figure 6.33 Introduction of pseudo method to represent a relationship of FF = 7

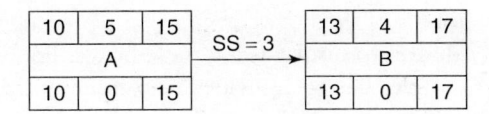

Figure 6.34 SS relationship with lead-lag factor of 3 days

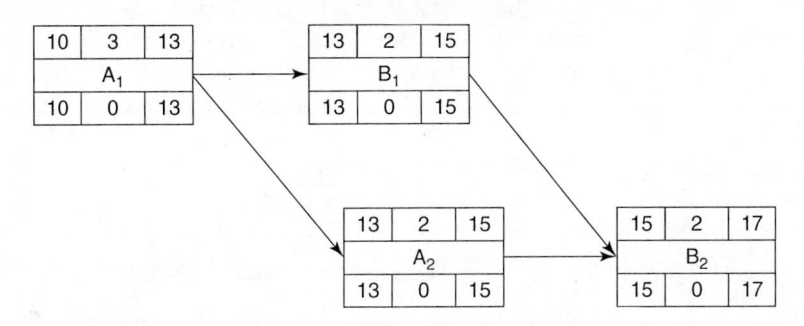

Figure 6.35 SS relationship with lead-lag factor of 3 days

Start-to-finish (SF)

Task *B* cannot finish until task *A* starts. The SF relationship is not that common in precedence network, but it has been included in the discussion for the sake of completeness. A SF relationship can be there with lead-lag factor, as shown in Figure 6.36, and this can also be represented by introducing a pseudo activity as shown in Figure 6.37.

The illustration of different relationships used in precedence network and their corresponding equivalent for conventional networks shows that all the types of relationships can be represented in conventional networks by splitting activities into multiple segments, though that will enlarge the arrow networks to a great extent. The precedence network, on the other hand, is smaller, easier to develop and to calculate, and clearer to interpret.

Precedence network resembles an AON diagram with activities on nodes or boxes and precedence relationship shown as arrow. However, precedence network without arrows is also possible. The numbering of activity follows rules similar to those followed in PERT and CPM. The time estimate for the activity could be one time estimate or three time estimate, though the three time estimate has to be converted to a single time before being used in precedence network.

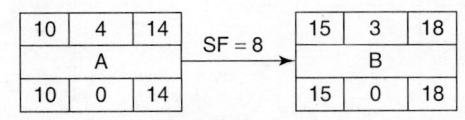

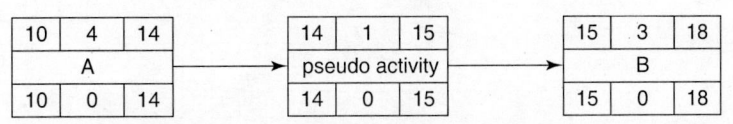

Figure 6.36 SF relationship with lead-lag factor of 8 days

Figure 6.37 SF relationship with lead-lag factor of 8 days

There could be many variants of the boxes or nodes in a precedence network, depending on the information the user desires to display using these boxes. For illustration, a typical box used for all the preceding examples has been divided into three horizontal parts, top, middle and bottom. The top is again divided vertically into three compartments, left, centre and right. Similarly, bottom portion is also divided into three departments, left, centre and right.

The top left compartment gives the earliest start time. The top centre compartment gives the activity duration. The top right compartment gives the earliest finish time. The middle portion gives activity description. The bottom left compartment gives the latest start time. The bottom centre compartment gives the total float. The bottom right compartment shows the latest finish time.

The process of forward pass and backward pass and the identification of critical path and critical activities are explained with respect to Figure 6.38, which is the precedence network for the example of retaining wall construction. Since the relationships that are used in Figure 6.38 are FS only, there is no difference in the computation procedure of forward pass and backward pass.

The forward pass gives the early start and early finish times of all the activities shown in the network, similar to the computations done in case of PERT and CPM. The computation is shown in Table 6.5 for a few activities for illustration. In the table, EST_{10} represents earliest start time of activity having activity code 10. EFT_{10} represents earliest finish time of activity having activity code 10. FS_{10-20} represents the lead lad factor of finish to start type. $FS_{10-20} = 0$ means activity with activity code 20 can start immediately after the completion of activity with activity code 10. The duration of activity with activity code is represented by d_{10}. Other representations in the table can be similarly interpreted.

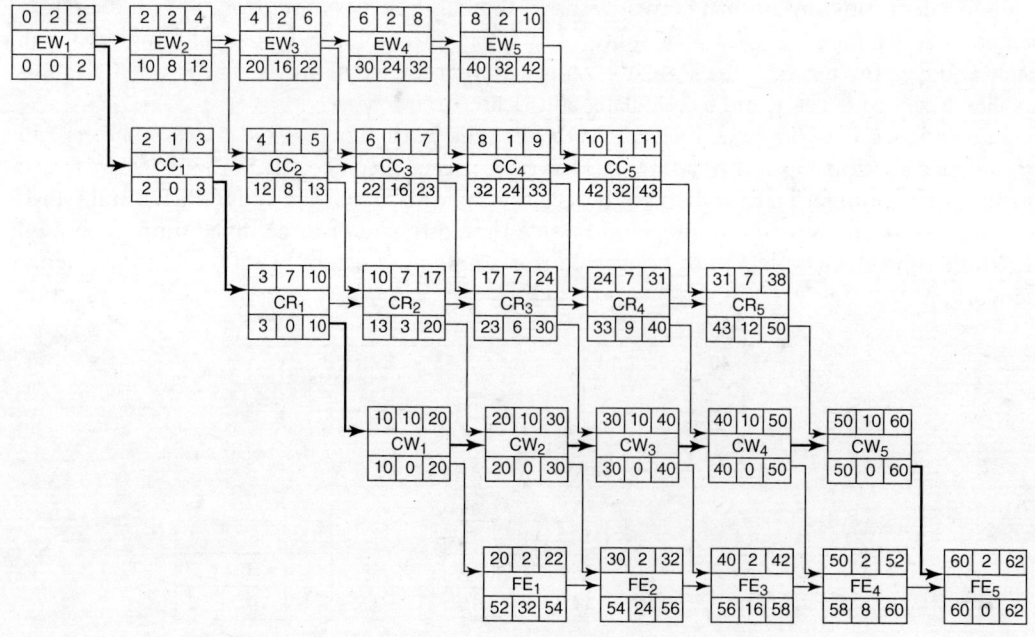

Figure 6.38 Example of precedence network for repetitive work

Table 6.5 Explanation of determination of early start and early finish activity times

Activity Code	Activity	Duration (days)	EST	Remarks	EFT	Remarks
10	EW_1	2	0	0 start activity, $EST_{10} = 0$	2	$EFT_{10} = EST_{10} + d_{10}$
20	EW_2	2	2	$EST_{20} = EFT_{10} + FS_{10-20}$ Hence, $EST_{20} = 2 + 0$	4	$EFT_{20} = EST_{20} + d_{20}$ Hence, $EFT_{20} = 2 + 2$
30	EW_3	2	4	$EST_{30} = EFT_{20} + FS_{20-30}$ Hence, $EST_{30} = 4 + 0$	6	$EFT_{30} = EST_{30} + d_{30}$ Hence, $EFT_{30} = 4 + 2$
40	EW_4	2	6	$EST_{40} = EFT_{30} + FS_{30-40}$ Hence, $EST_{40} = 6 + 0$	8	$EFT_{40} = EST_{40} + d_{40}$ Hence, $EFT_{40} = 6 + 2$
50	EW_5	2	8	$EST_{50} = EFT_{40} + FS_{40-50}$ Hence, $EST_{50} = 8 + 0$	10	$EFT_{50} = EST_{50} + d_{50}$ Hence, $EFT_{50} = 8 + 2$
60	CC_1	1	2	$EST_{60} = EFT_{10} + FS_{10-60}$ Hence, $EST_{60} = 2 + 0$	3	$EFT_{60} = EST_{60} + d_{60}$ Hence, $EFT_{60} = 2 + 1$
70	CC_2	1	4	$EST_{70} = Max(EFT_{20} + FS_{20-70}, EFT_{60} + FS_{60-70})$ Hence, $EST_{70} = 4 + 0$	5	$EFT_{70} = EST_{70} + d_{70}$ Hence, $EFT_{70} = 4 + 1$
.....	.....	.....	.....	.....	.....	.....
250	FE_5	2	60	$EST_{250} = Max(EFT_{200} + FS_{200-250}, EFT_{240} + FS_{240-250})$ Hence, $EST_{250} = 60 + 0$	62	$EFT_{250} = EST_{250} + d_{250}$ Hence, $EFT_{250} = 60 + 2$

The process is reversed for backward pass. Using the same logic but starting at activity FE_5, a late finish is determined. In majority of the cases, the late finish is taken to be the same as the early finish, that is, day 62, although it is possible to insert a higher number as the late finish. In that case, backward pass would not end at time period 0. In this example, the late finish of activity FE_5 has been taken same as the early finish. The late start is calculated by subtracting the activity duration from the late finish. The process is repeated, taking the lower value on the backward pass and adding a lead/lag factor until a late start for activity EW_1 is calculated. The backward computations for some of the activities are shown in Table 6.6 for illustration. In the table, LST_{250} represents latest start time of activity having activity code 250. LFT_{250} represents latest finish time of activity having activity code 250. Other representations in the table can be similarly interpreted.

Critical path(s) is (are) identified next. The critical path consists of activities with zero float logically linked between the start activity and the finish activity. Float is the difference between the late start and the early start. The critical path in this example is EW_1-CC_1-CR_1-CW_1-CW_2-CW_3-CW_4-CW_5-FE_5. It is interesting to note that critical activity has identical early start and late start. The critical path is shown with dark arrows in Figure 6.38. The bar chart resulting from the network of Figure 6.38 is shown in Figure 6.39.

The top bars in Figure 6.39 show the bar chart based on early start time, while the bottom bars represent the bar chart based on late start time. The breaks or splits in bars for some activities represent 'start' and 'stop' of those activities. For example, in the bar chart based on

Table 6.6 Explanation of determination of late finish and late start activity times

Activity Code	Activity	Duration (days)	LFT	Remarks	LST	Remarks
250	FE_5	2	62	$LFT_{250} = EFT_{250}$	60	$LST_{250} = LFT_{250} - d_{250}$
240	FE_4	2	60	$LFT_{240} = LST_{250} - FS_{240-250}$	58	$LST_{240} = LFT_{240} - d_{240}$
230	FE_3	2	58	$LFT_{230} = LST_{240} - FS_{240-230}$	56	$LST_{230} = LFT_{230} - d_{230}$
220	FE_2	2	56	$LFT_{220} = LST_{230} - FS_{230-220}$	54	$LST_{220} = LFT_{220} - d_{220}$
210	FE_1	2	54	$LFT_{210} = LST_{220} - FS_{220-210}$	52	$LST_{210} = LFT_{210} - d_{210}$
200	CW_5	10	60	$LFT_{200} = LST_{250} - FS_{200-250}$	50	$LST_{200} = LFT_{200} - d_{200}$
190	CW_4	10	50	$LFT_{190} = Min(LST_{240} - FS_{190-240}, LST_{200} - FS_{190-200})$	40	$LST_{190} = LFT_{190} - d_{190}$
180	CW_3	10	40	$LFT_{180} = Min(LST_{230} - FS_{180-230}, LST_{190} - FS_{180-190})$	30	$LST_{180} = LFT_{180} - d_{180}$
.....	.....	.....	.....	.....	.....	.....
10	EW_1	2	2	$LFT_{10} = Min(LST_{60} - FS_{10-60}, LST_{20} - FS_{10-20})$	0	$LST_{10} = LFT_{10} - d_{10}$

early start, the activity CC is not done on a continuous basis; instead, it has been done in five segments and, accordingly, the split in bar is shown in Figure 6.39.

In precedence network, forward and backward pass algorithms change if the network has other two types of relationships, namely FF and SF.

In the forward pass, an activity start time is calculated with FS and SS relationships and the finish time is calculated with FF and SF relationships, along with the activity duration. If the early start and the early finish so calculated differ by more than the activities duration, then the activity is split according to the rules outlined in the algorithm. Analogous procedure exists for the backward pass.

We take up the problem of retaining wall construction again and draw the network as given in Figure 6.40. However, now let us assume that the activity splitting is not allowed. In other

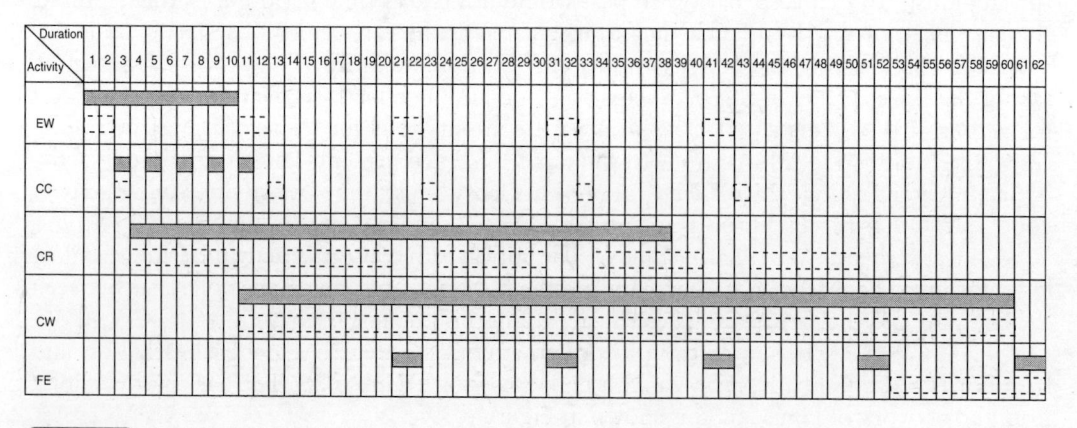

Figure 6.39 Bar chart resulting from the network of Figure 6.38

Table 6.7 Explanation of determination of early start and early finish activity times

Node	Activity	Duration	EST	Remarks	EFT	Remarks
10	EW	10	0	0 start activity, $EST_{10} = 0$	10	$EFT_{10} = EST_{10} + d_{10}$
20	CC	5	6	$EST_{20} = $ Max. of $[EST_{10} + SS_{10-20} = 0 + 2 = 2, EFT_{10} + FF_{10-20} - d_{20} = 10 + 1 - 5 = 6] = $ Max. of $(2, 6) = 6$	11	$EFT_{20} = EST_{20} + d_{20}$ Hence, $EFT_{20} = 6 + 5 = 11$
30	CR	35	7	$EST_{30} = $ Max. of $[EST_{20} + SS_{20-30} = 6 + 1 = 7, EFT_{20} + FF_{20-30} - d_{30} = 11 + 7 - 35 = -17] = $ Max. of $(7, -17) = 7$	42	$EFT_{30} = EST_{30} + d_{30}$ Hence, $EFT_{30} = 7 + 35 = 42$
40	CW	50	14	$EST_{40} = $ Max. of $[EST_{30} + SS_{30-40} = 7 + 7 = 14, EFT_{30} + FF_{30-40} - d_{40} = 42 + 10 - 50 = 2] = $ Max. of $(14, 2) = 14$	8	$EFT_{40} = EST_{40} + d_{40}$ Hence, $EFT_{40} = 14 + 50 = 64$
50	FE	10	56	$EST_{50} = $ Max. of $[EST_{40} + SS_{40-50} = 14 + 10 = 24, EFT_{40} + FF_{40-50} - d_{50} = 64 + 2 - 10 = 56] = $ Max. of $(24, 56) = 56$	66	$EFT_{50} = EST_{50} + d_{50}$ Hence, $EFT_{50} = 56 + 10 = 66$

words, EW, CC, CR, CW and FE activities once started in any of the five sections have to be completed in all respect.

The forward and backward pass computations are given in Table 6.7 and Table 6.8.

The bar chart resulting from the network of Figure 6.40 is shown in Figure 6.41. Incidentally, the bar charts based on early start of activities and late start of activities are similar(why??).

Table 6.8 Explanation of determination of late finish and late start activity times

Node	Activity	Duration	LFT	Remarks	LST	Remarks
50	FE	10	66	$LFT_{50} = EFT_{50}$	56	$LST_{50} = LFT_{50} - d_{50} = 66 - 10 = 56$
40	CW	50	64	$LFT_{40} = $ Min. of $[LFT_{50} - FF_{40-50} = 66 - 2 = 64, LST_{50} - SS_{40-50} + d_{40} = 56 - 10 + 50 = 96] = $ Min$[64, 96] = 64$	14	$LST_{40} = LFT_{40} - d_{40} = 64 - 50 = 14$
30	CR	35	42	$LFT_{30} = $ Min. of $[LFT_{40} - FF_{30-40} = 64 - 10 = 54, LST_{40} - SS_{30-40} + d_{30} = 14 - 7 + 35 = 42] = $ Min$[54, 42] = 42$	7	$LST_{30} = LFT_{30} - d_{30} = 42 - 35 = 7$
20	CC	5	11	$LFT_{20} = $ Min. of $[LFT_{30} - FF_{20-30} = 42 - 7 = 35, LST_{30} - SS_{20-30} + d_{20} = 7 - 1 + 5 = 11] = $ Min$[35, 11] = 11$	6	$LST_{20} = LFT_{20} - d_{20} = 11 - 5 = 6$
10	EW	10	10	$LFT_{10} = $ Min. of $[LST_{20} - FF_{10-20} = 11 - 1 = 10, LST_{20} - SS_{10-20} + d_{10} = 6 - 2 + 10 = 14] = $ Min.$[10, 14] = 10$	0	$LST_{10} = LFT_{10} - d_{10} = 10 - 10 = 0$

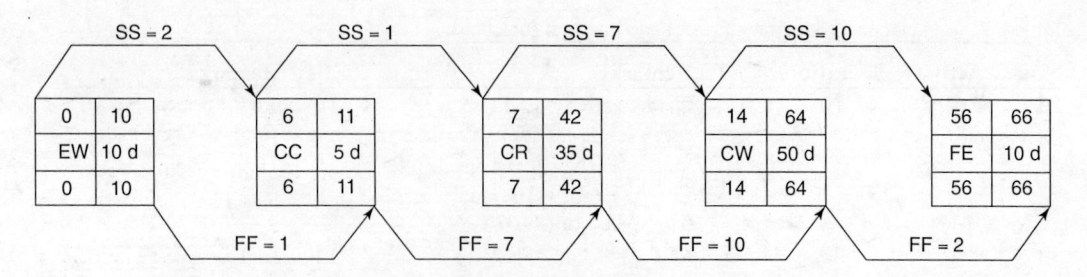

Figure 6.40 Revised network of retaining wall construction

EST	EFT
Activity	Duration
LST	LFT

It can be observed from the two bar charts (see Figure 6.39 and Figure 6.41) that the project duration has increased from 62 days, when activity splitting was allowed, to 66 days when activity splitting was not allowed.

The critical path obtained in a precedence network may not be a neat path (not connected clearly in a back-to-back sequence as in PERT/CPM). This has been illustrated in Solved Example 6.1.

There is an interesting feature of precedence network pertaining to critical activities. In conventional networks such as PERT and CPM discussed earlier, the increase (decrease) in critical activity duration would result in increase (decrease) in project duration. However, in case of precedence network, the increase in critical activity duration may not always result in increase in project duration. There could be three possibilities associated with increase

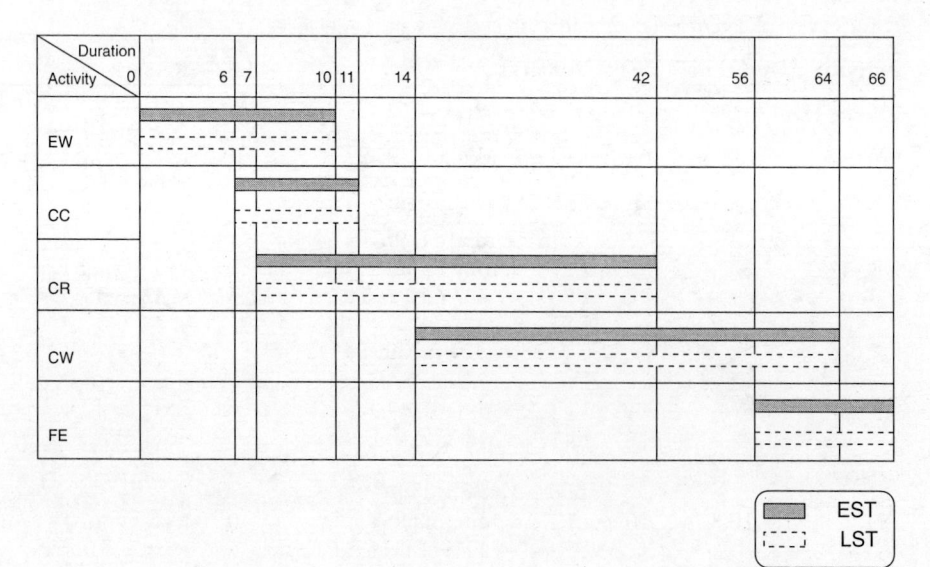

Figure 6.41 Bar chart resulting from the network of Figure 6.40

Table 6.9 Different types of critical activity

S. No.		Situation	Impact on project duration	Type of critical activity
1	a	Increase in critical activity duration	Project duration is lengthened	Such critical activities are known as 'normal critical activities'
	b	Decrease in critical activity duration	Project duration is shortened	
2	a	Increase in critical activity duration	No effect on project duration	Such critical activities are known as 'neutral critical activities'
	b	Decrease in critical activity duration	No effect on project duration	
3	a	Increase in critical activity duration	Project duration is shortened	Such critical activities are known as 'reverse or perverse critical activities'
	b	Decrease in critical activity duration	Project duration is lengthened	

(decrease) in duration of critical activity, with reference to the impact on project duration. These are shown in Table 6.9.

From Table 6.9, it is clear that three types of critical activities could be encountered in precedence network.

1. Normal critical activities: Increase (decrease) in critical activity duration would result in increase (decrease) in project duration.
2. Neutral critical activities: Increase (decrease) in critical activity duration would not result in any change in the project duration.
3. Reverse or perverse critical activities: Increase (decrease) in critical activity duration would result in decrease (increase) in project duration. In case an activity falls on the critical path and it is related with its predecessor in FF manner and with its successor in SS manner, then a decrease (increase) in the duration for such activity will increase (decrease) the project duration. Indeed, this is applicable when the activity interruption is not allowed.

A reduction in the overall project duration is also possible by reduction in the length of critical lead/lag factor.

The precedence network is becoming popular these days in the construction industry. Most of the latest software such as Primavera and MS Project, which are fast becoming a standard around the world, are based on this type of network (of course, Primavera and Microsoft Project can also produce bar charts from the precedence networks automatically, if it is desired). Some other features of precedence network are discussed next.

In precedence network, more than one relationship can be assigned between two activities. This is helpful in dealing with series activities. For example, let us assume that two activities, earthwork (EW) and cement concrete (CC), are to be executed in a project of retaining wall construction. The retaining wall is planned to be taken up in five sections. Earthwork and CC in each section require 2 days and 1 day, respectively. Thus, the total duration of earthwork and CC would be 10 days and 5 days, respectively. If it is planned to take up CC immediately after sec-

Sl no.	Activity	1	2	3	4	5	6	7	8	9	10	11	12
1	EW		▓	▓	▓	▓	▓	▓	▓	▓	▓		
2	CC				▓	▓	▓	▓					

Figure 6.42 Bar chart when activity splitting is not allowed

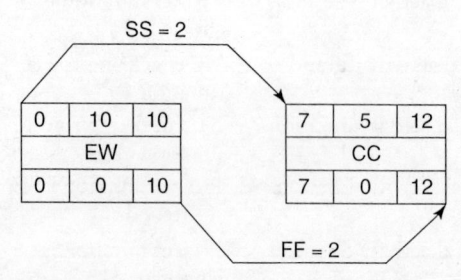

Figure 6.43 Illustration of dual relationship in precedence network

tion 1 is completed (i.e., a relation of SS = 2), the bar chart would be as shown in Figure 6.42, if activity splitting is not allowed. However, the bar chart shown in Figure 6.42 has a fallacy as CC cannot be completed until the entire earthwork is completed.

This situation can be avoided if we have more than one relationship between two activities. For example, we can have a relationship of SS = 2 and FF = 2, as shown in Figure 6.43, and by doing so, the above-mentioned fallacy can be avoided. The corresponding bar chart would be as shown in Figure 6.44 when activity splitting is not allowed. However, when the activity splitting is allowed, one of the possible bar charts could be as shown in Figure 6.45.

There is no requirement of keeping a single start and end node. In fact, there can be multiple start and end nodes in a precedence network, as shown in Figure 6.46. There are two start nodes, A and B, and two end nodes, F and G, in Figure 6.46.

Using precedence network, the early and late start schedules can be prepared for two cases—(1) when activity interruption (splitting) is not allowed, and (2) when activity interruptions are allowed. Further, in precedence network, the total float definition adopted (LS-ES or LF-EF) for networks such as PERT/CPM may or may not give same values. Whenever an activity is interrupted and the period of interruption differs in the early and late start schedules, the two methods of calculating total float will yield different results.

Sl no.	Activity	1	2	3	4	5	6	7	8	9	10	11	12
1	EW												
2	CC												

Figure 6.44 Bar chart when activity splitting is not allowed

Sl no.	Activity	1	2	3	4	5	6	7	8	9	10	11	12
1	EW												
2	CC												

Figure 6.45 Bar chart when activity splitting is allowed

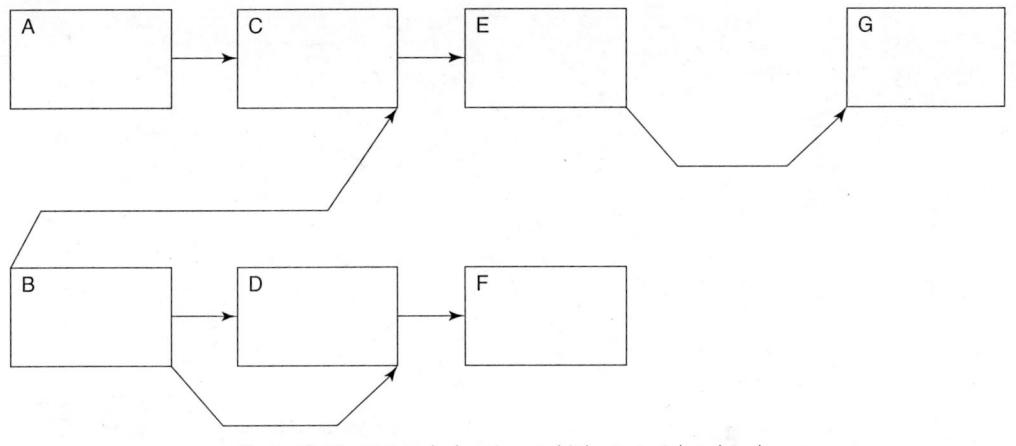

Figure 6.46 Network showing multiple start and end nodes

6.11 THE LINE-OF-BALANCE (LOB)

The line-of-balance technique of programming and control of projects was originated by the Goodyear Company in the early 1940s, and it was developed for the programming and control of both repetitive and non-repetitive projects by the US Navy in the early 1950s (Turban 1968, Johnston 1981, Lutz and Halpin 1992, all cited in Arditi et al. 2001).

The line-of-balance technique was first applied to industrial manufacturing and production control. The basic concepts of line-of-balance have been applied in the construction industry as a planning and scheduling method (Lumsden 1968, Khisty 1970, both cited in Arditi et al. 2001).

The line-of-balance technique assumes that the rate of production for an activity is uniform. In other words, the rate of production of an activity is linear. In line-of-balance, the time is usually plotted on horizontal axis, and units or stages of an activity on the vertical axis. The production rate of an activity is the slope of the production line and is expressed in terms of units per time.

The line-of-balance technique uses *man-hour estimate* and *optimum crew or gang size*. The line-of-balance diagram can be plotted once the man-hour estimates and optimum size of crew are determined by the planner based on his or her experience, or in consultation with the implementing agencies such as vendor or subcontractor.

The two oblique and parallel lines used in line-of-balance technique denote the start and finish times, respectively, of each activity in all the units from the first to the last. The slope of the lines is equal to the actual rate of output of the activity. This has been explained in detail in later sections.

The line-of-balance technique finds favour in the construction industry mostly for the execution of repetitive unit projects such as mass housing, high-rise buildings, tunnels and highways. A typical line-of-balance diagram is shown in Figure 6.47 for a housing project.

Consider an example to illustrate the construction of LOB for scheduling of a housing project, where 10 houses are to be constructed. For simplicity, let us assume that substructure, superstructure and finishing are the only three activities involved in the construction of a house. Further, let us assume that substructure, superstructure and finishing take 10, 20 and

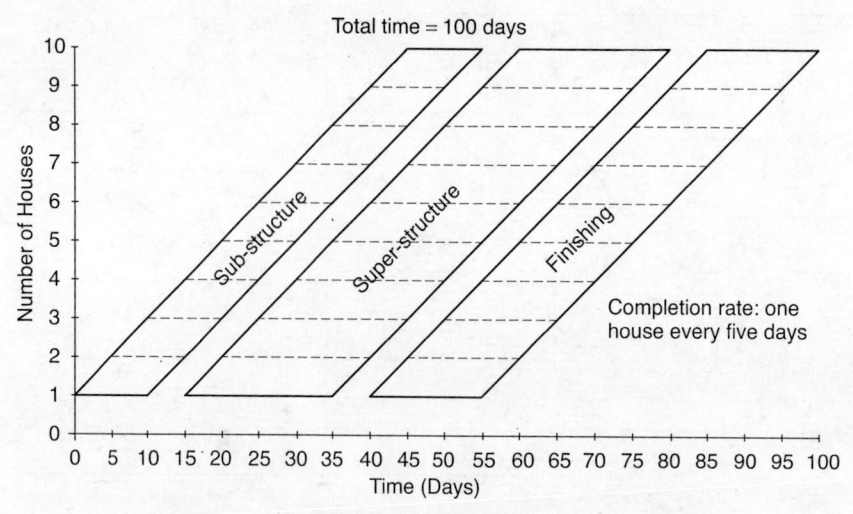

Figure 6.47 LOB diagram for scheduling of a housing project

15 days, respectively, for a single house. To provide for the margin of error in the time taken to complete each of these operations, the time buffer is usually placed in-between two activities. Let us assume the buffer in this case to be of five days. Thus, the construction of a single house needs 55 days to complete (10 + 5 + 20 + 5 + 15 = 55). That is, the first house is completed in all respect on the 55th day from the start. It is further assumed that now onwards, every 5th day one house is to be completed, i.e., second house on 60th day from start, third house on 65th day, and so on, and the last, or 10th, house gets completed on the 100th day. This is called the rate of construction, and in this case, the rate of construction is one house every five days. This rate is represented diagrammatically for all the three activities, and completion dates are drawn for each of the 10 houses. Thus, the substructure, superstructure and finishes of all houses are joined by sloping lines. These form a series of bands, which are the LOB schedules, for each activity. It can be seen that in this case, all the sloping lines are parallel to the handover line. Each activity is proceeding at the rate of one house per week, at the same rate as the handover schedule.

To provide a margin of error and to ensure that one activity does not interfere with another, time buffer of five days has been selected for each activity, though only as an example. In practice, each time buffer would probably be different. The construction manager selects the buffer according to his own experience and his own forecast of the risks, difficulties, or delays that may occur. For example, if in a particular locality carpenters were known to be in short supply, longer buffers would be inserted after activities involving carpenters. Conversely, only very small buffers may be required when little difficulty is expected in achieving planned gang size and output. The crew size is determined in such a way that production rate of all the activities become more or less uniform.

The buffer can be uniform or non-uniform depending on the situation. In practice, each time the buffer would probably be different. For example, if in a particular locality welders were known to be in short supply, longer buffers would be inserted after activities involving welders. Conversely, small buffers would be sufficient when not much difficulty is foreseen in mobilizing the required crew size.

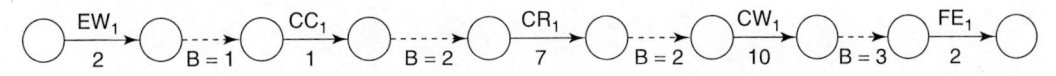

Figure 6.48 Logic diagram of construction of one segment of retaining wall

Arditi et al. (2002) list out some of the challenges associated with LOB scheduling:
- developing an algorithm that handles project acceleration efficiently and accurately,
- recognizing time and space dependencies,
- calculating LOB quantities,
- dealing with resource and milestone constraints,
- incorporating the occasional non-linear and discrete activities,
- defining a radically new concept of criticalness,
- including the effect of the learning curve,
- developing an optimal strategy to reduce project duration by increasing the rate of production of selected activities,
- performing cost optimization, and
- improving the visual presentation of LOB diagrams.

The LOB technique is highly effective in determining areas of weakness, and focusing the attention of the manager on items requiring more immediate attention, especially for repetitive works. However, the LOB has not found lasting popularity in construction industry, mainly due to the widespread availability of commercial software for network-based techniques and the non-availability of appropriate commercial software having capability to make LOB diagrams. An attempt to develop a computer application was made by Psarros (1987) but it was not free from flaws. Subsequently, Suh in 1993 developed a program called RUSS.

We consider the same example of retaining wall construction to illustrate how LOB can be used for scheduling of a project. In this example, 5 sides (each of 100 m length, called one segment) are to be constructed. Thus, there are five segments to be constructed. A step-by-step method for preparing a LOB is given in the following section.

Step 1: Construction of Network or Logic Diagram for One Segment

As a first step, usually a logic or network diagram for one of the many units to be produced is prepared and incorporated into the LOB schedule. The logic diagram for one segment of retaining wall is as given in Figure 6.48). The same notations have been used for this illustration also. For example, EW_1 represents the earthwork in segment one; CC_1 represents cement concrete in segment one; and so on. The durations are also assumed to be the same as for the previous illustrations.

The estimation of duration is dependent on the quantity of each of the activities and the associated crew. The quantities involved and the number of workers assumed are shown for all the activities associated with the construction of retaining wall, in Table 6.10.

It is assumed that there will be work for 12 hours each day. It is also assumed that the formwork, reinforcement and concreting would be done by the same set of workers.

Table 6.10 The calculation of duration of activities for retaining wall construction

Activity	Quantity for one segment	Productivity assumed	Man-hours per activity	No. of workers assumed	Theoretical duration (days)	Duration (t) days taken for LOB schedule
Earthwork	100 m^3	3 MH^1/m^3	300	13	1.92	2
Cement concrete	8 m^3	20 MH/m^3	160	13	1.03	1
Concrete raft	70 m^2 formwork	4 MH/m^2				
	2 t reinforcement	60 MH/t	880	11	6.67	7
	24 m^3 concrete	20 MH/m^3				
Concrete wall	500 m^2 formwork	4 MH/m^2				
	4.4 t reinforcement	60 MH/t	3,514	30	9.76	10
	62.5 m^3 concrete	20 MH/m^3				
Fencing	100 m	1.5 MH/m^2	150	7	1.79	2

Step 2: Buffer Estimation and 'Start' and 'End' Buffer Identification

The buffers are estimated in such a manner that the crew of one activity does not interfere with the crew of another activity. The buffers, B, for this illustration have been estimated and shown under the dotted arrows. For example, B = 1 means a buffer of one day between the earthwork (EW) and cement concrete activities.

It would be pertinent here to distinguish between start and end buffers. In case the rate of construction (speed) of a subsequent activity is faster than the activity under consideration, buffer is kept at the end. Conversely, if the rate of production of subsequent activity is slower than the activity under consideration, the buffer is placed at the start. For example, completion of earthwork is achieved in 2 days while the completion of cement concrete is achieved in 1 day itself. Naturally, the speed of the latter activity is faster than that of the former. Thus, in order to avoid the interference of the latter crews with the crews of earthwork activity, it is desirable to have the end buffer and derive the start of cement concrete as shown in Table 6.11. Now, considering the case of cement concrete and concrete raft, it is clear that the speed in the latter is slow (concrete raft is taking 7 days to complete)—thus requiring a start buffer.

Step 3: Computations of Start and Finish of First and Last Segments of Each of the Activities Involved

Using the concept of start and end buffers, the start and finish of the first segment and the fifth segment are computed for each of the activities involved in the construction of retaining wall. The computations are shown in Table 6.11.

Starting day of earthwork for segment 1 = 0
Duration of EW$_1$ = 2 days
Thus, finish of earthwork for segment 1 = 0 + 2 = 2
Start of earthwork for segment 5 = 0 + 4 × 2 = 8

The rate of construction of cement concrete (1 day) is faster than the rate of construction of earthwork (2 days). Thus, unless the start of the cement concrete is delayed, the crew of cement concrete will catch up with the earthwork crew and, thus, there will be interference among the two crews. In this case, it is prudent to have the buffer between

Table 6.11 Computations of line-of-balance schedule for the construction of retaining wall

Activity	Duration (t) Days	Time (T) (n−1) walls	Buffer days	Type of buffer	Start day of wall 1	Start day of wall 5
Earthwork	2	8	1	End	0	0 + 8 = 8
Cement concrete	1	4	2	Start	11 − 4 = 7	0 + 8 + 2 + 1 = 11
Concrete raft	7	28	2	Start	7 + 1 + 2 = 10	10 + 28 = 38
Concrete wall	10	40	3	End	10 + 7 + 2 = 19	19 + 40 = 59
Fencing	2	8			72 − 8 = 64	59 + 10 + 3 = 72

[1]MH = Man hour

the completion of earthwork on the last segment and the start of the cement concrete on the last segment. Hence, start day of cement concrete activity for segment 5 = start day of earthwork for segment 5 + duration of earthwork for segment 5 + buffer following earthwork activity = 8 + 2 + 1 = 11.

With this information, the start day of the cement concrete for segment 1 can be computed. This is equal to start of activity CC for segment 5 − T (time taken for cement concrete to get completed in 4 segments) = 11 − 4 = 7.

The rate of construction of concrete raft (7 days) is slower than the rate of construction of cement concrete (1 day). So, concrete raft may immediately start after the buffer, without the fear of interference with cement concrete activity. Thus, start day of concrete raft for first segment = start day of cement concrete for segment 1 + duration of cement concrete for segment one + buffer following cement concrete = 7 + 1 + 2 = 10. The start day of concrete raft for fifth segment = start day of concrete raft for segment one + T (duration for completing concrete raft in four segments) =10 + 28 = 38.

In this manner, the computations are completed. It may be noted that the start day of last segment is computed first, followed by the start day of first segment for an activity that has an end-buffer association with its succeeding activity, while the reverse manner is followed in case of start buffer.

Step 4: Drawing line-of-balance chart
The resulting start and finish times for each activity for all the segments are shown in Figure 6.49.

Step 5: Adjusting the schedule
It can be noted from Figure 6.49 that the rates of construction of different activities are not uniform. In other words, these are not 'balanced' and some improvement is desirable. In order to bring about a more 'balanced' schedule, a number of steps can be taken. For example, the crew size can be adjusted since we cannot make changes either in the quantity or in the assumed productivity. From the above figure, it can be seen that we can reduce the speed of cement concrete activity by reducing the number of crew. Also, we can try to increase the speed of the activities 'concrete raft' and 'concrete wall'. This can be achieved by increasing the speed of construction of these activities by increasing the number of crews.

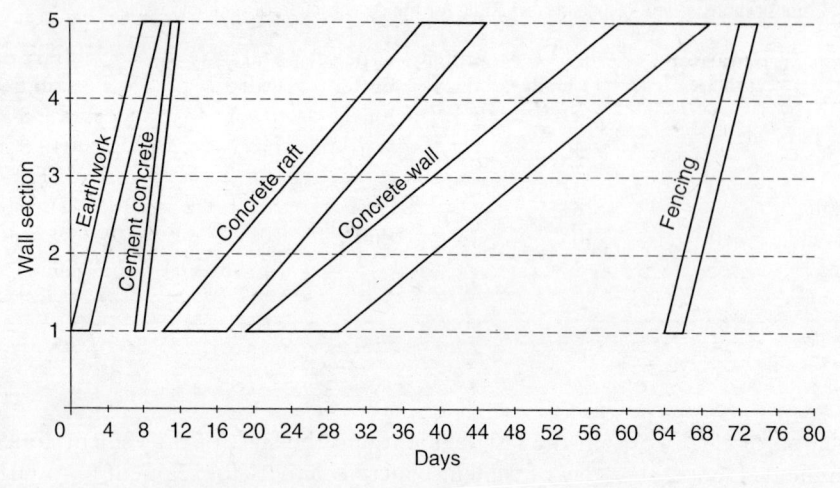

Figure 6.49 LOB schedule for construction of retaining wall

The ideal situation would be to aim for 'parallel scheduling', as shown in Figure 6.47 for the housing construction example, wherein the rate of construction of each of the activities is same. In our retaining wall example, it is also possible to achieve a parallel scheduling by decreasing the speed of cement concrete activity and increasing the speed of concrete raft and concrete wall activities to two days per segment, and thereby reducing the overall duration for the construction. However, such reduction in duration should be checked for its technological feasibility as well as its cost-saving potential. For example, if it is found that the extra cost incurred in increasing the crew size and thereby reducing the overall duration has more cost associated with it than the saving it generates by decrease in indirect cost, there is no point in such reduction in duration. Thus, there could be many possible schedules and the choice of a particular schedule will be dependent on the decision maker and the situation prevailing at that time. Needless to say, the computations for start and finish times of some of the activities may undergo change. The revised computations for start and finish times after adjusting the crew size are shown in Table 6.12. The overall completion time is now 32 days as against 74

Table 6.12 Revised computations of line-of-balance schedule for the construction of retaining wall

Activity	Duration (t) days	Time (T) (n − 1) walls	Buffer days	Type of buffer	Start day of wall 1	Start day of wall 5
Earthwork	2	8	1	End	0	0 + 8 = 8
Cement concrete	2	8	2	Start	11 − 8 = 3	0 + 2 + 8 + 1 = 11
Concrete raft	2	8	2	Start	3 + 2 + 2 = 7	7 + 8 = 15
Concrete wall	3	12	4	End	7 + 2 + 2 = 11	11 + 12 = 23
Fencing	2	8			30 − 8 = 22	23 + 3 + 4 = 30

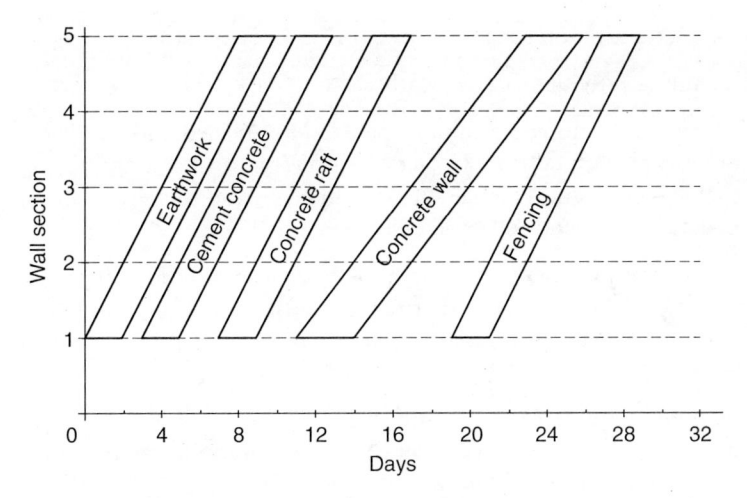

Figure 6.50 Revised line-of-balance schedule for the construction of retaining wall

days. Also, note that the buffer between the activities 'concrete wall' and 'fencing' has now been changed to 4 days from the previous buffer of 3 days. The revised LOB schedule based on these computations for our retaining wall example is shown in Figure 6.50. As explained earlier, there could be further improvements as well.

6.12 NETWORK TECHNIQUES ADVANTAGES

Having completed a discussion on the basic definitions above, the following paragraphs discuss some of the salient features of two of the most popular networks used in project planning—critical path method (CPM) and project evaluation and review technique (PERT). Network planning can be used for computing project completion time, identifying critical activities, etc. In fact, in PERT the uncertainty involved in completing an activity in a certain time can also be built in. It may be pointed out that not any network diagram can be called a PERT diagram (or chart) to distinguish it from a bar chart! Historically, PERT was developed for use in the aerospace industry and is most frequently used in R&D programmes. As a technique, PERT incorporates the uncertainty in the estimated time for completion of an activity by using the optimistic, mean and pessimistic times to define the likely time of completion of an activity, and then using the latter in further calculations to determine the times for project completion, etc.

As mentioned above, these networks can also be used for purposes such as resource allocation and monitoring. Indeed, in order to be effective, the network needs to be constantly updated (through feedback depending upon actual progress in different activities) and revised schedules drawn up. It may be remembered that critical activities identified at any stage may change as the project progresses, though simply a change in the actual (or likely) date of completion does not necessarily require reworking and redrawing the network. In other words, an important aspect of monitoring project progress is to assess any changes in the set of critical activities, and make appropriate adjustments in resource allocation, etc. Some of the advantages of network-based scheduling are highlighted below:

(a) Facilitates focusing attention on specific objectives

(b) The results of planning activities can be clearly seen

(c) Mathematical computation for determining parameters such as early and late start and finish dates can be easily carried out. In fact, the exercise can be computerized for large projects, and enable construction management to benefit from the higher speeds of calculation available, data storage and retrieval, etc.

(d) Activities on the critical path and those with float can be easily identified. This enables the manager to draw a list of activities that require closer attention, and take timely corrective action in case of slippage. In other words, 'management by exception' is made possible

(e) Facilitate identification of possible conflicts in the schedule of activities, in terms of sharing of resources, etc. As a corollary, resources can be dynamically reallocated taking account of emerging changes in the criticality of activities, etc.

(f) Facilitates drawing up of summaries, monitoring and review reports for consultants, clients, etc.

(g) Provide an overall graphic master plan of a complex project, clearly showing job interrelationships

(h) Helps in drawing up alternative plans of work—different alternatives can be examined by varying the pattern of resources allocation (which in turn changes the time estimates for the constituent activities)

(i) Cost-time studies can be carried out and an optimum project duration depending upon different levels of resource commitment can be found

REFERENCES

1. Arditi, D., Tokdemir O.B. and Suh K., 2002, 'Challenges in Line of Balance Scheduling', *ASCE Journal of Construction Engineering and Management*, 128(6), pp. 545–556.

2. Arditi, D., Tokdemir O.B. and Suh K., 2001, 'Effect of learning on line of balance scheduling', *International Journal of Project Management*, 19(5), pp. 265–277.

3. Burke, R., 1996, *Project management: Planning and control*, 2nd ed., West Sussex (UK): Wiley.

4. Callahan, M.T., Quackenbush D.G. and Rowings J.E., 1992, *Construction project scheduling*, New York: McGraw-Hill.

5. Chitkara, K.K., 1998, *Construction project management: Planning, scheduling and controlling*, New Delhi: McGraw-Hill.

6. Choudhary, S., 1988, *Project Management*, New Delhi: Tata McGraw-Hill.

7. Harris, R.B. and Loannou P.G., 1998, 'Scheduling projects with repeating activities', *Journal of Construction Engineering and Management*, ASCE, 124(4): 269–78.

8. Issues and Considerations in Project Management, available at http://www.maxwideman.com/issacons.

9. Johnston, D.W., 1981, 'Linear scheduling method for highway construction', *Journal of the Construction Division*, 107(CO2), pp. 247–261.

10. Khisty, C.J., 1970, 'The application of the "Line of Balance" technique to the construction industry', *Indian Concrete Journal*, July, pp. 297/300–319/320.

11. Lumsden, P., 1968, *The Line of Balance Method*, Tarrytown: Pergamon.

12. Lutz, J.D. and Halpin D.W., 1992, 'Analyzing linear construction operations using simulation and line of balance', Proceedings of Transportation Research Board, 71st Annual Meeting, Transportation Research Record 1351, Transportation Research Board, National Academy Press, Washington, D.C., pp. 48–56.

13. Nunnally, S.W., 1993, *Construction methods and management*, 4th ed., New Jersey (USA): Prentice-Hall.

14. Oberlender, G.D., 1993, *Project management for engineering and construction*, USA: McGraw-Hill.

15. Psarros, M.E., 1987, 'SYRUS: A program for repetitive projects', Masters thesis, Department of Civil Engineering, Illinois Institute of technology, Chicago.

16. Shtub, A., Bard J. and Globersons S., 1994, *Project management, engineering, technology and implementation*, international edition, London: Prentice-Hall.

17. Suh, K., 1993, A Scheduling System for Repetitive Unit Construction Using Line-of-Balance Technology, PhD thesis submitted at the Illinois Institute of Technology, USA.

18. Suhail, S.A. and Neale R.H., 1994, 'CPM/LOB: New methodology to integrate CPM and Line of Balance', *Journal of Construction Engineering and Management*, 120(3), pp. 667–684.

19. Turban, E., 1968, 'The line of balance scheduling highway maintenance projects', *Journal of Industrial Engineering*, 19(9), pp. 440–448.

20. Weist, J.D. and Levy F.K., 2005, *A Management Guide to PERT/CPM with GERT/PDM/DCPM and other Networks*, 2nd ed., New Delhi: Prentice Hall.

21. Wiest, J.D., 1981, 'Precedence Diagramming Method: Some unusual characteristics and their implications for project managers', *Journal of Operations Management*, 1(3), pp. 121–130.

SOLVED EXAMPLES

Example 6.1

For the network drawn in Figure Q6.1.1, perform the forward pass and backward pass. Calculate the early start, early finish, late start and late finish times for each of the activities. Find the critical activities and comment on the critical path(s).

| 212 | CONSTRUCTION PROJECT MANAGEMENT

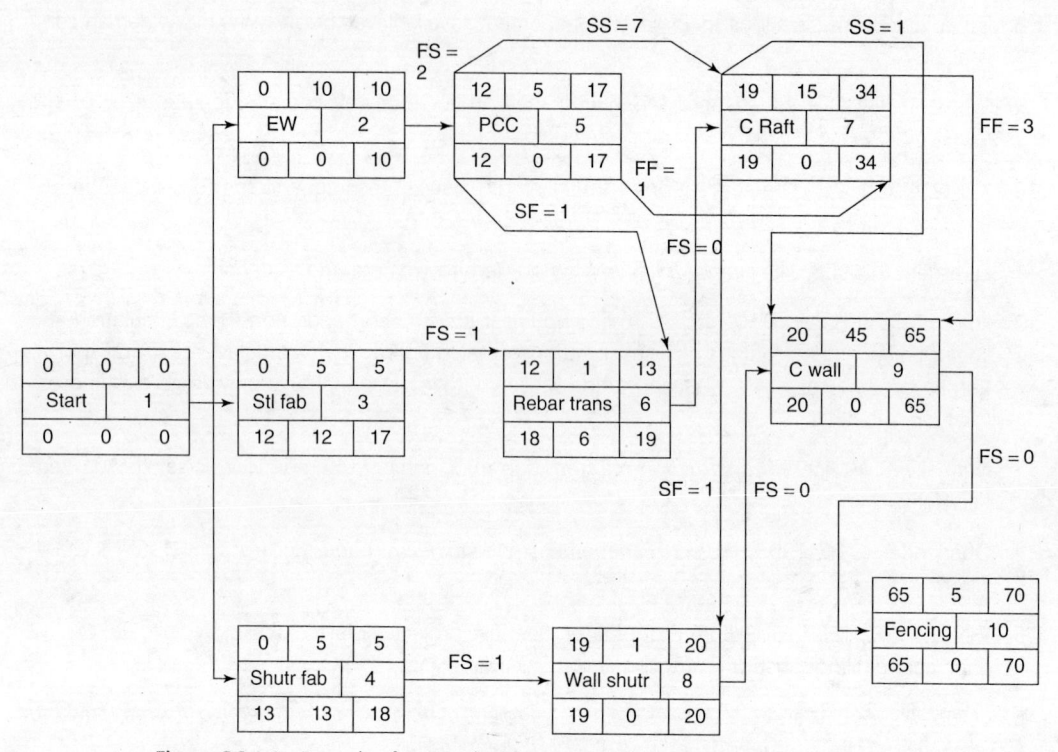

Figure Q6.1.1 Example of a precedence network with FS, FF, SS and SF relationships

EST	DUR	EFT
Activity	Node No.	
LST	TF	LFT

Solution

The computation of early start and late start times for all the activities of the network (Figure Q6.1.1) has been performed in Table S6.1.1, while the computation of late start and late finish times of all the activities have been performed in Table S6.1.2. It is evident from Table S6.1.1 and Table S6.1.2 that manual computations become complicated when there are relationships other than the finish to start in a precedence network. The use and interpretation of precedence networks with other than FS precedence relationship requires considerable care. The type of precedence relationship that occurs along the critical path must also be taken into account.

Total slack/float for each activity has been computed using the following relationships:

$$TF = LST - EST$$
$$TF = LFT - EFT$$

The total float values are given in Table S6.1.3.

Table S6.1.1 Forward pass explanation for the network of Figure Q6.1.1

Node	Activity	Duration (Dur$_i$)	EST	Remarks	EFT	Remarks
1	Start	0	0	0, Start Activity $EST_1 = 0$	0	$EFT_1 = EST_1 + dur_1$ $= 0 + 0 = 0$
2	EW	10	0	$EST_2 = EFT_1 + FS_{0-1}$ $= 0 + 0 = 0$	10	$EFT_2 = EST_2 + dur_2$ $= 0 + 10 = 10$
3	Steel fabrication	5	0	$EST_3 = EFT_1 + FS_{1-3}$ $= 0 + 0 = 0$	5	$EFT_3 = EST_3 + dur_3$ $= 0 + 5 = 5$
4	Shutter fabrication	5	0	$EST_4 = EFT_1 + FS_{1-4}$ $= 0 + 0 = 0$	5	$EFT_4 = EST_4 + dur_4$ $= 0 + 5 = 5$
5	PCC	5	12	$EST_5 = EFT_2 + FS_{2-5}$ $FS_{2-5} = 2$, $EST_5 = 10 + 2 = 12$	17	$EFT_5 = EST_5 + dur_5$ $= 12 + 5 = 17$
6	Rebar transport	1	12	$EST_6 =$ max. of $[EFT_3 + FS_{3-6} = 5 + 1 = 6$; and $EST_5 + SF_{5-6} - dur_6 = 12 + 1 - 1 = 12] =$ max. of $(6, 12) = 12$	13	$EFT_6 = EST_6 + dur_6$ $= 12 + 1 = 13$
7	Concrete raft	15	19	$EST_7 =$ max. of $[EST_5 + SS_{5-7} = 12 + 7 = 19$, $EFT_5 + FF_{5-7} - dur_7 = 17 + 1 - 15 = 3$, and $EFT_6 + FS_{6-7} = 13 + 0 = 13] =$ max. of $(19, 3, 13) = 19$	34	$EFT_7 = EST_7 + dur_7$ $= 19 + 15 = 34$
8	Wall shutter	1	19	$EST_8 =$ max. of $[EFT_4 + FS_{4-8} = 5 + 1 = 6$, and $EST_7 + SF_{7-8} - dur_8 = 19 + 1 - 1 = 19 =$ max. of $(6, 19) = 19$	20	$EFT_8 = EST_8 + dur_8$ $= 19 + 1 = 20$
9	Concrete wall	45	20	$EST_9 =$ max. of $[EFT_8 + FS_{8-9} = 20 + 0 = 20$, $EFT_7 + FF_{7-9} - dur_9 = 34 + 3 - 45 = -8$, and $EST_7 + SS_{7-9} = 19 + 1 = 20] =$ max. of $(20, -8, 20) = 20$	65	$EFT_9 = EST_9 + dur_9$ $= 20 + 45 = 65$
10	Fencing	5	65	$EST_{10} = EFT_9 + FS_{9-10}$ $= 65 + 0 = 65$	70	$EFT_{10} = EST_{10} + dur_{10}$ $= 65 + 5 = 70$

The critical paths for the network are 1-2-5-7-9-10 (see Figure S6.1.1) and 1-2-5-7-8-9-10 (see Figure S6.1.2).

As is evident from Figure S6.1.2, the critical path so obtained is not a neat path (not connected clearly in a back-to-back sequence as in PERT/CPM) since activity 8 is also critical.

The identification of different types of critical activities from the network (see Figure Q6.1.1) is left as an exercise for the readers.

Table S6.1.2 Backward pass explanation for Figure Q6.1.1

Node	Activity	Duration dur_i	LFT	Remarks	LST	Remarks
10	Fencing	5	70	Taking LFT_{10} same as EFT_{10}	65	$LST_{10} = LFT_{10} - dur_{10}$ $= 70 - 5 = 65$
9	Concrete wall	45	65	$LFT_9 = LST_{10} - FS_{9-10} = 65 - 0 = 65$	20	$LST_9 = LFT_9 - dur_9$ $= 65 - 45 = 20$
8	Wall raft	1	20	$LFT_8 = LST_9 - FS_{8-9}$ $= 20 - 0 = 20$	19	$LST_8 = LFT_8 - dur_8$ $= 20 - 1 = 19$
7	Concrete raft	15	34	$LFT_7 = $ Min of $[LFT_9 - FF_{7-9}$ $= 65 - 3, LST_9 - SS_{7-9} + dur_7$ $= 20 - 1 + 15,$ and $LST_8 -$ $SF_{7-8} + dur_7 = 20 - 1 + 15]$ $= $ Min of $(62, 34, 34) = 34$	19	$LST_7 = LFT_7 = dur_7$ $= 34 - 15 = 19$
6	Rebar transport	1	19	$LFT_6 = LST_7 - FS_{6-7}$ $= 19 - 0 = 19$	18	$LST_6 = LFT_6 - dur_6$ $= 19 - 1 = 18$
5	PCC	5	17	$LFT_5 = $ Min of $[LFT_7 - FF_{5-7}$ $= 34 - 1, LST_7 - SS_{5-7} + dur_5$ $= 19 - 7 + 5,$ and $LST_6 -$ SF_{5-6} $+ dur_5 = 18 - 1 + 5]$ $= $ Min of $(33, 17, 22 = 17$	12	$LST_5 = LFT_5 - dur_5$ $= 17 - 5 = 12$
4	Shutter fabrication	5	18	$LFT_4 = LST_8 - FS_{4-8}$ $= 19 - 1 = 18$	13	$LST_4 = LFT_4 - dur_4$ $= 18 - 5 = 13$
3	Steel fabrication	5	17	$LFT_3 = LST_6 - FS_{3-6}$ $= 18 - 1 = 17$	12	$LST_3 = LFT_3 - dur_3$ $= 17 - 5 = 12$
2	EW	10	10	$LFT_2 = LST_5 - FS_{2-5}$ $= 12 - 2 = 10$	0	$LST_2 = LFT_2 - dur_2$ $= 10 - 10 = 0$
1	Start	0	0	$LFT_1 = $ Min of $[LST_2 - FS_{1-2}$ $= 0 - 0, LST_3 - FS_{1-3} = 12-0,$ and $LST_4 - FS_{1-4} = 13 - 0]$ $= $ Min of $(0, 12, 13) = 0$	0	$LST_1 = LFT_1 - dur_1$ $= 0 - 0 = 0$

Table S6.1.3 Calculation of floats of activities

Node	EST	LST	Float (LST − EST)
1	0	0	0
2	0	0	0
3	0	12	12
4	0	13	13
5	12	12	0
6	12	18	6
7	19	19	0
8	19	19	0
9	20	20	0
10	65	65	0

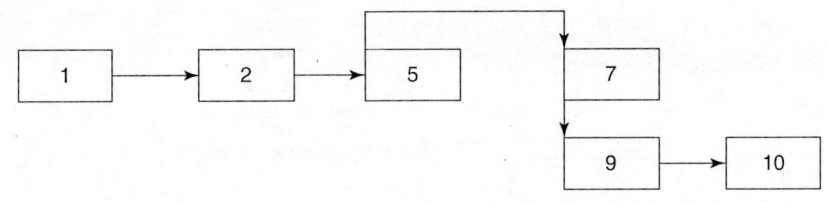

Figure S6.1.1 Critical path 1 for network of Figure Q6.1.1

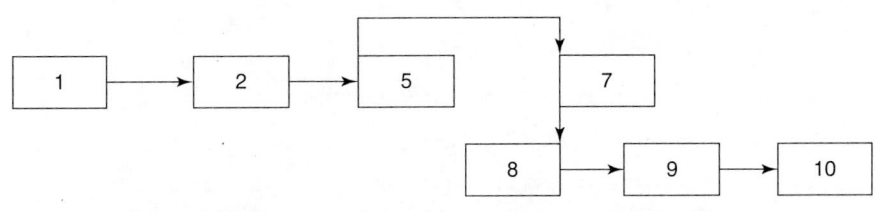

Figure S6.1.2 Critical path 2 for network of Figure Q6.1.1

REVIEW QUESTIONS

1. State whether True or False:
    a. Work-breakdown structure is a breakdown of project activity schedules from weekly to daily and from daily to hourly schedules for minute control.
    b. Work-breakdown structure is a breakdown of project into its components that enables meaningful organization of project scope.
    c. Gantt charts are useful in working out the period-to-period requirements of resources on a project.
    d. A Gantt chart can't identify 'critical' activities.
    e. Gantt chart is more effective than 'network diagram' for presenting project schedules to the company management.
    f. Arrow diagramming method of network representation can show different types of relationships such as start–start and finish–finish.
    g. If the early occurrence time and the late occurrence time for the starting and finishing events of an activity are same, the activity is always a critical activity.
    h. In activity on arrow (AOA) type of network diagram, the length of arrow has no relation to its time duration.
    i. AON networks can represent online finish–start type of relationships between activities.
    j. 'Slack' is the extra time available to perform an activity without affecting the project completion time.
    k. 'Float' on the activity indicates the duration by which total estimated duration of activity can be reduced while crashing.
    l. Float is the difference between 'time available to perform a task' and 'time required to perform the task'.

m. 'Total float' is the sum of 'free float' and 'independent float'.
n. PERT is used for scheduling product development projects.
o. PERT is an acronym for Project Estimation and Review Technique.
p. There is only a 50 per cent chance for completing a project in the scheduled time calculated using PERT model.
q. PERT is used when estimates for task durations can be worked out using past data.
r. Concept of 'crashing' is not applicable to PERT model.
s. CPM is an acronym for Cost and Profile Monitoring.
t. Critical path method aims at optimizing project duration at minimum cost.
u. Critical path is the shortest path in the project network.
v. Critical path is the longest path in the project network.
w. Critical path duration is the minimum time required to complete the project.
x. Critical path method uses 'three time estimation' for estimating task durations.
y. There can be only one critical path in a project network.
z. In precedence network, the activities are represented by arrows and the events are represented by nodes.
aa. Sometimes, it is necessary to introduce 'dummy' activities in precedence networks for maintaining relationship logic.
bb. Line-of-balance (LOB) technique is used to balance work at different workstations in a mass-production assembly line.
cc. Line-of-balance technique is useful for planning the construction of a number of similar structures.
dd. Line-of-balance charts are also known as Gantt charts after the name of the engineer who introduced the technique.

2. Short-answer type questions
   a. What is WBS? How is this helpful in planning, monitoring and controlling the programme of projects?
   b. Describe the process of drawing a Gantt chart for project schedule. What are the advantages and limitations of Gantt charts?
   c. How is the estimation of time duration of an activity different in PERT and CPM approaches? Under what situations are these approaches more suitable?
   d. Differentiate between CPM and PERT models for project scheduling.
   e. What is meant by line-of-balance technique, and what are its specific uses and advantages in project planning? Illustrate your answer with a suitable example.
   f. What do you understand by a dummy activity in an AOA network? In which situations does it become necessary to add a dummy activity in a network? Illustrate with suitable examples.
   g. Differentiate and compare between 'activity on arrow' network diagrams and 'activity on node' network diagrams.

CONSTRUCTION PLANNING | 217 |

h. What are the different steps involved in finding the probability of completing a project in the desired duration under certainty, using PERT approach?
i. Enumerate the advantages of using the AON network instead of the AOA network.
j. Explain the reasons for the need to develop the PERT methodology for scheduling and monitoring of projects. What is the process of review adopted in PERT?
k. Define 'critical path' in a CPM network.
l. How do you ensure timely completion of a project using CPM?
m. Does the critical path, once identified, remain the same till completion of project? If not, what are the reasons for its change?
n. What do you understand by the terms 'total float' and 'free float' in project schedule?
o. How are the floats useful in effective management of projects?
p. What do you understand by the 'three time estimates' method for network scheduling? Illustrate your answer by giving the appropriate formulae and the circumstances under which it is adopted.
q. What is the probability of completing the project within the total scheduled time calculated on the basis of 'three time estimate' method?
r. What are the advantages of the precedence network over the conventional 'arrow diagram' network representation for project schedules?
s. How do you indicate 'delays' and 'overlaps' between related activities?
t. Activity 'A' has estimated a duration of 18 days as a predecessor to activity 'B', which has an estimated duration of 25 days. Calculate the 'earliest start' for activity 'B' if the relationship between the two is:
   i. FF + 14 days
   ii. SS + 10 days
   iii. FS − 7 days
   iv. SF + 32 days
   v. FS + 9 days

3. Numerical Problems
   1. The mean time for project completion using PERT method of time estimation is 25 weeks with a standard deviation of 3 weeks. What is the duration beyond which the project completion is not likely to extend?
   2. A project network has five paths—P1, P2, P3, P4 and P5. The total activity duration computed along each of these paths are—22, 31, 20, 30 and 19 days, respectively. What is the minimum duration for project completion?
   3. The early occurrence time and the late occurrence time of the starting event of an activity are 10 and 12, respectively. The early occurrence time and the late occurrence time of the finishing event of the same activity are 22 and 26, respectively. If the estimated duration for the activity is 10 time units, what is the total float available for the activity?
   4. Two project activities 'A' and 'B' (B being dependent on A) have a finish–finish relationship with 8 days 'lag'. The estimated durations for the activities are 10 days and 15 days, respectively. If

activity 'A' is scheduled to start on the 12th day from the beginning of project, on which day can you schedule to start activity 'B'?

5. For the network shown in Figure Q6.5.1:
   (a) What are the critical path and the project duration?
   (b) If the estimated duration for activity 'C' is revised to 8 days, will the critical path remain the same as earlier?

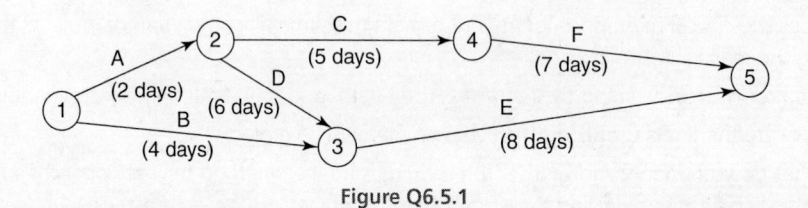

Figure Q6.5.1

6. Referring to the list of activities below (see Figure Q6.6.1), find out the total time for project completion, critical path, early and late start and finish dates, and total floats available on non-critical activities.

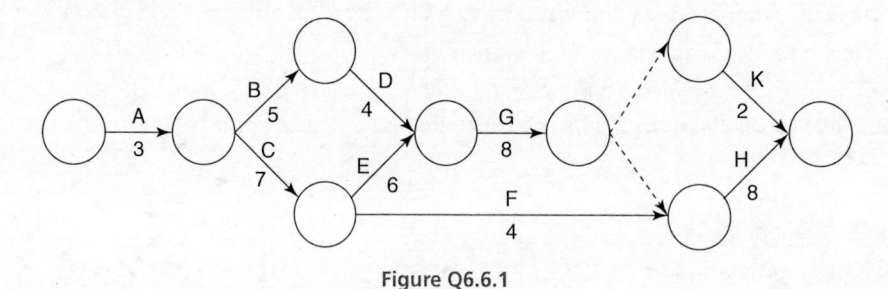

Figure Q6.6.1

7. For the activity table given in Table Q6.7.1, find the expected time for completion of project with 84.1 per cent probability.

Table Q6.7.1

Activity	Estimated activity duration		
	T_o	T_m	T_p
(1, 2)	3	6	15
(1, 6)	2	5	14
(2, 3)	6	12	30
(2, 4)	2	5	8
(3, 5)	5	11	17
(4, 5)	3	6	15
(6, 7)	3	9	27
(5, 8)	1	4	7
(7, 8)	4	19	28

8. Draw the network for the data given in Table Q6.8.1, pertaining to a small construction project. Compute the event times for all the activities. Find out the critical path. Also, calculate the three floats for the non-critical activities.

Table Q6.8.1

Activity	Duration (in weeks)	Depends upon
A	4	—
B	2	A
C	2	A, B, F
D	2	C, L
E	6	D, G, H
F	2	A
G	2	F
H	3	G, L
K	5	A
L	3	F, K

9. The average time for project completion using PERT method is 32 weeks, with a standard deviation of the longest path being 4 weeks. Using the Y value table for normal distribution curve, find the: (a) probability of completing the project in 32 weeks, (b) probability of completing the project in 36.6 weeks, and (c) probability of completing the project in 29 weeks.

10. Identify different paths in the network and the critical path for the network shown in Figure Q6.9.1. Also find the project duration. Each node shows activity name and their estimated duration.

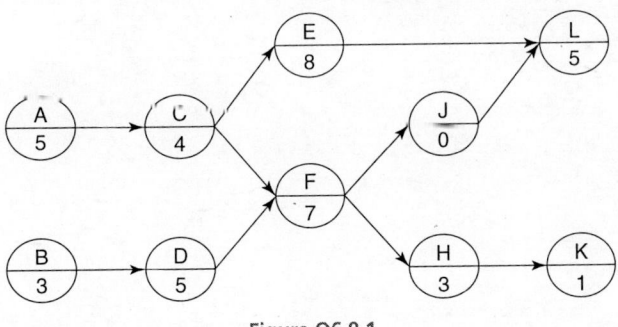

Figure Q6.9.1

11. For the network shown in Figure Q6.10.1, calculate the project duration. Also, identify the critical and non-critical activities. Find the floats in non-critical activities.

12. A small project has 12 activities—A, B, C, D... L. The relationship among these activities is given in the network shown in Figure Q6.11.1. The three estimates of duration for each of these activities are given in Table Q6.12.1. (1) Draw an activity on node (AON) network for the project. (2) Calculate the minimum expected completion time for the project. (3) Calculate the

Figure Q6.10.1

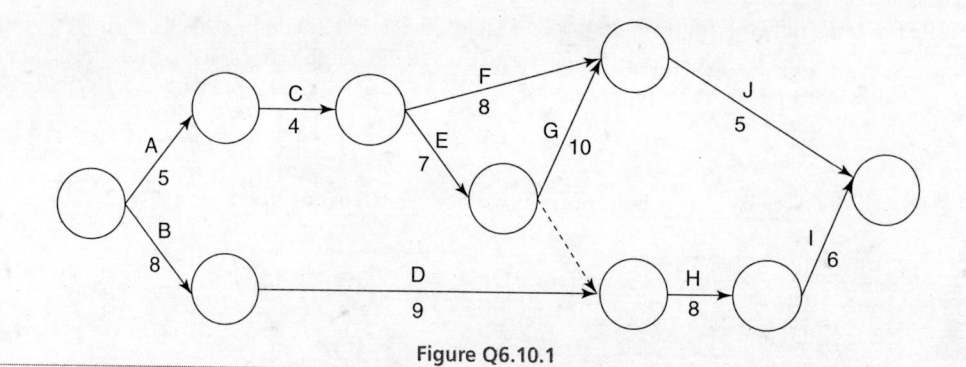

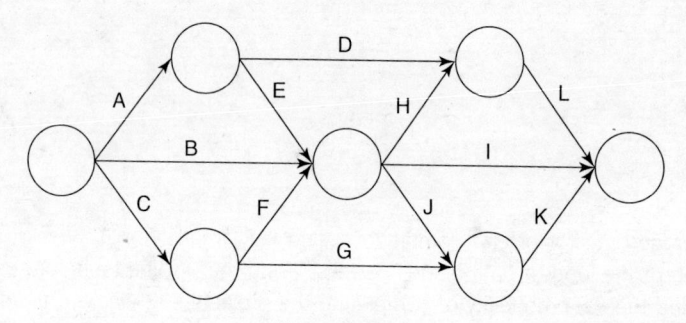

Figure Q6.11.1

Table Q6.12.1

Activity	Optimistic	Most likely	Pessimistic
A	2	3	4
B	5	6	7
C	5	6	7
D	3	4	5
E	2	3	4
F	3	4	5
G	8	10	18
H	5	6	7
I	7	11	15
J	2	3	4
K	3	4	5
L	7	11	15

probability that the project would be completed within 25 days. (4) Calculate the probability that the project cannot be completed within 30 days.

13. For the data given in Table Q6.13.1, find the (a) critical path and earliest time to complete the project, and (b) early and late start, early and late finish, total float and free float for all activities.

Table Q6.13.1

Activity	Duration (weeks)	Predecessor(s)	Successor(s)
A	3	—	B, C
B	4	A	D, E, F
C	6	A	E, F
D	4	B	G
E	5	B, C	J
F	5	B, C	H, I
G	5	D	K
H	3	F	L
I	8	F	K
J	9	E	L
K	7	G, I	—
L	6	H, J	—

14. A construction company engaged in undertaking small projects has recently been awarded a project. The project activities and the estimated time for their completion are listed in Table Q6.14.1, along with the information on immediate predecessors.

    (a) Construct a network for the project.
    (b) Determine the critical path and the project completion time.
    (c) What is the probability of completing the project in the completion time you have arrived at?

Table Q6.14.1

Activity	t_o	t_m	t_p	Immediate predecessor
A	8	10	12	—
B	6	7	9	—
C	3	3	4	—
D	10	20	30	A
E	6	7	8	C
F	9	10	11	B, D, E
G	6	7	10	B, D, E
H	14	15	16	F
I	10	11	13	F
J	6	7	8	G, H
K	4	7	8	I, H
L	1	2	4	G, H

(d) Determine the time interval within which the probability of completion of the project will be 90 per cent.

15. A new product is to be released on a deadline for which 25 days are left. The remaining activities in product launch with their interdependencies, the expected average times and the standard deviations are given in Table Q6.15.1. Draw a project network using AOA convention and find the probability of completing the project in time.

Table Q6.15.1

Activity	Preceding activities	Expected duration t_e	Standard deviation S_t
A	—	6	1.5
B	—	3	0.5
C	A	5	1.0
D	A	4	0.5
E	A	3	0.5
F	C	3	0.5
G	D	5	1.0
H	B, E, D	5	2.0
I	H	2	0.5
J	I, G, F	3.5	1.5

16. Convert the project network of Figure Q6.16.1 to activity on arrow (AOA) convention. Figures inside the node under activity name indicate duration in days.

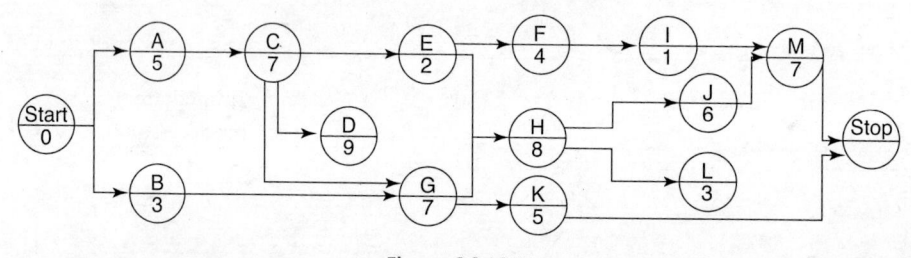

Figure Q6.16.1

17. You have to construct a building to set up your office. You own a plot of 100 square yards (30 feet × 30 feet) in a prime location. For the time being, you are in a position to construct ground floor alone due to financial constraints. So, you have decided to construct a simple RC structure. It consists of individual footing for 16 columns (columns are spaced every 10 feet in both directions) and a plinth beam at about 1 m from the existing road level. All the columns are going up to roof level, and you plan to have an RC roof slab and beam. The in-between space is constructed by using locally available bricks. There is earth filling up to plinth level, and after proper compaction of filled-up earth, flooring is laid. Other activities include

CONSTRUCTION PLANNING

plastering, painting, doors and windows, plumbing and electrical work. For your ready reference, the estimated duration for individual activities and their dependence are listed in Table Q6.17.1. How much time will it take to construct the building?

Table Q6.17.1 Activity details

Activity No.	Activity description	Duration (days)	Depends on
1	Site clearance	2	
2	Excavation in foundations	6	1
3	Foundations including plinth beam	10	2
4	Column casting up to roof level	10	3
5	Slab and beam of roof	21	4
6	Earth filling and flooring	10	5FS + 10d
7	Brickwork	10	5FS + 7d
8	Doors/windows	3	7FF
9	Plastering	10	7SS + 7d
10	Painting	7	9SS + 7d
11	Plumbing	7	7FF + 3d
12	Electrical works	7	11SS
13	Handover	1	10, 11, 12

7

Project Scheduling and Resource Levelling

Introduction, resource levelling, resource allocation, importance of project scheduling, other schedules derived from project schedules, network crashing and cost-time trade-off

7.1 INTRODUCTION

After the network calculations are made, it is important to present their results in a form that is easily understood by all the stakeholders involved with the project. Most of the times, the results of network techniques such as PERT/Critical Path Method/Precedence Diagram are plotted in the form of a Bar chart or a Gantt chart for its obvious advantages. A Gantt chart is plotted using either the early start time of activities or the late start time of activities. In the former case the chart is known as early Gantt chart, and in the latter case, late Gantt chart. The bar charts themselves are dependent on the ideal condition of availability of efficient resources as and when needed by an activity in a project. Due to the competition between activities for a particular resource, the demand may at times exceed the planned availability.

Through these charts, a range of information can be derived—for example, the total floats available in an activity, the critical activity, etc. As discussed earlier, these charts can also be used for plotting and comparing the planned progress versus the actual progress achieved. Further, these charts are useful for plotting the resources required on a day as well as for giving useful information on the cost aspects.

It may be recalled that while preparing the network diagram we concentrated mainly on the technological constraints (for instance, one activity cannot start until the other is over), and assumed that the resources are unlimited. Further, the resources can be mobilized on demand anytime we require them. Although such ideal conditions may prevail in some exceptional projects, in most of the real-life projects we face resource constraints. In real-life situations, the activity start times not only have to deal with technological constraints in the form of their precedence relationships, but also face the challenge of resource availability constraints. That is, we have limited quantity of resources at our disposal.

The planning for scheduling of activities, thus, has to account for the different constraints that may be imposed on the availability of any of these resources, and ensure that sudden changes in the requirement of these are avoided. For example, using an equipment at the outset and then at the end of the project should be avoided, as even idle (or idling) equipment carry a certain cost.

Similarly, sudden increase (or retrenchment) of labour should also be avoided!! Thus, effort has to be made to ensure that resources are allocated to activities in a manner that any changes in demands are gradual to the extent possible, and that the available resources are optimally used.

Resource constraints pose several operational problems in network-based scheduling. The concepts of float and criticality of activities start losing their normal meanings. For example, now the early start of an activity is governed not only by technological constraints but resource constraints as well. Hence, there cannot now be a single early start schedule as in the case without resource constraints. Similarly, the value of slack or float is no longer unique since it is dependent on early start and late start schedule.

The problem of scheduling activities so that none of the precedence relationships are violated and none of the resource availabilities are exceeded is a difficult task, and it could result into a number of combinations of possible early start and late start schedules. In literature, scheduling problems with limited resources is classified as a large combinatorial problem. A number of programmes based on some rules of thumb (also called heuristic) have been developed to solve such problems. These heuristic programmes for resource scheduling can broadly be classified under two categories: (1) resource levelling and (2) resource allocation.

7.2 RESOURCE LEVELLING

In resource levelling, the constraint is the fixed project duration. That is, the project must get completed by a fixed date. The attempt of such heuristic is to reduce peak requirement of resources and to smooth out period-to-period assignments. Such problems are also referred to as 'time limited resource considerations' problems. The assessment of resources is done using a resource-loading or resource-aggregation chart. We take an example to illustrate the concept of resource levelling.

Let us assume that there are a total of seven activities, A to G, in the example network (Figure 7.1). The duration of each of the activities is written below the arrow, while the resource requirement is shown in the bracket adjacent to the activity name. For example, the duration for activity A is three days and the resource required is two units. The early start and late start time of events or nodes are also shown in the network, from which the float available in a particular activity can be calculated and critical activities identified. The critical path of the network is 1-2-4-5-6 and it consists of activities A, C, E and G. The critical path is shown by bold arrows in the network.

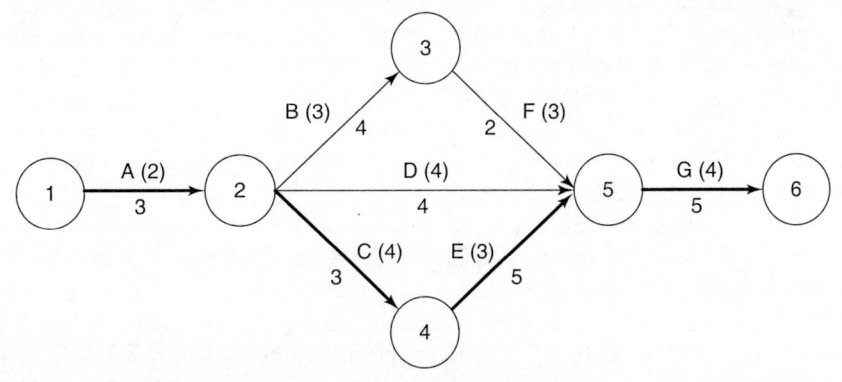

Figure 7.1 Network for resource leveling illustration

Table 7.1 Resource-loading table showing daily requirements of workers based on early-start order

Activity	Early start	Duration (days)	Resources	1	2	3	4	5	6	7	8	9	10	11	12	13	14	15	16
(1,2)	0	3	2	2	2	2													
(2,4)	3	3	4				4	4	4										
(2,3)	3	4	3				3	3	3	3									
(2,5)	3	4	4				4	4	4	4									
(4,5)	6	5	3								3	3	3	3	3				
(3,5)	7	2	3								3	3							
(5,6)	11	5	4												4	4	4	4	4
	Total			2	2	2	11	11	11	10	6	6	3	3	4	4	4	4	4

Resource levelling can, in principle, be carried out through the following steps:

1. The project network is prepared based on the data provided for each activity. Event times and activity times are computed as illustrated earlier; thus, total float is also computed for each of the activities.
2. The activities are ranked in order of their earliest start date (refer to Table 7.1). Activity (1, 2) has '0' as its earliest start time. Activities (2, 3), (2, 4) and (2, 5) all have their earliest start times on day 3, while activities (4, 5), (3, 5) and (5, 6) have their earliest start times on days 6, 7 and 11, respectively. The resources required on daily basis for each of the activities are summed up and shown in the form of a chart called resource-aggregation or resource-loading chart. Figure 7.2 shows a resource-loading chart based on the earliest start time of all the activities. The project takes a total of 87 man-days to complete, and the daily requirement varies from a minimum of 2 resources on days 1 to 3, to a maximum of 11 resources on days 4, 5 and 6.
3. Now, the activities are ranked in order of their latest start date (refer to Table 7.2). It may be noted that the latest start date of an activity is the latest time of the finish event less the duration. Thus, latest start times of activities (1, 2), (2, 4), (2, 3), (4, 5), (2, 5), (3, 5) and (5, 6) are on days 0, 3, 5, 6, 7, 9 and 11, respectively, in the ascending order. The resource-loading chart shown in Figure 7.3 is prepared based on the ascending order of latest start times of each activity. It can be noticed that the requirement of resources varies from a minimum of 2 to a maximum of 10 resources.
4. The two resource-loading charts obtained from steps 2 and 3 are compared. The two charts provide the two extreme arrangements of resource requirements. In the case where peaks and valleys are seen in the utilization pattern for a resource, the activities are manipulated by visual inspection and an acceptable resource requirement is found between the two extremes. The bottom line is to ensure continuous deployment of resources and to avoid large variations in the utilization pattern. One such compromise solution is shown

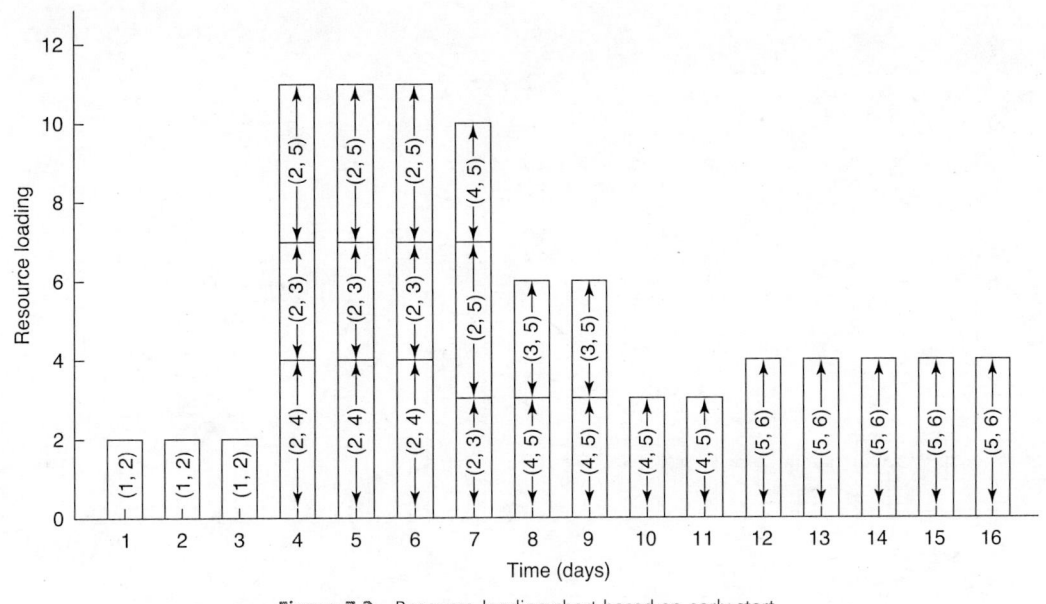

Figure 7.2 Resource-loading chart based on early start

in Figure 7.4 (also refer to Table 7.3). This figure has been obtained by allocating the resources in the following manner.

(1) The early start time of non-critical activity (2, 3) has been followed.
(2) The late start time of non-critical activities (2, 5), and (3, 5) have been followed.
(3) The critical activities (1, 2), (2, 4), (4, 5), and (5, 6) have not been disturbed.

Table 7.2 Resource-loading table showing daily requirements of workers based on late-start order

Activity	Late start	Duration (days)	Resources	1	2	3	4	5	6	7	8	9	10	11	12	13	14	15	16
(1,2)	0	3	2	2	2	2													
(2,4)	3	3	4				4	4	4										
(2,3)	5	4	3						3	3	3	3							
(4,5)	6	5	3								3	3	3	3	3				
(2,5)	7	4	4								4	4	4	4					
(3,5)	9	2	3										3	3					
(5,6)	11	5	4												4	4	4	4	4
	Total			2	2	2	4	4	7	6	10	10	10	10	4	4	4	4	4

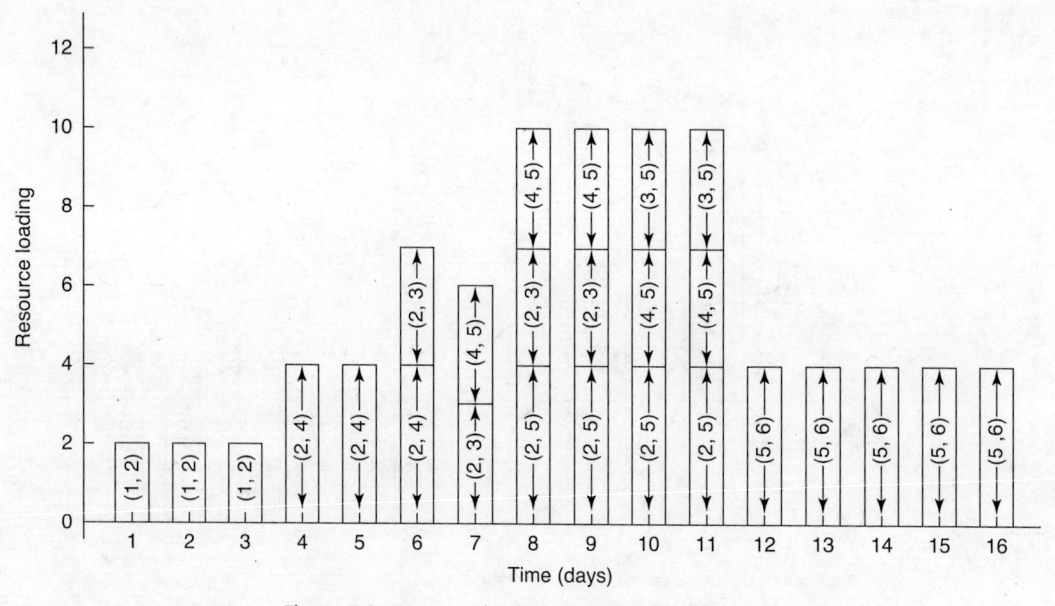

Figure 7.3 Resource-loading chart based on late start

The above manipulations have resulted in reducing the peak requirement (from 11 to 10) besides bringing a gradual change in resource deployment. One can also note that there is no change in the project duration of 16 days. This is also known as time-constrained leveling.

As can be understood from this example, it would be extremely difficult to employ this technique of visual examination for large problems. For simple problems the levelling exercise

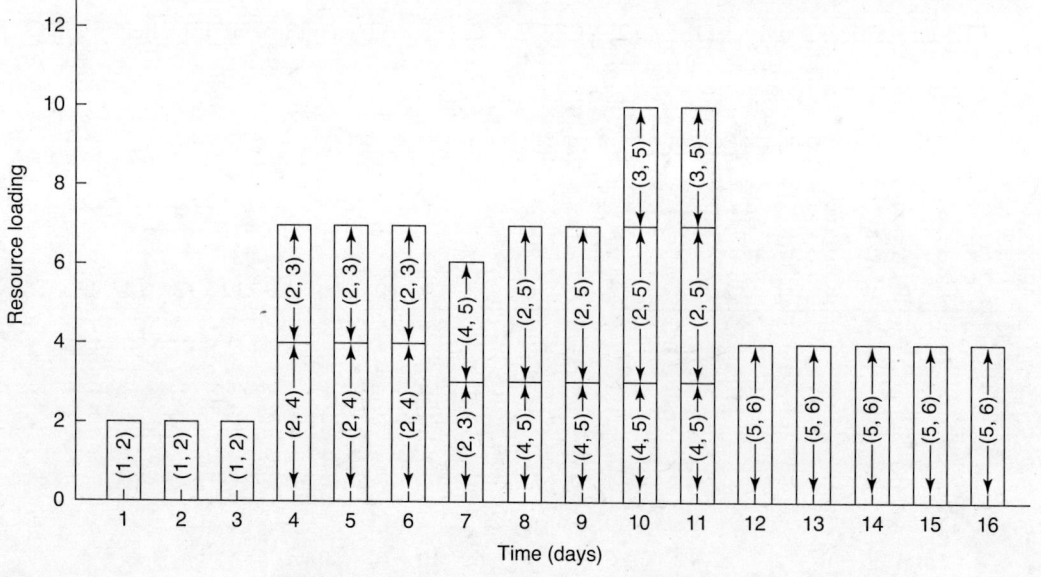

Figure 7.4 Resource-levelled chart (time constrained)

Table 7.3 Resource-loading table showing daily requirements of workers after levelling

Activity	Duration	Resources	1	2	3	4	5	6	7	8	9	10	11	12	13	14	15	16
(1,2)	3	2	2	2	2													
(2,4)	3	4				4	4	4										
(2,3)	4	3				3	3	3	3									
(4,5)	5	3							3	3	3	3	3					
(2,5)	4	4								4	4	4	4					
(3,5)	2	3										3	3					
(5,6)	5	4												4	4	4	4	4
Total			2	2	2	7	7	7	6	7	7	10	10	4	4	4	4	4

can be completed in one attempt, but for larger problems, the resource levelling cannot be carried out in a single step and is a largely iterative process. Computers can be employed to good advantage for levelling of resources under a time-constrained situation. A number of heuristics have been developed for this purpose and they are discussed elsewhere.

7.3 RESOURCE ALLOCATION

Here, availability of resources is a constraint. In other words, the resources have fixed limits. For illustrating resource allocation, we take the same problem (project network of Figure 7.1) that was used for resource levelling. The difference here is resource constraint. Let us take the maximum availability of resources as eight. In case it is not possible to resolve the resource over-allocation on a particular day, some activities may have to be delayed. The basic objective here is to find out which activities can be delayed and by how much, and finally to arrive at the shortest possible time to complete the project satisfying the resource constraints.

When the resource requirement on a particular day exceeds eight, we decide the priority of competing activities on a predefined set of rules. Let the activities with earliest start time get the first priority. In case there is a tie between two or more activities, the priorities are decided on the basis of float available in the activities. That is, the activity with the minimum float gets first priority. Further, non-critical activities may need to be rescheduled in order to free the resources for critical activities. Also, let's not stop an activity in-between once it has started.

The network of Figure 7.1 can be redrawn to time scale. This is shown in Figure 7.5.

Day 1
Only activity (1, 2) is scheduled for day 1 and only two resources are required, which is less than the maximum. Thus on day 1, requirement—2, availability—8, remaining—6.

Day 2
Activity (1, 2) continues. Requirement—2, availability—8, remaining—6.

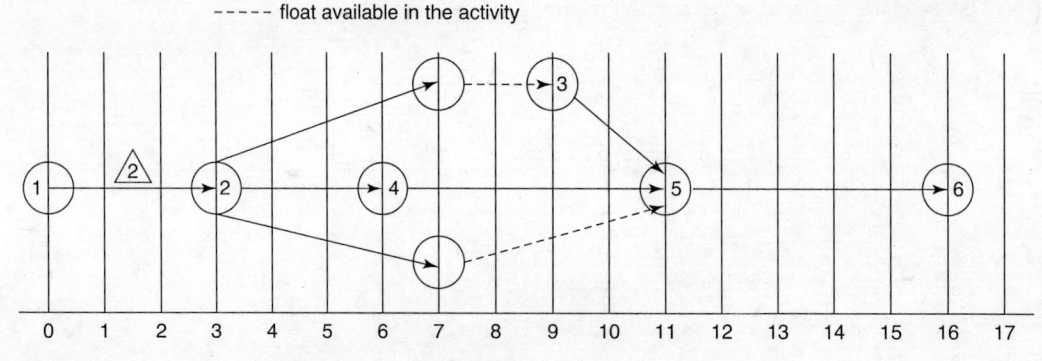

Figure 7.5 Network of Figure 7.1 redrawn to time scale

Day 3

Activity (1, 2) continues. Requirement—2, availability—8, remaining—6.

The time-scale network for days 1–3 is shown in Fig 7.6.

Day 4

The three activities (2, 3), (2, 4) and (2, 5) can be started on day 4 (see Fig 7.7). Thus, resource requirement would become 3 + 4 + 4 = 11, which is more than 8, the maximum limit. The decision rule 'activity with least float gets priority' for resource assignment comes into play here.

Activity (2, 4) lies on the critical path, i.e., float = 0, and hence, resources need to be allocated to it first. The activity (2, 4) requires 4 units of resources, while availability is 8; thus, resources remaining after assigning to (2, 4) are 4 (8 − 4 = 4). Next in queue is activity (2, 3) with float = 2 days. Thus, resources are assigned to (2, 3). The requirement of this activity is 3, while now the availability is 4; thus, 1 resource is left out after this allocation.

Activity (2, 5) is left out since it needs 4 resources but the available resource is just 1; hence, we need to postpone this activity. Also, the float of this activity is reduced to 3 now.

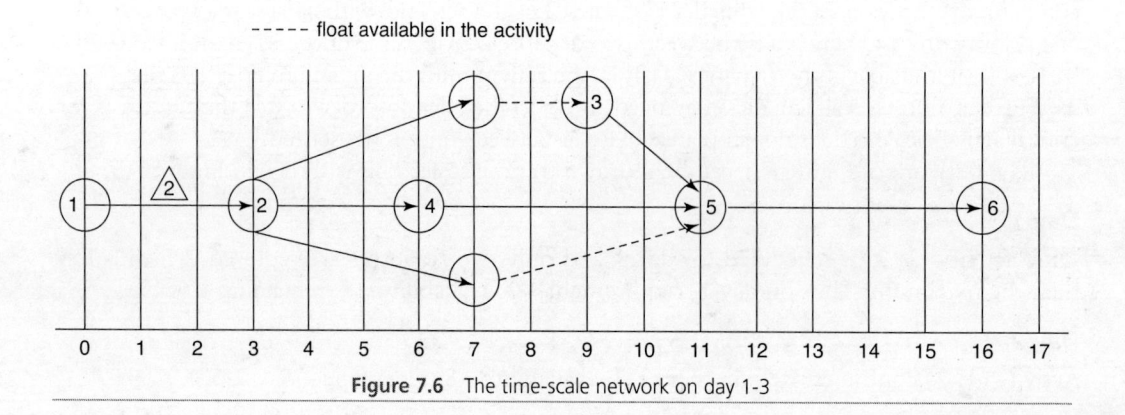

Figure 7.6 The time-scale network on day 1–3

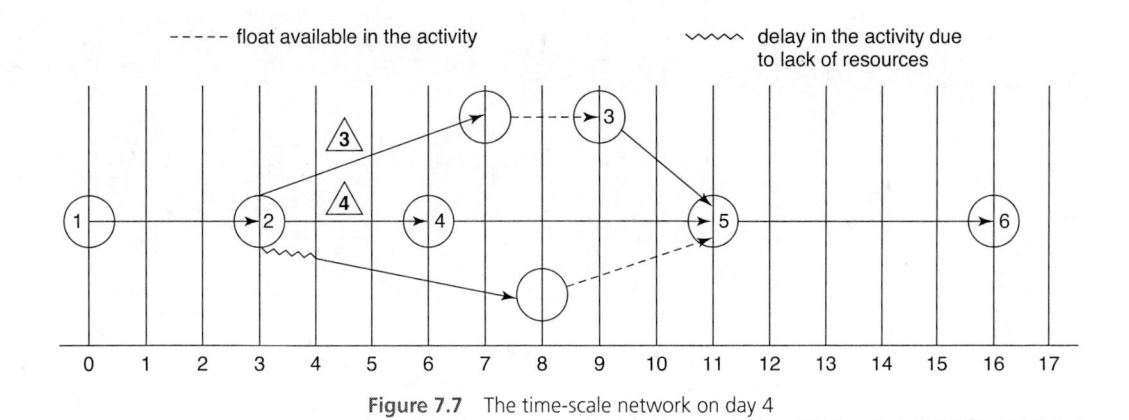

Figure 7.7 The time-scale network on day 4

Day 5

Since according to our 'decision rule', the activities in progress cannot be stopped in-between, the resources are allocated to them first. Thus, the resource assignment on day 5 is as below:

Activity (2, 4): requirement—4, availability—8, remaining—4
Activity (2, 3): requirement—3, availability—4, remaining—1
Activity (2, 5), again, has to be postponed because of the unavailability of resources. Also, the float of this activity is reduced to 2 now (see Fig 7.8).

Day 6

Based on the previous argument that the activities in progress cannot be stopped in-between, resources must be allocated to them first. We have the following resource assignment for day 6:

Activity (2, 4): requirement—4, availability—8, remaining—4
Activity (2, 3): requirement—3, availability—4, remaining—1
Activity (2, 5) has to be postponed because of unavailability of resources. The float remaining for this activity is 1 day (see Fig 7.9).

Day 7

Since activities in progress cannot be stopped in-between, resources must be allocated to them first. Thus, resource assignment for day 7 is as below:

Activity (2, 3): requirement—3, availability—8, remaining—5

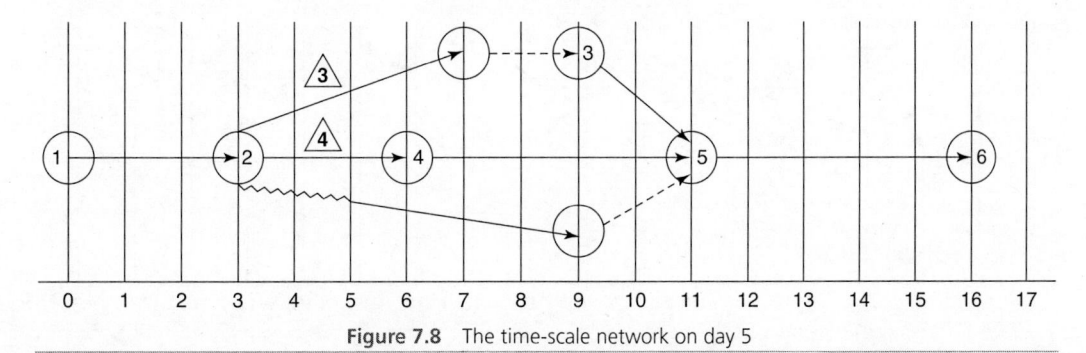

Figure 7.8 The time-scale network on day 5

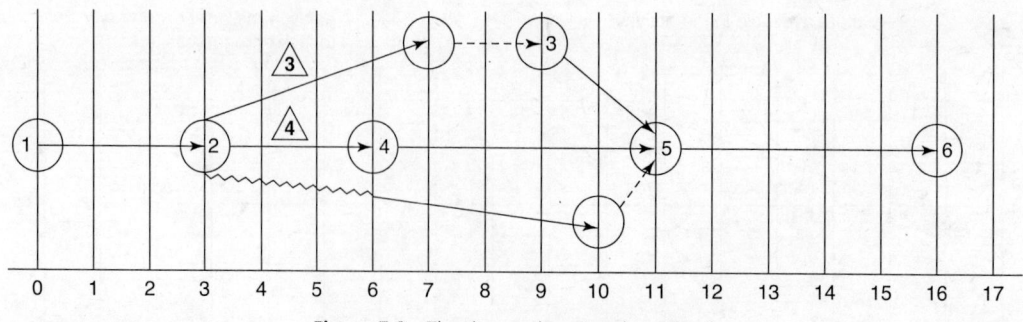

Figure 7.9 The time-scale network on day 6

Activities (4, 5) and (2, 5) both compete for the remaining resources. The activity (4, 5) requires 3 resources while activity (2, 5) requires 4 resources. Thus total requirement for these two activities is 7 as against the remaining 5 resources. According to the algorithm, the first preference would be given to activity (2, 5) since it has an earlier start time compared to activity (4, 5). This is despite activity (4, 5) being on the critical path.

Requirement for activity (2, 4)–4, availability—5, remaining—1

Activity (4, 5) has to be postponed because of the unavailability of resources. This will increase the project duration by one day (see Fig 7.10). Thus the project duration has become 17 days now. Also note the 1 day float available now for activity (3, 5).

Day 8
Activity (2, 5) continues. Requirement—4, availability—8, remaining—4

Activity (4, 5) can be taken up now. The resources required are 3 and the availability is 4. The remaining 1 resource remains unutilized.

Day 9
Activities (2, 5) and (4, 5) continue.

Activity (2, 5): Requirement—4, availability—8, remaining—4
Activity (4, 5): Requirement—3, availability—4, remaining—1

The time-scale network for days 8-9 is shown in Fig 7.11.

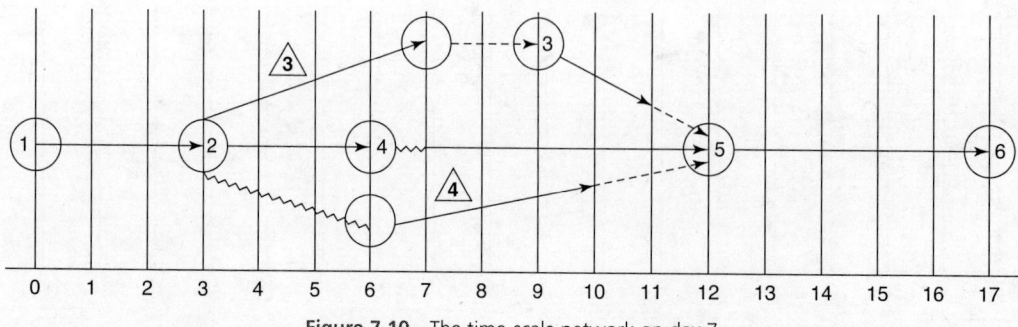

Figure 7.10 The time-scale network on day 7

Figure 7.11 The time-scale network on day 8–9

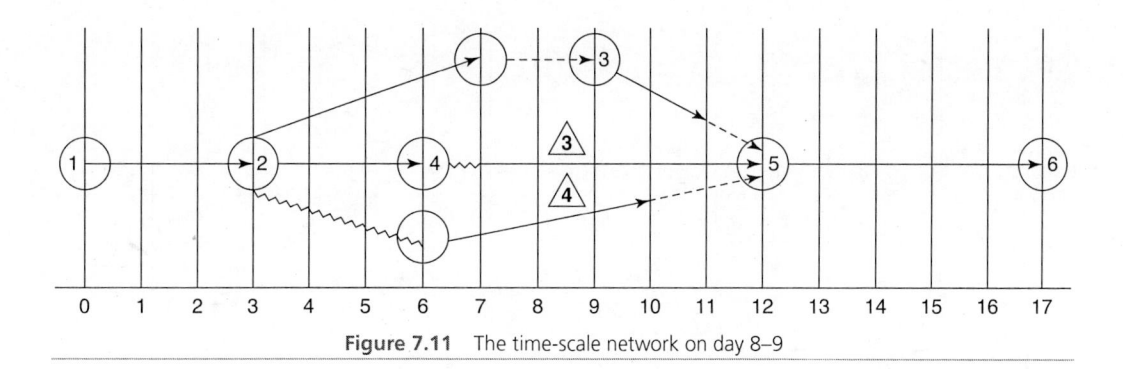

Day 10

Since activities in progress cannot be stopped in-between, resources must be allocated to them first.

Activity (2, 5): Requirement—4, availability—8, remaining—4
Activity (4, 5): Requirement—3, availability—4, remaining—1
Activity (3, 5) cannot be taken up on this day since it requires 3 resources as against the remaining 1 resource.

Thus it has to be delayed which would exhaust the 1 day float available with this activity (see Fig 7.12).

Day 11

Activity (2, 5) is completed.

Activity (4, 5) continues. Requirement—3, availability—8, remaining—5
Activity (3, 5) can be taken up now. Requirement—3, availability—5, remaining—2

Day 12

Activities (4, 5) and (3, 5) continue.

Activity (4, 5) : Requirement—3, availability—8, remaining—5
Activity (3, 5): Requirement—3, availability—5, remaining—2

The time scale network for days 11–12 is shown in Fig 7.13.

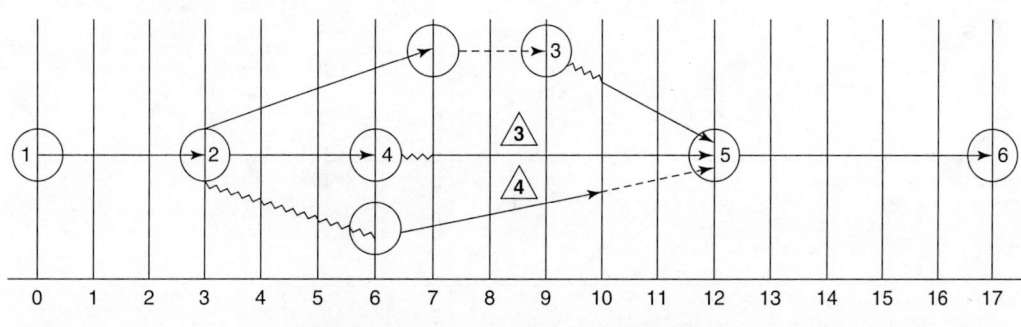

Figure 7.12 The time-scale network on day 10

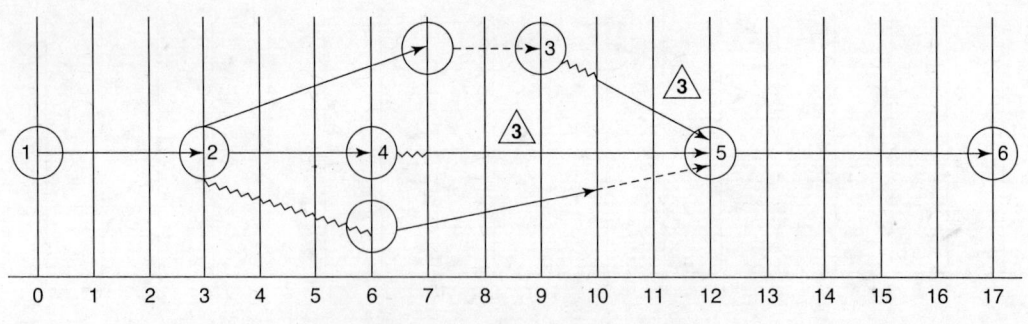

Figure 7.13 The time-scale network on day 11-12

Day 13
Only activity (5, 6) requires resource.

    Activity (5, 6): Requirement—4, availability—8, remaining—4

Day 14
Activity (5, 6) continues.

    Activity (5, 6): Requirement—4, availability—8, remaining —4

Day 15
Activity (5, 6) continues.

    Activity (5, 6): Requirement—4, availability—8, remaining —4

Day 16
Activity (5, 6) continues.

    Activity (5, 6): Requirement—4, availability—8, remaining —4

Day 17
Activity (5, 6) continues.

    Activity (5, 6): Requirement—4, availability—8, remaining —4

The time-scale network for days 13–17 is shown in Fig 7.14.

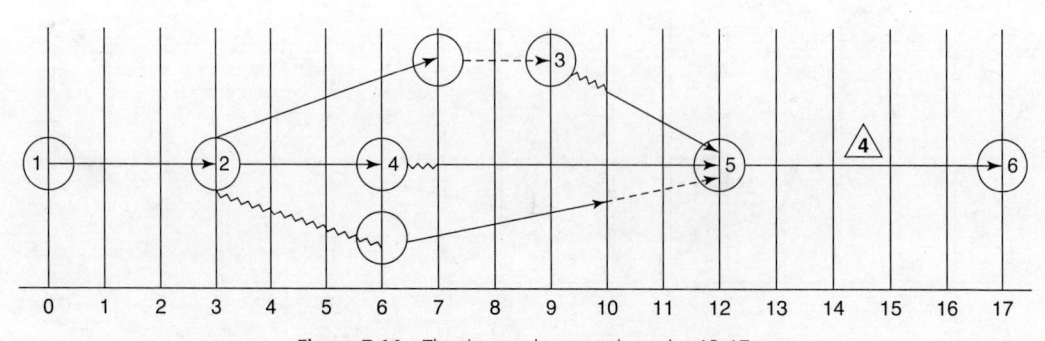

Figure 7.14 The time-scale network on day 13-17

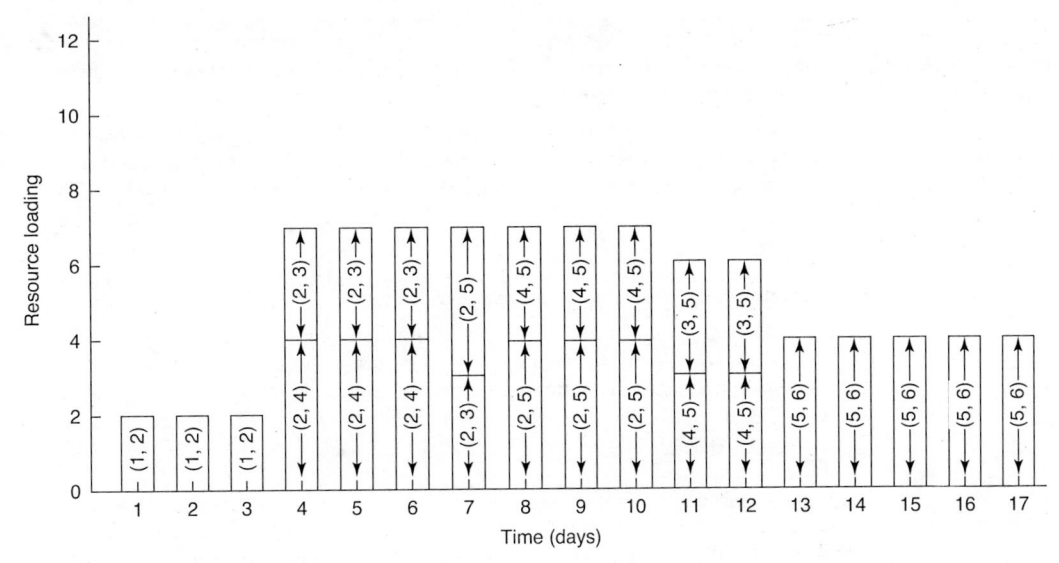

Figure 7.15 Resource-allocation chart on different days (non-time constrained /resource constrained leveling)

Thus, we observe that the project duration has been increased by 1 day from the original 16 days, with no resource constraint to 17 days with resource limit of 8. The resource requirement varies from 2 to 7. The resource-loading chart on different days (day 1 to day 17) for the total project duration is shown in Figure 7.15.

The steps illustrated above can be carried out manually only if the project is small. For large projects involving multiple resources, we need to look to computer programs that have been illustrated in Chapter 18.

The examples presented for resource levelling and resource allocation are that of a very simplistic nature. First of all, we have considered only a single type of resource used for all types of activities. Further, we have considered that the resource is interchangeable across different activities. Unfortunately, in real-life situations, both these assumptions do not hold true. In practice, we have to deal with different types of resources such as equipment, materials, labours, etc., and further, a single activity may require more than one type of resource. Such complicated problems are solved using computer programs and are discussed in detail in Chapter 18.

7.4   IMPORTANCE OF PROJECT SCHEDULING

The success of a project heavily depends on how effective the scheduling is and how tightly the project can be controlled. Poor scheduling can easily result in completion delays and cost overruns. These, in turn, result in claims and counter-claims, disagreements and disputes. The appropriate method should be used in planning and scheduling a project; management decisions should not depend on experience and intuition alone.

Preparing project schedule is extremely important due to a number of reasons. Project schedule is the basis of extracting a number of other schedules and information. Even at the tendering stage, project schedules need to be prepared although there may not be required information for preparing project schedules. Project schedules are helpful not only for a

Table 7.4 Project schedule preparation steps in general

Identification and classification of tasks	This is the first critical step wherein all individual tasks or activities required to complete the project are listed out.
Defining duration of a task	This is essentially the first iteration at estimating the time durations involved in completing the different tasks. All available information should be used to make an accurate estimate of the duration of each task, and depending upon the nature of the project and the activity, the estimated time may be given in hours, days, or months.
Task sequence	Activities in a construction project may appear to be independent, but often depend upon each other—for example, the painting in a house is done by an independent agency, but the work can commence only *after* the plastering of the walls is completed. Continuing with the example of house construction, there is obvious interdependence in electrical, plumbing and finishing works. Although it appears trivial, such issues of 'inherent' logic and 'predecessor' and 'successor' tasks should be sorted out at the outset to reduce the possibility of difficulties later on. In larger projects, it also helps to identify tasks that can be completed or carried out at the same time. Such tasks, or others that require careful handling, may be 'flagged' at the outset and special attention paid to their progress and any possible slippage. To that extent, it may be noted that an easy-to-understand depiction of relationships between different activities is an important function of the project planner. Different methods of such representations, such as a network diagram, are available and discussed in greater detail elsewhere.
Creating a calendar of events	Once the list of activities, their interdependence and the time involved in completing them is known from the steps above, a calendar of events can be easily created highlighting the milestone events.

contractor, but for the owners as well. In order to draw up a comprehensive schedule, the steps mentioned in Table 7.4 must be systematically followed.

For the contractors, project schedules help in timely mobilization of the required resources and help in identification of bottlenecks in the early stages of project duration. They also act as reference for comparison with actual progress, cost of construction, and profit margin at any given time during the course of monitoring. For the owner also, project schedule helps in understanding the requirement of finance to be mobilized. Also, with the project schedule in place, it is easy for the owner/client to monitor the project accordingly. The detailed steps carried out to prepare a project schedule by a contractor are listed below:

- Identify and study the scope of work from the contract document (tender drawings, bill of quantities, terms and conditions, etc.). Contractors normally do not have access to all the details needed to prepare an exhaustive schedule at the beginning, but that should not be an excuse for non-preparation of schedule for the project. In fact, contractors should try to extract as much information as possible, even by unofficially meeting the owner's representative and architects for the project. Owners try not to give information based on incomplete documentation on their part, for the fear of getting penalized at a later date.
- Decide on the construction methodology for completing the project within the given time schedule. The time schedule specified in the tender document mostly governs the construction methodology adopted by the contractor.

- The total project scope is finally brought down into work packages and activities. This breakdown into smaller activities helps the planner in estimating resource requirement and duration for completing the activities. The productivity norms as well as the experience of the planner are utilized to arrive at the estimates of duration and resource requirement for these activities.
- Finally, using the relationships existing among different activities in the form of a network for the project is prepared. The network can be prepared by using either PERT/CPM or the precedence network.
- From the network, a week-wise or month-wise project schedule is derived.
- In large projects involving a number of subprojects, a master project schedule is prepared in addition to preparing schedules for each of the subprojects.
- Finally, it should be kept in mind that these schedules need to be revised as and when any new information is available or at a fixed interval such as fortnightly or monthly.

7.5 OTHER SCHEDULES DERIVED FROM PROJECT SCHEDULES

From a project schedule, the following schedules can be prepared:

1. Invoice schedule
2. Cash inflow and cash outflow schedule
3. Staff schedule
4. Labour schedule
5. Material schedule
6. Specialized subcontractor schedule
7. Plant and equipment schedule
8. Working capital schedule
9. Estimation of direct and indirect costs

These are explained with the help of an example. Let us assume that a contractor has been awarded a contract for the expansion of an automobile factory. The contract value is five crore rupees and the contract duration is 10 months from the date of receipt of letter of intent (explained in later chapters). The bill of quantities for the project is given in Table 7.5.

The schedule of the project (in bar chart form) from the project network prepared by the contractor is given in Figure 7.16. The schedule shows the thirteen items from 1 to 13. The durations for each of the items are given in months. For example, earthwork is scheduled to take 6 months, from month 1 to month 6.

The figure also gives the break-up of quantities in percentage terms on a monthly basis. For example, the quantities of earthwork planned to be executed from month 1 to month 6 are 5 per cent, 15 per cent, 20 per cent, 25 per cent, 25 per cent and 10 per cent, respectively. That is, 5 per cent of 5,000 m^3 is planned for month 1, 15 per cent of 5,000 m^3 is planned for month 2, and so on. The percentage break-up is given for all the items, which can be interpreted in a similar manner.

Table 7.5 Bill of quantity for the expansion of automobile factory

S. No.	Item description	Unit of measurement	Quantity	Unit rate (₹)	Amount (₹ lakh)
1	Earthwork—all soils	m³	5,000	100	5
2	Concrete works	m³	2,000	4,000	80
3	Formwork	m²	14,000	300	42
4	Reinforcement works	t	250	45,000	112.5
5	Brickwork	m³	1,700	2,000	34
6	Plastering—all types	m²	20,000	75	15
7	Painting—all types	m²	5,000	100	5
8	Flooring—all types	m²	LS	LS	110
9	Waterproofing works	m²	LS	LS	12
10	Aluminium work	m²	300	4,500	13.5
11	Electrical work	Pkg	1	LS	35
12	Sanitary and plumbing works	Pkg	1	LS	20
13	Road works	m²	LS	LS	16
Total amount					**500**

S. No.	Item Description	Duration (months)	Time in months (1–10) — breakup
1	Earthwork – all soils	6	5%, 15%, 20%, 25%, 25%, 10% (months 1–6)
2	Concrete work	6	10%, 20%, 20%, 25%, 15%, 10% (months 2–7)
3	Form work	6	10%, 20%, 20%, 25%, 15%, 10% (months 2–7)
4	Reinforcement works	6	10%, 20%, 20%, 25%, 15%, 10% (months 2–7)
5	Brickwork	7	10%, 15%, 20%, 20%, 20%, 10%, 5% (months 3–9)
6	Plastering – all types	7	10%, 15%, 20%, 20%, 20%, 10%, 5% (months 4–10)
7	Painting – all types	7	10%, 15%, 20%, 20%, 20%, 10%, 5% (months 4–10)
8	Flooring – all types	7	10%, 15%, 20%, 20%, 20%, 10%, 5% (months 4–10)
9	Waterproofing works	2	50%, 50% (months 6–7)
10	Aluminium works	4	25%, 25%, 25%, 25% (months 5–8)
11	Electrical works	5	25%, 25%, 25%, 15%, 10% (months 6–10)
12	Sanitary and plumbing works	8	5%, 10%, 15%, 20%, 15%, 15%, 10%, 10% (months 3–10)
13	Road works	2	50%, 50% (months 6–7)

Figure 7.16 The bar chart and break-up of planned quantities of different items in percent

7.5.1 Preparing Invoice Schedule

Invoice is the amount of money realized by executing a particular activity or work. Accordingly, invoice value is the product of the quantities of work done each month for a particular activity and the corresponding rates. The rates are nothing but the quoted rates by the contractor. It must be kept in mind that the invoice schedule is the estimated billing amount on a monthly or weekly basis. The actual invoice may be exactly as per the planned one or it may be ahead or behind the estimated figure. This schedule also acts as the basis for monitoring progress of invoicing. There is a limitation, however, in using this schedule to monitor physical progress of a project, as completion dates of individual activities/milestones cannot be indicated herein.

The steps involved in the preparation of invoice schedule are as follows:

1. From the quantities of items/activities taken from bill of quantities, a month-wise or week-wise quantities to be executed every month is prepared corresponding to the project schedule. In Figure 7.16, the month-wise break-up of quantities (in percent terms) for each activity is shown above the bar corresponding to that activity. For example, for earthwork—all soils activity, the monthly break-up is 5 per cent, 15 per cent, 20 per cent, 25 per cent, 25 per cent and 10 per cent corresponding to months 1, 2, 3, 4, 5 and 6, respectively. One can also assume a uniform distribution of quantities for simplicity, although it may not be applicable in practice. For example, 833.33 m^3 (16.67 per cent of 5,000 m^3) of excavation may be assumed to get executed every month to complete the total quantity of 5,000 m^3 in 6 months.

2. Based on the quantities and unit rates (to be obtained from quoted rates) for each month, calculate invoicing amounts for all the activities. For large projects involving a number of items or activities, only major items need to be taken into account for this exercise. The remaining items can be clubbed together as other items. However, quantities for such items need not be mentioned. For projects having subprojects, a consolidated invoice schedule for the project can be prepared by combining the invoice schedules of subprojects.

3. As discussed elsewhere, one of the major problems that may be faced in the preparation of an invoice schedule is insufficient details in the drawings and bills of quantities available at the time of scheduling. For example, the item-wise split-up of quantities (in terms of concrete, shuttering, steel, etc., in columns, footings, floors) may not be available. In such cases, reasonable assumptions and past experience are utilized to arrive at the quantities.

4. The above steps are easy to carry out for an item rate contract. But for other types of contract such as lump-sum contracts, split-up of lump-sum price for each stage of payment will be taken into account since payment may not be based on quantities of work done.

5. The total invoicing figure will also include escalations as per accepted terms in the contract. Further, a distinction should be made between the invoice figure and the actual payment due. The entire bill amount is not payable to the contractor. There are only certain percentages of the total work done that are paid to the contractor. The rest of the money is kept either as retention money or as certain deductions in lieu of the advances drawn earlier by the contractor and certain taxes. The percentage payment and deduction applicable are as per the agreed terms of payment and should be taken into account in arriving at invoicing figures.

The month-wise break-up of quantities for all the activities are given in Table 7.6, and the invoice schedule is given in Table 7.7 for the case project.

Table 7.6 Distribution of quantity planned to be executed activity-wise on a monthly basis

Item description	M1	M2	M4	M5	M6	M7	M8	M9	M10
Earthwork—all soils	250	750	1,250	1,250	500				
Concrete works		200	400	500	300	200			
Formwork			1,400	2,800	3,500	2,100	1,400		
Reinforcement works			25	50	62.5	37.5	25		
Brickwork				255	340	340	340	170	85
Plastering—all types				2,000	3,000	4,000	4,000	4,000	2,000	1,000
Painting—all types				500	750	1,000	1,000	1,000	500	250
Flooring—all types										
Waterproofing works										
Aluminium work					75	75	75	75		
Electrical work										
Sanitary and plumbing works										
Road works										

7.5.2 Schedule of Milestone Events

Milestone events are very important in the proposed schedule of a project. Top management is more concerned with the scheduled completion of a milestone event. Some examples of milestone events are—completion of foundation, completion of superstructure, and completion of sanitary and plumbing works. Thus, from the schedule of the project, the details of milestone

Table 7.7 Estimated bill value or invoice value on weekly basis for the case project (values in ₹. lakh)

Item description	M1	M2	M3	M4	M5	M6	M7	M8	M9	M10
Earthwork—all soils	0.25	0.75	1.00	1.25	1.25	0.50				
Concrete works		8.00	16.00	16.00	20.00	12.00	8.00			
Formwork			4.20	8.40	8.40	10.50	6.30	4.20		
Reinforcement works			11.25	22.50	22.50	28.125	16.875	11.25		
Brickwork				3.40	5.10	6.80	6.80	6.80	3.40	1.70
Plastering—all types					1.50	2.25	3.00	3.00	3.00	1.50	0.75
Painting—all types					0.50	0.75	1.00	1.00	1.00	0.50	0.25
Flooring—all types					11.00	16.50	22.00	22.00	22.00	11.00	5.50
Waterproofing works								6.00	6.00		
Aluminium work						3.375	3.375	3.375	3.375		
Electrical work							8.75	8.75	8.75	5.25	3.50
Sanitary and plumbing works				1.00	2.00	3.00	4.00	3.00	3.00	2.00	2.00
Road works							8.00	8.00			
Total	0.25	24.20	52.30	68.25	92.55	92.60	85.375	50.525	21.95	12.00
Cumulative	0.25	24.45	76.75	145.00	237.55	330.15	415.525	466.05	488.00	500.00

S. No.	Description of milestone events	Estimated date of completion	Projected slippage in weeks	Description of impact of problems	Action to be taken	Person responsible	Latest estimated project completion date	Projected slippage in weeks
1	Completion of foundation							
2	Completion of superstructure							
3	Completion of sanitary and plumbing works							

Figure 7.17 Schedule of milestone events

events are separated and reported in the schedule of milestone events. The slippage, if any, and its financial implications are closely monitored, and the impact of slippage on other activities is also studied. The corrective actions to be taken to bring the slipped milestone event back on track are proposed. The person responsible for correcting the slippage is also identified in such reports. A typical format for preparing this report is given in Figure 7.17.

The schedule of milestone events has gained importance these days as any slippage from the scheduled completion date of a milestone event can become the cause for invoking 'liquidated damages' clause by the owner. Thus, from the perspective of a contractor, this schedule is very important.

7.5.3 Schedule of Plant and Equipment

In this schedule (see Table 7.8), the date of particular plant and equipment is also mentioned. The release date of a plant and equipment that is surplus at project site is mentioned as well. Finally, the time required for doing pre-dispatch maintenance is mentioned.

Table 7.8 Schedule of plant and equipment for the case project

S. No.	Plant and equipment description	Nos required	Requirement From	Requirement To	Remarks
1	Earth rammer of capacity 8 tonnes	2	M6	M7	Required for 2 months
2	Concrete mixer 500 l/5 m³ capacity	1	M2	M7	
3	Builders hoist	2	M3	M7	
4	Front-end dumper	1	M1	M10	
5	Theodolite with optical plummet	1	M1	M10	
6	Tilting/dumpy level	2	M1	M10	
7	Compression testing machine 100 t	1	M2	M7	
8	Bar-bending machine	1	M2	M7	
9	Bar-shearing machine	1	M2	M7	
10	Wood surface planner—horizontal	1	M2	M7	
11	Water pump centrifugal multistage	1	M2	M10	
12	Mobile diesel compressor	1	M2	M10	

Table 7.9 Schedule of project staff

S. No.	Staff category	Nos required	Requirement From	To	Remarks
1	Project manager	1	M1	M10	
2	Resident engineer	1	M1	M10	
3	Civil engineer	1	M2	M10	
4	Jr engineer—electrical	1	M6	M10	
5	Jr engineer—civil	1	M2	M7	For concreting activities
6	Jr engineer—civil	1	M6	M7	For road works
7	Foreman	1	M2	M7	For formwork activities
8	Foreman	1	M2	M7	For concreting activities
9	Accountant	1	M1	M10	
10	Time keeper	1	M1	M10	

7.5.4 Schedule of Project Staff

Depending on the invoicing to be done per month, a month-wise schedule for project staff is maintained (see Table 7.9). In this schedule, the date by which a particular staff is required is also mentioned. The release date of a particular category of staff that is surplus at project site is to be mentioned as well.

The number of staff is dependent on the nature of activity, the quantum of work, the working hours adopted by the organization, the proportion of work which the main contractor is planning to subcontract, etc. As discussed in Chapter 1, a large number of project staff is needed for performing different activities. Some of the staffs are specialized in some areas such as planning, billing, supervision, and so on.

7.5.5 Schedule of Labour Requirement

A typical format of labour requirement schedule is shown in Figure 7.18. From the split-up of quantity, labour requirement is calculated. The requirement of labour is estimated for each activity. The number of labourers is dependent on the quantity involved in an activity and the productivity of labourers. While the productivity aspect has been dealt with separately elsewhere in the book, it has been referred to for illustration here. Thus, in our example, the quantity of earthwork involved in month 1 is 833.33 m^3. If the productivity of labourers for this work is assumed to be 3 man hours per m^3, the total man hours required for this activity alone would be 2,500 man hours. If it is assumed that each day consists of

S. No.	Description (labour category)	Total man-months	Month 1	Month 2	Month 3	...	Month 6	Remarks
1	Unskilled							
2	Semi-skilled							
3	Skilled							

Figure 7.18 Schedule of labour requirement

Table 7.10 Schedule of materials requirement for the case project

S. No.	Description	Unit	Qty	M1	M2	M3	M4	M5	M6	M7	M8	M9	M10	Remarks
1	Cement	Bags	17,8251*		10	20	20	25	15	10				[2000 × 6.5 + 1700 × 1.25 + 20000 × 0.109] × 1.03
2	Sand—coarse, fine and filling	m^3	3,010											[2000 × 0.60 + 1700 × 0.66 + 20000 × 0.030] × 1.03
3	Aggregates—40, 20, 12/10, 6 mm and above 40 mm for roads	m^3	1,854											[2000 × 0.9] × 1.03
4	Bricks—75 class designation	Each	850,000											1700 × 500 including wastage
5	Reinforcement—TMT bars	t	258											250 × 1.03

[1]Cement is required for concreting, brickwork and plastering. For concrete the per-m^3 requirement is assumed to be approximately 6.5 bags; for one m^3 of brickwork (with 1:6 cement mortar) it is assumed to be 1.25 bags; and for plastering (1:4 cement mortar) it is assumed to be 0.109 bags per sqm. The exact requirement would come from the design mix. Wastage factor of 3 per cent is also added on top of this to get the total requirement.

10 working hours, and there are 25 working days, the number of workers required for this activity alone would be 10 (2500/10/25) for month 1. The requirement of workers for other months for the same activity is also computed in a similar manner.

A similar exercise is carried out for other activities also. The monthly requirement of labourers is obtained by summing up the requirement for each activity planned for a month. Care should be taken that computations for different types of workers (for example, unskilled, semi-skilled and skilled) are carried out separately. This way, one can find out the requirement for workmen of different categories and present the same in the form of a schedule.

7.5.6 Schedule of Materials Requirement

Based on the quantity of a particular item, the type and quantity of materials to be purchased are established. The schedule helps the project manager to assess the timing at which a particular material is desired for the project. The material requirement as a function of time shown month-wise in the schedule (could be any other measurement of duration such as week) is predetermined based on the bar chart of the project. The sample schedule of materials requirement is provided in Table 7.10 for the case project.

For example, if the material for the item brickwork is required to be calculated, we find that the number of bricks required in months 3, 4, 5, 6, 7, 8 and 9 are 85,000, 127,500, 170,000, 170,000, 170,000, 85,000 and 42,500, respectively. This is on an assumption that 500 bricks are needed for 1 m^3 of brickwork. Such predetermined schedules help a project

Table 7.11 Schedule of specialized agencies

S. No.	Item description	Unit	Qty	Date of start	Date of completion	Remarks
1	Waterproofing works			M7	M8	
2	Aluminium works			M5	M8	
3	Sanitary and plumbing works			M3	M10	
4	Road works			M6	M7	

manager in taking the procurement action well in time, thus realizing advantages such as discounts from the material supplier.

If, on the other hand, material is procured on an emergency basis, chances are that one may compromise on quality besides getting the material at higher prices.

7.5.7 Schedule of Specialized Agencies

In a project, a contractor may not be in a position to do all the activities in an economical manner. Further, for some of the activities in a project, he may not have the requisite expertise. In such cases, the contractor enters into a contractual arrangement with some specialized agencies called subcontractors or speciality contractors. As a first step, a list of activities for which the contractor wants to engage subcontractors is prepared. The scheduled dates of start and completion of such activities are noted. Let us assume that the contractor wants to engage subcontractors for activities 'waterproofing', 'aluminium work', 'sanitary and plumbing works', and 'road works'. The scheduled starts for these works are shown in Table 7.11, which has been replicated from Figure 7.16.

Looking at the schedule, the contractor would know the latest time to start the process of finalization of contractor for a particular activity. For example, if the contractor takes one month to finalize the deal with his subcontractors, he would start the process of subcontractor selection for 'waterproofing' at the start of month 6, while the selection process for 'aluminium work', 'sanitary and plumbing works', and 'road works' would be initiated at the start of months 4, 2 and 5, respectively.

There are other types of schedules that can be prepared based on the project schedule. These are schedule of direct costs, schedule of overheads, schedule of cash inflows and outflows, etc. Typical formats for all these schedules have been given in Figures 7.19 to 7.22. These have been described at appropriate places in detail.

7.5.8 Schedule of Direct Costs

S. No.	Item description	Unit	Total qty	Unit cost	Total cost	Month 1	...	Month 6	Remarks
1									
2									
3									
	Total								

Figure 7.19 Schedule of direct costs

7.5.9 Schedule of Overheads

S. No.	Description	Total amount	Month 1	Month 2	...	Month 6	Remarks
1	Salaries for staff						
2	Conveyance at site						
3	....						
	Total of overheads						
	Overheads cumulative total						
	Direct costs (cumulative)						
	Invoicing (cumulative)						
	Site contribution						

Figure 7.20 Schedule of overheads

The site contribution listed in Figure 7.31 is equal to invoicing minus overhead minus direct costs.

7.5.10 Schedule of Cash Inflow

S. No.	Description	Month 1	Month 2	...	Month 6	Remarks
1	Invoicing					
2	Progressive payments (A)					
3	Advance cumulative (B)					
4	Gross amount due (C) = (A + B)					
5	Less: Recoveries					
6	Advance (a)					
7	Cash retention (b)					
8	Income tax (c)					
9	Cement (d)					
10	Steel (e)					
11	Water Supply (f)					
12	Electricity (g)					
13	Other services (h)					
14	Total recoveries (D) = (a + b + c + d + e + f + g + h)					
15	Net amount due = (C − D)					
16	Add release of retention (X)					
17	Total inflow (E) = (C − D + X)					

Figure 7.21 Schedule of cash inflow

7.5.11 Schedule of Cash Outflow

S. No.	Description	Month 1	Month 2	...	Month 6	Remarks
1	Departmental labour (a)					
2	Subcontractor labour (b)					
3	Specialized contractor (c)					
	Material purchase (d)					
	Tools and tackles (e)					
	Plant and equipment (f)					
	Site overheads (g)					
	Temporary structures and site facilities (h)					
	Staff expenses (i)					
	Administrative expenses (j)					
4	Total outflow (F) = (a + b + c + d + e + f + g + h + i + j)					
5	Net inflow = (E − F)					
6	Cumulative inflow					

Figure 7.22 Schedule of cash outflow

7.6 NETWORK CRASHING AND COST-TIME TRADE-OFF

In the previous chapter, we discussed the computational procedures of critical path method. In this section, we discuss one of the application areas of this network technique. One of the assumptions in critical path method is that the duration of an activity can be reduced or crashed to a certain extent by increasing the resources assigned to it. As is known, the execution of an activity involves both direct costs and indirect costs. These terminologies are explained in detail elsewhere.

It is known that any reduction in duration (by increasing resources) of critical path activities can reduce the project duration and, thereby, enhance the possibility of reduction in project cost. However, as will be explained shortly, there is no point in attempting to crash all the activities by increasing the resources.

An activity can be performed at its normal or most efficient pace or it can be performed at higher speed. The duration associated with the former is called 'normal duration' and the duration associated with the latter is called 'crash duration'.

Some activities along the critical path sometimes need to be shortened in order to reduce the overall duration of the project. This leads to a decrease in the indirect expenses (due to decrease in duration), as shown in Figure 7.23, and an increase in the direct expenses (due to more mobilization of resources), as shown in Figure 7.24. As is evident from Figure 7.23 and Figure 7.24, the relationship between the cost of the job and the duration has been assumed to be linear. The steeper the slope of the line, the higher the cost of expediting the job at an earlier date.

The expediting of an activity to an earlier time is referred to as crashing. There are three cases that normally arise:

Figure 7.23 Indirect cost vs time

Figure 7.24 Direct cost vs time

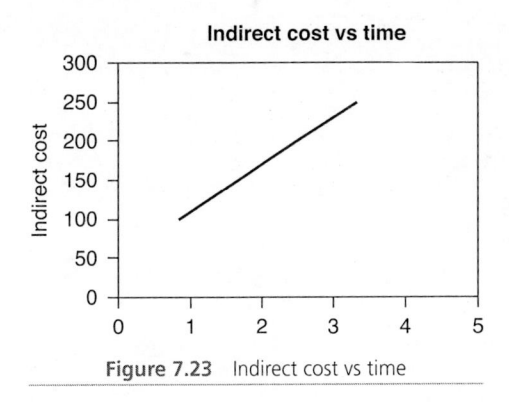

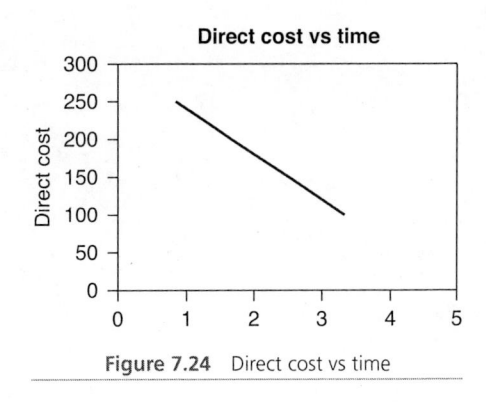

1. Line sloping down to the right—The steeper the slope, the higher the cost of crashing
2. Horizontal line—There would not be any cost of crashing
3. Vertical line—The activity cannot be shortened regardless of the extra resources applied to it

Figure 7.25 depicts the three possible cases.

Some important inferences drawn from Figure 7.25 are:

1. There is a point beyond which there can be no decrease in direct costs.
2. Slowing the project even further may only result in the increase in expenditure (last portion of the graph).

The point of the minimum cost of project is known as the optimum point. In order to find the optimum point, the project network is drawn based on the normal duration of the activities. This is the 'maximum length' schedule. The duration of the project is thus noted. It can be shortened by expediting jobs along the critical path. If the added cost of expediting the job is less than the saving in the indirect expenses which result from shortening the project, then a less expensive schedule can be found. New schedules are found as long as there is a reduction in the cost of the project.

At each step of the process, only the activities along the critical path are considered for crashing. The cost-time slope of each activity is examined and the activity with the least slope determined. This is the activity that can be shortened with the least increase in direct costs. If the increase in the direct costs is less than the savings in the indirect costs, then the activity is

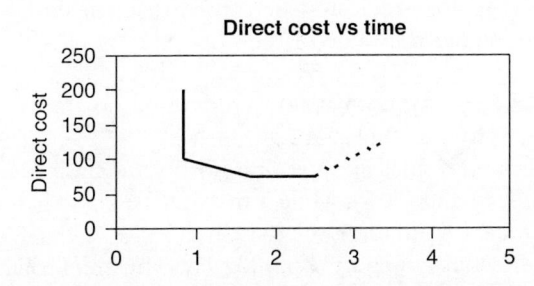

Figure 7.25 The three possible cases in 'direct cost vs time'

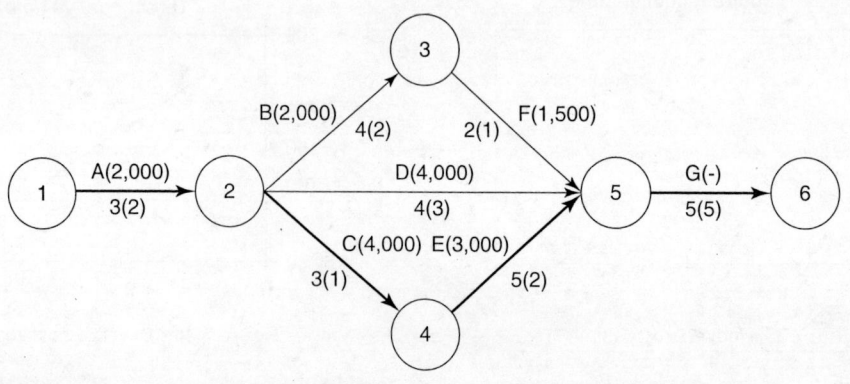

Figure 7.26 Network used for illustrating crashing

crashed up to the point where no further shortening of the schedule is possible (either because the activity duration cannot be reduced further or some other activity has become critical along a parallel path). The remaining critical activities are examined and the one with the flattest cost slope value is selected. The process is repeated until no further shortening of critical activities is possible or until the costs of such shortening would exceed the savings that result from reducing the project length.

We illustrate the process of crashing by the network shown in Figure 7.26. Suppose that the project consists of seven activities—A, B, C, D, E, F and G. As evident from the network, the critical path is 1-2-4-5-6 consisting of activities A, C, E and G, with project duration of 16 days. Please note that the 16 days duration is based on the normal duration.

The normal and crash duration of each activity is given under the arrow. The cost of crashing per day of all activities is given in the parentheses of all activities adjacent to the activity name. For example, the normal cost of activity A is 3 days while the crash duration is 2 days, and the crash cost per day is ₹ 2,000. The crash cost per day is given by the cost slope of an activity, which is calculated from the information on normal duration, crash duration, normal cost and crash cost. The cost slope has been computed from the expression—

$$\text{Cost slope} = \frac{(\text{Crash cost} - \text{Normal cost})}{(\text{Normal duration} - \text{Crash duration})} \quad (7.1)$$

In other words, the cost slope indicates the extra cost incurred by reducing the activity duration by one day. The higher the value of the slope, the more is the cost of crashing the activities.

The details of normal duration, normal cost, crash duration and crash cost of each of the activities of the project are given in Table 7.12.

As can be observed from this small example, it is possible to crash some activities such as (1, 2), (2, 3), (2, 4), (2, 5), (4, 5) and (3, 5), while it may not be possible for some activities such as (5, 6) (both normal and crash durations are same here).

Suppose that the indirect expenses are ₹ 6,000/day. Thus, the total project cost if all the activities are executed at their normal pace is given by:

Table 7.12 Details of activity cost and duration

Activity	Normal Duration	Normal Cost (₹)	Crash Duration	Crash Cost (₹)	Cost slope ₹/day
(1, 2)	3	5,000	2	7,000	2,000
(2, 3)	4	6,000	2	10,000	2,000
(2, 4)	3	9,000	1	17,000	4,000
(2, 5)	4	5,000	3	9,000	4,000
(4, 5)	5	7,000	2	16,000	3,000
(3, 5)	2	8,000	1	9,500	1,500
(5, 6)	5	20,000	5	20,000	–

Total Project Cost
= [*Normal cost of activities* (1, 2), (2, 3), (2, 4), (2, 5), (4, 5), *and* (3, 5)]
+ (*Indirect cost per day*) × *Duration of the project*)
= (5,000 + 6,000 + 9,000 + 5,000 + 7,000 + 8,000 + 20,000) + (6,000 × 16)
= (60,000) + (96,000) = 156,000.

It may be noted that the crashing exercise would have given the same result even if we had assumed the base cost to be equal to zero, as this was a constant in all the steps. Since every activity has to be performed to complete the project, a base cost of every activity has to be paid. These costs may be regarded as the 'fixed'.

To shorten the project duration, we have to shorten the duration of activities along the critical path (2-4-5-6). We observe that the activity (1, 2) is on the critical path and has the least slope (₹ 2,000/day), and hence, can be crashed first. This activity can be crashed by one day; thus, the project duration reduces by a day. Project duration has become 15 days now.

The project cost to complete in 15 *days*
= *cost to complete the project in* 16 *days* + *cost of crashing by a day*
− *saving in indirect cost*
= 156,000 + 2,000 − 6,000
= 152,000.

Thus, we have obtained a reduction of ₹ 4,000 and a reduction in duration by 1 day.

The next higher cost slope (₹ 3,000 per day) on the critical path is for activity (4, 5) and we can crash this activity to a maximum of 3 days (from a normal duration of 5 days to a crash duration of 2 days). However, we will crash it in two steps of one day each. With one day of crashing, the project duration will become 14 days now.

The project cost to complete in 14 *days*
= *the project cost to complete in* 15 *days* + *cost of crashing by a day*
− *saving in indirect cost*
= 152,000 + 3,000 − 6,000
= 149,000.

Table 7.13 Options available for further crashing

Option	Cost (₹/day)
C and B	4,000 + 2,000 = 6,000
C and F	4,000 + 3,000 = 7,000
E and B	3,000 + 2,000 = 5,000
E and F	3,000 + 1,500 = 4,500

The project duration, thus, reduces by another day and the cost has also decreased by ₹ 3,000 over the previous crash cost.

The activity (4, 5) is crashed again by a day. The project duration becomes 13 days now.

The project cost to complete in 13 days
   = the project cost to complete in 14 days + cost of crashing by a day
   − saving in indirect cost
   = 149,000 + 3,000 − 6,000
   = 146,000.

The project duration, thus, reduces by another day and the cost has also decreased by ₹ 3,000 over the previous crash cost.

There is still scope of crashing activity (4, 5). However, note that there are two critical paths 1-2-4-5-6 and 1-2-3-5-6, both of 13 days duration. The available options for crashing are given in Table 7.13. It may further be noted that the cost of crashing one day is the summation of individual cost slopes.

The lowest-cost slope option is given by activities E and F. The crash costs of these two activities combined together is ₹ 4,500. Thus, we crash activities E and F by 1 day, which makes the project duration equal to 12 days.

The project cost to complete in 12 days
   = the project cost to complete in 13 days + cost of crashing by a day
   − saving in indirect cost
   = 146,000 + 4,500 − 6,000
   = 144,500.

The project duration, thus, reduces by another day and the cost has also decreased by ₹ 1,500 over the previous crash cost.

Again, we have two critical paths 1-2-4-5-6 and 1-2-3-5-6, both with project duration of 12 days. The available option for crashing is activities C and B together. The crash cost would be equal to 4,000 + 2,000 = ₹ 6,000. The project duration reduces by another day and becomes equal to 11 days.

The project cost to complete in 11 days
   = the project cost to complete in 12 days + cost of crashing by a day
   − saving in indirect cost
   = 144,500 + 6,000 − 6,000
   = 144,500.

The project duration reduces by another day but there is no increase or decrease in the cost over the previous crash cost.

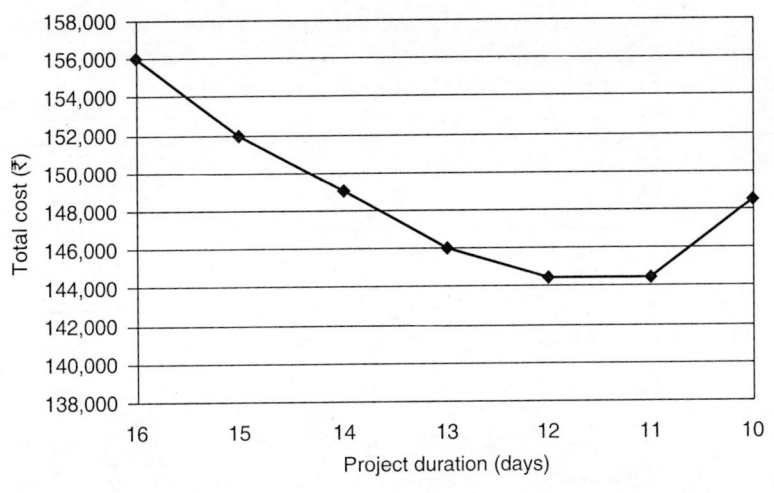

Figure 7.27 Time vs cost for the example problem

Now, we observe that we have 3 critical paths, 1-2-4-5-6, 1-2-3-5-6 and 1-2-5-6, each of 11 days duration. Our only available option for crashing is to crash activities B, C and D together by one day, and the cost of crashing the three activities together is equal to the sum of cost slopes of activities (2, 3), (2, 4) and (2, 5)—that is, ₹ 10,000. Thus, project duration becomes 10 days.

The project cost to complete in 10 *days*
$$= \text{the project cost to complete in } 11 \text{ days} + \text{cost of crashing by a day}$$
$$- \text{saving in indirect cost}$$
$$= 144{,}500 + 10{,}000 - 6{,}000$$
$$= 148{,}500.$$

The project duration reduces by another day but the cost has increased by ₹ 4,000 over the previous crash cost.

Thus, we have reached a stage where the decrease in duration is accompanied by a significant increase in the direct cost, forcing us to stop further crashing. If we combine all our results in a graph showing how project length affects the schedule costs, we obtain the curve as in Figure 7.27, which shows the minimum project cost corresponding to project durations of 11 days and 12 days.

It may be noted that the crashing exercise would have given the same result even if we had assumed the base cost to be equal to zero, as this was a constant in all the steps. Since every activity has to be performed to complete the project, a base cost of every activity has to be paid. These costs may be regarded as the 'fixed' or 'sunk' costs of the project. Neglecting the base cost of activities since these remain the same irrespective of crashing, base cost of activities has been assumed to be Re 0. This cost has to be paid for all activities and is independent of the crashing done.

The time-cost curve shown in Figure 7.27 helps the project manager to choose a suitable schedule for the project. The decision to select a particular duration does not depend on the cost alone, but is determined by a number of other criteria such as the considerations of safety of workers, the risks involved in a particular completion schedule, and so on. Thus, a project manager may not always go for the schedule associated with the least cost and he may choose to go for the moderate cost schedule after incorporating other factors.

REFERENCES

1. Weist, J.D. and F.K. Levy, 2005, *A Management Guide to PERT/CPM with GERT/PDM/DCPM and other Networks*, 2nd ed., New Delhi: Prentice Hall.

REVIEW QUESTIONS

1. State whether True or False:
   a. During resource levelling exercise, the critical activities are given priority for resource allocation.
   b. Time-cost trade-off analysis in CPM tries to optimize the project duration to minimize the total project cost.
   c. An activity with the highest cost slope is to be considered first for crashing.
   d. When there is more than one critical path in a network, crashing an activity on any one path will reduce the project duration.
   e. Resource levelling exercise uses activity floats to reschedule activities, without delaying project completion.

2. Short-answer type questions
   a. What do you understand by 'crashing' of an activity? How is it used for optimizing project cost? Explain with help of a diagram.
   b. Explain the step-by-step process of 'resource scheduling'. What do you understand by 'resource levelling' and how do you achieve it?
   c. Define 'crashing' and explain the basic assumptions that enable crashing in CPM. Why crashing cannot be attempted in PERT method?
   d. What is the need of resource levelling in any construction project?
   e. Differentiate between resource smoothening and resource levelling.
   f. What do you mean by project scheduling and why is it important?
   g. Discuss resource levelling and allocation.
   h. What is meant by 'project crashing'?
   i. Discuss time-cost trade-off.
   j. Discuss different types of schedules.
   k. Define direct cost and indirect cost.
   l. What are the different problems associated with project scheduling?

3. Numerical problems
   a. A construction company has been awarded a contract to construct a flyover in a city with a completion period of 18 months. The major activities in the project and the relationships among them, the normal and crash durations, and the corresponding normal and crash costs are given in Table Q7.1.1.

Table Q7.1.1 Data for Question 3a

Activity	Immediate predecessor(s)	Duration (months) Normal	Crash	Cost (in ₹) Normal	Crash
A	–	6	4	24,000	34,000
B	–	4	3	12,000	22,000
C	A	5	3	20,000	28,000
D	A	7	4	29,000	47,000
E	B	6	5	26,000	34,000
F	B	8	5	34,000	52,000
G	C, E	10	6	27,000	47,000
H	D, F	9	7	34,000	48,000

(i) If the project attracts a penalty of ₹ 10,000 per month for project completion beyond 18 months, find out the total project cost including penalty if the company does not 'crash' any activity.

(ii) Which of the activities should be crashed first in order to reduce the penalty payable and reduce the total cost?

b. The precedence requirements, normal and crash activity times, and normal and crash costs for a construction project are given in Table Q7.2.1. The overhead costs are ₹ 4 lakh per week.

(i) Find the critical path and time for completion under normal working conditions.

(ii) Find the optimum duration for the project to minimize the total project cost. What is the total project cost for the optimum duration?

c. Construction of a school building is scheduled to start from May 1, 2008. The details of different activities of the project are given in Table Q7.3.1. The budgeted costs shown in the table for different activities are to be spent uniformly over their duration. The activities start on the first day of the start month and finishes on the last day of the finish month. Draw a Gantt chart for project schedule and work out the monthly and cumulative cash flows for the project.

d. The details of activities for a small project are given in Table Q7.4.1. If the indirect cost for the project is ₹ 11.5 lakh per week, (a) what is the minimum time in which the project can be completed con-

Table Q7.2.1 Data for Question 3b

Activity	Immediate predecessor(s)	Duration (weeks) Normal	Crash	Cost (in ₹ lakh) Normal	Crash
A		4	2	10	11
B	A	3	2	6	9
C	A	2	1	4	6
D	B	5	3	14	18
E	B, C	1	1	9	9
F	C	3	2	7	8
G	E, F	4	2	13	25
H	D, E	4	1	12	18

Table Q7.3.1 Data for Question 3c

Activity	Duration	Start month	Finish month	Budgeted cost
A	4	May 08	Aug. 08	₹ 80 lakh
B	2	Sep. 08	Oct. 08	₹ 60 lakh
C	6	Sep. 08	Feb. 09	₹ 90 lakh
D	9	March 09	Nov. 09	₹ 45 lakh
E	2	May 08	June 08	₹ 50 lakh
F	5	July 08	Nov. 08	₹ 100 lakh
G	3	Nov. 08	Jan. 09	₹ 75 lakh
H	7	July 08	Jan. 09	₹ 70 lakh
I	10	Feb. 09	Nov. 09	₹ 150 lakh
J	1	Dec. 09	Dec. 09	₹ 40 lakh

sidering maximum possible crashing, and (b) what would be the total project cost if the schedule is crashed to the maximum?

e. A critical activity in a project is estimated to need 15 days to complete at a cost of ₹ 15,000. The activity can be expedited to complete in 12 days by spending a total amount of ₹ 27,000. If the indirect costs of the project are ₹ 3,000 per day, is it economically advisable to complete the activity early by crashing?

f. In a small project there are nine activities. The duration for each of these activities and the labour required to do them are given in Table Q7.6.1. The project must be completed within 27 days.

Table Q7.4.1 Data for Question 3d

Activity	Duration (weeks) Normal	Crash	Cost Normal	Crash
(1, 2)	6	3	₹ 40 lakh	₹ 70 lakh
(1, 3)	5	3	₹ 30 lakh	₹ 52 lakh
(2, 4)	2	1	₹ 60 lakh	₹ 84 lakh
(3, 4)	10	6	₹ 70 lakh	₹ 98 lakh
(2, 5)	3	2	₹ 45 lakh	₹ 63 lakh
(4, 5)	4	2	₹ 26 lakh	₹ 50 lakh

Table Q7.6.1 Data for Question 3f

Activities	Duration (days)	Labour
(1, 2)	4	2
(2, 3)	6	3
(2, 5)	9	4
(2, 4)	2	4
(3, 4)	3	3
(3, 7)	8	3
(5, 6)	10	2
(6, 7)	4	2
(4, 7)	2	1

Nevertheless, the contractor wishes to carry out some resource levelling/smoothing in order that there are no excessive peaks or troughs in his labour schedule. Prepare labour schedule based on early start and late start of activities, and by visual inspection indicate the adjustment you would make in activities in order to perform resource scheduling.

g. For the data given in Table Q7.7.1, perform resource levelling within the minimum project duration for the smallest number of workers if activities cannot be split.

Table Q7.7.1 Data for Question 3g

Activities	Duration (days)	No. of workers
(1, 2)	8	4
(1, 3)	7	10
(1, 5)	12	5
(2, 3)	4	6
(2, 4)	10	7
(3, 4)	3	5
(3, 5)	5	8
(3, 6)	10	7
(4, 6)	7	11
(5, 6)	4	3

8

Contractor's Estimation of Cost and Bidding Strategy

Contractor's estimation and bidding process, bidding models, determination of optimum mark-up level, bidding and estimation practices in Indian construction industry

8.1 CONTRACTOR'S ESTIMATION AND BIDDING PROCESS

The procedure of cost estimation explained in Chapter 4 is suitable mostly for small and medium projects, and that too, mostly from the owners' perspective. For a contractor, the stakes involved are very high. The estimates should be as close as possible to the actual cost for a contractor. If the estimates are on a lower side, the contractor, even if he is likely to win the bid, may have to bear monetary losses at the time of execution. If the estimates are on a higher side, the likelihood of winning the bid is very less.

The estimation and bidding process for a contracting organization is a systematic process (see Figure 8.1). These are discussed here briefly based mainly from the contractor's perspective.

8.1.1 Get Involved in Pre-qualification Process

For the contractor, the tendering process starts with the receipt of request for pre-qualification in case of selective (limited tender) bidding, while it may start at the onset of notice inviting tender in case of open bidding. These days, it is quite common for large projects to resort to selective bidding wherein the first stage is pre-qualification of contractors. An owner seeks information related to financial position, the availability of manpower, plant and equipment, and past experience of the contractor, usually in structured format. The contractor furnishes the desired information along with the supporting documents. The owner scrutinizes the documents received from a number of contractors. The owner shortlists a set of usually six to seven contractors based on the pre-decided selection criteria. These contractors are called pre-qualified bidders for the given project. The prequalification process was explained in detail in Section 5.4.1 of Chapter 5.

In case a contractor pre-qualifies in the pre-qualification process, he is intimated (by the owner) to collect the tender document at an appropriate time when the documents are ready.

CONTRACTOR'S ESTIMATION OF COST AND BIDDING STRATEGY | 257

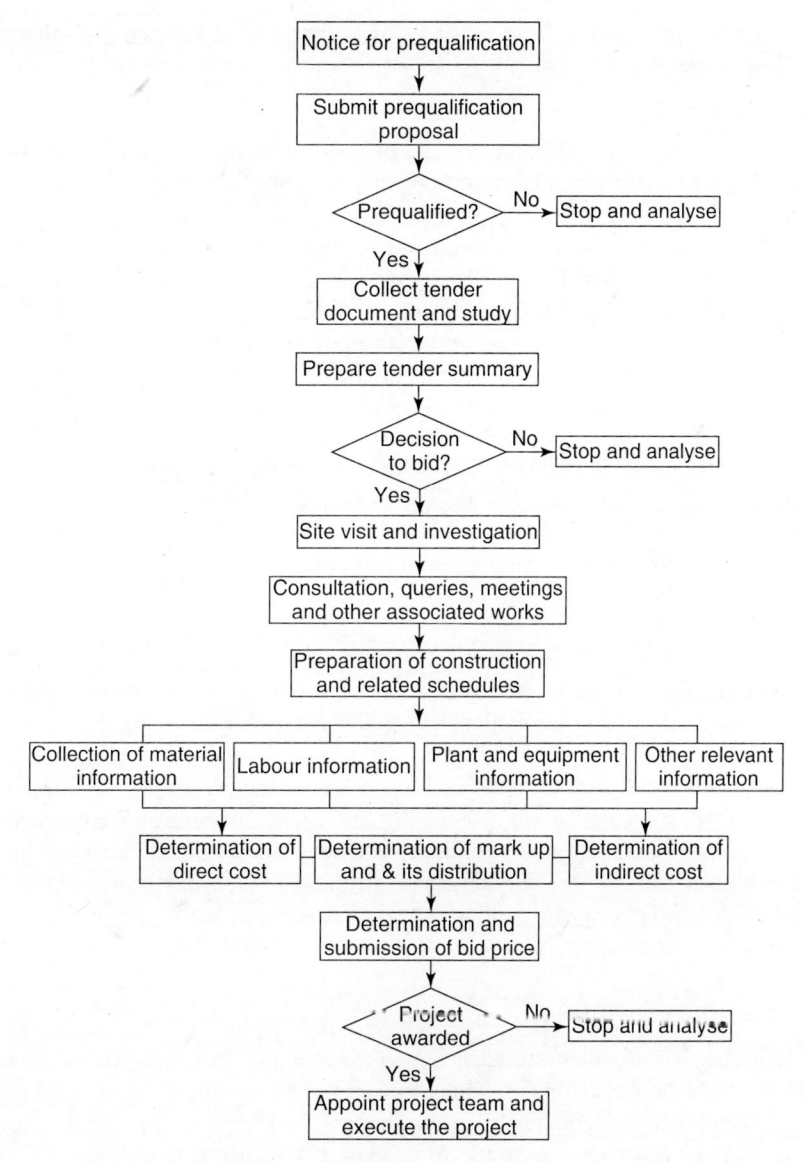

Figure 8.1 Estimation and bidding process from contractor's perspective

It may be noted that a contractor has an option not to bid for the project even if he has prequalified for a project. However, it is not a very common decision to not participate in the bid at this stage.

In case the contractor decides to take part in the bidding process, he sends a representative to collect the tender document. The owner charges some amount as cost of tender document before issuing the documents.

8.1.2 Study the Tender Document, Drawings and Prepare Tender Summary

Study of Documents

The contractor, on receipt of the tender document, studies the general conditions of contract, the special conditions of contract, the specifications, the bill of quantities and the drawings of the project. These documents are quite voluminous and usually the important points are noted down from the documents. Also, depending on the type of contract, some of the mentioned documents may not be there in the tender document. For example, in a lump-sum contract there may not be a bill of quantities. Also, in some contracts the preparation of drawings may be within the contractor's scope. In some contracts, contractor may be expected to develop the alternative design and the corresponding bill of quantities.

Study of Drawings

Although a set of drawings is usually issued to the contractors bidding for a particular project, along with other tender documents, in case the drawings are not issued, the same should be inspected from the architect's office. Any ambiguity in the drawings vis-à-vis specification and bill of quantities as well as any unclear aspect of drawings must be clarified by consulting the architect.

Preparing Tender at a Glance, or Tender Summary

A typical 'tender at a glance' report would contain—(1) general information related to client, consultant and architect; (2) commercial terms; (3) information on mobilization advance, plant and machinery advance, material advance, etc.; (4) information on taxes and duties; (5) information on terms of payment; (6) information on escalation clause, liquidated damages and bonus clauses; (7) information on arbitration and dispute resolution clauses; (8) information on insurance; (9) information on facilities to be provided at site; and (10) information on materials issued by the clients. The details of a typical tender at a glance report are given in Appendix 6. While preparing the tender at a glance, the emphasis is more on highlighting those points that have financial implications.

8.1.3 Decisions to Take

During the bidding process, the contractor has to make many decisions. To bid or not to bid itself is a big decision he has to make. The contractor has an option not to bid even at the stage when the tender document has been purchased. Further, the contractor has to decide on whether to bid independently or in a joint venture (JV) with some other contractor(s). In case of a joint venture, the contractor has to furnish the memorandum of understanding signed with the JV partner(s) to the client. The MOU specifies the roles, responsibilities and liabilities of each party.

At the time of bidding, the contractor also decides the item to be subcontracted and the extent to which the work is to be subcontracted. Once the items to be subcontracted are identified, the contractor mails enquiries to subcontractors and asks for their quote.

The contractor also needs to decide on the construction method to be adopted for the project at this stage. This is very important as there could be a number of possible ways to carry out a construction activity. The contractor has to choose the most economical method to perform an activity in order to be competitive. As an illustration, let us take the production of concrete as an activity. The concrete can be produced by various options including (1) utilizing the conventional mixer machine, (2) using batching plant, and (3) procuring ready mixed concrete. In a given situation, option 1 may be economical and under some other situations, option 2 or option 3 may be economical. The contractor has to analyse factors such as time constraint, space constraint and quality constraint in the project, and decide the most economical method for the given situation. Similarly, alternatives need to be generated for the different activities of the project, be it transporting of concrete, placing of concrete, or formwork, and the most economical alternative then chosen.

8.1.4 Arrange for Site Visit and Investigation

Site visit and its investigation are a prerequisite before estimating the rates for any activity. The visit is carried out mostly in the presence of the owner or his representatives. Site visit helps to understand the following aspects—(1) description of site; (2) positions of existing services; (3) description of ground conditions; (4) availability of labour in the vicinity of site; (5) assessment of security and law-and-order situation; (6) description of access to the site; (7) topographical details of the site; (8) availability of water and power; (9) facilities available for waste and excess earth disposal; and (10) description of any demolition works or temporary works to adjoining buildings.

It is advisable to carry a format of questionnaire that needs to be filled up during the site visit so that the information needed to prepare estimates is readily available. The site investigation helps to collect (1) general information related to site; (2) information on taxes, duties and tariffs; (3) information on laws/regulation; (4) meteorological information; (5) information related to access to project site by road, rail, air and water routes; (6) information on public utilities and services; (7) information on material availability and their rates; (8) information on site topography; (9) information on basic inputs for estimating material rates; (10) information on site facilities; (11) information on labour availability and their rates; (12) information on subcontractor availability and their rates for different items of works; and (13) information on availability of plant and machinery, and their rates. A detailed discussion on site investigation is provided in Appendix 7.

8.1.5 Consultation, Queries and Meetings, and Other Associated Works

Contractor's rate estimation is a collective process. The estimator has to get the inputs from a number of departments within the organization. He may have to consult purchase department, plant and equipment department, accounts department and legal department for updates on materials, plant and equipment, preparation of earnest money deposit (EMD), and examination of certain clauses. The contractor has to oversee various aspects:

- Arrange for earnest money deposit in the required mode of submission. In large organizations, usually there is a request letter that has to be addressed to the concerned accounts authority for getting the earnest money.
- Owners usually ask the contractor to furnish a proof that the official signing the tender documents is legally entitled to do so. For this, a copy of power of attorney on behalf of the person signing the documents is also demanded.

The estimator may require the services of designer and construction method engineer in case alternative designs and drawings are to be prepared for the project. In such cases, the estimator may also need to take off quantities.

In case there are ambiguities in the tender document issued by the owner, the same is noted and communicated to the client well in advance of the pre-bid meetings. Pre-bid meetings are held by the owner to clarify the queries raised by different contractors bidding for the project. In such meetings, representatives of project management firm/architects/service consultants are invited. It is desirable to send copies of the queries to the owner in advance so that he can prepare the relevant clarification. A detailed discussion on pre-bid meeting is given in Appendix 8.

8.1.6 Prepare Construction Schedule and Other Related Schedules

As repeatedly pointed out, it is of utmost importance to plan the different construction activities (the sequence and their timings) of the project. This ensures smooth running of the contract, without delays and hitches that would otherwise cause disorganization and consequent loss of time and money.

Although many details are not available at tendering and estimation stage, this should not prove to be a deterrent in preparing the schedule. A schedule, however approximate, must be prepared at the tendering stage. The importance of preparing the schedule has already been discussed at length in previous chapters.

It may be recalled that the construction schedule prepared for the project is helpful to derive certain other schedules also. These schedules include invoice schedule, cash inflow and outflow schedule, schedule of labour, schedule of material, schedule of plant and equipment deployment, schedule of specialized subcontractors' deployment, schedule of milestone events, schedule of direct costs and schedule of overheads. These schedules help the estimator when he is using the operational estimate procedure. The schedules also help in projecting the working capital requirement for the project. Further, these schedules help in estimating the overheads more judiciously. Certain financial charges and amount towards labour benefit can also be calculated from these schedules. Total man-hour requirement, return on capital employed, earned values, etc., are other aspects that can be calculated using these schedules.

8.1.7 Collect Information

Collect Information on Material

Calculate material requirement per unit of an item/activity Material constitutes about 55 per cent to 75 per cent in any typical project. Hence, any error in estimating the material cost may have a catastrophic result on the outcome of the project. While estimating the materials requirement for any item or activity of a project, one should clearly list out all the materials that are required for producing unit quantity of that item. It is a general tendency to go for thumb rules for calculation of material quantities, especially in the case of items like concrete. However, it is recommended that mix design be carried out for this and material quantities derived from the mix design. Further, all considerations of wastages, breakages and bulkages should be noted in estimating the material quantities. Materials productivity is defined as the quantity of work done per unit of materials—that is,

$$\text{Material productivity} = \frac{\text{Work units achieved}}{\text{Materials quantity consumed}} \quad (8.1)$$

Collect Information on Labour

Calculate labour requirement There may be a number of types of labourers present at any construction project site, such as departmental labour, plant operators and helpers, mates, masons, carpenter, fitters and subcontractor's labourers. They are also sometimes characterized as skilled, semiskilled and unskilled labourers. They are employed in different modes—job contracts; labour contracts; regular employment; project-based employment; daily-wage employment; casual employment, etc.

Labour costs are estimated based on the analysis of past performance data, study of published production norms, study of construction processes, thumb rules, and labour cost prevailing in the area where the project site is located. Labour rates should ideally be estimated considering productivity norms. For example, if activity A requires X man-hours to complete, then the labour rate shall be X multiplied by the labour standard hourly rate.

Labour productivity, however, will depend on a number of parameters including—complexity of work; repetition of work; quality control; mechanization; quality of supervision; management policies; resource management; work environment; climatic and weather conditions; and labour availability. Accordingly, suitable correction factors should be taken into account while estimating for the productivity norms.

The worker's productivity in mathematical terms can be defined as below:

$$Worker's\ productivity = Quantity\ of\ work\ done\ per\ man\ hour$$

$$= \frac{Qnantity\ of\ work\ done}{Man\ hour\ consumed} \qquad (8.2)$$

As can be expected, labour productivity varies from project to project. Some of the typical causes of low labour productivity are workers' low morale, poor pre-work preparation by supervisors and directional failures of the project management.

Some of the ways through which labour productivity can be enhanced are—reducing unproductive time by constantly reviewing and minimizing causes contributing to unproductive time; replacing labour with appropriate equipment where economically feasible; substituting inefficient working tools with appropriate and efficient tools; improving method of executing work and remembering that there is always a better way of doing a task; improving working conditions; and employing competent supervisors.

Standard productivity norms are established and well documented by several agencies such as, Central Public Works Department (CPWD), and MES, as well as big construction companies. It must be pointed out here that standard productivity is applicable under ideal conditions, whereas actual productivity is far less due to various factors, as listed above. In a typical building project, the efficiency varies from 60 per cent to 70 per cent.

Standard hourly rates are derived keeping in view factors such as basic wages, bonus, employee provident fund (EPF) and family pension, retrenchment, notice pay and insurance. The basic wages vary from state to state, and before estimating the labour rates, the minimum wages to be paid to labourer should be checked with the minimum wages list specific to a state. Presently, bonus to the labourer is paid at a rate of 20 per cent of the total wages earned by the labourer. EPF and family pension are contributed by the employer at the rate of 12 per cent of total wages earned, and if the labourer has worked for more than 240 days, he is entitled for 15 days' pay for retrenchment and notice pay equivalent to 26 days' wages. These statutory requirements must be cross-checked every time the labour standard hourly wages are estimated.

In addition, there are other factors that may have to be considered, such as accommodation cost; mobilization and demobilization; canteen and medical facilities; sports and entertainment, and incentives. If the construction project is located in a foreign country, then depending on the country, one may have to incorporate recruitment expenses, airfare and social security costs also in order to arrive at the standard hourly wages.

Collect Information on Plant and Equipment
Identify the plant and equipment needed for completion of the project within the scheduled duration and get information on their output The output of mechanical plant is an important variable. The productivity of mechanical plant and equipment is defined as the quantity of work done per equipment hour—that is,

$$Equipment\ productivity = \frac{Work\ units\ achieved}{Equipment\ hour\ consumed} \qquad (8.3)$$

Analysis of plant and machinery performance reports from different project sites and consulting manufacturer's technical literature are important factors for deciding on plant and equipment cost. Depending on the project, a suitable correction factor should be applied to arrive at the productivity figure.

8.1.8   Determining Bid Price

Schematically, the tender price or the bid price can be mathematically represented as the sum of total cost and mark-up. The total cost consists of direct costs and indirect costs—the cost of material, labour and plant and equipment could be taken under the former head, while expenses on items such as personnel recruitment, training, and research and development are essentially indirect costs. Several definitions for these are available in literature. According to one of the more widely accepted definitions, a direct cost of an activity is physically traceable to the activity in an economic manner, and thus, is a cost not incurred if the activity is not performed.

Indirect costs, on the other hand, are business costs other than direct costs and are not physically traceable to the activity, and may be incurred even if the activity is not performed. Some of the salient features of direct and indirect costs are briefly discussed in the following paragraphs.

A schematic representation of the different components of a bid price is given in Figure 8.2. From the figure, it follows that,

$$\begin{aligned}
Bid\ price &= Total\ cost + Mark\ up\ amount & (8.4) \\
Bid\ price &= Direct\ cost + Indirect\ cost + Mark\ up\ amount & (8.5) \\
Direct\ cost &= Labour\ cost + Plant\ and\ equipment\ cost + Crew\ cost + Materials\ cost \\
&\quad + Subcontractors'\ cost & (8.6) \\
Indirect\ cost &= Project\ overheads + Common\ plant\ and\ equipment\ cost \\
&\quad + Common\ workmen\ cost & (8.7) \\
Mark\ up\ amount &= Profit + Contingency + Allowances\ for\ risks + General\ overheads & (8.8)
\end{aligned}$$

```
                          Bid price
                             |
              ┌──────────────┴──────────────┐
          Total cost                   Mark up amount
              |                              |
      ┌───────┴───────┐                      ├──── Profit
  Direct cost    Indirect cost               ├──── Contingency
      |               |                      ├──── Allowances for risks
      ├── Labour      |                      └──── General overhead
      ├── Equipment   ├──── Project overhead
      ├── Crew        ├──── Common plant and equipment
      ├── Materials   └──── Common workmen
      └── Subcontractors
```

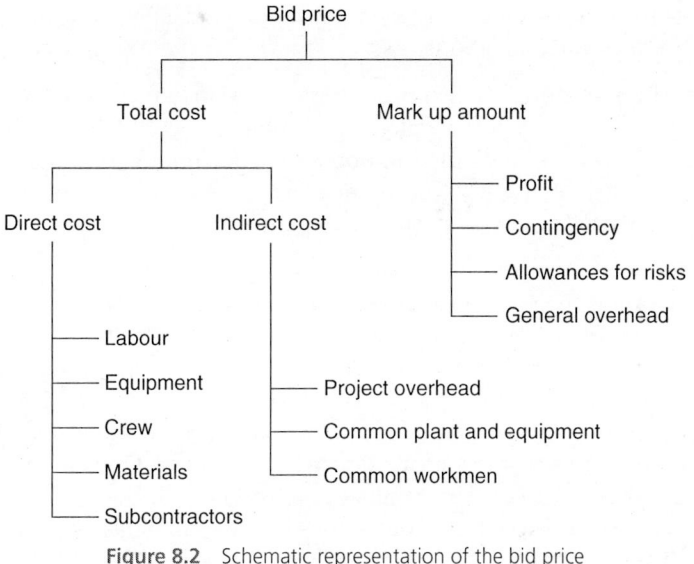

Figure 8.2 Schematic representation of the bid price

Although, strictly speaking, the mark-up could also be considered a part of indirect costs, it is discussed separately to facilitate better understanding and highlight its importance in a construction project. In fact, some authors consider the sum of project overheads and general overheads as indirect costs. In other words, the indirect costs can be said to be made up of general and project overheads, in the sense that the latter are directly related to the project under consideration and the former are more in the nature of overheads incurred at the head office of a company.

In this text, however, we will treat general overheads as part of mark-up.

Estimate Direct Cost

As mentioned above, direct costs are physically related to the actual performance of the activity, and cannot be incurred in the event that the activity is not actually performed. In a construction project, the costs of material, labour, etc., are some of the important components of direct cost, as discussed below.

Cost of material   This includes the cost of all materials that actually go into the building, like cement, stone and reinforcement steel, including the actual rate of the material, excise duty, sales tax, packing and forwarding expenses, freight charges, octroi and entry taxes.

Cost of labour   This component includes the cost incurred on the labour employed at the site. It consists of not only the actual basic wage paid to the employee, but also bonus payments that may be made from time to time, and costs towards welfare schemes such as pension, retrenchment, insurance, health benefits and notice pay. It may be borne in mind that local laws may provide for mandatory contributions by the employer towards schemes providing social security and retirement benefits for employees. At times, companies also provide housing to their employees with or without arrangements for preparing food, and also incur a cost at the time of their mobilization and demobilization, and all costs incurred towards such arrangements are also counted as labour cost.

Cost of plant and equipment This covers the cost incurred towards owning and operating the equipment required for a particular activity. It could be taken to consist of cost incurred in mobilization, owning and operating the equipment, cost of the fuel/oil/lubricants involved, cost of repair of equipment and maintaining the inventory of spare parts, operators and drivers costs, statutory expenses, personnel carriers expenses, and demobilization costs.

Subcontractor cost As mentioned earlier, the main contractor quite often takes help of specialized subcontractors for some of the items such as anti-termite treatment, waterproofing, woodwork and aluminium works. For these items/activities, the main contractor invites quotation from a number of subcontractors in a similar manner as that followed by the owner, though the process may not be that rigorous. The contractor makes a comparative statement of the bids received from the subcontractors for an item/activity. Out of the quotations received for an item/activity, the contractor selects the most responsive bid judiciously. The contractor notes down any 'exclusions' in the selected subcontractor's bid and loads this into the subcontractor's bid. This is termed as subcontractor's cost for that item.

Crew cost Crew is a small group of men (worker, supervisor, etc.) and machineries. Sometimes, the direct cost may be computed on the basis of crew cost directly instead of computing the labour and equipment costs separately. Examples of crew are excavation crew (may consist of 2–3 workers, an excavator and 2–3 dumpers), piping crew (may consist of 2–3 pipe layers and a couple of pipe-laying machines), welding crew (may consist of 1 welder, a couple of helpers, and a welding and grinding machine), and so on. Thus, in order to compute the cost of a crew, we need to add the cost of men and machineries as explained earlier. Knowing the requirement of number of crews to be employed for a particular activity, and the cost to employ one crew, we can determine the crew cost associated with that activity.

Estimate Indirect Costs

Indirect costs are also known as overheads. Several expenses incurred at the project site cannot be attributable to a single activity and, thus, they are kept under indirect costs or project overheads. In fact, there could always be an indirect component even with material costs, labour costs, plant and equipment costs, or with subcontractor costs. For example, as far as indirect material cost is concerned, it can be further classified into purchase cost and inventory-carrying cost. Under the former, there could be expenses on account of purchase or expenses of the procurement department at office or site, stationery cost, material-receiving cost, inspection, warehousing and bill-processing cost. On the other hand, inventory-carrying cost consists of cost of funds blocked, storage charges, material-handling charges, insurance, obsolescence, breakage, damages, pilferage, and the cost towards staff and labour engaged in the stores.

It is almost impossible to make watertight compartments between direct and indirect costs. Broadly, indirect costs include project overheads and common workmen, plant and equipment costs. These are briefly discussed below.

Project overheads The following paragraphs give a more detailed description of some of the major heads of expenditure that may be involved in a project.

Salaries and benefits This covers salaries of project staff including engineers, supervisor and other administrative personnel posted at a construction site, and also a part of the salaries of staff indirectly related to the site. The latter could include staff involved in maintaining logistics, supporting staff, and staff engaged in warehousing and maintenance of equipment. The salaries of personnel involved in temporary facilities construction, operation, and maintenance and dismantling are usually excluded from this head. The gross salaries are calculated inclusive of the basic salary and allowances towards house rent, leave travel, bonus, medical assistance, etc.

Now, for purposes of estimation, two components are involved as far as salaries are concerned—the manpower required in terms of man-months for the personnel at different levels, and the gross salaries involved. It may be noted that these parameters give only the actual cost, and the organization needs to add a mark-up to these in an appropriate manner before arriving at a final figure.

For making a quick analysis, let us take a project for which the contractor has estimated his bid price as ₹ 250 lakh. The contractor has set a productivity norm for his staff to be 5 lakh per staff per month. Hence, the total staff month required would be 250 lakh/5 lakh = 50. In order to calculate the average staff month, we need to know the duration of the project. Let the duration be 10 months for this example. Hence, on an average, 5 staff are required. For calculating the salaries and benefits paid to the staff, we need to know what are the monthly benefits and the annual benefits. The annual benefits are distributed on a pro rata basis, while the monthly benefits are added as they are. For example, let us assume that monthly salary and benefits given to a staff are ₹ 25,000 and annual benefits amount to ₹ 60,000. Hence, the cost that will be taken for calculating the staff cost for the example project would be ₹ 60,000 × 10/12 = ₹ 50,000, since the project duration is only 10 months.

Insurance cost Different kinds of risk are associated with a construction project as far as a contractor is concerned, and to cover these, appropriate insurance schemes are worked out with insuring agencies. Some of the insurance policies could be contractors all risk (CAR) policy, workmen compensation, group insurance for staff, site and camp establishment, cash in transit, materials at site, and export credit guarantee (ECGC). The cost of the premiums involved is clubbed under this expenditure head.

Financing cost In order to be able to mobilize and carry out a job, a contractor needs to arrange for operating funds from financial institutions, banks, etc., and these loans attract interest. Apart from such interest payments, cost is also incurred on account of preparing a bid bond and a performance bond, advance and payment guarantee, retention guarantee, letter of credit establishment charges, guarantee for temporary workers, bank service charges, credit for advances, work insurance, social security, unforeseen fines and penalties, etc., and the financing cost should be appropriately accounted for when estimating the total expenditure involved in a project.

Progress photographs and video A monthly sum (usually a lump sum) is usually set aside for such expenses, on the basis of past experience and the total expenditure estimated on the basis of the duration of the project.

Conveyance cost To facilitate travel of personnel posted at sites, vehicles are often locally hired, and the cost is estimated on the basis of the number of vehicles required (which, in turn, depends on the size of the project, the duration of the project, the hierarchy of project staff, etc.). The conveyance vehicles could be jeeps, cars, vans, or buses. These vehicles may be either owned by the contractor himself or taken on rental basis. In any case, the cost for keeping these vehicles for a month should be multiplied with the number of months the vehicles are proposed to be kept at site, in order to get the conveyance cost.

Travel and transfer costs The expenses incurred in travelling by staff either from closing project site or from home office to the new project sites are charged to the new site. Accordingly, this cost is estimated based on the total staff strength estimated for the proposed project.

Visits of headquarter personnel Personnel from the head office often visit the site for meetings, inspection, or coordination. Depending upon the size and duration of the project, the total cost for such trips can be estimated using a unit cost for such a visit from past experience. Apart from the travel itself, such expenses could include stay, transport, entertainment, applicable taxes and other incidental expenses.

Temporary site installations and facilities As part of a project, temporary site installations such as a site office, workshops, stores, fencing of equipment, warehouse and storage areas, sanitary facilities, first-aid facilities, furniture and fittings, temporary roads, car parks, and even residential units and messing facilities for the staff need to be set up, and such expenses are held under this head. As far as estimates for such facilities are concerned, it need not be very accurate and could be based on the plinth area rate estimates, with the area itself being determined on the basis of the size and duration of the project. It may also be noted that for large projects, contractors usually prefer to construct temporary site accommodation near the project site itself, provided land is available, in which case the costs of leasing land as well as of construction also need to be taken into account. However, in the case of rented accommodation being leased, the cost should be accounted for under 'accommodation' and not this head.

Client's and consultant's requirements At times, a contractor is required to provide certain facilities to the client and consulting organizations in terms of office space, office staff, furniture and fittings, telephone, stationery, operating and maintenance, transport, etc., which need to be kept in mind separately.

Utilities Appropriate arrangements for water, power and sewage facilities need to be made in connection with the project office and the residential area (if set up). Depending upon the conditions of the contract, some of these may be provided free of cost or at a predetermined rate, though usually the contractor is required to set up his own distribution/collection network. In the case of water, a contractor may choose to set up his own borewell and pump sets, and the cost towards running these pumps and the distribution cost through temporary pipelines need to be accounted for. In the case of power also, there are two possibilities—either the power is made available at a single point on a chargeable basis by the owner, or the contractor is required to make his own arrangement by either buying it from external sources or installing his own generator sets. In both cases, the total cost on this account should be considered.

Taxes and duties Construction activities attract different taxes and duties. Some examples of taxes are works contract tax, excise duty, customs duty, income tax, sales tax, service tax and value-added tax. The relevant taxes and duties need to be checked for the construction project and suitably accounted for.

Miscellaneous expenses Apart from the expenses listed above, some of the other expenses, which could be estimated either on a lump-sum basis or on the basis of the unit rates, are:

a. administrative expenses at the site
b. general field expense consisting of operating costs such as stationery, documentation, and communication facilities such as telephone, fax and postage
c. small tools and consumables including fuel and lubricants
d. cost for mobilization and demobilization of personnel, equipment, temporary facilities, small tools, consumables, camp structure and other facilities
e. establishment expenses including the cost incurred on field supervisors, technical staff and workers, materials management staff and workers, plant and machinery management staff and workers, administration staff, workers and managers
f. technical management costs for designs and drawings, project consultancy, data processing, site laboratory and testing, technical library, subcontracted workers' expenses, on-site surveying, site clearance
g. safety and protective equipment costs
h. costs incurred towards payment of certain professional fees, government liaison fees
i. sundry expenses, catering expenses, laundering and camp-operating costs, etc.

Common plant and equipment Some common plant and equipment that are difficult to be apportioned to an activity directly are kept under the head of 'common plant and equipment'. The cost towards these plant and equipment is uniformly distributed across all the items in the bill of quantities. Some examples of common plant and equipment are diesel generators and water pumps.

Common workmen Expenses associated with workmen used for different activities in a manner that their role cannot be directly attributed to any single activity, are uniformly distributed across all the items of bill of quantities. Such workmen may include safety steward, quality control technicians, survey helpers, peons, security men and workmen employed in accounting office, time office, stores, technical office, etc.

It may be noticed that the costs mentioned above under indirect costs may vary from project to project and from contractor to contractor.

8.1.9 Analysis of Rates

Having identified the different cost components associated with labour, material, plant and equipment, subcontractor and overheads, the analysis of rates can be performed using any one of the following methods or a combination thereof.

Operational Estimating

The operational estimating procedure is quite common, especially with contracting organization, as it fits well with the construction schedule. In civil engineering projects, it is quite common to have resources not working on a continuous basis, due to any number of reasons such as lack of front availability, inclement weather, and so on. The operational estimation method recognizes and takes into account the idle time of resources, especially the plant and equipment. The method is explained in a step-by-step manner.

Table 8.1 Extract of bill item from bill of quantity

Earthwork	Unit	Quantity	Rate	Amount
Earthwork in soft soil from a depth of 0 m to 1.5 m	m³	10,000	—	—
Earthwork in dense soil from a depth of 1.5 m to 3.0 m	m³	15,000	—	—
—	m³	8,000	—	—
—	m³	7,000	—	—
Total scope of work		40,000		

- To start with, the items falling within a group in a bill of quantities are identified and the total scope is determined. For example, under earthwork items in the bill of quantity, there may be items such as those given in Table 8.1.
  Here, earthwork is considered as an operation with a total scope of work equal to 40,000 m³.

- The time available for the total scope of work of this operation is read out from the construction schedule prepared at the time of tendering. For this example, let us assume that duration of operation 'earthwork' is four months.

- The estimator would select the resources (see Table 8.2) to be used in the operation and use the duration of the activities and the unit cost of each resource to produce the total cost of the operation. This total cost has to be assigned back to each bill item.

Table 8.2 Estimation of resources for earthwork operation

Particulars	Remarks/explanation
Total scope	40,000 m^3
Duration	4 *months*
Disposal	1 *km lead*
Average excavation per day	$\dfrac{40,000}{4 \times 26} \approx 400 \, m^3$, Considering 4 Sundays in a month
Accounting for Peak excavation (25% more)	$400 \times 1.25 = 500 \, m^3$
Type of Plant and Equipment	Excavator (L&T Poclain CK 90), 50 m³/h, 10 hours working, considering 2 hrs overtime Tippers, 6 m³ capacity
Computation for no. of tippers	*Loading time from Poclain* = 7 *buckets* × 1.2 *min* = 8.4 *min* *Travel time to cover* 1 *km* × 2 *at* 15 *km/h* = $\dfrac{2 \times 60}{15} = 8 \, min$ *Tipping time* = 1 *min* *Turning/manoeuvring time* = 3 *min* *Total time required per cycle* = 8.4 + 8 + 1 + 3 = 20.4 *min*
No. of tippers required	$= \dfrac{(Time\ to\ load + Time\ to\ haul)}{Time\ to\ load} = \dfrac{8.4 + 12}{8.4} = 2.4$ say 3

Having calculated the type and number of equipment required, the operational estimating process of owning cost (see Table 8.3), power and fuel cost (see Table 8.4), and labour cost (see Table 8.5) are explained for the earthwork operation.

Equipment-Owning Cost

Table 8.3 Illustration of equipment-owning cost

Description	No.	Hire charges (₹/month)	Months	Total (₹)
Excavator	1	125,000	5	625,000
Dumpers	3	50,000	5	750,000
Subtotal				1,375,000
Spares and maintenance @25% of above				343,750
Mobilization of P&M—Poclain (L.S.)				50,000
Mobilization of tippers (L.S.)				60,000
Total P&M owning cost				1,828,750

Operating Fuel and Power for Equipment

Table 8.4 Illustration of operating fuel and power cost for the required resources

Description	No.	Months	Days	Hours	Fuel/Power	Cost (₹/l)	O&L[1]	Amount (₹)
Poclain	1	4	26	10	13 l	38	1.3	667,888
Tippers	3	4	26	10	5 l/h	38	1.3	770,640
Total cost for operating fuel								1,438,528

[1] Oil and lubricants

Operating Labour for Equipment

Table 8.5 Illustration for operating labour for the required resources

Description	No.	Months	Days	Wage (₹/day)	Amount (₹)
Excavator operator	1	4	26	220	22,880
Helpers for bucket cleaning	2	4	26	180	37,440
Dumper drivers	3	4	26	200	62,400
Subtotal					122,720
Add overtime charges @50%					61,360
Total for operating labour for equipment					184,080

Table 8.6 Summary of owning cost, fuel/power cost and operational labour cost for the required resources

Description	Amount (₹)
P&M owning cost	1,828,750
Fuel/power cost	1,438,528
Operating labour	184,080
Total	3,451,358
Cost per m³ ₹ 3,451,358/40,000 = ₹ 86.28 say	86 per m³

Summary The results of operational estimating for owning cost, power and fuel cost, and labour cost are summarized in Table 8.6. The total cost is ₹ 3,451,358 for the bill of quantity corresponding to 40,000 m3.

- The method of assignment is open to the estimator, but the simplest way is to apportion the costs pro rata with the item quantity. That is, for the given extract of bill of quantity where the total earthwork quantity is 40,000 m³ contained in 4 different bill items of varying quantity, the rate would be ₹ 86.28/m³ (₹ 3,451,358 ÷ 40,000, where the total cost of the plant owning + power and fuel + labour = ₹ 3,451,358, and the combined quantity = 40,000 m³).

The link between operational estimating, planning and bill of quantities is that the estimator would group together the bill items they would want to estimate as an operation. This group of bill items would be represented in the construction programme as one or more activities, and so, the construction programme would provide the duration of the operation.

In general, conventional bill of quantities are not designed to suit operational estimating.

Unit Rate Estimating

The unit rate estimating method is most commonly used for building works, especially where the traditional bill of quantities is issued by the owner along with the other tender documents.

The method involves selection of the required resources (such as labour, material, and plant and equipment) and selection of their output or usage rates. For the selection of resources and their associated output or usage rates, the estimator uses his experience or the contracting company's past record. Some companies maintain a data bank of output or usage rates for a large number of jobs executed in the past.

Thus, for each of the selected resources the following arithmetic calculation is performed in unit rate estimating method:

$$\text{Unit rate} = \text{The output or usage rate } p \text{ of the resources} \times \text{The unit cost } c \text{ of the resources} \quad (8.9)$$

The output rate is expressed as work quantity per hour, and the usage rate is the time taken to do a fixed quantity of work. Thus, by definition, the unit of output rate would be m³/h or m²/h or m/h, depending on the unit of measurement used for the item, and the unit of usage rate would be h/m³ or h/m² or h/m, depending on the unit of measurement.

The unit cost of resources could be in terms of either ₹/h or ₹/m³ (or ₹/m², and so on, depending on the unit of measurement). Thus, by combining the output or usage rate p of

Table 8.7 Unit rate estimating method for derivation of material rates for 1 m³ of cement concrete

Material	Usage rate 'p' qty/m³	Unit cost 'c' (₹/unit) including wastage	Total cost (₹)
Coarse aggregate	0.880 m³	1000/m³ (5% wastage assumed)	880.00
Fine aggregate	0.440 m³	1000/m³ (5% wastage assumed)	440.00
Cement	6.6 bags	225/bag (3% wastage assumed)	1,485.00
Total			2,805.00

the resources with the cost of the resources c, the cost per unit for an item can be obtained and the sum of the unit cost for all the resources will be the estimate of the total direct cost for the item.

For example, the unit rate estimating method for earthwork in soft soil can be obtained from the estimate of output/usage rate and unit cost. Assuming that the output of labour in such soil is 3 man-hours per m³ and assuming the cost of 1 man-hour to be ₹ 20, the direct cost of 1 m³ earthwork would be 3 h/m³ × 20 ₹/h = ₹ 60. An illustration of hourly labour wage determination is given in the 'solved examples' section. Another illustration of unit rate estimating is given in Table 8.7 and Table 8.8 for concrete producing and placing in a project. The unit rate estimation has been used for material and labour cost estimation. The tables are self-explanatory.

It is also possible that for the direct cost estimation of a particular item, the estimator can use the combination of unit rate estimation and operational estimating. Similarly, the estimate of indirect cost can be performed using the combination of operational estimating and unit rate estimating.

Combined Operational Estimating and Unit Rate Estimating

In some cases, it may so happen that some component of the total cost—for example, plant and equipment component—is worked out following operational estimating procedure, and the other components such as labour and material are worked out based on unit rate estimating. In such cases, both the costs are combined to get the unit rate for the specific item. Let us assume that the operational estimating gives a rate of ₹ x/m³ for concrete operation, which is made up of 10 different bill of quantity items totalling to 100,000 m³ of concrete. Thus, ₹ x/m³ cost of plant would be set against each item, together with the labour and material costs computed based on unit rate estimating procedure and illustrated under unit rate estimating. The process is illustrated in 'solved examples' section in detail.

Table 8.8 Unit rate estimating method for labour for producing and placing of 1 m³ of cement concrete

Labour	Output/usage	Cost per hour (₹)	Total cost (₹)
Head mason	0.4 h/m³	27.50	11.00
Mason	1.6 h/m³	27.50	44.00
Mazdoor	25.6 h/m³	22.50	576.00
Bhisti	4.8 h/m³	25.00	120.00
Labour cost per m³			751.00

8.1.10 Fix Mark-up

Mark-up Definition

From Equation 8.8, it is clear that mark-up is the sum of profit, contingency, allowances for risk, and general overheads. It can be expressed either (1) in terms of some percent of total cost TC or (2) in terms of some percent of bid price B explained latter. In the second case, it is also referred to as 'off-top'.

Profit This is a reward for carrying out the business and thus taking the risks involved. Risky business carries more profit, and vice versa.

Contingency and allowances for risks Construction projects are a risky proposition full of uncertainties and risks. In spite of different details known at the tendering stage, there are uncertainties and risks pertaining to timely completion, budget escalation, site conditions, soil characteristics, labour and material availability, and so on. In order to safeguard against these eventualities, contractors keep certain contingency provisions. Indeed, the higher the uncertainties involved in a project at the time of tender, the higher is the contingency provisions.

General overheads Several expenses incurred at the corporate headquarters of a company cannot be directly traced to any particular project and, accordingly, such expenses are charged from the running projects on a pro rata basis. Such expenses include those incurred towards research and development, publicity and advertisement, the cost of unsuccessful bids that necessarily need to be distributed over other projects, recruitment of personnel, and salaries of security personnel at the head office, and are termed as 'general overheads'. The 'general overheads' for a construction company vary between 2 per cent and 3 per cent of the contract value, depending on factors such as turnover, staff strength, nature of expenses incurred at head office, and the number of projects in hand.

It is clear that for a contracting agency, since the direct and indirect costs are actually incurred, the bid price (the cost at which it accepts to carry out the job) needs to be higher to provide a certain profit, to cover general overheads, and also to cover risks involved. These three components are collectively referred to as 'mark-up', which a contracting agency adds to the cost at the time of bidding.

The direct and indirect costs have been discussed in some detail above, and the effort here is directed to better understand some of the factors on which the mark-up depends.

At the outset, it should be relatively clear that the actual direct and indirect costs of a project should not normally vary too much from one contracting agency to the other, especially if they are comparable in stature (in terms of turnover, number of employees, etc.). Thus, any difference in competitors' bids to carry out a particular job largely depends on their selected mark-ups. In principle, the mark-up should be such that it allows the maximum margin of profit, while at the same time keeping the bid low enough for the job to be awarded. This makes the mark-up selection a complex issue and a function of many parameters, as is brought out in the following paragraphs. It may be borne in mind that companies usually do not disclose their strategies towards deciding their mark-up and, therefore, scientific studies based on actual data are relatively few.

Factors Affecting Mark-up

Factors affecting the mark-up could be related not only to the project involved, but also to the contracting agency and the client, besides factors such as uncertainty and risk. As far as factors related to the project are concerned, the size, location, duration, complexity, cost, construction method and season of the work could be cited as factors affecting the mark-up. Similarly, for the same project, a contracting agency could have a different mark-up depending upon the intensity of competition (number of bidders), the ongoing commitments of the contractor, and the client involved. Independent of the factors cited so far, the mark-up is likely to depend on the past relationship of the contracting agency with the client and the contracting agency's corporate strategy in terms of willingness to build a long-term relationship with the client, as also on the contracting agency's attitude towards risk. The following factors affecting the mark-up have been compiled from past studies, namely from the work of Park (1964); Broemser (1968); Benjamin (1969); Miller and Starr (1969); deNeufville et al. (1977); Ahmad and Minkarah (1988); Shash (1998); Chua and Li (2000); and Jha (2004):

- Number of competitors and the intensity of competition
- Size, cost and intensity of the project
- Type of project—buildings, infrastructure projects, etc.
- Duration of the project
- Location of the project
- Season in which the work is done
- Degree of hazard and difficulty associated with the project
- Name of owner/consultant and designers, and time available for bid preparation
- Labour availability and productivity
- Material availability and costs
- Percent of the work which is to be subcontracted and the bids of subcontractors
- Insurance cost and fringe benefits
- Availability of supervisory talent
- Method of performing the work
- Uncertainty in estimate and historic profit
- The current and forecasted economic conditions
- The contractor's risk attitudes

According to a survey conducted among the top 400 contractors identified by *Engineering News Record*, the following factors emerged in the order of importance towards influencing the mark-up decision:

1. Degree of hazard
2. Degree of difficulty
3. Type of job
4. Uncertainty in estimate
5. Historical profit

6. Current workload
7. Risk of investment
8. Rate of return
9. Owner
10. Location

It may be noted that in the above study, most contractors participating were from the industrialized world. In a similar exercise carried out in India (Jha 2004), the following emerged as parameters most likely to affect the mark-up:

1. Previous experience with the clients/consultants as regards decision and payment
2. Degree of hazard associated with the project
3. Project complexity
4. Number of agencies/statutory bodies involved in approval process
5. Bureaucratic delays

A comparison of the two lists above is an indictment of the level of professionalism in a developing country and clearly shows the apprehension of a contracting agency towards working in a system where multiple clearances are required from regulatory agencies, or where there are delays in obtaining material, drawings, etc., from the client, or where there are delays in terms of clearance of bills, etc. Interestingly, it was also found that highly critical parameters may not always result in increase in mark-up. In other words, a prudent contractor in a developing country could prefer to handle critical parameters more carefully during execution rather than loading (increasing) his rate, a possibility that could eventually adversely affect his chance of winning the contract. It can also be seen that there is little commonality among the top-ranked parameters in India and the more industrialized countries, clearly highlighting the importance of local conditions and their role in deciding the mark-up.

Distribution of Mark-up

The distribution of mark-up is an important decision that a contractor has to make. We take up the example discussed in Chapter 7 for illustration of mark-up distribution. It may be recalled that total value of the project was ₹ 500 lakh and the markup amount was ₹ 50 lakh. Thus, the total cost to the contractor for this project works out to be ₹ 450 lakh. The mark-up in terms of percentage works out to be 10 per cent of the value, or 11.11 per cent of the total cost. There are different ways to distribute the mark-up amount of ₹ 50 lakh. Although it does not affect the total bid price, the timing of cash receipt plays an important role in easing the negative cash flow that is usually encountered in the beginning phases of a project.

Table 8.9 shows the three ways in which the total mark-up has been distributed. In the first case, the total mark-up is distributed uniformly across all the activities of the project. Thus, it can be seen that the mark-up percentage of 11.11 per cent is uniformly applied to all the items of the project, totalling to ₹ 50 lakh.

In the second case, the activities that are likely to be taken up early in the project carry a higher mark-up percent. For example, the activities earthwork—all soils, concrete works, formwork, reinforcement and brickwork carry mark-up of 22.22 per cent, 13.89 per cent, 18.52 per cent, 12.84 per cent and 15.34 per cent, respectively, which are relatively high compared

Table 8.9 Illustration of uniform, front and back loading of mark-up

				Mark-up distribution (total mark-up amount = ₹ 50 lakh)					
				Uniform loading		Front loading		Back loading	
S. No.	Item description	Bid amount (₹ lakh)	Total cost (₹ lakh)	Mark-up amount (₹ lakh)	Mark-up %	Mark-up amount (₹ lakh)	Mark-up %	Mark-up amount (₹ lakh)	Mark-up %
(1)	(2)	(3)	(4)	(5)	(6)	(7)	(8)	(9)	(10)
1	Earthwork—all soils	5.00	4.50	0.50	11.11	1.00	22.22	0.25	5.56
2	Concrete works	80.00	72.00	8.00	11.11	10.00	13.89	6.00	8.33
3	Formwork	42.00	37.80	4.20	11.11	7.00	18.52	3.00	7.94
4	Reinforcement works	112.50	101.25	11.25	11.11	13.00	12.84	7.00	6.91
5	Brickwork	34.00	30.60	3.40	11.11	4.70	15.36	2.00	6.54
6	Plastering—all types	15.00	13.50	1.50	11.11	1.50	11.11	3.05	22.59
7	Painting—all types	5.00	4.50	0.50	11.11	0.50	11.11	1.00	22.22
8	Flooring—all types	110.00	99.00	11.00	11.11	7.00	7.07	14.00	14.14
9	Waterproofing works	12.00	10.80	1.20	11.11	1.20	11.11	1.20	11.11
10	Aluminium work	13.50	12.15	1.35	11.11	0.50	4.12	2.50	20.58
11	Electrical work	35.00	31.50	3.50	11.11	1.00	3.17	6.00	19.05
12	Sanitary and plumbing works	20.00	18.00	2.00	11.11	1.00	5.56	1.00	5.56
13	Road works	16.00	14.40	1.60	11.11	1.60	11.11	3.00	20.83
	Total	500.00	450.00	50.00	100.00%	50.00	100.00%	50.00	100.00%

to the uniform mark-up of 11.11 per cent. This is also known as 'front-end rate loading' and has an effect of improving the cash flow in the early stages of the contract. It may be noted that the total mark-up is still ₹ 50 lakh.

In the third case, the mark-up percent is higher for the items that are planned to be completed in the final or later stages of the project. This is referred to as 'back-end rate loading'. It can be noticed from Table 8.9 that the activities such as plastering—all types, painting—all types, flooring—all types, aluminium work and electrical work have mark-up of 22.59 per cent, 22.22 per cent, 14.14 per cent, 20.58 per cent and 19.05 per cent, respectively, which are relatively higher than the uniform value of 11.11 per cent, although the total mark-up amount still remains at ₹ 50 lakh. The back-end rate loading is adopted when inflation rate is higher than the interest rate, so that the contractor can recover the amount on account of price escalation.

8.1.11 Computing Bid Price

After the estimation of direct cost and indirect cost, and the finalization of mark-up percent, it is not very difficult to compute the bid price for a project. For the purpose of illustrating the computation of bid price, let us assume that direct cost of a project is DC and the indirect cost is IC. Thus, total cost TC is given by

$$TC = DC + IC \qquad (8.10)$$

It may be recalled that mark up is applied either (1) in terms of some per cent of total cost TC or (2) in terms of some percent of the bid price B. The computation of bid price for both the situations are given below:

For the 1st case, that is when the mark up is expressed in terms of some percent of the total cost TC, the bid price is computed as:

$$B = TC + \frac{\text{mark up (\%)}}{100} \times TC \qquad (8.11)$$

Assuming that mark-up is 10 per cent of the total cost TC, the bid price B is given as:

$$B = TC + \frac{10}{100} \times TC \Rightarrow B = 1.10 \times TC$$

For the 2nd case, that is when the mark up is expressed in terms of some percent of the bid price B, the bid price is computed as:

$$B = TC + \frac{\text{mark up (\%)}}{100} \times B$$

$$\Rightarrow B = \frac{TC}{\left(1 - \frac{\text{mark up (\%)}}{100}\right)} \qquad (8.12)$$

Assuming that mark-up is 10 per cent of the bid price B, the bid price B is given as:

$$B = TC + \frac{10}{100} \times B \Rightarrow B = \frac{TC}{0.9} = 1.11 \times TC$$

The multiplication factor CO, for the direct cost of individual items would be given as:

$$CO = \frac{B}{L + M + P} \qquad (8.13)$$

where L, M and P are the labour, material and plant and equipment costs for all the activities of the project. The expressions for computation of L, M and P are given below:

$$L = \sum_{i=1}^{n} l_i \qquad (8.14)$$

$$M = \sum_{i=1}^{n} m_i \qquad (8.15)$$

$$P = \sum_{i=1}^{n} p_i \qquad (8.16)$$

In the above expressions, n is total activity/items in the project.

CO is also referred to as *cover on LMP*. The product of CO and $(L + M + P)$ amount is entered under the rate column of bill of quantities, which when multiplied with the total quantity of that item gives the amount corresponding to this item. A similar process is repeated for all the items and the total sum, i.e., bid price B, is obtained.

Also, knowing the bid price B and total cost TC, mark up (in terms of percent) can be obtained as given below:

(1) Mark up in terms of percent of total cost TC

$$\text{mark up (\%)} = \left(\frac{B}{TC} - 1\right) \times 100\% \qquad (8.17)$$

(2) Mark up in terms of percent of bid price B, referred to as off-top is given as:

$$\text{off-top (\%)} = \left(1 - \frac{TC}{B}\right) \times 100\% \qquad (8.18)$$

As mentioned earlier, mark up obtained in the 2nd case is referred to as '*off-top percent*' in some construction companies.

8.1.12 Submit Bid

After obtaining the bid price, an offer letter containing the final price and discounts on overall rates, if applicable, is prepared mentioning the validity of the offer in the letter. Any other terms and conditions, assumptions and deviations from the contract conditions are also mentioned in the offer letter. Sometimes, the owners ask for an unconditional offer. Under such circumstances, it is advisable to avoid putting any condition in the offer letter as that may lead to rejection of the bid. Other details and documents as desired by the owner are also submitted along with the offer letter. The entire document is submitted either in one packet or in two packets, technical bid and financial bid, as desired by the owner.

8.1.13 Post-submission Activities

The contractor, should he be invited, also needs to attend techno-commercial discussions. Based on these discussions, sometimes the contractor is asked to revise his bid price. In case the contractor is awarded the project, the estimating engineer compiles all tender documents giving details of the entire tendering process, including various correspondences with the owner, consultant, architect, vendor and subcontractor. The document so prepared is forwarded to the person in charge of executing the project. In case the bid is lost, all the compiled documents are analysed once again, and are kept in record for future references.

8.2 BIDDING MODELS

From the point of view of a contractor, the bidding process represents a game where he would like to 'win' a bid at a price that gives him the maximum profit. It is clear that the probability of winning a bid decreases as a contractor tries to increase his profit margin. Some of these issues have been discussed at some length in the section above on 'mark-up'. Further, the probability for a contractor to win a bid depends not only on his mark-up, etc., but also on the approach adopted by the competitors. It should also be noted that in present-day construction, the criteria laid down by the client for evaluation of bids become relevant to the bidding process. At times, a client may choose to evaluate a bid on criteria such as the contractor's record on quality in executing the job and schedule compliance, apart from consideration of the price quoted.

A suitable model based on probability to explain the bidding phenomenon in a construction project has yet eluded researchers, though initial efforts in the direction were made as early as 1956 by Friedman. A brief discussion on some of the models used is included here for the

```
                           Bidding models
                                 |
        ┌────────────────────────┼────────────────────────┐
   Game theory-based       Statistics-based         Cash flow-based
                                 |
                    ┌────────────┴────────────┐
              Expected monetary         Expected utility
                value-based               value-based
```

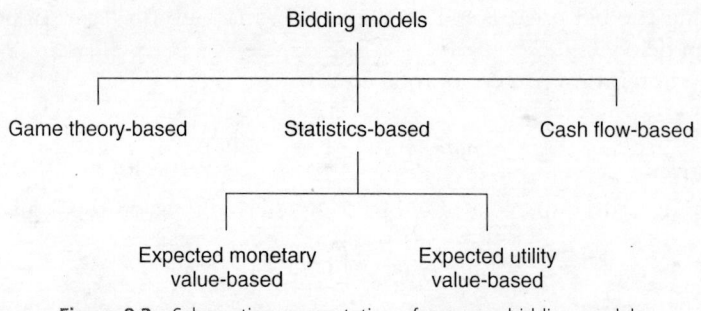

Figure 8.3 Schematic representation of common bidding models

sake of completeness, and a rigorous treatment of concepts of probability and statistics in the bidding process has been avoided.

A schematic summary of some of the common models is shown in Figure 8.3. The following paragraphs also explain some of the salient features of these models.

8.2.1 Game Theory Models

In these models, the probability distribution from which the opponent's costs are drawn is assumed to be known to both players. In other words, the objective of maximizing the expected profit in this problem implicitly assumes that both competitors have the same utility function or that they are risk-indifferent. Now, the game theory postulates a rational and intelligent opponent, whose interests are completely opposed to those of the other 'players'. The objective of the game theory approach is to find a pair of strategies that represent an equilibrium solution. In view of the several limiting assumptions, models based on the game theory have not been found to be of much practical value in the construction industry.

8.2.2 Statistical Bidding Strategy Models

All statistical bidding strategy models aim at assisting decision-makers in selecting an optimal mark-up level for a bid. For this, they rely mainly on historical data collected over a period of time, and the inherent assumption in all the models is that 'the competitor is of a consistent nature whose behaviour can be described by a probability distribution'. A successful bidding model is one that does not result in 'too cheap on a bid'. At times, a classification of statistical bidding strategy models into the following three categories of decision problems is presented in literature:

1. Decision-making under certainty
2. Decision-making under risk
3. Decision-making under uncertainty

It should, however, be noted that as far as the construction industry is concerned, 'decision-making under risk' is considered most relevant, where statistical analysis is carried out on the assumption that the contracting organization has enough historical data to estimate the probability of being a low bidder at various mark-up levels.

Yet another classification of the statistics-based models is given in terms of expected monetary value and expected utility value, as discussed briefly in the following paragraphs.

Expected Monetary Value Models

Friedman can be considered as the father of these models, which are based on the principle that profit maximization is the single objective function. The following assumptions were laid down for these models, and later became the basis of the successive models also.

- The objective of the contractor is to maximize his expected profit, i.e., his expected monetary value, in every bidding opportunity.
- The contractor would like to win the contract, even at a loss, in order to keep the production going.
- The contractor would like to minimize the competitor's profit.
- The actual cost of fulfilling a contract is a random variable and the ratio of the contractor's true cost to his estimated cost is represented by a random variable

Given their importance in terms of understanding of the bidding process, the statistical models proposed by Friedman (1956), Gates (1967) and Park (1962) are briefly discussed below. It should be noted that the basic assumption of all calculations is that a relationship exists between the tender sum and the 'probability' or 'chance' of winning the contract. Thus, the aim of the probabilistic models turns out to simply be able to express the relationship in a quantitative manner.

Friedman's model Friedman in 1956 attempted to develop an expected value model describing the bidding situation with probabilistic formulations. He adopted the approach in which the 'value' that a contractor associates with a potential profit on a job is synonymous with the monetary value of the profit. Friedman's (1956) model caters to the following cases:

1. All of the competitors on a particular job are known.
2. The exact number of competitors is known, but their identity is not known.
3. Both the number and identity of the competitors are unknown.

For the first case, Friedman concluded:

Probability of winning against a number of known competitors

$$= \text{Probability of beating competitor } A \\ \times \text{Probability of beating competitor } B \\ \times \text{Probability of beating competitor } C \\ \times \text{Probability of beating competitor } D \ldots \ldots \ldots, \text{etc.} \quad (8.19)$$

For the second case, he concluded:

Probability of winning against a number of unknown competitors for a given mark-up = (Probability of beating one typical competitor)^n (8.20)

The analysis of the third case is more complex and is beyond the scope of this text. A close observation of the above model shows that too much emphasis is placed on the number of bidders for a particular bid. Since the model utilizes the power relationship, it results into an abnormally low probability when the number of competitors increases. Taking lead from this, Gates proposed his bidding model on more or less similar assumptions.

Gates' model Gates, too, recognizes that the true cost of performing the work is a random variable at the time the bid is prepared. As per Gates (1967), the probability of winning against 'n' known competitors for a given mark-up

$$p = \frac{1}{\left[\frac{(1-p(A))}{p(A)} + \frac{(1-p(B))}{p(B)} + \frac{(1-p(C))}{p(C)} + \cdots + 1\right]} \qquad (8.21)$$

and the probability of winning against 'n' unknown competitors is given by

$$p = \frac{1}{n \times \left[\frac{(1-p(typ))}{p(typ)}\right] + 1} \qquad (8.22)$$

where p(typ) is the probability of beating a typical competitor.

Gates' model clearly indicates that a contractor who is competitive should be able to obtain 'his share' of the work and, hence, win approximately one job in seven against six competitors, if it is assumed that he has a 50 per cent chance of beating each one individually.

In the models suggested by Friedman and Gates, we find that in order to find the probability of winning associated with any mark-up percentage, we need to either (1) calculate the probability of beating individual contactors such as contractor A, B, or C, or (2) calculate the probability of beating a typical contractor. The two situations are described below.

Calculating probability of beating an individual competitor In order to calculate the probability of beating an individual competitor X by a contractor C, the step-by-step process is given below:

1. From the bids submitted on past contracts, the bid price of competitor X is obtained. It is usual for the owner to announce the bid price of all the competitors in a particular project.

2. Knowing the bid price, B, of competitor X on a project, the ratio B/TC for that project can be found out, where TC is the estimated cost of contractor C on that project. The B/TC ratio for a project would indicate the mark-up used by competitor X [recall equation (8.17): $mark\ up\ (\%) = \left(\frac{B}{TC} - 1\right) \times 100\%$]. For example, let us assume that the bid price B of competitor X is ₹ 1,000 lakh and the estimated cost, TC, of contractor C for this project is ₹ 900 lakh.

   Thus, B/TC for this particular project would be $= \frac{1000}{900} = 1.1111$, which corresponds to a mark-up $(\%) = \left(\frac{1000}{900} - 1\right) \times 100\% = 11.11\%$. Indeed, the inherent assumption here is that the costs of both competitor X and contractor C are almost equal.

3. Thus, contractor C will have a large set of data on B/TC and a number of such occurrences (frequency) corresponding to each B/TC ratio. For illustration, let us assume that contractor C has a record of 36 past bids in which he has met competitor X. The details are given in Table 8.10.

4. A histogram is now plotted between B/TC and the number of occurrences, as shown in Figure 8.4. The histogram shows that there was one instance when competitor X had quoted below the estimated cost of contractor C. There were two instances when competitor X had used a mark-up of 5 per cent. Likewise, the histogram can be interpreted.

CONTRACTOR'S ESTIMATION OF COST AND BIDDING STRATEGY | 281

Table 8.10 Details of bids involving C and X

S. No.	X's bid price, B, to C's estimated cost, C, (B/TC)	Mark-up % of X's bid	Number of occurrences (frequency), n
(1)	(2)	(3)	(4)
1	0.95	−5.00	1
2	1.05	5.00	2
3	1.10	10.00	3
4	1.15	15.00	4
5	1.20	20.00	5
6	1.25	25.00	6
7	1.30	30.00	5
8	1.35	35.00	4
9	1.40	40.00	3
10	1.45	45.00	2
11	1.50	50.00	1
		Total Σ n	36

5. If we have a sufficiently large data set (preferably 30 or more as in this case), we can compute the mean µ and standard deviation σ of the B/TC of all the bids and approximate these with a normal distribution curve as shown in Figure 8.5. Please note that the mean µ and standard deviation σ of the B/TC for the given example are 1.25 and 0.13 respectively. Now, using the properties of normal distribution, we can derive a wealth of infor-

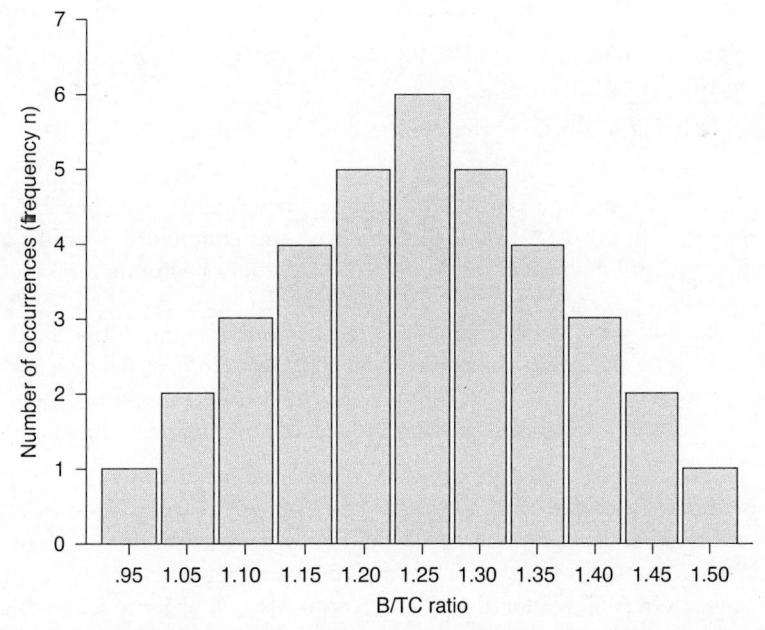

Figure 8.4 Histogram of B/C ratios of 36 past contracts in which contractor C competed with X

Std. dev = .13
Mean = 1.25
N = 36.00

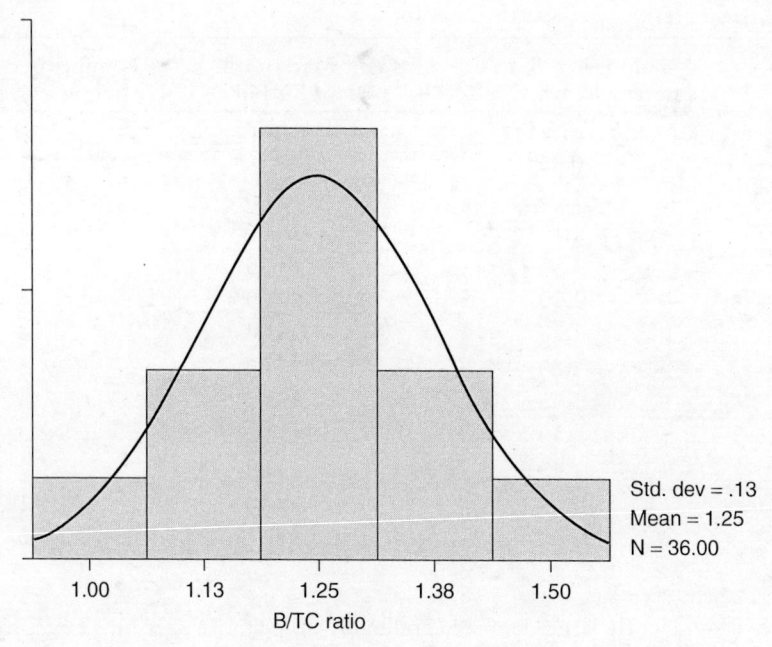

Figure 8.5 Normal distribution approximation of histogram shown in Figure 8.4

mation regarding the particular competitor, which can be helpful in formulating a suitable bidding strategy against this competitor.

For example, we can derive answers to questions such as:

1. What is the mark-up percent that will likely result in a 60 per cent or, say, 90 per cent probability of beating this competitor?
2. What is the probability of beating this competitor if contractor C bids at 10 per cent mark-up level?

The μ and σ characterize the competitor's behaviour. For example, a small value of σ for the B/TC ratio against a competitor would indicate that the competitor is using a consistent mark-up policy and it would be easier for contractor C to apply a winning mark-up against this competitor.

The record of μ and σ can be maintained for most frequent competitors. Thus, contractor C would know that if he is bidding at a mark-up level of, say, 10 per cent, he has a 40 per cent chance of beating competitor A, a 50 per cent chance of beating competitor B, and so on. This information can be put in the relevant bidding models to compute the probability of winning a bid.

Calculating probability of beating a typical contractor The situation described above has some limitations. First of all, contractor C would be more interested in knowing his chances to win the bid rather than knowing that he has a 50 per cent chance of beating competitor X, a 40 per cent chance of beating competitor Y, and so on. Also, in order to apply the probability of beating a particular competitor X, contractor C has to have a large set of data. There may be a situation wherein contractor C has very little past record against a particular competitor.

Worse still, contractor C may be facing some competitor for the first time. In such a situation, it is advisable to carry out the analysis based on the record of the lowest bidder in past projects. After all, contractor C would be interested in beating the lowest competitor at any point of time. The behaviour of a 'typical contractor' is assumed to be close to the behaviour of the lowest competitor. In other words, the μ and σ values of a typical contractor are obtained from the historical records of B/TC ratios of the lowest competitor. Now, B stands for the bid price of the lowest competitor and TC is the estimated cost of contractor C.

For this purpose, bid data of 21 construction projects were collected (Jha 2004). Due to confidentiality of the subject, only one construction company (we will denote this company by C) agreed to share the data on bids in which this company had participated. The bid details of the 21 cases studied, the price quoted by C in these 21 cases, the total cost estimated by C in each of these cases, and the quoted price of the lowest bidder are presented in Table 8.11. From the table, it can be seen that C has been a winner in three cases and the loser in eighteen.

Table 8.11 Details of bids involving C

S. No.	C's price	C's cost	Lowest price B	Lowest bidders' bid to C's cost, B/TC	Mark-up % of lowest bidder	Probability of beating lowest bidder at given mark-up %
(1)	(2)	(3)	(4)	(5)	(6)	(7)
1	812	731	578	0.79	−20.93	= 20/21 × 100 = 95.24
2	4,917	4,425	3,694	0.83	−16.52	= 19/21 × 100 = 90.48
3	3,800	3,420	3,050	0.89	−10.82	= 18/21 × 100 = 85.71
4	1,632	1,469	1,314	0.89	−10.55	= 17/21 × 100 = 80.95
5	1,481	1,333	1,226	0.92	−8.03	= 16/21 × 100 = 76.19
6	1,681	1,513	1,414	0.93	−6.54	= 15/21 × 100 = 71.43
7	2,153	1,938	1,799	0.93	−7.17	= 14/21 × 100 = 66.67
8	585	527	501	0.95	−4.93	= 13/21 × 100 = 61.90
9	617	555	532	0.96	−4.14	= 12/21 × 100 = 57.14
10	1,301	1,171	1,144	0.98	−2.31	= 11/21 × 100 = 52.38
11	3,146	2,831	2,820	1.00	−0.39	= 10/21 × 100 = 47.62
12	3,280	2,952	2,978	1.01	0.88	= 9/21 × 100 = 42.86
13	564	508	518	1.02	1.97	= 8/21 × 100 = 38.10
14	3,127	2,814	2,875	1.02	2.17	= 7/21 × 100 = 33.33
15	2,050	1,845	1,927	1.04	4.44	= 6/21 × 100 = 28.57
16	6,675	6,008	6,392	1.06	6.39	= 5/21 × 100 = 23.81
17	1,401	1,261	1,346	1.07	6.74	= 4/21 × 100 = 19.05
18	3,034	2,731	2,949	1.08	7.98	= 3/21 × 100 = 14.29
19	1,294	1,165	1,294	1.11	11.07	= 2/21 × 100 = 9.52
20	921	829	921	1.11	11.10	= 1/21 × 100 = 4.76
21	3,490	3,141	3,490	1.11	11.11	= 0/21 × 100 = 0.00

It can be further observed from Table 8.11 that price difference of C with respect to the lowest bidder has ranged from −20.93 per cent to 11.11 per cent. On comparison of the lowest bid prices with C's total cost estimates, it is found that the lowest bidder has bagged the contract at −20.93 per cent to 11.11 percent over C's total cost estimate. In eleven of the cases, the lowest bidder has submitted his bid at a price that is lower than even the total cost of C. It is not logical to assume that the lowest bidder will be quoting below his total cost estimates. This large difference is definitely due to the difference in total cost estimate of the organizations. Thus, the 'expected monetary value models', in which it is assumed that cost estimates of the competitors are at par, do not fit in here. The histogram corresponding to the data of Table 8.11 is shown in Figure 8.6, and the probability of beating the lowest bidder by C at different mark-up percentages is shown in Figure 8.7.

Using this approach, contractor C need not compute the individual probabilities of beating different competitors and then combine them to get the probability of winning. Instead, the probability of beating a typical bidder is directly read off from the cumulative frequency curve, such as given in Figure 8.7. It must be noted that one needs to have sufficient data in order to get a stable distribution. In case a sufficient number of past records is not available, the mean μ and standard deviation σ of a typical bidder can be assumed to be 1.06 and 0.065, respectively. Also, in case there is no information on number of bidders for a project, it can be assumed to be around six.

Comparison between Friedman's model and Gates' model According to Rickwood (1972), Friedman's model is found to be more correct when the cost estimates of different competitors are nearly the same and the difference in bid price is mainly due to the difference in mark-up. On the other hand, Gates' model gives more accurate results when mark-ups

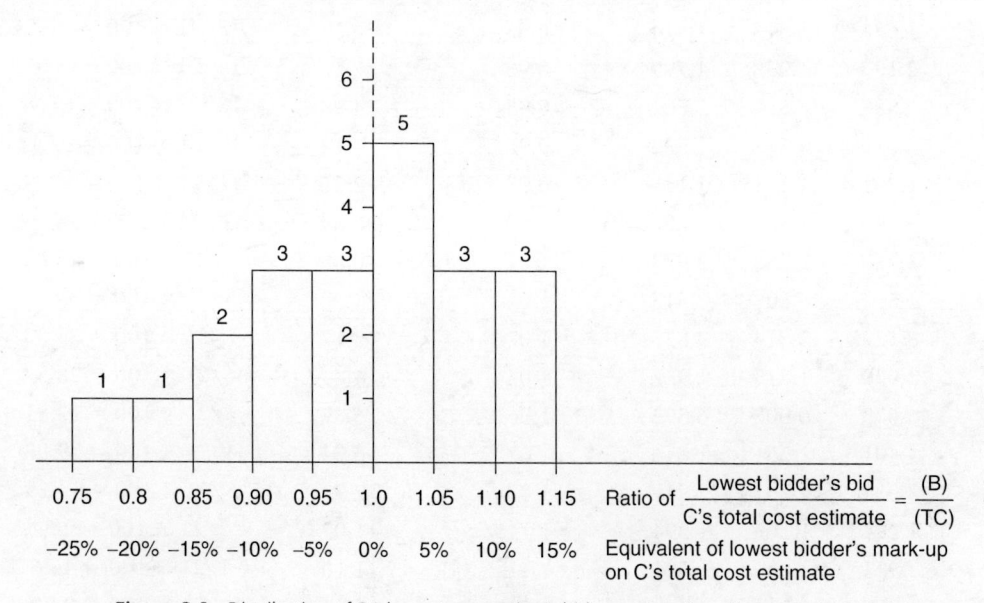

Figure 8.6 Distribution of 21 lowest competitor's bids compared to C's cost estimates

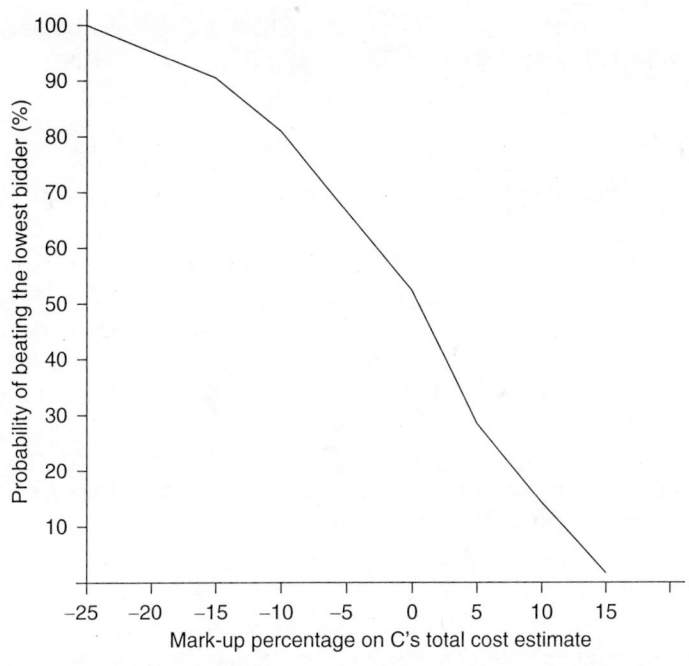

Figure 8.7 Probability of beating the lowest bidder vs mark-up

used by competitors are nearly the same and the difference in bid price is mainly on account of differences in cost estimates. Some researchers suggest taking the weighted average of the mark-ups obtained from Friedman's and Gates' models. Friedman's model, in most cases, determines a lower optimum mark-up than that of Gates'. In this sense, Gates' model is more optimistic as it assumes that one can still win the bid at a high mark-up. With Freidman's and Gates' models being viewed as pessimistic and optimistic, respectively, a moderate bidding strategy would be to assign equal weight to the mark-ups obtained using Friedman's and Gates' models, and thus consider the average of their optimum mark-ups.

Park's model Park (1962) presented a simplified version of Friedman's model wherein uncertainty associated with the true cost of executing the project was not considered. He suggested that both the number of bidders and the size of the job have some influence on the optimal mark-up. Park opined that when the number of bidders are more, say 10, then the chances for earning profit by the lowest bidder are quiet low. Park suggested the following relationship between number of bidders and mark-up:

$$\left[\frac{N_1}{N_2}\right]^x = \frac{M_2}{M_1} \tag{8.23}$$

Where N_1 and N_2 represent the number of bidders; M_1 and M_2 are the respective optimal mark-ups; and x is a constant, which is generally taken to be in the range of 0.5 to 0.8. Park also related the direct job cost to the optimal mark-up of another job with a different estimated direct job cost, by the following relation:

$$\left[\frac{C_1}{C_2}\right]^x = \frac{M_2}{M_1} \quad (8.24)$$

Where C_1 and C_2 are the estimates of the respective direct job costs and x is a constant in the range 0.15 to 0.30.

Bernoulli (1968) first critiqued the validity of the 'expected monetary value' and opined that though the price of any item is dependent only on the thing itself and is equal for everyone, the utility may be dependent on the particular circumstances of the person making the mark-up decision. He further reinforced his argument by placing the example of a pauper and a rich man both getting, say, ₹ 1,000. Although both are getting the same amount, the utility of ₹ 1,000 to a pauper may be more than it would be to the rich man. Many a time, the decision taken by a construction company cannot be explained by the expected monetary considerations alone, and under such situations, the 'expected utility value' models can give a better explanation of the rationale behind a particular decision.

Expected Utility Value Models

If a contractor were only interested in the accumulation and possession of wealth, he would preferably sell the company and retire once he 'had made it'. Shrewd investment of the proceeds from such a sale would provide a comfortable and probably more secure source of income. However, one is rarely guided by such instincts alone. Instead, there are certain other motivating factors such as the thrill of managing and completing a complex project, a sense of power and importance, and recognition by one's peers. Utility can be defined as the description of an individual's wealth preferences. Understandably, utility value would be unique for every individual and would be dependent on the current business status and the available opportunities. Mathematically, the utility functions are characterized under three classes—risk-indifferent, risk-averse and risk taker or gambler. Howard (1968) and Carr (1977) have made important contributions in the development of 'expected utility value models'.

8.2.3 Cash Flow-Based Models

The basic approach in all the cash flow-based models is the evaluation of effective cash flow characteristics of a project and the assumption of a rate of return during the bid stage. Fondahl and Bacarreza (1972) developed a cash flow model for the calculation of mark-up assuming a fixed minimum rate of return. The cash flows and mark-up calculations were made using a computer. They defined mark-up as:

$$\text{Mark-up (percent)} = \left(\frac{\text{Present worth of disbursements}}{\text{Present worth of receipts}} - 1\right) \times 100 \quad (8.25)$$

In the model developed by Fondahl and Bacarreza (1972), the minimum required rate of return and the selection of an appropriate rate of return are to be decided by the decision-maker. Farid (1981) also developed the Fair and Reasonable Mark-up (FaRM) pricing model based on cash flow approach.

8.3 DETERMINATION OF OPTIMUM MARK-UP LEVEL

A step-by-step process of optimum mark-up determination is delineated below:

1. To start with, a mark-up percent varying usually between 1 per cent and 20 per cent of total cost *TC* is assumed. This is increased in small increments of one per cent to start with, and is reduced as and when the mark-up is approaching the optimum level.

2. The mark-up amount corresponding to a given mark-up percent, *M*, is computed using the expression:

$$\text{Mark up amount} = \text{Bid price} - \text{Total cost} = B - TC \quad (8.26)$$

$$\Rightarrow \text{Mark up amount} = TC + \frac{\text{mark up (\%)}}{100} \times TC - TC = \frac{\text{mark up (\%)}}{100} \times TC \quad (8.27)$$

3. Probability to beat different competitors, say $x_1, x_2, x_3, \ldots x_n$, is computed from the past history of competitors, as explained earlier.

4. Using Friedman's model, the probability to win at mark-up level *M* is calculated.

5. Expected mark-up amount is determined using the expression:

$$\begin{aligned}&\text{Expected mark up amount}\\&= \text{Mark up amount for a given mark up percent } M\\&\times \text{Probability of winning at the mark up percent } M\end{aligned} \quad (8.28)$$

6. The mark-up percent is increased to $(M + 1)\%$ now, and the expected value of mark-up amount is calculated again as described in steps 3 to 5.

7. Based on the above values of mark-up percent and the expected mark-up amount, a curve is plotted for mark-up percent versus expected mark-up amount.

8. The optimum mark-up percent is read from the above plot. Let this be M_1.

9. Steps 1 to 8 are repeated for Gates' model also, and optimum mark-up is read from the plot. Let this be M_2.

10. The average mark-up $\left(\frac{M_1+M_2}{2}\right)$ can be taken as the optimum mark-up.

An example to illustrate the above process of optimum mark-up determination is given in the 'solved examples' section.

8.4 BIDDING AND ESTIMATION PRACTICES IN INDIAN CONSTRUCTION INDUSTRY

In this section, the current bidding and estimation practices for establishing the various components (viz. labour, material, plant and equipment, overhead and mark-up) adopted by the Indian construction industry are discussed. The findings presented here are based on a questionnaire survey conducted by Jha (2000) among 12 large and medium Indian construction companies. Analyses of responses confirm that despite the advancement in computational facilities, arbitrariness and discretionary approach in estimation and bidding process still prevail. The mark-up ranges prevailing in different fields of civil engineering projects are discussed as well, based on the analysis of the questionnaire survey.

8.4.1 Prevailing Estimation Practices

To a question on estimation practices adopted by the organizations, 82 per cent of the respondents confirm use of first principle by their firms, i.e., they analyse rates of individual items using the labour, material and plant costs for individual items. Only 18 per cent of the respondents accept referring to a standard schedule of rates, along the lines of Delhi Schedule of Rates (DSR[2]), and they update the rates from time to time for their future references. The only organizations that do not use the first principle of estimating are either the government or multinational firms dealing mostly in electrical and telecommunication field. However, all respondents agree that when they are asked to quote with reference to DSR for any private/public works, they quote with reference to DSR and their quote percentages (as in September–October 2000) range from −5 per cent to 30 per cent over DSR 1997. These quoted percentages are not project-specific (in terms of referring to complexity, size, location, etc.), but are dependent on the bidder organization's internal policies.

8.4.2 Use of Statistical/Mathematical Tools in Estimation

To a question on 'use of statistical/mathematical tools by the estimation engineers' of the respondent organizations, 55 per cent respondents confirm use of these tools for estimation purpose and for judging their chances of winning a bid. In order to verify the consistency among multiple responses received from one organization, checks are carried out across the responses to assess the actual practice of the organization. The checks reveal that there is complete agreement among all responses received from one organization.

8.4.3 Breakup of Mark-up

Through personal interviews and the literature survey, it is established that the mark-up is quite a complex phenomenon and is a function of various parameters. To assess the mark-up under individual components of work, a question was formed and it was assumed that the mean value of responses would indicate the expected mark-up on these components. The details of individual mark-up percentage applied to various components such as labour, material, plant and machinery, overhead and subcontractor prevailing within the construction industry are presented in Table 8.12. The actual estimation of individual components is discussed in subsequent paragraphs.

Table 8.12 Details of mark-up percentage

Components significantly contributing in deciding mark-up	Labour component	Material component	Plant and equipment component	Overhead component	Subcontractor component
Mean mark-up %	17	14	12	11	13
Standard deviation	7	5	8	6	4

---

[2] A periodical publication of CPWD, a central government organization that contains the schedule of rates prevailing during that time.

8.4.4 Labour Cost Estimation

For assessing the labour cost for any item, engineers rely mainly on the prevailing labour rates in the vicinity of sites and the productivity norms as published by various research organizations. In addition, rule of thumb or the past experience of the estimator is used. To assess the percentage of people using the above method of analysis of rates for various items of work, a question is asked. The response to this question indicates that a good 54 per cent of the respondents rely on productivity norms, while 42 per cent base their analysis on experience/thumb rule.

8.4.5 Plant and Equipment Cost Estimation

On a similar question on 'estimation of plant and equipment costs', depreciating cost of plant and equipment is considered to be the most significant factor, and a majority of the respondents (58 per cent) prefer this method. However, 42 per cent of the respondents use the concept of fixed-hire charges for estimating plant and machinery cost.

8.4.6 Dealing with Uncertainties

In response to the question on 'dealing with uncertainties in costs', a majority of the respondents (68 per cent) seem to take care of this by applying a correction factor. A small percentage (18 per cent) takes care of the uncertainty factor by adjusting the mark-up. Some respondents also adopt means other than the ones listed in the questionnaire for tackling this problem, while some others feel that the uncertainty gets covered in the escalation clause.

8.4.7 Average Range of Mark-up

Since the mark-up is a confidential point for any contracting organization, direct response to this point is extremely difficult. Therefore, a five-point ordinal scale for the response was designed by Jha (2000). Each point represents a range of mark-ups, such as 1 for mark-up range of 0 per cent–5 per cent, 2 for mark-up range of 5 per cent–10 per cent, 3 for mark-up range of 10 per cent–15 per cent, 4 for mark-up range of 15 per cent–20 per cent, and 5 for mark-up range of more than 20 per cent. However, the interpretation of responses is made with reference to the midpoint of the range values. For example, the response value of 3 for the range 10 per cent–15 per cent is considered as equivalent to 12.5 per cent mark-up; response value of 4 for the range 15 per cent–20 per cent is considered as equivalent to 17.5 per cent mark-up; and so on.

The statistics of responses to 'mark-up in different business sectors' are presented in Table 8.13. Column 2 presents the mean value of mark-up as interpreted from the midpoint of each response. On close observation of this table, it can be safely concluded that as the complexity of job increases and the job orients towards specialization, the mark-up range also tends to increase. It is also observed that increased competition in traditional areas like buildings, roads, large industrial projects and factories tends to pull the mark-up level down.

8.4.8 Mark-up Distribution

In response to the question on mark-up distribution, varied responses were received. Industry professionals prefer distribution of total mark-up amount in a mixed way. About 43 per cent of the respondents achieve this by evenly distributing the mark-up across all the items. Five per cent of the respondents front-load it, while 33 per cent achieve this distribution by unevenly

Table 8.13 Statistics of responses to mark-up percentage in different business sectors

Business sector	Mean value of mark-up derived from the range of response points	Sample size	Std deviation	Variance
Buildings	10.00%	16	4.80%	23.30%
Factories	11.56%	16	3.75%	14.10%
Roads	10.83%	12	5.36%	28.80%
Bridges	13.75%	10	4.29%	18.40%
Large industrial projects	10.44%	17	4.69%	13.20%
Piling job	12.95%	11	4.15%	17.30%
Tunnelling job	14.31%	11	5.13%	26.40%
Jetties	13.21%	11	1.89%	3.60%

loading it into different items. Nineteen per cent of the respondents confirm that they do not follow any fixed policy in this regard and depending on the situation, they take any of the above measures to achieve this goal. None of the respondents seems to distribute the mark-up by back-loading, and this is an expected outcome in Indian conditions since the problem of inflation is not so acute in this country.

8.4.9 Mark-up Range

The mark-up range that is operated by the various respondents for different types of projects is given in Table 8.14.

8.4.10 Summary and Conclusion from the Study

The study reveals that the primary mode of cost estimation is the 'first principle'. The survey also shows that no uniform method exists for the estimation of direct-cost items, viz. plant cost and sundry charges. These are organization-specific and depend on the guidelines prevailing in the organization.

Labour cost estimation is highly subjective and is based mainly on the experience of the decision-maker. Statistical and mathematical techniques are not much in use in the Indian construction industry.

Table 8.14 Percentage of projects with a given mark-up range in different types of projects

Mark-up range	Building	Factories	Roads	Bridges	Large industrial projects	Piling	Tunnelling	Jetties
0–5%	13%	—	—	—	6%	—	—	—
5–10%	38%	31%	67%	10%	47%	18%	18%	—
10–15%	43%	63%	8%	60%	35%	64%	46%	86%
15–20%	—	—	17%	20%	6%	9%	18%	14%
20–25%	6%	6%	8%	10%	6%	9%	18%	—

There is no uniformity of approach when plant and equipment costs are estimated for tendering purpose. This is also true while dealing with uncertainties in the cost items. In both these cases, a variety of methods are employed.

It is also observed that a contracting organization does not apply equal mark-up percentage on different components of direct and indirect costs, viz. labour, material, plant and machinery, subcontractor and overhead components.

Although there seems to be an increasing trend of mark-up level as the work becomes more specialized, the mark-up percentage differences observed between two business sectors with different expertise are not statistically significant. The prevailing mark-up percentage ranges from 10 per cent to 15 per cent for all business sectors. Similarly, project duration and project size also do not seem to influence mark-up decision, which is established through the insignificant differences between means of responses obtained for mark-up values for a variety of project durations and sizes. The mark-up range for this class also remains between 10 per cent and 15 per cent.

REFERENCES

1. Ahmad, I. and Minkarah, I., 1988, 'Questionnaire survey on bidding in construction.' *Journal of Management in Engineering*, ASCE, 4(3), pp. 229–243.

2. Al-Harbi, K.M., Johnston, D.W. and Fayadh, H., 1994, 'Building Construction Detailed Estimating Practices in Saudi Arabia', *Journal of Construction Engineering and Management*, ASCE, 120(4), pp. 774–784.

3. Benjamin, N.B.H., 1969, 'Competitive Bidding for Building Construction Contracts.' *Technical Report* No. 106, The Construction Institute, Department of Civil Engineering, Stanford University, Stanford.

4. Bernoulli, D., 1968, 'Expositions of a new theory of the measurement of risk', In *Utility Theory: A book of readings,* A.L.Page (Editor), New York: John Wiley and Sons.

5. Broemser, G.M., *Competitive bidding in the Construction Industry*, Ph.D. thesis, Graduate School of Business, Stanford University, Stanford, California, 1968.

6. Carr, R.I., 1989, Cost Estimating Principles, *Journal of Construction Engineering and Management*, ASCE, 115(4), pp. 545–551.

7. Chua, D.K.H. and Li, D., 2000, 'Key Factors in Bid Reasoning Model', *Journal of Construction Engineering and Management*, ASCE, 126(5), pp. 349–357.

8. de Neufville, R., Hani, E.N. and Lesage, Y., 1977, 'Bidding models: effects of bidders' risk aversion', *Journal of the Construction Division*, ASCE, 103(1), pp 57–70.

9. Farid, F., 1981, Fair and Reasonable Mark-up (FaRM) Pricing Model: a present value approach to pricing of construction contracts, Doctoral thesis submitted at The Graduate College, University of Illinois at Urbana Champaign.

10. Fondahl, J.W. and Bacarreza, R.R., 1972, Construction Contract Markup Related to Forecasted Cash Flow, Technical Report No.161, Construction Industry Institute, Department of Civil Engineering, Stanford University, Stanford.

11. Friedman, L., 1956, 'A competitive bidding strategy.' *Operations Research*, 4. pp. 104–112.

12. Gates, M., 1967, 'Bidding Strategies and Probabilities,' *Journal of Construction Division,* ASCE, paper 5159, 93, pp. 75–107.
13. Harris, F. and R. McCaffer, 2005, *Modern Construction Management,* 5th ed., Blackwell Publishing.
14. Hegazy, T., 2002, *Computer-based construction project management,* New Jersey: Prentice Hall.
15. Howard, R., 1968, 'The foundations of decision analysis', IEEE transactions on systems science and cybernetics, SSC4(3), pp. 211–219.
16. Jha, K.N., 2000, *Analysis of Mark-up in Indian Construction Contracts,* M.Tech. thesis, Department of Civil Engineering, Indian Institute of Technology Delhi.
17. Jha, K.N., 2004, Factors for the success of a construction project: An empirical study, Doctoal thesis submitted to the Department of Civil Engineering, Indian Institute of Technology Delhi.
18. Miller, D.W. and Starr, M.K., 1969, *Executive Decisions and Operations Research,* 2nd Edition, Englewood Cliffs: Prentice Hall.
19. Park, W.R., 1962, 'How Low to Bid to Get Both Job and Profit', Engineering News Record, June 20, pp. 122–123.
20. Park, W.R., 1964, Profit Optimization through Strategic Bidding, ASCE Bulletin, 6(5), December.
21. Rickwood, A.K., 1972, An investigation into the tenability of Bidding Theory and Techniques, M.Sc. project Report, Loughborough University.
22. Shash, A.A., 1998, 'Bidding Practices of Subcontractors in Colarado', *Journal of Construction Engineering and Management,* ASCE, 123(3), pp. 219–225.

SOLVED EXAMPLES

Example 8.1

Derive the rate from contractor's perspective for one m³ of excavation of earth (a) in normal soil where the depth of excavation varies from 0 m to 0.5 m, and (b) in a trench excavation where the depth of excavation varies from 0 m to 1.5 m. Analyse the rates for manual excavation.

Solution

(a) The working of rate for open excavation in normal soil, from 0 m to 0.5 m depth, is given in Table S8.1.1. It can be observed that contractor has productivity data from his past experience.

Table S8.1.1

Item	Productivity norms[3]	Rate	Remarks
Labour	2.16 man-hours @₹ 22.5/h[4]	= 2.16 × 22.50 = ₹ 48.60/m³	See footnote

---

[3] One should be cautious in using the productivity norms as they depend on a number of conditions discussed elsewhere in the book.

[4] Appropriate man-hour rates should be checked as it also varies from state to state. It may also vary from organization to organization.

(b) The working of rate for trench excavation in normal soil, from 0 m to 1.5 m depth, is given in Table S8.1.2. It can be observed that contractor has productivity data from his past experience.

Table S8.1.2

Item	Productivity norms	Rate	Remarks
Labour	3 man-hours @₹ 22.5/h	67.5	See footnote
Slope and working space 15%	15% of above	10.12	For large excavation, the exact slope can be calculated from construction methods drawing
Dewatering and shoring	Being considered as 1% of cost of labour	0.68	On a lump sum, though, it is possible to calculate this separately
	Total per m³	78.30	Say ₹ 78/m³

It should be noted here that the productivity norms vary from one site to another depending on a number of conditions. Similarly, the hourly rate for different classes of labour also varies from place to place and is dependent on a number of factors. These have been explained elsewhere.

Example 8.2
The total concrete quantity to be poured in a project is specified to be 100,000 m³ of M15 grade, and it is to be completed in 16 months, although the total project duration is 20 months. The contractor needs to estimate the cost of 1 m³ of M15 grade concrete.

Solution
As described earlier, the total cost of concreting would be the sum of the material cost, the labour cost, the plant and equipment cost, and the overhead cost. However, we are not calculating overhead cost for this problem. In real life, overhead costs are distributed on a pro rata basis across all the items of the project.

The plant and equipment cost and labour cost calculations are given in Part A, while material cost computation is given in Part B.

Part A Plant and equipment cost and associated labour cost
The cost calculation of plant and equipment involves calculation of type and number of plant and equipment, followed by the calculation of their owning cost, operating fuel and power cost, operating labour cost, labour cost for placing concrete, labour cost for curing, etc. These are explained below:

Estimate of plant and equipment The first step in such an exercise is to calculate the daily or hourly concrete production required to achieve the total concreting activities in the given duration. This is shown in Table S8.2.1.

Please note that for per-day quantity calculation, 26 days have been taken into account. This is on account of four non-working days (Sunday or any other day) in a month. Also, a multiplying factor of 1.5 has been considered since it is not practical to assume that work will always progress based on the plan. In construction projects, it is usual to start at lower production rates and then slowly attain a peak. Hence, in order to take care of peak concrete production requirement, a multiplying factor is used.

Table S8.2.1 Estimate of hourly concrete production

Total duration	= 20 months
Concrete duration	= 16 months
Concreting per day	= 10000/(16 × 26) = 240.38 m^3
Peak concreting 1.5 times × 240.38	= 360.57 m^3
Concrete per hour	= 360.57/8 hrs = 45 m^3

The type and number of plant and equipment requirement are worked out as given in Table S8.2.2. In case power supply is not available at the project location, the provision of diesel generator set should be considered appropriately.

The calculation of owning cost, operating fuel and power cost, operating labour cost, labour cost for placing concrete, and labour cost for curing are given in Tables S8.2.3, S8.2.4, S8.2.5, S8.2.6 and S8.2.7, respectively. The summary of plant and equipment cost and associated labour cost is provided in Table S8.2.8.

Part B Material cost for M15 grade concrete

The material cost for M15 grade concrete is shown in Table S8.2.9

The total cost per m^3 for M15 grade concrete is estimated as shown in Table S8.2.10

Example 8.3
Illustration of labour cost calculation

The labour can be of unskilled, semiskilled and skilled categories. Sometimes, it is required to calculate the average man-day rate for labour. In such cases, it is normal to assume a weighted

Table S8.2.2 Type and number of plant and equipment requirement

Type	No.	Remarks
Batching plant = 30 m^3 capacity	= 2	
Wheel loaders	= 2	
Transit mixers	4/Batching plant (depends on the lead, type of element, mode of placing (pump/manual, etc.)	
Mixer machines 10/7	= 5	Can be other than 5 depending up on the situation
Tough riders	= 3	
High lift pump 10 HP	= 3	
Curing pump	= 4	
Concrete pump with boom placer	= 1	
Weight batchers	= 5	
Vibrators	= 15	
Builders hoist	= 5	
Tower cranes (40% apportioning to concrete)	= 3	

Table S8.2.3 Plant- and equipment-owning cost (in ₹)

Description	No.	Hire charges/month (₹)	Months	Total (₹)
Batching plant	2	250,000	17	8,500,000
Wheel loader	2	46,000	17	1,564,000
Transit mixer	8	18,600	17	2,529,600
Mixer machine	5	2,000	17	170,000
Concrete pump	1	125,000	10	1,250,000
Weight batchers	5	1,000	17	85,000
Builders hoist	5	15,000	17	1,275,000
Vibrators	15	500	17	127,500
Tough riders	3	3,600	15	162,000
High-lift pump	3	10,000	17	510,000
Curing pump	4	1,700	17	115,600
Tower cranes (only 40%)	3	525,000	15	23,625,000
Subtotal				39,913,700
Spares and maintenance cost @20%				7,982,740
Batching plant/tower crane foundation/rails				150,000
Assembling and dismantling				600,000
Mobilization of P&M				1,000,000
Total for P&M-owning cost				49,646,440

Table S8.2.4 Plant- and equipment-operating fuel and power

Description	No.	Months	Days	Hours	Fuel (l) / Power (units)	Cost (₹)	O&L	Amount (₹)
Batching plant	2	16	26	8	64 units	6	1	2,555,904
Wheel loader	2	16	26	8	12 l	38	1.3	3,945,677
Transit mixer	8	16	26	8	12 l	38	1.3	15,782,707
Mixer machine	5	16	26	6	6 l	38	1.3	3,699,072
Builders hoist	5	16	26	6	15 units	6	1	1,123,200
Tough riders	3	16	26	6	1 l	38	1.3	369,907
Vibrators	10	16	26	5	1.5 units	6	1	187,200
Concrete pump	1	9	26	6	60 units	6	1	505,440
High lift pump	3	12	26	4	10 units	6	1	224,640
Curing pump	4	16	26	4	5 units	6	1	199,680
Tower cranes (40% for concrete)	3	14	26	6	45 units	6	1	707,616
Total operating fuel/power								29,301,043

Table S8.2.5 Plant- and equipment-operating labour

Description	No.	Months	Days	Wage (₹)	Amount (₹)
Batching plant	4	16	26	220	366,080
Transit mixer/wheel loader	10	16	26	220	915,200
Mixer machine	5	16	26	180	374,400
Vibrators	10	16	26	180	748,800
Builders hoist	5	16	26	220	457,600
Pumps	4	16	26	220	366,080
Concrete pump	1	12	26	220	68,640
Tower cranes (40% for concrete)	3	12	26	350	327,600
Tough riders	5	12	26	200	312,000
Subtotal					3,936,400
Overtime @50%					1,968,200
Total					5,904,600

Table S8.2.6 Concrete-placing labour

Description	No.	Months	Days	Wage (₹)	Amount (₹)
Mason	15	16	26	200	1,248,000
F masons	15	16	26	180	1,123,200
Carpenters	10	16	26	200	832,000
Fitters	10	16	26	200	832,000
Mate	5	16	26	200	416,000
Helpers	30	16	26	180	2,246,400
Subtotal					6,697,600
Overtime charges @50%					3,348,800
Total for placing labour					10,046,400

Table S8.2.7 Labour cost for curing

Description	No.	Months	Days	Wage (₹)	Amount (₹)
Pump operator	4	16	30	200	384,000
Helpers/coolies	10	16	30	180	864,000
Sunday wages					
Operators	4		75	200	60,000
Helpers/coolies	10		75	180	135,000
Total for curing labour					**195,000**
Miscellaneous					
Admixtures 50,000 m^3 × ₹ 30					1,500,000

Table S8.2.8 Summary of plant and equipment cost and associated labour cost

Description	Amount (in ₹)
Plant- and machinery-owning cost	49,646,440
Fuel/power cost	29,301,043
Operating labour	5,904,600
Placing labour	10,046,400
Curing labour	195,000
Miscellaneous	1,500,000
Total	96,593,483
Cost per m³	965.93

Table S8.2.9 Material cost for 1 m³ of M15 grade concrete

Description	Cost	Remarks
Cement 5.8 bags × ₹ 225	= ₹ 1,305	The quantity of cement, sand and aggregate are to be calculated from mix design or from standard norms. Rates to be verified.
Sand 0.6 m³ × ₹ 700	= ₹ 420	
Aggregate 20 mm 0.9 m³ × ₹ 900	= ₹ 810	
Total per m³	= ₹ 2,535	

Table S8.2.10 Summary of total cost for 1 m³ of M15 grade concrete

Description	Cost	Remarks
Plant and equipment and labour for 1 m³	= ₹ 966	From A
Material cost for 1 m³ of M15 grade concrete	= ₹ 2,535	From B
Miscellaneous labour for pouring 1 m³	= ₹ 25	Based on existing labour rates
Grand total cost for 1 m³ of M15 grade concrete	= ₹ 3,526	Excludes overhead cost

average on the assumption that for every skilled workman, there would be 2 semiskilled workmen and 4 unskilled workmen. The basic minimum wages across the country for these categories of labourers vary. For example, the rates payable in Delhi would be different from the rates payable in, say, Uttar Pradesh and Rajasthan. A typical calculation is illustrated in Table S8.3.1.

Table S8.3.1 Illustration of labour cost computation

S. No.	Description	Skilled	Semiskilled	Unskilled
1	Basic rates	220	200	180
2	PF @12.0%	26.4	24	21.6
3	Tools/tackles and minor consumable @7.5%	16.5	15	13.5
4	Overtime @25%	55	50	45
5	Total	317.9	289	260.1

(Continued)

Table S8.3.1 Continued

S. No.	Description	Category Skilled	Semi-skilled	Unskilled
6	Overhead and profit @12.5%	39.74	36.125	32.51
7	Grand total	357.64	325.125	292.61
8	Weighted average man-day cost, considering one skilled worker for every two semi-skilled workers and four unskilled workers	₹ 311.19; say, ₹ 311 [(1 × 357.64 + 2 × 325.125 + 4 × 292.61)/7]		
9	Hourly rate considering 8 hours of working time	311/8 = 38.875; say, ₹ 39/h		

Example 8.4 Illustration of material cost calculation

Material cost calculation is the easiest among all the cost calculations. Some illustrations are provided in Table S8.4.1, Table S8.4.2 and Table S8.4.3.

Table S8.4.1 Calculating cost of cement OPC-43 Grade

S. No.	Description of item	Unit	Quantity	Cost/Unit	Amount (₹)
1	Basic rate of material ex-cement factory price inclusive of all taxes	t	1.00	4,500	4,500.00
2	Transportation to site	t	1.00	1,000	1,000.00
3	Inter-carting and handling at site (assuming 4 handling altogether @₹ 2/bag)	t	1.00	160	160.00
4	Total (1 + 2 + 3)	t	1.00		5,660.00
5	Wastage @2.5%	t	1.00		141.50
6	Credit from selling empty bags @₹ 2 per bag	t	1.00	40	40.00
7	Total cost per MT (4 + 5 − 6)	t	1.00		5,761.50
8	Cost per bag (7/20) considering 50 kg per bag	Bag	1.00		288.075; say, 288/bag

Table S8.4.2 Calculating cost of steel reinforcement (Fe 415)

S. No.	Description of item	Unit	Quantity	Cost/Unit	Amount (₹)
1	Basic rate of material	t	1.00	29,000	29,000.00
2	Transportation to site	Incl.	1.00		
3	Inter-carting and handling at site	t	1.00	300	300.00
4	Wastage	t	3%	29,000	725.00
5	Deduct for scrap	t		12,000	
6	Cost per ton				30,025.00

Table S8.4.3 Calculating cost of fine sand

S. No.	Description of item	Unit	Quantity	Cost/Unit	Amount (₹)
1	Basic cost at site	m^3	1.00	750	750
2	Add for screening	m^3	1.00		
3	Wastage	m^3	8.00%	750	60
4	Cost per m^3				810

Example 8.5 Illustration of plant and equipment cost

A contractor wants to estimate the equipment cost for undertaking earthwork involving a quantity of 40,000 m^3. The contractor has derived the following requirement for equipment for this work.

Equipment required	Capacity	Nos required	Duration for which required
Poclain CK 90	50 m^3/h	1	5 months (10 hrs working considering 2 hrs overtime)
Dumpers/tippers	6 m^3 capacity	3	5 months (10 hrs working considering 2 hrs overtime)

Calculate the equipment cost per unit m^3 of earthwork for the contractor.

Solution

Plant and equipment cost consists of owning and operating cost, the cost of fuel, oil and lubricants, cost of spare parts, and cost towards minor and major repairs. The option for a contractor is either to own the equipment or to hire them on rental. The costs of the above equipment are calculated under three cost heads, owning cost, operating fuel and operating labour, as shown in Tables S8.5.1, S8.5.2 and S8.5.3, respectively.

Table S8.5.1 Equipment-owning cost

S. No.	Description	Nos	Hire charge/ month (in ₹)	Months	Total (in ₹)
1	Poclain	1	125,000	5	625,000
2	Dumpers	3	50,000	5	750,000
3	Subtotal (1 + 2)				1,375,000
4	Spares and maintenance (3 × 0.25)			@25%	343,750
5	Mobilization of Poclain				50,000
6	Mobilization of tippers @₹ 20,000 per number for 3 numbers				60,000
7	Total owning cost				1,828,750

Table S8.5.2 Equipment-operating fuel cost

Description	Nos	Months	Days	Hrs	F/P	Cost	O&L	Amount (in ₹)
Poclain	1	4	26	10	13 l	38	1.3	667,888
Tippers	3	4	26	10	5 l/h	38	1.3	770,640
Total cost for operating fuel								**1,438,528**

Table S8.5.3 Equipment-operating labour

Description	No.	Months	Days	Wage (₹)	Amount (₹)
Poclain operator	1	4	26	220	22,880
Helpers for bucket cleaning	2	4	26	180	37,440
Dumper drivers	3	4	26	200	62,400
Subtotal					**122,720**
Add overtime charges @50%					61,360
Total for P&M operating labor					**184,080**

Table S8.5.4 Summary

Description	Amount (in ₹)
P&M-owning cost	1,828,750
Fuel/power cost	1,438,528
Operating labour	184,080
Total	3,451,358
Cost per m³ 3,451,358/40,000 = 86.28	86 per m³

Example 8.6 Illustration of subcontractor cost calculation

A contractor is evaluating the bids of 4 subcontractors for a given item in a project. The details of final price and exclusion/inclusion in the bid price of the 4 subcontractors are given in Table S8.6.1. The contractor has estimated that the charges towards electricity, should it be provided to the subcontractor would cost ₹ 25,000. The contractor has further estimated that the charges towards plant and machinery, should it be provided to the subcontractor would cost ₹ 75,000. The tax applicable for the given item is at present 12%. Help the contractor award the bid to the most eligible subcontractor.

Table S8.6.1 Data for Example 8.6

Item description	Subcontractor 1	Subcontractor 2	Subcontractor 3	Subcontractor 4
Price (₹)	550,000	520,000	475,000	490,000
Tax included	Yes	No	Yes	Yes
Electricity included	Yes	Yes	No	No
Plant and machinery included	Yes	Yes	No	No

Solution

The computation details of the total bid price which would be the cost to the contractor for engaging the subcontractor is shown in Table S8.6.2. It is clear that awarding the given item to Subcontractor 1 would have least cost liability for the contractor. Thus, Subcontractor 1 is most eligible for this item.

Table S8.6.2 Data for Example 8.6

Item description	Subcontractor 1	Subcontractor 2	Subcontractor 3	Subcontractor 4
Price (₹)	550,000	520,000	475,000	490,000
Tax @12% of above in case it is excluded (₹)	—	62,400	—	—
Electricity charges -add ₹ 25,000 if it is excluded from the bid (₹)	—	—	25,000	25,000
Plant and machinery charges -add ₹ 75,000 if it is excluded from the bid (₹)	—	—	75,000	75,000
Total bid amount = the cost implication for the contractor (₹)	550,000	582,400	575,000	590,000

Example 8.7

For a project, the schedule of quantities to be executed month-wise along with the total quantities are given in Figure Q8.7.1. Compute the average monthly labour requirement and the associated cost towards labour benefit.

Sl No.	Item description	Quantity	1	2	3	4	5	6	7	8	9	10
1	Earthwork-all soils	5,000	250	750	1,000	1,250	1,250	500				
2	Concrete-PCC & RCC	2,000		200	400	400	500	300	200			
3	Centring & shuttering	14,000		1,400	2,800	2,800	3,500	2,100	1,400			
4	Reinforcement & structural steel	250		25	50	50	63	38	25			
5	Brickwork	1,700			170	255	340	340	340	170	85	
6	Plastering-all types	20,000				2,000	3,000	4,000	4,000	4,000	2,000	1,000
7	Painting-all types	5,000				500	750	1,000	1,000	1,000	500	250
8	Aluminium work	300					75	75	75	75		

Figure Q8.7.1 Schedule of quantities for Example 8.7

Solution

This is achieved in two steps—(1) calculating total man-hours and, hence, total labour days required for executing the entire project, and (2) calculating labour benefits.

1. Calculating total man-hours and, hence, total labour days required for executing the entire project

In order to calculate the total man-hours needed to complete the project, we need to know (1) the month-wise schedule of quantities to be executed and (2) the productivity norms for executing the different activities. In this example, we are given the information on the first part. For the second part, we may refer to the appropriate productivity norm. The following values of productivity are obtained.

The monthly labour requirement has been calculated on the assumption of 10 working hours every day (8 normal hours and 2 overtime hours) and 26 workdays every month. Thus, for month one, the labour requirement for earthwork would be calculated thus:

$$\text{Labour required} = \frac{(\text{quantity scheduled for the month} \times \text{productivity norm})}{(\text{working hours per day} \times \text{working days in a month})}$$

$$= \frac{250 \times 3}{10 \times 26} = 2.88, \quad \text{say 3 workers}$$

Similarly, the labour required for concreting in month 2 would be

$$\text{Labour required} = \frac{(\text{quantity scheduled for the month} \times \text{productivity norm})}{(\text{working hours per day} \times \text{working days in a month})}$$

$$= \frac{200 \times 18}{10 \times 26} = 13.85, \quad \text{say 14 workers}$$

In addition to the above labour requirement, the project would also require some miscellaneous workers such as electrician, mechanic, security men, store helper, office peon, computer operator, and typist. For calculating the labour benefits, the total number of such miscellaneous workers is also added in the average labour requirement. Let us assume that the total miscellaneous workers are estimated to be 9 every month for the total project duration of 10 months.

Hence, total number of labours required on the project month-wise would be as given in Table S8.7.2 and Figure S8.7.1). The average requirement of labour would be = (12 + 60 + 112 + 131 + 165 + 120 + 94 + 44 + 25 + 17)/10 = 780/10 = 78, and the total man-hours for the project would be = 78 × 10 × 26 = 20,280. The total man-hours required for a project is an important data used for knowing the quantum of works involved in a project. It is another way of speaking about a project. We talk about a project in terms of one million labour hours, one-and-a-half million labour hours, and so on. If we talk about projects in terms of cost or capacity, we cannot get a very clear picture of the magnitude of a project, while if we talk about projects in terms of labour hours it is more relevant due to its wider applicability.

The average monthly requirement of workers is useful for calculating the labour benefits, which ultimately form part of the indirect costs. The calculation of labour benefits is shown below:

Table S8.7.1 Solution details for Example 8.7

S. No.	Item	Unit	Quantity	P[5]	M1	M2	M3	M4	M5	M6	M7	M8	M9	M10
1.	Earthwork	m³	5,000	3	3	9	12	15	15	6				
2.	Concreting	m³	2,000	18		14	28	28	35	21	14			
3.	Formwork	m²	14,000	4		22	43	43	54	32	22			
4.	Reinforcement	t	250	60		6	12	12	15	9	6			
5.	Brickwork	m³	1,700	12			8	12	16	16	16	8	4	
6.	Plastering	m²	20,000	1				8	12	16	16	16	8	4
7.	Painting	m²	5,000	2				4	6	8	8	8	4	2
8.	Aluminium work	m²	300	9					3	3	3	3		
9.	Monthly requirement for project activities				3	51	103	122	156	111	85	35	16	8

Table S8.7.2 Computation of labour requirement

Month	1	2	3	4	5	6	7	8	9	10
Labour requirement from Table S8.7.1	3	51	103	122	156	111	85	35	16	8
Miscellaneous workers (9 assumed)	9	9	9	9	9	9	9	9	9	9
Total for the month	12	60	112	131	165	120	94	44	25	17

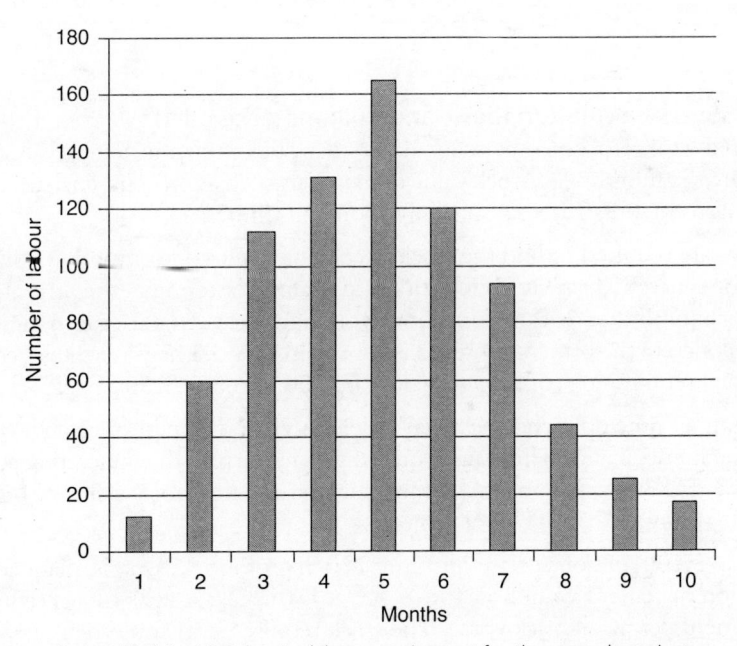

Figure S8.7.1 Month-wise labour requirement for the example project

---

[5] *P* indicates productivity norms in man-hours assumed by the contractor. As discussed elsewhere, this may vary from contractor to contractor.

2. Calculating labour benefits

The average number of labour calculated for the above problem is 78. Let us assume that the average wages paid is ₹ 125 per day. The details of computations for labour benefits are shown in Tables S8.7.3 and S8.7.4.

Table S8.7.3 Calculation details for labour benefits

S. No.	Description	Value (₹)	% of wages	Amount (₹)
1	Bonus (20% of the total wages including overtime)	804,375	20%	160,875
2	EPF and family pension	14,062,500	12%	843,750
3	ESI (applicable only when the proposed project is inside a running factory)		6.25%	
4	Subtotal			1,004,625

Table S8.7.4 Calculation for retrenchment and notice pay

S. No.	Description	No. of workmen	Days payable	Wage per day (₹)	Amount (₹)
5	Retrenchment	24	15	125	45,000
6	Notice pay	24	26	125	78,000
7	Subtotal				123,000
	Total labour benefit (4 + 7)				1,127,625

Example 8.8 Stock calculation (basic and bulk material part)

This is illustrated with respect to the schedule of quantities shown in Figure Q8.7.1.

We calculate the month-wise stock value of materials such as cement, coarse aggregate, fine aggregate, reinforcement, structural steel, plywood and timber.

- Suppose we are required to find the stock of cement required for month 2. Now, cement is required for concreting activity, brickwork activity and plastering in the project. Concreting planned for month 2 = 200 m^3, and brickwork and plastering are nil. Assuming that for one m^3 of concrete the cement required is approximately 6.5 bags (each bag containing 50 kg), the total requirement of cement would be 200 × 6.5 = 1,300 bags.
- If the lead time (time difference between order placed and receipt of cement) for cement is 15 days, the number of cement bags required for supporting the concreting activity would be 1,300 × 15/30 = 650 bags. In monetary terms, the stock value = 650 × 225 = ₹ 146,250 (assuming ₹ 225/bag of cement).
- For month 3, 400 m^3 of concreting and 170 m^3 of brickwork is planned. The total consumption of cement would be 400 × 6.5 + 170 × 2.2 = 2,974 bags (considering 2.2 bags of cement per m^3 of brickwork). The stock requirement would be 2,974 × 15/30 = 1,487 bags, assuming 15 days' lead time. In monetary terms, it is equivalent to 1,487 × 225 = 334,575. In this manner, the month-wise stock of cement in monetary terms can be calculated.

- For calculation of stock of coarse aggregate and fine aggregate, the material coefficients for unit quantity can be referred to and the stock value obtained assuming appropriate lead time.
- For formwork material such as staging material (props, cup-lock system, framing material, etc.) and sheathing material (plywood and timber), we need to know the formwork area scheduled to be achieved for the month, the cycle time for formwork activity and the repetition possible for the formwork materials. A detailed discussion can be found elsewhere in the text.

Example 8.9 Stock calculation (temporary structure part)

A contractor has to construct a number of temporary structures at site including site office, cement godown, store, client's office, architect's office, maintenance workshop, carpentry workshop, canteen, labour hutment and site barricading. A substantial portion of money goes into constructing these temporary structures and they are also treated as part of stock at site. An illustration for providing such stock value is shown in Table S8.9.1.

Table S8.9.1 Stock calculation for temporary structure

S. No.	Description	Unit	Quantity	Unit rate	Amount (₹)
1	Site office–within site	m²	77.5	2,500	193,750
2	Cement godown	m²	60.0	1,250	75,000
3	Store	m²	60.0	1,250	75,000
4	Client's office	m²	60.0	4,500	270,000
5	Architect's office	m²	35.0	4,500	157,500
6	Maintenance workshop	m²	40.0	1,250	50,000
7	Carpentry shop	m²	40.0	1,250	50,000
8	Canteen	m²	40.0	750	30,000
9	Labour hutment for 500	m²	13,500	100	1,350,000
10	Site barricading	m²	1,000	235	235,000
11	Computers	No.	3	50,000	150,000
12	Client's facilities	LS			1,000,000
	Subtotal				3,398,750
					Say, ₹ 34 lakh

It is usual to distribute the total cost of temporary construction on a pro rata basis depending on the invoicing achieved per month.

Example 8.10 Stock calculation

For a particular construction project, the month-wise stock of basic and bulk materials (in monetary terms) is given in Table Q8.10.1. The contractor's estimated cost of construction of temporary structure is ₹ 34 lakh. The month-wise estimated invoice (in terms of percentage of contract value) is also given. Calculate the average monthly stock value for the project.

Table Q8.10.1 Stock values (₹ lakh)

S. No.	Description	1	2	3	4	5	6	7	8	9	10
	Invoice	2%	10%	15%	20%	15%	15%	10%	7%	4%	2%
1	Cement	1	2	5	8	6	6	3	2	1.75	0.5
2	Coarse aggregate	0.4	0.6	1.4	2.5	2	1.75	1	7.5	5	0.4
3	Fine aggregate	0.4	0.6	1.4	2.5	2	1.75	1	7.5	5	0.4
4	Reinforcement	4	6	8	10	8.5	6.25	4	2.5	1	0.6
5	Plywood	0.4	0.8	1	0.8	7	0.65	0.5	0.35	0.3	0.1
6	Timber	0.4	0.8	1	0.8	7	0.65	0.5	0.35	0.3	0.1

Solution

We split the stock of temporary structures (₹ 34 lakh obtained through Example 8.9) month-wise on pro rata basis of invoice values. Thus, the contribution of stock of temporary structures for month 1 will be 2% of ₹ 34 lakh = ₹ 0.68 lakh; for month 2 it will be 10 per cent of ₹ 34 lakh = ₹ 3.4 lakh; and so on. These values are added in the total month-wise stock of basic and bulk materials, as shown below:

S. No.	Description	1	2	3	4	5	6	7	8	9	10
	Invoice	2%	10%	15%	20%	15%	15%	10%	7%	4%	2%
A	Stock of temporary structure	0.68	3.4	5.1	6.8	5.1	5.1	3.4	2.38	1.36	0.68
B	Cement	1	2	5	8	6	6	3	2	1.75	0.5
C	Coarse aggregate	0.4	0.6	1.4	2.5	2	1.75	1	7.5	5	0.4
D	Fine aggregate	0.4	0.6	1.4	2.5	2	1.75	1	7.5	5	0.4
E	Reinforcement	4	6	8	10	8.5	6.25	4	2.5	1	0.6
F	Plywood	0.4	0.8	1	0.8	7	0.65	0.5	0.35	0.3	0.1
G	Timber	0.4	0.8	1	0.8	7	0.65	0.5	0.35	0.3	0.1
H	Stock of basic and bulk materials H = (B + C + D + E + F + G)	6.6	10.8	17.8	24.6	32.5	17.05	10	20.2	13.35	2.1
I	Total stock value (₹, in lakh) I = A + H	7.28	14.2	22.9	31.4	37.6	22.15	13.4	22.58	14.71	2.78
J	Cumulative stock	7.28	21.48	44.38	75.78	113.38	135.53	148.93	171.51	186.22	189.0

Thus, average monthly stock = cumulative stock/duration = 189.0 /10 = 18.9 lakh.

Example 8.11 Average outstanding and unadjusted advance calculation

For the given data, compute the average outstanding and unadjusted advance.

Table Q8.11.1 Data for Example 8.11

Month	1	2	3	4	5	6	7	8	9	10
% invoice (₹ lakh)	2	10	15	20	15	15	10	7	4	2

1. Contract value = ₹ 30 crore. The projected percentage invoicing every month is given in Table Q8.11.1
2. Mobilization advance = 5% of contract value, to be recovered in 5 equal instalments beginning 3rd month
3. Retention money = 10% of billed amount (invoice), payable fully one month after the completion of project
4. Contractor raises the bill for the month on the last day of that month
5. Owner delays the payment by one month from the date of submission of bill

Solution

The computation of month-wise invoice, cumulative invoice, cumulative retention, cumulative outstanding and cumulative unadjusted advance is shown in Table S8.11.1. From the table, we can find the average invoice, average outstanding and average unadjusted advance.

Table S8.11.1 Solution details for Example 8.11

	Month	1	2	3	4	5	6	7	8	9	10	11
1	Month	1	2	3	4	5	6	7	8	9	10	11
2	% invoice (of contract value)	2	10	15	20	15	15	10	7	4	2	
3	Invoice value (₹ lakh)	60	300	450	600	450	450	300	210	120	60	
4	Cumulative invoice	60	360	810	1,410	1,860	2,310	2,610	2,820	2,940	3,000	
5	Retention (10% of invoice)	6	30	45	60	45	45	30	21	12	6	
6	Cum retention	6	36	81	141	186	231	261	282	294	300	
7	Payment due	54	324	729	1,269	1,674	2,079	2,349	2,538	2,646	2,700	300
8	Payment received		54	324	729	1,269	1,674	2,079	2,349	2,538	2,646	2,700
9	Outstanding	54	270	405	540	405	405	270	189	108	54	
	Cumulative outstanding	54	324	729	1,269	1,674	2,079	2,349	2,538	2,646	2,700	
10	Unadjusted advance	150	150	120	90	60	30	0	0	0	0	0
11	Cumulative unadjusted advance	150	300	420	510	570	600	600	600	600	600	

$$\text{Average invoice} = \frac{\text{Cumulative invoice}}{\text{Duration}} = \frac{3,000}{10} = Rs.\ 300\ lakh$$

$$\text{Average outstanding} = \frac{\text{Cumulative outstanding}}{\text{Duration}} = \frac{2,700}{10} = Rs.\ 270\ lakh$$

$$\text{Average unadjusted advance} = \frac{\text{Cumulative value of unadjusted advance}}{\text{Duration}} = \frac{600}{10} = Rs.\ 60\ lakh$$

Example 8.12 Working capital computation

For a given project, the contractor has estimated that he would be required to keep an average stock (basic and bulk materials, and pro-rated temporary structure stock) equivalent to ₹ 20 lakh. Further, it is also estimated that the average outstanding and average unadjusted advance would be ₹ 270 lakh and ₹ 60 lakh, respectively. Estimate the average monthly working capital required for the project.

Solution
Given,

Average monthly stock = ₹ 20 lakh
Average outstanding = ₹ 270 lakh
Average unadjusted advance = ₹ 60 lakh

We have,

Average working capital
= Average monthly stock + Average outstanding − Average unadjusted advance
= 20 + 270 − 60 = ₹ 230 lakh

Thus, average working capital = ₹ 230 lakh

Example 8.13 Average fixed asset (AFA) computation

A project requires a total of four equipment. The approximate book value of equipment on arrival at site is given (see Table Q8.13.1). The contracting company follows a straight-line method of depreciation, and rate of depreciation is one per cent every month. Calculate the average fixed asset (AFA) employed by the contracting company for the project assuming that mixer machine, weight batcher and builders hoist are required for 10 months, and dewatering pump is required for 6 months only.

Table Q8.13.1 Data for Example 8.13

S. No.	Description	Estimated book value on arrival at site (in ₹)	Nos	Requirement
1	Mixer machine	112,000	1	For 10 months
2	Weight batcher	68,000	1	For 10 months
3	Builders hoist — 2t	166,000	1	For 10 months
4	Dewatering pump	267,000	1	For 6 months

Solution

For finding the average fixed asset, we need to know the type of equipment asset and their required numbers. Further, we need to know the opening and closing book values of these assets. From these book values, the average book value for each of the assets is computed. This is shown in Table S8.13.1.

Table S8.13.1 Solution details for Example 8.13

S. No.	Description	Opening book value	Closing book value	Average of opening and closing book values
1	Mixer machine	112,000	Book value at the end of 10 month = 112,000 − 10% of 112,000 = 100,800	(112,000 + 100,800)/2 = 106,400
2	Weight batcher	68,000	Book value at the end of 10 month = 68,000 − 10% of 68,000 = 61,200	(68,000 + 61,200)/2 = 64,600
3	Builders Hoist—2t	166,000	Book value at the end of 10 month = 166,000 − 10% of 166,000 = 149,400	(166,000 + 149,400)/2 = 157,700
4	Dewatering pump	267,000	Book value at the end of 6 month = 267,000 − 6% of 267,000 = 250,980	(267,000 + 250,980) = 258,990

Average monthly fixed asset for the overall project duration of 10 months = 106,400 + 64,600 + 157,700 + 258,990 × 6/10 = 106,400 + 64,600 + 157,700 + 155,394 = ₹ 484,094

Example 8.14 Illustration of AFE calculation

For a given project, the contractor has estimated that he would be required to keep an average stock equivalent to ₹ 20 lakh. Further, it is also estimated that the average outstanding and average unadjusted advance would be ₹ 270 lakh and ₹ 60 lakh, respectively. Estimate the average funds likely to be employed for this project if it is known that the contractor would be required to keep an average fixed asset of ₹ 50 lakh.

Solution

Given,

Average monthly stock = ₹ 20 lakh
Average outstanding = ₹ 270 lakh
Average unadjusted advance = ₹ 60 lakh
Average fixed asset = ₹ 50 lakh

We have,

Average funds employed (AFE) = average working capital + average fixed asset
= (average monthly stock + average outstanding − average unadjusted advance) + average fixed asset
= (20 + 270 − 60) + 50
= 230 + 50 = ₹ 280 lakh

Example 8.15 Bid price and cover computations

For a given project, the contractor has estimated the direct cost and indirect cost of labour, material, plant and specialized contractors (see Table Q8.15.1). The contractor plans to apply an off top of 15 per cent, 10 per cent, 10 per cent and 5 per cent on labour, material, plant and subcontractor costs, respectively. Further, the contractor has kept contingency amounts of ₹ 5 lakh, ₹ 20 lakh, ₹ 1 lakh and ₹ 35 lakh for labour, material, plant and subcontractor costs, respectively. Calculate the bid price and cover to be applied on labour, material, plant and subcontractor costs.

Table Q8.15.1 Data for Example 8.15

Description	Labour	Material	Plant	Subcontractor
Direct cost	100	700	50	1,000
Indirect cost	80	35	5	100
Off top (%)	15	10	10	5
Contingencies	5	20	1	35

Solution
The computation is shown in Table S8.15.1

Table S8.15.1 Solution details for Example 8.15

Description	Labour	Material	Plant	Subcontractor	Total	Remarks
Direct cost	100	700	50	1,000	1,850	(A)
Indirect cost	80	35	5	100	220	(B)
Total 1	180	735	55	1,100	2,070	C = (A) + (B)
Off top (%)	15	10	10	5		(D) Given
Off top	31.76	81.67	6.11	57.89	177.43	(E) = (F) − C
Total sales of civil	211.76	816.67	61.11	1,157.89	2,247.43	F = (C)/[1 − (D/100)]
Contingencies	5	20	1	35	61	(G) Given
Total 2	216.76	836.67	62.11	1,192.89	2,308.43	(H) = (F) + (G)
Cess @1% on sales	2.19	8.45	0.63	12.05	23.32	(I) = (J) - (H)
Sales price	218.95	845.12	62.74	1,204.94	2,331.75	(J) = (H)/(1−(Cess%/100))
Cover	2.19	1.21	1.25	1.20	1.26	(K) = (J)/(A)

Example 8.16
For a building project, the estimated costs of civil, electrical, fire-fighting, and plumbing and sanitary works are ₹ 3,500 lakh, ₹ 280 lakh, ₹ 150 lakh and ₹ 125 lakh, respectively. The cost of common plant and equipment has been estimated to be ₹ 10 lakh. The contractor has further estimated his overheads to be ₹ 400 lakh. There is no escalation clause in the contract and the duration of the project is 30 months. Thus, the contractor has also estimated the likely escalation in cost over the total duration of the project. The escalation is estimated to be ₹ 50 lakh. If the contractor desires to operate at 10 per cent markup on his bid price, what would be the bid price of the contractor given that the liability of contractor towards works contract tax is ₹ 80 lakh? Also, find the multiplication factor (cover) through which the direct cost of bid items is to be multiplied in order to get the bid price.

Solution

The computation of bid price and the cover is shown in Table S8.16.1. The explanation is given under 'remarks' column.

Table S8.16.1 Solution details for Example 8.16

			All values in ₹ (lakh)		Remarks
A	Direct cost				
	a1	Civil works	3,500		Given
	a2	Electrical	280		Given
	a3	Fire-fighting	150		Given
	a4	Plumbing and sanitary	125		Given
	a5	Subtotal		4,055	a5 = a1 + a2 + a3 + a4
B	Indirect cost				
	b1	Overheads	400		Given
	b2	Common plant and equipment	10		Given
	b3	Subtotal		410	b3 = b1 + b2
C	Escalation		50		Given
	c1	Subtotal		50	
D	Total cost			4,515	D = a5 + b3 + c1
E	Mark-up 10% on price (off-top)			502	E = F − D
F	Price			5,017	F = D/(1 − 10/100) = D/0.9
G	Works contract tax			80	Given
H	Bid price			5,097	H = F + G
I	Multiplication factor, i.e., cover on direct cost (L + M + P)			1.25	I = H/a5

Example 8.17 Calculation of Return on Capital Employed (ROCE) and EVA

The bid price quoted by a contractor for a construction project is ₹ 3,000 lakh, consisting of ₹ 2,300 lakh as his total cost and project overhead, and ₹ 100 lakh as his general overhead. The taxes are levied at a rate of 40 per cent on the gross profit. The contractor has estimated the average funds employed (AFE) for the project to be ₹ 500 lakh. The project duration is 16 months. Assume cost of capital at 15 per cent. Calculate the ROCE and EVA for the project.

Solution

The computations for ROCE and EVA are shown in Table S8.17.1 and are self-explanatory.

Table S8.17.1 Solution details for Example 8.17

S. No.	Description	Computation	Remarks
1	Bid price	₹ 3,000 lakh	Given
2	Total cost including project overheads	₹ 2,300 lakh	Given
3	General overhead	₹ 100 lakh	Can be between 1.5% and 3% of sales depending on the contractor
4	Gross profit before taxes (PBIT)	₹ 600 lakh	₹ 3,000 lakh − ₹ 2,300 lakh − ₹ 100 lakh
5	Taxes @40%	₹ 240 lakh	Currently at 38%
6	Net earnings	₹ 360 lakh	PBIT − taxes
7	Net earnings per annum	₹ 360 lakh × 12/16 = ₹ 270 lakh	Net earnings per annum × 100/duration of the project
8	ROCE	₹ 270 lakh × 100/ ₹ 500 lakh = 54%	Net earnings per annum × 100/AFE
9	EVA	(₹ 360 lakh − ₹ 500 lakh × 16 × 15) ÷ (100 × 12) = ₹ 260 lakh	EVA = net earnings − capital charges on AFE Capital charges on AFE = AFE × project duration × interest charges/12

Example 8.18

The bid data of 12 projects on which contractor C competed with competitor X is collected (Jha 2000) and presented in Table Q8.18.1. The estimated cost for each of the 12 projects is shown in the table. Find the probability of beating competitor X at different mark-up percentages.

Table Q8.18.1 Details of bids involving C and X

S. No.	C's bid price (₹ lakh)	C's cost (TC) (₹ lakh)	X's bid price (B) (₹ lakh)	X's bid to C's total cost B/TC	Mark-up % of X (in terms of % of C's cost) = $\left(\dfrac{B}{TC} - 1\right) \times 100\%$
(1)	(2)	(3)	(4)	(5)	(6)
1	3,618	3,289	3,083	0.94	−6.26
2	3,564	3,240	3,083	0.95	−4.85
3	921	838	798	0.95	−4.77
4	617	561	538	0.96	−4.10
5	1,600	1,455	1,404	0.96	−3.51
6	2,153	1,957	1,935	0.99	−1.12
7	1,481	1,346	1,395	1.04	3.64
8	3,280	2,982	2,978	1.00	−0.13
9	1,404	1,276	1,397	1.09	9.48

10	1,401	1,274	1,398	1.10	9.73
11	92,500	84,091	95,000	1.13	12.97
12	1,750	1,591	2,000	1.26	25.71

Solution

The bid data from Table Q8.18.1 has been converted into a histogram shown in Figure S8.18.1, which shows the number of bids and the B/TC ratio. Finally, the histogram has been converted into a probability distribution curve shown in Figure S8.18.2. From this, the probability of beating competitor X at different mark-up percentages can be read.

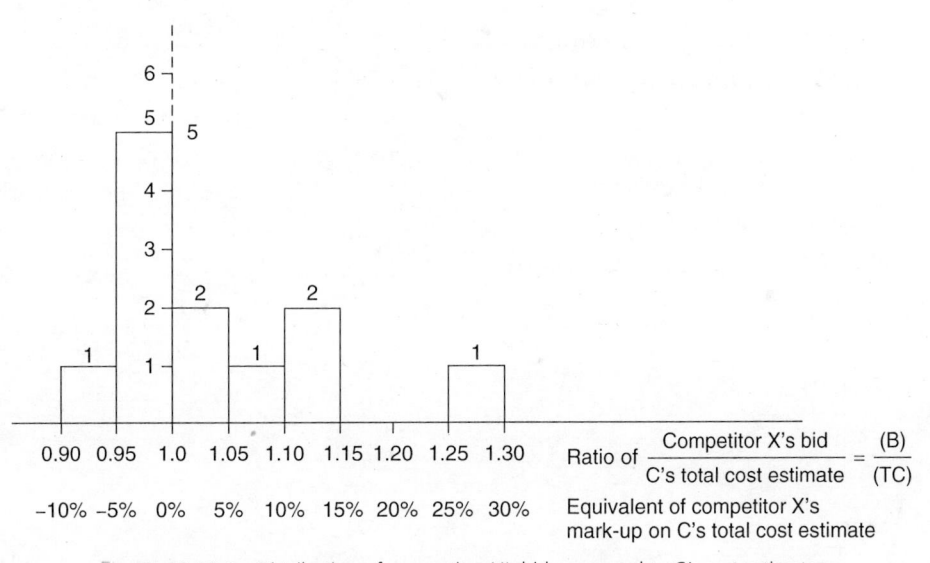

Figure S8.18.1 Distribution of competitor X's bid compared to C's cost estimates

Ratio of $\dfrac{\text{Competitor X's bid}}{\text{C's total cost estimate}} = \dfrac{(B)}{(TC)}$

Equivalent of competitor X's mark-up on C's total cost estimate

Example 8.19 Optimal mark-up calculation

The histogram shown in Figure Q8.19.1 is the compilation of past behaviour in 15 bids of a typical contractor against you as a contractor. In the histogram, B/TC ratio indicates $\dfrac{\text{Competitor's bid price}}{\text{Your estimated total cost}}$

a. Based on the given behaviour, what is the mark-up value that this competitor uses on average? What is the probability of winning against this competitor if you use a mark-up of 12 per cent?

b. In a new project with an estimated cost of ₹ 50,000,000, what is your optimum mark-up strategy against four typical competitors using Friedman's model? What is the expected mark-up amount at optimal mark-up?

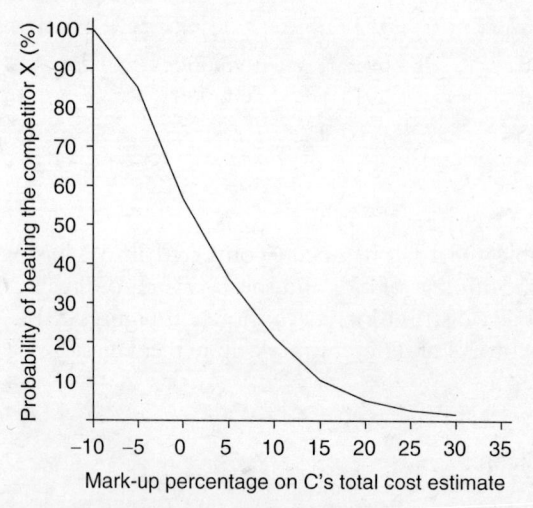

Figure S8.18.2 Probability of beating competitor X vs mark-up

c. In a new project with an estimated cost of ₹ 50,000,000, what is your optimum mark-up strategy against four typical competitors using Gates' model? What is the expected mark-up amount at optimal mark-up?
d. What would be your suggested mark-up level?

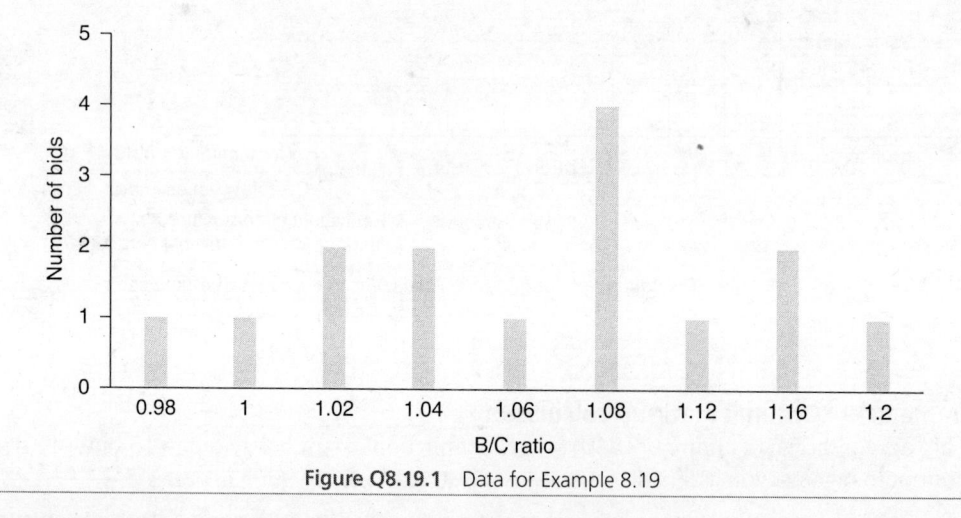

Figure Q8.19.1 Data for Example 8.19

Solution

a. In order to find the probability of winning against this competitor, we need to know the mean and the standard deviation of the bid-to-total cost ratio (B/TC). The mean of the B/TC ratio is 1.07467 and the standard deviation is 0.0630.

Thus, Z at 12 per cent mark-up level $= \dfrac{(1.12 - 1.07467)}{0.0630} = 0.7195$

Probability value can be read from standard normal distribution table given in Appendix 5 for $Z = 0.7195$, which is 0.7642. Hence, the probability of winning against the competitor at 12 per cent mark-up $= 1 - 0.7642 = 0.2358$ (23.5 per cent).

b. The computation method of optimal mark-up level is shown in Table S8.19.1.

Table S8.19.1 Computation details for Example 8.19b

Mark-up %	x	z	Probability from standard normal distribution table	Probability of winning the bid with only one typical competitor	Final probability of winning the bid with 4 typical competitors	Expected mark-up amount (₹)
	$x = 1 +$ (mark-up %/100)	$z = (x - \mu)/\sigma$	y	$p(typ) = 1 - y$	$P_{final} = p(typ) \times p(typ) \times p(typ) \times p(typ)$	$E = ₹ 50,000,000 \times (x - 1) \times P_{final}$
0%	1.00	−1.19	0.11799	0.88201	0.6052	0.00
1%	1.01	−1.03	0.15235	0.84765	0.5162	258,123.17
2%	1.02	−0.87	0.19279	0.80721	0.4246	424,570.77
3%	1.03	−0.71	0.23918	0.76082	0.3351	502,600.14
4%	1.04	−0.55	0.29108	0.70892	0.2526	505,144.28
5%	1.05	−0.39	0.34771	0.65229	0.1810	452,585.96

Here, Z is calculated using the expression $Z = \frac{x-\mu}{\sigma}; x = 1 + \frac{mark\ up\ \%}{100}$. Thus, for mark-up percentage of 1%, x would be 1.01 and Z approximately −1.03. Similarly, other values of Z can be calculated for different values of x.

The probability value corresponding to $Z = -1.03$ can be read out from standard normal distribution table, which is 0.15235. Thus, probability of winning the bid when only single competitor is there $= 1 - 0.15235 = 0.84765$.

Since there are four typical competitors, the probability of winning the bid from Friedman's model is given by $P_{final} = 0.84765 \times 0.84765 \times 0.84765 \times 0.84765 = 0.5162$. The expected mark-up amount at this mark-up level would be $E = 50,000,000 \times (x - 1) \times P_{final} = 50,000,000 \times (1.01-1) \times 0.5162 = ₹ 258,123$ (the value shown in table varies slightly from this value due to rounding-off error).

Similarly, the expected value of mark-up amount can be computed at different mark-up percentages. In the table, the computations for mark-up level up to 5 per cent have been performed. It may be noted that expected mark-up amount has been increasing initially up to 4 per cent and then started decreasing.

Hence, according to Friedman's model, the optimal mark-up that should be applied is 4 per cent. The results are also shown graphically in Figure S8.19.1. In fact the exact mark-up value works out to be 3.53% (verify using spreadsheet!!!). All these computations can be performed quite easily in a computer spreadsheet.

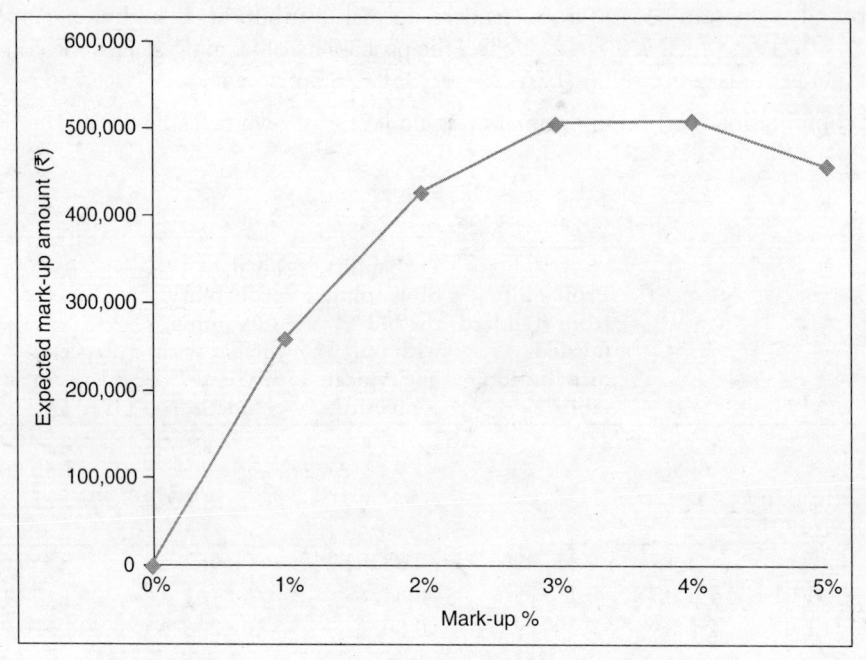

Figure S8.19.1 Determination of optimal mark-up using Friedman's model

c. The computation method of optimal mark-up level using Gates' model is shown in Table S8.19.2. Computation steps till column 5 in the table are similar to those used in mark-up determination using Friedman's model. The final probability (see column 6 in Table S8.19.2) is computed using the expression $P_{final} = \frac{1}{\left[n\left(\frac{1-p(typ)}{p(typ)}\right)+1\right]}$. The value of n in the problem is 4.

Table S8.19.2 Computation details for Example 8.19c

Mark-up %	x x = 1 + (mark-up %/100)	z z = (x − μ)/σ	Probability from standard normal distribution table y	Probability of winning the bid with only one typical competitor $p(typ) = 1 - y$	Final probability of winning the bid with 4 typical competitors $P_{final} = 1/[n(1 - p(typ))/(p(typ)) + 1]$	Expected mark-up amount (₹) E = ₹ 50,000,000 × (x − 1) × P_{final}
0%	1.00	−1.19	0.11799	0.88201	0.6514	0.00
1%	1.01	−1.03	0.15235	0.84765	0.5817	290,874.56
2%	1.02	−0.87	0.19279	0.80721	0.5114	511,422.96
3%	1.03	−0.71	0.23918	0.76082	0.4430	664,459.40
4%	1.04	−0.55	0.29108	0.70892	0.3784	756,888.57
5%	1.05	−0.39	0.34771	0.65229	0.3193	798,148.75
6%	1.06	−0.23	0.40796	0.59204	0.2662	798,651.71
7%	1.07	−0.07	0.47048	0.52952	0.2196	768,558.85

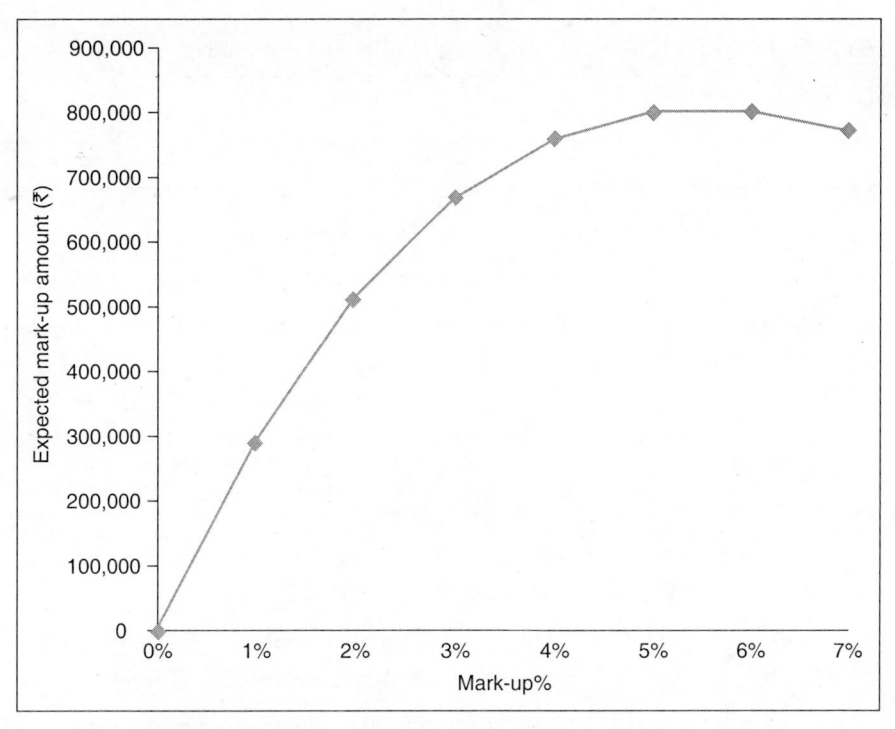

Figure S8.19.2 Determination of optimal mark-up using Gates' model

Thus, for a mark-up value of one percent, the final probability of winning the bid against 4 typical com-petitors, $P_{final} = \frac{1}{4 \times \frac{(1-0.84765)}{0.84675}+1} = 0.5817$ and the expected mark-up amount, as before, would be $E = 50,000,000 \times (x - 1) \times P_{final} = 50,000,000 \times (1.01-1) \times 0.5817 = ₹ 290,850$ (the value shown in table varies slightly from this value due to rounding-off error).

Similarly, the expected value of mark-up amount can be computed at different mark-up percentages. In the table, the computations for mark-up level up to 7 per cent have been performed. It may be noted that expected mark-up amount has been increasing initially up to 6 per cent and then started decreasing.

Hence, according to Gates' model, the optimal mark-up that should be applied is 6 per cent. The results are also shown graphically in Figure S8.19.2. In fact the exact mark-up value works out to be 5.50% (verify using spreadsheet!!!). All these computations can be performed quite easily in a computer spreadsheet.

d. The optimal mark-up obtained using Friedman's model = 3.53 per cent, and the optimal mark-up using Gates' model = 5.50 per cent. Thus, a moderate mark-up that should be used $= \frac{(3.53+5.50)}{2} = 4.515\%$, in order to get the optimal expected mark-up amount.

REVIEW QUESTIONS

1. State whether True or False:
   a. In restrictive bidding, the first stage is pre-qualification of contractors.
   b. Earnest money deposit is nothing but a type of security deposit by contractor.
   c. Materials productivity is the ratio between work units and materials quantity.
   d. Workers' productivity is measured by quantity of work done per man-hour.
   e. Bid price in general includes project overheads, common labour cost, plant and equipment cost, labour cost, material cost, subcontractor cost, general overheads, profits and allowances for risks.
   f. Indirect cost subsumes cost of material, cost of labour, cost of plant and equipment, subcontractor cost and crew cost.
   g. Direct costs are nothing but all the project overheads.
   h. Project overheads subsume salaries and benefits, insurance cost, financing cost, conveyance cost, travel and transfer cost, taxes and duties.
   i. Operational estimating and unit rate estimating are two important ways to analyse the rates.
   j. Mark-up includes the sum of general overheads, provisions for risks, and profit margin.
   k. Bidding models are generally based on game theory, cash flow and statistics (expected monetary value and expected utility value).

2. Calculate material cost per m² for Kota stone.

3. Derive the rate of one cubic metre of M20 and M25 grade concrete. The estimated quantities for different constituents of one m³ of concrete are listed in the given table for your reference.

Table Q3.1 Mix proportion of concrete by wt (kg/m³)

Grade	Water	Cement	Fine aggregate	Coarse aggregate	
M20	195	350	661	436	655
M25	200	380	626	435	652

4. A 10-month project involves execution of some items. The contractor estimates the project value to be ₹ 30 crore. The estimated invoice percentage per month is given in Table Q4.1.

Table Q4.1 Data for Question 4

Month	1	2	3	4	5	6	7	8	9	10
Invoice	1%	5%	10%	12.5%	12.5%	15%	20%	20%	3%	1%

CONTRACTOR'S ESTIMATION OF COST AND BIDDING STRATEGY | 319 |

Payment conditions are such that contractor estimates an average outstanding duration of 45 days. The mobilization advance is 10 per cent of the contract value, which is non-interest-bearing. The recovery of the mobilization advance starts from the 3rd running bill and the entire recovery must be completed in the 6th running bill.

The stock proposed to be maintained at site during the contract period has also been estimated by the contractor. For the first two months and the ninth month, it is estimated to be 25 lakh, while from 3rd month to 8th month it is estimated to be 50 lakh.

Calculate the return on capital employed by the contractor and find the EVA to the contractor for the project.

5. For the data given in Table Q5.1, calculate the return on capital employed and the EVA for the project.

Table Q5.1 Data for Question 5

S. No.	Description	Amount
1	Bid price	₹ 31 crore
2	Total cost	₹ 22 crore
3	General overheads of the contractor	3% of bid price
4	PBIT	?
5	ROCE	?
6	EVA	?

Corporate tax rate is 38 per cent. Average working capital employed by the contractor is ₹ 5.5 crore. The cost of capital is 15 per cent.

6. Calculate the average monthly stock from the data given in Table Q6.1

Table Q6.1 Data for Question 6

Month	1	2	3	4	5	6	7	8	9	10
Stock value in ₹ (lakh)	25	25	50	50	50	50	50	50	50	25

7. For the data given in Table Q7.1, calculate the average outstanding duration for the contractor.

Table Q7.1 Data for Question 7

Month	1	2	3	4	5	6	7	8	9	10
Outstanding amount in ₹ (lakh)	30	170	390	540	580	665	860	940	440	60

8. Contractor X is developing his bidding strategy against competitor Y. Table Q7.1 shows the details of B/C ratio (Y's bid to X's cost) for the 10 previous bids in which contractor X has competed against contractor Y.

Table Q8.1 Data for Question 8

B/C	Number of bids
1.04	2
1.08	4
1.12	1
1.16	2
1.20	1
Total	10

a. Based on the bidding behaviour, what is the mark-up value that this competitor uses on average? What is the probability of winning against this competitor if you use a mark-up of 14 per cent?

b. In a new project with ₹ 5 crore estimated cost, what is your optimum mark-up strategy against four typical competitors using Friedman's model? What is the expected profit at optimal mark-up?

9. The previous record of a contractor's bidding encounters against three competitors is:

Job No.	Contractor's cost estimate (₹ crore)	Bid price of competitors (₹ crore)		
		A	B	C
1	8.5	10.5	11	9.5
2	16	21	18	16
3	7	–	–	9
4	20	24	–	22

Using Friedman's and Gates' models, determine the mark-up needed to optimize the expected profit in the following cases, and comment on the results:

a. Bidding against A, B and C in a new job with estimated total cost of ₹ 50 lakh

b. Bidding against A, B and C in a new job with estimated total cost of ₹ 25 crore

c. Bidding against six typical competitors with behaviour close to that of competitor B

Comment on the impact of project size in (a) and (b) on the estimated optimum mark-up.

10. Your company is very keen on winning a job for which you submitted a bid of ₹ 550 lakh. Your cost estimate for the job is about ₹ 500 lakh. After the bid opening, your company and two others are selected for final negotiation with the owner. At the meeting, you are asked if you are willing to reduce your final bid. You are also told that there is a high chance of additional work being added, in the order of about 20 per cent of the original volume of work. Based on this information, how much will you go down in your bid?

11. The previous record of a contractor's bidding encounters against four competitors is given in Table Q11.1.

Table Q11.1 Data for Question 11

Job No.	Contractor's cost estimate (₹ lakh)	Bid price of competitors (₹ lakh)			
		A	B	C	D
1	775	950	850	875	–
2	1,000	–	1,000	1,100	–
3	650	750	700	825	775
4	600	–	800	700	–
		Hospitals	Townships	Office building	Institutional building

Using Friedman's and Gates' models, determine the mark-up needed to optimize expected profit in the following cases, using an estimated cost of ₹ 200 lakh:

a. Bidding against A, B and C in a new job
b. Bidding against six unknown competitors in an office-building project
c. You are bidding for a project that involves building several townhouses. Companies A, B and C are also interested in bidding for that job, in addition to two unknown bidders.

12. Why is site investigation important for proper bidding? How will the issues of improper site investigation be resolved contractually? Give adequate examples.

9

Construction Equipment Management

Introduction, classification of construction equipment, factors behind the selection of construction equipment, earthwork equipment, concreting equipment behind, hoisting equipment, plant and equipment acquisition, depreciation, depreciation and taxation, methods of calculating depreciation, example of depreciation calculations for equipment at a site, the effect of depreciation and tax on selection of alternatives, evaluating replacement alternatives, advanced concepts in economic analysis

9.1 INTRODUCTION

Construction equipment are one of the very important resources of modern-day construction, especially in infrastructure projects. Such projects utilize equipment for most of the works including earthmoving operations, aggregate production, concrete production and its placement, and so on. In fact, one cannot think of any major construction activity without the involvement of construction equipment. There are different types of construction equipment suitable for different activities in a construction project. The choice of construction equipment defines the construction method, which in a way leads to the determination of time and cost for the project. In order to select the right equipment to perform a specific task at the least cost, it is essential to know the features of a construction equipment including its rate of production and the associated cost to operate the equipment. In this chapter, major types of equipment have briefly been introduced. For a contractor, the decision to employ a particular type of equipment is influenced by the contract he enters with the owners, the site condition and the location of the project site. Specifications indicating the capabilities and the different features of construction equipment are provided by manufacturers.

The factors to be considered for the selection of some important equipment have been discussed. An effort has also been made in this chapter to cover topics such as financial issues involved in purchase of equipment, issues involved in choosing an equipment, accounting for assets generated, and considerations in computation of depreciation of equipment, using primarily a case-study approach. Certain aspects of implications of the above factors on tax calculation are also covered. It should be understood that the calculations, etc., shown here are purely illustrative, and simplified to only bring out the features intended.

9.2 CLASSIFICATION OF CONSTRUCTION EQUIPMENT

Construction equipment have been classified in various ways.

1. The classification could be on the basis of the function an equipment performs—for example, material-loading function, material-transporting function, etc. In the functional classification, equipment are grouped into (a) power units, (b) prime movers, which make it possible for the equipment to operate, (c) tractors, (d) material-handling equipment and (e) material-processing equipment.

2. It could be on the basis of the operation in which the equipment is used. For example, under this classification we have—(a) equipment used for loosening and moving the materials found in their natural state (equipment under this class include compressors, pumps, excavators, earthmoving equipment, trenchers, and conveying and hauling equipment); (b) equipment used for processing the materials, such as aggregate, concrete and asphalt production equipment; (c) equipment used for transporting the processed materials; and (d) equipment used for placing finished materials.

3. Another basis is the purpose, whether general-purpose or special-purpose equipment. For example, in general-purpose equipment we can have earthwork equipment, material-hoisting equipment, concreting equipment, etc., while in special-purpose equipment we can have bridge-construction equipment, marine equipment, coffer dams and caissons equipment, large-diameter pipe-laying equipment, and so on. Classification based on the purpose could be company-specific. For example, if a construction company is into tunnel construction, the tunnel-boring machine could be the general-purpose equipment for this company, while it may be a special-purpose equipment for a company that is mostly into building construction. A detailed discussion on classification of construction equipment can be seen in *Construction Equipment Handbook* by Day and Benjamin (1991). The classification helps in identifying equipment and their spare parts, besides helping in accounting for the cost related to the equipment. Most companies maintain alphanumeric codes to classify equipment and their spare parts.

9.3 FACTORS BEHIND THE SELECTION OF CONSTRUCTION EQUIPMENT

The selection of the appropriate construction equipment is an important part of job planning. The contractor has many different options to choose from, which makes the selection even more complicated. A planner has to choose the alternative that provides the best value from a cost and schedule perspective.

Selection of equipment for construction projects generally involves two classes of factors or considerations—hard factors and soft factors. Examples of hard factors include technical specifications of the equipment, physical dimensions of the site and constructed facility, and cost calculations. As can be noticed, hard factors are tangible in nature. On the other hand, soft factors are mostly intangible, qualitative and informal in nature. Some examples of soft factors are safety considerations, company policies regarding purchase/rental, market fluctuations and environmental constraints. The soft factors influence decision-making to a large extent.

A number of researchers have worked to develop a method to assist in equipment selection and, consequently, different models have evolved over the years. For example, there are optimization model, graphics model and database-centred models to select tower cranes. Different expert systems have also been developed to assist in equipment selection. Finally, there are artificial intelligence-based models. Harris and McCaffer (2001) have developed their equipment selection model based on multi-attribute decision-making. Dynamic programming-based models have also been developed for equipment selection. A number of computer programs have been developed to assist in the equipment-selection process.

Similarly, commercial software is available that offers solutions for the selection and location of a specific crane model and for lift planning (Meehan 2005). Examples are Compu-Crane and LPS (NCI 2006), Cranimation and Tower-Management (Cranimax 2006), LiftPlanner (LiftPlanner Software 2006) and MéthoCAD (Progistik 2006). These software packages, commonly used by engineering and construction firms, address mainly the technical aspects of crane location and lift planning.

In the following paragraphs, we discuss some of the factors that govern the equipment-selection decision. It may be difficult to come up with a set of decision variables that are applicable for the selection of all types of equipment used in construction, and hence, only a few common decision variables are discussed. The specific decision variables for a particular class of equipment selection are discussed subsequently.

9.3.1 Economic Considerations

The economic considerations such as owning costs, operating labour costs and operating fuel costs of equipment are most important in selection of equipment. Besides, the resale value, the replacement costs of existing equipment, and the salvage value associated with the equipment are also important. The economics of equipment selection and replacement issues are dealt with in detail in later parts.

9.3.2 Company-specific

The selection of equipment by a company may be governed by its policy on 'owning' or 'renting'. While emphasis on 'owning' may result in purchase of equipment keeping in mind the future requirement of projects, the emphasis on 'renting' may lead to putting too much focus on short-term benefits. This may explain the situation of a construction company opting for two 30 m^3/h batching plant owned by the company and currently idle, instead of the required one 45 m^3/h batching plant that it can get on rent. It is clear that the company wants to deal with the 30 m^3/h batching plant keeping future projects in mind, even though going for the 45 m^3/h batching plant option at present may have been economical.

Further, if the company project forecast says that there will be a considerable number of projects involving a particular type of construction equipment, say X, for the next couple of years, then the decision would be to buy only this type of equipment. A construction company specializing in a particular type of construction, such as tunnelling, will have a tendency to procure only those equipment that are used in tunnel construction. The equipment-selection decision also depends on the amount of outsourcing the company does to execute its projects. A company opting to subcontract a majority of their project work will tend to keep low equipment asset.

9.3.3 Site-specific

Site conditions—both ground conditions as well as climatic conditions—may affect the equipment-selection decision. For example, the soil and profile of a site may dictate whether to go for a crawler-mounted equipment or a wheel-mounted equipment. If there is a power line at or in the vicinity of site, one may go for a fixed-base kind of equipment rather than a mobile kind of equipment. Similarly, climatic conditions such as the presence of strong winds, visibility level and noise level may affect equipment-selection decision.

Further, the access leading to the site may also affect the decision. Heavy traffic congestion near a site may lead to a decision to produce the concrete at site and, hence, selecting a concrete batching plant rather than relying on ready-mix concrete. Similarly, if there is only a narrow road leading to the site, it may have a bearing on type of transportation equipment selected for the site.

9.3.4 Equipment-specific

Construction equipment come with high price tags. While it may be tempting to go for the equipment with low initial price, it is preferable to opt for standard equipment. Such equipment are manufactured in large numbers by the manufacturers, and their spare parts are easily available, which would ensure minimum downtime. Besides, they can also fetch good salvage money at the time of their disposal.

It is a general tendency to go for such equipment that can bring in 'uniformity' in the type of equipment that are already available with the company. For example, a company would like to go for a uniform type of engines for different machines such as excavators, dumpers and tractors.

The size of equipment selected is also an important consideration. Although the unit production cost may be cheaper for equipment of large size, it is also true that large equipment require correspondingly larger sizes of matching equipment. Thus, downtime in one primary unit may lead to downtime in dependent equipment also. A trade-off between unit production costs versus size must be obtained, and as far as possible, equipment of similar sizes should be selected for the project.

Besides the above factors, the versatility (whether it can perform more than one function) of equipment, the adaptability for future use, the past experience with the equipment, and the interaction with other equipment are also to be noted.

9.3.5 Client- and Project-specific

The owner/client in a certain project may have certain preferences that are not in line with the construction company's preferred policies as far as equipment procurement is concerned. The schedule, quality and safety requirements demanded of a particular project may in some cases force the company to yield to the demands of the client.

9.3.6 Manufacturer-specific

A construction company may prefer to buy equipment from the same manufacturer again and again, and that too from a specific dealer. This may be to bring in uniformity in the equipment fleet possessed by the company or because the company is familiar with the working style of the manufacturer and the dealer. Long association may not only result in cheaper price, but it also ensures prompt services by the manufacturer with regard to the company.

9.3.7 Labour Consideration

Shortage of manpower in some situations may lead to a decision in favour of procuring equipment that is highly automated. Further, the selection of equipment may be governed by the availability or non-availability of trained manpower. The company may not be inclined to select some sophisticated equipment if it finds that there is limited availability of manpower to operate the same.

In the following sections, we briefly describe some common construction equipment that can be seen in most of the construction sites these days.

9.4 EARTHWORK EQUIPMENT

A brief description of the equipment used in earthwork is given in Table 9.1 and they are shown in Figure 9.1.

The selection of earthmoving equipment is mainly dependent on the quantities of material to be moved, the available time to complete the work, the job conditions, the prevailing soil types, the swell and compaction factors, etc. The job conditions include factors such as soil conditions, availability of loading and dumping area, accessibility of site, traffic flows and weather conditions at site (Alkass and Harris 1988). The type of supervision and the length of working day and shift work are also important for the selection of a particular type of earthmoving equipment. The decision to choose between a track-mounted earthmoving equipment and a wheel-mounted one depends to a large extent on the prevailing job conditions.

In order to plan the number of earthwork equipment needed, the planner first determines the following:

1. The suitable class of equipment for earthwork—for example, if the soil to be excavated is loose and marshy, and bulk excavation is involved in the project, one may opt for a dragline. The planner needs to be familiar with different classes of equipment used in earthwork and their suitability of working in different conditions
2. The appropriate model of equipment within that class—for a given class of equipment, be it an excavator, a scraper, or a dragline, manufacturers usually come up with different models with different performance characteristics such as payload of bucket and speeds obtainable on different soil resistances. For example, draglines come in different capacities ranging from 0.38 m^3 to 3.06 m^3; scrapers in capacities ranging from 8 m^3 to 50 m^3; and so on
3. The number of equipment needed for the project to carry out the given quantity—once the model with suitable payload capacity is decided, the only thing we need to determine is the cycle time
4. The number of associated equipment required to support the main equipment

9.5 CONCRETING EQUIPMENT

Selection of concreting equipment (see Table 9.2) can be complicated and difficult. The decision will involve many issues that have to be analysed. The following factors mentioned by Alkass et al. (1993) are noteworthy:

Table 9.1 Earthwork equipment

Description	Purpose	Application	Remarks	Capacity
Backhoe	For excavation below the ground (lower elevation)	Cutting of trenches, pits, etc., levelling and loading	Suitable for heavy positive cutting	Struck bucket capacity 0.38 m^3 to 3.25 m^3
Shovel or front shovel	For excavation above its own track or wheel level	For cutting and for loading	Suitable for heavy positive cutting in all types of dry soil	Struck bucket capacity 0.38 m^3 to 3.25 m^3
Dragline	For bulk excavation in loose soils below its own track level	For canals and pits excavation, cutting and desilting of ditches	Suitable for loose soils, marshy land and areas containing water	0.38 m^3 to 3.06 m^3
Clamshell or grab	For deep confined cutting in pits, trenches	Such as shafts, pits, wells	Consists of a hydraulically controlled bucket suspended from a lifting arm	0.38 m^3 to 3.06 m^3
Dozers	For moving earth up to a distance of about 100 m, shallow excavation, and acting as a towing tractor and pusher to scraper machines	Clearing and grubbing sites, excavation of surface earth, and maintaining roads	Can be track-mounted or wheel-mounted	Blade capacity 1.14 m^3 to 6.11 m^3
Roller compactor	For compaction of earth or other materials	Used for large works of highways, canals and airports	Comes in different varieties such as (1) smooth-wheeled roller, (2) vibratory roller, (3) pneumatic-tyred roller, and (4) sheep-foot roller	For (1) 8–10 t (2) 4–17 t (3) 11–25 t (4) 2.5–11.5 t Capacity can be increased by ballasting
Scraper	For site stripping and levelling, loading, hauling and discharging over long distances	Comes in different varieties such as towed scrapers, two-axle scrapers and three-axle scrapers	Best suited for haul distances varying between 150 m and 900 m	Sizes varying from 8 m^3 to 50 m^3
Dumper	For horizontal transportation of materials on and off sites	Suitable for hauling on softer sub-grades. Large-capacity dumpers are used in mines and quarries	Comes in different varieties with front tipping, side tipping, or elevated tipping arrangements	Load capacity of 1 t to about 80 t; 20 t is common for small dumper
Grader	For spreading fill and fine-trimming the sub-grade. Grader performs a follow-up operation to scraping or bulldozing	Grading and finishing the upper surface of the earthen formations and embankments	Graders usually operate in the forward direction	

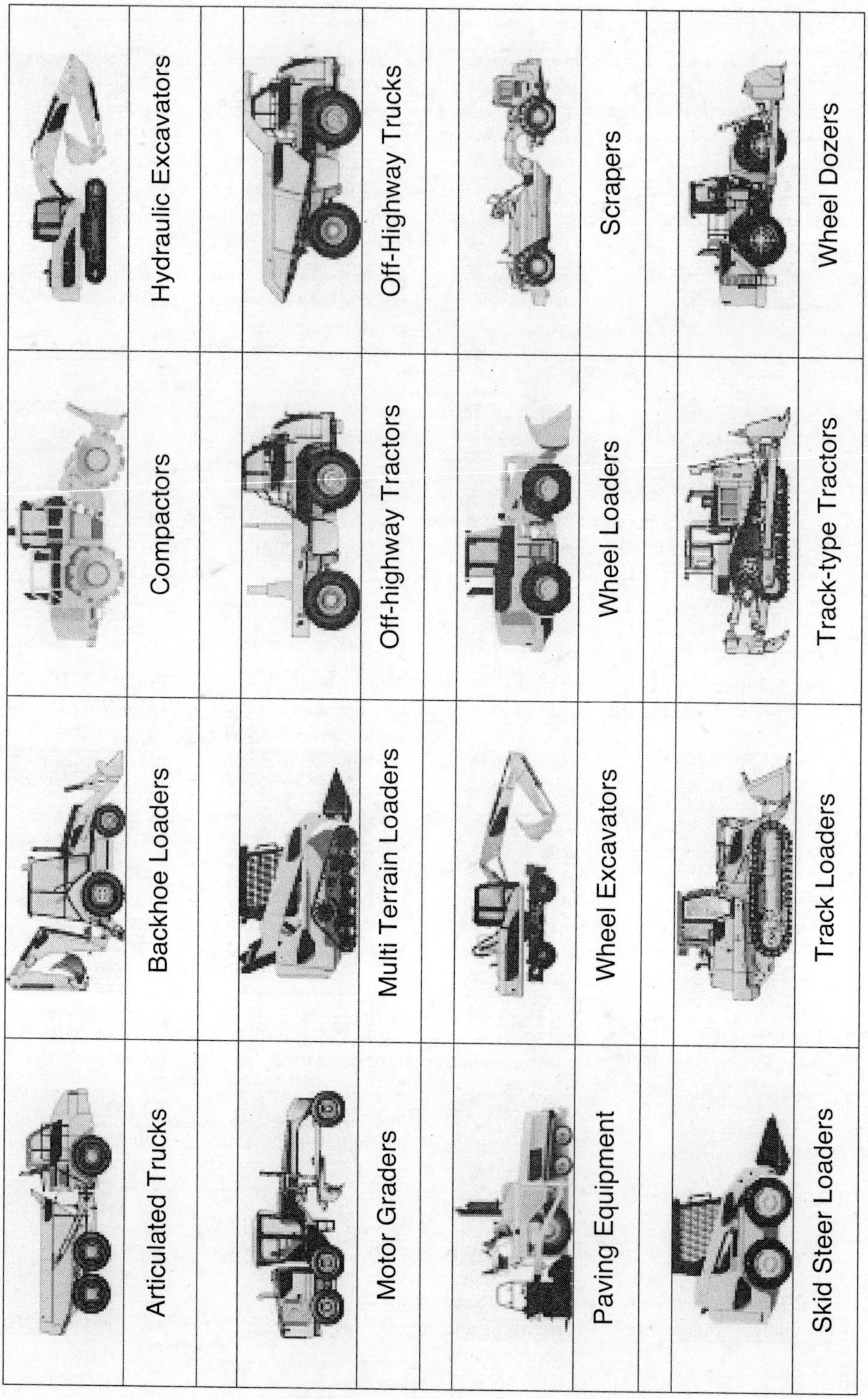

Hydraulic Excavators	Off-Highway Trucks	Scrapers	Wheel Dozers
Compactors	Off-highway Tractors	Wheel Loaders	Track-type Tractors
Backhoe Loaders	Multi Terrain Loaders	Wheel Excavators	Track Loaders
Articulated Trucks	Motor Graders	Paving Equipment	Skid Steer Loaders

Figure 9.1 Some common earthwork equipment

Source: Caterpillar website

Table 9.2 Concreting equipment

Description	Purpose	Capacity	Remarks
Concrete batching and mixing plant	For weighing and mixing large quantity of concrete constituents	20 m^3/h to 250 m^3/h	Comes in different varieties such as stationary and mobile batching plants
Concrete mixers	For mixing small quantities of concrete constituents	Capacity could be up to 200 l/batch for small mixers, and between 200 and 750 l/batch for large mixers	Comes in different varieties such as non-tilting drum type and tilting drum type
Concrete transit mixers	For transporting concrete from batching plant	3 m^3 to 9 m^3	Capacity also depends on the permissible axle load
Concrete pumps, static or portable	For horizontal and vertical transportation of large volumes of concrete in short duration	30 m^3/h for ordinary construction; can be in excess of 120 m^3/h for specialized construction	Direct acting pumps—an output of up to 60 m^3/h through 220 mm delivery pipes. Concrete can be easily pumped up to a distance of 450 m horizontally or 50 m vertically
			Squeeze pumps—an output of up to 20 m^3/h through 75 mm delivery pipes. Concrete can be easily pumped up to a distance of 90 m horizontally or 30 m vertically

- Site characteristics such as boundary conditions, equipment manoeuvre, provision of temporary roads, noise limitations and other restrictions
- Equipment availability—local availability of equipment, whether the contractor owns that equipment
- Continuity of operation
- Effect of permanent work
- Weather conditions
- Temporary works
- Time restrictions
- Concrete specifications

Concrete-mixing equipment selection will depend on factors such as the maximum and the total output required in a given time frame, the method of transporting the mixed concrete and the requirement of discharge height of the mixer. Concrete-placement equipment selection depends on factors such as the capacity of the vehicle, the output of the vehicle, the site characteristics, the weather conditions, the operator's efficiency, the rental costs, and the temporary haul roads. It is important that the system that guarantees minimum costs be selected.

9.6 HOISTING EQUIPMENT

9.6.1 Hoists

Hoists are a means of transporting materials or passengers vertically by means of a moving level platform. Materials hoists come in basically two forms—static and mobile models. The static version consists of a mast or tower with the lift platform either cantilevered from the small section mast or centrally suspended with guides on either side within an enclosing tower. Mobile hoists usually do not need tying to the structure unless extension pieces are fitted, in which case they are treated as cantilever hoists.

Passenger hoists, like the materials hoist, can be driven by petrol, diesel, or electric motor and can be of a cantilever or enclosed variety. The cantilever type consists of one or two passenger hoist cages operating on one side or both sides of the cantilever tower; the alternative form consists of a passenger hoist cage operating within an enclosing tower. Tying-back requirements are similar to those needed for materials hoist.

9.6.2 Cranes

A crane is a device used for hoisting and placing materials and machinery. It is used to facilitate handling of materials such as formwork, reinforcement, pipes and structural steel items. The use of cranes has greatly increased in the construction industry due mainly to the need to raise the large and heavy prefabricated components often used in modern structures. Cranes are available in different ranges and the choice of a crane is dependent on factors such as the weight of the materials and machineries to be lifted, the overall dimension, the distance to which they are to be placed, and the prevailing site condition. Construction cranes can be broadly classified under two major categories—tower cranes and mobile cranes.

Selecting between a mobile crane and a tower crane is one of the decisions faced by the planner. This is followed by the selection of an appropriate model of mobile or tower crane. These decisions are influenced by a number of factors. The additional cost associated with a tower crane would include the cost of design and construction of a suitable foundation, the cost towards erection and dismantling, the cost of providing gye ropes, etc. The additional advantage of using tower cranes is that they provide an almost noise-free working environment, and they can also work under high wind speed. Shapiro et al. (1991) point out that 'some of the tower cranes can operate in winds up to 72.5 km per hour, while mobile cranes are designed for winds of about 32.3 km/h in conjunction with rated loads, which does not include an allowance for wind on the lifted load.' Use of tower crane also enables the site personnel to open the job so that any area of the project can be constructed at the same time as any other area.

Tower Crane

In appearance most of the tower cranes look alike, but they are not of same type. Various types of tower cranes are available in the market and it is important to choose the right tower crane for a project. The tower cranes offer several advantages over conventional cranes and it is required to be familiar with the different features of the tower crane. The manufacturer's catalogue offers useful advice such as the lifting capacities at different radii (see Figure 9.2). Tower cranes are available in different types, sizes and capacities. The lifting capacity is one of the important selection parameters besides length of reach (radius), maximum hook height above ground, and crane positioning.

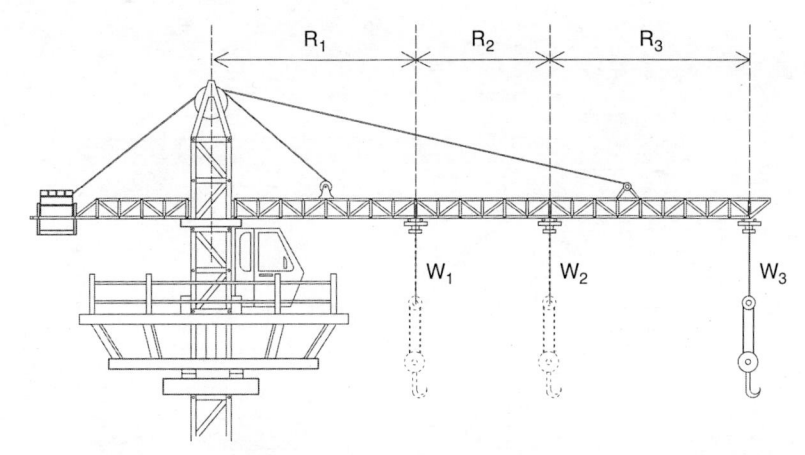

Figure 9.2 Manufacturer's way of providing lifting capacity at different radii

Tower cranes have distinct advantages over conventional lattice boom crawlers or truck cranes because the boom or jib looms high above the work site. The tower crane's jib can place its load anywhere within the radius of operation without interfering with the structure over which it swings. In addition, the operator can either be on crane or control the crane remotely using instrumentation located on the building structure, while enjoying an excellent view of the load and its surroundings at all times.

The use of tower crane has become synonymous with medium- to high-rise construction, be it for buildings or for bridges, dams, cooling towers, etc. Tower cranes can be fixed or can be mounted on rails. In the latter case, it is referred to as travelling tower crane. Sites with multiple tower cranes in operation are very common these days, especially for fast-track construction projects.

Tower cranes can be classified into top-slewing tower cranes and bottom-slewing tower cranes. The main differences are in the manner in which they are erected and dismantled, and in the maximum lifting height.

Top-slewing tower cranes Top-slewing cranes (see Figure 9.3) require a longer duration to erect and dismantle, but they can be erected virtually up to any height. These cranes are suitable for medium-to high-rise construction projects where they are needed for longer durations.

The main parts of a typical top-slewing crane are—undercarriage, mast, operator's cabin, slewing ring, jib and counterjib. These cranes are usually of stationary type, resting on the concrete or structural steel foundations. In the travelling tower crane, modular concrete blocks are used in the base to provide stability. The crane travels on heavy wheeled bogies mounted on a wide-gauge rail track. The travelling tower crane provides better site coverage. For additional stability, counter-weights in the form of modular concrete blocks are also provided in the counterjib of the crane.

Top-slewing tower cranes have a mast (also called tower) with modular, lattice-type sections. The mast is erected with the help of another lifting device. The crane rises in height by addition of sections one by one. Tower cranes also have provision of telescopic climbing frame through

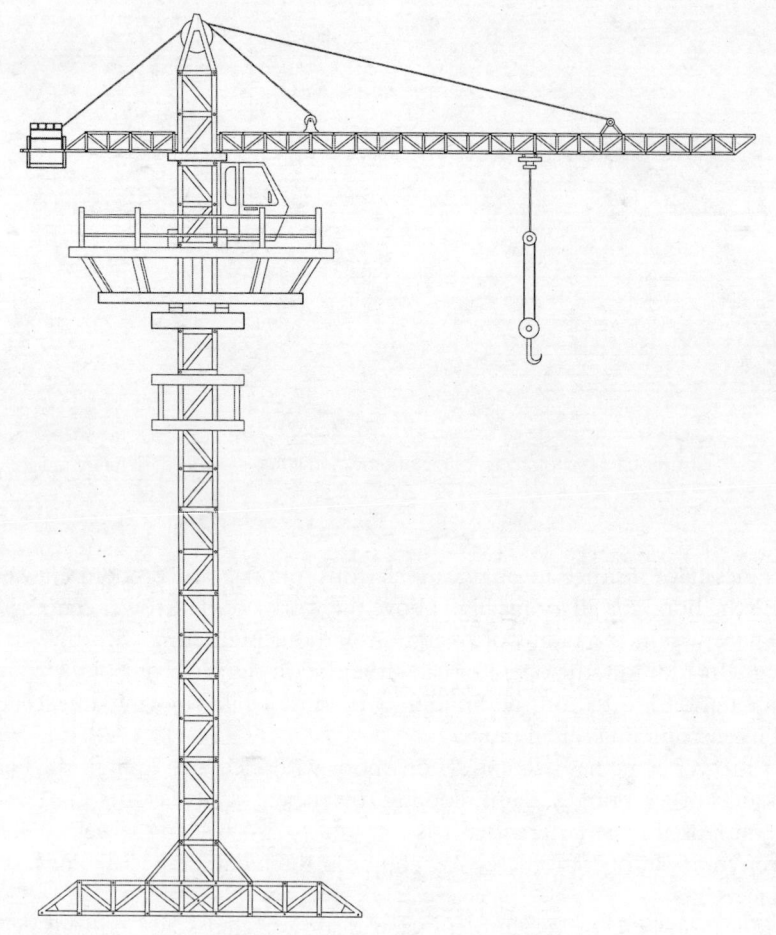

Figure 9.3 Tower crane (top-slewing)

which new sections are inserted and height of the crane is raised. Slewing ring provides the rotating mechanism of the crane and is located at the top of the mast just below the boom. In the top-slewing tower crane, only the jib, tower top and operator's cabin rotate.

Bottom-slewing tower cranes Bottom-slewing cranes (see Figure 9.4) have height limitations although they can be erected and dismantled very quickly—the reason they are known to be self-erecting and fast-erecting. These are suitable for low-rise construction and for short-term assignments.

Bottom-slewing tower cranes also consist of undercarriage, slewing ring, slewing platform, mast and jib. It has a slewing ring located near the base and both the mast and the jib rotate relative to the base. Since virtually the entire crane rotates, anchoring the crane with some fixed support is not possible. This is the reason for the shorter height of a bottom-slewing crane relative to a top-slewing crane.

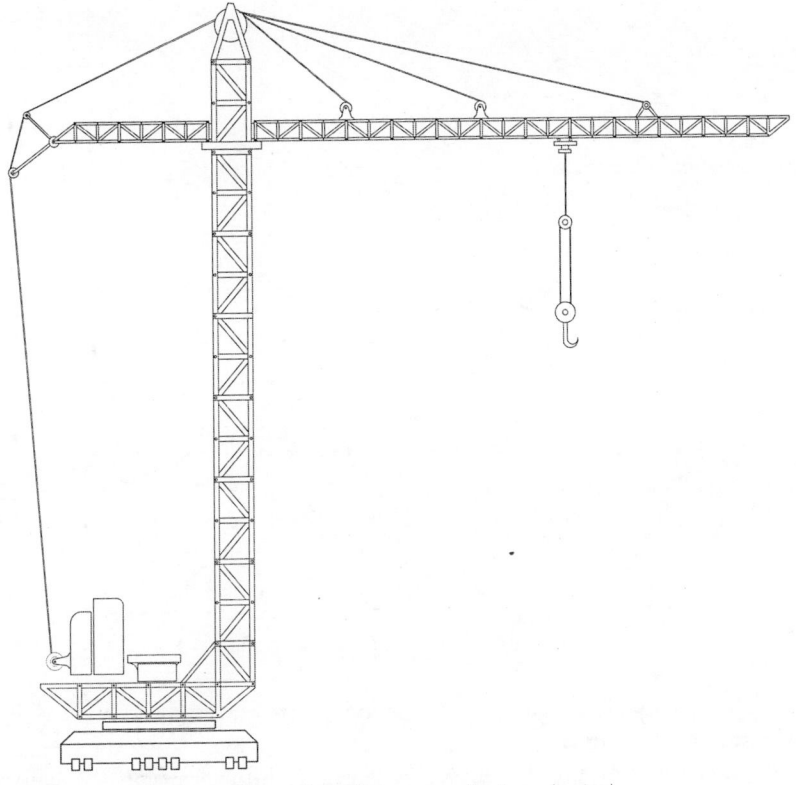

Figure 9.4 Tower crane (bottom-slewing)

Mobile Crane

The mobile cranes have their own prime movers and, thus, are capable of moving freely on the project sites. Mobile cranes come in a wide variety of designs and capacities. The cranes typically consist of an undercarriage, an engine to propel the movement, a boom at an inclination, and an operator's cabin.

The boom could be either a lattice-type steel boom or a telescopic-type boom. The lattice boom has the advantage of simplicity in design and, consequently, lower cost. In order to extend or shorten the length of the lattice boom, the lifting operation is stopped and the extension pieces (also called fly jibs) are added or removed. The telescopic boom has the advantage of having a boom that can be adjusted in length even during the lifting operation, which makes it faster when compared to lattice boom structure.

Mobile cranes can be either crawler-mounted or wheel-mounted depending on the arrangement of the undercarriage. Also, in contrast to tower cranes, mobile cranes can be used for purposes other than lifting such as excavation, pile driving and demolition, if fitted with suitable attachments.

Mobile cranes are characterized by the maximum lifting capacities. The lifting capacities vary at different radii and inclinations. In all type of cranes, the shorter the lifting radius, the greater will be the lifting capacity. The small-capacity machines have a fixed boom or jib length,

Figure 9.5 Lattice boom truck-mounted

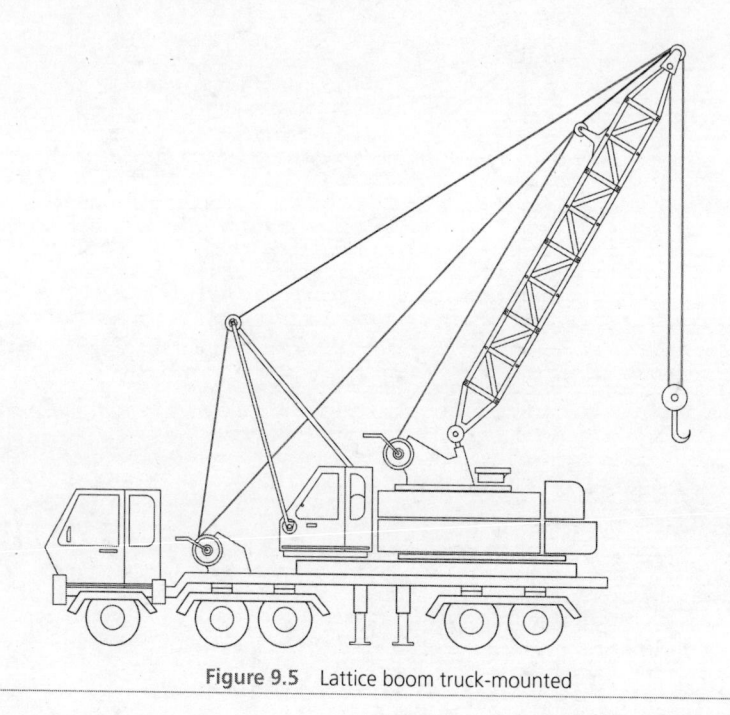

whereas the high-capacity cranes can have a sectional lattice boom or a telescopic boom to obtain various radii and lifting capacities. The mobile cranes are known for their heavy load-lifting capacity and even a smaller mobile crane can have lifting capacities equal to or more than a tower crane.

Mobile cranes can be classified into four major categories—(1) truck crane; (2) crawler crane; (3) rough-terrain crane; and (4) all-terrain crane. These are explained briefly in the following paragraphs:

Truck cranes The term 'truck crane' comes from the fact that the entire superstructure of the crane consisting of boom, engine, counterweight and operator's cabin is mounted on a special truck. Truck cranes are suitable for short-term lifting assignments only and should not be considered if the lifting assignment extends for longer duration. The booms in these types of cranes could be either of lattice type (see Figure 9.5) or of telescopic type (see Figure 9.6). Telescopic boom cranes are very popular because of the short time period required to prepare the crane for use upon arrival on site, making them ideally suitable for short hire periods. Truck cranes can travel with ease on rough terrain as well as on public roads. The cranes can travel between sites at speeds of up to 48 km/h, which makes them very mobile, but to be fully efficient they need a firm and level surface from which to operate.

Crawler cranes Crawler cranes (see Figure 9.7) have better manoeuvrability and can be quickly relocated at different locations on a project site. However, for shifting to another project site, these cranes require a truck or a trailer. The inability to be shifted from one site to another on its own is one of the main disadvantages of the crawler crane. Nevertheless, these cranes can work in difficult ground conditions. The cranes normally have lattice boom but are also available with the telescopic boom option.

CONSTRUCTION EQUIPMENT MANAGEMENT | 335 |

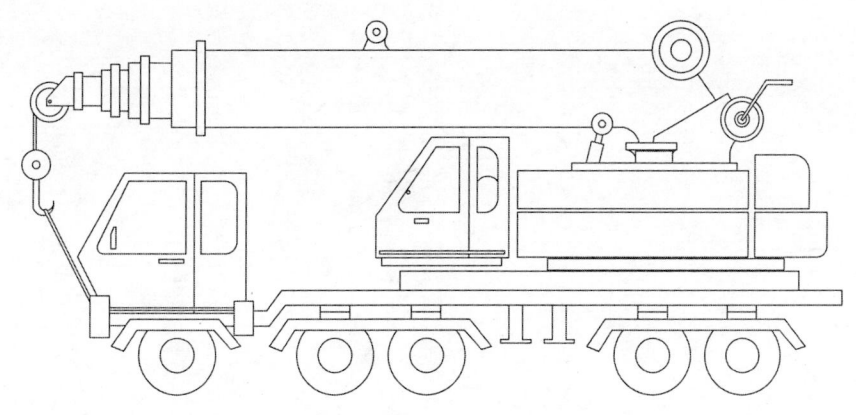

Figure 9.6 Telescopic boom truck-mounted

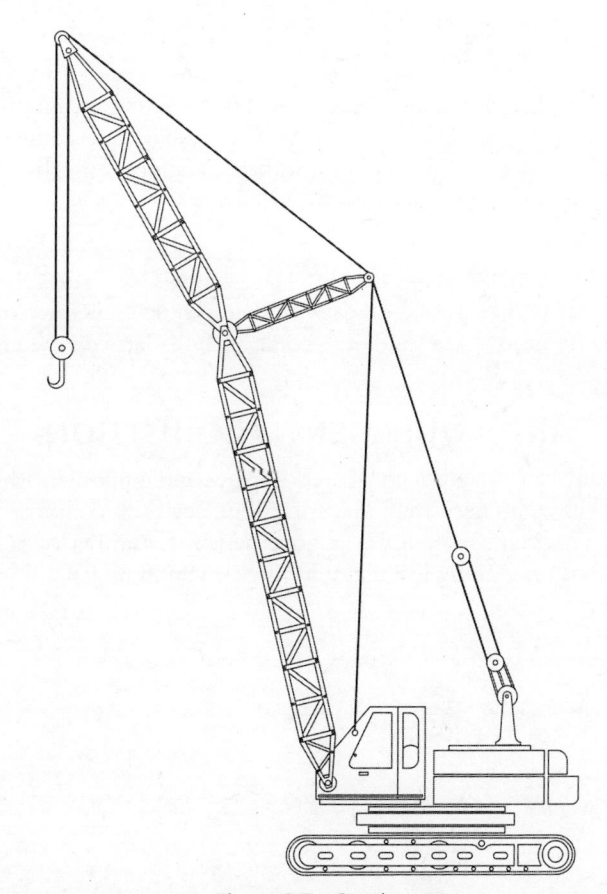

Figure 9.7 Crawler crane

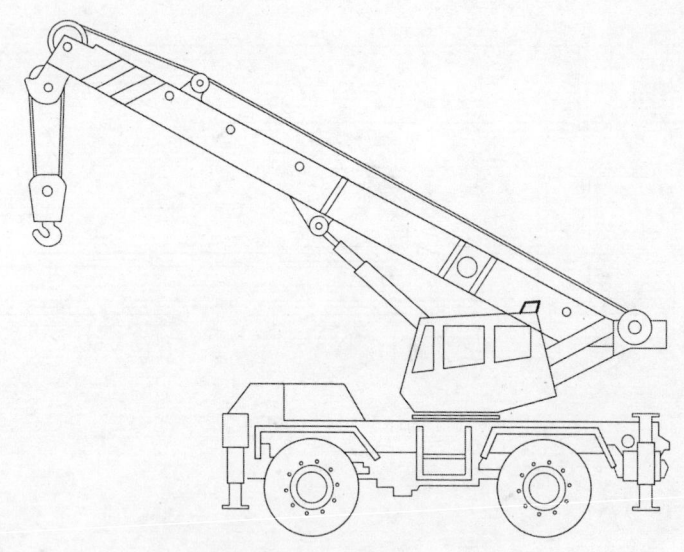

Figure 9.8 Rough-terrain mobile crane

Rough-terrain cranes Rough-terrain cranes (see Figure 9.8) have high mobility and can work best if combined with the operation of a tower crane. These cranes are very good in rough terrain conditions and where frequent relocation of cranes is needed. They are available with both the lattice- and telescopic-type booms. The rough-terrain cranes can manage grades even up to 35°.

All-terrain cranes As the name suggests, all-terrain cranes (see Figure 9.9) can manoeuvre through rough terrain as well as public roads with ease. These cranes are technologically the most advanced of all the mobile cranes. They come in both lattice-type and telescopic-type boom options.

9.7 PLANT AND EQUIPMENT ACQUISITION

A construction company can acquire a construction plant and equipment through (1) cash or outright purchase, (2) hire purchase and (3) leasing. The decision to choose between purchasing and renting is an important one, and it is governed by a number of factors. A simplified form of this type of decision-making is illustrated through an example.

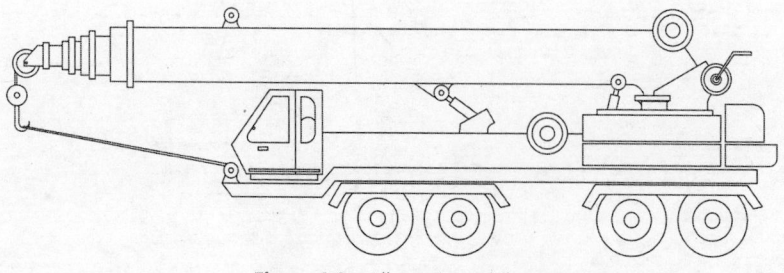

Figure 9.9 All-terrain mobile crane

In cases where the company has surplus funds available, it can invest in cash purchase of plant and equipment provided such investment guarantees a return more than the minimum attractive rate of return of the company. The purchase of equipment raises the asset of the company and, in fact, this entitles the company for certain tax incentives. The title (ownership) of the plant and equipment is with the company in the purchasing option.

In the hire purchase option, the company enters into an agreement with the financer of the plant and equipment. The financer receives the specified rental from the company using the equipment during the entire agreement period. The title of the equipment lies with the financer during this agreement period. Upon the completion of agreement period of equipment, the title is transferred in the name of the company against some nominal amount of money. Through the hire purchase option, the company is able to avoid large capital investment though it may be paying higher interest and, thus, higher rental.

In the leasing option, the company pays the lease rentals to the lessor (one who owns the equipment) for the use of equipment for a specified period. The title of the equipment never gets transferred to the company using the equipment. These days, a number of business houses are in the business of leasing construction equipment and, consequently, different variants of leasing with a number of interesting features are appearing on a regular basis.

9.8   DEPRECIATION

The following is an illustrative example to highlight the concept of depreciation—X purchases a truck, T, for use in his business, and for some reason if he wants to sell it even the very next day, it is unlikely that he will get a buyer who will buy it for the price he (X) paid. In other words, the truck starts to lose value as soon as it is driven out of the dealership. Continuing with the example, as soon as it is bought, the truck becomes an operational asset for X, and has a certain value in the account books of the company. Now, the annual loss in the value of the truck (on account of depreciation), too, should be appropriately reflected in the account books, even though no physical transaction is taking place (that reduces the value of the truck), i.e., the book value of the truck should change every year that it continues in X's ownership. Finally, a stage is reached when the truck may stop running and has no value to the business, and may be sold, scrapped, or written off.

In a general sense, 'depreciation' discusses the different issues related to this gradual change (reduction) in the value of an asset. Formally, there are a number of definitions for depreciation, although each one conveys similar meanings. For example, according to one definition, 'depreciation is the allocation of the cost of an asset over a period of time for accounting and tax purposes.' Another definition discusses depreciation in terms of a decline in the value of an asset due to general wear and tear, deterioration, or obsolescence. It may be noted, however, that not all assets depreciate with time. Land is one such example—it does not wear out like vehicles or equipment, and should be handled separately.

Although the above example of X buying a truck was drawn for a single asset, in general a company 'owns' various assets in the form of buildings, office furniture, machinery and equipment, etc., all of which are acquired at a certain value, and they depreciate over time. In the end, they cease to be of any monetary value, and can only be scrapped and a certain value extracted (salvaged) from the scrap.

From the point of view of bookkeeping, depreciation is considered as an expense and is listed as such (as an expense) in the firm's financial statement. It may be noted that this

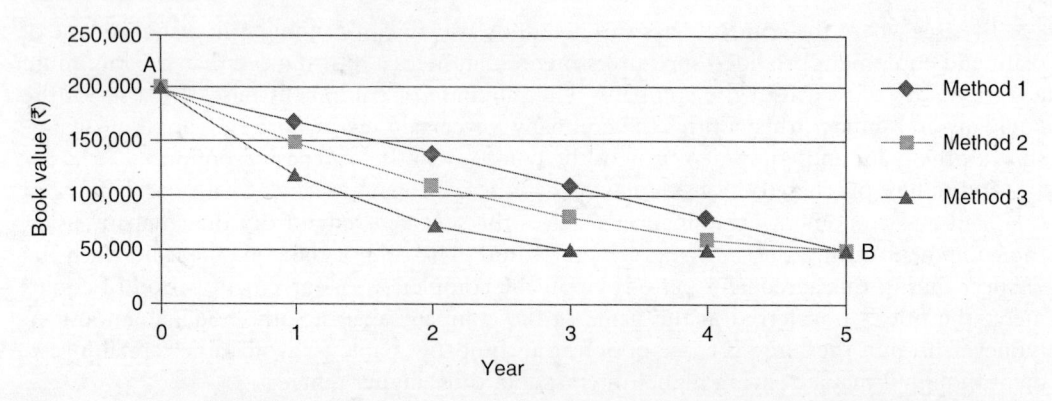

Figure 9.10 Schematic representation of depreciation in the value of an asset

'expense' is only technical in nature and the funds are not really transferred out of the system. As far as the idea of the final value is concerned, it is referred to as the salvage value. Now, depending upon various factors, an asset may take different periods of time to depreciate to its salvage value, and therefore, the concept of 'service life' also needs to be borne in mind. A more rigorous treatment of these terms is taken up elsewhere.

Figure 9.10 is a schematic representation of the depreciation process, and shows how the 'value' of an asset changes from the initial value at the time the asset is acquired (point A) to the final salvage value at the end of the useful life (point B). Depending on the nature of the variation taken between A and B (as schematically shown by methods 1, 2 and 3), the annual depreciation cost and the book value (calculated by subtracting the total depreciation from the initial cost) of the asset can be quantitatively determined. It should be borne in mind that it is entirely possible for a firm to continue to have (own) assets that have zero book value. In other words, the actual service life (the period during which an asset continues to be functionally useful) and the book life (the period during which the asset depreciates to its final value) could be different. Obviously, at the end of the book life of an asset, no depreciation can be charged.

Our discussion here largely focuses on the commonly used methods for calculating depreciation. However, since most regulatory and tax systems allow tax breaks on (the expenditure for) depreciation, the issue needs more than a cursory attention, and other related topics such as the implication on taxation, replacement cost, etc., are also briefly discussed. To facilitate understanding and highlight the importance of the subject, a brief discussion of the latter topics is taken up first in the following paragraphs.

9.9   DEPRECIATION AND TAXATION

As mentioned above, depreciation can be looked upon as an expenditure on which tax benefits are often available. The following illustrative example has been included to show the close relationship between depreciation expenses and tax liability.

Example 1

Given that the gross income is ₹ 200,000, and assuming that (i) depreciation expenses are fully deductible from gross income and (ii) the income tax is 25 per cent of the taxable income, determine the tax liability in cases when (a) there are no depreciation expenses, and (b) there is an expense of, say, ₹ 40,000 towards depreciation.

Table 9.3 Illustrative example showing the effect of depreciation expenses on tax liability

	Case I				Case II		
Gross income (a)	Depreciation expense (b)	Net income (c) = (a) − (b)	Tax payable @25% of (c)	Depreciation expense (b)	Net income (c) = (a) − (b)	Tax payable @25% of (c)
200,000	Nil	200,000	50,000	40,000	160,000	40,000

Table 9.3 summarizes the total and taxable incomes, and the tax liability in the two cases. It can be clearly seen that there is a net saving of ₹ 10,000 in taxes for the conditions given. Now, having explained the implication of depreciation expenses on the tax liability, another related issue is how should the total amount of depreciation be distributed over the service life of the asset. Before discussing the commonly used methods, let us look at another simple illustrative example.

Example 2

Let the total depreciation involved in an asset be ₹ 50,000 over a two-year period. Assuming a gross income of ₹ 100,000 in each year, and taxes payable at a rate of 20 per cent of the net income, compare the tax liabilities in the following cases:

Case I: Depreciation amounts in the first and second years are equal (₹ 25,000)
Case II: Depreciation amounts in the first and second years are ₹ 40,000 and ₹ 10,000, respectively

Table 9.4 showing the calculations is largely self-explanatory. It can be seen that the total tax paid over the two-year period is the same (30,000). However, in Case I a sum of ₹ 15,000 is paid each year, whereas the amounts in the two years in Case II are ₹ 12,000 and ₹ 18,000, respectively. Now, considering the time value for money, payment of taxes in a manner suggested in Case II is beneficial on two counts—(a) more funds remain available at the end of year 1, considering that depreciation expenses are only a bookkeeping exercise, and (b) less tax is paid in year 1, and additional funds are available, at least for some time. It is for these reasons that accountants prefer to use methods that allow for faster initial depreciation in order to maximize the returns on tax savings.

It is in this backdrop that regulatory measures are required to define a framework for calculations of depreciation rates and service lives of assets. While a detailed treatment of such frameworks is outside the purview of the discussion here, some of the basic methods for depreciation calculations are explained in the following paragraphs.

Table 9.4 Illustrative example showing effect of depreciation pattern on tax liability

			Case I			Case II		
Year	Gross income (a)	Depreciation expense (b)	Net income (c) = (a)−(b)	Tax payable [20% of (c)]	Depreciation expense (b')	Net income (c') = (a)−(b')	Tax payable [20% of (c')]
1	100,000	25,000	75,000	15,000	40,000	60,000	12,000
2	100,000	25,000	75,000	15,000	10,000	90,000	18,000

9.10 METHODS OF CALCULATING DEPRECIATION

From the qualitative framework given in Figure 9.10, it is clear that in order to deal with the issues related to depreciation quantitatively, any method should appropriately account for the following:

(a) purchase price or initial cost of the asset (P),
(b) economic life or recovery period allowed for the asset (N), and
(c) the salvage value of the asset (E).

Some of the more commonly used methods are discussed next.

9.10.1 Straight-line Method

Straight-line depreciation is considered to be the most common method of calculating depreciating assets, and is based on the assumption that the asset loses an equal amount of value each year. This annual depreciation is calculated by subtracting the salvage value of the asset from the purchase price, and then dividing this number by the estimated useful life of the asset.

Thus, computation of annual depreciation expense here requires three numbers—the initial cost of the asset, its estimated useful life, and the salvage value at the end. For example, a truck purchased for ₹ 500,000 and expected to have an economic life of ten years, and not having any salvage value, has a total depreciation expense of ₹ 500,000, which is reported over the ten-year period at the rate of ₹ 50,000 every year. As far as the book value of the truck is concerned, starting from ₹ 500,000, it decreases by ₹ 50,000 every year and becomes zero over a ten-year period.

Mathematically, straight-line depreciation amount calculation can be represented as:

$$R_m = \text{Rate of depreciation} = \frac{1}{N} \qquad (9.1)$$

$$D_m = \text{Annual depreciation} = (P - E) \times R_m \qquad (9.2)$$

$$BV_m = \text{Book value at } EOY_m \text{ (end of year m)} = BV_{m-1} - D_m \qquad (9.3)$$

Example 3

For an asset having an initial cost of ₹ 2 lakh, and a salvage value of ₹ 50,000 at the end of an economic life of 5 years, determine the annual depreciation and the book value at the end of each year during the economic life of the asset.

Taking P, N and E to be ₹ 200,000, 5 years and ₹ 50,000, respectively, the rate of depreciation, the total depreciation and the annual depreciation can be calculated as follows:

(a) rate of depreciation (R_m) = 1/N = 0.2
(b) total depreciation (P − E) = (200,000 − 50,000) = 150,000
(c) annual depreciation (D_m) = (P − E) × R_m = 150,000 × 0.2 = 30,000

Using these results, Table 9.5 summarizes the change in the book value of the asset over the five-year period.

Table 9.5 Calculation of depreciation using straight-line method

Year (m)	Opening book value (BV_{m-1})	Annual depreciation (D_m)	Closing book value (BV_m)
0	0	0	200,000
1	200,000	30,000	170,000
2	170,000	30,000	140,000
3	140,000	30,000	110,000
4	110,000	30,000	80,000
5	80,000	30,000	50,000

9.10.2 Sum of Years Digit Method

One of the commonly used methods of calculating depreciation, it leads to a higher rate of depreciation initially which reduce progressively. As illustrated in Example 2 above, such a strategy could be useful to obtain greater tax benefits in the early years of the economic life of the asset. Mathematically, the steps in the method can be summarized as follows:

1. For an asset having an economic life of N years, determine the sum of years (SOY)

$$SOY = (1+2+3+\cdots+N) = \frac{N \times (N+1)}{2} \qquad (9.4)$$

2. Determine the annual rate of depreciation (R_m)

$$R_m = \frac{N-(m-1)}{SOY} \qquad (9.5)$$

3. Determine the total depreciation involved (D)

$$D = \text{Total depreciation} = (P - E) \qquad (9.6)$$

4. Determine the annual depreciation for the year m (D_m)

$$D_m = (P - E) \times R_m \qquad (9.7)$$

5. Determine the book value of the asset for the year m (BV_m)

$$BV_m = BV_{m-1} - D_m \qquad (9.8)$$

Example 4

Using the data in Example 3, tabulate the annual depreciation and the book values of the asset using the sum of digits method.

Although the details of the simple computations have been left out, Table 9.6 shows the annual depreciations and the opening and closing book values of the asset using the sum of digits method.

From Table 9.6, it can be seen that though the total depreciation is the same in the SOY and straight-line methods (i.e., ₹ 150,000), the former method gives higher amounts in the first and second years (₹ 50,000 and ₹ 40,000, respectively). This reduces the book value of the asset, and also gives higher tax benefits. Of course, the book value at the end of the useful life (₹ 50,000) is the same in both cases.

Table 9.6 Calculation of depreciation using sum of years digit method

Year (m)	Opening book value (BV_{m-1})	Annual rate of depreciation (R_m)	Annual depreciation (D_m)	Closing book value (BV_m)
0	0	0	0	200,000
1	200,000	5/15	50,000	150,000
2	150,000	4/15	40,000	110,000
3	110,000	3/15	30,000	80,000
4	80,000	2/15	20,000	60,000
5	60,000	1/15	10,000	50,000

9.10.3 Declining Balance Method

In this method, the depreciation for a given year is calculated on the basis of the undepreciated balance (instantaneous book value), rather than the original cost. Further, the method does not take into account any salvage value of the asset. The rate of depreciation, R_m, in this case can be represented by a factor M/N, where M is a constant and N is the service life of the asset. Frequently, M is taken as 2 and the rate of depreciation as 2/N. In such a case, the method is also referred to as 'double declining balance (DDB) method'. It may be noted that the rate of depreciation in this case is twice that of the straight-line method, with the additional condition that the salvage value is taken as zero. However, in the case of a given finite salvage value, this value becomes the lower bound of the book value. Mathematically, the formulation for the double declining balance method can be represented as:

$$R_m = \frac{2}{N} \tag{9.9}$$

$$D_m = BV_{m-1} \times R_m \tag{9.10}$$

$$BV_m = BV_{m-1} - D_m \tag{9.11}$$

Example 5
Using the data in Example 3, tabulate the annual depreciation and the book values of the asset using the double declining balance method.

In this case, given that the economic life of the asset is 5 years, the rate of depreciation (R_m) can be taken as (2/N = 2/5 = 0.4). Now, this factor is applied to determine the annual depreciation (D_m), and the book value of the asset is allowed to reduce till such time as the value does not go below the salvage value (E). While details of the computations have been left out, the final table showing the annual depreciations and the opening and closing book values of the asset using the double declining balance method is given in Table 9.7.

It can be seen from Table 9.7 that the decline in the value of the asset is even more steep here compared to the sum of years method, and that the value falls to the salvage level at the end of three years itself, whereas the economic life of the asset itself is taken as five years. It may also be noted that the book value of an asset never becomes zero when the depreciation is determined using the double rate declining balance method.

CONSTRUCTION EQUIPMENT MANAGEMENT | 343 |

Table 9.7 Calculation of depreciation using double declining balance method

Year (m)	Opening book value (BV_{m-1})	Annual rate of depreciation (R_m)	Annual depreciation (D_m)	Closing book value (BV_m)
0	0	0	0	200,000
1	200,000	0.4	80,000	120,000
2	120,000	0.4	48,000	72,000
3	72,000	0.4	28,800 (22,000) #	43,200 (50,000) #
4	50,000	0.4	ZERO #	50,000 #
5	50,000	0.4	ZERO #	50,000 #

#: The value of the asset is not allowed to fall below the salvage value, and the amounts for depreciation are adjusted accordingly. Thus, in this case, the D_m calculated may be taken to give an upper bound of the allowable depreciation, which should be adjusted according to the book and salvage values.

In view of the above, regulatory and tax authorities sometimes permit a taxpayer to switch from double declining balance method to straight-line depreciation method at any point in the tax life. Now, this switch could be advantageous when the depreciation using the DDB method becomes less than that determined by the straight-line method—in other words, when the ratio of difference between the undepreciated balance and the predicted tax salvage to the remaining tax life exceeds the depreciation determined by the DDB method, i.e.,

$$\frac{(\text{Undepreciated balance} - \text{predicted salvage value})}{\text{Remaining tax life}} \geq \text{Depreciation using DDB method} \quad (9.12)$$

The following example illustrates the concept of this switch point. It may be mentioned that the switch point normally lies between (N/2 + 1) and N, where N is the period.

Example 6
Assume the purchase cost and salvage value of an equipment with a tax life of 20 years to be ₹ 10 lakh and ₹ 1 lakh, respectively. Initially, depreciation is calculated using the double declining balance method, and the details are shown in Table 9.8. It can be seen that the depreciation value calculated using this method is greater than the [(undepreciated balance − predicted tax salvage)/remaining tax life] till the 15 year. However, for year 16, depreciation calculated using double declining balance method is ₹ 20,589, which is less than that calculated using the straight line at that point. Hence, in year 16, a switch to the straight-line method (from the double declining balance method) may be made. This is also shown graphically in Figure 9.11. Compared to the thumb rule of N/2 + 1 and N, as given above, it may be noted that the switch point found here (16) is, indeed, between these values (11 and 20, respectively).

9.10.4 Sinking Fund Method
The idea of this method is basically to have enough funds to be able to replace the asset at the end of its service life. To achieve this, a fixed sum is set aside from revenue each year and taken

| 344 | CONSTRUCTION PROJECT MANAGEMENT

Table 9.8 Calculating depreciation using double declining balance method and switch-point location

Year (m)	Opening book value (BV_{m-1})	Annual rate of depreciation (R_m)	Annual depreciation (D_m)	Closing book value (BV_m)	(Undepreciated balance– predicted tax salvage)/ Remaining tax life
0	1,000,000	0.0	0	1,000,000	45000
1	1,000,000	0.1	100,000	900,000	42105
2	900,000	0.1	90,000	810,000	39444
3	810,000	0.1	81,000	729,000	37000
4	729,000	0.1	72,900	656,100	34756
5	656,100	0.1	65,610	590,490	32699
6	590,490	0.1	59,049	531,441	30817
7	531,441	0.1	53,144	478,297	29100
8	478,297	0.1	47,830	430,467	27539
9	430,467	0.1	43,047	387,420	26129
10	387,420	0.1	38,742	348,678	24868
11	348,678	0.1	34,867	313,810	23757
12	313,810	0.1	31,381	282,429	22804
13	282,429	0.1	28,243	254,186	22027
14	254,186	0.1	25,419	228,767	21461
15	228,767	0.1	22,877	205,890	21178
16	**205,891**	**0.1**	**20,589**	**185,302**	**21325**
17	184,713	0.1	18,471	166,242	21325
18	163,535	0.1	16,354	147,182	21325
19	142,356	0.1	14,236	128,120	21325
20	121,178	0.1	12,118	109,060	21325

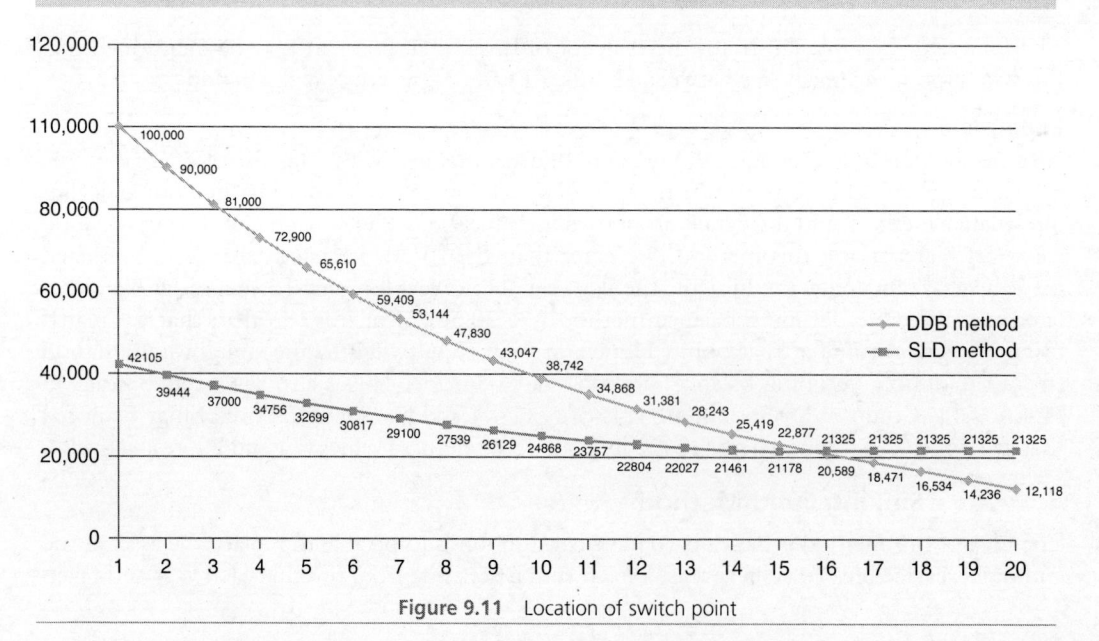

Figure 9.11 Location of switch point

to be invested with compound interest throughout the life of the asset, such that after successive instalments the sum accumulates to produce the original purchase price of the asset less its salvage value. Thus, the method works in terms of a 'sinking fund factor', which is determined on the basis of the initial and salvage values of the asset, the service life, and a rate of (compound) interest. This factor can be determined using simple interest formulae, and also special tables developed for the purpose, discussed elsewhere.

Example 7
For the data given in Example 3, taking a rate of interest (i) to be 10 per cent, determine the sinking fund factor and the book value of the asset at different points in time.

Given that the initial cost is ₹ 200,000 (P), the service life is 5 years (N), and the salvage value is ₹ 50,000 (E), taking the rate of interest as 10 per cent (i), we need to determine the quantum of equal instalments, which if invested at the rate of 10 per cent will yield a sum of ₹ 1.5 lakh (i.e., P − E = 200,000 − 50,000) at the end of a five-year period. Using available tables, the factor is seen to be 0.1638. Thus, the annual depreciation can be taken as ₹ 24,570 (i.e., 0.1638 × 150,000). Using this value, Table 9.9 summarizes the details of the changes in the book value of the asset.

It can be seen from Table 9.9 that because the amount set aside is assumed to be generating compound interest, the total depreciation in the method (24,570 × 5 = 122,850) is less than the estimate arrived at from a simple P − E (which is 150,000) type of formulation. Also, as a corollary to this, the asset has a book value higher than the salvage value at the end of the useful service life.

9.10.5 Accelerated Depreciation

Accelerated depreciation is a method that allows faster write-offs than the straight-line method. It provides a greater 'tax shield' effect than straight-line depreciation, and for this reason, companies with large tax burdens might prefer to use the accelerated depreciation method, even if it reduces the income shown on financial statement. Accelerated depreciation methods are popular for writing off equipment that might be replaced before the end of its useful life since such equipment might become obsolete(e.g., computers).

One example of an accelerated depreciation method is the modified accelerated cost recovery system (MACRS), which is adopted in the USA to calculate the depreciation of assets placed in service after 1986. This system classifies each kind of assets by its useful life of 3 years,

Table 9.9 Using the sinking factor method to monitor changes in the book value of assets

EOY	Fixed depreciation	Interest	Net depreciation (D_m)	BV
0	0	0		200,000.00
1	24,570	0	24,570.00	175,430.00
2	24,570	2,457.00	27,027.00	148,403.00
3	24,570	2,702.70	29,729.70	118,673.30
4	24,570	2,972.97	32,702.67	85,970.63
5	24,570	3,270.27	35,972.94	~50,000.00

Table 9.10 Asset category and the applicable rates of depreciation

Year	3 years	5 years	7 years	10 years
1	33.33	20.00	14.29	10.00
2	44.45	32.00	24.49	18.00
3	14.81	19.20	17.49	14.40
4	7.41	11.52	12.49	11.52
5		11.52	8.93	9.22
6		5.76	8.92	7.37
7			8.93	6.55
8			4.46	6.55
9				6.56
10				6.55
11				3.28

5 years, 7 years, 10 years, 15 years and 20 years, and it does not consider the salvage value. In other words, the minimum permissible life period for tax computation for different kinds of assets is laid down. This restricts the permissible depreciation expenses that may be charged for an asset and prevents a tax payee from claiming indiscriminate tax benefits in the initial life of the asset. Table 9.10 gives a few examples of the applicable rate of depreciation for prescribed minimum service lives.

In order to calculate the depreciation using MACRS, the original book value is multiplied with the percentages shown against each year for the particular class of asset. It should be noted that for 3-year class asset, depreciation percentages are given till 4 years; for 5-year class asset the percentages are given for 6 years; and so on. For an example of a 5-year class asset of original book value of ₹ 200,000, the depreciation calculation is given in Table 9.11.

9.11 EXAMPLE OF DEPRECIATION CALCULATIONS FOR EQUIPMENT AT A SITE

The discussion so far has been limited to the commonly used methods for calculating depreciation and a simple comparison of these methods in terms of how the book value

Table 9.11 Calculation of depreciation using MACRS method for 5-year class asset

Year (m) (A)	Opening book value (BV_{m-1}) (B)	Annual rate of depreciation (R_m) (C)	Annual depreciation (D_m) (D) = (C)*P	Closing book value (BV_m) (E) = (B) – (D)
0	0	0	0	200,000
1	200,000	20.00% = 0.20	40,000	160,000
2	160,000	32.00% = 0.32	64,000	96,000
3	96,000	19.20% = 0.192	38,400	57,600
4	57,600	11.52% = 0.1152	23,040	34,560
5	34,560	11.52% = 0.1152	23,040	11,520
6	11,520	5.76% = 0.0576	11,520	0

Table 9.12 Calculation of depreciation calculations for equipment at a site

S. No.	Description (A)	Nos (B)	Month (C)	Opening book value as on 31.12.2005 (D)	Opening value for total (E)	Dep. rate (F)	Dep. per month* (G)	Total dep. (H)
1	Mixer machine	13	4	73,360	953,680	4.75%	443	23,036
2	Tippers—6 m^3	2	12	315,054	630,108	11.31%	5,664	135,936
3	Vibrators	13	4	20,704	269,152	5.15%	137	7,124
4	Builders hoist—2t	4	13	109,230	436,921	4.75%	657	34,164
5	Compressor	1	13	242,375	242,375	5.15%	1,609	20,917
6	Compression testing machine—100 t	4	13	118,805	475,220	4.75%	716	37,232
7	Concrete pump	2	15	6,114,175	12,228,350	4.75%	36,951	1,108,530
8	Earth rammer	1	21	30,654	30,654	11.31%	556	11,676
9	Woodcutting and planner	1	8	21,865	21,865	4.75%	131	1,048
10	Level	1	21	289,915	289,915	4.75%	1,754	36,834
11	DG set—65KVA	1	13	203,555	203,555	5.15%	1,352	17,576
12	Welding generator	1	13	48,275	48,275	5.15%	322	4,186
	Total depreciation							1,438,259

* As mentioned in the text, the rate of depreciation is included only for the sake of completion, and cannot be directly used to determine the actual depreciation involved (shown in column G). The latter is based on the initial cost (and the salvage value, depending upon the method used), and these values have not been provided here. In other words, for the sake of illustration, the values in column G have been provided independently in this example.

changes, etc. Now, a large company continues to buy different pieces of individual equipment at different points in time at different costs, and they are used at different sites for varying periods of time. Thus, for the purpose of bookkeeping and the expenses incurred at a particular site on account of depreciation of equipment used at that site, the actual depreciation expenses incurred at any site need to be determined. In this context, a month or a quarter could be taken as single units for calculating depreciation expenses. The following example has been included to illustrate the concept of this integration of depreciation expenses at a single site.

Table 9.12 gives details of the number(s) of the different equipment mobilized at the site (columns A and B), the number of months an equipment is likely to be required (column C), and the opening book value of each of the equipment (column D). It may be noted that the value in column D could be different for the individual pieces of the equipment, though in this example they have been taken to be identical. For example, though each mixer machine

Table 9.13 Problem statement

Description	Equipment A	Equipment B
Initial cost	₹ 12.5 lakh	₹ 7.5 lakh
Operating cost	₹ 3 lakh	₹ 5.5 lakh
Economic life	5 years	5 years
Corporate tax	50%	50%
Salvage value	0	0

may have different opening values, to simplify calculations, all mixers have been taken to have the opening value of ₹ 73,360, and the total opening value for the equipment in column E has been arrived at by multiplying the corresponding numbers in columns B and D. Further, columns F and G give the rate of depreciation and the amount of depreciation for individual equipment, respectively, and the total depreciation in column H can be calculated by multiplying the corresponding values in columns B, C and G. It should be noted that the rate of depreciation given in column F is included only for the sake of completion, and cannot be directly used to determine the actual depreciation involved (column G), as the latter is based on the initial cost (and the salvage value, depending upon the method used), and these values have not been provided here. This exercise is carried out for all the equipment to be used, and the total depreciation in column H is determined as the sum of all values. This value for the present example can be worked out as ₹ 1,438,259.

9.12 THE EFFECT OF DEPRECIATION AND TAX ON SELECTION OF ALTERNATIVES

In the previous section, we discussed the different methods of depreciation. In the following paragraphs, we discuss the effect of depreciation and taxes on selection of alternatives. For this, let us look at the problem statement given in Table 9.13.

Equipment A

The before-tax cash-flow diagram for equipment A is shown in Figure 9.12.

For getting the post-tax cash-flow diagram, the relevant computations are given in Table 9.14 along with the explanation.

The post-tax cash-flow diagram corresponding to equipment A is as shown in Figure 9.13.

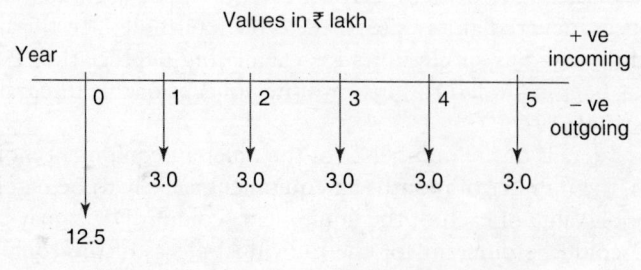

Figure 9.12 Pre-tax cash-flow diagram for equipment A

CONSTRUCTION EQUIPMENT MANAGEMENT | 349

Table 9.14 Computations for post-tax cash-flow for equipment A

S. No.	Description	Amount	Explanation
(1)	Operating cost	300,000	Given
(2)	Depreciation using straight-line method	250,000	Depreciation per year = (1,250,000 − 0) ÷ 5
(3)	Total taxable amount	550,000	(1) + (2)
(4)	Corporate tax @50%	275,000	(4) × 0.5
(5)	Net cost per annum	25,000	(1) − (4)

Equipment B

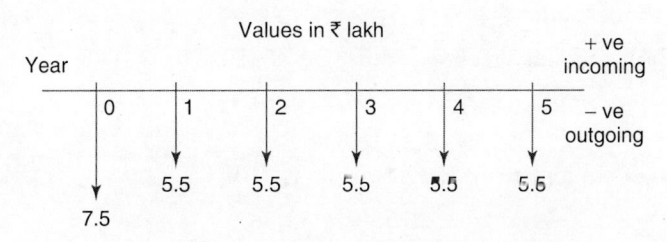

The before-tax cash-flow diagram for equipment B is shown in Figure 9.14.

For the post-tax cash-flow diagram, the relevant computations are given in Table 9.15. The post-tax cash-flow diagram is shown in Figure 9.15.

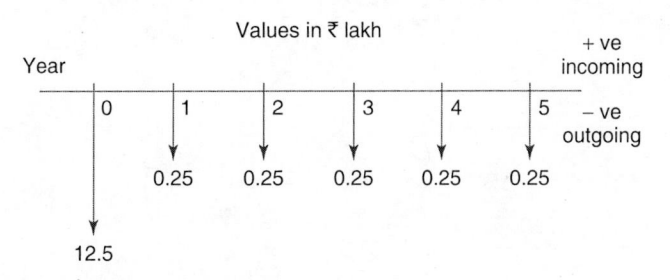

Figure 9.13 Post-tax cash-flow diagram for equipment A

Figure 9.14 Pre-tax cash-flow diagram for equipment B

Table 9.15 Computations for post-tax cash-flow for equipment B

S. No.	Description	Amount	Explanation
(1)	Operating cost	550,000	Given
(2)	Depreciation	150,000	Depreciation per year = (750,000 − 0) ÷ 5
(3)	Total taxable amount	700,000	(1) + (2)
(4)	Corporate tax @50%	350,000	(3) × 0.5
(5)	Net cost per annum	200,000	(1) − (4)

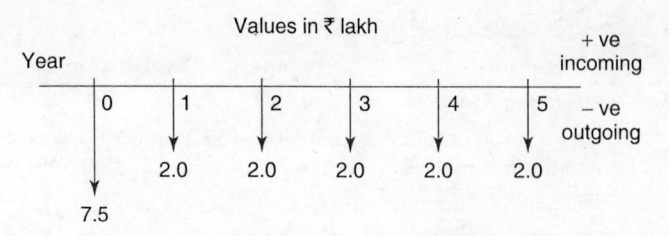

```
                    Values in ₹ lakh
Year                                              + ve
                                                  incoming
      0    1    2    3    4    5
                                            - ve
                                            outgoing
           2.0  2.0  2.0  2.0  2.0
      7.5
```

Figure 9.15 Post-tax cash-flow diagram for equipment B

Once the cash-flow diagram after tax is prepared, we can use any of the methods discussed earlier (present worth, annual worth, or incremental rate of return) to select the alternative.

(a) **Present worth method:**

Net present worth of equipment A:

$$\text{NPW} = -12.5 - 0.25(P/A, 10\%, 5)$$
$$= -12.5 - 0.25 \times 3.7908$$
$$= -13.45 \text{ lakh}$$

Net present worth of equipment B:

$$\text{NPW} = -7.5 - 2(P/A, 10\%, 5)$$
$$= -7.5 - 2 \times 3.7908$$
$$= -15.08 \text{ lakh}$$

Hence, choose equipment A as its net present worth is more (in other words, it costs less).

(b) **Annual worth method:**

$$\text{Annual worth of equipment A} = -12.5 \times (A/P, 10\%, 5) - 0.25$$
$$= -12.5 \times (0.2638) - 0.25$$
$$= -3.55 \text{ lakh}$$

$$\text{Annual worth of equipment B} = -7.5 \times (A/P, 10\%, 5) - 2$$
$$= -7.5 \times (0.2638) - 2$$
$$= -3.98 \text{ lakh}$$

Hence, choose equipment A as its cost is less.

(c) **Incremental rate of return method:**

Preferring equipment A over B, and considering post-tax cash-flow diagrams of equipment A and B, the incremental cash-flow diagram is shown in Figure 9.16.

Assume $i = 10\%$

$$\text{NPW} = -5 + 1.75 \times (P/A, 10\%, 5)$$
$$= -5 + 1.75 \times (3.7908)$$
$$= 1.633 \text{ lakh}$$

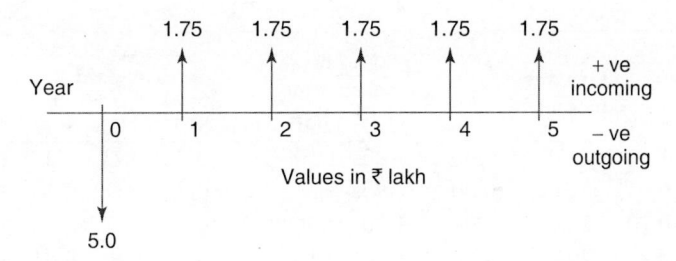

```
                    1.75  1.75  1.75  1.75  1.75
                     ↑     ↑     ↑     ↑     ↑            + ve
        Year                                              incoming
              ┌─────┬─────┬─────┬─────┬─────┐
              0     1     2     3     4     5            − ve
                                                         outgoing
                         Values in ₹ lakh
              ↓
             5.0
```

Figure 9.16 Incremental cash-flow diagram—preferring equipment A over B

Assume i = 15%

$$\text{NPW} = -5 + 1.75 \times (P/A, 15\%, 5)$$
$$= -5 + 1.75 \times (3.3522)$$
$$= 0.866 \text{ lakh}$$

Assume i = 20%

$$\text{NPW} = -5 + 1.75 \times (P/A, 20\%, 5)$$
$$= -5 + 1.75 \times (2.9906)$$
$$= 0.233 \text{ lakh}$$

Assume i = 25%

$$\text{NPW} = -5 + 1.75 \times (P/A, 25\%, 5)$$
$$= -5 + 1.75 \times (2.68928)$$
$$= -0.294 \text{ lakh}$$

By interpolation, i = 21.53%

Since this is more than MARR = 10 per cent, it will be preferable to choose equipment A. Thus, all the three methods give the same result—that of choosing equipment A.

We take another example to illustrate the effect of method of depreciation on the cash flow and, thereby, on its present worth, annual worth, or rate of return. We assume that some equipment has been purchased for ₹ 3.5 lakh, and its accounting life and service life happen to be five years. The predicted salvage value is nil. We consider the three cases—(1) when the data on income alone is given, (2) when the data on expense alone is given, and (3) when both the income and expense data are given. For each of the three cases, we discuss the implication of depreciation (using straight-line method, double declining balance method, and sum of years digit method) and taxes.

Case 1: Income alone is given

Table 9.16 shows the computation of depreciation using the three methods—straight-line (SLD), double declining balance (DDB) and sum of years digit (SOY). It may be noted that a switchover has been made from double declining balance method of depreciation to straight-line method of depreciation in year 4. The gross income for all the five years is assumed to be ₹ 2.5 lakh and expenses are assumed to be nil for all these five years.

For calculating the net taxable income, we subtract the depreciation and expenses from the gross income. The total deduction in case of straight-line depreciation would be (0.7 + 0), and

Table 9.16 Case 1: Income alone is given

Year	Depreciation SLD	Depreciation DDB	Depreciation SOY	Gross income	Expense	Net taxable income SLD	Net taxable income DDB	Net taxable income SOY	Tax paid SLD	Tax paid DDB	Tax paid SOY	Net income SLD	Net income DDB	Net income SOY
1	0.70	1.40	1.17	2.50	0.00	1.80	1.10	1.33	0.72	0.44	0.53	1.78	2.06	1.97
2	0.70	0.84	0.93	2.50	0.00	1.80	1.66	1.57	0.72	0.66	0.63	1.78	1.84	1.87
3	0.70	0.50	0.70	2.50	0.00	1.80	2.00	1.80	0.72	0.80	0.72	1.78	1.70	1.78
4	0.70	0.38	0.47	2.50	0.00	1.80	2.12	2.03	0.72	0.85	0.81	1.78	1.65	1.69
5	0.70	0.38	0.23	2.50	0.00	1.80	2.12	2.27	0.72	0.85	0.91	1.78	1.65	1.59
								Total	3.60	3.60	3.60			

the net taxable income = 2.5 − (0.7 + 0) = 1.8 lakh. If the tax rate is assumed to be 40 per cent on the net taxable income, we get the total tax component as 1.8 × 40/100 = 0.72 lakh; thus, the net income for year 1 would be 2.5 − 0.72 = 1.78 lakh.

Similar calculations are performed for the other two methods of depreciation. We find that although the total tax implication is same (₹ 3.6 lakh) in all the three cases of depreciation, the timing of tax payment varies. For the faster method of calculating depreciation such as double declining balance method, the tax paid in the initial years (₹ 0.44 lakh in year 1, ₹ 0.66 lakh in year 2, and so on) is less and the net income post-tax is more when compared to slower methods of depreciation such as straight-line method (₹ 0.72 lakh for year 1 to year 5). This is precisely the reason why construction companies, if given a choice, would opt for the faster method of depreciation.

Case 2: Operating expense alone is given

In this case (see Table 9.17), let us assume that there is no data provided on the gross income made from operating the equipment. Also, let us assume that the deductible expense to operate the equipment is ₹ 1 lakh annually for all the five years. Thus, for straight-line method of depreciation, the total expense would be the sum of depreciation for the year and the expense, that is, 0.7 + 1.0 = 1.7 lakh. The net taxable income would be 0 − (0.7 + 1.0) = −1.7 lakh. The tax at a rate of 40 per cent on net taxable income would be −1.70 × 40/100 = −0.68 lakh. The negative sign in tax indicates the resultant saving in the tax on account of the admissible expenses. Thus, for year 1, the post-tax net expense would be total annual expense minus saving in tax on account of depreciation and annual expense, that is, 1.0 − 0.68 = 0.32 lakh.

Table 9.17 Case 2: Operating expense alone is given

Year	Depreciation SLD	Depreciation DDB	Depreciation SOY	Gross income	Expense	Net taxable income SLD	Net taxable income DDB	Net taxable income SOY	Tax SLD	Tax DDB	Tax SOY	Post-tax net expense SLD	Post-tax net expense DDB	Post-tax net expense SOY
1	0.7	1.4	1.17	0	1.0	−1.70	−2.40	−2.17	−0.68	−0.96	−0.87	0.32	0.04	0.13
2	0.7	0.84	0.93	0	1.0	−1.70	−1.84	−1.93	−0.68	−0.74	−0.77	0.32	0.26	0.23
3	0.7	0.50	0.70	0	1.0	−1.70	−1.50	−1.70	−0.68	−0.60	−0.68	0.32	0.40	0.32
4	0.7	0.38	0.47	0	1.0	−1.70	−1.38	−1.47	−0.68	−0.55	−0.59	0.32	0.45	0.41
5	0.7	0.38	0.23	0	1.0	−1.70	−1.38	−1.23	−0.68	−0.55	−0.49	0.32	0.45	0.51
								Total	−3.40	−3.40	−3.40			

CONSTRUCTION EQUIPMENT MANAGEMENT | 353

Table 9.18 Case 3: Income and expense both are given

Year	Depreciation SLD	Depreciation DDB	Depreciation SOY	Gross income	Expense	Net taxable income SLD	Net taxable income DDB	Net taxable income SOY	Tax paid SLD	Tax paid DDB	Tax paid SOY	Net income SLD	Net income DDB	Net income SOY
1	0.7	1.40	1.17	2.5	1.0	0.80	0.10	0.33	0.32	0.04	0.13	2.18	2.46	2.37
2	0.7	0.84	0.93	2.5	1.0	0.80	0.66	0.57	0.32	0.26	0.23	2.18	2.24	2.27
3	0.7	0.50	0.70	2.5	1.0	0.80	1.00	0.80	0.32	0.40	0.32	2.18	2.10	2.18
4	0.7	0.38	0.47	2.5	1.0	0.80	1.12	1.03	0.32	0.45	0.41	2.18	2.05	2.09
5	0.7	0.38	0.23	2.5	1.0	0.80	1.12	1.27	0.32	0.45	0.51	2.18	2.05	1.99
								Total	1.60	1.60	1.60			

Similar calculations are performed for all the years using all the three methods of depreciation. It can be observed that post-tax expenses come down drastically due to tax saving on account of admissible expenses.

Case 3: Income and expense both are given

Let us assume gross income and expense to be ₹ 2.5 lakh and ₹ 1 lakh, respectively, on an annual basis(see Table 9.18). The net taxable income for year 1 using straight-line depreciation would be 2.5 − (0.7 + 1.0) = 0.8 lakh. The tax to be paid at the rate of 40 per cent on taxable income would be 0.8 × 40/100 = 0.32 lakh. Thus, net income for year 1 would be 2.5 − 0.32 = 2.18 lakh. Similarly, calculations for all the five years can be performed using all the three methods of depreciation, and the net income can be found as well.

Here also, we find that total tax implication remains same (₹ 1.60 lakh) although the timing of tax payment varies.

9.13 EVALUATING REPLACEMENT ALTERNATIVES

Apart from the decision involving selection of new equipment, a construction manager also needs to time his purchase as well as decide what to do with the older equipment. As would be expected, any equipment has a service life beyond which it does not make sense economically to continue using the same equipment. Replacement of an existing productive asset becomes necessary because of:

(a) Physical or mechanical impairment

(b) Technological obsolescence

(c) Uneconomic (increasing) maintenance and/or operating cost

(d) Inadequacy

Replacement analysis always involves comparison of an existing asset with a new one. The existing asset is termed as 'defender' and the new one is known as 'challenger' in replacement analysis.

The defender would have been purchased some years ago at some cost—say, Po. Today, if it is traded in or sold because of a consideration to buy a challenger, the value obtained will be lower—say, P. The term (Po−P) is called the sunk cost. The sunk cost is never taken into account in engineering economic analysis.

The term P is known as 'trade in value' and is equal to the market value at which the equipment can be sold. The term P is not to be confused with salvage value E, as the equipment with P value can still serve for some more time. Salvage value is the value that is obtained when a productive asset has served its life.

The term 'life' has many variants such as economic life, service or useful life, ownership life and accounting life. The economic life is the period over which the given equipment has its lowest uniform equivalent annual cost. In other words, economic life is the period that will be terminated when a new piece of equipment has a uniform annual cost lower than the cost of keeping the equipment for one or more years. Service or useful life is the period over which the equipment will produce useful service. The economic life of an asset can be smaller than its useful life, due to the increase in maintenance cost component with its age.

The period until the equipment is sold or otherwise disposed of is known as ownership life, and the period selected by the accountant over which the equipment will be depreciated in the books of the company is known as the accounting life. The accountant normally uses a period equal to the useful life, although when permitted by tax rules, he may accomplish a fast write-off of the equipment.

When comparing a defender with a challenger, it is advisable to consider the problem from the viewpoint of a third party that has a choice of buying the defender or the challenger. This is also known as the 'third-party concept' in considering replacement decision. Thus, it is not required to subtract the 'trade in value' from the initial cost of the challenger, and it is also not desirable to add sunk cost of the existing asset to the challenger. The following example will illustrate the concept clearly.

Example 8

An equipment that was purchased at a cost of ₹ 22,000 four years ago is considered for replacement against a challenger whose cost is ₹ 24,000. The existing equipment can be traded in today at ₹ 6,000, and if kept on for another 6 years, it will have a salvage value of ₹ 2,000. The annual maintenance cost of the existing asset is ₹ 7,000. The challenger has an annual operating cost of ₹ 3,500, and its salvage value is ₹ 3,000 at end of year 10, i = 15%

Here, Po = ₹ 22,000 and P = ₹ 6,000; and E = ₹ 2,000 at the end of 6 years from now.

Thus, the equivalent annual cost of existing asset (defender)

$$= [6,000 - 2,000 \times (P/F, 15\%, 6)] \times (A/P, 15\%, 6) + 7,000 = 8,356.77$$

Please note that the sunk cost (Po − P) = 22,000 − 6,000 = 16,000 is not used anywhere in the analysis.

Equivalent annual cost of challenger

$$= [24,000 - 3,000 \times (P/F, 15\%, 10)] \times (A/P, 15\%, 10) + 3,500$$
$$= 8,135.40$$

Since the equivalent annual cost of challenger is less than that of defender, it would be economical to replace the existing equipment with the new equipment. Thus replacement is desirable.

The terms used frequently in replacement analysis include depreciation, replacement cost, investment cost, maintenance and repair cost, downtime and loss of productivity cost, obsolescence cost, sunk cost and life of the equipment.

The different steps involved in replacement analysis are summarized below.

Step 1 Calculate depreciation and replacement cost. The method of computation of depreciation is already explained and, thus, can be obtained on hourly basis for an assumed salvage value depending on trends in the market and technology. The replacement cost is to be assumed for a particular type of equipment for different years.

Step 2 Calculate investment cost for the equipment for each year of the analysis period. For example, the investment cost can be assumed to be some percent of the value of the equipment at the beginning of a year.

Step 3 Forecast the maintenance and repair cost based on past experience and records available for similar types of equipment.

Step 4 Forecast the downtime and loss of productivity cost. Downtime cost is the product of percentage downtime and given value of operating cost. If other dependent equipment is also idling due to breakdown of this equipment, downtime costs of dependent equipment need to be considered—for example, idling of trucks due to downtime of loader.

Step 5 Calculate the obsolescence cost, which is the product of the obsolescence factor and the given operating cost. The obsolescence factor could be taken as zero just at the time of purchase, which goes on increasing as and when the equipment gets older.

Step 6 Calculate total costs per year, which is the sum of depreciation and replacement cost, investment cost, maintenance and repair cost, downtime and loss of productivity cost, and obsolescence cost. All the cost computations can be performed on hourly basis assuming the total working time of equipment in any year.

Step 7 Plot the total cost per year for different years to find out the appropriate time of replacement and economic life. One may notice that the total cost per year reduces for initial years and then starts increasing. The time from where the total costs start increasing is the ideal time to replace the equipment.

9.14 ADVANCED CONCEPTS IN ECONOMIC ANALYSIS

9.14.1 Sensitivity Analysis

In all our preceding discussions, we had arrived at a particular decision (in the form of acceptance or rejection of a proposal, and the selection of one alternative among different possible alternatives) in a given situation, assuming that the estimates or the expected value for different variables such as initial cost, receipts, disbursements, interest rate, life of the assets, salvage value, etc., were accurate and constant. Unfortunately, in real-life situations this is not the case. Barring a few variables, such as initial cost of the asset, the rest are all estimates or forecasts that may prove to be wrong on most of the occasions. The life of the asset could be longer or shorter than our estimate; the interest rate could be higher or lower than the assumed value; and so on. The salvage value may be more or less than the assumed value.

We may like to know what will happen to the net present worth associated with a particular investment alternative when certain variables like incomings (receipts) value or outgoings (disbursement) value vary from the expected value. Sensitivity analysis is, thus, aimed at studying the impact of change in the value of variable(s) on the economic decision in a particular situation. In a sense, it aims to answer 'what if'. For example, what will happen if the annual disbursement value increases by 10 per cent or 20 per cent from the current value? Will it turn the positive present worth into negative? Will it change the earlier decision?

The changes (increase or decrease in the assumed values) in the variable values may or may not lead to reversal of our earlier decision. If even a slight change in one variable leads to a reversal of decision—from, say, acceptance of one alternative to its rejection—we say that the variable is highly sensitive. On the other hand, even if a large change in one variable does not change the decision, we say that the variable is not sensitive or is insensitive.

The sensitivity analysis is also aimed at identifying the sensitivity of a particular variable. Once the identification has been made of variables in categories such as highly sensitive, less sensitive, or insensitive, the management can focus their attention on the highly sensitive variables. That is, for such variables they can dedicate more energy and effort in preparing their estimate. In some situations, external help in the form of engagement of consultants can also be thought of.

Sensitivity analysis is basically a non-probabilistic technique, as it does not consider the probability of occurrence associated with the variation in variables. There could be different forms of sensitivity analysis. These are depicted in Figure 9.17.

In its simplest form, sensitivity analysis for a single alternative can be performed and the impact (on decision) of change in a single variable can be studied. Then, there could be simultaneous changes in two variables corresponding to a single alternative, which can be studied. Finally, for a single alternative we can study the changes in more than two variables at a time.

Sensitivity analysis can be performed for more than one alternative also. Here also, we can see the impact of variation of one variable on the decision. There can be cases in which we would like to change two variables at a time or more than two variables at a time, and see their impact on the decision arrived at earlier by assuming all the other variables as fixed or constant.

Sensitivity analysis can be performed with any method of evaluation of alternatives—for example, present worth analysis, annual cost or worth analysis, or internal rate of return method of analysis. Also, the analysis can be performed at different stages of a project either

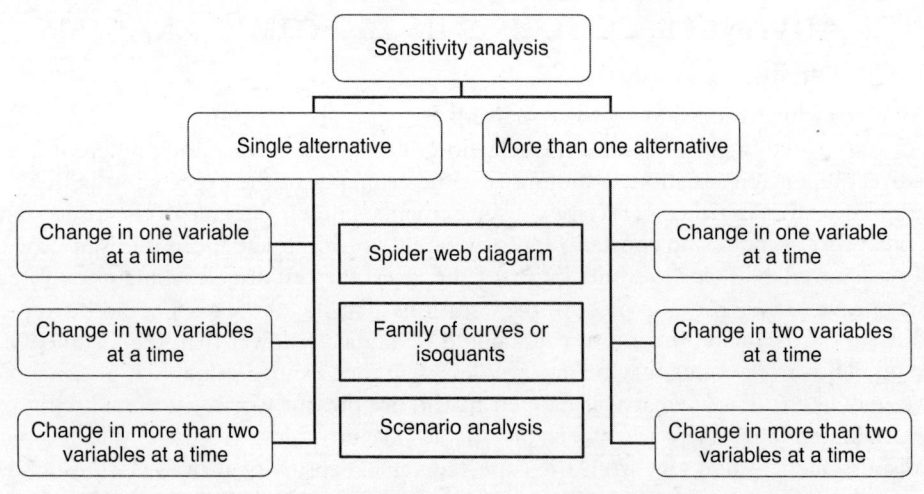

Figure 9.17 Different forms of sensitivity analysis and the methods of presentation of results

Table 9.19 Problem data for illustrating sensitivity analysis

Description	Alternative 1
Acquisition cost (first cost)	₹ 500,000
Incomings	₹ 100,000 every year for 10 years
Outgoings	₹ 5,000 every year for 10 years
Salvage value	₹ 50,000
Interest rate	12%
Service life	10 years

with pre-tax cash flow or post-tax cash flow. However, it is preferable to perform sensitivity analysis with post-tax cash flow. It is customary to show the results of sensitivity analysis in the form of sensitivity graphs.

The different forms of sensitivity analysis and the presentation of their results are illustrated with the help of an example. Consider the following alternative to acquire a new piece of equipment. The acquisition cost, incomings, outgoings, salvage value, interest rate and service life associated with the alternative are provided in Table 9.19.

We illustrate the sensitivity analysis for alternative 1 with variation in one variable at a time. As a first step, we find the net present worth of alternative 1 with the given values of variables.

Thus, net present worth
$$= -500{,}000 + 100{,}000(P/A, 12\%, 10) - 5{,}000(P/A, 12\%, 10) + 50{,}000(P/F, 12\%, 10)$$
$$= 52{,}869.40$$

Let us assume that the estimate of incoming value goes wrong and instead of ₹ 100,000 it is ₹ 90,000. Thus, the new net present worth keeping all other variables same would be
$$= -500{,}000 + 90{,}000(P/A, 12\%, 10) - 5{,}000(P/A, 12\%, 10) + 50{,}000(P/F, 12\%, 10)$$
$$= -3632.80$$

We find that the decision taken on net present worth would get reserved.

The above analysis can also be performed in terms of percentage variation in incoming values. Let the percentage variation in incomings be x. We can find the percent change in incomings at which the net present worth becomes zero. In this example, if the incoming value decreases by 9.35 per cent, the net present worth turns out to be negative. The value of incoming at which net present worth is zero is known as *breakeven point*. The sensitivity graphs showing the effect of changes in incomings on net present worth both in terms of absolute values and percentages are shown in Figure 9.18 (see the line corresponding to incomings variable).

A similar analysis can be performed by changing other variables one at a time, and the combined sensitivity graphs are shown in Figure 9.18. The slope of the sensitivity line indicates the sensitivity of a particular variable. The steeper the slope, the more sensitive the variable is, and the milder the slope, the less sensitive the variable. Thus, it can be seen from the figure that variables such as incomings and interest rates are very sensitive, while variables such as outgoings and salvage value are less sensitive.

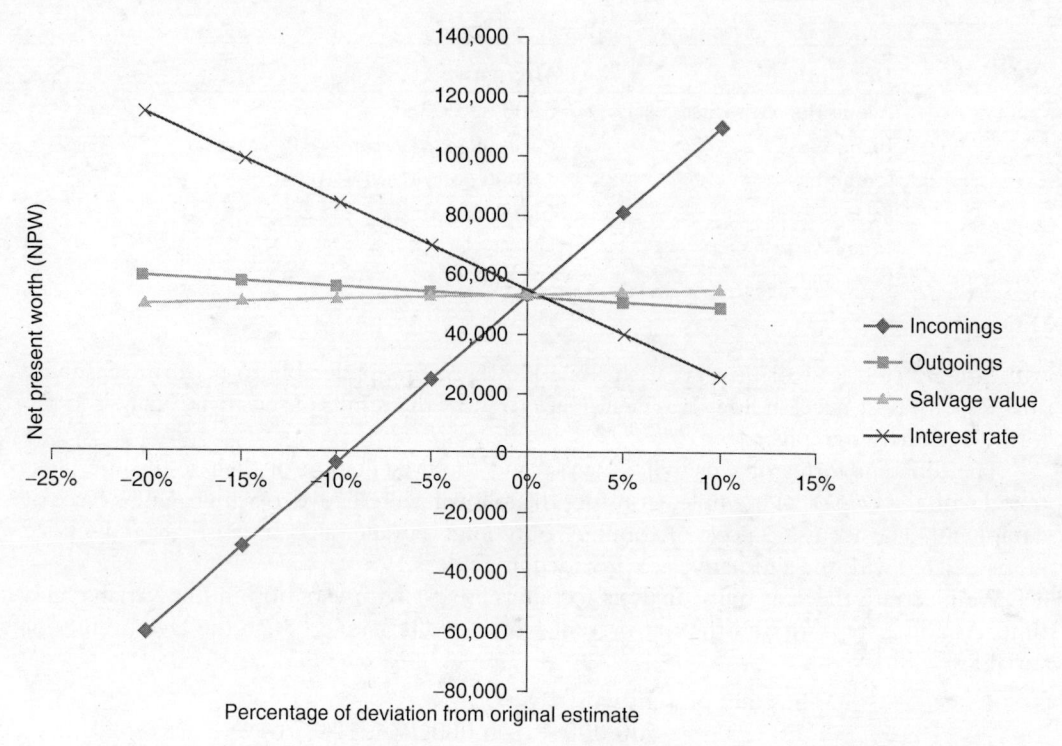

Figure 9.18 Sensitivity graph showing the effect of changes in different variables on net present worth

Next, we take the variables 'incomings' and 'service life' together, and study their sensitivity. In order to perform the sensitivity analysis of two variables simultaneously, for example, incomings (R) and life of the equipment (N), we write the expression of net present worth as:

$$NPW = -500,000 + R \times (P/A, 12\%, N) - 5,000 \times (P/A, 12\%, N) + 50,000 \times (P/F, 12\%, N)$$

There are two ways in which we can show the results of the sensitivity analysis. These are—(1) family of curves and (2) isoquants. The two representations are shown in Figure 9.19 and Figure 9.20, respectively.

In family of curves representation, one variable is kept fixed while another variable is varied. For example, service life N is fixed at some value, N1, and net present worth is computed for different values of incomings R. This process is repeated for N = N2 (say).

In isoquants, an indifference line is plotted by varying two variables, say, service life N and incomings R, at a time while keeping all the other variables fixed. For drawing the indifference line, first we try to get points, say, (N1, R1), (N2, R2), and so on, at which the net present worth is zero. At these points, the decision-maker is indifferent to this proposal, that is, he is neither for nor against this proposal. The indifference line is obtained by joining together the points. Thus, the indifference line divides the acceptance and rejection zones.

Conducting scenario analysis is the best approach when performing sensitivity analysis involving changes in more than two variables at a time. In scenario analysis, a number of

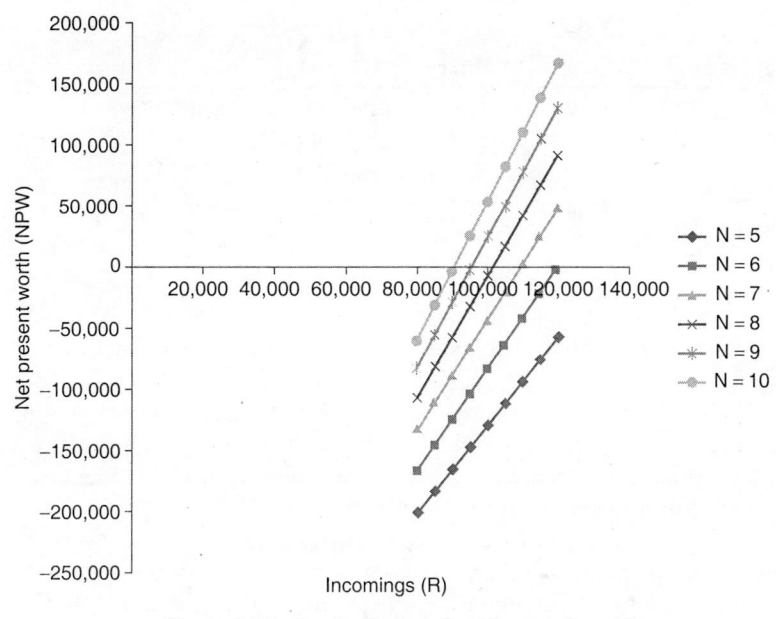

Figure 9.19 Family of curves for the example problem

scenarios such as best scenario, normal scenario and worst scenario (in some texts, these scenarios are referred to as less favourable estimate, objective estimate and more favourable estimate) are generated. The objective of such scenario analysis is to get a feel of what happens under the most favourable and the most adverse configuration of key variables. For example, the best, the normal and the worst scenarios for the previous example could be as given in Table 9.20.

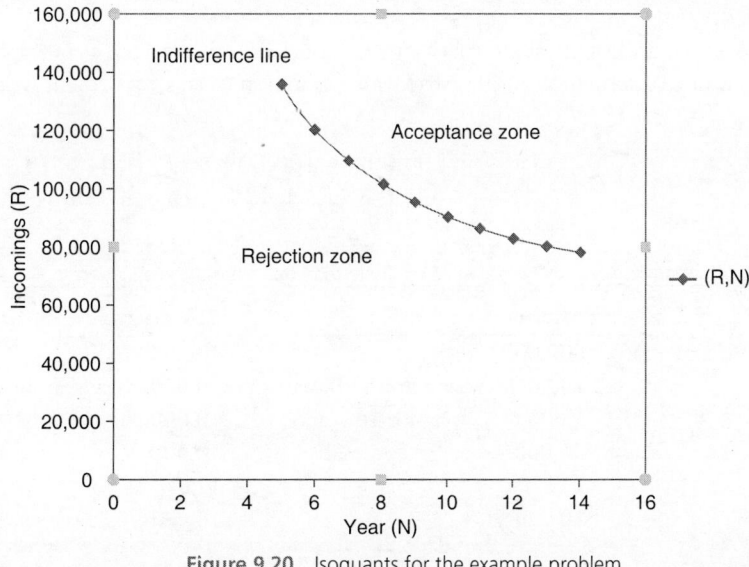

Figure 9.20 Isoquants for the example problem

Table 9.20 Data for illustrating scenario analysis

	Worst	Normal	Best
Incomings	85,000	100,000	115,000
Outgoings	7,000	5,000	4,000
Salvage value	25,000	50,000	60,000
Interest rate	15%	12%	10%
Service life	8 years	10 years	12 years

Now, corresponding to each of the above scenarios, the net present worth can be computed. Based on the net present worth values for each of the scenarios, the decision-maker would be in a better position to take a decision. For example, if the net present worth corresponding to the worst scenario is a large negative value, and the net present worth corresponding to the normal and best scenarios are low positive value and moderate positive value, respectively, the decision would be not to acquire the asset.

Scenario analysis is considered superior to sensitivity analysis because it considers variations in more than two key variables together. However, scenario analysis has some limitations—(1) it assumes that there are a few well-delineated scenarios, which may not be true in many cases; and (2) there is a huge requirement of data to perform scenario analysis. For example, even for six input variables we would require $6 \times 3 = 18$ input data altogether corresponding to the best, normal and worst scenarios.

Sensitivity analysis can also be performed when more than one alternative is involved. Here also, we can vary one variable, two variables, or more than two variables at a time. However, in this text we discuss the variation in one variable only. For other cases, the reader can refer to advanced texts on construction economics. For illustrating the sensitivity analysis corresponding to changes in one variable, we take up two alternatives (see Table 9.21) in which the information on acquisition cost, incomings, outgoings, salvage values, interest rate and service life are provided.

As a first step, we find the best alternative by any of the methods such as net present worth, annual worth, or rate of return. We will restrict our discussion to the net present worth method.

Alternative 1:-

$$NPW = -500,000 + 100,000 \times (P/A, 12\%, 10) \\ - 5,000 \times (P/A, 12\%, 10) + 50,000 \times (P/F\,12\%, 10)$$

Table 9.21 Data for illustrating sensitivity analysis for a 'more than one alternative' situation

Description	Alternative 1	Alternative 2
Acquisition cost (first cost)	500,000	400,000
Incomings	100,000 every year for 10 years	80,000 every year for 10 years
Outgoings	5,000 every year for 10 years	10,000 per year for 10 years
Salvage value	50,000	40,000
Interest rate	12%	12%
Service life	10 years	10 years

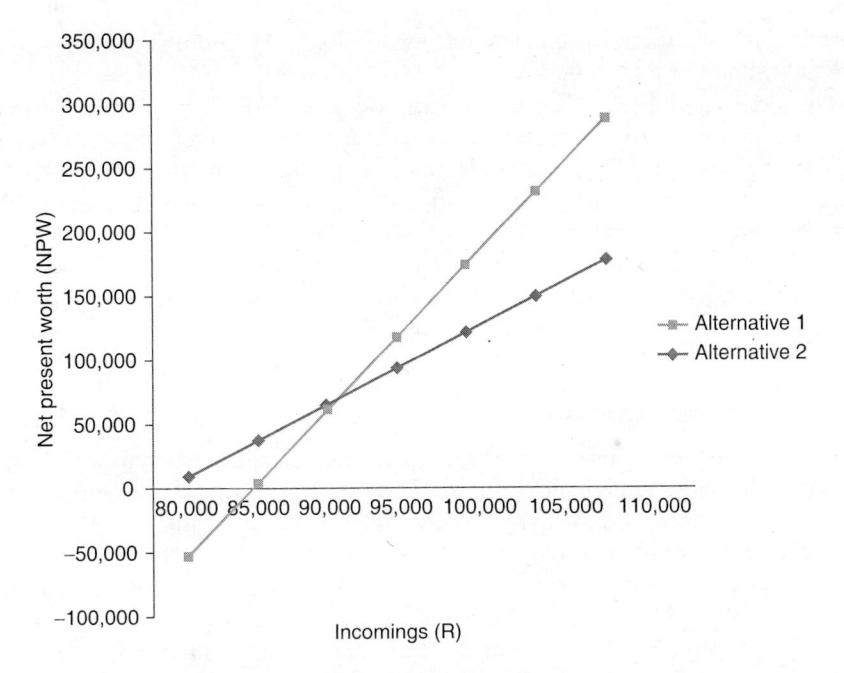

![Figure 9.21 Sensitivity analysis for more than one alternative]

Figure 9.21 Sensitivity analysis for more than one alternative

$$= -500{,}000 + 100{,}000 \times 5.6502 - 5{,}000 \times 5.6502$$
$$+ 50{,}000 \times 0.3220$$
$$= +52{,}869$$

Alternative 2:-

$$NPW = -500{,}000 + 90{,}000 \times (P/A, 12\%, 10)$$
$$- 5{,}000 \times (P/A, 12\%, 10) + 50{,}000\, (P/F, 12\%, 10)$$
$$= -500{,}000 + 90{,}000 \times 5.6502 - 5{,}000 \times 5.6502$$
$$+ 50{,}000 \times 0.3220$$
$$= +8{,}394$$

The net present worth of alternative 1 = 52,869.00
The net present worth of alternative 2 = 8,394.00

Now, let us change each of these variables one by one. For example, consider the changes in the variable 'incoming'. In case the 'incoming' of alternative 1 changes to 90,000 from the existing 100,000, the new net present worth of alternative 1 changes to −3,663. We can find that if the incoming value reduces to less than 92,128.6, the decision is reversed. This is shown graphically in Figure 9.21. Such analysis addresses questions such as—At what value of incomings is alternative 1 preferred to alternative 2? At what service life of the assets is alternative 1 preferred to alternative 2?

The main limitation of sensitivity analysis is that it is non-probabilistic in nature. One may recollect that for none of the cases we considered the likelihood of occurrence of a particular variable value. It merely shows us what happens to the net present worth, the annual worth, or the rate of return when there is a change in some variable(s), without providing any information on the likelihood of the changes.

However, despite its inherent deficiencies, sensitivity analysis offers certain benefits to the decision-maker—(1) it shows how robust or vulnerable a particular alternative is with reference to changes in the value of different variables; and (2) it enables the decision-maker to distinguish the sensitive variables from the insensitive variables, and thereby to focus attention on making the estimate of sensitive variables. Commonly, in sensitivity analysis only one variable is changed at a time, which may not reflect the real-world situation since variables tend to move together. There is subjectivity involved in sensitivity analysis, and thus, may lead one decision-maker to accept a proposal while another may reject it.

9.14.2 Breakeven Analysis

Breakeven analysis is another way of performing sensitivity analysis wherein we are more concerned about finding the value (called the breakeven point) at which the reversal of decision takes place. It is in contrast to sensitivity analysis, where not much emphasis is placed on finding this breakeven value. In sensitivity analysis, we ask what will happen to the project if the invoice or billing declines, if the costs increase, or if something else happens. However, we may also be interested in knowing how much should be produced and sold at a minimum to ensure that the project does not 'lose money'. Such an exercise is called *breakeven analysis* and the minimum quantity at which loss is avoided is called the breakeven point. The breakeven analysis is also referred to as *cost-volume-profit* analysis.

In order to illustrate the concept of break-even analysis, we take a simple example wherein a ready-mix concrete (RMC) manufacturer wants to find out the minimum production of concrete which will just be enough to recover its total cost incurred in a particular month. The total cost (TC) incurred in a month is the sum total of its indirect cost (IC) and direct cost(DC). The indirect costs in this example are those costs that are incurred irrespective of concrete production taking place or not. However, the direct costs are proportional to the volume or quantity of production. A detailed discussion on direct and indirect costs is provided elsewhere in this text.

$$\text{The Total Cost } TC = IC + DC = IC + n \times UDC \qquad (9.13)$$

The relation between DC and UDC is given by the following expression.

$$DC = n \times UDC \qquad (9.14)$$

From the above expression, it is clear that UDC is the direct cost for one unit of production.

Let the sales price fixed by the RMC supplier be P per unit of concrete sold. If the quantity of concrete sold is n units, the revenue R would be computed by the expression:

$$R = n \times P \qquad (9.15)$$

Gross profit Z for the period would be defined as:

$$Z = R - TC = n \times P - IC - n \times UDC = n \times (P - UDC) - IC \quad (9.16)$$

The net profit Z_1 after taking taxes into account is given by:

$$Z_1 = Z \times (1 - t) \quad (9.17)$$

Where t is the tax rate.

Breakeven point is defined as one where profit equals zero. In order to determine the concrete quantity n at which the RMC seller just recovers its total cost, we equate total cost to revenue.

Thus, at breakeven point, Total cost TC = Revenue R. (Note that at this point, profit $Z = 0$.)

At break even, we have:

$$TC = R$$
$$\Rightarrow IC + n \times UDC = n \times P$$

The quantity produced at breakeven point is denoted with B
Thus,

$$IC + B \times UDC = B \times P$$
$$\Rightarrow B = \frac{IC}{(P - UDC)} \quad (9.18)$$

The denominator in the above expression $(P - UDC)$ is also known as 'contribution'. For n less than B, the RMC seller is making losses, while for n greater than B, the RMC seller is making profits.

The result of breakeven analysis is shown on the *breakeven chart*. The chart contains direct cost line and indirect cost line besides the total cost line. It also has the revenue line. The point of intersection of the total cost and the revenue line is known as breakeven point. The ordinate corresponding to this intersection point gives the breakeven quantity of concrete to be produced in order to just recover the total cost incurred by the RMC seller. An illustration of breakeven chart is given for an example problem.

Example
Let us assume that a ready-mix concrete company sells RMC for a price of ₹ 3,500/m³. The fixed cost of the company for the production is ₹ 40,000/month and the variable cost associated with per-m³ production of RMC is ₹ 1,500. It is desired to determine the breakeven quantity.

The breakeven point can be found out from the breakeven chart (Figure 9.22) drawn for the above example. The chart is self-explanatory.

Companies try to lower the breakeven point by resorting to different means such as (1) increasing the sales price, (2) reducing the total cost of production, or (3) increasing the quantity of production. These measures are adopted either in isolation or in combination.

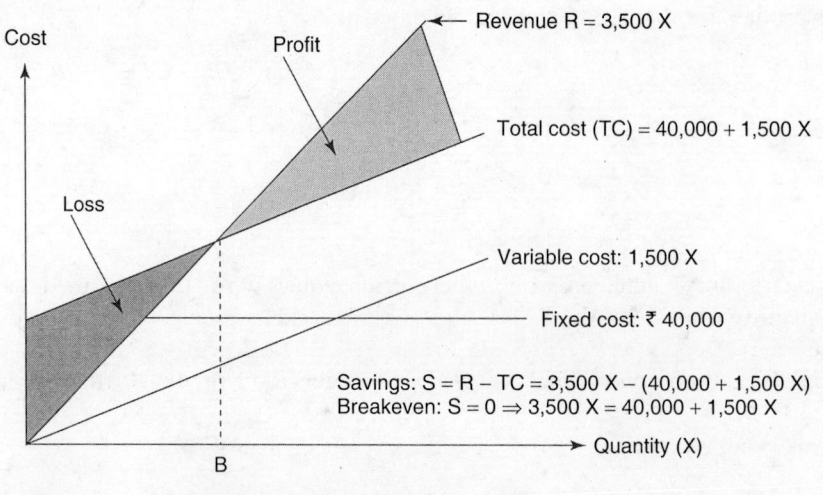

Figure 9.22 Breakeven chart for the example problem

Breakeven analysis is useful in situations involving 'make or buy' decision, 'lease or purchase' decision, and so on. The breakeven analysis performed here is an illustration of linear breakeven analysis with the following inherent assumptions:

1. Income is only from the productions under study. In the previous example, it was assumed that the ready-mix concrete-manufacturing company is into concrete production only.
2. Whatever quantity is produced is sold out.
3. The per-unit direct cost, indirect cost and sales price associated with the production are constant over the study period. These are also constant over the quantity produced.

In real-life situations we may not observe linearity, and to model such situations we have non-linear breakeven analysis. Interested readers can refer to advanced texts on construction economics for more details on the subject.

In non-linear analysis, the cost and revenue do not increase or decrease in linear fashion with increase or decrease in production level. Based on the information on cost, revenue and associated production level (n), one can fit a curve and get an expression of P in terms of n, or C in terms of n. These curves can then be used to compute the marginal cost and the marginal revenue.

Marginal cost is the additional cost incurred by the company to produce one extra unit of product. For example, suppose that the cost to produce 100 m³ of concrete is ₹ 200,000 and the cost to produce 101 m³ of concrete is ₹ 201,000. Here, the marginal cost would be equal to $(201,000 - 200,000)/(101 - 100) = $ ₹ 1,000.

Marginal revenue is the additional money realized by selling one extra unit of product. For example, if the revenue raised by selling 100 m³ of concrete is ₹ 250,000 and by selling 101 m³ of concrete the revenue is ₹ 252,000, the marginal revenue is $(252,000 - 250,000)/(101 - 100) = $ ₹ 2,000. Using the concepts of marginal cost and marginal revenue, and the principle of calculus, one can determine the production level at which the firm would be able to maximize its profit.

9.14.3 Incorporating Probability in Cash Flows

In our preceding discussions, we initially assumed that cash flows are certain, which was later relaxed and we assumed that variations in cash flows and other variables are possible. In fact, at a time we addressed the single variable changes, two variables changes and multiple variables changes for single alternative, and more than single alternative cases; however, those discussions were based on deterministic approach. We never considered the probabilities of occurrence of a given cash flow at a given period. In the subsequent sections, we would like to bring in the concepts of probability in our analysis. Our approach of illustrating the concepts would be primarily on the bases of few examples. First we discuss some commonly used terms in the context of alternative evaluation under risk and then we move on to specifics of aggregated cash flows situation.

9.14.4 Determining Expected Values, Variance, Standard Deviation, and Coefficient of Variation

The following terms are used very frequently in the context of alternative evaluation under risk especially when probability of occurrence of different variables is accounted for. They are (1) expected value, (2) variance, standard deviation, and coefficient of variation. These terms are explained with respect to a simple example.

Assume that a company considers investment in two proposals: (1) I_1—Investing in ready-mix concrete plant and (2) I_2—Investing in an infrastructure bond floated by National Highways Authority of India. Market report, prepared by the Market Research Department, has come up with the forecast of market situation. The report says that chances of booming construction market (extremely favorable demand) for the RMC product is 25%, while the chances of normal demand situation and recession in construction market (extremely unfavourable demand situation) are 50% and 25% respectively. The expected return obtained in each of the scenarios for the two investment plans have been estimated and is produced in Table 9.22.

The expected return $E(X)$ is simply a weighted average of all returns. The weights are probabilities of occurrence of different states. Mathematically, expected return is computed using the following expression.

$$E(X) = \sum_{i=1}^{n} R_i \times P_i \qquad (9.19)$$

Where $E(X)$ is expected return, R_i = Return for the ith probability, and P_i = Probability of occurrence of that return.

Thus, the expected value of returns from ready-mix concrete plant (I_1) is

$$E(X)_{I_1} = 40 \times \frac{25}{100} + 20 \times \frac{50}{100} + (-10) \times \frac{25}{100} = 17.5\%$$

Table 9.22 Percentage return for the two investment plans

Investment plan	Boom (25%)	Normal (50%)	Recession (25%)
Ready mix concrete plant (I_1)	40%	20%	−10%
Infrastructure bond (I_2)	20%	15%	10%

And, the expected value of returns from infrastructure bond option is

$$E(X)_{I_2} = 20 \times \frac{25}{100} + 15 \times \frac{50}{100} + 10 \times \frac{25}{100} = 15\%$$

It can be noticed that although the expected return in the ready-mix concrete plant option is higher, the range of possible return varies from as low as (–)10% to as high as 40%. The range or deviation of return in this case is 50% [40–(–10)]. In case of infrastructure bond option, the expected return is 15% [lower than the ready-mix concrete plant (I_1) option] but has a lower range of variation in return. The return from infrastructure bond option varies from 10% to 20%, thus a range or deviation of 10%. This range of variation indicates the risk associated with an investment. Higher is the range or deviation, higher is the risk involved. In this sense the ready-mix concrete plant (I_1) is more risky than infrastructure bond option.

Risk is measured by the following terms:

Variability of an investment proposal is measured by variance σ^2 and standard deviation σ commonly. In some cases, coefficient of variation is also used. The term variance σ^2 is defined as below:

$$\sigma^2 = \sum_{i=1}^{n} [X_i - E(X)]^2 P(X_i) \qquad (9.20)$$

The standard deviation is the positive square root of the variance and is defined as below:

$$\sigma = \sqrt{\sigma^2} = \sqrt{\sum_{i=1}^{n} [X_i - E(X)]^2 P(X_i)} \qquad (9.21)$$

In the context of the current example, $E(X)$ for the ready-mix concrete plant option $E(X)_{I_1}$ is 17.5, and thus, variance associated with this option is

$$\sigma_1^2 = (40 - 17.5)^2 (0.25) + (20 - 17.5)^2 (0.5) + (-10 - 17.5)^2 (0.25)$$
$$= 318.75$$

The variance associated with infrastructure bond option is

$$\sigma_2^2 = (20 - 15)^2 (0.25) + (15 - 15)^2 (0.50) + (10 - 15)^2 (0.25)$$
$$= 12.50$$

Assuming a hypothetical investment option wherein the returns corresponding to booming, normal, and recession market conditions are 1,000%, 800%, and 600% respectively. One can compute the expected return to be

$$E(X)_{I_3} = 1,000 \times \frac{25}{100} + 900 \times \frac{50}{100} + 600 \times \frac{25}{100} = 850\%$$

The variance associated with this hypothetical investment plan is

$$\sigma_3^2 = (1{,}000 - 850)^2 (0.25) + (900 - 850)^2 (0.50) + (600 - 850)^2 (0.25)$$
$$= 22{,}500$$

It may be noticed that variance associated with the hypothetical example is far too high when compared to the first two proposals— ready-mix concrete plant and infrastructure bond option. However, it would be wrong to say the hypothetical example a risky proposal because of the high variance associated. In fact, when the expected values differ to a large extent among the competing investment options, a term called coefficient of variation is used to measure the risk associated with different options. Coefficient of variation is defined as below:

$$C.V. = \frac{\sigma}{E(X)} \quad (9.22)$$

The coefficients of variation associated with the three alternatives are given below:

C.V. for ready-mix concrete plant option $= \dfrac{17.85}{17.50} = 1.02$

C.V. for infrastructure bond option $= \dfrac{3.54}{15} = 0.24$

C.V. for hypothetical investment option $= \dfrac{150}{850} = 0.18$

As it can be noticed in terms of risk, the hypothetical investment plan is less risky than the remaining two plans. This could not have been found out had risk been categorized based on variance or standard deviation values alone. Thus, whenever there are large differences in the magnitude of expected values of different alternatives, coefficient of variation should be used as a measure of risk measurement.

9.14.5 Determining Expected Values and Variance in Aggregated Cash Flows

Let's take an investment option where X_0 is to be invested now and the expected return at time $t = 1, 2, 3, \ldots$, and n are $X_1, X_2, X_3, \ldots$, and X_n respectively. The corresponding cash flows diagram is shown in Figure 9.23.

It is further assumed that the cash flow at a given time is independent of the cash flows at other periods. In other words, $X_1, X_2, X_3, \ldots$, and X_n are not dependent on each other. Under such circumstances, the expected value of the proposal is computed as explained below:

In all our preceding discussions, we had assumed that cash flows at $t = 0, 1, 2, 3, \ldots, n$ were fixed and certain to occur. The net present worth under such assumption is given by the following expression:

$$NPW = X_0 + \frac{X_1}{(1+i)} + \frac{X_2}{(1+i)^2} + \frac{X_3}{(1+i)^3} + \ldots + \frac{X_n}{(1+i)^n} \quad (9.23)$$

$$= X_0 + X_1(P/F, i, 1) + X_2(P/F, i, 2) + X_3(P/F, i, 3)$$
$$+ \ldots + X_n(P/F, i, n) \quad (9.24)$$

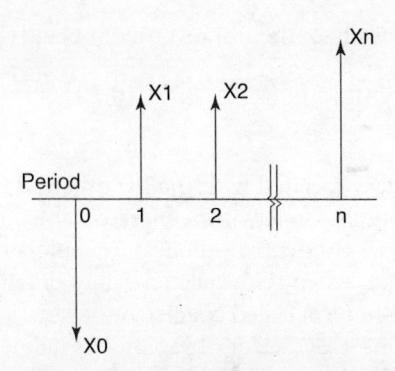

Figure 9.23 Cash flow diagram assumed for computation of expected value and variance

When the cash flows (X_i) are considered a random variable, the expected value for the net present worth can be found out using the following expression:

$$EV(NPW) = \mu_0 + \frac{\mu_1}{(1+i)} + \frac{\mu_2}{(1+i)^2} + \frac{\mu_3}{(1+i)^3} + \ldots\ldots\ldots + \frac{\mu_n}{(1+i)^n} \quad (9.25)$$

Alternatively,

$$EV(NPW) = \mu_0 + \mu_1(P/F,i,1) + \mu_2(P/F,i,2) + \mu_3(P/F,i,3) \quad (9.26)$$
$$+ \ldots\ldots\ldots + \mu_n(P/F,i,n)$$

The variance is given by the following expression:

$$Var(NPW) = \sigma_0^2 + \frac{\sigma_1^2}{(1+i)^2} + \frac{\sigma_2^2}{(1+i)^4} + \frac{\sigma_3^2}{(1+i)^6} + \ldots\ldots\ldots + \frac{\sigma_n^2}{(1+i)^{2n}} \quad (9.27)$$

Alternatively,

$$Var(NPW) = \sigma_0^2 + \sigma_1^2(P/F,i,2) + \sigma_2^2(P/F,i,4) + \sigma_3^2(P/F,i,6) \quad (9.28)$$
$$+ \ldots\ldots + \sigma_n^2(P/F,i,2n)$$

As mentioned earlier, our discussion was limited to independent cash flow for a single investment. More complex situation could be corresponding to cases having correlation between different cash flows which are beyond the scope of this text.

Example 9

Let's assume that an investment proposal is being evaluated. The proposal requires an investment of ₹ 60,000. The chances of getting a return of ₹ 15,000, 30,000, or 40,000 at the end of year one are 0.25, 0.50, and 0.25 respectively. Similarly, the chances of getting a return of ₹ 15,000, 30,000, or 40,000 at the end of year two are 0.50, 0.25, and 0.25 respectively. And finally getting a return of ₹ 15,000, 30,000 or ₹ 40,000 at the end of year three are 0.25, 0.25, and 0.5 respectively. The question is whether we should invest in this proposal? Let the interest rate be 10%.

CONSTRUCTION EQUIPMENT MANAGEMENT | 369

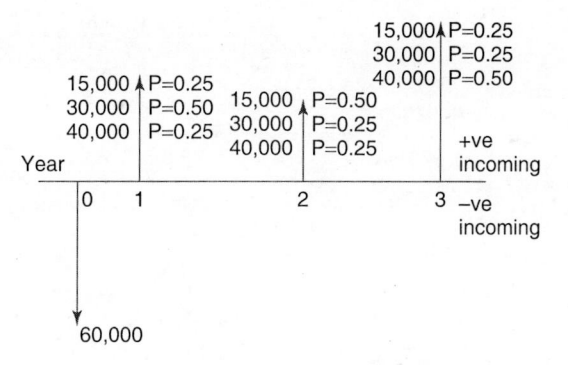

Figure 9.24 Cash flow diagram for the example problem

As before, the first step in solving such problems is to draw a cash flow diagram. This is shown in Figure 9.24:

In the next step we calculate the expected values of returns at the end of year one, two, and three. For example, The expected values of returns at the end of year one, two, and three $E(X)_1$, $E(X)_2$, and $E(X)_3$ respectively will be:

$$E(X)_1 = \mu_1 = 15,000 \times 0.25 + 30,000 \times 0.50 + 40,000 \times 0.25 = 28,750$$
$$E(X)_2 = \mu_2 = 15,000 \times 0.50 + 30,000 \times 0.25 + 40,000 \times 0.25 = 25,000$$
$$E(X)_3 = \mu_3 = 15,000 \times 0.25 + 30,000 \times 0.25 + 40,000 \times 0.50 = 31,250$$

The variance associated with cash flows at the end of each year can be computed as below:

$$\sigma_1^2 = (28,750 - 15,000)^2 (0.25) + (28,750 - 30,000)^2 (0.50) + (28,750 - 40,000)^2 (0.25)$$
$$= 7,96,87,500$$

Thus, $\sigma_1 = 8926.78$

$$\sigma_2^2 = (25,000 - 15,000)^2 (0.50) + (25,000 - 30,000)^2 (0.25) + (25,000 - 40,000)^2 (0.25)$$
$$= 11,25,00,000$$

$$\sigma_3^2 = (31,250 - 15,000)^2 (0.25) + (31,250 - 30,000)^2 (0.25) + (31,250 - 40,000)^2 (0.50)$$
$$= 10,46,87,500$$

Thus, $\sigma_2 = 10,606.60$, and $\sigma_3 = 10,231.69$

The expected values and variances are shown in the revised cash flow diagram as shown in Figure 9.25:

The expected value of net present worth of the above proposal is given by,

$$EV(NPV) = -60,000 + \frac{28,750}{1.10} + \frac{25,000}{1.10^2} + \frac{31,250}{1.10^3}$$
$$= 10,276.11$$

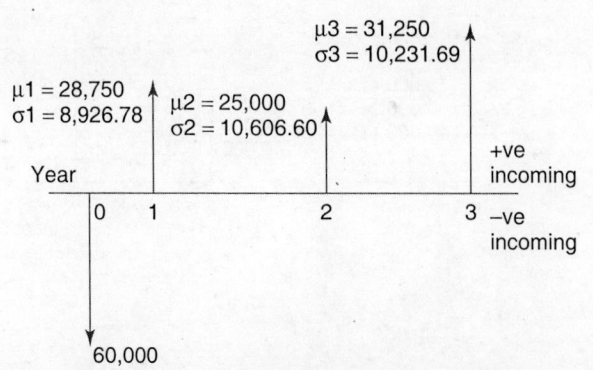

Figure 9.25 Cash flow diagram showing the expected values and variance of returns

The variance of aggregated cash flows is given by,

$$Var(PW) = \sigma_0^2 + \frac{\sigma_1^2}{(1+i)^2} + \frac{\sigma_2^2}{(1+i)^4} + \frac{\sigma_3^2}{(1+i)^6}$$

Here, $\sigma_0 = 0$ (the cash flow at time $t = 0$ is certain, and hence variance is zero). $\sigma_1 = 8926.78$, $\sigma_2 = 10,606.60$, and $\sigma_3 = 10,231.69$

Thus,
$$Var(PW) = 0^2 + \frac{(8926.78)^2}{(1+0.10)^2} + \frac{(10606.60)^2}{(1+0.10)^4} + \frac{(10,231.69)^2}{(1+0.10)^6}$$
$$= 20,17,89,698.60$$

And the standard deviation of the aggregated cash flow = 14,205.27

Using the expected value and the standard deviation of the aggregated cash flows, we can determine the risk associated with the investment proposal. The risk here is defined as the chances of making losses that is net present worth less than zero. The computation of risk is done using the properties of normal distribution.

$$Z = \frac{x - \mu}{\sigma}$$

$x = 0$, $\mu = 10,276.11$, $\sigma = 14,205.52$

Thus, $Z = -0.72$

From, Appendix 5, the probability corresponding to $Z = 0.72$ is 0.2672. Thus, the probability that the given aggregated cash flows will fetch a negative net present value, and thereby, losses is 0.50 − 0.2672 = 0.2378.

In terms of percentage, there is 23.78% chance that this investment option if pursued will run into losses. This is the advantage of using probabilistic analysis in our discussion. In the earlier cases, the moment we were getting a positive net present worth we were accepting the proposal. We were not bothered to look beyond. Now we understand that even though we are getting a positive net present worth still we may run into losses and the chances of running into losses is 23.78%. This is also shown pictorially in Figure 9.26.

The hatched portion in the Figure 9.26 shows the probability of making losses which is 23.78%

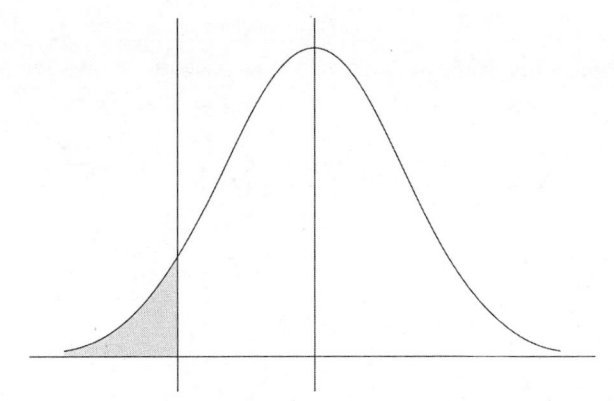

Figure 9.26 Normal distribution showing probability of making losses

Example 10

We take another example to illustrate the above concepts. Let's take an investment plan in which ₹ 200,000 is to be invested now. The chances that it would fetch ₹ 80,000, ₹ 110,000, or ₹ 150,000 at the end of year one are 0.3, 0.4, and 0.3 respectively. The return at the end of year two corresponding to first year end return of ₹ 80,000 is ₹ 40,000, ₹ 100,000, or ₹ 180,000 with associated probabilities of 0.2, 0.6, and 0.2 respectively. For first year end return of ₹ 110,000, the corresponding return at the end of year two are ₹ 100,000, ₹ 150,000, or ₹ 160,000 with associated probabilities of 0.3, 0.4, and 0.3 respectively. Finally for a first year end return of ₹ 150,000, the probable return at the end of year two is ₹ 100,000, ₹ 200,000 or ₹ 240,000 with associated probabilities of 0.1, 0.8, and 0.1. If the interest rate is 12%, what is the expected return and the associated risk of the investment proposal? What is the chance that above investment will run into losses?

Solution

Such problems can be solved using the following two methods.

Method I:

$$NPW = -200,000 + 0.3 \times \left[\frac{80,000}{1.12} + \frac{0.2 \times 40,000 + 0.6 \times 100,000 + 0.2 \times 180,000}{1.12^2}\right]$$

$$+0.4 \times \left[\frac{110,000}{1.12} + \frac{0.3 \times 100,000 + 0.4 \times 150,000 + 0.3 \times 160,000}{1.12^2}\right]$$

$$+0.3 \times \left[\frac{150,000}{1.12} + \frac{0.1 \times 100,000 + 0.8 \times 200,000 + 0.1 \times 240,000}{1.12^2}\right]$$

$$= -200,000 + [46,301 + 83,291 + 86,575]$$

$$= ₹ 16,167$$

Method II

The information provided in the problem can be best represented through a decision tree as shown in Figure 9.27 and the problem can be solved using the concepts of joint probability.

372 | CONSTRUCTION PROJECT MANAGEMENT

Table 9.23 Joint probability, NPW, and weighted NPW for various paths

Path	Joint probability of the path	NPW of the path	Weighted net present worth
I	0.06	−96,684	−5,801
II	0.18	−48,852	−8,793
III	0.06	14,923	895
IV	0.12	−22,066	−2,648
V	0.16	17,793	2,847
VI	0.12	25,765	3,092
VII	0.03	13,648	409
VIII	0.24	93,367	22,408
IX	0.03	1,25,255	3,758
		Total:	₹ 16,167

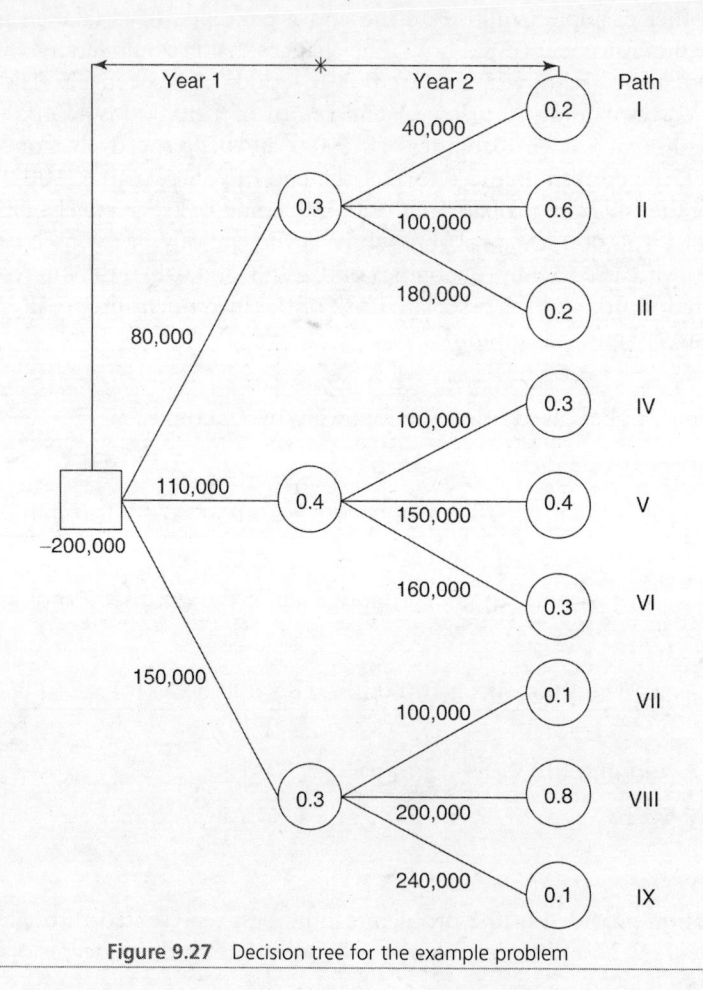

Figure 9.27 Decision tree for the example problem

Table 9.24 NPW in ascending order and cumulative probability

Path number	NPW arranged in ascending order	Cumulative probability
I	−96,684	0.06
II	−48,852	0.24
IV	−22,066	0.36
VII	13,648	0.39
III	14,923	0.45
V	17,793	0.61
VI	25,765	0.73
VIII	93,367	0.97
IX	125,255	1.00

There are a total of nine paths in the decision tree. For each path we can compute the joint probability. For example, for path 1, the joint probability is 0.3 × 0.2 = 0.06. In other words, there is a probability of 0.06 that the return at the end of first and second year will be ₹ 80,000 and ₹ 40,000 respectively. Similarly for path 2, the joint probability of having a return of ₹ 80,000 and ₹ 100,000 at the end of year one and two respectively is 0.18 (0.3 × 0.6). The joint probabilities of other paths are shown in Table 9.23. In the Table 9.23, the third column shows the net present worth of the path. For example, net present worth of path 1 is (−) 96,684 which is obtained from the following computation.

$$\text{Net present values of path 1} = -200,000 + \frac{80,000}{1.12} + \frac{40,000}{1.12^2} = (-)96,684.$$

In a similar manner, the net present values of other paths have been calculated. The sum of net present values of all the paths put together presents the net present value of aggregated cash flow.

In the next step, PW of different paths are arranged in ascending manner and cumulative probability is worked out. This is shown in Table 9.24.

Results of above table are sometimes presented in a diagram as shown in Figure 9.28 in the risk analysis. Such diagrams are also known as investment risk profile. Such diagrams are plotted with net present worth on x-axis and cumulative probability on y-axis. Looking at the diagram, it is very easy to understand the risk associated with the proposal.

It is clear that for drawing the investment risk profile, the net present worth is arranged in ascending order first and they are plotted against cumulative probability values.

The investment risk profile, if coupled with the aspiration level of investor, gives rise to another diagram which is referred to as Acceptable Investment Diagram (AID). In such diagrams, the rates of return are plotted on x-axis while the possibility that an investment will exceed a given rate of return is plotted as y-axis. The rejection zone is bound from (1) x-axis, (2) y-axis, (3) a line joining x-axis with payoff coefficient, (4) a line joining y-axis with loss coefficient, and (5) a line known as aspiration level which joins loss coefficient and payoff coefficient. The term 'loss coefficient' is the probability required by the decision maker that an investment's rate of return exceeds a minimum percentage while 'payoff coefficient' is the probability desired by the decision maker that an investment's rate of return will exceed an attractive level.

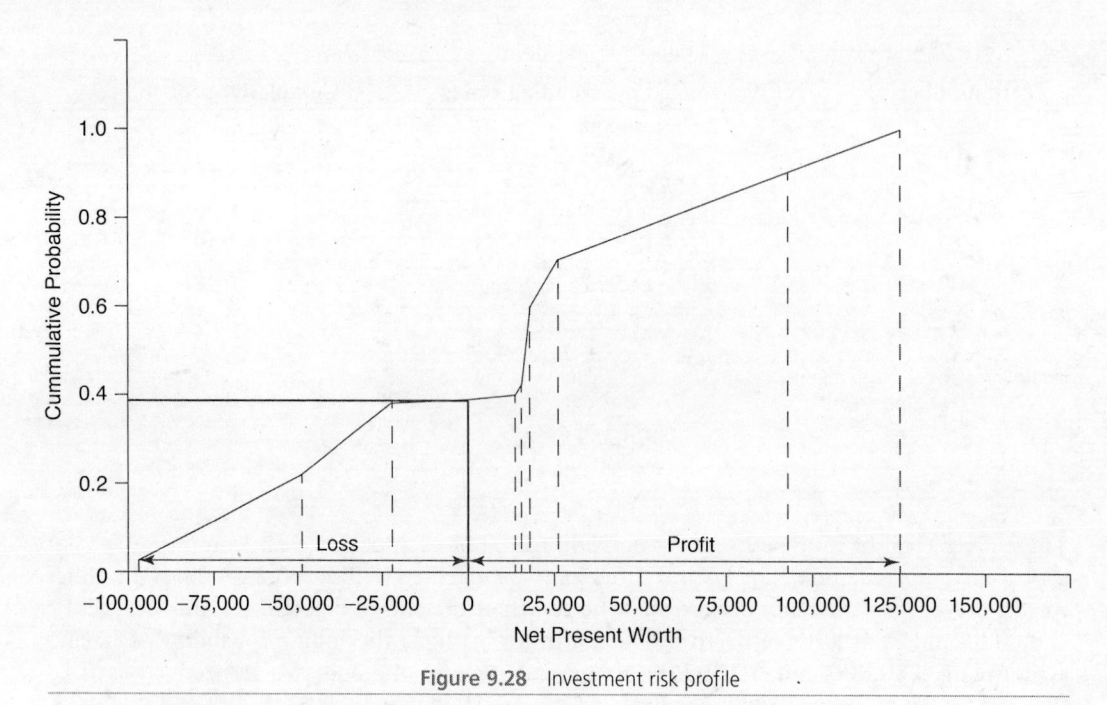

Figure 9.28 Investment risk profile

Let's assume that a decision maker aspires to invest in only those proposals where the chances of making losses are not more than 10% and the chances of making 20% or more return is 40%. While the former data would be helpful in locating loss coefficient (see point P in Figures 9.29 and 9.30) the latter would be helpful for locating payoff coefficient (see point Q in Figures 9.29 and 9.30). With these two points known, the rejection zone can easily be plotted. The investment risk profile of an investment alternative falling in the rejection zone

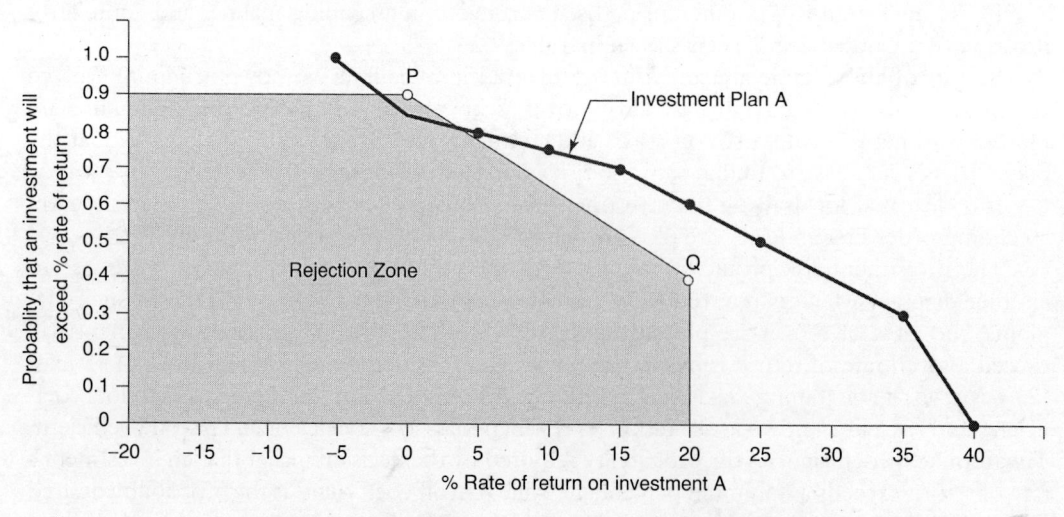

Figure 9.29 Acceptable investment diagram (AID) for investment plan A

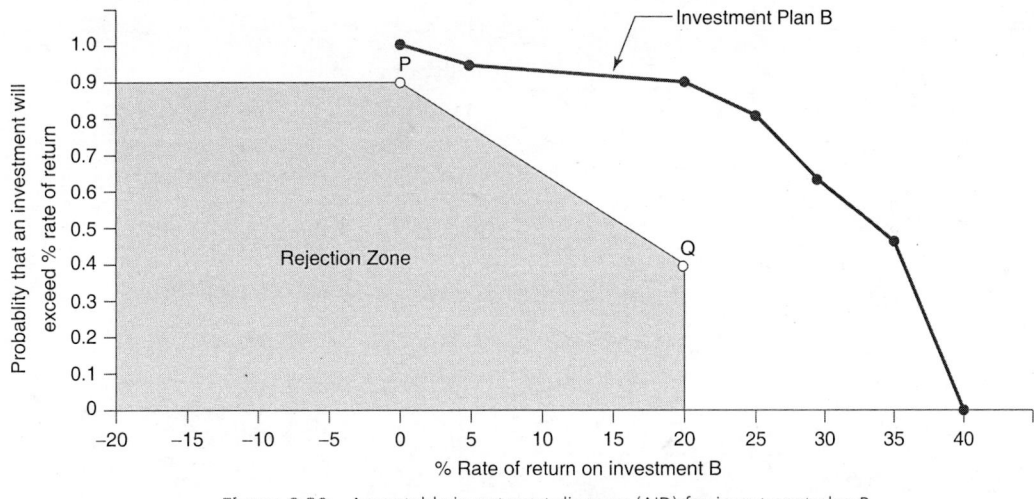

CONSTRUCTION EQUIPMENT MANAGEMENT | 375

Figure 9.30 Acceptable investment diagram (AID) for investment plan B

will be rejected by the decision maker. On the other hand, if the investment risk profile for an investment is beyond the boundary of rejection zone, the decision maker would opt for that investment alternative.

The two situations are shown in acceptable investment diagrams for two hypothetical investment plans A and B. In one case (see Figure 9.29), the investment plan A has failed to meet the investor's aspiration. In other words, the investment risk profile enters into rejection zone and thus would be rejected by the decision maker for the given set of aspiration. In the other case (see Figure 9.30), the investment plan B is well outside the rejection zone and thus would be an acceptable proposition.

9.14.6 Selection of Alternatives Based on Expected Value and Variance of Returns

In the previous section, the process of finding the expected value, variance, and drawing of investment risk profile were explained. When it comes to selecting appropriate alternative among more than one proposal, the one with the maximum expected value is chosen. Also, given two proposals with equal expected returns, the proposal with the least variance is chosen. In case of selecting between 'high expected value and high variance' and 'low expected value and low variance' appropriate tradeoff between return and variability is required. In some cases we may like to opt for 'most probable future' and 'aspiration level of investor' decision criteria. The evaluation methods are explained with the help of an example.

Let's assume there are three investment proposals being considered. Although each proposal requires equal investment of ₹ 1,000, the returns and the chances of their occurrence vary. The returns and their probabilities for the three proposals (P_1, P_2, and P_3) are shown in Table 9.25.

In solving such problems, the first step is to find out the expected return and the variance associated with the proposal. Other steps are also explained.

$$E(X)_1 = \mu_1 = -100 \times 0.10 + 0 \times 0.20 + 1,150 \times 0.30 + 1,300 \times 0.25 + 1,500 \times 0.15 = 885$$

Table 9.25 Return and their probabilities for the three proposals

Return on an investment of ₹ 1,000	Probability of given return		
	Proposal 1 (P_1)	Proposal 2 (P_2)	Proposal 3 (P_3)
–100	0.1	0.05	0.25
0	0.2	0.05	0.1
1,150	0.3	0.3	0.2
1,300	0.25	0.4	0.3
1,500	0.15	0.2	0.15

$$\sigma_1^2 = (885-(-100))^2 \times (0.10) + (885-0)^2 \times (0.20) + (885-1{,}150)^2 \times (0.30)$$
$$+ (885-1{,}300)^2 \times (0.25) + (885-1{,}500)^2 \times (0.15) = 3{,}74{,}525$$

Hence, σ_1 = ₹612

Similarly, the expected values and standard deviation of the other proposals can be worked out. These are shown in the following table:

Table 9.26 Expected return and standard deviation of the three proposals.

	Proposal 1 (P_1)	Proposal 2 (P_2)	Proposal 3 (P_3)
Expected return	₹ 885	₹ 1,160	₹ 820
Standard deviation σ	₹ 612	₹ 1,752	₹ 1,523

Based on the principle of maximization of expected value, the choice between proposal P_1 and proposal P_3 is simple. It can be observed that for proposal P_1, the expected value is higher than the proposal P_3 and the standard deviation is also less. Thus, between proposal P_1 and proposal P_3, the proposal P_1 is a better choice. However, when it comes to the selection between proposal P_1 and proposal P_2, the choice is difficult to make. This is because for proposal P_2, the expected value and standard deviation both are higher than proposal P_1. In such cases suitable tradeoff between return and variability is to be made. Alternatively, we can use the following methods.

1. **Most probable future:** According to this criterion, the investment that has the greatest return for the most probable future is preferred. It is assumed that the future with the highest probability of occurrence is certain. In other words, most probable future is the return corresponding to the maximum probability. In the example, for proposal P_1, the maximum probability (p = 0.3) is for a return of ₹ 1,150. Similarly, for proposals P_2 and P_3, the most probable future is for a return of ₹ 1,300 (p = 0.4) and ₹ 1,300 (p = 0.3) respectively (see Table 9.27). Although, the return corresponding to proposals P_2 and P_3 are same, a greater probability of occurrence for proposal P_2 makes it the most preferred choice. Thus, based on the most probable future criterion, alternative P_2 is the most preferable.

Table 9.27 Return at most probable future for the three alternatives

Alternatives	Return at most probable future, ₹
P_2: $p = 0.40$	1,300
P_3: $p = 0.30$	1,300
P_1: $p = 0.25$	1,150

Table 9.28 Probability of return for the given aspiration level (Minimum return of ₹ 1,150)

Proposal	Probability of return of ₹ 1,150 or more
Proposal 1 (P_1)	0.30 + 0.25 + 0.15 = 0.70
Proposal 2 (P_2)	0.30 + 0.40 + 0.20 = 0.90
Proposal 3 (P_3)	0.20 + 0.30 + 0.15 = 0.65

Table 9.29 Probability of return for the given aspiration level (return ≥0)

Proposal	Probability to make no losses (i.e. return ≥0)
Proposal 1 (P_1)	0.20 + 0.30 + 0.25 + 0.15 = 0.90
Proposal 2 (P_2)	0.05 + 0.30 + 0.40 + 0.20 = 0.95
Proposal 3 (P_3)	0.10 + 0.20 + 0.30 + 0.15 = 0.75

2. **Aspiration level:** This is based on the minimum amount that will satisfy the decision maker. Assuming an aspiration level of ₹ 1,150, the probability of return of ₹ 1,150 or more can be computed for the three proposals as given in Table 9.28.

The decision in this case would be to opt for Proposal 2 (P_2)

If the aspiration is to make no losses, i.e., return ≥0, the probability of return ≥ 0 can be computed for the three proposals as given in Table 9.29.

The decision in this case would be to opt for Proposal 2 (P_2)

From the preceding discussions, it is evident that the decision made is dependent on the decision criteria adopted by the decision maker.

SOLVED EXAMPLES

Example 9.1

Equipment X is purchased at a cost of ₹ 350,000. The income generated by this equipment is ₹ 250,000 annually for next five years. Calculate the rate of return post-tax for this deal. Assume (a) straight-line method of depreciation, (b) sum of years digit method of depreciation, and (c) double declining balance method of depreciation. You can use straight-line method of depreciation in case there is a need of switching over in double declining balance method. Assume tax rate at present is 50 per cent. Prepare cash-flow diagram, both pre-tax and post-tax, for each of the cases described.

Solution

The purchase cost of the equipment X is ₹ 350,000, and the equipment produces an annual income of ₹ 250,000 before taking into account depreciation and taxes. Thus, pre-tax cash-flow diagram would be as shown in Figure S9.1.1.

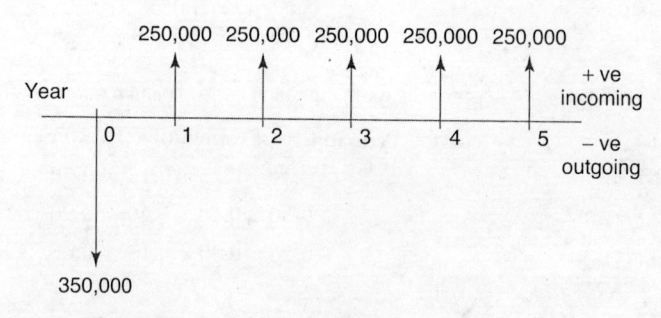

Figure S9.1.1 Pre-tax cash-flow diagram for Example 9.1(a)

(a) We calculate the depreciation of the equipment using the straight-line method (see Table S9.1.1). Column 1 of Table S9.1.1 shows the period in year, column 2 shows the book value at the beginning of the year, column 3 shows the depreciation as calculated using straight line method and column 4 shows the book value of the equipment at the end of year.

Table S9.1.1 Straight-line depreciation method

m	BV_{m-1} (₹)	Depreciation (₹)	BV_m (₹)
1	350,000	(350,000 − 0)/5 = 70,000	350,000 − 70,000 = 280,000
2	280,000	(350,000 − 0)/5 = 70,000	280,000 − 70,000 = 210,000
3	210,000	(350,000 − 0)/5 = 70,000	210,000 − 70,000 = 140,000
4	140,000	(350,000 − 0)/5 = 70,000	140,000 − 70,000 = 70,000
5	70,000	(350,000 − 0)/5 = 70,000	70,000 − 70,000 = 0

In order to generate the post-tax cash-flow diagram, we will take into consideration depreciation and tax. The post-tax annual income is computed as given in Table S9.1.2.

Table S9.1.2 Computation details for generating post-tax cash-flow

A	Income before tax and depreciation	₹ 250,000
B	Depreciation (constant for all the 5 years)	₹ 70,000
C	Total taxable income (A − B)	₹ 180,000
D	Tax @50% (C × 0.5)	₹ 90,000
E	Income after tax (A − D)	₹ 160,000

The post-tax cash-flow diagram is as shown in Figure S9.1.2.

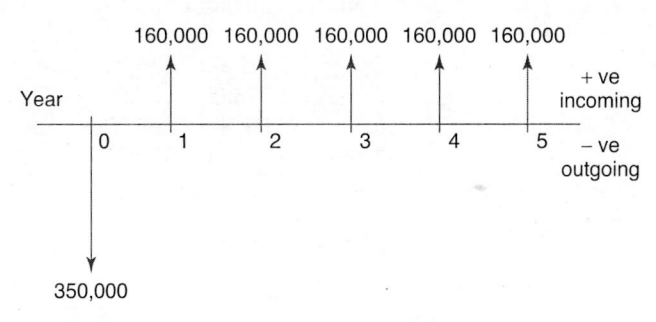

Figure S9.1.2 Post-tax cash-flow diagram for Example 9.1(a)

Based on the post-tax cash-flow diagram, it is easy to use the concept of rate of return learnt in Chapter 3 and derive the post-tax return.

The post-tax return is obtained by solving the following equation using trial and error.

$NPW = 0$,

$\Rightarrow 160{,}000 \, (P/A, i, 5) - 350{,}000 = 0, \, i = 36.36\%$

(b) We calculate the depreciation of the equipment using the sum of years digit method (see Table S9.1.3). Column 1 shows the period in year; Column 2 shows the book value at the beginning of the year; Column 3 shows the rate of depreciation; Column 4 shows the depreciation as calculated using sum of years digit method; and Column 5 shows the book value of the equipment at the end of the year. The sum of years, N, in this case = 15 (5 × 6/2). The rate of depreciation would vary every year. For example, in year 1 it would be 5/15, and in year 2 it would be 4/15. Finally, in year 5 it would be equal to 1/15.

Table S9.1.3 Depreciation using sum of years digit method

m	BV_{m-1}	R_m	Depreciation D_m	BV_m
1	350,000	5/15	(5/15) × (350,000 − 0) = 117,000	350,000 − 117,000 = 233,000
2	233,000	4/15	(4/15) × (350,000 − 0) = 93,000	233,000 − 93,000 = 140,000
3	140,000	3/15	(3/15) × (350,000 − 0) = 70,000	140,000 − 70,000 = 70,000
4	70,000	2/15	(2/15) × (350,000 − 0) = 47,000	70,000 − 47,000 = 23,000
5	23,000	1/15	(1/15) × (350,000 − 0) = 23,000	23,000 − 23,000 = 0

The rate of return calculation using sum of years method of depreciation is as shown in Table S9.1.4.

Table S9.1.4 Computation details for generating post-tax cash-flow

	Year	1	2	3	4	5
A	Income before tax and depreciation	250,000	250,000	250,000	250,000	250,000
B	Depreciation	117,000	93,000	70,000	47,000	23,000
C	Total taxable income (A − B)	133,000	157,000	180,000	203,000	227,000
D	Tax @50% (C × 0.5)	66,500	78,500	90,000	101,500	113,500
E	Income after tax (A − D)	183,500	171,500	160,000	148,500	136,500

The post-tax cash-flow diagram is as shown in Figure S9.1.3.

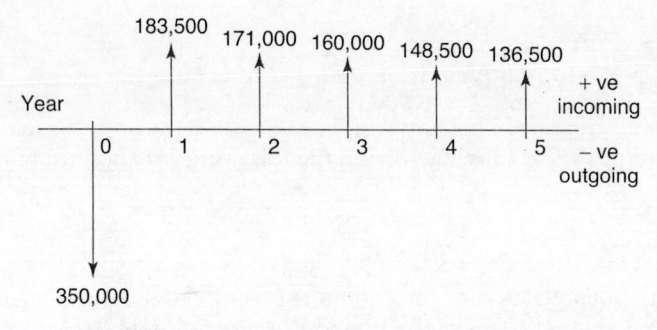

Figure S9.1.3 Post-tax cash-flow diagram for Example 9.1(b)

The post-tax return is obtained by solving the following equation using trial and error.

$$NPW = 0$$
$$\Rightarrow 183,500 \times (P/F, i, 1) + 171,000 \times (P/F, i, 2) + 160,000 \times (P/F, i, 3) + 148,500 \times (P/F, i, 4) + 136,500 \, (P/F, i, 5) - 350,000 = 0$$
$$\Rightarrow i = 38.40\%$$

(c) We calculate the depreciation of the equipment using the double declining balance method (see Table S9.1.5). Column 1 shows the period in year; Column 2 shows the book value at the beginning of the year; Column 3 shows the rate of depreciation; Column 4 shows the depreciation as calculated using double declining balance method; and Column 5 shows the book value of the equipment at the end of the year. The rate of depreciation for years 1, 2 and 3 is 2/N = 2/5. A switch has been made in year 4 from double declining balance method to straight-line method. This is because the depreciation calculated using double declining balance method (2/5 × 75,600 = 30,240) is less than the depreciation calculated using straight-line method (37,800).

CONSTRUCTION EQUIPMENT MANAGEMENT | 381

Table S9.1.5 Depreciation using double declining balance method

m	BV_{m-1}	R_m	Depreciation D_m	BV_m
1	350,000	2/5	$(2/5) \times (350,000) = 140,000$	$350,000 - 140,000 = 210,000$
2	210,000	2/5	$(2/5) \times (210,000) = 84,000$	$210,000 - 84,000 = 126,000$
3	126,000	2/5	$(2/5) \times (126,000) = 50,400$	$126,000 - 50,400 = 75,600$
4	75,600	1/2	$(1/2) \times (75,600 - 0) = 37,800$	$75,600 - 37,800 = 37,800$[1]
5	37,800	1/2	$(1/2) \times (75,600 - 0) = 37,800$	$37,800 - 37,800 = 0$

[1] Note the change in method of depreciation from double declining balance method to straight-line method.

The rate of return calculation using double declining balance method of depreciation is as shown in Table S9.1.6.

Table S9.1.6 Computation details for generating post-tax cash-flow

	Year	1	2	3	4	5
A	Income before tax and depreciation	250,000	250,000	250,000	250,000	250,000
B	Depreciation	140,000	84,000	50,400	37,800	37,800
C	Total taxable income (A − B)	110,000	166,000	199,600	212,200	212,200
D	Tax @50% (C × 0.5)	55,000	83,000	99,800	106,100	106,100
E	Income after tax (A − D)	195,000	167,000	150,200	143,900	143,900

The post-tax cash-flow diagram is as shown in Figure S9.1.4.

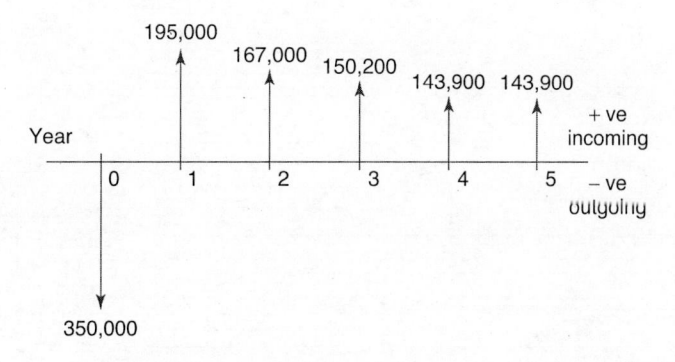

Figure S9.1.4 Post-tax cash-flow diagram for Example 9.1(c)

The post-tax return is obtained by solving the following equation using trial and error.

$$NPW = 0$$
$$\Rightarrow 195,000 \times (P/F, i, 1) + 167,000 \times (P/F, i, 2) + 150,200 \times (P/F, i, 3) + 143,900 \times (P/F, i, 4) + 143,900 \times (P/F, i, 5) - 350,000 = 0$$
$$\Rightarrow i = 38.87\%$$

Example 9.2

The purchase price of an earthmoving equipment is ₹ 10 lakh. The predicted resale value of the equipment, the operating, and the maintenance cost, and repair cost are given in Table Q9.2.1. The cost of capital is 15 per cent. Calculate the optimum replacement age.

Table Q9.2.1 Data for Example 9.2

Year	1	2	3	4	5	6	7	8
Predicted resale value (₹)	900,000	850,000	800,000	700,000	600,000	400,000	300,000	200,000
Operating and maintenance cost (₹)	30,000	40,000	45,000	50,000	65,000	75,000	85,000	95,000
Repair cost (₹)	10,000	20,000	25,000	30,000	35,000	45,000	55,000	70,000
Sum of operating and maintenance cost and repair cost (₹)	40,000	60,000	70,000	80,000	100,000	120,000	140,000	165,000

Solution

The computations involved in the solution of this problem are shown in Table S9.2.1 and are self-explanatory.

From Figure S9.2.1, it is clear that the equipment has its lowest EAC in year 3. Thereafter, the cost starts increasing. Thus, the ideal time to replace the equipment is at the end of three years.

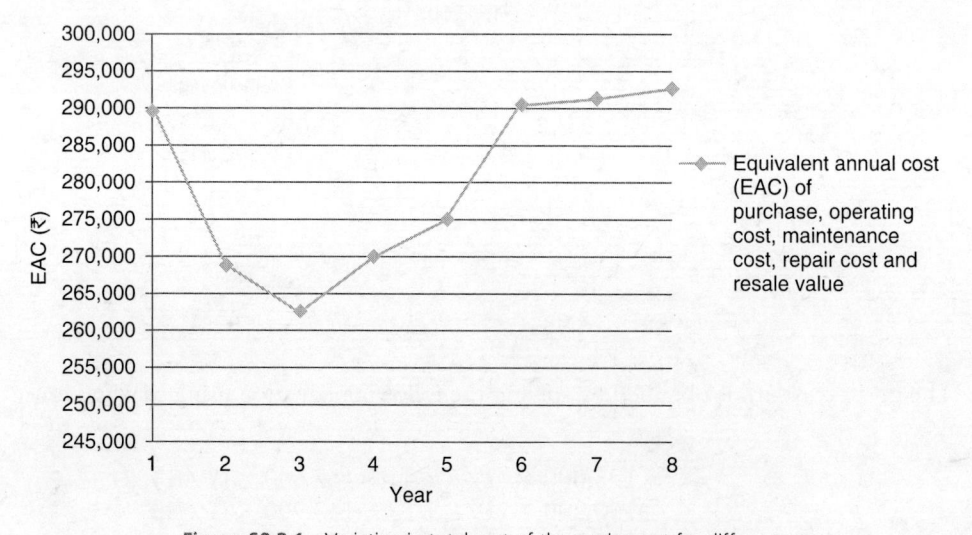

Figure S9.2.1 Variation in total cost of the equipment for different years

Table S9.2.1 Computation details for EAC

Year (1)	(A/P, i,n) (2)	EAC of purchase price (3)=P×(2)	sum of operating cost and maintenance and repair cost (4)	(P/F, i,n) (5)	Present worth of operating and maintenance and repair cost (6)=(4)×(5)	Cumulative sum of present worth of operating and maintenance and repair cost (7)	EAC of present worth of operating and maintenance and repair costs (8)=(7)×(2)	Resale value (9)	Present worth of resale value (10)=(9)×(5)	EAC of present worth of resale value (11)=(10)×(2)	EAC of purchase, operating and maintenance and repair costs, and resale value (12)=(3)+(8)−(11)
1	1.1500	1,150,000	40,000	0.8696	34,784.00	34,784.00	40,001.60	900,000	782,640.00	900,036.00	289,965.60
2	0.6151	615,100	60,000	0.7561	45,366.00	80,150.00	49,300.27	850,000	642,685.00	395,315.54	269,084.72
3	0.4380	438,000	70,000	0.6575	46,025.00	126,175.00	55,264.65	800,000	526,000.00	230,388.00	262,876.65
4	0.3503	350,300	80,000	0.5718	45,744.00	171,919.00	60,223.23	700,000	400,260.00	140,211.08	270,312.15
5	0.2983	298,300	100,000	0.4972	49,720.00	221,639.00	66,114.91	600,000	298,320.00	88,988.86	275,426.06
6	0.2642	264,200	120,000	0.4323	51,876.00	273,515.00	72,262.66	400,000	172,920.00	45,685.46	290,777.20
7	0.2404	240,400	140,000	0.3759	52,626.00	326,141.00	78,404.30	300,000	112,770.00	27,109.91	291,694.39
8	0.2229	222,900	165,000	0.3269	53,938.50	380,079.50	84,719.72	200,000	65,380.00	14,573.20	293,046.52

Purchase Price P = 1,000,000

REFERENCES

1. Alkass, S., Aronian, A. and Moselhi, O., 1993, 'Computer-Aided Equipment Selection for Transporting and Placing Concrete', *Journal of Construction Engineering and Management*, 119(3), pp. 445–65.
2. Alkass, S. and Harris, F., 1988, 'An expert system for earthmoving equipment selection in road construction', *Journal of Construction Engineering and Management*, 114(3), pp. 426–440.
3. Amirkhanian, S.N. and Baker, N.J., 1992, 'Expert system for equipment selection for earth-moving operations', *Journal of Construction Engineering and Management*, 118(2), pp. 318–331.
4. Arditi, D. et al., 1997, 'Innovation in construction equipment and its flow into the construction industry', *Journal of Construction Engineering and Management*, 123(4), pp. 371–378.
5. Caterpillar, http://www.caterpillar.com.
6. Collier, C.A. and Jacques, D.E., 1984, 'Optimum equipment life by minimum life-cycle costs', *Journal of Construction Engineering and Management*, 110(2), pp. 248–265.
7. Cranimax, 2006, http://www.cranimax.com.
8. Day, D.A. and Benjamin, N.B.H., 1991, *Construction Equipment Guide*, 2nd ed., New York: John Wiley & Sons, Inc.
9. Douglas, J., 1975, 'Past and future of construction equipment—Part III', *Journal of Construction Division*, 101(CO4), pp. 699–701.
10. Elazouni, A.M. and Basha, I.M., 1996, 'Evaluating the Performance of Construction Equipment Operators in Egypt', *Journal of Construction Engineering and Management*, 122(2), pp. 109–114.
11. Gates, M. and Scarpa, A., 1975, 'Optimum Size of Hauling Unit', *Journal of the Construction Division*, ASCE, 101(CO4), pp. 853–860.
12. Gates, M. and Scarpa, A., 1975, 'Optimum Size of Equipment', *Journal of the Construction Division*, ASCE, 101(CO4), pp. 861–867.
13. Harris, F. and McCaffer, R., 2005, *Modern Construction Management*, 5th ed., Blackwell Publishing.
14. Klump, E.H., 1975, 'Past and future of construction equipment—Part II', *Journal of Construction Division*, 101(CO4), pp. 689–698.
15. Lift Planner Software, 2006, http://www.liftplanner.net.
16. Meehan, J., 2005, 'Computerize to organize: Lift planning', *Cranes Today*, 369, 50.
17. NCI, 2006.
18. Peurifoy, R.L. and Schexnayder, C.J., 2002, *Construction Planning, Equipment and Methods*, New York: McGraw Hill.
19. Prognistik, 2006.
20. Schexnayder, C.J. and David, S.A., 2002, 'Past and Future of Construction Equipment—Part IV', *Journal of Construction Engineering and Management*, 128(4), pp. 279–286.

21. Selinger, S., 1983, 'Economic Service Life of Building Construction Equipment', *Journal of Construction Engineering and Management*, 109(4), pp. 398–405.
22. Singh, B., Singh, S.P. and Singh, B., 2004, 'Some issues related to pumping of concrete', *The Indian Concrete Journal*, September, pp. 41–44.
23. Shapiro, H.I., Shapiro, J.P. and Shapiro, L.K., 1991, *Cranes and Derricks*, 2nd ed., New York: McGraw Hill.
24. Touran, A. et al., 1997, 'Rational equipment selection method based on soil conditions', *Journal of Construction Engineering and Management*, 123(1), pp. 85–88.

REVIEW QUESTIONS

1. State whether True or False:
   a. The choice of construction equipment defines the construction method, which in a way leads to the determination of time and cost for the project.
   b. Construction equipment can be classified on the basis of functional aspects, operational aspects, and purpose (general or special).
   c. The two broadly important factors for selection of construction equipment are hard factors and soft factors.
   d. Hard factors for equipment selection are intangible in nature, whereas soft factors are tangible and quantifiable in nature.
   e. Some of the decision variables affecting selection of construction equipment are—economic considerations, company-specific, site-specific, equipment-specific, client- and project-specific, manufacturer-specific and labour-specific.
   f. Straight-line method, sum of years digit method, and declining balance method are the methods to select construction equipment.
   g. Replacement of an existing productive asset becomes necessary due to physical or mechanical impairment, technological obsolescence, uneconomic increasing maintenance cost, and functional inadequacy.
   h. Sensitivity analysis is a non-probabilistic technique and it is aimed at identifying the sensitivity of a particular variable.
   i. Breakeven analysis is equivalent to cost-volume-profit analysis.
   j. (Price - direct cost) is known as contribution per unit.
   k. Breakeven point = indirect cost/contribution.
   l. Companies try to lower breakeven point by decreasing sales price, increasing total cost of production, or decreasing quantity of production.
2. What are the advantages and disadvantages of using machines in construction in a country like India?

3. Enumerate various types of equipment required for (a) earthwork, (b) concrete and (c) material hoisting.

4. Describe various types of lifting cranes. State the various uses of tower crane in construction industry.

5. A building costing ₹ 1,600,000 is expected to have a 35 years life with 25 per cent salvage value. Calculate the depreciation charge for years 4, 9, 18 and 26 using the

   a. Straight-line method

   b. Sum of years method

   c. Double declining balance method

6. An asset having a first cost of ₹ 1,150,000 is expected to have a life of 12 years with a salvage value of ₹ 100,000. In what year does the depreciation charge by the straight-line method first exceed the depreciation charge allowed by the

   a. sum of years method and

   b. double declining balance method?

7. Table Q7.1 gives details of the equipment used at a construction site. Complete the table and determine the total expenses for depreciation of equipment at the site.

Table Q7.1   Details of the equipment used at a construction site

S. No.	Description	Nos	Month	Opening book value as on 31.12.2005	Opening value for total	Depreciation rate	Depreciation per month	Total depreciation
1	Batching plant—20 cum/hr	1	10	1,197,450		5.15%	7,940	
2	Front-end loader	1	10	1,116,742		11.31%	20,094	
3	Transit mixers	3	10	1,282,895		4.75%	7,754	
4	Tough rider	2	6	167,069		11.31%	2,997	
5	Dewatering pump	1	4	174,635		4.75%	1,057	
6	Curing pump—multi-stage type	1	10	64,595		4.75%	392	
7	Tower crane	1	6	7,279,250		5.15%	48,281	
8	Theodolite	1	10	598,265		4.75%	3,614	

8. Check regulations published by the accounting authorities for specific rules regarding depreciation and methods of calculating depreciation for various types of assets.

9. Two pumps can be used for pumping a corrosive liquid. A pump with a brass impeller costs ₹ 40,000 and is expected to last for three years. A pump with a stainless steel impeller will cost ₹ 95,000 and lasts for five years. An overhaul costing ₹ 15,000 will be required after 2,000 operating hours of the brass option, while an overhaul of ₹ 35,000 is estimated for the stainless steel

CONSTRUCTION EQUIPMENT MANAGEMENT | 387

option after 9,000 hours. If the operating cost of each pump is ₹ 25/hour, how many hours/year must the pump be required to justify the purchase of an expensive pump? (Use interest rate of 10 per cent/year.)

10. A construction company engaged in constructing sewage treatment plant is considering the economics of furnishing an in-house water-testing lab instead of sending samples to independent labs for analysis. If the lab is furnished so that all the tests can be conducted in-house, the initial cost will be ₹ 1,250,000. A technician will be required at a cost of ₹ 650,000/year. The cost of power, chemicals, etc., will be ₹ 250 per sample. If it is partially furnished, then initial cost will be ₹ 500,000 and a part-time technician at ₹ 250,000 per year. The cost of in-house analysis is ₹ 150/sample, but since all tests cannot be conducted by the company, outside testing will be required at the cost of ₹ 1,000 per sample. If the company elects to continue with the present condition of outside testing, the cost will be ₹ 1,750 per sample. If the lab equipment will have a useful life of 12 years and the company's MARR is 10 per cent/year, how many samples must be tested each year in order to justify (1) the complete lab system and (2) the partial lab system? If the company expects to test 175 samples per year, which alternative among the three should be selected?

11. An engineer is trying to decide between two ways to pump concrete up to the top floors of a seven-storey building under construction. Plan 1 requires the purchase of equipment costing ₹ 300,000 and costing between ₹ 20 per ton and ₹ 35 per ton to operate with the expected cost of ₹ 25 per ton. The equipment is able to pump100 tons per day. If purchased, the equipment will last for 5 years, have no salvage value, and can be used for 50 to 100 days per year. Plan 2 is an equipment-leasing option and is expected to cost ₹ 125,000 per year for equipment, with an optimistic cost of ₹ 90,000 and a pessimistic value of ₹ 160,000 per year. In addition, a ₹ 250 per-hour labour cost will be incurred for operation of the leased equipment.

    Plot the equivalent uniform annual cost of each plan versus total annual operating cost and lease cost at i = 12 per cent. Determine which plan should be selected for use of

    a. 50 days per year and
    b. 100 days per year.

12. The purchase price of a small diesel generator set is ₹ 10 lakh. The operating costs based on the annual average estimated hours of operation are ₹ 40,000 in the first year and ₹ 60,000 in the second year, rising by ₹ 15,000 each year thereafter. The resale value of the diesel generator set can be assumed to be as predicted in table Q12.1. The cost of capital is 10 percent. Calculate the optimum replacement age.

Year	1	2	3	4	5	6	7
Predicted resale value (₹ lakh)	9	8	7.5	6	4	2.5	1

13. In a project, a contractor needs an earth-excavating equipment, which he has the option of buying either from manufacturer M/s XYZ Equipment or from M/s EFG Equipment. M/s XYZ makes it under brand name A and M/s EFG under brand name B. For your convenience, assume that technically, there is no difference between them, and the decision has to be based purely on financial

considerations. Some of the information that may be required in order to be able to make the decision is summarized in Table Q13.1.

Table Q13.1 Data for Question 13

Item	A	B
Cost of new equipment	₹ 500,000	₹ 400,000
Service life (years)	6	6
Salvage value at the end of year 6	₹ 200,000	₹ 150,000
Annual operating disbursement consisting of operating and maintenance costs for the first three years	₹ 250,000	₹ 275,000
Annual operating disbursement consisting of operating and maintenance costs for the last three years	₹ 300,000	₹ 325,000

Analyse and recommend which equipment you are going to buy. You can assume a minimum rate of return of 15 per cent per annum. Repeat the calculation for other rates of return also and see if a change in rate of return can change your decision.

14. You want to invest ₹ 1,00,000/-. You have two proposals in hand, (a) to invest in a stock and (b) to invest in an infrastructure project. The rate of return of return on the investment depends on the state of economy of the market. The chances of occurrence of each state of economy and the probable return for the two investment options are given below. Find the expected returns from both the proposals.

State of economy	Probability of this state occurring	Rate of return on investment in stocks	Rate of return on investment in infrastructure
Boom	0.3	100%	20%
Normal	0.4	15%	15%
Recession	0.3	(−)70%	10%

Hint: Expected return for both options is 15%.

15. The cash flows for an investment of ₹ 6,000 for three years are shown in the following Table along with the probabilities of their occurrences. If the discount rate is 10% what is the net present value of the proposal. What is the probability of making losses in this investment plan?

Year 1 Cash flow	Probability	Year 2 Cash flow	Probability	Year 3 Cash flow	Probability
₹ 1,500	0.25	₹ 1,500	0.50	₹ 1,500	0.25
₹ 3,000	0.50	₹ 3,000	0.25	₹ 3,000	0.25
₹ 4,000	0.25	₹ 4,000	0.25	₹ 4,000	0.50

Hint: NPV = ₹ 1,027

16. A machine can be designed with various degrees of reliability. The cost of loss resulting from a failure is ₹ 25,000. At the extra costs shown below, various designs can be built to reduce the chance of failure. The higher cost designs will also create higher annual disbursements for insurance, property taxes, and certain maintenance and repair costs. The economic life is 5 years with zero salvage and the minimum require rate of return is 20%.

Investment for reliabilty, ₹	Operating disbursement per year, ₹	Reliability of the machine
0	0	0.000
5,000	300	0.500
10,000	600	0.700
15,000	900	0.850
20,000	1,200	0.940
25,000	1,500	0.990
30,000	1,800	0.995
35,000	2,100	0.999

10

Construction Accounts Management

General, principles of accounting, accounting process, construction contract revenue recognition, construction contract status report, limitations of accounting, balance sheet, profit and loss account, working capital, ratio analysis, funds flow statement

10.1 GENERAL

As has been mentioned at several places, construction projects involve consumption of resources in terms of manpower, equipment, materials and, of course, time. Now, *money* could serve as a common denominator to reduce all these components to a level-playing field and compare costs of projects, profits (or loss) to the contractor, and so on. In the case of construction projects, which often take several years to complete, issues related to the time value of money, proper planning for cash flow, etc., also acquire tremendous significance. It needs to be borne in mind that transactions are taking place at different times, and it stands to reason that the financial position is viewed as on a certain cut-off date.

Given the uncertainties and the nature of the transactions, certain basic procedures and conventions have evolved in the field of accounting. This chapter aims at introducing basic concepts of accounting and equips a construction engineer to handle day-to-day accounting issues. Effort has been made to avoid details of accounting procedure and complexities, which are beyond the scope of the present text.

In principle, accounting can be defined as a system of recording, collecting, summarizing, analysing and presenting information about an organization or a project in monetary terms. The findings of this exercise are usually presented in the following forms:

a. balance sheet,
b. profit and loss account,
c. a statement of changes in financial position.

When drawing up these statements, it should be borne in mind that they often become the basis for crucial decision-making. In fact, financial statements are of interest to owners and management of companies, potential investors, financial institutions, credit-rating agencies, employees, statutory and regulatory authorities, and competitors. Thus, effort is made to present various facts in an organized manner and at times the analysis is left to the user, who may like to better understand different aspects of the report. For example, whereas the investor would be primarily interested in knowing the prospects of his investment in the company, the tax authorities would be interested in knowing the tax liabilities of the company; the employees in the company may like to know about the stability and profitability of the company, while the creditors may be interested to know whether the amount owing to them would be paid when due; further, an analyst may be interested to know the extent of reliance on external source (debt) in relation to internal source (equity), and the management may be more keen to measure the firm's liquidity, financial flexibility, profit-generation ability and debt-payment ability.

10.2 PRINCIPLES OF ACCOUNTING

In a manner of speaking, accounting can be called the language of business, and like any language, it aims at effective communication of business transactions. To that extent, it is very important for any accountant, who is an exponent of the accounting language, to keep a true and accurate record of transactions as and when they take place. It is important to note that the accounts should be maintained in a manner that the financial position of a company, project, etc., on a particular date can be easily and accurately ascertained. Now, if this exercise is carried out with sufficient rigour, it should not be difficult to determine the 'profit' or 'loss' during a particular period.

Though a purely theoretical treatment has been avoided, this section presents a simple treatment of basic issues related to accounting, profit and loss account, and balance sheet, and other relevant information required to analyse this information and make logical conclusions.

Here, it is important to keep in mind that accounting is largely based on the following assumptions:

a. Transactions is recorded from the viewpoint of the person for whom the accounts are maintained. Consider a transaction involving transfer of, say, X tonnes of cement worth ₹ Y lakh from a cement supplier's godown to a contractor's batching plant. Now, the books of the cement supplier should reflect the 'receipt' of ₹ Y lakh and the 'depletion' of the available stock of cement by X tonnes. On the other hand, as far as the books of the contractor are concerned, they should show an increase (augmentation) of cement stock by X tonnes and an 'expense' of ₹ Y lakh.
b. The establishment is going to exist forever.
c. Provide for all prospective losses, but do not account for prospective profits.
d. Sum of debits is always equal to the sum of credits. In other words, the total funds owned are equal to the total funds owed.
e. Costs and revenues are taken into account or accrued as and when they are incurred or earned, rather than when they are paid or received. This is also known as the principle of accruals.
f. It is very important that identical or very similar items are treated in the same way in different accounting periods and in different sets of account. This is known as the principle of consistency.

While a detailed treatment of the legal and regulatory framework for accounting is out of the scope of the present text, it may be noted that in India, accounting procedures are governed by the provisions of The Companies Act (1956), Income Tax Act (1961), accounting standards and guidance notes issued by the Institute of Chartered Accountants of India (ICAI), and the Securities and Exchange Board of India (SEBI) listing agreement.

10.3 ACCOUNTING PROCESS

The profit and loss account and the balance sheet are two important financial documents. This section is aimed at understanding the process of preparation of these two documents. The entire exercise of preparation of profit and loss account and balance sheet is carried out involving the following seven steps.

Step 1 Initiation of transaction

A construction company enters into a number of transactions (in a broad sense, a transaction is an event involving money) on a daily basis. The accountants come into action when transactions take place. However, no transaction is recorded unless there is documentary evidence of the same. The documents could be in the form of an invoice, a cash voucher, etc. The documents need to be preserved for crosschecking the accounts as well as for auditing purposes.

Step 2 Keeping records of transactions in a chronological manner

The transactions mentioned above could be cash transactions or non-cash transactions. For cash transactions a cashbook is maintained, while non-cash transactions are recorded in journals chronologically. The term 'journal' derives from the fact that this is the book of first entry. A cashbook, on the other hand, keeps record of all cash transactions in chronological manner and shows the cash balance at any point of time. Depending on the nature of cash transactions, entries in cashbook are recorded on either debit or credit side. The explanation of entering some transactions on debit side and some on credit side is given later in the text. Entries in cashbook have serial numbers $C_1, C_2, C_3, \ldots$ and so on, which are used as cross-reference in the T-accounts, also called ledgers.

The journal records both the debit and credit sides of non-cash transactions. It is mandatory these days to record transactions following double-entry procedures. The principle of double entry is that for every debit there is an equal and corresponding credit. In this system (double entry), the bookkeeping convention requires debits to be defined as resource flowing into the accounts (i.e., the account received), credits being resource flowing out (i.e., the account given).

Conventionally, the debit side is written first, followed by the credit side. For reference purpose, the journal has a provision for writing ledger folio (LF). By convention, the entries in the journal are recorded as J_1, J_2, etc., instead of simply 1, 2, etc. A typical ledger is shown in Table 10.1.

Table 10.2a and Table 10.2b give a summary of the conventions usually adopted to record transactions in a cashbook and a journal. It may be noted from the tables that in cases where there is an increase (or decrease) in the assets, the entry is recorded as debit (or credit). Simi-

Table 10.1 A typical journal

Date	S. No.	Account	LF	Debit	Credit
25/3/2008	J_1	Debtor		300,000	
		Sales			300,000

Table 10.2a Conventions used for recording debit transactions

Item description	Recorded under
Increase in asset	Debit
Decrease in liability	Debit
Increase in expense	Debit
Decrease in income	Debit

Table 10.2b Conventions used for recording credit transactions

Item description	Recorded under
Decrease in asset	Credit
Increase in liability	Credit
Decrease in expense	Credit
Increase in income	Credit

larly, an increase in liability or in income is recorded on the credit side, and an increase in expense is recorded on the debit side.

Step 3 Making adjustment entries

Some transactions such as wages paid to labourers and depreciation in the fixed asset (owned by a company) are not recorded on a daily basis, and an appropriate entry is made only at the time of preparing financial statements such as balance sheet and profit and loss account. Such entries are known as adjustment entries, and made in the journal only at the time of closing the accounts.

Step 4 Recording transactions in T-account or ledger

As and when the entries in cashbook and journal become substantial in number, they are posted in relevant T-accounts (so-called due to their T-shape) or ledgers. For some organizations, posting may be done on a daily basis, while for others it may be on a weekly basis. In a ledger, the debit entries are recorded on the left-hand side and the credit entries on the right-hand side. A typical T-account is shown in Table 10.3.

Step 5 Preparing initial trial balance

Each of the T-account or ledger mentioned earlier is closed at the time of preparing the initial trial balance, which is a prerequisite for preparing the financial statements of profit and loss account and balance sheet. For this, the debit or credit balances from the T-accounts are carried over in the initial trial balance.

Step 6 Preparing profit and loss account and closing of T-accounts

The trial balance as prepared above becomes the source and basis for preparing the profit and loss account for a period and the balance sheet as of any date. For preparing the latter from the trial balance, the items related to revenue and expenditure are extracted. This combined with adjustment entries form the basis of profit and loss account. The adjustment entries now consist of stocks of raw materials, stock of work in progress, and stock of finished goods, besides

Table 10.3 A typical T-account

Debit			Credit		
Reference	Head	Amount	Reference	Head	Amount
T_1	Building	100,000			

Table 10.4 Data for illustrating contract revenue

S. No.	Description	Amount (₹ lakh)
1	Contract value	500.0
2	Original estimated cost of the contractor to complete the contract	450.0
3	Billed (invoice) to date	350.0
4	Payments received to date	315.0
5	Costs incurred to date	333.0
6	Estimated (forecasted) costs to complete	135.0
7	Costs paid to date	300.0

other entries such as purchases made, interest paid, income taxes paid, depreciation and wages incurred for the period for which profit and loss statement is desired.

Step 7 Preparing the balance sheet

Preparation of a final trial balance sheet precedes the preparation of balance sheet as on a particular date for a company. The final trial balance includes items related to assets and liabilities from the initial trial balance, and also includes the balances from the profit and loss accounts. A detailed discussion on balance sheet is given elsewhere in the book.

10.4 CONSTRUCTION CONTRACT REVENUE RECOGNITION

The construction industry is characterized by the long-term nature of projects, the involvement of several uncertainties, and the frequent change orders observed in projects. This calls for a slightly different approach in accounting processes, especially when it comes to recognizing the revenue. There are different methods used to demarcate revenue recognition. These are

1. Cash method of revenue recognition
2. Straight accrual method of revenue recognition
3. Completed contract method of revenue recognition
4. Percentage of completion method of revenue recognition

These methods are explained with the help of illustrative examples. Let us begin with the given information about a construction company (see Table 10.4).

10.4.1 Cash Method of Revenue Recognition

In this method, the revenue for the project at any point of time is calculated by subtracting costs paid to date from the payments received to date.

$$Revenue = payment\ received\ to\ date - costs\ paid\ to\ date \quad (10.1)$$

For the data provided in Table 10.4, revenue to date = ₹ 315.0 lakh − ₹ 300.0 lakh = ₹ 15.0 lakh

10.4.2 Straight Accrual Method of Revenue Recognition

In this method, the revenue for the project at any point of time is calculated by subtracting costs incurred to date (even if the cost is not yet paid) from the billed (invoice) to date (even if the billed amount is not yet realized).

$$Revenue = billed\ (invoice)\ to\ date - costs\ incurred\ to\ date \qquad (10.2)$$

For the data provided in Table 10.4, revenue to date = ₹ 350.0 lakh − ₹ 333.0 lakh = ₹ 17.0 lakh

10.4.3 Completed Contract Method of Revenue Recognition

In this method, the revenue is recognized only when the project has been completed. For the example project, the revenue would be zero since the project has not yet been completed. In cases where the contracts extend to several years, the company will not be in a position to show any revenue, and thereby profits, during the interim period. Obviously, for a long-duration construction project, this method has drawbacks.

10.4.4 Percentage of Completion Method of Revenue Recognition

In the first step, the percentage completion of the contract is evaluated. This is determined using the following relationship:

$$Percentage\ completion = \frac{Costs\ incurred\ to\ date}{(Costs\ incurred\ to\ date + Estimated\ cost\ to\ complete)} \times 100 \qquad (10.3)$$

In the second step, the revenue for the project is calculated by multiplying the contract value with the percentage completion obtained in the first step.

$$Revenue = Percentage\ complete \times Contract\ value \qquad (10.4)$$

For the example project, the revenue recognized using this method would be ₹ 500.0 lakh × ₹ 333 lakh × 100%/(₹ 333 lakh + ₹ 135 lakh) = ₹ 500 lakh × 71.15% = ₹ 355.75 lakh. Gross profit to date for this case would then be ₹ 355.75 lakh (revenue recognized) − ₹ 333.0 lakh (cost incurred to date) = ₹ 22.75 lakh.

In view of the high uncertainty that is characteristic of the construction industry, some construction companies do not record profits on projects till the billing reaches 50 per cent of the original contract value. Till such time, the cost is accumulated and the total cost incurred during the year is considered as 'invoiced sales' for the year. When the billing crosses 50 per cent of the original contract value, the revenue from the project is recognized but even then, a contingency amount varying from 1 per cent to 5 per cent is kept for unforeseen reasons while calculating the revenue and, thereby, the profit. Obviously, the provision for contingency is lowered towards the advanced completion percentage.

10.5 CONSTRUCTION CONTRACT STATUS REPORT

As discussed elsewhere, a construction company executes a number of construction contracts in a given financial year. Each of these construction contracts may not be getting completed in a given financial year. While some contracts might start and get over in a financial year, some others

Table 10.5 Data needed for preparing contract status report

Name of the project	Contract value (₹)	Cost to date (₹)	Total amount billed to date (₹)	Estimated cost to complete the balance work (₹)	Current estimated total cost (₹)
(1)	(2)	(3)	(4)	(5)	(6) = (3) + (5)
Project 1	1,500,000	600,000	800,000	850,000	1,450,000
Project 2	3,000,000	1,350,000	1,200,000	1,100,000	2,450,000
Project 3	2,000,000	900,000	1,200,000	1,000,000	1,900,000
Project 4	2,000,000	950,000	1,000,000	1,150,000	2,100,000

may be in different stages of completion. For example, some contracts may be completed about 30 per cent; some others may be 50 per cent complete; and the rest may be about 80 per cent complete. Contract status report, also known as work-in-progress schedule, is a summary of each of the contracts being executed by the company which are not yet completed. This report is usually prepared by each of the project managers responsible for the mentioned contract. The report from each of the contract is compiled at head-office level. Table 10.5 illustrates the procedure of preparing contract status report.

For preparing the contract status report, the data required by the project manager includes contract value, cost to date, estimated cost to complete the balance work, and total amount billed to date. In case the contract value is revised due to change order, the revised contract value is to be noted. The source of 'cost to date' is the project cost report discussed in Chapter 12. The estimated total cost of completion is obtained by adding 'cost to date' and 'estimated cost to complete the balance work'. The estimated profit or loss, then, can be obtained by subtracting 'estimated total cost of completion' from the 'contract value'.

$$Percent\ completed = \frac{Cost\ to\ date}{Estimated\ total\ cost\ of\ completion} \quad (10.5)$$

The revenue recognized till date can be computed based on any one of the methods explained in the previous section, depending on the accounting norms followed by the company. In most of the cases, the percentage completion method of revenue recognition is adopted. Keeping in mind the accounting principles discussed earlier, the proportional profit (based on the percentage completion) is reported in case of projects showing profit, while for projects showing loss, the entire loss is projected in the status report. Alternatively, the revenue can also be computed by adding cost to date and total profit on pro-rated basis, or total loss, as the case may be. The computations are shown in Table 10.6.

For completion of the discussion, we introduce the concepts of over-billing and under-billing. These terms are discussed in detail elsewhere.

$$Over\ billing = Billed\ to\ date - Revenue \quad (10.6)$$

$$Under\ billing = Revenue - Billed\ to\ date \quad (10.7)$$

Based on the above two formulaes, columns (11) and (12) have been computed in Table 10.6.

Table 10.6 Data needed for preparing contract status report

Name of the project	Estimated profit (loss) at completion (₹)	Percentage completion	Total profit (loss) recognized till date (₹)	Amount earned to date (revenue) (₹)	Over-billings (₹)	Under-billings (₹)
	(7) = (2) − (6)	(8) = (3)/ (6) × 100	(9) = (8) × (7) for profit only	(10) = (2) × (8) or (3) + (9)	(11)	(12)
Project 1	50,000	41.38%	20,690	620,690	179,310	
Project 2	550,000	55.10%	303,061	1,653,061		453,061
Project 3	100,000	47.37%	47,368	947,400	252,600	
Project 4	(100,000)	45.24%	(100,000)	904,800	95,200	

10.6 LIMITATIONS OF ACCOUNTING

Although accounting is a powerful tool to monitor financial transaction and evaluate the financial health of an organization, there are some inherent limitations that should be kept in mind when using the tool and arriving at any conclusion. The first such limitation that can be easily understood is that only transactions with money value can be recorded in the normal accounting procedures. Thus, generation (or loss) of goodwill cannot be kept track of. Similarly, records can be maintained only considering the original value. It is clear that the value of a plot of land, an equipment, or a building can be clearly defined only at the time of purchase (or sale), and it is difficult to assign any value in the intermediate period. In fact, an intermediate valuation of stock in terms of market price or cost price, whichever is lower, always involves the personal judgement of an accountant, which should be avoided in a scientific treatment and is also the bone of contention in a large proportion of disputes involving taxation, etc.

10.7 BALANCE SHEET

The balance sheet is one of the most important and fundamental financial statements of a firm. As the term suggests, the balance sheet provides information about the overall financial standing/position of a firm, giving a clear picture of its assets and liabilities among other things. In other words, a balance sheet presents a summary of what the company owns and what the company owes. Now, given the fact that transactions take place at different times (dates), it is obvious that neither assets nor liabilities are really permanent (though, as stated above, the establishment is assumed to be permanent in drawing up financial statements). Such a picture can at best be a snapshot—relevant for a particular point of time, say March 31, 2005, or December 31, 1990. Obviously, such a financial position is true for only a particular point in time, and bound to be different on the preceding or the following day. It may be mentioned here that the actual selection of the time and date of the balance sheet is more or less a matter of convention and practice, and it is more important to clearly understand the concept of the snapshot view of the overall financial status.

Example 1

Consider the transactions given in Table 10.7.

The balance sheet for the above transactions as on March 31, 2004, is shown in Table 10.8.

Table 10.7 Transaction details for example problem

Date	Detail	Remarks
19-03-04	A company ABC is created and a bank account for the firm opened	
20-03-04	Mr M transfers an amount of ₹ 500,000 to the company account	
25-03-04	An expense of ₹ 200,000 is made to M/s XYZ for purchase of equipment E1	Balance in bank reduced to ₹ 300,000
25-03-04	Equipment E1 is delivered and installed in the premises of ABC	

The balance sheet shown in Table 10.8 as on March 31, 2004, reflects a liability of ₹ 500,000 (which M/s ABC owes to Mr M) and assets in the form of a bank balance of ₹ 300,000 (the balance after funds for the equipment were transferred out), as well as equipment worth ₹ 200,000 that M/s ABC now owns. It may be pointed out that in any balance sheet, the sum of assets and liabilities should always match—this, in fact, is a cardinal principle of accounting. It may be noted that in this illustrative example, the dates for transfer of funds and installation of equipment have been kept identical to avoid the complication of an 'apparent mismatch' in the intervening period when the funds would have been transferred out and the equipment is not yet in place.

Traditionally, the liabilities comprised the left-hand side of the balance sheet and the assets comprised the right-hand side. The modern layout of balance sheet is linear and it gives details of liabilities beneath assets.

Figure 10.1 shows a typical balance sheet, and it can be seen, even without completely understanding the nuances or the strict definitions of the words used, that the information contained is critical from the point of view of understanding the financial health of the organization.

A close look at the balance sheet would reveal that assets and liabilities are arranged in the order of their liquidity. As far as listing the assets in the balance sheet is concerned, they can be listed in order of either liquidity promptness or fixity. Liquidity promptness refers to the ease with which an asset can be converted into cash (to repay a liability, etc.). In other words, the more liquid assets occupy the bottom position under asset heading. On the other hand, the amounts falling due within one year, such as bank loans, creditors and debentures, occupy the top position under liabilities heading.

Table 10.8 Balance sheet as on 31 March 2004

Assets (owned)					
Equipment	0	+	200,000	=	200,000
Cash	500,000	−	200,000	=	300,000
Total assets					500,000
Liabilities (Owed)					
Capital	0	+	500,000	=	500,000
Total liabilities					500,000

Balance sheet as on 31 March 2007

Description	As on 31.03.2007	As on 31.03.2006
₹ crore schedule	₹ crore	₹ crore
SOURCES OF FUNDS		
Share capital A	25.63	25.63
Reserves and surplus B	878.45	864.18
Total shareholders' funds	904.08	889.81
Loans		
(a) Secured C	482.48	196.39
(b) Unsecured D	1,068.58	999.51
Total loans	1,551.06	1,195.90
Deferred tax liability	85.54	67.73
Total	2,540.68	2,153.44
APPLICATION OF FUNDS		
Fixed assets		
(a) Gross block E	1,101.19	772.81
(b) Less: Depreciation	355.03	280.75
(c) Net block	746.16	492.06
(d) Items awaiting completion or commissioning	151.27	107.43
Total fixed assets	897.43	599.49
Investments F	228.64	126.47
Current assets, loans and advances		
A. Current assets G		
(a) Inventories	1,738.61	1,030.72
(b) Sundry debtors	0.54	2.80
(c) Cash and bank balances	208.37	1,006.00
(d) Other current assets	11.09	0.08
	1,958.61	2,039.60
B. Loans and advances H	347.61	199.10
Total current assets, loans and advances	2,306.22	2,238.70
Less: Current liabilities and provisions		
Current liabilities I	843.90	773.98
Provisions	47.71	37.24
Total current liabilities and provisions	891.61	811.22
Net current assets	1,414.61	1,427.48
Total	2,540.68	2,153.44

Figure 10.1 Balance sheet

It should also be noted that while a balance sheet is a very terse, matter-of-fact statement, which should in itself be complete, additional information is appended to better explain some of the features and facilitate better understanding. Such additional documents could be auditors' and/or director's report, other schedules forming part of accounts, (one such schedule

SCHEDULE G

Current Assets	Current year (₹ crore)	Previous year (₹ crore)
Inventories (as valued and certified by the management)		
(a) Stores, spares and embedded goods, at cost	299.09	169.15
(b) Fuel, at cost	5.05	1.91
(c) Materials in transit, at cost	3.82	1.28
(d) Work in progress		
Uncompleted contracts and value of work done	1,462.85	913.56
Less: Advances received/other recoveries	89.15	79.54
Less: Related site-mobilization expenses	112.87	67.07
	1,260.83	766.95
Add: Retention money	169.82	91.43
	1,430.65	858.38
Total	1,738.61	1,030.72

Figure 10.2 Schedule G

referred to as schedule 'G' giving details of current assets is given in Figure 10.2) notes explaining significant accounting policies, related parties' disclosure, etc.

Some of the common terms appearing in a typical balance sheet are discussed in the following sections. These are discussed under two broad sections—liabilities and assets.

10.7.1 Liabilities

Technically, liabilities represent the funds made available (to the company) by an external agency—be it owner(s), shareholders, or financial institutions. In other words, liabilities represent the funds owed (by the organization) to someone, and arise on account of borrowings by the company. Obviously, all liabilities need to be repaid at some point of time or the other, and depending upon the duration over which the funds are available, liabilities are often classified as (a) long term and (b) current. Usually, a period of one year is used as a benchmark to differentiate these liabilities, as briefly discussed below.

Long-term Liabilities

These liabilities arise in cases when the funds are available for more than one year. Debentures, bonds, mortgages and loans secured from financial institutions and commercial banks may be regarded as the source of long-term borrowings (or liabilities). Such liabilities are to be repaid (to the creditors) either as a single lump-sum payment at the time of maturity of the deposit, loan, or debenture, or in instalments over the life of the loan.

Current Liabilities

As against long-term liabilities, current liabilities are obligations payable in a short period, usually within the accounting period. It may be noted that liabilities due during the operating cycle (discussed elsewhere) of the organization are also deemed current liabilities. Such liabilities include accounts payable, taxes payable, accrued expenses, deferred income and short-term bank credit. It should be borne in mind that in some cases existing current liabilities can be cleared (liquidated) through the creation of additional current liabilities. Taking a short-term loan from one source to pay off a loan of another creditor is an example of such a transaction.

10.7.2 Assets

Although the term 'asset' is widely understood and used in common exchange of ideas, in business accounting for a resource to be considered as an asset of a company, it needs to satisfy the following conditions:

a. it must be valuable—a resource is valuable in the event that it is either cash or it is convertible into cash, or it can provide future benefits to the operations of the firm

b. it should be owned by the company in the legal sense, and merely possession or control is not sufficient for the resource to be considered as an asset

c. it must be acquired at a measurable money cost. In the event that an asset is not acquired for cash/ promise to pay cash, the test is what the asset would have cost, had cash been paid for it

Fixed Assets

As the name suggests, such assets are fixed and cannot be easily physically moved. They are usually acquired to be retained in business on long-term basis to produce goods and services, and not to be sold off easily. To that extent, fixed assets can also be looked upon as long-term assets, and are expected to remain with the company for periods longer than one accounting period. Very often, a viewer may use the extent and nature of these assets to judge the financial soundness of a company in terms of its future earnings, revenue, or profits. Fixed assets can be further classified into tangible and intangible fixed assets.

Tangible fixed assets Tangible fixed assets are those that have physical existence and contribute directly to generate goods and services. Land, buildings, plant, machinery, furniture, etc., are examples of tangible fixed assets. Such assets are shown at the cost they were acquired in the balance sheet, and this cost is distributed or spread over their useful life. Now, this yearly charge is referred to as depreciation. Thus, the amount of the tangible fixed assets shown in the balance sheet every year declines to the extent of the amount of depreciation charged in that year, and by the end of the useful life of the asset, it equals only to the salvage value, if any. It may be noted that the salvage value signifies the amount realized by the sale of the asset at the end of its useful life. The subject of different models for depreciation is very important to construction companies, in view of the large amounts of equipment often owned by them, and is taken up for a more detailed discussion elsewhere in the text.

Intangible fixed assets These assets do not directly generate goods and services, and in a manner of speaking, reflect the 'rights' of the firm. Such assets include patents, copyrights, trademarks and goodwill. Just as patents confer exclusive rights to use an invention, copyrights relate to production and sale of literary, musical and artistic works, and trademarks represent exclusive rights to use certain names, symbols, labels, designs, etc. It may be noted that intangible fixed assets are also written off over a period of time.

Current assets

Current assets comprise liquid assets of short duration. These assets are used in the operating cycle of the firm. Some examples of current assets are cash and bank balances, debtors, stock, etc. These assets will usually change their forms, time and time again, during the course of a year. Such assets in the ordinary and natural course of business move onward through the various processes of production, distribution and payment of goods until they become cash or equivalent, by which debts may be readily and immediately paid. Money used to purchase these assets is called floating or circulating capital.

Out of the given examples of current assets, cash is the most basic current asset. Bank balances are deposits made in the banks. Debtors are amount due to an entity or firm from customers buying on credit. In case of a construction company, debtors represent money that clients owe to the construction firm for construction services rendered by the company. Stocks consist of construction material for use in construction projects in case of construction companies. A more general definition of stock would be the value of inventory consisting of raw materials, work in progress, and finished goods.

10.8 PROFIT AND LOSS ACCOUNT

The profit and loss account is also known as the income and expenditure statement, the statement of earnings, and the statement of operations. This is a statement of the incomes and expenditures (or revenues and expenses) of the company during an accounting period, usually a year. Information on the following heads is commonly found in a profit and loss account:

- Gross profit
- Details of costs incurred
- Income from other investments of the company
- Interest details
- Outstanding expenses
- Prepaid expenses
- Income received in advance
- Outstanding income
- Depreciation
- Provision for bad and doubtful debt, etc.

A typical profit and loss account for a major construction company is shown in Figure 10.3.

As can be noticed, profit and loss account is an important financial statement that helps to ascertain the profit/loss of any particular business during a particular period of time. The statement, in essence, is the scoreboard of the performance of the organization in terms of profitability.

Some of the common terms appearing in a typical profit and loss account are discussed in the following paragraphs:

Revenues

Revenues are incomes accrued by sale of goods/services/assets or by the supply of the organization's resources to others. In other words, revenue refers to the overall value received from its customers or users. Interest, dividend, rent, etc., from the investments of the organization in other organizations, banks, etc., are also sources of revenue.

Profit and loss account for the year ended 31 March 2007

Schedule ₹ crore	Current year (₹ crore)	Previous year (₹ crore)
INCOME		
Income from operations J	2,394.50	2,028.14
Less: Company's share in turnover of integrated joint ventures	36.88	41.16
	2,357.62	1,986.98
Company's share in profit/(loss) of integrated joint ventures	24.38	3.09
Other income K	19.88	6.14
	2,401.88	1,996.21
EXPENDITURE		
Construction expenses L	1,837.03	1,601.11
Employees' remuneration and welfare expenses M	208.68	131.42
Office and site establishment expenses N	96.65	71.57
Interest O	61.97	41.39
Depreciation	79.66	52.45
	2,283.99	1,897.94
PROFIT BEFORE TAX AND EXCEPTIONAL ITEM	117.89	98.27
EXCEPTIONAL ITEM		
Transfer of development rights	—	40.00
PROFIT BEFORE TAX	117.89	138.27
Provision for current tax	18.19	12.00
Provision for deferred tax	17.81	0.68
Provision for fringe benefit tax	2.61	2.65
PROFIT BEFORE EARLIER YEARS' TAX	79.28	122.94
Add/(Less): Provision for earlier years' tax (refer note No. III (8) of Schedule Q)	(42.52)	1.86
PROFIT AFTER EARLIER YEARS' TAX	36.76	124.80
Add/(Less): Transferred from General Reserve No. 2	42.52	—
NET SURPLUS AFTER TAX	79.28	124.80
Add: Balance brought forward from last year	118.36	74.66
Add: Transferred from debenture redemption reserve	2.42	2.17
AMOUNT AVAILABLE FOR APPROPRIATION	200.06	201.63
Less: Appropriations:		
(a) Proposed dividend	19.22	17.94
(b) Tax on proposed dividend	3.27	2.52
(c) Debenture redemption reserve	8.54	2.81
(d) General Reserve No. 1	15.00	15.00
(e) General Reserve No. 2	—	45.00
	46.03	83.27

Balance carried to balance sheet	**154.03**	118.36
Basic EPS ₹		
— before earlier years' tax	**3.09**	5.36
— after earlier years' tax	**1.43**	5.44
Diluted EPS ₹		
— before earlier years' tax	**2.68**	5.35
— after earlier years' tax	**1.13**	5.44

The annexed notes (Schedule Q) form an integral part of the accounts.

Figure 10.3 Profit and loss account

Expenses

As against revenue, which can be considered as the inflow of funds into an organization, expenses cause an outflow of funds. Indeed, expenses should be made with the object of generating revenues! For example, in a construction company, expenses need to be made towards procuring materials, payment to workers and staff, procuring and installing plant and machinery (either through purchase or hire), and so on. To facilitate proper monitoring and control, expenses are classified under material cost, labour cost, plant and equipment cost, other costs such as cost toward fuel and power, insurance, etc. General and administrative expenses comprise salary of staff, rent of the organization premises, welfare, etc.

Net income/profit

Quite simply, the difference between the total revenues and the total expenses is termed as net income or profit. Now, a part of this net profit is retained to augment the reserves and surplus (retained earnings), and the remaining is used for paying dividends to shareholders.

10.9 WORKING CAPITAL

There has been considerable growth in the construction industry in the recent past, which has led to proliferation of small, medium and large construction companies. This has introduced stiff competition amongst all types of companies with different financial structures to win contracts for lesser profit margins. Companies with different financial structures have different working capital requirements; however, it is important that a company has sufficient working capital or access to funds to meet its short-term obligations. Too little may result in avoidable consequences such as cash-flow problems and failure to pay suppliers, labour, subcontractors, etc., on time. Ultimately, a business with insufficient working capital may be forced to cease trading even if it continues to report profit. Similarly, too much of working capital would make the funds idle, thereby hitting the profitability of the company. Working capital management, thus, seeks to arrive at a balance between liquidity and profitability, which are inversely related and, hence, lead towards opposite directions. Further, it aims at ascertaining the optimum level of cash and inventory that can be maintained and the optimum credit that should be allowed.

Stated simply, working capital or gross working capital refers to the current assets. When the term 'net capital' is used, it basically refers to the difference in current asset and current liability. It is said that working capital is to business what blood is to the body. This explains why it is important to manage the working capital effectively.

Moreover, working capital is costly and by suitably managing it, the profitability of business can be improved. Management of working capital means management of its key components, i.e., cash, debtors, inventory and creditors. The various factors that affect working capital include the nature of the business, the nature of raw materials used, the process technology used in the business, the nature of finished goods, the degree of competition in the market, and the paying habits of customers.

10.9.1 Need for Working Capital

The objective of financial decision-making is to maximize the shareholders' wealth. To achieve this, it is necessary to generate sufficient profits. The extent to which profits can be earned will naturally depend upon the magnitude of sales, among other things. A successful sales programme is, in other words, necessary for earning profits by any business enterprise. However, sales do not convert into cash instantly; there is invariably a time lag between the sale of goods and the receipt of cash. There is, therefore, a need for working capital in the form of current assets to deal with the problem arising out of the lack of immediate realization of cash against goods sold. Therefore, sufficient working capital is necessary to sustain sales activity. Technically, this is referred to as the operating or cash cycle and is explained in detail in the following section.

10.9.2 Operating Cycle

The operating cycle can be said to be at the heart of the need for working capital. The continuing flow from cash to suppliers, to inventory, to accounts receivable, and back into cash is what is called the operating cycle, which in fact is a continuous process. In other words, the term 'operating cycle' refers to the length of time necessary to complete the process of movement from cash to inventory (raw materials, consumables, other stocks), to finished goods, to receivables, and to cash again. A typical operating cycle is shown in Figure 10.4 and is explained in the context of a construction company.

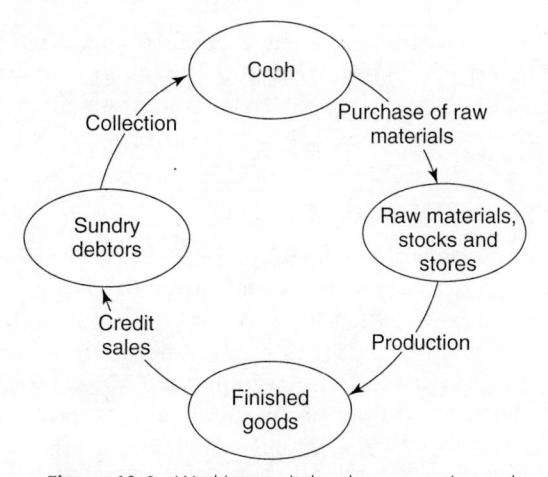

Figure 10.4 Working-capital cycle or operating cycle

A construction company is awarded a contract. The company may or may not receive mobilization advance. However, the company gets monthly payment for the work done during a particular month, and thus, cash is realized. The cash so obtained is spent in procuring inventory (materials, consumables and other stocks). The inventory is utilized in the production process of some construction activities of the contract. The inventories may be procured either in cash or on credit. In the later case, the supplier of inventory is treated as creditors of the company. The cash is also utilized in paying labour wages and equipment supplier, and in meeting overhead expenses. Upon completion of billing cycle, the company raises the bill for different activities carried out (it is equivalent to finished goods in case of manufacturing) during the billing period. Till the time payment is received against the bill, it is understood as credit sales and the client is regarded as debtor to the company. The retention and the other deductions from the certified bill are considered as outstanding and are treated as debtors. Further, the receipt of payment against the bill completes the operating cycle.

The provision for taxes, bad debts and expenses towards general overheads also needs to be considered in the cycle. A company operating profitably is generally cash-surplus. Failure to do so makes the company short of cash and eventually leads to its downfall.

If it were possible to complete the above cycle instantaneously, there would be no need for current assets (working capital). But since it is not possible, the firm is forced to have current assets. Since cash inflows and cash outflows do not match, firms have to necessarily keep cash or invest in short-term liquid securities so that they will be in a position to meet obligations when they become due.

Business activity does not come to an end after the realization of cash from customers. For a company the process is continuous, and hence the need for a regular supply of working capital. The magnitude of working capital required will not be constant, but will fluctuate. The changes in the level of working capital occur for three basic reasons—(1) changes in the level of sales and/or operating expenses, (2) policy changes and (3) changes in technology. To carry on with the business, a certain minimum level of working capital is necessary on a continuous and uninterrupted basis. For all practical purposes, this requirement will have to be met permanently as is the case with other fixed assets. This requirement is referred to as permanent or fixed working capital.

Any amount over and above the permanent level of working capital is temporary, fluctuating, or variable working capital. This portion of the required working capital is needed to meet fluctuations in demand consequent upon changes in production and sales as a result of seasonal changes.

10.9.3 Components of Working Capital

There are several components of working capital, as shown in Figure 10.5. As can be observed, these are broadly classified under current assets and current liabilities.

Current assets include cash/bank balances, inventories, sundry debtors, and loans and advances. These assets will usually change their forms, time and time again, during the course of a year. Such assets in the ordinary and natural course of business move onward through the various processes of production, distribution and payment of goods, until they become cash or equivalent, by which debts may be readily and immediately paid.

Current liabilities are all those liabilities that are falling due within a period of 12 months. The liabilities consist of sundry creditors, advances received from customers, overdraft facilities

CONSTRUCTION ACCOUNTS MANAGEMENT

Figure 10.5 Components of working capital

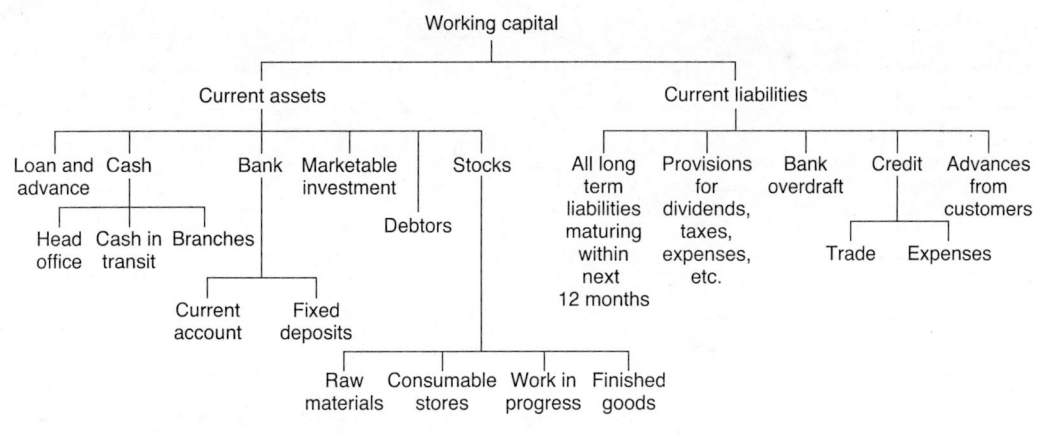

*Fixed deposits: To the extent to which immediate loans can be raised

(unsecured) and provisions. The provisions are mainly proposed dividends, corporate taxes, miscellaneous owings by the company, etc. In order to minimize the requirement of working capital, contracting organizations attempt to avail the maximum credit facility extended by suppliers. However, contracting organizations should meet the payment commitments made to suppliers on the due date.

The components of working capital in the context of a construction company are given below.

Current Assets

A Stock at sites/main depot
B Outstanding (including retention)
C Work in progress (WIP)
D Other current assets
E Excess cost over invoicing
F Work done unbilled
G Cash and bank balance

Current Liabilities

H Unadjusted advance (material, mobilization and plant)
I Vendor credit
J Other current liabilities

$$Net\ working\ capital = A + B + C + D + E + F + G - H - I - J \qquad (10.8)$$

10.9.4 Determination of Working Capital

It is evident from Figure 10.5 that there are a number of components of working capital and monitoring all of them is no easy task. Hence, some construction companies monitor only a few important components such as stock of materials, as well as customer outstanding from

Table 10.9 Data to illustrate different approaches in managing working capital

S. No.	Description	Conservative approach	Average approach	Aggressive approach
1	Sales (₹ lakh)	150	150	150
2	Earnings before interest and tax (EBIT) (₹ lakh)	15	15	15
3	Fixed asset (₹ lakh)	50	50	50
4	Current assets (₹ lakh)	60	50	40
5	Total assets (3 + 4) (₹ lakh)	110	100	90
6	Ratio of CA to FA (4/3)	1.2:1	1:1	0.8:1
7	Return on investment (EBIT/total asset)	13.60%	15.00%	16.70%

among current assets and advances provided by the customer or owner from among current liabilities. A budgetary norm is fixed for each of these three components and they are monitored accordingly. In order to calculate the net working capital, the revised formula then becomes:

$$Net\ working\ capital = A + B - H \tag{10.9}$$

The proportion of current asset to fixed asset depends on the psyche of the manager dealing with the working capital. An aggressive manager would prefer less of current assets than fixed assets. On the other hand, a conservative manager would stress more on the current assets than the fixed assets. An average manager would follow the intermediate path. The proportion of current asset to fixed asset impacts the return on assets employed and is evident from the given example (see Table 10.9). The example shows that for the same sale amount of ₹ 150 lakh, the variation in the ratio of current to fixed asset (from 1.2:1 to 0.8:1) results into the return on asset varying from 13.6 per cent to 16.7 per cent depending on the approach of the manager.

10.9.5 Financing Sources of Working Capital

One important issue in working capital management is to decide on the sources of finances for the working capital. The finances may be arranged from short-term sources or long-term sources. Arranging short-term finances are less costly compared to long-term finances. Here also, the proportion in which short-term finances and long-term finances are arranged is dependent on the manager's approach.

The three common approaches adopted in working capital finance are—matching approach, aggressive approach and conservative approach. In the matching approach, the permanent part of working capital is financed through long-term funds and the variable part through short-term funds. In the conservative approach, the permanent part of working capital as well as part of the variable working capital is financed through long-term funds, thereby avoiding risk but at the cost of profitability.

In the aggressive approach, the variable part of the working capital as well as part of the permanent working capital are financed through short-term funds, thereby involving a

Table 10.10 Different approaches in financing working capital

S. No.	Description	Matching approach	Conservative approach	Aggressive approach
1	Current assets (₹)	50,000	50,000	50,000
2	Fixed assets (₹)	50,000	50,000	50,000
3	Total assets (₹)	100,000	100,000	100,000
4	Short-term credit at 10% (₹)	25,000	10,000	40,000
5	Long-term credit at 12% (₹)	25,000	40,000	10,000
6	Current ratio (₹)	2:1	5:1	5:4
7	EBIT (₹)	15,000	15,000	15,000
8	Less interest (₹)	5,500	5,800	5,200
9	EBT (₹)	9,500	9,200	9,800
10	Tax @30% (₹)	2,850	2,760	2,940
11	Net income after tax (₹)	6,650	6,440	6,860
12	Return on equity considering a debt ratio of 50%	13.30%	12.88%	13.72%

relatively high degree of risk. In summary, in conservative approach the major sources of finances are long-term, while in aggressive approach the major sources of finances are short-term.

Depending on the approach, the mix of short-term and long-term finance has a bearing on the return. This is illustrated through an example (see Table 10.10).

It is evident from Table 10.10 that the interest liability changes depending on the approach of the manager. For example, interest in aggressive approach works out to be ₹ 5,200, while it is ₹ 5,800 for the conservative approach. This is hitting the return on equity. In aggressive approach, the return on equity works out to be 13.72 per cent, while it is 12.88 per cent in case of conservative approach. This difference could be even more pronounced if the difference in interest rates of short-term loan and long-term loan widens. This is the reason the determination of proportion of short-term and long-term finances demands appropriate attention. Having understood the importance of this, we now see the various options available for short-term and long-term finances of working capital. The common sources of finance are shown in Figure 10.6 and are described briefly in the following two sections.

Sources of Short-term Finance

The sources for short-term finance could be broadly classified in two categories—spontaneous and negotiated. While the first type does not require any formal arrangement on the part of the company availing them and come up automatically in the normal course of business, the negotiated category requires formal arrangements and is largely obtained from commercial banks. While trade credit comes under the spontaneous type, bank finances and some other sources such as commercial papers fall under the negotiated category. The sources of short-term finance are discussed briefly in the following paragraphs.

```
                          Sources of finance
                         ┌────────┴────────┐
                  Short term source    Long term source
                   ┌─────┴─────┐          ├── Retained earning
              Negotiated   Spontaneous    ├── Debentures ──┬── Ordinary
Other sources ──┬── Bank finances         ├── Shares ──────┼── Preference
                │    Trade credit         ├── Merchant bank└── Right issue
   ├── Commercial     ├── Over draft      ├── Industrial and commercial finance
   │   papers         ├── Cash credit     │   corporation
   ├── Factoring      ├── Purchase or     ├── Clearing bank loan facility
   ├── Forfaiting     │   discounting of bills
   └── Intercorporate ├── Letter of credit└── Government grant
       deposits       └── Working capital loan
```

Figure 10.6 Sources of finance

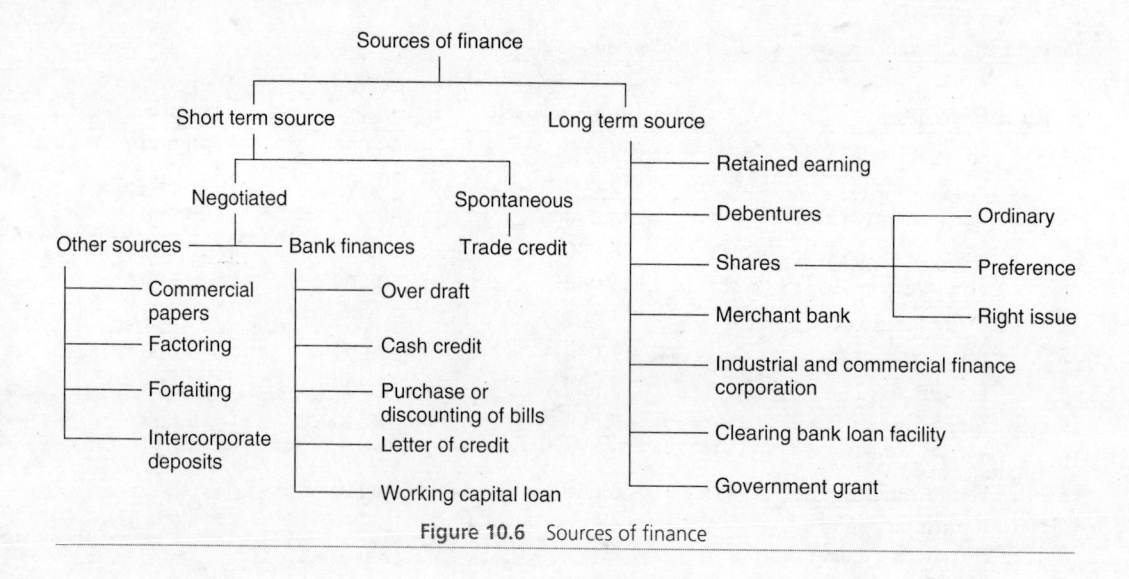

Trade credit Trade credit is very common in construction industry as it comes with fewer restrictions and is known for its simplicity and flexibility. Suppliers, vendors and subcontractors extend credit to the construction company. This form of financing does have benefits, but at the same time, it involves certain implicit and explicit costs. The cost of paying late in the form of reduced credit standing of the company is an example of implicit cost, while the cost of foregoing discount is an example of explicit cost.

Bank finance Banks are one of the important sources of finance. Bank finances could be of different types—overdraft facility, cash credit, purchase or discounting of bills, letter of credit, working capital loan, etc. Credits extended by banks could be either unsecured or against some collateral. The collaterals could be in the form of hypothecation, pledge, mortgage, or lien. The interest rate depends on whether interest is paid upfront or at maturity.

Other sources The other sources of short-term funds include commercial papers, forfaiting, factoring and inter-corporate deposits.

Commercial paper (CP) The concept of commercial paper (CP) originated in the USA on the pattern of short-term notes. Blue-chip companies in need of short-term funds could issue CP. In India, it came into existence in 1987. The maturity of CP lies between 15 days and less than one year.

Factoring Factoring is a process whereby a firm sells its receivables for cash to another firm known as the factor, which specializes in their collection and administration. In a broader sense, factoring is a type of bill discounting. The factor buys the client's receivables and controls the credits, and collects the same at maturity. In factoring, the seller of the goods informs the factor about an order received by the seller. The factor, in turn, evaluates the customer's (buyer of the goods) credit worthiness and approves the deal. The approval means that the factor has purchased the accounts receivables and now it is his responsibility to provide coverage for any bad debt loss. Of course, the factor offers this service for a certain commission.

Forfaiting This is an extension of factoring in the area of international trade. In forfaiting, an exporter discounts the bill with an agency known as the forfaitor. Once the export deal is finalized, the exporter intimates the forfaitor. The forfaitor then scrutinizes the deal. Once the forfaitor is satisfied, it quotes the discount rate and if it is agreeable to the exporter, the deal is signed.

Sources of Long-term Finance

The period involved in long-term finance is between 5 years and 10 years. Such finances are usually arranged to start a new business or to expand the existing business. The interest rates are usually high in view of the larger risks involved. Also, it is difficult to get finance on long-term basis, especially for construction companies. Some typical sources of long-term finance are:

1. Retained earnings
2. Clearing bank loan facility
3. Shares
4. Merchant bank
5. Industrial and commercial finance corporation
6. Debentures
7. Government grants

Retained earnings Retained earnings are those portions of earnings (usually between 30 per cent and 80 per cent of earnings after tax) of a company which are ploughed back (reinvested) into the business. These are also sometimes referred to as internal equity. The retained earnings are an important source of long-term finance.

Debentures Debentures are loans made to the company at fixed interest rates repayable at a set time. A company's reputation plays an important role in raising this type of loan. Debentures can be secured by mortgage on the firm's property as well. In case of liquidation or bankruptcy of the firm, those holding debentures can exercise their right almost ahead of all creditors.

Shares Equity shares are issued to the promoters of a company at the formation stage of the company. Subsequently, the company issues shares privately to the promoters' relatives, friends, business partners, employees, etc. Once the company grows, it raises capital from the public in the form of issue of equity shares.

Rights issue In rights issue, existing shareholders are offered new shares in exchange for their present shares, at some discount. The left-out shares (those not taken by existing shareholders) are put up for sale.

Merchant bank Merchant banks also provide loan at some fixed interest charges, which are negotiable. They usually attract higher interest charges. The loan is flexible in the sense that capital and interest repayment can be arranged to suit the company's (the borrower) future cash-flow position.

Loans Long-term loans are difficult to raise by construction companies, especially for new entrants in the business. This is because of the high risk involved in such loans for long terms. However, companies are able to raise long-term loans on higher interest rates. The new entrants first have to develop a good market reputation before realizing such loans.

Financial instruments Financial instruments such as bonds, floating-rate notes, bills of exchange, and promissory notes are also increasingly used for raising long-term finance for a company.

Venture capital These funds are an important source of finance for new companies. However, venture capital investors are specific to a business and they are very small in number.

10.10 RATIO ANALYSIS

Mathematically, ratio is the representation of one quantity (say, X) as a proportion of another (say, Y), and expressed as X/Y. To calculate the ratio, X is divided by Y, and the quotient (Z) multiplied by 100, in case the result is to be expressed as a percentage. Now, in obtaining Z some of the absolute value of X is lost, and the quotient is representative of the value only in relation to the value of Y.

The underlying principle of ratio analysis lies in the fact that it makes related information comparable. For instance, the fact that the net profits of an organization are, say, ₹ 10 lakh, provide very little basis for evaluation of the information in terms of performance. It is only when viewed against a target profit, profit achieved in previous years, percentage of profit in terms of the total sales, total assets, etc., that the figure of ₹ 10 lakh begins to make sense, and helps the management make strategic decisions. Now, in the context of business accounting, ratios acquire special significance as interest focuses not only on absolute values of parameters such as revenue, sales, etc., but also on their value in relation to others, as clearly brought out in the given example. The example clearly highlights the importance of ratios as an aide to business analysis, and the effort in this section is focused on providing an overview of some of the commonly used ratios in normal business accounting.

Table 10.11 shows the data for two organizations over a two-year period.

The following are some of the observations that can be made on the basis of the data presented in Table 10.11:

a. Whereas the sales for both A and B were identical in the first year, they grew at different levels for the two companies. The sales of A grew by 50 per cent, while B recorded a growth of only 25 per cent.
b. In the first year, profits of A and B are 25 per cent and 10 per cent of the respective sales values, and these numbers became 33 per cent and 20 per cent in the next year.

Table 10.11 Data for sales and profit for organizations A and B

Item	Organization A Year 1	Organization A Year 2	Organization B Year 1	Organization B Year 2
Sales (₹ lakh)	200	300	200	250
Profit (₹ lakh)	50	100	20	50

c. Whereas the profit of A doubled in the second year, B managed a two and a half times growth in profits.

In the above example, since both sales and profits were in the same units (rupees), it was possible to express the results in terms of percentages.

While the example highlights the importance and power of using ratios in analysis of business data, it should be remembered that computing ratios does not really add any new information, and only reveals a relationship in a more meaningful way. At the same time, pure ratios (or percentages) are also quite meaningless. For example, a statement that a company invests one per cent of its revenue in research and development does not communicate anything about the actual investments in research and development. Thus, in a manner of speaking, ratios and absolute numbers serve to complement each other as ways of presenting information, and ratio analysis should only be considered as an important tool in business analysis and decision-making.

It is clear from the above discussion that ratio analysis is basically useful when a comparison is involved—that of a number with another number, or that of a ratio with another ratio. Generally, three types of comparisons are commonly used in business—trend ratios, inter-firm comparison, and comparison with standards or plans. Trend ratios involve a comparison of ratios of a firm over time—that is, present ratios are compared with past ratios (for the same firm). A comparison of the profitability of a firm over a period, say, 1981 through 1985, is an illustration of trend ratio. Trend ratios are indicative of a general direction of change years. In inter-firm comparison, the subject of investigation is comparison of ratios of a firm with those of others in a similar line of business at the same time, or for the industry as a whole. A comparison of the market share or revenues of Hutch and Airtel would be an inter-firm comparison in the telecommunication sector. A comparison of investment in research and development by the telecommunication and construction industries could be looked upon as another example of inter-firm comparison. Finally, the comparison of ratios of a typical organization in a particular sector such as construction with that of a leading construction organization could be an example of comparison with standards. Also, if the comparison is carried out for different ratios against a target value set by management of an organization, it would be an example of 'comparison with plan'.

In the following paragraphs, discussion will be largely confined to different aspects of the ratios that are commonly used in the industry to study financial performance and integrity liquidity ratio, capital structure ratio, profitability ratio, activity ratio and supplementary ratio.

10.10.1 Liquidity Ratios

Liquidity in financial terms refers to the ease with which assets can be converted to cash to pay off debts, meet an expense, etc. To that extent, cash in hand has the highest liquidity, followed by, say, bank deposits in savings account, fixed deposits, etc. Naturally, then, given the time and complexities involved in selling land, or such other assets, they have the lesser liquidity. The liquidity ratio, therefore, is a measure of the ability of a firm to meet its short-term obligations and reflects its short-term financial strength. Depending on how the parameters involved are calculated, the following variants of liquidity ratios are commonly used:

a. Current ratio
b. Acid test/quick/liquid ratio

Current Ratio

$$\text{Current ratio} = \frac{\text{Current assets}}{\text{Current liabilities}} \quad (10.10)$$

Current ratio is a measure of short-term financial liquidity and indicates the ability of the firm to meet its short-term liabilities. A current ratio in excess of 1.3:1 is generally considered satisfactory for construction organizations (Schexnayder and Mayo 2004). A current ratio of less than 1:1 would certainly be undesirable in any industry, as some safety margin is required to protect the creditors' interest at least.

Acid Test of Liquidity or Quick Ratio

$$\text{Acid test ratio} = \frac{\text{Quick assets}}{\text{Current liabilities}} \quad (10.11)$$

The 'quick assets' include only the cash and bank balances, short-term marketable securities, and receivables. The acid ratio test is a more rigorous measure of a firm's ability to service short-term liabilities. The ratio is also widely accepted as the best available measure of the liquidity position of a firm. The recommended value of acid test ratio is in excess of 1.1:1 for a construction organization (Schexnayder and Mayo 2004).

10.10.2 Capital Structure Ratios

When a firm wishes to make a long-term borrowing, it needs to commit to

- pay the interest regularly, and
- repay the principal when it is due [in instalment(s) at due dates, or in a single payment at the time of maturity].

Now, for long-term creditors, both the criteria are important and they often base their judgement on the (long-term) financial soundness on the capital structure ratios, which reflect the long-term solvency of a firm. It is quite evident that the above two criteria are quite different but mutually dependent, and are examined using slightly different capital structure ratios. It may be borne in mind that usually two types of capital structure ratios are used—leverage ratios and coverage ratios.

Leverage Ratio

The source document for computing leverage ratio is the balance sheet. Some of the important leverage ratios are given below.

Debt equity ratio This ratio is computed to better understand the financial structure and integrity of a firm, as it reflects the relative contribution of creditors and owners of business in its financing. Now, depending on the relationship between creditor's claims and owner's capital, there can be several variants of the debt equity ratio as given below.

i. $\text{Debt equity ratio } DE_1 = \dfrac{\text{Long term debt}}{\text{Shareholders' equity}} \quad (10.12)$

In the above expression, long-term debt excludes the current liabilities and shareholders' equity includes both ordinary and preference capital, besides past accumulated profits (reserves and credit balance of the profit and loss account). The shareholders' equity so defined is also referred to as the 'net worth', and the debt equity ratio computed on this basis may also be called debt to net worth ratio.

ii. $\text{Debt equity ratio } DE_2 = \dfrac{\text{Total debt}}{\text{Shareholders' equity}}$ \hfill (10.13)

In the above expression, the total debt is the sum of the short-term and long-term debts.

Neither a very high nor a very low debt equity ratio is desirable, and each firm needs to determine an optimum level where the debt and equity are properly balanced.

Debt asset ratio This ratio relates the total debt to the total assets of the firm, and is defined as follows:

$$\text{Debt asset ratio} = \dfrac{\text{Total debt}}{\text{Total assets}} \quad (10.14)$$

In the above, the total debt comprises long-term debt and current liabilities, whereas the total assets consist of all assets minus fictitious assets (i.e., total debt plus shareholders' equity).

Coverage Ratio

This ratio is calculated from the profit and loss account. Coverage ratio measures the relation between what is 'normally' available from operations of the firms and the outsiders' claims. Some of the important coverage ratios are given below.

Interest coverage ratio

$$\text{Interest coverage ratio} = \dfrac{\text{Earnings before interest and tax (EBIT)}}{\text{Interest}} \quad (10.15)$$

This ratio measures the debt-servicing capacity of a firm. It shows how large the EBIT is, compared to the interest charges. The ratio, like the interest coverage ratio, reveals the safety margin available to the preference shareholders, and, in principle, a higher value of the ratio is considered better from the point of view of shareholders.

Dividend coverage ratio

$$\text{Dividend coverage ratio} = \dfrac{\text{Earnings after tax (EAT)}}{\text{Preference dividend}} \quad (10.16)$$

This ratio is a measure of the ability of a firm to pay dividend on preference shares that carry a stated rate of return.

10.10.3 Profitability Ratios

The importance of profits to any commercial firm cannot be overstressed—the entire exercise of proper management is directed to maximize profitability without compromising on other parameters such as quality and safety. In a certain manner, profitability is also a measure of efficiency and indicates public acceptance of the product. Moreover, profits provide funds for repaying debts incurred while financing the project and mobilizing resources. The profitability ratios have been defined to establish quantitative measures of the profitability of a firm, and are of interest to owners, management, creditors and regulatory bodies. These ratios can be determined on the basis of either sales or investments. In the former class, the ratios commonly used are the (gross or net) profit margin and the expenses (or operating) ratio. On the other hand, ratios such as those defined in terms of return on assets, capital employed, and equity of shareholders are based on investment. Some of these ratios are discussed below.

Profit Ratios

These ratios are a measure of the relation between profits and sales of a firm. Now, accounting procedures differentiate between gross, net and operating profits, and the profitability ratios based on sales could be defined as gross profit ratio, operating profit ratio and net profit ratio.

Gross profit ratio

$$Gross\ profit\ ratio = \frac{Gross\ profits}{Sales} \times 100 \qquad (10.17)$$

A relatively low gross profit ratio can obviously be seen as a signal requiring careful and detailed analysis of the factors involved. On the other hand, a high gross profit ratio is often regarded as a sign of good management, though in this case, too, the factors responsible should be carefully analysed.

Net profit ratio

$$Net\ profit\ ratio = \frac{Net\ profits}{Sales} \times 100 \qquad (10.18)$$

The net profit is calculated after deducting all operating and non-operating expenses, including taxes and interest. A high net profit margin ensures adequate return to the owners and is also indicative of the ability of a firm to withstand adverse conditions on account of a decline in sales, increase in cost of production, or fall in demand, at least for some time. Obviously, a low net profit margin has quite the opposite implications.

Expenses Ratios

These ratios constitute another methodology of representing profitability ratios, and are based basically on the expenses, rather than the profits. Simply speaking, the expenses ratio is the complementary component of the profit ratio, gross as well as net. For instance, if the profit ratio is deducted from 100 per cent, the resultant is nothing but the expense ratio. Of course, alternatively, the profit ratio can be obtained by subtracting the expenses ratio from 100 per cent. In principle, expense ratios are computed by dividing the expenses by the sales. Like in the case of profit-based ratios discussed above, there are different variants of expenses ratios. Some of these are discussed below.

Operating ratio

$$Operating\ ratio = \frac{Total\ operating\ cost}{Sales} \times 100 \qquad (10.19)$$

The operating ratio is an indicator of the total operating cost incurred for every hundred rupees of sales, and reflects the efficiency of cost management in controlling the operational cost. The total operating cost is usually taken as the sum of the cost of the goods sold and the operating expenses. A high operating cost alerts the management to the necessity of adopting appropriate measures to reduce cost.

Cost of goods sold ratio

$$\text{Cost of goods sold ratio} = \frac{\text{Cost of goods sold}}{\text{sales}} \times 100 \qquad (10.20)$$

The cost of goods sold ratio shows the percentage share of sales consumed by cost of goods sold (and, therefore, what proportion is available for expenses involved in selling and distribution, as well as financial expenses consisting of interest and dividends).

Profitability Ratios Based on Investment

So far, the discussion has been confined to understanding the profit generated in terms of sales, whether from the point of view of profit generated or from the standpoint of expense incurred. The following paragraphs discuss profitability from yet another perspective, namely, that of investment. It may be noted at this stage that, generally speaking, three kinds of investments are often referred to, namely, assets, capital employed and shareholders' equity. Based on each of these, three broad categories of returns on investment (commonly referred to as ROIs) have been identified—return on assets, return on capital employed, and return on shareholders' equity.

Return on assets (ROA)

$$\text{Return on assets (ROA)} = \frac{\text{Net profit after taxes and interest}}{\text{Total assets}} \times 100 \qquad (10.21)$$

The assets may be defined as total assets or simply the tangible assets (i.e., total assets minus intangible assets like goodwill, patents and trademarks, as well as fictitious assets).

Return on capital employed (ROCE)

$$\text{Return on capital employed (ROCE)} = \frac{\text{Net profit after taxes and interest}}{\text{Net capital employed}} \qquad (10.22)$$

This ratio indicates how efficiently the long-term funds are being used. The term 'net capital' refers to total assets minus current liabilities, or alternatively, it is the sum of the long-term liabilities and the owner's equity.

Return on shareholders' equity

$$\text{Return on shareholders' equity} = \frac{\text{Net profit after taxes}}{\text{Total shareholders' equity}} \times 100 \qquad (10.23)$$

The ratio measures the return on the owners' funds (investments) in a firm. Sometimes, the shareholders' equity is also referred to as 'net worth'. The ratio is indicative of how profitably the owners' funds have been utilized by the firm.

10.10.4 Activity Ratios

This set of ratios used in business accounting is concerned with measurement of efficiency in asset management. These ratios, at times also called efficiency ratios, measure the efficient employment of assets by relating the assets to sales and/or cost of goods sold. An activity ratio may, therefore, be defined as a test of the relationship between sales and/or cost of goods sold and the various assets of a firm. Depending upon the various types of assets, different types of activity ratios can be defined. Some of the important and commonly used activity or efficiency ratios are discussed in greater detail in the following paragraphs.

Inventory (or Stock) Turnover Ratio

$$Inventory\ turnover = \frac{Cost\ of\ goods\ sold}{Average\ inventory} \qquad (10.24)$$

In the above expression, the cost of goods sold is taken as the sum of the opening stock and the manufacturing stock, including purchases less the closing stock (inventory). Average inventory is the average of the opening and the closing inventory. A high inventory turnover ratio is generally considered better than a lower ratio since it implies good inventory management, but a very high ratio calls for a careful analysis.

Receivables (Debtors) Turnover Ratio

$$Debtors\ turnover = \frac{Credit\ sales}{Average\ debtors} \qquad (10.25)$$

In the above expression, the term 'average debtor' is the average of the opening and closing balance of debtors. The ratio is an indirect indication of the rapidity with which receivables or debtors can be converted into cash.

Assets Turnover Ratio

There are many variants of this ratio. The expressions for computing some of these ratios are given below:

$$Total\ assets\ turnover = \frac{Cost\ of\ goods\ sold}{Total\ assets} \qquad (10.26)$$

$$Fixed\ assets\ turnover = \frac{Cost\ of\ goods\ sold}{Fixed\ assets} \qquad (10.27)$$

$$Current\ assets\ turnover = \frac{Cost\ of\ goods\ sold}{Current\ assets} \qquad (10.28)$$

$$Working\ capital\ turnover\ ratio = \frac{Cost\ of\ goods\ sold}{Working\ capital} \qquad (10.29)$$

It may be noted that the numerator in all the cases is the same (total cost of goods sold), though the denominator is different depending upon the definition of the 'assets' used. It may be further noted that in the discussion here, the total and fixed assets are not inclusive of depreciation. The higher the above ratio, the more efficient is the management and utilization of the assets, and vice versa.

10.10.5 Supplementary Ratios

Besides the four broad heads of ratios discussed, there are numerous other ratios that could be useful for analyses of firms. It is not the intention to cover all of them here. However, some useful ratios related to productivity and ratios related to periods of credit offered by the firm and periods of credit enjoyed by the firm are described briefly under supplementary ratios.

Periods of Credit

The period of credit offered by the firm is defined as:

$$\text{Period of credit offered by the firm} = \frac{\text{Debtors}}{\text{Turnover}} \times 12, \quad \text{if the ratio is preferred in months} \tag{10.30}$$

For the value in weeks and days, the 12 in the expression should be replaced by 52 and 365, respectively.

The period of credit enjoyed by the firm is defined as:

$$\text{Period of credit enjoyed by the firm} = \frac{\text{Creditors}}{\text{Purchases}} \times 12, \quad \text{if the ratio is preferred in months} \tag{10.31}$$

For the value in weeks and days, the 12 in the expression should be replaced by 52 and 365, respectively.

The above ratios indicate credit control of the firms. The ratios indicate whether the firm is applying a sensible policy with respect to both its creditors and debtors.

Productivity Ratios

Many variants of productivity ratios are possible depending on the types of numerator and denominator considered. The productivity ratios are essentially the ratio of output to a given input. Some of the commonly used productivity ratios are listed below and these are self-explanatory.

- Turnover to number of employees
- Turnover to subcontractors
- Turnover to plant and equipment
- Profit to number of employees
- Plant and equipment value to number of employees

The summary of ratios discussed in the preceding sections is given in pictorial form in Figure 10.7.

10.11 FUNDS FLOW STATEMENT

The funds flow statement of a company is a useful financial statement that summarizes changes in the availability of funds over a period of time, and thus also serves as a very important supporting document to the balance sheet, or the profit and loss statement. In general, 'funds' in

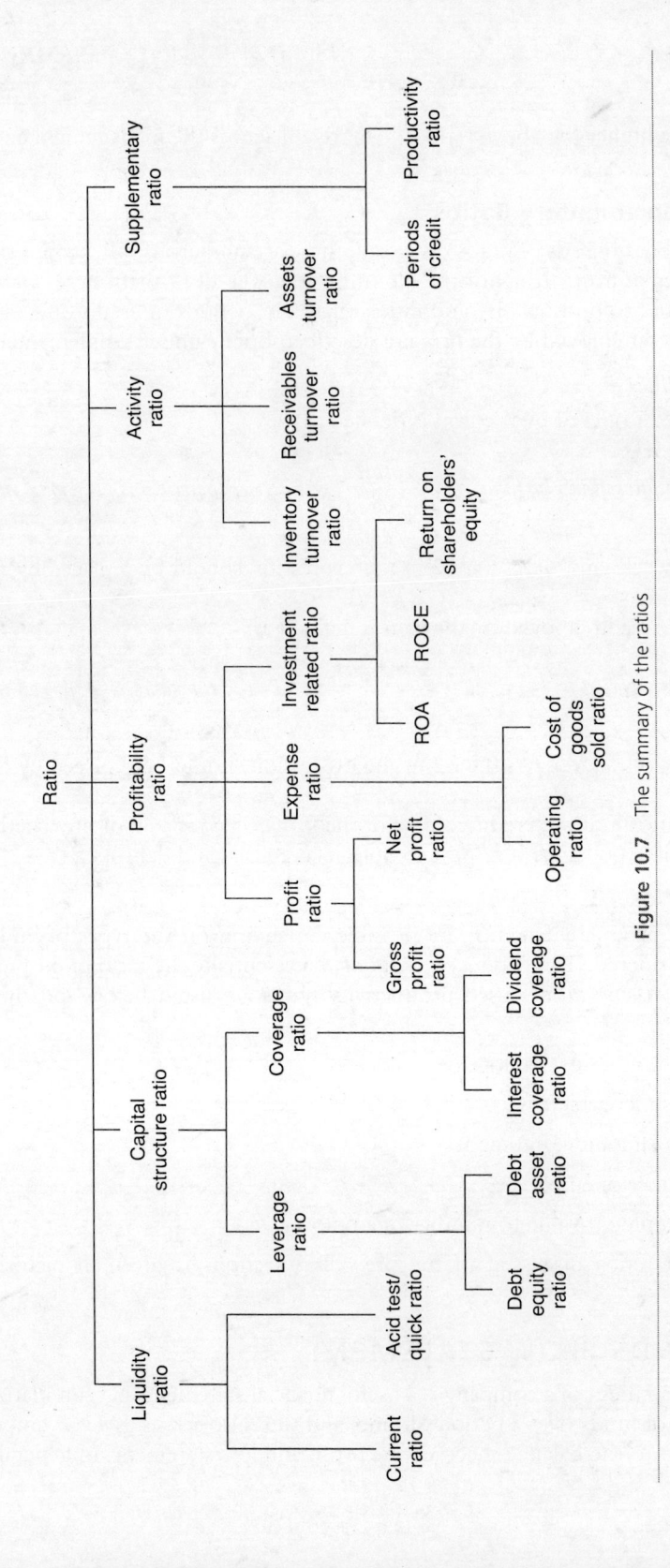

Figure 10.7 The summary of the ratios

this context are simply taken as the financial resources (working capital) available at a certain point in time—and are, therefore, taken as the excess of current assets over current liabilities, and represented as follows:

$$\text{Working capital (WC)} = \text{Current assets (CA)} - \text{Current liabilities (CL)} \qquad (10.32)$$

Changes in funds are described as flow of funds—an increase as inflow and decrease as outflow. Being so, increase/inflow in funds would be considered as source and decrease/outflow as use. Therefore, funds flow statement depicts factors responsible for changes in working capital during a given period. In other words, its purpose is to show wherefrom working capital originates and whereto the same goes during an accounting period. Accordingly, it portrays the sources and applications of working capital and highlights the basic changes in the resources and financial structure of a firm between its two consecutive balance-sheet dates. When funds are identified with working capital, the funds flow statement records the reasons for changes in working capital. Thus, funds flow statement may be defined as a statement that records sources from which funds have been obtained and uses to which they have been applied. If a transaction causes an increase in working capital, it is termed as a source of funds, and if it causes a decrease in working capital, it is termed as application or use of funds. Obviously, in the preparation of this statement, transactions affecting working capital are to be traced out. For this purpose, let us formulate some general rules so as to ascertain which transactions give rise to a source or use of working capital and which do not.

10.11.1 Changes in Working Capital or Funds

It has been emphasised elsewhere that any transaction causes some 'movement' of funds—and given the definition of working capital, it should be clear that a transaction affects working capital if it results in an increase (or decrease) in CA without a corresponding increase (or decrease) in the CL. Similarly, a transaction that causes increase (or decrease) in CL without a corresponding increase (or decrease) in the CA also affects the working capital.

As a corollary to this, a transaction does not affect working capital if it results in an increase (or decrease) in the CA followed by a corresponding increase (or decrease) in the amount of CL, or an increase (or decrease) in the amount of CL followed by a corresponding increase (or decrease) in the amount of CA. Besides these more or less arithmetical conditions, it should be noted that as discussed elsewhere in greater detail, the aggregate current assets and current liabilities of a company are essentially a sum total of several components. Thus, the working capital is not affected if there is only a change in the composition of current assets (or liabilities), i.e., an increase in one component of CA (or CL) accompanied by a decrease in the other component of CA (or CL). Similarly, any change in the liabilities also does not affect the working capital.

Increase in working capital can be brought about in a variety of ways. The most straightforward contribution could come from profits from ongoing business operations. Then, there could be other sources of 'income', such as sale of fixed assets, raising long-term borrowings, or raising share capital. All these become source of funds for the company.

Similarly, losses in business operations, purchase of fixed assets, payment of dividends, etc., result in a decrease in the working capital and need to be shown under 'use', 'application', or 'outflow' of funds.

Table 10.12 Sources and applications of funds statement

	(₹ lakh)	(₹ lakh)
Sources of funds		
Profit before taxes		2,000
Add for depreciation[1]		800
Total generated from operations		2,800
Sale of fixed assets		30
		2,830
Applications of funds		
Dividends paid	500	
Taxes paid	350	
Purchases of fixed assets	1,320	2,170
		660
Increase/decrease in working capital		
Decrease in stocks	60	
Decrease in work in progress	60	
Increase in debtors	370	
Increase in creditors	−50	
Increase in cash balances	220	
		660

[1] It may be recalled that depreciation does not involve movement of fund.

10.11.2 Determining Funds Generated/Used by Business Operations

As mentioned earlier, the profit and loss account of a company provides a summary of the transactions related to expenses incurred and revenues earned during a particular period. This information needs to be appropriately analysed using the principles outlined above to understand the implications as far as the working capital is concerned. Table 10.12 gives an illustrative example of such a classification. In fact, to that extent, the funds flow statement is only a summary of some of the transactions already presented in the profit and loss statement, prepared with a view to highlight the sources and application of funds along with increase/decrease in working capital for a particular period.

10.11.3 Application of Funds Flow Statement

A funds flow statement can be used for analysis of facts during a given period and also for effective financial planning for the future. An increase in working capital clearly involves higher interest costs, besides risks of inventory accumulations, receivables being locked up, maintaining more (idle) cash than required, and so on. The information provided by funds flow statement is crucial to the analysts of the firm, both internal and external, besides being useful to the creditors.

The analysts can get appropriate answers to several important financial questions. For example, analysis of the funds flow statement yields valuable information on subjects such as how the profits for a period were used, why the dividends (to the shareholders) were paid only at a certain rate, how dividends could still be given despite losses encountered in a period, and so on.

External analysts can also find a lot of important facts and figures in the funds statement of a company. For example, the statement clearly shows the extent of reliance on external sources (debt) in relation to internal (equity) sources, the major sources used to finance major non-current asset expansion like plant and equipment, the major source of funds used for repayment of long-term debentures or preference shares, and so on. Indeed, such disclosures to a trained reader reveal the financing and investment activities of a company. In fact, creditors can also use a funds flow statement to examine the mechanics of how the funds for activities such as expansion are being raised.

The funds flow statement also helps to better understand the effect of various financing and investment decisions on working capital in future. For example, in case the implementation of a certain decision results in excessive or inadequate working capital, steps need to be taken to improve the situation or review the decision itself. Consider a situation where the working capital position is expected to deteriorate—a decision can be made to raise funds by borrowing or by issuing new equity shares.

REFERENCES

1. Agrawal, R. and Sriniwasan, R., 2005, *Accounting Made Easy*, New Delhi: Tata McGraw-Hill.
2. Schexnayder, C.J. and Mayo, R.E., 2004, *Construction Management Fundamentals*, Singapore: McGraw-Hill.
3. Sharan, V., 2005, *Fundamentals of Financial Management*, New Delhi: Pearson Education.

REVIEW QUESTIONS

1. i. Net working capital means
    a. Current assets—current liabilities
    b. Fixed assets—fixed liabilities
    c. Current assets—non-tangible assets
   ii. An aggressive approach to current asset financing leads to
    a. Greater profitability
    b. Lower profitability
    c. No impact on profitability
   iii. The longer the operating cycle—
    a. The larger the size of current assets
    b. The smaller the size of current assets
    c. The size of current assets remain unchanged

iv. Mortgage involves
   a. Moveables as collateral
   b. Immovables as collateral
   c. No collateral
v. Commercial paper is issued by
   a. Any company
   b. Only blue-chip companies
   c. Only government companies
vi. Retained earning is liability. (True or False)
vii. Goodwill is tangible asset. (True or False)
viii. Factoring and forfaiting are sources of long-term finance. (True or False)
ix. T-account is also called ledger. (True or False)
x. Outstanding is part of current liability. (True or False)

2. Discuss the principles of accounting.
3. What is meant by initial trial balance? Why is it important to prepare an initial trial balance?
4. What are the different entries in a balance sheet?
5. What is meant by assets and liabilities? Give examples of assets and liabilities in the context of a construction organization.
6. Write short notes on T-account, cashbook, and profit and loss account.
7. What do you mean by working capital? Why is working capital management important?
8. What is operating cycle?
9. Differentiate between short-term and long-term finance.
10. Discuss different ratios and their relevance.
11. For the set of data given in Table Q11.1, calculate the revenue using the following methods of revenue recognition—(1) cash method of revenue recognition, (2) straight accrual method of revenue recognition, (3) completed contract method of revenue recognition, and (4) percentage of completion method of revenue recognition.

Table Q11.1 Data for Question 11

S. No.	Description	Amount (₹)
1	Contract amount	2,000,000
2	Original estimated cost	1,800,000
3	Billed to date	1,500,000
4	Payments received to date	1,300,000
5	Costs incurred to date	900,000
6	Forecasted costs to complete	850,000
7	Costs paid to date	800,000

CONSTRUCTION ACCOUNTS MANAGEMENT | 425 |

12. Table Q12.1 shows inter-firm comparison of construction companies. Calculate the current ratio and the net working capital for each of the companies and comment on them.

    Table Q12.1 Inter-firm comparison

Particulars	ABC	CDE	EFG	HIJ
Working capital as % of sales	28.94%	6.30%	27.33%	28.26%
Unadjusted advance (₹ lakh)	560	174	169	29
Unadjusted advance as % of backlog	6.28%	6.09%	5.05%	2.54%
Gross current assets (₹ lakh)	2,704	442	539	256
Gross current assets as a % of sales	70.18%	56.81%	79.62%	56.51%
Total liabilities and provisions (₹ lakh)	1,589	392	354	128
Total liabilities and provisions as a % of sales	41.24%	50.39%	52.29%	28.26%
Current ratio				
Net working capital				

13. From the transaction details of a construction company, M/s *New Construction Limited*, given in Table Q13.1, prepare the profit and loss account and the balance sheet. Assume all transactions to be cash transactions. Also, assume depreciation of machinery to be zero in this financial year.

    Table Q13.1 Transaction details of M/s *New Construction Company*

Transaction	Reference	Description
1	T_1	Purchased materials for ₹ 6,566
2	T_2	Paid salaries ₹ 668
3	T_3	Paid for telephone charges ₹ 1,691
4	T_4	Interest on loan paid ₹ 177
5	T_5	Tax paid ₹ 77
6	T_6	Sold goods for ₹ 9,614
7	T_7	Borrowed loan from Citibank ₹ 3,176
8	T_8	Purchased machinery for ₹ 3,176

14. Determine the working capital required for the business from the data given in Table Q14.1.
15. Write short notes on (a) Percentage of completion method, and (b) Completed contract method of project accounting.
16. What are the basic concepts and features behind preparing profit and loss account and balance sheet?

17. Working capital management assumes considerable significance in construction management. What are the concepts and strategies behind effective working capital management?

Table Q14.1 Data for Question 14

Particulars	₹ (lakh)
General reserves	50
Cum depreciation	20
Short term loans	30
Work in progress	40
Finished goods	10
Debtors	30
Sales	100
Commercial papers	15
Rental deposits	5
Other advances made	15
Equity capital	200
Fixed assets	150
Raw materials	25
Creditors	30
Advance received	10
Overdraft	50
Provision for tax	5
Cash in bank	5
Advance tax	10

11

Construction Material Management

Introduction, material procurement process in construction organization, materials management functions, inventory management

11.1 INTRODUCTION

The term 'materials' is used in a general sense and refers to the whole range of goods and services that are purchased or otherwise procured from sources outside the organization and are used or processed in order to provide finished products for sale. One can notice a fundamental difference in materials management in a factory-like situation and a construction project-situation. In a factory situation, products are standardized, the manufacturing location and process are permanent, and, as such, long-term planning is possible. Further, increase in input costs of materials can be passed on to the customer through increase in sales price of the end products. However, as has been repeatedly pointed out in the text, each construction project is unique in content and nature of execution. Project locations are transient and temporary, and the cost is predetermined at the start of the project, which calls for utmost care in material procurement in order to ensure that cost provisions are not overrun. Materials account for a large fraction of overall construction project cost.

A typical construction project requires a variety of materials. Materials in a project are broadly classified under capital equipment, construction machineries, and consumables. Some companies classify their materials under two broad categories—capital items and revenue items. Plant and machinery, vehicles, office equipment, land and buildings are placed under capital items, while non-capitalized items, heavy tools and tackles, small tools, consumables, electrical items, construction materials, special/one-time items, and spares are designated under revenue items.

The importance of materials management can be gauged from the fact that in any typical building project the share of material costs is about 55 per cent; labour costs, about 25 per cent; and POL (petrol, oil and lubricants), overheads, tax components and profits, about 5 per cent each. Thus, materials management occupies an important place in construction management and in recent times has received considerable attention. This is due to the following major factors:

- First, a worldwide recession has had the impact of reducing sales volumes and revenues, and this has forced management to reconsider how best to lower the levels of inventory in order to maintain margins (by reducing interest and obsolescence costs).
- Secondly, the changes in manufacturing philosophy, specifically the growth in just-in-time (JIT) applications, have reduced the need for inventory as an insurance buffer within the overall logistics activity.
- Thirdly, many companies have realized that a greater return on investment (ROI) can be obtained by developing the core business, and that investment in working capital items, such as inventory and debtors, returns far less in comparison.
- Fourthly, developments on the information technology (IT) front have provided a potential tool to reduce the inventory.

Materials management is integrated functioning of the different sections of a company dealing with the supply of materials and other related activities, so as to obtain maximum coordination and optimum expenditure on materials. The scope of materials management is vast as it begins with 'award of contract' and ends with material resting at its point of use. Typically, the objectives of materials management are to:

- minimize material cost;
- procure and provide material of desired quality when required;
- reduce investment tied in inventories for use in other productive purposes and to develop high inventory turnover ratios;
- purchase, receive, transport and store material efficiently and to reduce the related costs;
- cut down costs through simplification, standardization, value analysis, import substitution, etc.;
- modify paper-work procedure in order to minimize delays in procuring material; and
- train personnel in the field of materials management in order to increase operational efficiency.

In summary, materials management aims to provide the right item of the right quality, in the right quantity, at the right price, at the right place, and at the right time for ensuring uninterrupted execution of works.

Empirical evidence strongly suggests that materials management activities are either not being practiced or being practiced ineffectively in the construction industry. 'Not being able to get material' and 'not being able to get the right tool' are commonly heard complaints on the site. In fact, research indicates that as much as one-third of the worker's time is spent waiting for the right materials or tool to perform a construction task. 'Unavailability of materials and tools' may act as a major demotivator to workers' performance on a typical construction site. The result is reduced productivity and delay in project. It is the responsibility of the construction manager to make sure that the right material and the right tool are available to the worker.

Muehlhausen (1987) points out some of the reasons for poor materials management practices adopted by construction contractors:

- Senior management of construction firms do not recognize the impact that materials management has on the cost-effectiveness of their project operations.
- Personnel performing related materials management activities have not been properly selected and trained.
- Computerized systems related to materials management activities have not been properly selected, designed, or used to provide needed information for management and control.

The problem of low worker productivity and unreasonable construction project cost due to unavailable materials at the time and place of need will continue until construction managers have learnt how to manage the flow of materials to and through the construction site.

It can be appreciated that one of the areas of great concern today is the effective utilization of our planet's limited resources. This has led to a renewed understanding of the need for new and efficient construction materials, and the re-use and recycling of materials from existing facilities.

Reconstruction and replacement of an ageing infrastructure will place increased emphasis on the development of new and/or improved materials for use in pavements, bridges, water and electrical distribution facilities, buildings and other structures. Management skills and analytical tools are needed to effectively utilize the multitude of resources required for the completion of a construction project.

In this chapter, various issues related to materials management have been highlighted. These include material procurement process in construction organization, different functions of materials management, management of inventory, inventory-related costs, functions of inventory, classification of inventory systems, selective inventory control, and inventory models. It is envisaged that by following these principles, one may be able to manage the materials function effectively within the construction industry.

11.2 MATERIAL PROCUREMENT PROCESS IN CONSTRUCTION ORGANIZATION

In the context of a large construction organization, having countrywide domestic as well as overseas operations, the set-up for performing materials management function can be quite big. For an organization that has regional offices throughout the country and a head office in a major metropolitan of the country, the materials management department is normally headed by a person of vice president cadre and he or she is supported by a general manager. The general manager is overall in charge of central materials department and regional materials department. The responsibilities of central materials department are:

- to procure high-value materials centrally for different regions;
- to arrange for their transport to respective regions;
- to exercise inventory control;
- to enter into rate contract for frequently used construction materials such as cement, steel and plywood at organization level;
- to develop computerized procedures;
- to procure capital equipment and to arrange for their maintenance;

- to supply raw materials to company's manufacturing units, if any, and to cater to material needs of overseas projects;
- to import construction materials depending on the requirement of projects;
- to gather management information system (MIS) reports and their analyses;
- to standardize and codify the construction materials; and
- to dispose of the waste and scrap materials.

The regional materials department arranges for different construction materials for different projects falling in their jurisdiction. The department arranges for the transport of materials to different projects, manages inventory at regional level, enters into rate contract for the projects falling in the region, assists projects to develop computerized materials management procedures, and arranges for the maintenance of the capital equipment being utilized in the region.

A number of people are involved in the materials procurement process. For example, in a large construction organization, the project manager is by and large responsible for materials management. The planning engineer and the materials engineer are responsible for preparing the materials schedule, sending the requisition for the materials, monitoring and control of materials consumption at projects, comparing the tender provision and actual cost of materials, and so on. For selection of sources, especially for procurement of materials of natural origin such as aggregate and stones, the quality control engineer is responsible. The quality engineer is responsible for sampling and testing of materials received at project sites as also for checking their conformance with the specification. The stores in-charge is responsible for follow-up with the vendors, receiving and issuing the right materials, and so on.

After the contract is awarded to a construction organization, the planning engineer in association with the materials engineer prepare a materials schedule, and lists out the important materials and the departments responsible for their procurement. In a large organization having countrywide and overseas operations, materials are purchased at head-office level, regional- or branch-office level and project-site levels. The materials engineer in consultation with the planning engineer fills up the requisition for desired materials giving all the details such as quantity required, schedule of procurement, specification of materials, drawings, list of approved manufacturers/suppliers, and so on. The requisition in most of the cases needs approval from higher authorities at the project site or regional office, and is forwarded to the procurement or materials department, which may be located at regional office, head office, or project site itself, as the case may be.

At the regional- or head-office level, it is ascertained whether the requested material pertains to (a) rate contract items, (b) standard price-list items, or (c) other items. In case of rate contract items, purchase order is released as per the existing rate contract, and for standard price-list items the purchase order is placed with agreed discount. For other items, suitable vendors are identified, enquiries are sent with all the details received from project sites, and price offers are received from suppliers. The price offers are compared and evaluated; a structured negotiation takes place with a few suppliers; and the purchase order is issued to the most eligible supplier. The materials supplied by the supplier is dispatched to the project sites, which inspect, receive and dispatch the material receipt note to the concerned office from where the materials have been procured. Payment is released to the vendor after the materials receipt note has been received from the project sites.

11.3 MATERIALS MANAGEMENT FUNCTIONS

The functions involved in materials management are described below.

11.3.1 Materials Planning

The site-planning engineer is responsible for this function. Normally, purchase of any material is initiated by requisitioning for it. In order to have proper control on material purchase, it is a good idea to have some specific engineer approving the material requisition. The site-planning engineer makes a purchase requisition in consultation with the construction manager, following which the material is procured by administrative people (mainly stores people). Materials schedule should be the basis for raising material requisitions. Materials planning involves identifying materials; estimating quantities; defining specifications; forecasting requirements; and locating right sources for procurement.

11.3.2 Procurement

Before procurement, the site-planning engineer along with the materials engineer survey the local market and identify the items that are to be procured locally and those that are to be procured from the head office under centralized procurement system.

Local procurement should be kept to the minimum and, as far as possible, limited to non-engineered items and consumables. The advantages and disadvantages of centralized and local purchasing are given in Table 11.1.

For items that are required from head office (centralized purchasing), an authorized purchase request is sent to head materials manager, who will coordinate with the site materials engineer and organize for the items. Head materials manager and site materials engineer ensure that procurement is as per the terms and conditions of the contract. Early and late procurement of materials can be the two approaches to material procurement. The advantages and disadvantages of these approaches are given in Table 11.2 and Table 11.3.

11.3.3 Custody (Receiving, Warehousing and Issuing)

The main documents used are—inward register, material receipt note (MRN), delivery challan (DC), indent, dispatch covering note, outward register, loan register, repair register, and plant

Table 11.1 Advantages of centralized and local purchasing (George Stukhart 1995)

Centralized	Local
• Low unit price due to large volume of purchase. For some items there can be an agreement on rate for the country as a whole, valid for a particular period during which the rates are fixed irrespective of general increase observed in the same item.	• Faster response to material requirement for the project.
• Smoother purchasing action due to well-laid-out process.	• Better project control and regulated expense towards purchasing expenses. In centralized procurement, the indirect cost towards purchasing cost may be debited in wrong proportion to a particular project.
• Specialized person having better market knowledge involved in the process.	
• Dealing with regular suppliers and, hence, better negotiated price.	

Table 11.2 Advantages and disadvantages of early procurement

Advantages	Disadvantages
• Materials availability is assured and, hence, work will not suffer. Proper quality and price of materials can be assured as there is time to look around and shop.	• Materials may be stolen or may deteriorate during storage. Materials will require space for storage, which may also be needed for other uses. They need to be guarded and accounted for. Money locked up in the purchase of materials does not get interest and other works may suffer due to this.

and machinery movement register. The functions of warehousing cover receipt, inspection, storage, issue/dispatch, materials accounting, valuation and insurance.

11.3.4 Materials Accounting

The main purpose of materials accounting is monitoring inflow and consumption of raw materials. Material accounting involves materials stock accounting; materials issues and returns accounting; monthly stocktaking of selected materials; and materials wastage analysis.

Material wastage analyses aims at finding the causes of wastages and rectifying them. Wastage during procurement can result from various factors—buying materials of wrong specifications; buying more than the actual requirements; unnecessary buying of items to cater for unrealistic and unforeseen eventualities; untimely buying of short-life materials; improper and unnecessary handling of materials; and wastage while transportation. Some other reasons of wastage of materials are breakages and damages during handling; lack of pre-work preparation and coordination; inferior quality of materials; improper accounting and poor storekeeping; negligent and careless attitude of the supervisor; unforeseen circumstances like accidents, fire, etc.; high rate of deterioration due to long storage at the place of work; over-issues from the central stores and failures to return unused surplus materials; and thefts and pilferage.

11.3.5 Transportation

Construction materials used at project sites undergo considerable movement right from their point of origin to storage point, to the actual point of consumption. In some construction companies, materials are first received in central stores, from where they are dispatched to the project stores located at project sites and then finally to the workplace. The construction materials could be in the form of raw materials such as aggregate, sand and plywood, or they may be in semi-finished form such as mixed mortar, mixed concrete and dressed stones. In general,

Table 11.3 Advantages and disadvantages of late procurement

Advantages	Disadvantages
• All the disadvantages of early procurement are avoided.	• All the advantages of early procurement are lost. If guaranteed delivery dates are available, it may be satisfactory to arrange for deliveries to be made a week in advance of the starting date of the work.

raw materials procurement is within the scope of materials management team, while the procurement of semi-finished material comes under the purview of construction team. Proper care should be taken while planning for transportation of raw and semi-finished materials so as to avoid any adverse effect on the characteristics or performance of materials.

11.3.6 Inventory Monitoring and Control

Inventory may be defined as 'usable but idle resource'. If resource is some physical and tangible object such as materials, then it is generally termed as stock. Thus, stock and inventory are synonymous terms, though inventory has wider implications. It is very important to have a check on the inventory level. Scientific inventory management is an extremely important problem area in the materials management function. Materials account for more than half the total cost of any business, and organizations maintain a huge amount of stocks, though much of this could be reduced by following scientific principles. Inventory management is highly amenable to control. Across industries, there is a substantial potential for cost reduction through inventory control. Inventory being a symptom of poor performance, one can reduce inventories through proper design of procurement policies by reduction in the uncertainty of lead times by variety reduction and in many other ways.

The financial implications of inventory as an element of the company's assets must be understood properly. The following reasons could be used to justify inventory:

- Improves customer service
- Permits purchase and transportation economies, the argument here being based on the notion that both product procurement and transportation costs will be reduced if lot sizes are large
- Hedges against price changes: One can observe the tendency to hoard commodities in anticipation of price rise just before the budget (in the months of January/February in the Indian context)
- Protects against demand and lead-time uncertainties: When there are uncertainties in the customer demand patterns and the suppliers' replenishment lead times, it is preferable to invest in 'safety stock' so that services are maintained at acceptable levels to customers
- Hedges against contingencies: While there are now fewer labour disputes in most economies, fires, floods and other exogenous variables can create problems. It is argued that these will be minimized if stockholding is increased

There are a number of inventory control mechanisms such as ABC analysis (based on value of consumption), FSN analysis (based on movement from store), and VED analysis (based on necessity). Considering the importance of inventory monitoring and control, we discuss these in detail later in the text.

11.3.7 Materials Codification

Codification is an important function in materials management, especially for construction companies where thousands of different items are used all along the project duration. There are different systems of codification followed by construction organizations—namely, numeric, alphanumeric, and colour codification. As the term suggests, in numeric codification numbers are used, while in alphanumeric system both letters and numbers are used. Colour

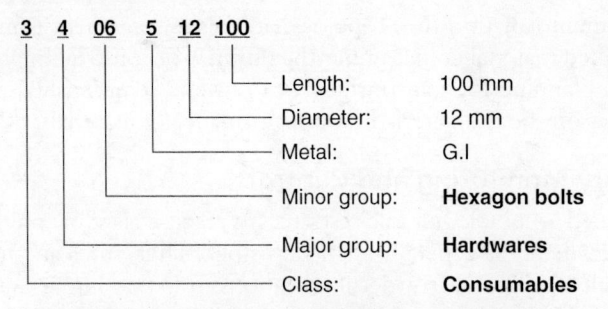

```
3 4 06 5 12 100
              │  │  │
              │  │  └── Length:        100 mm
              │  └───── Diameter:      12 mm
              └──────── Metal:         G.I
           └─────────── Minor group:   **Hexagon bolts**
        └────────────── Major group:   **Hardwares**
    └────────────────── Class:         **Consumables**
```

The description for material code '3406512100' therefore is:
'**GI make hexagon bolt of 12 mm diameter and 100 mm length**'

Figure 11.1 Illustration of material codification

codification uses different colour schemes to codify different items. An example of numeric codification could be—3 4 06 5 12 100 (see Figure 11.1), where the first digit represents the class of material (for example, 0 may mean capital items; 1, heavy tools and tackles; 2, small tools; 3, consumables; 4, electrical items; 5, construction materials; 6, special/one-time items; 7, spares; 8, reserved for future use; and 9, miscellaneous). The second digit may represent a major group of materials; the third set of digits may represent a minor group; and so on.

An example of alphanumeric codification used normally for equipment and spare parts could be 9 C B 6 2 0007. Here, the first digit could represent class (for example, spares); the second space, equipment category (for example, concreting equipment); the third space, equipment name (for example, batching machine); the fourth space, the make of equipment (for example, Schwing Stettor); the fifth space, the equipment model (for example, 56 m^3); and the sixth space, the serial-number coding for parts.

A proper system of material codification serves the following purposes:

- proper identification of items by all departments concerned
- avoiding use of long description of items
- avoiding duplicate stocks under different descriptions
- material accounting and control
- ensuring receipt and issue documents are posted in appropriate records
- helps in mechanization of records

11.3.8 Computerization

Computers are being used increasingly in the application of construction materials management. They find wide application in almost all the functions of materials management, including:

- forecasting the prices of materials based on past data and analysing past trends
- planning of different materials—using the construction schedule, one can work out the materials schedule quite easily using computers. This aspect was also discussed in Chapter 7. The quantity of different materials required for a project can also be worked out using computers

- developing the specification of materials—one can refer to past specifications stored in the computer and make suitable adjustments depending on the requirement of the project
- purchasing of materials—it is now possible to float enquiries for different materials online as well as invite bids from different suppliers online
- preparing a comparative statement and finalizing the supplier
- inventory control—computers can play an important role in decision-making as far as economic order quantity is concerned. An inventory analysis using different methods illustrated elsewhere can be easily performed using computers

11.3.9 Source Development (Vendor Development)

Source development is a continuous activity. As a policy, for every major item, more than one source should be identified. The vendor's performance should be evaluated regularly and only satisfactory vendors should be encouraged.

Managing vendors or suppliers is an important issue. It is not an easy task to manage a large number of vendors. Thus, it is better to have only a manageable number of vendors. It is also necessary to develop a long-term relationship with the vendors. It has become imperative and strategically important to manage vendors by considering them as partners in the business. Initiatives such as vendor-managed inventory (VMI) can play an important role in containing and managing the costs. To trace new sources of supply and to develop cordial relations with them in order to ensure continuous, reliable and quality material supply at reasonable rates, has become an important item in the agenda of professional materials managers.

11.3.10 Disposal

Every year, old and used items that are not economical to use have to be disposed off in a planned way. For this, the quantity and quality of materials to be disposed off should be assessed. The reusable items from the scrap can be retained for further use. The remaining scrap is disposed off either by selling it to some scrap dealer or through the process of floating enquiry, collecting quotation, and awarding to the highest bidder. Normally, while disposing off scraps, payment is collected in advance from the scrap buyer.

11.4 INVENTORY MANAGEMENT

It can be recalled that inventory is usable but idle resource. The problem of inventory management is of maintaining an adequate supply of something to meet an expected demand pattern for a given financial investment. This could be raw materials, work in progress, finished products, or spares and other indirect materials. Inventory is one of the indicators of management effectiveness on the materials management front. Inventory turnover ratio (annual demand/average inventory) is an index of business performance. A soundly managed organization will have higher inventory turnover ratio, and vice versa. Inventory management deals with the determination of optimal policies and procedures for procurement of commodities. Since it is quite difficult to imagine a real work situation in which the required material will be made available at the point of use instantaneously, maintaining inventories becomes almost necessary. Thus, inventories could be described as 'necessary evil'.

11.4.1 Inventory-Related Cost

Inventory related cost has following broad cost components.

1. Cost of carrying inventories (holding cost)
2. Cost of incurring shortages (stock-out cost)
3. Cost of replenishing inventories (ordering cost)

The three types of costs are the most commonly incorporated costs in inventory analysis, though there may be other cost parameters relevant in such an analysis including inflation and price discounts. These are explained in the following sections.

Cost of Carrying Inventories (Holding Cost)

This is expressed as ₹/item held in stock/unit time. This is the opportunity cost of blocking material in the non-productive form as inventories. Some of the cost elements that comprise carrying cost are cost of blocking capital (interest rate); cost of insurances; storage cost; and cost due to obsolescence, pilferage, deterioration, etc. It is generally expressed as a fraction of value of the goods stocked per year. For example, if the fraction of carrying charge is 20 per cent per year and a material worth ₹ 1,000 is kept in inventory for one year, the unit carrying cost will be ₹ 200/year. It is obvious that for items that are perishable in nature, the attributed carrying cost will be higher.

Cost of Incurring Shortages (Shortage Cost)

It is the opportunity cost of not having an item in stock when one is demanded. It may be due to lost sales or backlogging. In the case of backlogging (or back ordering), the order is not lost but is backlogged, to be cleared as soon as the item is available on the stock. In lost sales, the order is lost. Both cases are characterised by demand, penalty cost, emergency replenishment, loss of goodwill, etc. This is generally expressed as ₹/order.

Cost of Replenishing Inventories (Ordering or Set-up Cost)

This is the amount of money and efforts expended in procurement or acquisition of stock. It is generally called ordering cost. This cost is usually assumed to be independent of the quantity ordered, because the fixed-cost component is generally more significant than the variable component. Thus, it is expressed as ₹/order.

11.4.2 Functions of Inventory

As mentioned earlier, inventory is a necessary evil. It is necessary because it aims at absorbing the uncertainties of demand and supply by 'decoupling' the demand and supply sub-systems. Thus, an organization may be carrying inventory for the following reasons:

- Demand and lead-time uncertainties necessitate building of safety stock (buffer stock) so as to enable various sub-systems to operate somewhat in a decoupled manner. It is obvious that the larger the uncertainty of demand and supply, the larger will have to be the amount of buffer stock to be carried for a prescribed service level.
- Time lag in deliveries also necessitates building of inventories. If the replenishment lead times are positive, then stocks are needed for system operation.

- Cycle stocks may be maintained to get the economics of scale so that total system cost due to ordering, carrying inventory and backlogging is minimized. Technological requirements of batch processing also build up cycle stocks.
- Stocks may build up as pipeline inventory or work-in-progress inventory due to finiteness of production and transportation rates. This includes materials actually being worked on or moving between work centres, or materials in transit to distribution centres and customers.
- When the demand is seasonal, it may become economical to build inventory during periods of low demand to ease the strain of peak period demand.
- Inventory may also be built up for other reasons such as quantity discounts being offered by suppliers, discount sales, anticipated increase in material price, and possibility of future non-availability.

Different functional managers of an organization may view inventory from different points, leadings to conflicting objectives. This calls for an integrated systems approach to planning of inventories so that conflicting objectives can be scrutinized to enable the system to operate at minimum total inventory-related costs—both explicit costs such as purchase price as well as implicit costs such as carrying, shortage, transportation and inspection costs.

11.4.3 Inventory Policies

In this section we discuss (a) Lot size reorder point policy, (b) Fixed order interval scheduling policy, (c) Optional replenishment policy, (d) Two-bin system. In practice, there may be other policies that may be special cases of the policies mentioned or may be a combination of these policies. Some of the factors affecting the choice of an inventory policy are—the nature of the problem; the usage value of an item; and situational parameters. It is desirable to first select an operating policy before determining optimal values of its parameters.

Lot Size Reorder Point Policy

The inventory status is continuously reviewed and as soon as the inventory level falls to a prescribed value called 'reorder point', a fresh replenishment order of fixed quantity known as EOQ is placed. This is one of the very classical types of inventory policies. The decision variables for the design of policy are Lot size and reorder point.

Figure 11.2 shows the typical stock balance under this type of inventory policy. The solid line in this figure represents the actual inventory held with a finite lead time, while the broken line shows the ideal situation if no lead time existed.

Fixed Order Interval Scheduling Policy

The time between the consecutive replenishment orders is constant. A maximum stock level (S) is prescribed and the inventory status is reviewed periodically with a fixed interval (T). At each review an order of size O is placed, which takes the stock on hand plus an order equal to the maximum stock level. The order quantity could vary from period to period. In this policy, when the level of stock on hand is high at review, a smaller-size replenishment order is placed.

The decision variables for the design of policy are S, the maximum stock level, and T, the review period

Figure 11.3 shows the typical stock balances under the fixed reorder cycle policy.

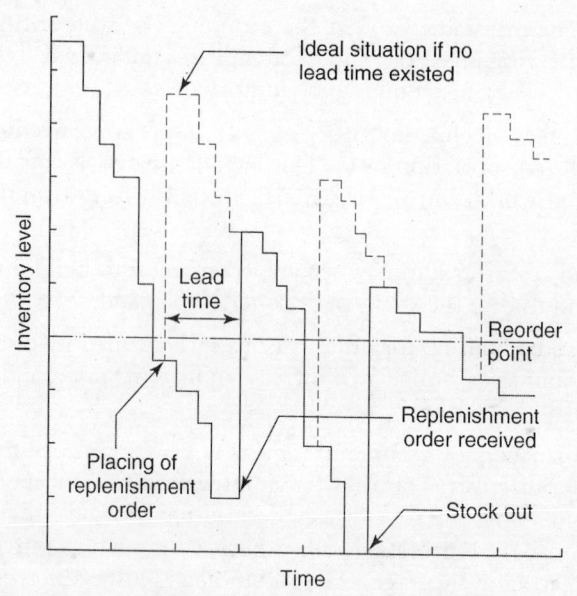

Figure 11.2 Typical stock balance under Lot Size Reorder Point Policy

Optional Replenishment Policy

This policy is also known as (s, S) policy. The status of stock is periodically reviewed and maximum stock level (S) and minimum stock level (s) are prescribed. At the time of review, if the stock on hand is less than or equal to s, an order of size Q is placed so that stock on hand plus the

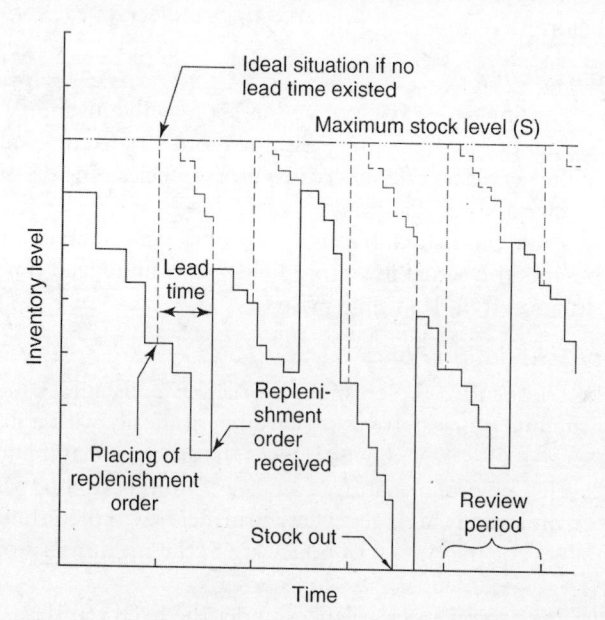

Figure 11.3 Typical stock balance under Fixed Order Interval Scheduling Policy

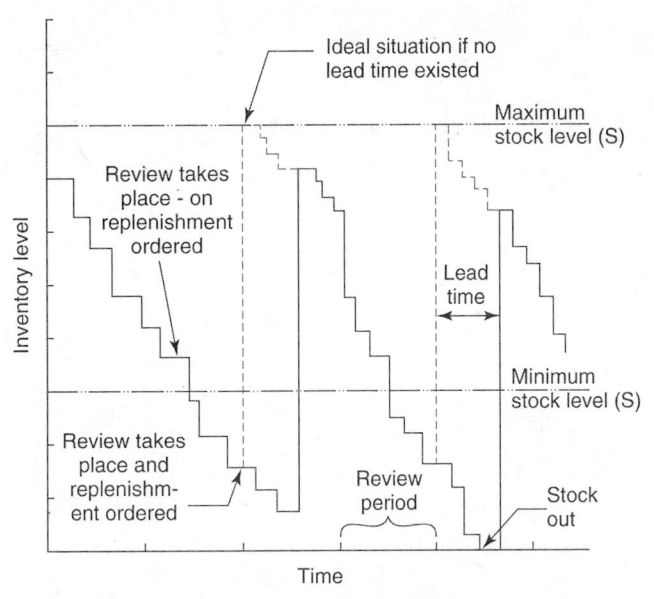

Figure 11.4 Typical stock balance under (s,S) policy

order equals the maximum stock level S. If stock on hand at review is higher than s, no order is placed and the situation is reviewed at the next review period.

The decision variables for the design of policy are S, s and T.

Figure 11.4 shows the typical stock balance under this policy.

Two-bin System

This system is simple to operate and easy to understand. There are two bins kept full of items. Item from the first bin are used first. The moment the first bin is exhausted, an order is placed for items and the second bin acts as buffer or safety cushion. Figure 11.5 shows all the stages of this system in a schematic manner which is self explanatory.

11.4.4 Selective Inventory Control

A list of items used in any typical construction can be found in Appendix 10. The appendix lists out the materials related to a typical construction project, in addition to commonly used small tools and equipment. These tools are used by craftsmen and skilled labour such as mason, carpenter, steel fitter, painter, welder, plumber and electrician. In addition, the appendix lists out general stores materials and administration and safety materials.

As is evident, there is a large variety of items stocked by a project site and applying scientific inventory control for all these items is neither feasible nor desirable. Since applying inventory control across all items may render the cost of inventory control more than its benefits, it may prove to be counter-productive.

Inventory control has to be exercised selectively. Depending upon the value criticality and usage frequency of an item, we may have to decide on an appropriate type of inventory

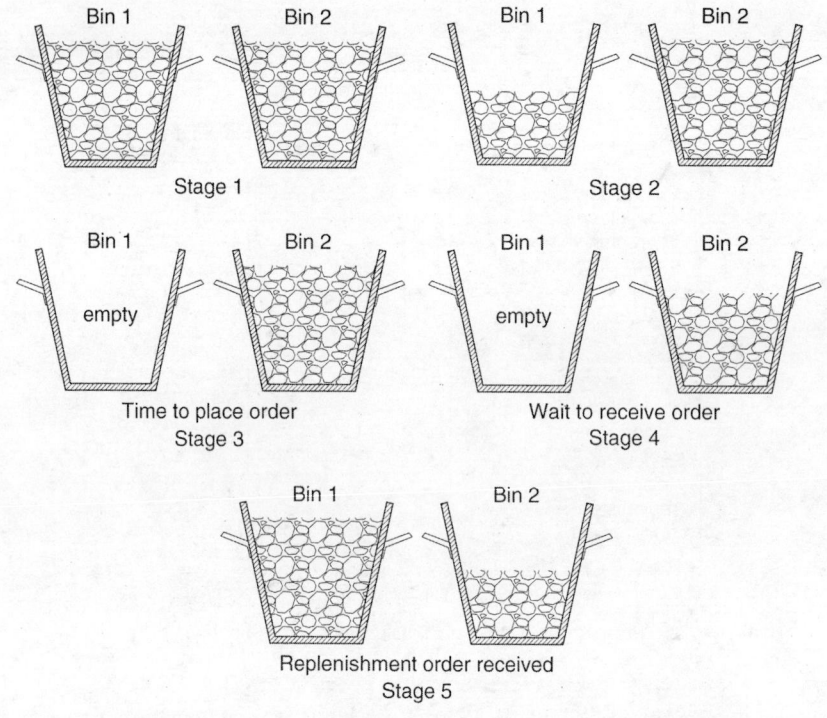

Figure 11.5 Schematic representation of different stages in Two bin system

policy. Selective inventory control, thus, plays a crucial role so that we can apply our limited control efforts more judiciously to the more significant group of items. In selective control, items are grouped in a few discrete categories depending upon value, criticality and usage frequency. Table 11.4 shows some of the ways to make such groupings. This type of grouping may well form the starting point for scientific inventory management in an organization.

ABC Analysis

This is based on Pareto's Law, which says that in any large group there are 'significant few' and 'insignificant many'. For example, only 20 per cent of the items may be accounting for 80 per cent of the total material cost procured by a construction organization. Here, the 20 per cent constitute the 'significant few' that require utmost attention.

To prepare an ABC-type curve, we may follow a simple procedure:

1. Different materials required for the project are identified and their estimated quantities worked out. The quantity estimate could be on the basis of either annual consumption or the project's total requirement.
2. The unit rates of materials are estimated.
3. The usage values for each of the materials are obtained by multiplying the estimated quantities and their unit rates. These values are converted into percentage of total annual usage cost or total project cost, as the case may be.

Table 11.4 Some commonly adopted inventory-control policies

Name of the policy	Expansion	Basis of classification	Remarks
VED	Vital, essential and desirable	It is based on the criticality of the item, which are classified in three categories.	Classification depends on the consequences of material stock-out when demanded.
FSN	Fast, slow and normal	It is based on consumption rate of the inventory.	It is helpful in controlling obsolescence.
HML	High, medium and low	It is based on unit price of material.	It is mainly used to control the inventory of purchased material.
XYZ	Value of balance stocks very high	It is used for classifying materials in storage.	Its main use is in review of inventory.
SDE	Scarce, difficult and easy to obtain	It is based on the level of difficulty in the procurement of inventory.	It is useful in lead-time analysis and decision related to the procurement of purchasing strategies.
GOLF	Government, ordinary, local and foreign	It is based on the inventory.	It is useful for decision related to the procurement strategy.
HML	High, medium and low price	It is based on the prices of materials.	It is useful for delegating the purchasing responsibilities.

4. The percentage usage cost for each of the materials is arranged in the descending order of their ranking, starting with the first rank, i.e., highest to lowest usage value. The cumulative percentage usage value is also calculated.

5. A curve as shown in Figure 11.6 is plotted, and points on the curve at which there are perceptible sudden changes of slopes are identified. In the absence of such sharp points, cut-off points corresponding to the top 10 per cent and the next 20 per cent or so are marked as a general indicator of A, B and C type of materials.

6. According to an empirical approach, 'A' class items account for about 70 per cent of the usage value, 'B' class items for about 20 per cent of the usage value, and 'C' class items for about 10 per cent of the usage value. In terms of numbers, 'A' class items constitute about 10 per cent of total items, 'B' class items about 20 per cent of total items, and 'C' class items about 70 per cent of total items. These percentages are indicative only and can vary depending on a number of factors.

Upon classification of materials into A, B and C types, suitable inventory policies can be decided. Corresponding to each type of materials, the implications on inventory policy are mentioned below:

Item type 'A' The salient features are:
- accurate forecast of quantities needed
- involvement of senior level for purchasing
- ordering is on requirement basis

Figure 11.6 Illustration of ABC analysis

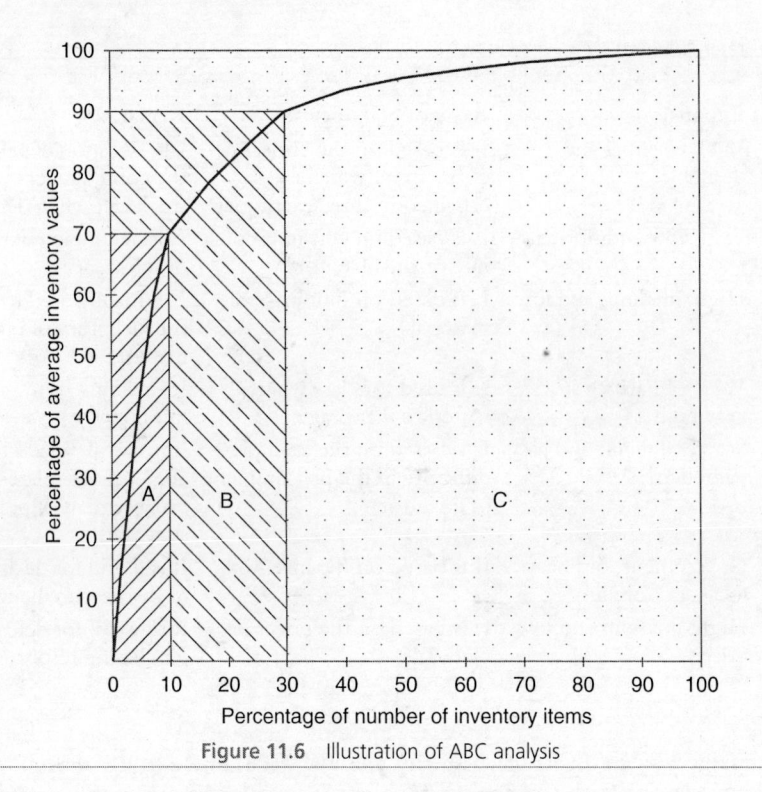

- enquiries for procurement need to be sent to a large number of suppliers
- strict degree of control is required, preferably monitoring on a weekly basis
- low safety stock is needed

Item type 'B' The salient features are:
- approximate forecast of quantities needed
- requires involvement of middle level for purchasing
- ordering is on EOQ basis
- enquiries for procurement need to be sent to three to five reliable suppliers
- moderate degree of control required, preferably monitoring on a monthly basis
- moderate safety stock needed

Item type 'C' The salient features are:
- no need of forecasting; even rough quantity estimate is sufficient
- junior-level staff is authorized to order purchase
- bulk ordering is preferred
- quotations from even two to three reliable suppliers are sufficient

- a relatively relaxed degree of control is sufficient, and monitoring can be done on a quarterly basis
- adequate safety stock can be maintained

VED Analysis

This analysis attempts to classify items into three categories depending on the consequences of material stock-out when demanded. As stated earlier, the cost of shortage may vary according to the seriousness of such a situation.

Thus, the items are classified into V (vital), E (essential) and D (desirable) categories. Vital items are the most critical having extremely high opportunity cost of shortage and must be available in stock when demanded. Essential items are quite critical with substantial cost associated with shortage and should be available in stock by and large. Desirable group of items do not have very serious consequences if not available when demanded, but these can be stocked items.

Obviously, the percentage risk of shortage with the vital group of items has to be kept quite small, thus calling for a high level of service. With 'essential' category we can take a relatively higher risk of shortage, and for 'desirable' category, even higher. Since even a C-class item may be vital or an A-class item may be desirable, we should carry out a two-way classification of items grouping them in nine distinct groups as A-V, A-E, A-D, B-V, B-E, B-D, C-V, C-E and C-D. We can then determine the aimed service level for each of these nine categories and plan for inventories accordingly.

Vital group comprises those items for the want of which the production will come to a stop—for example, power in the factory. Essential group features those items for whose non-availability the stock-out cost is very high. Desirable group contains items whose non-availability causes no immediate loss of production; the stock cost involved is very less and their absence may only cause minor disruption in the production for a short time.

The steps used for classifying materials as vital, essential and desirable are given below:

Step 1 Factors such as stock-out case, lead time, nature of items, and sources of supply are identified and considered for VED analysis.

Step 2 Assign points or weightages to the factors according to the importance they have to the company, as shown above.

Step 3 Divide each factor into three degrees and allocate points to each degree.

Step 4 Prepare categorization plan to provide basis for classification of items—for example, items scoring between 100 and 160 can be classified under desirable items; items between 161 and 230 can be classified under essential items; and items between 231 and 300 can be classified under vital items.

Step 5 Specify the degree and allocate weightages to all the factors.

Step 6 Evaluate and find the final score for every item, and specify the type of item.

FSN Analysis

Not all items are required with the same frequency. Some materials are required quite regularly, some are required very occasionally, and yet some others may have become obsolete and might not have been demanded for years together. FSN analysis groups them as fast-moving, slow-moving and non-moving (dead stock), respectively. Inventory policies and models for the three categories have to be different. Most inventory models in literature are valid for the fast-moving items exhibiting a regular movement (consumption) pattern. Many spare parts come under the slow-moving category, which have to be managed on a different basis. For non-moving dead stock, we have to determine optimal stock disposal rules rather than inventory provisioning rules. Categorization of materials into these three types on value and critical usage enables us to adopt the right type of inventory policy to suit a particular situation.

'F' items are those items that are fast-moving—i.e., in a given period of time, say, a month or a year, they have been issued a number of times. However, 'fast-moving' does not necessarily mean that these items are consumed in large quantities.

'S' items are those items that are slow-moving—in the sense that in the given period of time they have been issued in a very limited number.

'N' or non-moving items are those that are not at all issued for a considerable period of time.

Thus, the stores department, which is concerned with the moving of items, would prefer to classify items in the categories F-S-N, so that they can manage, operate and plan stores activity accordingly. For example, for efficient operations it would be necessary that fast-moving items are stored as near as possible to the point of issue, for these to be issued with minimum of handling. Also, such items must be stored at the floor level, avoiding higher heights. Thus, if the items are slow-moving or issued once in a while in a given period of time, they can be stored in the interior of the stores and even at greater heights because handling of these items becomes rare. Further, it is necessary for the stores in-charge to know about non-moving items for various reasons mentioned below:

1. Non-moving items mean unnecessary blockage of money which affect the rate of returns of the company.

2. Non-moving items also occupy valuable space in the stores without any usefulness and, therefore, it becomes necessary to identify these items and find reasons for their non-moving status. If justified, recommendation may be made to top management for their speedy disposal so that company operations are performed efficiently.

To some extent, inventory control can be exercised on the basis of FSN analysis. For example, fast-moving items can be controlled more severely, particularly when their value is also high. Similarly, slow-moving items may not be controlled and reviewed very frequently since their consumption may not be frequent and their value may not be high.

11.4.5 Inventory Models

There are a number of computer-based analytical inventory models available (such as economic order quantity [EOQ] model), most of which are able to generate economic purchase orders, shipping orders, delivery notes and invoices. Most models claim to improve management control by reducing inventory-holding costs without loss of customer service. The basic philosophy behind these models is to use a trade-off analysis by comparing the cost of inventory holding versus the cost of ordering. We are discussing one of the most popular inventory models in the following section.

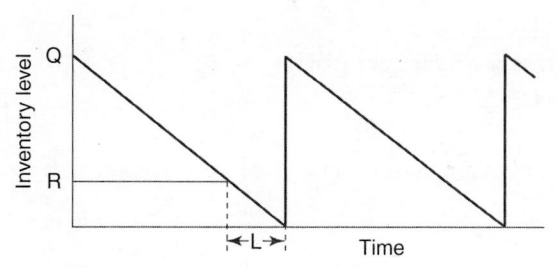

Figure 11.7 Inventory behaviour under EOQ model

Economic Order Quantity (EOQ) Model

The EOQ model provides answers on how much to order. Figure 11.7 shows the behaviour of EOQ model. The reorder point R and the quantity to be ordered, Q, are shown in the figure, as is the lead time L. The ordered quantity derived from this model is known as economic order quantity, EOQ.

It is usually less expensive to purchase (and transport) or produce a bunch of material at once than to order it in small quantities. If orders for large quantities are specified, there will be fewer orders placed. For purchasing, this means that quantity discounts and transportation efficiencies may be realized. The other side of the coin, however, is that larger lot sizes result in more inventory, and inventory is expensive to hold. EOQ model attempts to specify a balance between these opposing costs. This aspect is shown graphically in Figure 11.8, where it is clear that there is a decrease in cost associated with increase in order quantity, while there is increase in cost with increase of inventory.

The total cost is given by the sum of inventory-carrying cost and ordering cost.

$$\text{Total cost } TC = \text{Ordering cost} + \text{Carrying cost} \tag{11.1}$$

The following notations are used to develop the EOQ model:

D = Demand rate; unit/year
A = Ordering cost; ₹/order

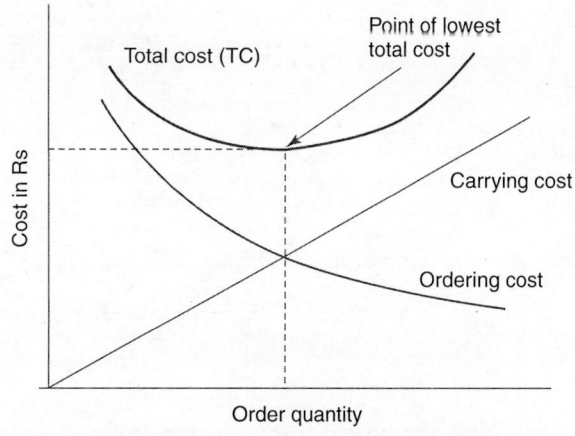

Figure 11.8 Total cost curve—EOQ model

C = Unit cost; ₹/unit of item
I = Inventory-carrying charges per year
H = Annual cost of carrying inventory/unit item
Q = Order quantity; number of units per lot

It is assumed that demand is at a uniform rate. Thus, the average inventory required would be $\frac{(0+Q)}{2} = \frac{Q}{2}$ throughout the year.

The total number of orders placed would be $\frac{D}{Q}$ per year.

Order cost per year = Number of orders placed per year × Cost per order

$$\text{Ordering cost per year} = \frac{A \times D}{Q} \quad (11.2)$$

$$\text{Carrying cost per year} = \frac{\text{Order quantity} \times \text{Unit cost of item} \times \text{Annual cost to carry}}{2}$$

$$\text{Carrying cost per year} = \frac{C \times I \times Q}{2} = \frac{H \times Q}{2} \quad (11.3)$$

where

$$H = C \times I \quad (11.4)$$

Using the notations mentioned above, we can write the expression of TC as:

$$TC = \frac{A \times D}{Q} + \frac{H \times Q}{2} \quad (11.5)$$

For optimum Q, one needs to find the particular value of Q which will minimize total cost. This can be done by differentiation, and one gets:

$$EOQ = \sqrt{\frac{2 \times \text{Order cost} \times \text{Demand}}{\text{Inventory carrying cost}}} \quad (11.6)$$

$$EOQ = \sqrt{\frac{2 \times A \times D}{I \times C}} \quad (11.7)$$

Some of the observations that are clear from the above expressions are:
a. The more the demand per year, the larger the order quantity.
b. The higher the order cost, the larger the order quantity.
c. The more expensive the item, the smaller the order quantity.
d. The higher the carrying cost, the smaller is the order quantity.

The derivation of EOQ is based on a number of assumptions such as:
- Demand is deterministic and continuous at a constant rate.
- The process continues infinitely.
- No constraints are imposed on quantities ordered, storage capacity, budget, etc.
- Replenishment is instantaneous (the entire order quantity is received all at one time as soon as the order is released).
- All costs are time-invariant.
- There are no shortages of items.
- The quantity discounts are not available.
- There is negligible or deterministic lead time.

The above assumptions mean that there is no uncertainty and one is able to predict the demand and the constancy of cost parameters such as item cost, carrying rate, or ordering cost. In real life, however, all these assumptions may not hold good.

In case the lead time is varying, one has to keep safety stock or buffer stock. Normally, the service level that is expected will decide the safety stock. If one keeps a very small quantity as safety stock, there is a danger that stock-out may occur. On the other hand, large safety or buffer stock may result in large inventory-carrying cost. Thus, safety stock will be decided based on the service level desired. The following relations are important and should be noted.

a. *Reorder Point = Demand or usage per period × Lead time* (11.8)

b. *Average Inventory Carried* $= \dfrac{Order\ Quantity}{2} + Safety\ Stock$ (11.9)

c. *Transit Inventory = Days in transit × Inventory carried per day* (11.10)

d. The higher the lead time, the more will be the safety stock, and vice versa. In general, the safety stock varies with the square root of lead time (assuming all other factors as constant). For example, if the lead time for an item is reduced by a factor of 4, then the safety stock will be reduced by a factor of 2 (i.e., ($\sqrt{4}$)).

Effect of Uncertainty in Demand

Generally, demand is never uniform all throughout the year. In case, the demand has a mean D_m and standard deviation σ_d, the reorder point is expressed as given below:

$$Reorder\ point = D_m \times Lead\ Time + Z \times \sigma_d \times \sqrt{Lead\ Time} \qquad (11.11)$$

Where, Z is standard normal variate for a given service level. In order to find the value of Z, one can use normal distribution table provided in Appendix 5 of the text. For ready reference, values of Z for some commonly used service levels are given in the Table 11.5. The service level is the probability of having material in stock when demand of this material occurs in a construction project.

Table 11.5 Values of Z for different service levels

Service level	90%	92%	94%	95%	96%	98%	99%
Z	1.29	1.41	1.56	1.65	1.75	2.05	2.33

SOLVED EXAMPLES

Example 11.1

A shop dealing in construction goods has seven different items in its inventory. The average number of units of each of these items held in the store along with their unit costs is given in Table Q11.1.1. The shopkeeper has decided to employ ABC inventory system. Classify the items in A, B and C categories.

Table Q11.1.1 Data for Example 11.1

Item	Average number of units	Average cost per unit in inventory (in ₹)
1	10,000	121.50
2	10,000	100.00
3	24,000	14.50
4	16,000	19.75
5	60,000	3.10
6	50,000	2.45
7	30,000	0.50

Solution

From Table S11.1.1, it is clear that total number of units stored in the shop is 200,000 and the cost is equal to ₹ 32.025 lakh. The total cost of inventory is worked out by summing up the cost of storing the given numbers of all the seven items. The percentage share of each item and cost are calculated as shown in Table S11.1.1. From these two values, cumulative percentage values are worked out.

Table S11.1.1 Computation details for Example 11.1

Item	Units	% of total	Cumulative % of total item	Unit cost (₹)	Total cost (₹ lakh)	% of total cost	Cum. % of total cost
1	10,000	5.00	5	121.50	12.15	37.9	37.9
2	10,000	5.00	10	100.00	10.00	31.2	69.2
3	24,000	12.00	22	14.50	3.48	10.9	80.0
4	16,000	8.00	30	19.75	3.16	9.9	89.9
5	60,000	30.00	60	3.10	1.86	5.8	95.7
6	50,000	25.00	85	2.45	1.225	3.8	99.5
7	30,000	15.00	100	0.50	0.15	0.5	100.0
Total	200,000	100.00			32.025	100.00	

A graph as in Figure S11.1.1 is drawn between cumulative percentage of cost and cumulative percentage of numbers. From the graph, it is clear that about 70 per cent cost is consumed in 10 per cent of the inventory items. These are for items 1 and 2. Thus, items 1 and 2 are Class

CONSTRUCTION MATERIAL MANAGEMENT | 449 |

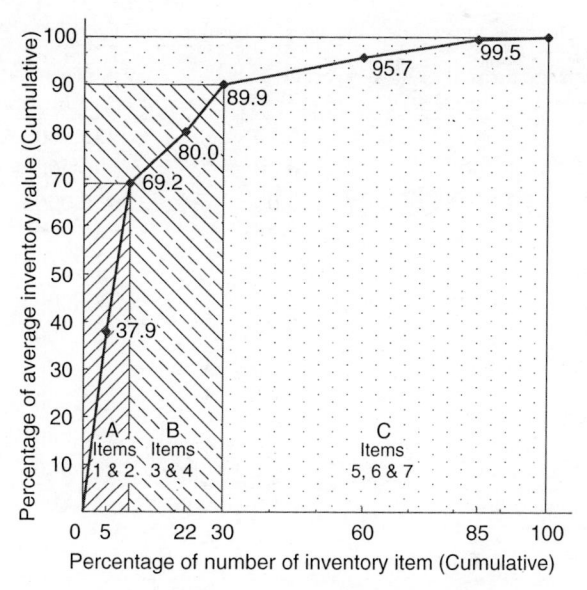

Figure S11.1.1 Cumulative percentage of cost and cumulative percentage of numbers for Example 11.1

'A' items. Similarly, items 3 and 4 have a cost share of about 20 per cent (89.9 per cent − 69.2 per cent) and an inventory share of 20 per cent. Thus, items 3 and 4 are Class 'B' items. Finally, we can see that items 5, 6 and 7 belong to Class 'C' as their cost share is about 10 per cent (100 per cent − 89.9 per cent) and inventory share is 70 per cent.

Example 11.2

A construction company stores various items (see Table Q11.2.1) in the central stores. The average annual consumption and cost per unit of items stored are given. Classify the items using ABC analysis.

Table Q11.2.1 Data for Example 11.2

Name of the item	Average annual consumption (No.)	Average cost per unit (₹)
a	5,000	45.00
b	1,000	90.00
c	2,000	225.00
d	4,000	11.25
e	50	300.00
f	6,000	62.50
g	2,000	67.50
h	4,000	18.75
i	50	375.00
j	250	105.00
k	200	187.50
l	50	150.00

Table S11.2.1 Computation details for average annual cost of consumption (Example 11.2)

Name of the item	Average annual consumption (No.)	Average cost per unit (₹)	Average annual cost of consumption (₹)	Ranking
a	5,000	45.00	225,000	3
b	1,000	90.00	90,000	5
c	2,000	225.00	450,000	1
d	4,000	11.25	45,000	7
e	50	300.00	15,000	11
f	6,000	62.50	375,000	2
g	2,000	67.50	135,000	4
h	4,000	18.75	75,000	6
i	50	375.00	22,500	10
j	250	105.00	22,500	9
k	200	187.50	37,500	8
l	50	150.00	7,500	12

Solution

The computations of average annual cost consumption of different items have been performed in Table S11.2.1. The ranking based on the average annual cost of consumption has also been performed.

Now, we prepare Table S11.2.2 in the descending order of average annual cost of consumption. The average annual cost in percentage of total annual cost is computed. The cumulative value of this percentage is also computed.

A graph as in Figure S11.2.1 is drawn. From the graph, it is clear that about 70 per cent cost is consumed by items c, f & a. Thus, items c, f & a are Class 'A' items. Similarly, items g, b & h have a cost share of about 20 per cent. Thus, items g, b & h are Class 'B' items. Finally, we can see that items d, k, j, i, e & l belong to Class 'C' as their cost share is about 10 per cent.

Table S11.2.2 Computation details for A,B,C classification (Example 11.2)

Rank in descending order of cost	Name of the item	Average annual cost (₹)	Annual cost (percentage)	Cumulative cost (%)	Category
1	c	450,000	30	30	A
2	f	375,000	25	55	A
3	a	225,000	15	70	A
4	g	135,000	9	79	B
5	b	90,000	6	85	B
6	h	75,000	5	90	B
7	d	45,000	3	93	C
8	k	37,500	2.5	95.5	C
9	j	26,250	1.75	97.25	C
10	i	18,750	1.25	98.5	C
11	e	15,000	1	99.5	C
12	l	7,500	0.5	100	C

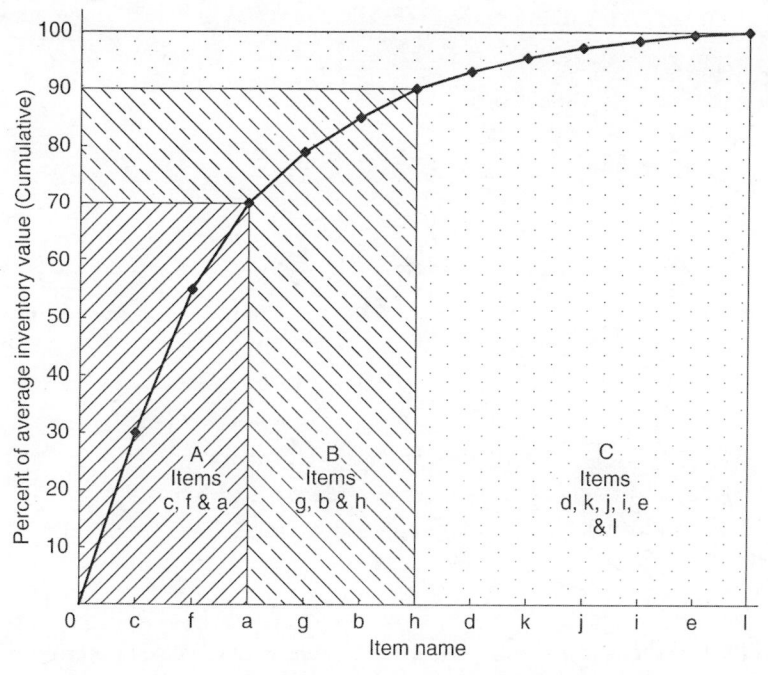

Figure S11.2.1 Illustration of A, B, C Classification for example 11.2

Example 11.3
A construction material trading company receives a total of 200 t as annual demand for steel reinforcement. The annual cost of carrying per-unit t of reinforcement is ₹ 2,000, and the cost to place an order is ₹ 25,000. What is the economic order quantity?

Solution
The economic order quantity is computed using the expression

$$EOQ = \sqrt{\frac{2 \times \text{Order cost} \times \text{Demand}}{\text{Inventory carrying cost}}}$$

$$EOQ = \sqrt{\frac{2 \times 25,000 \times 200}{2000}}$$

Thus, economic order quantity = 70.7 t per order.

Example 11.4
A construction company purchases 10,000 bags of cement annually. Each bag of cement costs ₹ 200 and the cost incurred in procuring each lot is ₹ 100. The cost of carrying is 25 per cent. What is the most economic order quantity? What is the average inventory level?

Solution

Given, Unit item cost (C) = ₹ 200
Ordering cost = ₹ 100 per order
Annual usage = 10,000 units
Carrying rate = 25%
To find, EOQ

$$EOQ = \sqrt{\frac{2 \times A \times D}{I \times C}}$$
$$= \sqrt{\frac{2 \times 100 \times 10{,}000}{0.25 \times 200}}$$
$$= \sqrt{(40{,}000)}$$
$$= 200 \text{ units}$$

Hence, the order quantity will be 200 units and on an average, 200/2 = 100 units in inventory will be carried.

Example 11.5

For Example 11.4, if the lead time of procuring cement is two weeks, determine the reorder point.

Solution

The lead time given is two weeks. So, the weekly usage is 10,000 units/52 weeks = 192 units. The reorder point will be = 192 units × 2 weeks = 384 units. That is, place an order for cement whenever the stock reaches the level of 384 units. The order quantity will be 200 units.

REFERENCES

1. Chitkara, K.K., 2006, *Construction Project Management: Planning, Scheduling and Controlling*, 10th reprint, New Delhi: Tata McGraw-Hill.

2. Gopalkrishnan, P. and Sundaresan, P., 1996, *Materials Management: An Integrated Approach*, New Delhi: Prentice-Hall of India.

3. Mohanty, R.P. and Deshmukh, S.G., 2001, *Essentials of supply chain management*, New Delhi: Phoenix Publishers.

4. Muehlhausen, F.B., 1987, 'Materials management model for construction management curriculum development', ASC Proceedings of the 23rd Annual Conference, Purdue University, West Lafayette, Indiana, USA.

5. Naddor, E., 1986, *Inventory Systems*, New York: Wiley.

6. Stukhart, G., 1995, *Construction Material Management*, Marcel Dekker.

REVIEW QUESTIONS

1. State whether True or False:
   a. Construction materials account for a large fraction of overall construction project cost.
   b. Materials are classified broadly into capital items and revenue items.
   c. Functions involved in material management are—materials planning, procurement, custody (receiving, warehousing and issuing), materials accounting, transportation, inventory monitoring and control, and materials codification.
   d. Costs related to inventory are holding cost, stock-out cost and ordering cost.
   e. Commonly used inventory-control policies are ABC analysis, VED analysis and FSN analysis.
   f. EOQ inventory model aims to provide an answer to how much to order—i.e., it fixes the trade-off between ordering cost and carrying cost.
2. Discuss the importance of inventory management in material management.
3. What are the different functions of material management?
4. What are the advantages and disadvantages of centralized and local purchasing?
5. What are the advantages and disadvantages of early and late procurement?
6. What are the benefits of proper classification of materials?
7. Discuss the role of vendor management in material management.
8. What are the inventory-related costs? What are the functions of inventories?
9. How do we classify different inventory systems?
10. Discuss different inventory-control policies.
11. Differentiate between ABC analysis, VED analysis and FSN analysis.
12. Why are inventory models needed? Discuss EOQ model.
13. Collect consumption data for 100 different items for an organization and classify these into an ABC framework. List these items in a two-way classification, ABC and VED, and identify the number of items belonging to each of these nine distinct groups.

12

Project Cost and Value Management

Project cost management, collection of cost-related information, cost codes, cost statement, value management in construction, steps in the application of value engineering, description of the case, value-engineering application in the case project

12.1 PROJECT COST MANAGEMENT

A project consists of a number of activities. Each of these activities consumes resources. Resources cost money. We can take certain steps through which we can control the costs of these activities. Project cost management is all about controlling cost of the resources needed to complete project activities. Apart from these controllable costs, there are certain aspects over which we do not have any control. These are called uncontrollable costs and they are the subject matter of risk management, taken up separately elsewhere in the book. The subject of project cost management can be taken up in four broad steps, as explained below.

12.1.1 Resources Planning Schedules

This aspect was discussed in the previous chapters. The objective here is to prepare different resources schedule such as labour and staff schedule, material schedule, plant and equipment schedule, and subcontractor or specialist's schedule. As mentioned earlier, these schedules show the quantity requirement of each of these resources either on a weekly basis or on a monthly basis. The basis for preparing these schedules is the project time schedule.

12.1.2 Cost Planning

Once the resource requirement is obtained, the estimate to complete each of these can be prepared based on the unit cost of the resources and the total units of the resources required. This process is called cost planning and it is a must for project cost management. The essential components of a cost plan are the project schedule and estimates. We have already discussed the project scheduling aspect in Chapter 7. The estimation aspect was covered in Chapters 4 and 8. It can be recalled that the estimate of a project is progressively developed as the project gets underway. To start with, a rough estimate based on previous projects of similar size and nature is prepared at the conceptual stage. This is gradually refined and finally a detailed estimate is prepared when the scope gets clearer and a detailed design is ready.

Cost planning aims at ascertaining cost before many of the decisions are made related to the design of a facility. It provides a statement of the main issues, identifies the various courses of action, determines the cost implications of each course, and provides a comprehensive economic picture of the project. The planner and the quantity surveyor should be continuously questioning whether it is giving value for money or whether there is any better way of performing the particular function.

At the detailed design stage, final decisions are made on all matters relating to design, specification, construction and others. Every part of the facility must be comprehensively designed and its cost checked. Where the estimated cost exceeds the cost target, either the element must be redesigned or other cost targets reduced to make more money available for the element in question—through all this, the overall project cost limit must remain unaltered.

The planner and the estimator should be continuously examining the cost aspects throughout the design process, and keep asking certain typical questions—'Is a particular feature, material, or component really giving value for money?' 'Is there a better alternative?' 'Is a certain item of expenditure really necessary?' Cost planning establishes needs, sets out the various solutions and their cost implications, and finally produces the probable cost of the project while maintaining a sensible balance between cost on one hand and quality, utility and appearance on the other.

12.1.3 Cost Budgeting

Cost budgeting is the process of allocating the overall cost estimate to individual work items of the project. Work items are groups of similar activities taken from bill of quantities. It is not necessary to go into each item of bill of quantities since that would require too much of the planning engineer's efforts without commensurate results. In practice, similar activities such as excavation in open might be clubbed with excavation up to depth of 1.5 m, and excavation from 1.5 m to 3.0 m depth under one common head, excavation work. Similarly, concreting of different grades given in the bill of quantities can be grouped under one common head 'concrete'. Although the rates quoted by the contractor may be different for different grades of concrete, it will be a huge task to allocate the cost to individual grades of concrete since that would require creating large numbers of cost codes. The same result can be achieved if the work items are created based on some weighted average techniques.

12.1.4 Cost Control

The objective of cost control is to ensure that the final cost of the project does not exceed the budgeted or planned cost. Project cost control can be seen as a three-step process:

1. Observe the cost expended for an item, an activity, or a group of activities.
2. Compare it with available standards. The standard could be a predefined accepted cost estimate (ACE) or it could be the tender estimate.
3. Compute the variance between the observed and the standard, communicating any warning sign immediately to the concerned people so that timely corrective measures can be taken.

The initial stages of the project such as conceptual and design stages offer the maximum possibility for influencing the final project cost. Thus, regular and close monitoring is needed during these stages of the project. The details of cost control aspects are discussed in Chapter 16.

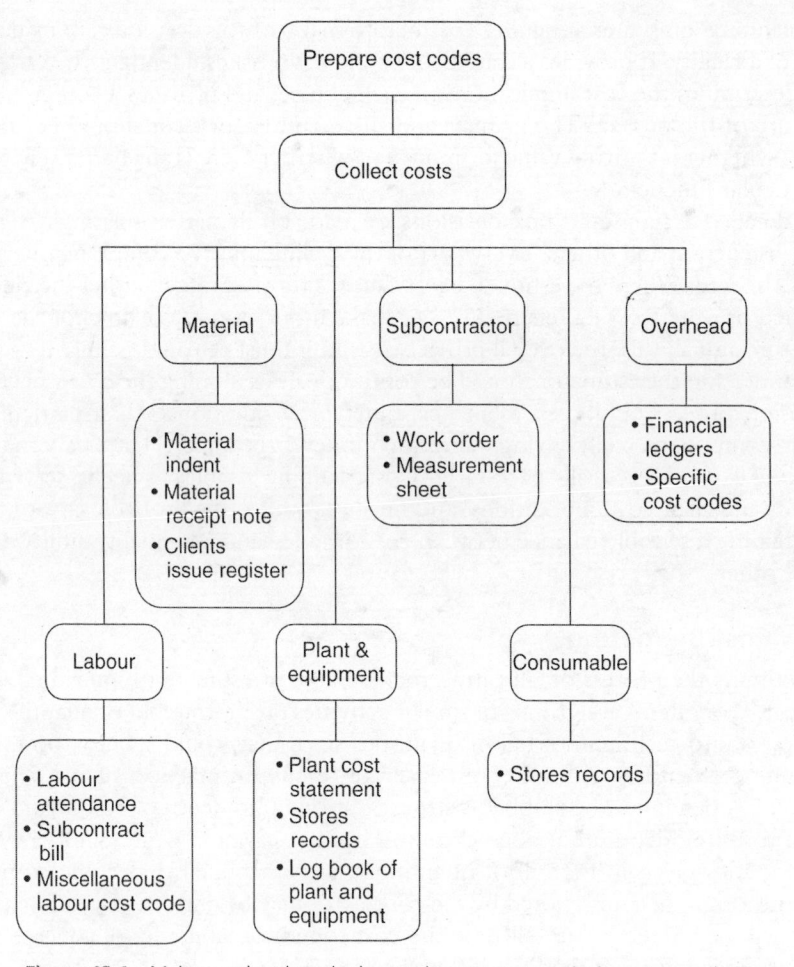

Figure 12.1 Major cost heads and relevant documents to assist in cost compilation

12.2 COLLECTION OF COST-RELATED INFORMATION

In this section, we discuss the collection of actual cost figures or data and the different reports that are prepared to collect such data. The cost of an item or an activity is collected essentially under the following broad heads (refer to Figure 12.1):

1. Labour costs, which include departmental labour and subcontract labour
2. Material cost
3. Plant and equipment cost
4. Subcontractor cost
5. Consumables cost
6. Overhead cost

Project name			Time office in-charge		
Name of the worker	Category	Cost code	In-time	Out-time	Signature

Figure 12.2 A typical format for recording daily labour attendance

In some cases, there may be further split-up of the above cost heads in terms of direct expense and indirect expense. For example, the total labour cost would be the sum of direct labour expenses and indirect labour expenses.

There are certain standard reports maintained at sites to record the above-mentioned costs, as illustrated in Figure 12.1.

12.2.1 Labour Cost

As mentioned earlier, in a project there can be two types of labour—one that is directly employed by the main contractor and the other employed by the subcontractor of the main contractor. While the former is referred to as departmental labour, the latter is called subcontractor labour. There are different costing processes adopted for these two categories of labour.

While the source document for collection of departmental labour cost is the daily labour attendance and allocation form (see Figure 12.2), the source data for knowing the subcontract labour cost is the subcontract bill. In addition, there is a certain component of miscellaneous labour that cannot be associated with any activity cost codes; they are termed as miscellaneous labour and find their place in overhead cost.

Normally, two accounts are maintained, one each for departmental labour and subcontractor labour. Daily labour attendance is maintained mentioning the relevant cost code.

It is possible that a worker may be used for more than two works having different cost codes. In such a situation, it is advisable to keep the worker's expense under only one cost code for that particular day. While preparing the labour attendance sheet, it is a good practice to check the cost codes against each such worker. The labour cost for each cost code is determined by multiplying the number of days in each cost code with the average man-day rate for that month. Average man-day rate is worked out from the total gross wages divided by the total man-days worked in that particular month. To keep the calculation simple, although the overtime amount paid to workers is taken in the gross wages, overtime hours are not considered in the denominator while calculating the number of days.

The cost code mentioned in the wage sheet is also summarized in labour cost allocation on a monthly basis to arrive at the total number of man-days worked for each cost code. It is a good practice to check whether the total of cost code-wise man-days and the total man-days worked are matching.

Notice pay, retrenchment compensation and employer's contribution to provident fund for labour are part of indirect labour costs, and these go under the project's indirect costs.

12.2.2 Material Cost

Materials used at construction project sites are basically of two types:

1. Client's supply: This is issued by the client for the execution of project. The owners may issue it on free-issue basis or on chargeable basis. While the former will not have any effect on cost statement, for the second case it is imperative that total quantity of the supplied material consumed is allocated cost code-wise and pricing done at agreed rates. Necessary provisions for wastage and inventory-carrying cost (like cost of storage sheds, handling charges, etc.) should be made for both categories of materials.

2. Own purchase: The materials that a construction company purchases could be of the following types:

   (i) *Bulk or basic materials*

   Items such as aggregates and sand can be termed as bulk or basic materials. For costing purpose, the total quantity consumed, including wastage, needs to be allocated cost code and priced on periodic weighted average rate (PWAR) basis.

   (ii) *Heavy tools*

   For such items, indents are prepared each month for write-off as suitable percent of cost on written-down rate basis to cover wear-and-tear and allocated to relevant cost codes. If any item is rendered unserviceable, the entire residual value is written off.

   (iii) *Scaffolding/staging materials*

   For these items, indents are to be prepared for rentals at suitable rates of value of materials cost per month and allocated to relevant cost codes.

   (iv) *Temporary structures/installation*

   Some common temporary structures at a typical site would be contractor's office, client's and consultant's office, workshops, labour colonies, etc. For costing purposes, proportionate cost to invoicing done till each month-end is written off. Necessary credit for possible/realistic salvage value is also sometimes considered.

   (v) *Small tools*

   Small tools are charged fully at the time of first issue. In other words, the value of the small tools is written off at the time of issue by stores and charged to relevant cost code.

The source document for collection of material cost is the indent, which is shown in Figure 12.3.

In a large organization, material can be procured either locally at project site or from the central depot of the organization through centralized procurement. The moment any material is received at a project site, site stores personnel prepare the material receipt note, commonly referred to as MRN, and the total cost of the material is booked in this project by the central procurement department. The cost allocation at site is done by looking at the cost code mentioned in the indent received from project engineers.

Suppose a project engineer needs a particular material. He will raise the material indent in which details such as item description, brief specification, unit of measurement, quantity, rates and cost code of the work head for which this material is needed are required to be filled up. The project engineer fills up the indent form leaving aside the rate part, which is usually filled up by the stores personnel. Care should be taken to fill the cost code correctly. Wrong entry of

Project name

S. No.	Description	UOM	Quantity	Cost code	Rate	Amount	Remarks

Requisition by
Authorized by
Storekeeper
Planning engineer

Figure 12.3 A typical indent format for requisition of material used in projects

cost codes can give an erroneous result while reconciling the material cost for a particular work head. This is the reason indent-raising authority is given only to a few people at site, and moreover, these entries are usually crosschecked by the planning engineer on a day-to-day basis.

Construction companies follow certain guidelines while compiling materials cost. No material cost is debited unless material receipt note is prepared at site. The value of goods purchased is based on purchase journal. The issue of material to the site is made on the basis of periodic weighted average rate. In case of transfer of materials from one site to another, debits are raised at site at PWAR based on acknowledged copy of delivery challan. For certain items of work (like concrete, formwork, or excavation) where there are subdivisions and for which collection of cost for each of the subdivisions may be difficult, the sub-items should be clubbed together under one single code.

Soon after the monthly bill is submitted, the cost statement, indicating the cumulative cost of each item under the relevant cost code booked till the date of the monthly bill, will be prepared. The cost statement will contain the split-up cost for each item of work (under corresponding code)—i.e., amount spent on labour (departmental and subcontract), materials (basic materials and consumables and spares), and overheads. Plant costs will be taken from plant cost statements (hire charges + operating costs) and included into work items as far as possible. Corresponding quantities of work done will be taken from the monthly bill to be included in JCR.

The costs in the cost statement may include amounts spent on portions of work in progress but not billed, as also amounts spent on certain enabling works like staging for formwork, precasting yard, partly fabricated structures, erection tools and tackles, which may be used for the balance work. It may, therefore, be necessary to remove from the cost all such amounts so as to assess the true cost of the quantities invoiced. The planning/billing engineer will assess the quantum of such costs that are to be removed from the total costs incurred till then. He may apportion such costs in proportion to the quantity billed or compute these through any other rational means. The amounts thus removed from the costs will be carried over as 'deferred expenses' and are to be included in 'estimates to complete'.

Similarly, care should be taken to include accrued expenses (provision for expenses) into the above cost. Accrued expenses are those expenses for which works were already carried out but payment not yet made. Thus,

$$\text{job status to date cost} = \text{actual cost incurred} - \text{deferred expenses} + \text{accrued expenses} \qquad (12.1)$$

Site accountant will keep track of all transactions that are likely to take place outside his site, concerning his project, and maintain a register of such items for preparation of the cost statement. In case of delay in receipt of debit advices for such items, he should estimate the cost based on reasonable assumptions. Since all transactions take place as a result of actions taken by the site or actions known to have been taken on behalf of the site, sufficient information would be available for the site accountant to compute such costs.

12.2.3 Plant and Equipment Cost

The components of plant and equipment cost include hire charges, labour, fuel and oil, spares, running and maintenance cost, one-time installation and erection cost, and dismantling cost. While some of the plants and equipment can be owned by the company directly, other plants may be rented from agencies. Even when some plants are owned by the company, they have a practice of debiting hire charges as if these were taken from outside agencies. This is done to know the exact usage cost of plant and equipment in a particular work cost head. Generally, for each of the plant and equipment, the companies maintain an asset code. Plant and equipment cost statement (contains hire charges + operating costs) is prepared for each of these asset codes being utilized at project site. From the log book of each of these plant and equipment assets, the cost is ploughed back to the relevant cost codes of activities for which these assets have been utilized. The operator costs are collected from the labour cost details discussed earlier. Expenses towards fuel, oil and spares are collected from the details available through stores department.

12.2.4 Subcontractor Cost

The source document for calculating subcontractor cost on a project is work order issued to the subcontractor by the main contractor and the periodic measurement bills. Relevant cost codes should be mentioned against each item of work in a subcontractor bill.

The subcontractor labour cost is recorded from relevant cost codes for which the particular subcontractor is engaged, through the subcontractor bill for the month. One can also mention the cost code for the subcontractor labour in the work order or the measurement sheet. Issues of materials to the subcontractor by the main contractor are done on a chargeable basis and are normally compiled under a separate cost code. Recovery from subcontractor's bill is made at predetermined intervals and is appropriately considered in the costing.

12.2.5 Overhead Cost

The overhead cost can be directly taken from the ledgers and grouped under fewer cost codes, averting costing of individual vouchers at sites. Once the basic elements of cost are collected for the period, some of the costs can be reallocated to work items and overheads, if necessary. Also, the general overheads need to be allocated to all projects being undertaken by the company in proportion to the contract value of the project. The expenses related to staff that can be directly identified to 'specific projects' shall be taken to the respective jobs. The common staff expenses and other administration expenses are allocated to all the projects being undertaken by the contracting organization in proportion to the invoicing of each of the projects.

It can be observed from the above discussion that cost collection for a large project is a big exercise that involves a number of people such as planning engineer, billing engineer, plant and equipment staff, storekeeper and accountants. Planning engineer is supposed to coordinate the process. It is expected that all the involved people do their part in time so that project cost is compiled at regular intervals and made use of in an effective manner.

12.3   COST CODES

These codes are designed based on the nature of the activity for which a particular cost is incurred. Cost codes are allocated by the planning department depending on the nature of activity at site. The cost codes should be such that these are easily compatible with the bill of quantity. The number of cost codes should be neither too large nor too small.

The basic point is that all the expenses incurred in and for the project should be recorded in one of the cost codes, no matter how small or how large the number of cost codes for a project. Some companies maintain additional cost codes for staging and shuttering, temporary structures, operating cost of plant and equipment, and installation cost of batching plants and quarry.

While on the subject of cost code, another point that merits attention is the comparison of cost with some standard such as accepted cost estimate. Harris and McCaffer (2005) suggest adoption of **coarse-grained system** that describes no more than 15 cost codes. They quote the study by Fine which finds the following interesting results.

If there are 30 cost heads in a project, about 2 per cent items are misallocated, while for a project with 200 cost heads about 50 per cent items are misallocated, and for a project with 2,000 cost heads, about 2 per cent of items are correctly allocated.

The activity codes are widely circulated to the authorized project staff so that correct allocation of costs to different cost heads is achieved. These allocations are further checked centrally by the project-planning engineer on a daily basis.

12.4   COST STATEMENT

Costing is a method of collecting expenses at the job site under various heads of accounts called cost codes. The cost codes are to be finalized at the beginning of the job and communicated to all section heads at the job site and all staff who are empowered to authorize the indents.

In preparing the cost statement, the following system has to be adopted:

1. Allot cost codes: This has already been discussed
2. Collect (a) labour costs, (b) subcontractor costs, (c) materials cost, (d) consumables, (e) plant costs, and (f) overheads cost. The method of collecting the costs and the documents through which these are collected have already been explained
3. Determine provision for expenses
4. Determine deferred expenses
5. Compile above costs in the given formats, finalize total costs code-wise, and arrive at project costs

The cost statement will reflect both the cost during the month and the cumulative costs for the job. The application of cost statement is in the preparation of project cost report, which has been discussed later. Typical formats of cost statements are given in Figure 12.4 to Figure 12.7.

The source of plant and equipment charges is the plant and equipment department. Time office and stores maintain the records of labour cost and material cost, respectively, while the voucher payment is used to know the other expenses concerning a particular plant and equipment.

12.5 VALUE MANAGEMENT IN CONSTRUCTION

The concept of value management (VM), also known as value analysis (VA) or value engineering (VE), evolved during World War II. It is a systematic approach for obtaining value for the money spent. VE is one of the most effective techniques known to identify and

All figures in ₹

Asset code	Description	Plant and equipment hire charges	Labour cost	Material cost	Other expenses	Total
xyz123	Batching plant	1,505,600	273,075	661,301	12,295	2,452,271
abc243	Jeep	27,000	29,943	74,787	5,751	137,481
abw195	Tipper	224,100	53,027	169,746	5,814	452,687
		1,756,700	356,045	905,834	23,860	3,042,439

Figure 12.4 Cost statement of plant and equipment for a given month

All figures in ₹

Cost code	Description	Labour cost	Subcontract cost	Material cost	Plant cost	Overheads	Total
	Bid items						
2010	Excavation	92,233	1,081,975	590,604	0	0	1,764,812
2100	Reinforcement	126,809	7,380,561	46,472,033	0	0	53,979,403
	Infrastructure						0
1501	Labour colony	123,861	1,191,575	761,994	0	0	2,077,430
1505	Approach road		29,907				29,907
	Overheads						
1020	Conveyance	52,193	1,062,573			513,511	1,628,277
1130	Taxes					1,480,085	1,480,085
		395,096	10,746,591	47,824,631	0	1,993,596	60,959,914
Source of charges:		Time office	Subcontract records	Stores		Voucher payment	

Figure 12.5 Cost statement—without plant cost allocation

PROJECT COST AND VALUE MANAGEMENT | 463 |

All figures in ₹

Cost code	Description	Labour cost	Subcontract cost	Material cost	Plant cost	Overheads	Total
	Bid items						
2010	Excavation	92,233	1,081,975	590,604			1,764,812
2100	Reinforcement	126,809	7,380,561	46,472,033	2,452,271		56,431,674
	Infrastructure						
1501	Labour colony	123,861	1,191,575	761,994	452,687		2,530,117
1505	Approach road		29,907				29,907
	Overheads						
1020	Conveyance	52,193	1,062,573		137,481	513,511	1,765,758
1130	Taxes					1,480,085	1,480,085
		395,096	10,746,591	47,824,631	3,042,439	1,993,596	64,002,353
	Source of charges:	Time office ↓	Subcontract records ↓	Stores ↓	Cost statement of plant and equipment ↓	Voucher payment ↓	

Figure 12.6 Integrated cost statement

All figures in ₹

Description	Labour cost	Subcontract cost	Material cost	Plant cost	Overheads	Total cost	Expenses provision	Deferred expenses	Net Cost
	A	B	C	D	E	F = A + B + C + D + E	G	H	I = F + G − H
Bid items	219,042	8,462,536	47,062,637	2,452,271	0	58,196,486	5,000,000	0	63,196,486
Infrastructure	123,861	1,221,482	761,994	452,687	0	2,560,024	0	1,280,012	1,280,012
Overheads	52,193	1,062,573	—	137,481	1,993,596	3,245,843	0	0	3,245,843
Total	395,096	10,746,591	47,824,631	3,042,439	1,993,596	64,002,353	5,000,000	1,280,012	67,722,341
Source of charges:	Time office ↓	Subcontract records ↓	Stores ↓	Cost statement of plant and equipment ↓	Voucher Payments ↓				

Figure 12.7 Cost statement summary

eliminate unnecessary costs in product design, testing, manufacturing, construction, operations and maintenance. VE involves answering the question—'What else will accomplish the function of a system, process, product, or component at a reduced cost?' (Dell'Isola 1982)

The core of value management lies in the analysis of function and is concerned with the elimination or modification of anything that adds cost to an item without adding to its function. In value engineering, 'function' is that which makes the product work or sell, and accordingly, we have 'work' functions and 'sell' functions. All functions can be divided into two levels of importance—basic and secondary. The basic function is the primary function of a product or a service, while the secondary functions are not directly accomplishing the primary purpose but play a supporting role and provide additional benefits.

Some other related terms used in value engineering are worth, cost and value. Worth refers to the least cost required to provide the functions that are required by the user of the finished project. Worth is established by comparison, such as comparing it with the cost of its functional equivalent. Cost is the total amount of money required to obtain and use the functions that have been specified. Value is the relationship of worth to cost as realized by the owner, based on his needs and resources in any given situation. The ratio of worth to cost is the principal measure of value.

$$Value\ index = \frac{Worth}{Cost} = \frac{Utility}{Cost} \qquad (12.2)$$

Thus, value may be increased by—(1) improving the utility with no change in cost, (2) retaining the same utility for less cost, and (3) combining improved utility with less cost. The situation in which worth and cost are equal represents 'fairness of deal', while the situation in which worth of a project is more than the cost paid represents a situation of 'good bargain'. The situation in which worth is less than the cost paid for a project represents 'poor value'. An optimum value is obtained when all utility criteria are met at the lowest overall cost.

Value engineering can be applied to any phase of a construction project, though the best results may be expected during the initial stage of a project. VE has its best chance for success in the integrated design–build organization. Beginning at the pre-design stages, when the potential for value engineering is the highest, the owner meets his entire integrated project team and provides inputs. As the integrated team begins work, multiple facets of value delivery are hypothesised, tested and enacted continuously. Team members across different functional lines consider the project in a holistic sense.

When applying value engineering, all expenditures relating to construction, maintenance, replacement, etc., are considered and through the use of creative techniques and the latest technical information regarding new materials and methods, alternate solutions are developed for the specific functions. It is claimed that application of value engineering can result in a saving of about 15 per cent–20 per cent of the construction costs.

Construction is one among many types of project-based production systems—shipbuilding, movie-making, software engineering, product development and all forms of work-order system. The aim is to provide the required functions to the user and ensure that anything else is eliminated, thereby reducing unnecessary costs. So, the moot question is—how are value engineering (VE) principles relevant to the construction industry?

VE principles are dovetailed with the construction type of temporary production system where production starts only after the user defines its needs specifically, as part of the exercise of eliminating wastages. As such, each activity of production is targeted towards fulfilling user requirements. Such is not the case with the manufacturing or product development industry, where in many cases a product's required function is decided by the producer and not by the user, whether it is the case of mobiles, laptops, automobiles, or a number of other products where the customer has to pay for many additional features which may not be user-specific and may even be undesirable to him. In the manufacturing industry, the scope for modification or elimination of the undesirable features of a product to a particular customer is limited, while the construction industry offers unlimited scope for alterations specific to the user.

It is an irony that construction projects, whether big, small, or of national importance, get delayed or often have long gestation in many countries. Also, there are large variations in the final cost of constructed facilities and the originally proposed cost of the projects. The delays and the cost overrun could be due to many reasons, but ultimately these lead to decline in the value to a great extent. There is, therefore, a need to move from the traditional method of construction to alternative approaches that can deliver fast-track constructions. The 'design and build' approach is an alternative. It may be noted that the last few years have witnessed a growing trend in the adoption of 'design and build' construction contracts.

In the 'design and build' approach, a particular entity forges a single contract with the owner to provide for architectural/engineering design services and construction. Here, the designer works directly under the project manager, unlike in the traditional design–bid–build projects where a designer works under the owner and there is no contact between the project manager and the designer. The 'design and build' approach has been found to be extremely beneficial when applied to fast-track projects in different countries. Fast-track construction is generally based on the norms for the industrial plant structures where the design and the construction work can go on simultaneously leading to reduction in the project-cycle time. Further, single-point responsibility eliminates the scope for blaming others for delay, cost overrun, value loss, etc. The integrated design–build team focuses not on getting money out of the project, but on putting value into the project for the duration of the facility's useful life.

In the subsequent sections, we discuss the applicability of value engineering at different stages of a design–build project through a real-world case study. For this, first we discuss the different steps in the application of value engineering.

12.6 STEPS IN THE APPLICATION OF VALUE ENGINEERING

As mentioned earlier, the different steps described here are for the application of VE in design-and-build projects. With few modifications, however, the steps can be applied to different types of construction projects. The schematic representations of these steps are given in Figure 12.8.

1. Within the scope of design-and-build industrial projects, different plant and non-plant structures are compared with respect to their cost. Normally, only a few structures would be responsible for a very high cost. Accordingly, only these structures merit attention and application of VE, subject to the constraints of time and resources.
2. Further, within a given structure, various elements such as foundation and superstructure construction, and material aspects are analysed cost-wise. Here also, only the elements with high cost implications are explored and analysed from the viewpoint of VE.

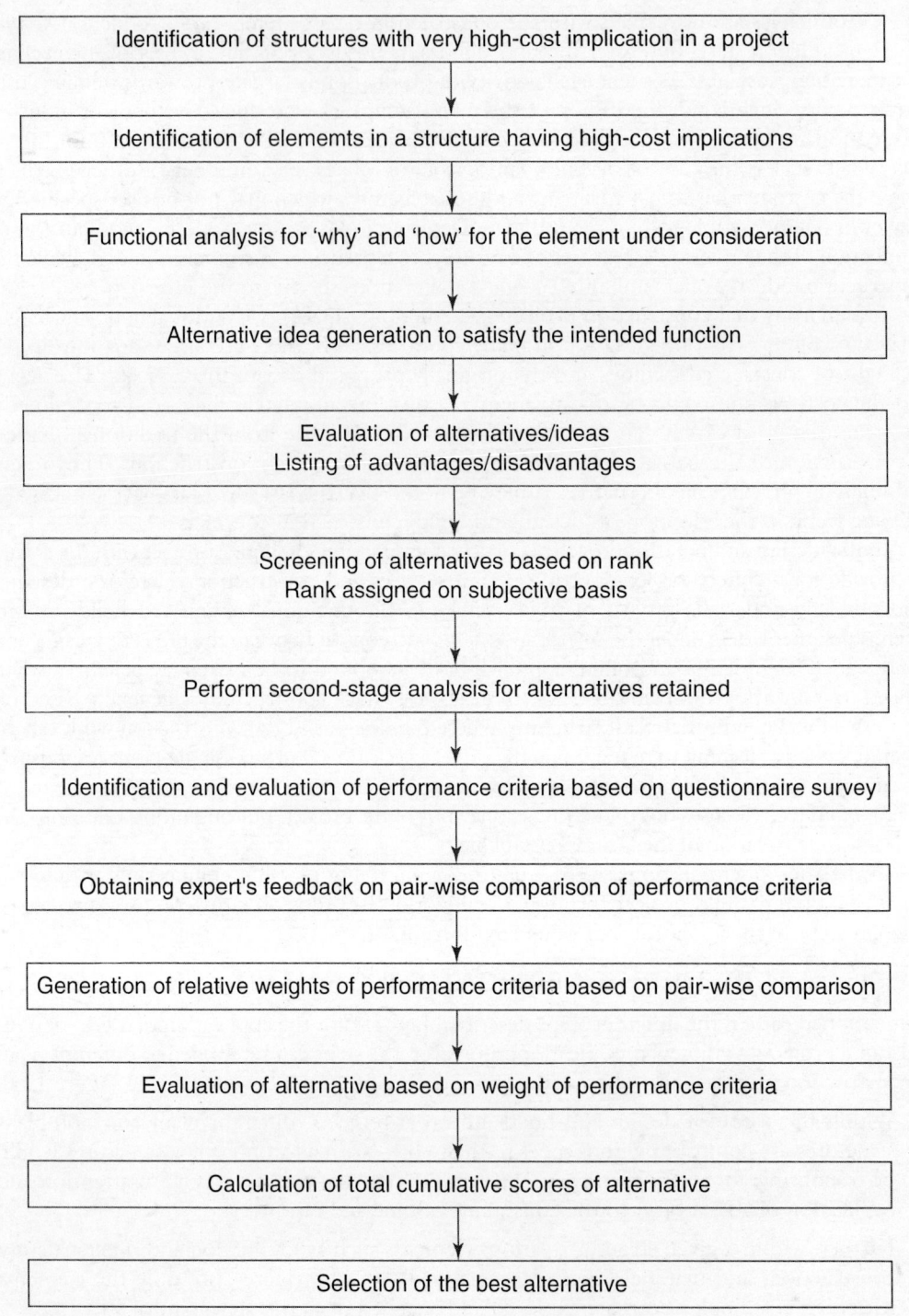

```
┌─────────────────────────────────────────────────────────────────────────┐
│  Identification of structures with very high-cost implication in a project │
└─────────────────────────────────────────────────────────────────────────┘
                                    ↓
┌─────────────────────────────────────────────────────────────────────────┐
│  Identification of elememts in a structure having high-cost implications │
└─────────────────────────────────────────────────────────────────────────┘
                                    ↓
┌─────────────────────────────────────────────────────────────────────────┐
│  Functional analysis for 'why' and 'how' for the element under consideration │
└─────────────────────────────────────────────────────────────────────────┘
                                    ↓
┌─────────────────────────────────────────────────────────────────────────┐
│  Alternative idea generation to satisfy the intended function           │
└─────────────────────────────────────────────────────────────────────────┘
                                    ↓
┌─────────────────────────────────────────────────────────────────────────┐
│  Evaluation of alternatives/ideas                                        │
│  Listing of advantages/disadvantages                                     │
└─────────────────────────────────────────────────────────────────────────┘
                                    ↓
┌─────────────────────────────────────────────────────────────────────────┐
│  Screening of alternatives based on rank                                 │
│  Rank assigned on subjective basis                                       │
└─────────────────────────────────────────────────────────────────────────┘
                                    ↓
┌─────────────────────────────────────────────────────────────────────────┐
│  Perform second-stage analysis for alternatives retained                 │
└─────────────────────────────────────────────────────────────────────────┘
                                    ↓
┌─────────────────────────────────────────────────────────────────────────┐
│  Identification and evaluation of performance criteria based on questionnaire survey │
└─────────────────────────────────────────────────────────────────────────┘
                                    ↓
┌─────────────────────────────────────────────────────────────────────────┐
│  Obtaining expert's feedback on pair-wise comparison of performance criteria │
└─────────────────────────────────────────────────────────────────────────┘
                                    ↓
┌─────────────────────────────────────────────────────────────────────────┐
│  Generation of relative weights of performance criteria based on pair-wise comparison │
└─────────────────────────────────────────────────────────────────────────┘
                                    ↓
┌─────────────────────────────────────────────────────────────────────────┐
│  Evaluation of alternative based on weight of performance criteria       │
└─────────────────────────────────────────────────────────────────────────┘
                                    ↓
┌─────────────────────────────────────────────────────────────────────────┐
│  Calculation of total cumulative scores of alternative                   │
└─────────────────────────────────────────────────────────────────────────┘
                                    ↓
┌─────────────────────────────────────────────────────────────────────────┐
│  Selection of the best alternative                                       │
└─────────────────────────────────────────────────────────────────────────┘
```

Figure 12.8 Schematic diagram depicting different steps in the application of VE in design-and-build project

3. The above step is followed by the construction of a FAST[1] diagram, which assists in breaking up a large problem and helps us to orderly identify the basic and secondary functions of the element under consideration. The 'why' and 'how' questions are closely linked to each other logically.

4. All the possible alternatives/ideas are then generated to cater to the functions as identified in the previous step, through a brainstorming session. Let us assume that for some element the alternatives generated are identified as x_1, x_2, x_3, x_4 and x_5.

5. The identified alternatives/ideas are then evaluated in two stages. The first stage comprises a rough screening process. This is done by enumerating advantages and disadvantages associated with each of the alternatives. These are ranked based on the subjective assessment of the evaluators. The evaluation is done not merely by counting the number of advantages or disadvantages, but also by taking into account the strength or importance of a particular advantage or disadvantage associated with a particular alternative. Some of the seemingly unattractive alternatives are dropped here itself and removed from the subsequent analysis. The ideas retained at this stage are taken to the next stage. For example, let us assume that out of the five alternatives x_1 to x_5 generated earlier, two alternatives x_1 and x_2 are rejected at this stage. In that case, the alternatives x_3, x_4 and x_5 only will be taken in the next step of analysis.

6. Now, in order to select the best alternative/idea from the remaining ones (x_3, x_4 and x_5), performance criteria are identified through literature review or interaction with experts. In this study, the identification and preference of each of the performance criteria was carried out using a questionnaire survey. For each of the performance criteria, respondents were asked to give a rating on a five-point scale in which '5' represented 'extremely important', '4' represented 'major important', and '3', '2' and '1' represented 'important', 'minor important' and 'slightly important', respectively. The performance criteria were then ranked based on the mean values of the responses. The criterion with the highest mean value was rated as the first rank and the criterion with the lowest mean value was rated as the last. Intermediate values were assigned depending on the mean values of the responses.

7. Evaluation matrix was then used for obtaining the weights of each performance criterion. Each criterion was assigned a letter of the alphabet and then compared with the other criteria based on the preference of the owner/designer for each particular project. The importance of one criterion in relation to another can be measured in terms of a four-point scale in which '4' represents 'major preference', '3' represents 'medium preference', '2' represents 'minor preference', and '1' represents 'slight preference'. After all the comparative evaluations are made, the raw scores of each criterion are totalled by summing up the assigned letters in the matrix. After this, the raw scores are adjusted to a scale of 1–10,

---

[1]FAST provides a systematic approach to assure correct basic function identification. It graphically and visually displays each sub-function, supporting function, critical-path function and basic function. It provides for identification of high-cost areas. A FAST diagram may appear similar to a PERT network and a work simplification flowchart. However, FAST is function-oriented and not time- or operation-oriented. A critical path of all functions required to make the product/service work or sell is one result of a FAST diagram. FAST presents a Christmas-tree breakdown of all functions involved. The tree is visualized as lying in the horizontal position. The apex shows the basic function, the trunk shows the critical-path function, and the limbs or branches show the secondary and aesthetic functions connected to the critical path. The roots of the tree can be visualized as the reason for being of the study.

wherein 10 is assigned to the criterion with the highest raw score and the other criteria are adjusted accordingly, which finally gives us the weights of the criteria (Hammond and Hassanani 1996).

8. Then, the alternatives x_3, x_4 and x_5 are evaluated against each of the performance criteria. It is assumed that all the alternatives that have survived meet the minimal needs or basic functions of the owner or the user. The scoring system used for this purpose is a five-point scale in which the extremes '1' and '5' represent 'poor' and 'excellent', respectively. The ranks were given by experts for each of the alternatives x_3, x_4 and x_5 for their corresponding performance criterion.

9. The ranks of each alternative are multiplied by the corresponding weights of the criteria, and thereby the resulting scores are calculated. The total scores are thus known, through summation of the resulting scores, for each alternative.

10. The alternative with the highest score is thus selected to cater to the desired function. The same process is adopted for all the elements in a particular structure.

The application of the above methodology is illustrated through a live case study discussed in the subsequent section.

12.7 DESCRIPTION OF THE CASE

The real-world design-and-build construction project that we have chosen for our case study is an industrial building of a reputed glass manufacturer (the client). A leading Indian construction company was awarded the contract for civil and structural work for integrating float-glass plant as a design-and-build lump-sum contract. The project is located in the northern parts of India. The initial contract value of the design and construction project was ₹ 92 crore. This industrial project was chosen as our case project since its construction was focused mainly on satisfying the functional requirements, while the aesthetic requirements were not given much attention except in the case of a few buildings in the plant such as administrative building. Hence, VE becomes a potent tool for realizing maximum benefits.

The scope of work for the glass plant consisted of formulating the architectural/structural/services designs and preparation of working drawings, preparation of the required plans/drawings for approval from the necessary and appointed authorities, and execution of the works in accordance with the drawings and contractual specifications.

The glass plant under construction consisted of more than thirty plant and non-plant structures. Some of the major structures in the project were batch plant, furnace, float bath, annealing lehr, cold end and warehouse. Based on the Pareto Law, it has often been noticed that only about 20 per cent of the items constitutes nearly 80 per cent of the cost. This was found to be true for this plant as well. Only five structures—namely, warehouse, annealing lehr, float bath, furnace and main office building—constituted about 76 per cent of the total cost (see Figure 12.9). As suggested in VE, the main focus should be restricted to the structures having higher saving potential. Further, out of the five structures mentioned, in-depth VE analysis was done only for the warehouse, mainly due to constraints of time and human resources. The cost break-up for the civil items showed that the items contributing to most of the total cost were pile foundation, flooring, granular sub-base (GSB), fabrication and fit-up structural steel (refer to Figure 12.10).

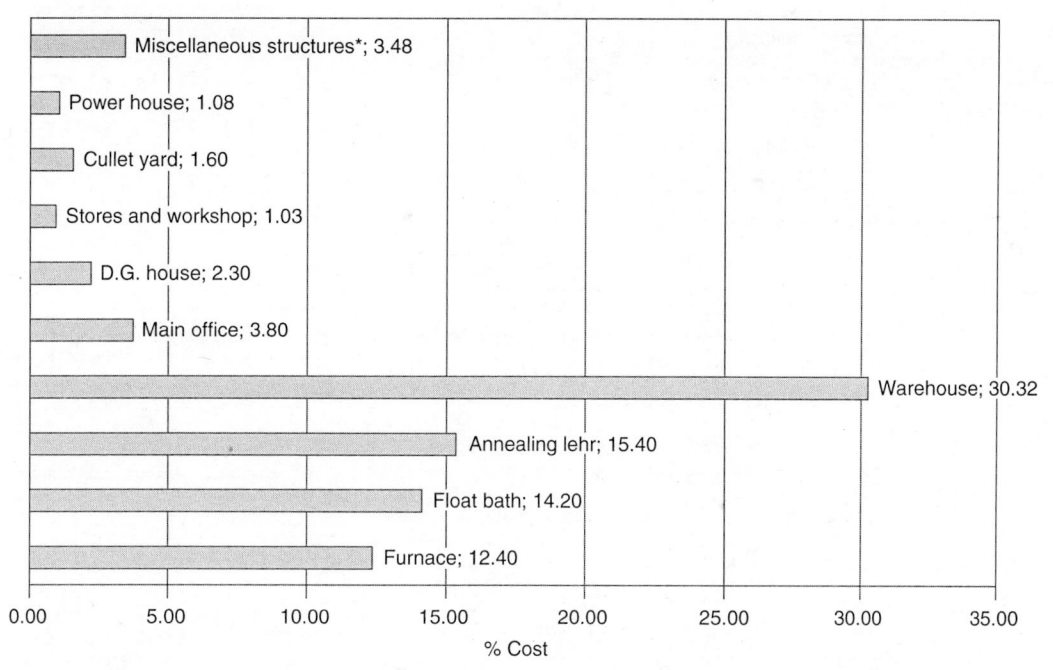

Figure 12.9 Break-up of cost of different structures for the case project

Bar chart (% Cost):
- Miscellaneous structures*: 3.48
- Power house: 1.08
- Cullet yard: 1.60
- Stores and workshop: 1.03
- D.G. house: 2.30
- Main office: 3.80
- Warehouse: 30.32
- Annealing lehr: 15.40
- Float bath: 14.20
- Furnace: 12.40

* Miscellaneous structures contain the following

Structures	Cost (%)
Fork/ box/ pallet yard	0.70
LPG yard	0.30
Scrap yard	0.30
Fabrication yard	0.30
Weigh bridge	0.18
Cooling water	0.80
Boiler room	0.20
Canteen	0.70

As seen from the cost break-up of the warehouse (Figure 12.10), the cost of the foundation forms a major part of its total cost. It is taken up first for the VE study as it has the maximum saving potential.

12.8 VALUE-ENGINEERING APPLICATION IN THE CASE PROJECT

As mentioned, the cost of foundation was a major element in the case project. Accordingly, this has been taken up for VE application. Besides, VE has been considered on flooring system, superstructure construction, material selection for sub-base, etc.

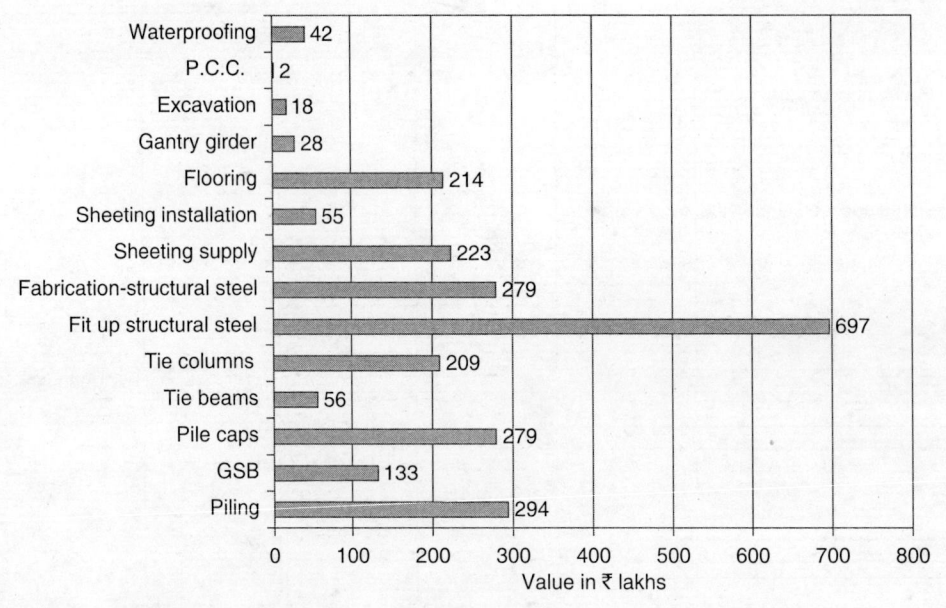

Figure 12.10 Break-up of cost of different items/elements in warehouse structure

Item	Value in ₹ lakhs
Waterproofing	42
P.C.C.	2
Excavation	18
Gantry girder	28
Flooring	214
Sheeting installation	55
Sheeting supply	223
Fabrication-structural steel	279
Fit up structural steel	697
Tie columns	209
Tie beams	56
Pile caps	279
GSB	133
Piling	294

12.8.1 Foundation Design

In the case project, the soil stratum at the warehouse site had a low bearing capacity such that the spread foundations were found to be infeasible and uneconomical, which, in turn, necessitated the deep foundation such as piles. The bearing capacity of the soil as obtained from the plate-load test was 8 t/m^2. The groundwater table was at an average depth of 2.5 m. In the warehouse alone, there were, in all, 134 pile caps resting over two piles running 22 m deep on an average.

A FAST diagram was drawn in order to identify the basic and secondary functions (see Figure 12.11). Some of the functions of the foundations identified are—transfer load, compact soil, resist settlement, and avoid liquefaction. The functions thus identified helped in thinking up new alternative ideas. The alternative solutions for the foundation of warehouse are:

1. Strap footings
2. Concrete piers—a pair of columns (each of size 400 mm × 500 mm) carrying a load of 600 kN each, resting on a pile cap of dimension 2.5 m × 1.5 m supported by two piles running 22 m deep
3. Deep vibratory compaction (Baker, undated)
4. Vibrated stone columns (Farel and Taylor 2004)
5. Rammed aggregate piers (RAP) as per the details given in Majchrzak et al. (2004)

To explain the last mentioned in detail, reinforce the soil with four rammed aggregate piers 10 m deep of 750 mm diameter, with total area equal to 30 per cent of the footing area. Crushed aggregates of 20 mm–37.5 mm are used for the RAP. Conventional isolated footing is designed on top of RAP. Aggregates are rammed using 10 kN–20 kN hydraulic hammers. From the design

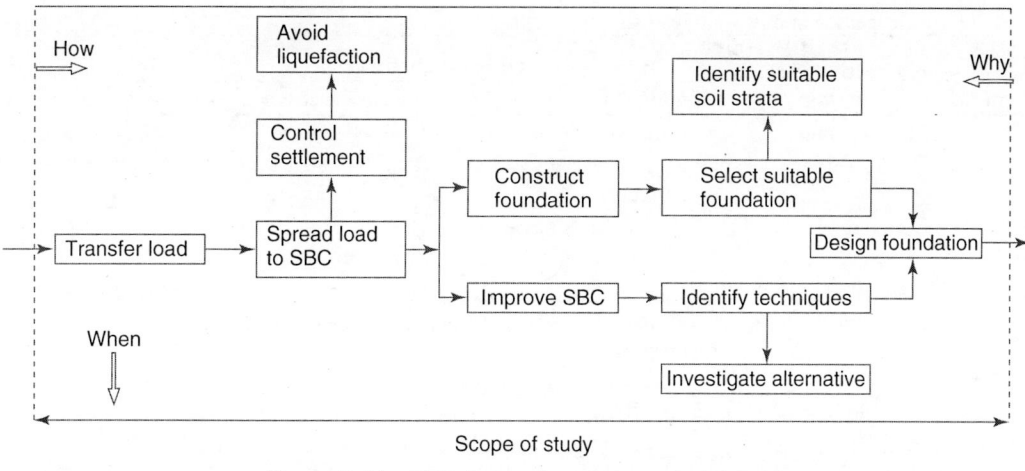

Figure 12.11 FAST diagram for warehouse foundation

calculations for RAP, it is found that four RAPs of 750 mm diameter and 10 m depth increase the bearing capacity of the soil from 70 kN/m² to 250 kN/m², and are sufficient to substitute 22 m deep pile foundation. The settlement as found from the previous case studies is within 8 mm–20 mm, which is less than the permissible settlement of 50 mm for the warehouse. The productivity of piles and RAPs per foundation is the same for the pile driving and RAP installation equipment, respectively, and so, the time to construct will be almost the same.

The advantages and disadvantages corresponding to each of the above alternatives for the foundation are enumerated in Table 12.1. The top three alternatives—namely, vibrated stone column (rank 1), rammed aggregate piers (rank 2), and deep vibratory compaction method (rank 3)—were selected for further analysis.

In order to arrive at the best alternatives, nine performance criteria for evaluating the alternatives were identified with the help of experts. In order to know the preferences for these nine criteria, a questionnaire survey was undertaken in which the respondents were asked to evaluate the criteria on a five-point scale. The responses obtained from the 20 respondents are shown in Table 12.2.

The mean value of each criterion and the corresponding rank are shown in Table 12.3. For illustration, let us take cost as the criterion. Twelve respondents have rated it as 'extremely important', five have rated as 'major important' and three respondents have rated it as 'important'. Thus, the mean value for 'cost' criterion works out to be $(12 \times 5 + 5 \times 4 + 3 \times 3) \div 20 = 4.45$. In a similar manner the mean values of other criteria have been worked out. The criterion with the maximum mean value (incidentally, it is for the cost criterion) is ranked one and so on.

Furthermore, a pair-wise comparison, i.e., matrix evaluation, is done to get the weight of each performance criterion. This is shown in Table 12.4. For pair-wise comparison, one criterion is taken at a time and is compared with the remaining criteria on a four-point scale, as mentioned earlier. For illustration, let us take the entry A-2 in row 1 and column 1 of Table 12.4. Here, the entry A-2 means that the criterion A has minor preference over criterion

Table 12.1 Comparison of qualitative features of various alternatives

Item no.	Creative idea listing	Advantages	Disadvantages	Idea rating
A1	Strap footings	a. Controls settlement b. Simple to design and construct	a. High cost b. Time-consuming	5
A2	Concrete piers	a. Good for soils having shallow hard strata	a. Complexity increases with greater depth	4
A3	Deep vibratory compaction	a. Simple and cost-effective for soils having less than 15 per cent fines b. Improves soil by compaction	a. Bearing capacity can only be doubled b. Stiffness of soil remains low	3
A4	Vibrated stone columns	a. Reduces sub-structure cost b. Improved densification c. Acts as sand drain	a. Stiffness is less than RAP b. Lower resistance to lateral forces	1
A5	Rammed aggregate piers (RAP)	a. Reduces complexity and cost b. Improves the soil radially c. Lowers settlement	a. Design is complex for high uplift pressure b. Soil is not compacted	2

B. The other entries in the table can be similarly interpreted. Then, the raw scores of each criterion are totalled. For example, the raw score of criterion A works out to be 20 (2 + 3 + 1 + 2 + 3 + 4 + 4 + 1). Similarly, these scores are calculated for all the criteria. The raw scores are finally adjusted to a scale of 1–10, with 10 being assigned to the criterion with the highest raw score. The other criteria are adjusted accordingly. The assigned score for each criterion is shown in Table 12.5, which also shows the ranks assigned by the experts on a five-point scale for each criterion for the three alternatives, viz. deep vibratory compaction, vibrated stone

Table 12.2 Evaluation of criteria governing foundation selection

S. No.	Criteria	Extremely important	Major important	Important	Minor important	Slightly important
A	Cost	12	5	3	0	0
B	Soil compaction	3	6	4	5	2
C	Construction time	2	6	4	5	3
D	Liquefaction control	7	5	3	3	2
E	Uplift resistance	5	4	10	1	0
F	Lateral resistance	2	4	6	4	4
G	Stiffness	0	3	6	6	5
H	Constructability	1	3	5	9	2
I	Soil settlement	10	5	5	0	0

Table 12.3 Evaluation criteria and their relative weights

S. No.	Criteria	E(x)	Rank
A	Cost	4.45	1
B	Soil compaction	3.15	5
C	Construction time	2.95	6
D	Liquefaction control	3.60	4
E	Uplift resistance	3.65	3
F	Lateral resistance	2.80	7
G	Stiffness	2.35	9
H	Constructability	2.60	8
I	Soil settlement	4.25	2

Table 12.4 Pair-wise comparison matrix for performance criteria evaluation

	B	C	D	E	F	G	H	I
A	A-2	A-3	A-1	A-2	A-3	A-4	A-4	A-1
	B	B-2	D-1	E-1	B-1	B-3	B-3	I-2
		C	D-3	E-2	F-1	C-2	H-1	I-2
			D	D-1	D-2	D-3	D-3	I-1
				E	E-1	E-3	E-3	I-2
					F	F-2	F-1	I-3
						G	G-1	I-3
							H	I-3

4 - Major preference
3 - Medium preference
2 - Minor preference
1 - Slight preference

Table 12.5 Analysis matrix

S. No.	Criteria	Assigned weights	Deep vibratory compaction Rank	Deep vibratory compaction Score	Vibrated stone column Rank	Vibrated stone column Score	Rammed aggregate piers Rank	Rammed aggregate piers Score
A	Cost	10.0	4	40.0	4	40.0	3	30.0
B	Soil compaction	4.5	3	13.5	4	18.0	3	13.5
C	Construction time	1.0	4	4.0	3	3.0	3	3.0
D	Liquefaction control	6.5	1	6.5	2	13.0	4	26.0
E	Uplift resistance	5.0	1	5.0	3	15.0	4	20.0
F	Lateral resistance	2.0	2	4.0	3	6.0	5	10.0
G	Stiffness	0.5	2	1.0	3	1.5	4	2.0
H	Constructability	0.5	4	2.0	3	1.5	3	1.5
I	Soil settlement	8.0	2	16.0	4	32.0	4	32.0
	Total scores			92.0		130.0		138.0

column and rammed aggregate piers. For example, in respect of the cost criterion, the experts have given a rating of 4, 4 and 3, respectively, to the three alternatives. Similarly, for the soil compaction criterion, the experts have assigned the ratings 3, 4 and 3, respectively, to the three alternatives.

The ratings given by the experts for each criterion corresponding to an alternative are multiplied by the weight of the criterion obtained earlier to get the score. The individual scores of each criterion are added to get the total score of a particular alternative. It is observed that the total scores obtained for the three alternatives are 92.0, 130.0 and 138.0 respectively.

Hence, it can be concluded that under the given circumstances, the alternative of providing rammed aggregate piers will give the best result. Without going through the VE process, the actual alternative selected was the arrangement of pile and pile caps for the foundation. It would be interesting to see the difference in costs of the actually implemented alternative and the best possible alternative as suggested by the VE process.

The original design consisting of a pair of piles and pile caps supporting the columns costs ₹ 103,420, whereas the cost of proposed isolated footings with RAP is ₹ 49,866, which is less than half the cost of the original design. The figure of ₹ 49,866 has been arrived at by adding the cost of RAP (which includes the excavation, the cost of aggregates, and the owning and operating cost of equipment for RAP) and the cost of constructing spread footing (which includes the cost of excavation and the cost of reinforced cement concrete including formwork). The total saving with the proposed modification, thus, works out to be ₹ 7,176,236 for the warehouse structure alone.

12.8.2 Flooring System

The flooring for the warehouse consists of grade slab, which is designed for the load of 60 kN/m². The presence of high groundwater table at an average depth of about 2.5 m may result in building up of upward water thrust on the grade slab. The total plinth area of the warehouse was 25,950 m² (with length = 250.00 m and breadth = 103.80 m).

As per the specification of the existing design, the layer below the flooring consisted of granular sub-base (GSB) of compacted thickness of 225 mm in single layer, with specified graded stone metal as per the relevant Indian Standards specification, including conveying of material to the site and spreading in uniform layers on prepared surface, watering and compacting with vibratory roller having minimum 80 kN–100 kN static weight and at OMC (Optimum Moisture Content) to achieve desired density including all material, labour, machinery with all lead, lifts, etc. With an area of 25,950 m² and allowing for some deductions, the total quantity of the sub-base material was found to be 5,618.31 m³.

The main function of GSB provided under grade slab is to transfer load from the grade slab to the soil beneath, and at the same time provide for dissipation of upward water thrust due to the presence of high water table. Suitable drainage system consisting of perforated pipes is provided for this purpose (Arm 2003). At the same time, the GSB material should have sufficient strength so that it does not get crushed during compaction, as the finer crushed particles will try to clog the air voids and help in building the water pressure.

Applying the same methodology as adopted for the foundation system, the functional evaluation for the GSB has been done and various alternatives have been proposed:

1. MSWI bottom ash (Maria 2001)
2. Recycled concrete (Prasad 2001)

Table 12.6 Comparison of qualitative features of various alternatives for sub-base

S. No.	Creative idea listing	Idea evaluation — Advantages	Idea evaluation — Disadvantages	Idea rating
1	MSWI bottom ash	(a) Particles are well graded as that of sandy gravel (b) Good bearing strength (c) Low deformation under load than the gravel	(a) Non-availability (b) High porosity leads to leaching (c) Organic content must be less than two per cent (d) Performance is impaired by presence of high amount of incombustible material	2
2	Recycled concrete	(a) Can be graded as per requirements (b) Low cost	(a) Impurities must be kept to a minimum (b) Crushing and grading requires mechanical set-up	3
3	Air-cooled blast furnace slag (AcBFS)	(a) Available near steel-plant sites for free (b) High bearing strength as compared to sandy gravels (c) High durability (d) Organic content is minimum	(a) Use limited to local areas as transporting them over a long distance may be costly	1
4	Coal fly-ash stabilized bases	(a) Fly ash when added to lime gives cementative properties and can be used to stabilize weak soils (b) The thickness of lime fly-ash soil layer for use as sub-base or base course is designed in accordance with IRC:37, 1984, with a minimum thickness of 150 mm	(a) The thickness of the stabilized layer can be determined by the CBR method, which requires testing (b) The percentages of addition of fly ash and lime in the soil can be determined by the test of the specimens, which is cumbersome and should be done by a reputed authority	4
5	Quarry waste	(a) Screening of mining can be utilized as a filler material in bases (b) Marble slurry can be utilized (c) Huge environmental hazards material can be beneficially utilized	(a) Screening are non-uniform in grading and can vary across different sites (b) Filter press has to be used to generate value-added product from marble slurry	5

   3. Air-cooled blast furnace slag (AcBFS)
   4. Coal fly-ash stabilized bases (Mudge 1971)
   5. Quarry waste

As already discussed, the advantages and disadvantages corresponding to each of the five alternatives were listed out for preliminary screening process. These are provided in Table 12.6 and are self-explanatory.

After the preliminary screening, the alternatives 'coal fly-ash stabilized bases' and 'quarry base' were dropped and the remaining three alternatives were taken up for subsequent

Table 12.7 Pair-wise comparison of performance evaluation criteria

	B	C	D	E	F	G	H	I	J	K	L
A	A-1	A-1	A-2	A-4	A-3	A-2	A-3	A-3	A-3	A-4	A-2
	B	C-1	B-1	B-3	B-2	B-1	B-2	B-3	B-3	B-3	B-3
		C	C-1	C-3	C-3	C-2	C-2	C-3	C-3	C-4	C-2
			D	D-3	D-2	D-1	D-2	D-2	D-3	D-3	D-1
				E	F-2	G-3	H-2	I-1	J-1	E-1	L-2
					F	G-2	H-1	F-1	F-1	F-2	L-1

4 - Major preference G   G-1 G-2 G-2 G-3 G-1
3 - Medium preference   H   H-1 H-2 H-2 L-1
2 - Minor preference   I   I-1 I-2 L-2
1 - Slight preference   J   J-1 L-2
                                      K   K-2

analysis. These chosen alternatives were evaluated using the same procedure as described for the foundation system. The matrix evaluation and the relative weights obtained for the performance criteria are presented in Tables 12.7 and 12.8, respectively.

It can be observed from the tables that the maximum raw score has been obtained for the 'availability' criterion, and accordingly, it has been assigned a weight of 10. Other criteria have also been assigned weights in a similar proportion. The information on the assigned weight and the rank of each criterion for each alternative is provided in Table 12.9, and on this basis the total score for each alternative is calculated. According to the results obtained from the analysis matrix, the most viable substitute for sub-base material is found to be AcBFS with a total score 193.2, followed by the recycled crushed aggregates with a score of 158.2, and MSWI bottom ash with the least score of 126.7. AcBFS has been used at some places on an experimental basis and has shown very good performance. AcBFS and clean crushed concrete

Table 12.8 Evaluation criteria and their relative weights

S. No.	Criteria	Raw score	Assigned weight
A	Availability	28	10.0
B	Resilient modulus	21	7.5
C	Bearing capacity	24	8.6
D	Durability	17	6.1
E	Resistance to freezing/thawing	1	0.4
F	Particle size distribution	6	2.1
G	Permeability	14	5.0
H	Dry density	8	2.9
I	% Air voids	4	1.4
J	Residual water content	2	0.7
K	Organic content	2	0.7
L	Deformation under load	8	2.9

Table 12.9 Analysis matrix for material selection

S. No.	Criteria	Assigned weights	MSWI bottom ash Rank	Score	AcBFS Rank	Score	Recycled crushed aggregates Rank	Score
A	Availability	10.0	1	10.0	4	40.0	2	20.0
B	Resilient modulus	7.5	3	22.5	4	30.0	4	30.0
C	Bearing capacity	8.6	3	25.8	5	43.0	4	34.4
D	Durability	6.1	4	24.4	3	18.3	3	18.3
E	Resistance to freezing/thawing	0.4	3	1.2	3	1.2	3	1.2
F	Particle size distribution	2.1	4	8.4	4	8.4	4	8.4
G	Permeability	5.0	2	10.0	4	20.0	3	15.0
H	Dry density	2.9	3	8.7	4	11.6	4	11.6
I	% Air voids	1.4	2	2.8	3	4.2	2	2.8
J	Residual water content	0.7	4	2.8	3	2.1	4	2.8
K	Organic content	0.7	2	1.4	4	2.8	3	2.1
L	Deformation under load	2.9	3	8.7	4	11.6	4	11.6
	Total scores			126.7		193.2		158.2

could be used as embankment fill, as a capping layer and as a sub-base. Their properties are best utilized in a sub-base.

Comparative cost calculation for the originally designed material and the proposed material has been made and summarized in the results. The original design consisting of GSB costs ₹ 150/m² (this is the sum of ₹ 110/m² cost of material and ₹ 40/m² for cost of labour and plant and machinery). The cost of suggested material AcBFS works out to be ₹ 96/m² (₹ 56/m² for material cost and ₹ 40/m² towards labour and plant and machinery). The cost saving in this item for the warehouse alone works out to be ₹ 1,349, 340.

12.8.3 Precast vs in-situ Construction

Warehouse is a cast-in-situ RCC framed structure with a number of columns and tie beams at different levels. The roof is covered with sheeting material supported on steel trusses. Considering the intended functions of the complete warehouse itself, the possible alternatives were generated. Some of the alternatives satisfying the functional requirements have been proposed and they are analysed. Although there could be a large number of possible alternatives such as pre-cast construction, construction using structural steel, etc., this study compared the original alternative with the pre-cast construction alternative. Here also, the analysis focused only on replacing a few cast-in-situ members with pre-cast construction. The structural elements that have been analysed in the study are gantry beams, tie beams, rib slabs and roof truss sheathing beams. The basic assumptions made for cost comparison between cast-in-situ and pre-cast construction take into account the saving in construction cost as well as the saving in construction time. The saving potential of pre-cast construction is estimated for the pre-cast gantry girder, the truss sheathing beam, the tie beam (+3.25 m), and the tie beam at truss level in 174, 83, 48 and 83 elements, respectively. The cost comparison for these elements for both the alternatives show that there is a saving potential of ₹ 247,828 if the mentioned cast-in-situ elements are replaced with the pre-cast elements. Although the amount is not significant, the time saved in using pre-cast construction would be of interest. It was found that replacing the cast-in-situ elements with pre-cast construction could save about 49 days of construction time, considering a cycle time of 21 days for the cast-in-situ gantry girder and 14 days for the cast-in-situ tie beams and truss sheathing beam. The time saving has been converted into monetary value and it works out to be ₹ 1,025,000 from the expression used by Warszawski (2000). The total saving, thus, works out to be ₹ 1,272,828 for the elements under consideration in the warehouse structure alone.

12.8.4 Discussion of Results

Value engineering study for the foundation implies that various alternatives available for the design of the system need to be explored at the time of design itself, and designers should not try to optimise their own productivity without considering the cost implications of their design, since in many cases design flaws contribute to about 50 per cent of the value loss. In our study, it has been found that poor selection of the foundation system has resulted in a loss of value to a similar extent. The foundation unit that could have been constructed at ₹ 49,866 without compromising on the intended function has been performed at a cost of ₹ 103,420. According to our findings, the pile foundation gave a poor value of 0.48 (49,866 ÷ 103,420) for RAPs.

It has been observed in many cases that contractors have better knowledge regarding the availability and cost of local materials, as compared to the designer, and are in a better position to provide feedback to the design team about the possible alternatives. The applicability for the intended function can be verified by the designers. In this respect, design-and-build projects offer an ideal opportunity. However, as has been found in this study, lack of VE application has resulted into loss in value to the tune of ₹ 1,349,340 in just one item of one structure.

Although the cost comparison of cast-in-situ and pre-cast construction has resulted in a marginal saving in construction cost, the possible saving in time has been neglected due to lack of VE application during design stage, thereby resulting in a considerable loss in value.

While VE application can be done in all the phases of a project, in the case study we have illustrated its applicability primarily in the design stage. Also, it should be noted that the cost-saving potential diminishes as time progresses from commencement to completion. Therefore, it is desirable to apply it as early as possible in a project. Looking at the increasing trend of adoption of design-and-build projects vis-à-vis traditional projects, it is further desirable to apply VE techniques for avoiding loss of value to the different stakeholders of a project.

REFERENCES

1. Arm, M., 2003, 'Mechanical Properties of Residues as Unbound Road Materials', PhD thesis, Stockholm, Sweden.

2. Baker, Hayward, Undated, Vibro System, available at. http://www.HaywardBaker.com.

3. Dell'Isola, A.J., 1982, *Value Engineering in the Construction Industry*, 3rd edition, New York: Van Nostrand Reinhold.

4. Farell, T. and Taylor, A., 2004, 'Rammed Aggregate Pier Design and Construction in California—Performance, Constructability and Economics', *SEAOC Convention Proceedings*.

5. Hammad, A.A. and Hassanain, M.A., 1996, 'Value Engineering in the Assessment of Exterior Building Wall System', *Journal of Architectural Engineering*, September, p. 115.

6. Kaufman, J.J., 1990, *Value Engineering for the Practitioner*, 3rd edition, North Carolina State University, Raleigh, N.C.

7. Majchrzak, M., Lew, M., Sorensen, K. and Farrell, T., 2004, 'Settlement of Shallow Foundations Constructed over Reinforced Soil: Design Estimates vs Measurements', Proceedings of the Fifth International Conference on Case Histories in Geotechnical Engineering, Paper No. 1.64, New York, April 13–17.

8. Maria, I., Enric, V., Xavier, Q., Marilda, B., Angel, L. and Feliciano, P., 2001, 'Use of Bottom Ash from Municipal Solid Waste Incinerator as a Road Material', Paper No. 37, In International Ash Utilization Symposium, Centre for Applied Energy and Research, University of Kentucky.

9. Mudge, A.E., 1971, Value Engineering: A Systematic Approach, New York: McGraw-Hill.

10. Prasad, M.M., 2001, *Utilization of Industrial Waste Byproduct in Road Construction*, IRC (19) 94.

11. Parker, D.E., 1985, *Value Engineering Theory*, The Lawerence D. Miler Value Foundation, Washington, D.C.

12. Warszawski, A., 2000, *Industrial and Robotics in Building*, New York: Harper & Row Publishers.

REVIEW QUESTIONS

1. State whether True or False:
   a. Four broad steps involved in project cost management are—resource planning schedule, cost planning, cost budgeting and cost control.
   b. The cost of an activity is collected under these broad heads—labour costs, material costs, plant and equipment costs, subcontractor costs, consumable costs, and overheads.
   c. Cost planning aims at ascertaining cost before many of the decisions are made related to design of a facility.
   d. Cost budgeting is the process of allocating the overall cost estimate to individual work items of the project.
   e. The objective of cost control is to ensure that the final cost of the project does not exceed the budgeted or planned cost.
   f. Labour cost accounting is done under heads of departmental labour and subcontractor labour.
   g. The two types of materials used in construction project are client-supplied and owner-purchased.
   h. The job status to date cost is given by actual cost incurred—deferred expenses and accrued expenses.
   i. Plant and equipment cost = hire charges + operating cost.

2. What are the different steps taken for project cost management? Discuss in brief.

3. What do you mean by project overheads?
4. What are the critical aspects that should be considered as vital while cost planning?
5. Differentiate between cost budgeting and cost planning?
6. Discuss in brief three steps involved in cost control and why cost control is important?
7. Why is the cost code important?
8. How is the cost statement prepared? Discuss its importance in brief.
9. Visit a construction site and collect the cost information on various cost heads as given in the text.
10. Study cost statement of any significant project being executed around you.
11. What is value engineering? In what stage of the project value engineering can provide maximum advantages?
12. Discuss the various steps involved in the application of value engineering.
13. Draw FAST diagram for (a) plastering of wall, (b) flooring, (c) roof slab construction.
14. Clearly explain (a) worth, (b) cost, and (c) value.
15. Apply the value engineering in the context of (a) road construction project, and (b) building project.

13
Construction Quality Management

Introduction, construction quality, inspection, quality control and quality assurance in projects, total quality management, quality gurus and their teachings, cost of quality, ISO standards, CONQUAS—construction quality assessment system, audit, construction productivity

13.1 INTRODUCTION

The quality of construction is one of the matters of great concern with most civil engineering constructions. Collins (1996) while describing quality as the world's oldest documented profession reports that poor quality can have far-reaching consequences. The following statement recorded during the reign of a Babylonian king is worth mentioning:

If a builder constructed a house but did not make his work strong with the result that the house which he built collapsed and so caused the death of the owner of the house, the builder shall be put to death.

Existing laws might prevent such harsh penalties in the present scenario, but the consequences may rather be in terms of loss in productivity, additional expenditures by way of rework and repair, re-inspection and retest in the short term. In the long term, poor quality can hurt reputation, and if the company continues in the same way it might have to close its shop for want of new projects. If a number of construction companies of a country start neglecting the quality aspects in their projects, this also starts reflecting on the reputation of the country.

There is a great difference between the quality of constructed facility and the samples that are submitted for testing. For example, quality of concrete is checked from the cube that is cured very well, but the actual constructed facility might not have received adequate curing. Similarly, there may be defects like faulty slopes in any flooring and roofing works, leading water to stagnate; there could be dampness on walls due to some bad brick used or bad workmanship; or there may be deliberate saving on material quality.

According to a conservative estimate, the cost of poor quality could be as high as 200 per cent to 300 per cent of the cost of construction in the early ages of works in some of the cases, which is a severe drain on our precious resources. Contractors are asked to rectify the defects, but many times they are irreparable and users are forced to take over the facility even with a long list of defects, as *fait accompli*. The user then suffers throughout the life of the structure by way of attending to recurring defects from time to time, or he sacrifices his comforts and keeps quiet and ends up paying much more cost than what contractor would have paid to do the right quality of work.

Construction quality has not got enough attention also due to the policy of awarding the project on bid price alone (lowest bidder system). The inherent assumption in adopting such a policy is that given the same set of drawings and specifications, all contractors including the contractor with the lowest bid price will produce similar construction quality. Unfortunately, this is not often the case. The lowest bidder reeling under cutthroat price tries to cut corners and starts compromising on quality. The majority among such contractors tend to neglect the complaints of the customer and produce bad quality of work. In the short run, such contractors do rise very fast, but in the long run they ultimately fail. On the other hand, there are some contractors who consistently make attempts to produce good-quality work. In the short run such contractors may seem to be on the losing side; in the long run, however, they prove to be quite successful and, in fact, some customers engage them even by paying extra price (premium).

Quality has widely been recognized as a distinctive competency that can be used by business to increase profitability and market share. The recognized success of Japanese firms with low cost, high quality, reliable and innovative products has had considerable impact on the western attitude to quality, particularly in forcing them to rethink their belief that quality is expensive (Love et al. 1995). While this concept is well understood in the manufacturing industry, the same cannot be claimed for the construction industry.

Customer satisfaction, which is of paramount importance in quality management, deals with life-cycle cost, that is, after-sales service or the cost to users. Incidentally, in the construction industry, while the supplier (constructor) is responsible for any defects up to defect liability period or six months to one year from the date of completion, there is no provision that can hold the supplier responsible for defects. Many times, the customer (user) takes possession of the constructed facility much after the defect liability period is over and spends a huge sum of money in rectification, which may at times be irreparable too. Also, most customers are not so demanding with regard to the supplier (constructor/developer) because of their ignorance of their rights.

13.2 CONSTRUCTION QUALITY

13.2.1 Definition of Quality

The term 'quality' has many connotations when used by different stakeholders. Some of the ways in which 'quality' has been defined are given in Box 13.1:

Box 13.1 Quality definitions

It is the **fitness for purpose**.
It is **conformance to specification**.
It is about **meeting or exceeding the needs of the customer**.
It is **value for money**.
It is **customer satisfaction/customer delight**.
It is **doing it right the first time and every time**.
It is **reduction of variability**.

No matter what definition we follow for quality, it becomes very complex when we try to put it into actual practice. For a user, quality is nothing but satisfaction with the appearance, performance and reliability of the project for a given price range.

The term 'quality' is often associated with products that are costly; however, it does not mean that products of low price cannot be of good quality. If the product meets the stated and unstated (intended) requirement of the customer, it can still be called a quality product. Most of the works in 'quality' are reported from the manufacturing industry, while very few are reported in the context of construction quality.

In the context of construction, suppose the contractor provides (i) a 110 mm thick RCC slab or (ii) a 95 mm thick RCC slab, against the customer's requirement of 100 mm thick RCC slab. Can these be considered quality products? Suppose the owner has desired that M25 grade of concrete be used in the slab, and the contractor has provided M25 grade concrete but the concrete that was supposed to have been received by the customer on Monday reached him on Tuesday. Has the contractor done a quality job?

How do we define construction quality? Is it the quality of materials being used in construction, the quality of workmanship, or the fulfilment of the end user's ultimate requirements? Schexnayder and Mayo (2004) extend the definition of construction quality beyond just 'supplying the right materials' and add that construction quality is also about finishing the project safely, on time, within budget, and without claims and litigation.

Suppose a contractor has not used the specified material in the specified quantity and he is still able to achieve the strength and serviceability requirement. Can this product be called a quality product? In another example, suppose the materials specified have been followed properly for the construction of a roof slab, but the slope that should have been provided as per the requirement is not provided. Can this qualify as a quality product? As a final example, suppose that out of 100 components in a construction project, about 10 components are not as per the requirement. Can this be called a quality project?

Most of the questions posed above can be addressed reasonably well if quality is assumed to be 'sticking to specification'—in other words, fulfilling the promises (contractual obligation) or delivering what has been promised. In construction, the promises are made by agreeing to 'do something' or 'not do something' in the form of signing the contract document.

Specifications, bill of quantities, general and special conditions of contract, drawings, etc., are part of the contract document. In the context of quality, specifications and contract drawing play the most important role. It would be virtually impossible to achieve the desired quality if the specifications and drawings for a project are unclear and confusing. Thus, specifications and drawings should try to capture the requirement or need of the customer in clear terms. These two documents should spell out clearly the materials to be used, the manufacturers, the guidelines for ensuring quality of materials at the time of purchase, the work method, the sequence to be followed, the tools and equipment to be used, the information on line and levels, and so on.

The relevant standards governing material and workmanship must be clearly specified. More importantly, qualitative terms describing the material-selection process or workmanship should be avoided as far as possible because this may bring in subjectivity and make the predefined quality a debatable issue. In summary, the objective of construction quality is to ensure that the constructed facility will perform its intended function.

According to International Organization for Standardization (ISO), quality is an inherent characteristic (distinguishing feature). Some of these characteristics are obtained from the stated, implied, or obligatory needs. Let us assume that the need of a particular concrete slab is to attain 20 N/mm^2 strength at 28 days of casting, in which case 'strength' becomes one of the quality characteristics for the concrete slab. The quality characteristic is not restricted to products such as concrete slab, but can be extended to include the process of construction, person, machine, the company as a whole, and so on. Each of the entity can have different quality characteristics.

Total quality in the construction industry can be defined as a measurable process of continuous improvement that is focused on the needs and expectations of the customer. Success requires a partnership characterized by input, involvement, commitment and action from owners, contractors, architects, engineers, subcontractors and suppliers (Deffenbaugh 1993).

13.2.2 Evolution of Quality

Modern quality control techniques were developed in the United States in the 1920s. In 1924, Shewart applied statistical quality control (SQC) in manufacturing in which statistical techniques are combined with conventional quality control methods. In 1928, H.F. Dodge and H.G. Roming published a theoretical consideration of stochastic statistics applied to sampling inspections.

In the early 1940s, a number of developments took place in the use of sampling tables for acceptance inspection. During World War II, the need for strict quality control became a necessity due to increased production of war materials. The quality control techniques used by Department of Defense (DOD) propelled their suppliers also to adopt such techniques. Quality control techniques and statistical analysis techniques in particular have advanced greatly since that time. After World War II, the emphasis was on promotion of quality control techniques on the part of suppliers and quality assurance on the part of inspection agencies of DOD.

In 1969, Feigenbaum used the term 'total quality' for the first time and it referred to wider issues such as planning, organization and management responsibility. Feigenbaum applied the statistical quality control techniques of Shewart to cover the whole company and not just one operation of a company. During this period itself, Ishikawa explained the term 'total quality control' and interpreted it to be 'company-wide quality control'. He emphasised the need for all employees, from top management to workers, to study and participate in quality control. Thus, in the 1960s and the early part of 1970s, the keywords in quality were 'zero defect' and 'total quality control', or TQC.

The TQC route was religiously followed in Japanese industries including in construction. TQC was not only applied to ensure the product quality but was also used to ensure management effectiveness. This saw a rise in the operation of Japanese companies worldwide. Japanese products were synonymous with the term 'quality'. The Japanese concept of TQC has come to be known as total quality management (TQM) in modern quality parlance (Burati, Jr, et al. 1991).

The 1980s were a time of intense global competition and a country's economic performance and reputation for quality was made up of the reputations and performances of its individual companies and products/services. Countries were striving to take lead in producing quality products at cheaper price. Quality management got a major thrust and it was now regarded as a key variable in the competitive positioning of firms and in ensuring market share. Quality gurus such as Deming, Crosby and Juran contributed a lot in the development of the quality field. The emphasis now shifted from the traditional approach of making the product—tracking the mistake by inspection and correcting it—to 'making it right the first time' and 'zero defect'.

During the 1990s, ISO 9000 became the internationally recognized standard for quality management systems. ISO standards specify the requirements for documentation, implementation and maintenance of a quality system. TQM became part of a much wider concept that addresses overall organizational performance and recognizes the importance of processes. People started realizing the benefits of total quality management.

In the 21st century, TQM has developed in many countries and it has helped organizations in achieving excellent performance, particularly in customer and business results. The quality movement is now gradually moving towards the much wider 'business excellence' or 'Excellence' model.

13.3 INSPECTION, QUALITY CONTROL AND QUALITY ASSURANCE IN PROJECTS

Quality standards obtained from modern construction projects have not kept pace with developments in technology and management in construction industry. Recurring incidents of faulty design and construction have caused untold damage and loss of life and property. Economic and legal implications of construction failures are nothing compared to the human lives that are lost and the permanent or temporary physical, mental and psychological suffering. Construction quality can be affected by:

- Whether a clear set of design and drawings is available—sometimes the confusion in design and drawings may show up in poor quality of construction
- Whether a clear, well-laid-out and unambiguous set of specifications is available
- Whether a clearly defined quality-control methodology exists
- Whether there has been usage of proper materials, workers and equipment during the construction processes

A major cause of controversy in quality control is delegation of responsibility and authority pertaining to quality assurance. Traditionally, designers as the agents of the owner are responsible for ensuring compliance as per specifications (design specifications). The responsibility of supervising the contractor's performance can be delegated to the designer's field staff—the clerk of works. He is responsible for seeing that the work is performed according to specifications, but he has no power to enforce compliance.

To alleviate (overcome) the problem of responsibility and authority in quality control, a *linear responsibility chart* (LRC) that describes all the persons within the quality control programme, their responsibilities, authority and interrelationships relative to quality control tasks is proposed. LRC is facilitated by the quality control matrix, which clearly defines the quality control requirements and the quality control methods. The two charts form the basis for developing the quality control programme.

A well-defined quality control programme should be established for each project and the organization structure of the programme should be very explicit. For an efficient quality control system, it is essential to develop and encourage cooperation among the participants so as to minimize the adversary relationships the traditional contract methods tend to generate. Designers should work closely with contractors to meet project quality objectives. Situations in which the designers 'police' the contractor's performance do alienate them from the latter.

Quality control is the responsibility of the entire project team (including owner). Quality control inspections should be done with the motive of encouraging and ensuring good workmanship rather than to catch culprits. Specifications should be realistic and these must consider natural variations in workmanship. In spite of the diverse factors influencing construction quality control, it remains a possible goal.

The process entails a system approach that makes participants in the construction process work together as a team. By taking better care and offering incentives for good workmanship, quality construction can be obtained. Legislation and regulations can ensure and safeguard the health, safety and welfare of the public by securing better construction quality.

13.3.1 Inspection

Inspection usually entails checking the physical appearance of an item against what is required. Activities such as measuring, examining, testing and gauging one or more characteristics of a product or service and comparing this with specified requirements are part of inspection.

It is generally a non-destructive qualitative observation such as checking performance against descriptive specifications and, thus, it could be subjective in nature. In some cases, gauges or machines may be required to do some simple measurements or examinations. Collecting concrete cube samples and testing them for quality interpretation is one of the most common examples of inspection in concrete construction operation. The three common levels of inspection are—(1) at the time of receiving the raw materials, parts, assemblies and other purchased items; (2) at the time of processing; and (3) final inspection prior to acceptance of product. The process of inspection is undertaken in a structured way using checklists. One such sample checklist is shown in Figure 13.1. Needless to say, the checklist helps to compare the characteristics required out of a product vis-à-vis what is inbuilt in the product.

S. No.	Description of formwork inspection items	Compliance		Remarks
		Yes	No	
1.0	Whether the approved design, drawing and specifications available for the formwork system?			
2.0	Are fall prevention/protection measures provided to all voids and exposed edges?			
3.0	Is suitable access and egress provided to the formwork areas?			
4.0	Is the formwork firmly supported on base plate, ground or supporting structure with good holding condition?			
5.0	Whether the working platform of adequate width (min 450 mm) has been provided?			
6.0	Is safety net provided if the height for work is more than 6 meters?			
7.0				
8.0				
.. ..				
.. ..				

Figure 13.1 A sample checklist for formwork inspection

13.3.2 Quality Control (QC)

Oakland (1995) defines 'quality control' as essentially the activities and techniques employed to achieve and maintain the quality of a product, process, or service. It involves a monitoring activity, but also concerns finding and eliminating causes of quality problems so that the requirements of the customer are continuously met. According to ISO, quality control is defined as a set of activities or techniques whose purpose is to ensure that all quality requirements are being met. In order to achieve this purpose, processes are monitored and performance problems are solved. Thus, quality control describes those actions that provide the means to control and measure the characteristics of an item, process, or facility against the established requirements. Quality control is basically the responsibility of the production personnel. A typical quality control programme would consist of defining quality standard, defining procedures for the measurement of attainment of that standard, execution of the procedures to determine probable attainment or non-attainment of the standard, and the power to enforce and maintain the defined standard as measured according to the defined procedure.

In the context of construction, quality control is administered by the contractors or by the specialist consultants such as consulting engineers or testing laboratories. Construction quality control entails performing inspection, test, measurement and documentation necessary to check, verify and correct the quality of construction materials and methods. Primary objectives of construction quality control are to produce a *safe, reliable and durable* structure so that the owner gets the best value for his investment.

The construction industry does not abide by a formal quality control programme as do the construction-related industries. Quality control on some projects could be haphazard and inconsistent. Because of heterogeneity, it is impossible to employ a uniform approach to check quality standards of construction work. Three major quality control methods commonly used on construction projects are:

- Inspection
- Testing
- Sampling

Techniques used vary from subjective evaluation to objective assessment of quality attained. The type adopted depends on the characteristics of construction activities or systems being examined and the degree of certainty desired. While all the methods may be feasible, not all of them are applicable on a particular activity. It is necessary that the methods to be used are as defined in the contract documents, to eliminate any confusion. Because of the nature of construction work, absolute compliance with specifications is impractical. The objective of quality assurance examination is to determine the degree of compliance with contract quality standards. A realistic approach is to first establish a *minimum quality standard* that will be the basics of acceptance or rejection. Appropriate quality control methods can thereafter be used to judge if variations are within the acceptable tolerances. Best results are obtained if quality control is consistent and the techniques used are appropriate.

13.3.3 Quality Assurance (QA)

According to Oakland (1995), quality assurance is broadly the prevention of quality problems through planned and systematic activities (including documentation). These will include the establishment of a good quality management system, the assessment of its adequacy, the audit of the operation of the system, and the review of the system itself. According to ISO, quality assurance is defined as a set of activities whose purpose is to demonstrate that an entity (such as product, processes, person, department and organization) meets all quality requirements. QA activities are carried out in order to inspire the confidence of both customers and managers, that all quality requirements are being met.

In the context of construction, quality assurance activities include all those planned and systematic administrative and surveillance functions initiated by project owner or regulatory agents to enforce and certify, with adequate confidence, compliance with established project quality standards to ensure that the completed structure and/or its components will fulfil the desired purposes *efficiently, effectively and economically*. The increase in complexity in a project has further increased the need for more efficient QA measures to ensure compliance with contract specifications.

Quality assurance programmes encompass the following:

- Establishing the procedure for defining, developing and establishing quality standards in design, construction and sometimes the operational stages of the structure and/or its components
- Establishing the procedure to be used to monitor, test, inspect, measure and perform current and review activities to assure compliance with established quality standards, with regard to construction materials, methods and personnel
- Defining the administrative procedure and requirements, organizational relationships and responsibilities, communications and information patterns, and other management activities required to execute, document and assure attainment of the established quality standards

It can be commonly observed that engineers/contractors use the term QA and QC interchangeably, which is not correct. While QA is a construction management process, QC is a sampling or inspection process. The focus in quality assurance is on defect prevention, while the focus in quality control is on defect detection once the item is constructed. In fact, it can be said that quality control is an element of a quality assurance programme.

13.4   TOTAL QUALITY MANAGEMENT

According to Oakland (1995), TQM is a way of planning, organizing and understanding each activity that depends on each individual at each level. Ideas of continuous learning allied to concepts such as empowerment and partnership, which are facets of TQM, also imply that a change in behaviour and culture is required if construction firms are to become learning organizations.

This is a complete management philosophy that permeates every aspect of a company and places quality as a strategic issue. Total quality management is accomplished through an integrated effort among all levels in a company to increase customer satisfaction by continuously improving current performance. TQM is a management-led approach applicable in all the operations of a company and the responsibility of ensuring quality is collective. The philosophy of TQM is one of prevention rather than defect detection. According to Pheng and Teo (2004), TQM is a way of thinking about goals, organizations, processes and people to ensure that the right things are done right the first time. It is an approach to improving the competitiveness and effectiveness, and flexibility of the whole organization. The essential elements of TQM are:

- Management commitment and leadership
- Training
- Teamwork
- Statistical methods
- Cost of quality
- Supplier involvement

It is believed that adoption of TQM by construction companies will result in higher customer satisfaction, better quality products and higher market share. However, adoption of TQM requires a complete turnaround in the corporate culture and management approach, as compared to the traditional way of top management giving orders and employees merely obeying those.

Construction, being different from manufacturing and other industries, has many unique problems that cause hindrances in adoption of TQM. Some of the major problems identified are:

1. lack of teamwork
2. poor communication
3. inadequate planning and scheduling

The causes identified for the above problems are:

1. no team-building exercises at the inception of projects
2. lack of understanding of team members' expectations
3. little or no team-oriented planning and scheduling

13.5 QUALITY GURUS AND THEIR TEACHINGS

Deming, Juran and Crosby are some of the world-famous quality gurus. All of them have come out with their own ideas and concepts on quality. These are briefly discussed below.

13.5.1 Deming

Deming modified the 'plan, do, check, act' (PDCA) cycle originated by Shewart. He named this as PDSA (plan, do, study, act) cycle. Conceptually, PDSA cycle, now also known as Deming cycle, is one of the problem-solving methods. The cycle is shown in Figure 13.2.

As the name suggests, PDSA cycle suggests preparation of a plan of things to be done as a first step, followed by its execution (doing whatever has been planned) as a second step. In the third step, results of the plan during execution are studied. Issues regarding the execution exactly as per plan and any variations are studied during this step. Finally, in the fourth step, the results are checked by actually identifying what went according to plan and what did not follow the plan. Using this insight, a revised and improved plan is worked out and the entire process is repeated.

Figure 13.2 The PDSA cycle

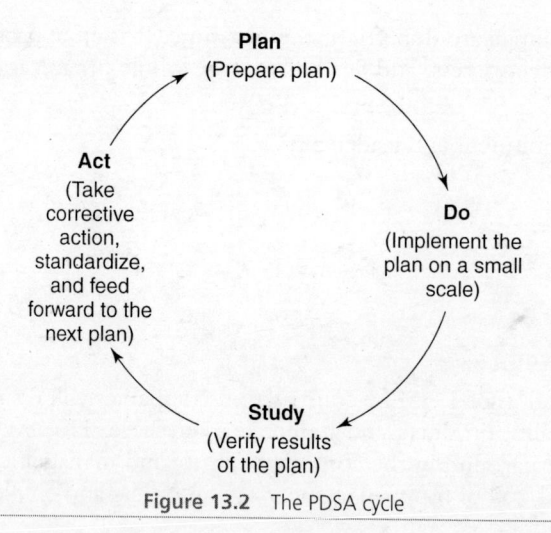

Deming's fourteen points provide a theory for management to improve quality, productivity and competitive positions. These points were first presented in his book *Out of the Crisis* and are briefly presented below:

1. The management should create and publish a statement of the aims and purposes toward improvement of product and service of the company (for example, to become competitive and stay in business, and to provide jobs). It should be accessible to all employees. The management must constantly demonstrate their commitment to the statement.
2. Top management and employees must learn and adopt the new philosophy. In a new economic age, management must awaken to the challenges and take on leadership for change.
3. Cease dependence on inspection to achieve quality. Eliminate the need for inspection on a mass basis by building quality into the product in the first place.
4. End the practice of awarding business on the basis of price tag. Instead, minimize total cost. Move towards a single supplier for any one item, on a long-term relationship of loyalty and trust.
5. Improve constantly the system of production and service, to improve quality and productivity, and thus constantly decrease cost.
6. Institute training on the job.
7. Teach and institute leadership. The aim of supervision should be to help people and machines to do a better job. Supervision of management is in need of overhaul, as is supervision of production workers.
8. Drive out fear, and create trust so that everyone may work effectively for the company.
9. Break down barriers between departments. Optimize the aims and purposes of the company, and the efforts of teams, groups and staff areas. People in research, design, sales and production must work as a team, to foresee problems of production and in use that may be encountered with the product or the service.

10. Eliminate slogans, exhortations and targets for the workforce asking for zero defects and new levels of productivity. Such exhortations only create adversarial relationships, as the bulk of the causes of low quality and low productivity belong to the system and, thus, lie beyond the power of the workforce.
11. Eliminate numerical quotas for production. Instead, learn and institute methods for improvement. Learn the capabilities of processes, and how to improve them.
12. (a) Remove barriers that rob the hourly worker of his right to pride of workmanship. The responsibility of supervisors must be changed from looking at sheer numbers to quality. (b) Remove barriers that rob people in management and in engineering of their right to pride of workmanship. This means abolishment of the annual or merit rating and of management by objective.
13. Institute a vigorous programme of education and self-improvement for everyone.
14. Put everyone in the company to work to accomplish the transformation. The transformation is everyone's work.

13.5.2 Juran

Joseph Juran developed the idea of the quality trilogy—quality planning, quality control and quality improvement. Juran emphasised the necessity for management at all levels to be committed to the quality effort with hands-on involvement. Juran recommended project improvements based on return on investment to achieve breakthrough results. He concentrated not only on the end customer, but identified other external and internal customers as well. According to him, quality is 'fitness of use'.

Juran suggested his trilogy (see Figure 13.3) for process improvement. The trilogy has three components—planning, control and improvement. The planning part consists of:

- establishing the goals and identification of both external and internal customers
- identification of needs of both external and internal customers, and translating these into deliverables that are understandable by the organization and its suppliers

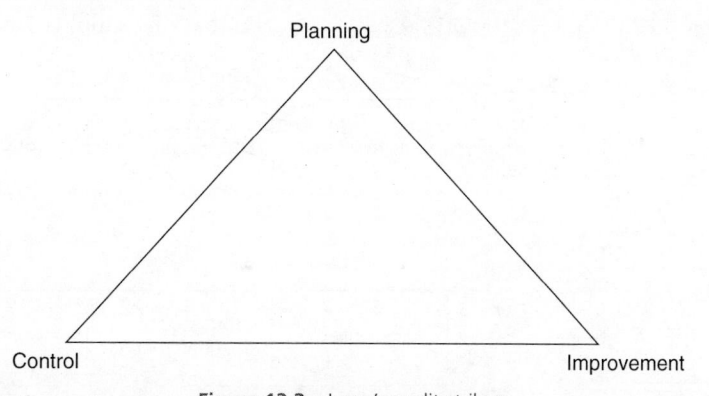

Figure 13.3 Juran's quality trilogy

- developing the product and/or service according to the needs of the customer, at an optimum cost
- developing the processes capable of producing the product and/or service
- putting plans into operation

The control part consists of:

- determining the variables to be controlled and the means to measure these variables
- setting goals for the control
- measuring actual performance
- comparing the actual performance with the goals set during planning
- acting on the difference between actual performance and performance goals

The improvement part consists of:

- attaining a performance level that is higher that the current level of performance
- identifying the different improvement measures and adopting an effective problem-solving method
- implementing the solution of the problem found during this stage to the quality planning process

The process thus adopted is repeated starting from setting a fresh goal.

Juran is also known for his 'triple-role concept' (see Figure 13.4) of quality involving customer, processor and supplier. In the context of construction, according to the 'triple-role concept' it can be said that each stakeholder at every level (for example, corporate level, region or branch level, department level, section level and individual level) has three roles to play—that of supplier, processor and customer. It may be recalled that some of the stakeholders in a construction project are architect/engineer, owner and contractor. The construction process works when the owner communicates his requirement to the architect/engineer, a step that converts the requirement into the form of plan and specifications, which when provided to the contractor are realized in the form of a constructed facility.

Now, according to the 'triple role' concept, the architect/engineer plays the role of a customer to the owner when he receives the requirement from the owner, that of a processor when he processes the design for the proposed facility, and that of a supplier when he supplies

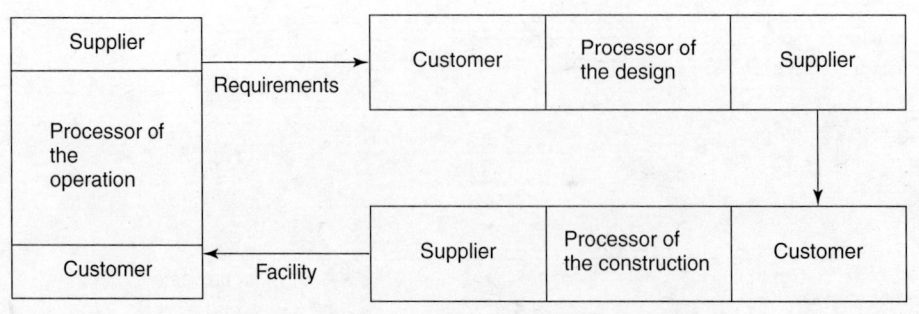

Figure 13.4 'Triple role' concept as applied in construction

the plan and specifications to the contractor to convert these into reality. The contractor plays the customer to the architect/engineer when he receives the plan and specifications from him (architect/engineer), the processor when he processes the actual construction, and the supplier when he hands over the constructed facility to the owner. The higher the satisfaction of each of the stakeholders in the construction process, the better the project quality.

13.5.3 Philip Crosby

Philip Crosby is known for his concepts of 'do it right first time' and 'zero defects'. He believed that doing it right the first time is less expensive than the cost of detecting and correcting the non-conformities. He defined quality as conformance with requirements that the company itself has established for its products, based directly on customer needs. He emphasised prevention management in every area. The four absolutes of quality management according to Philip Crosby are:

- Quality is conformance with requirements.
- Prevention of non-conformance is the objective, not appraisal.
- The performance standard is 'zero defects', not 'that's close enough'.
- Measurement of quality is the cost of non-conformance.

13.6 COST OF QUALITY

Construction projects are capital-intensive and cost of quality acquires a great significance. According to Juran, the cost of quality can be considered in terms of economics of the conformance quality. The quality cost breakdown shown in Figure 13.5 is based on the work of Feigenbaum (1983), who first described the concept in 1956.

From Figure 13.5, it is clear that

$$Quality\ costs = Quality\ control\ costs + Failure\ costs \tag{13.1}$$

Where, $\quad Quality\ control\ costs = Prevention\ costs + Appraisal\ costs \tag{13.2}$

and $\quad Failure\ costs = Internal\ failure\ costs + External\ failure\ costs \tag{13.3}$

The prevention costs used in the above equation refer to the cost of quality control activities undertaken before and during production. In other words, prevention cost is the cost of efforts undertaken to prevent failures. The appraisal cost is given by the costs incurred for quality control or quality assurance after production—for example, the costs of inspection, testing and examination to assess that the specified quality is being maintained.

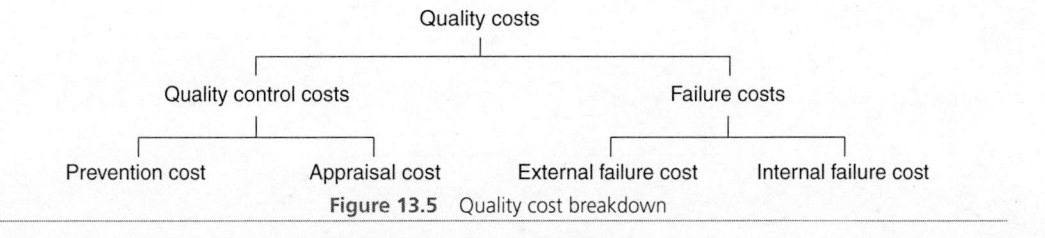

Figure 13.5 Quality cost breakdown

The internal failure cost is the cost resulting from a product or a service failing to meet the quality requirements—for example, warranties and return, liability costs, product recall cost, and direct cost or allowances.

The external assurance costs include:

- Costs relating to the demonstration and proof/objective evidence to customers
- Cost of testing by recognized, independent testing bodies for quality assurance provisions, demonstration and assessments
- Cost of independent assessment/third-party agency performing a detailed and in-depth study of company's QA activities

The prevention and appraisal costs being optional, they have also been referred to as discretionary (Blank and Solarzano 1978) or controllable costs (Besterfield 1979). A failure refers to the non-achievement of requirements. The relationship between cost and quality level is also shown pictorially in Figure 13.6 and is self-explanatory.

For illustration, let us take an example of a leaking roof. The costs associated with this will be as described below.

Conformance Cost

Prevention cost: This includes costs of various components that help to prevent the defect, such as preparation of specification and work procedure for construction joint preparation, stripping time of formwork, and training of staff and workmen.

Appraisal cost: This includes costs of all such activities that result in the cost of checking whether it is right. In this case, it will include checking the concrete-making materials against agreed specification, inspection before placing concrete, the calibration and maintenance of equipment used for testing of concrete, and the assessment and approval of all suppliers.

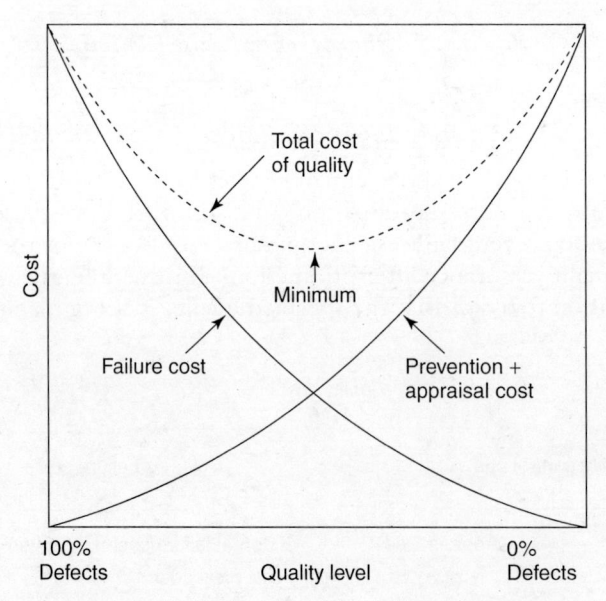

Figure 13.6 Cost versus quality level — classic view (adapted from Brown and Kane 1984)

Non-conformance Cost

This has two main components: internal failure cost and external failure cost. Together, they indicate the cost of getting it wrong.

Internal failure cost: This includes cost due to wastage of materials, grouting the roof slab by waterproofing compounds, and re-examination of works that have been rectified.

External failure cost: This includes repairing and servicing the defective parts, replacement of flooring components including transportation cost, cost associated with handling and servicing of customer complaints, and the impact on reputation and image which impinges directly on future prospects of sale.

Organizations in the construction industry spend money for prevention and appraisal, but the magnitude of these costs is very less when compared to the total cost of the project. It is widely believed that if the prevention and appraisal costs are more—that is, the cost of conformance is more—the failure or the non-conformance cost will be less. Typically, while the conformance costs are controllable variables, the non-conformance costs are the resultant variable.

The cost of quality includes direct and indirect costs associated with labour, materials and equipment used in quality management activities and for correcting deviations (Davies et al. 1989). Joseph Juran is widely credited with making the earliest references to losses due to poor quality of products and services. His broad-based, general application of a 'cost of quality' philosophy represented a considerable expansion of the existing body of knowledge when he first introduced it in 1951. Since that time, many significant contributions have been made by a number of prominent authors. Juran, along with Armand Feigenbaum, was primarily responsible for the writings that led to the development of current 'cost of quality' concepts (Feigenbaum 1961, Juran and Gryna 1988, Bajpai and Willey 1989, and Companella 1990).

Many quality experts, including W. Edwards Deming, Philip Crosby and Genachi Taguchi, have put forth numerous ideas and principles related to cost of quality (Logothetis 1999). Juran refers to two types of quality costs—control costs, which include the prevention and appraisal categories, and failure costs, which include both internal and external failure categories (Juran and Gryna 1988). American quality consultant Philip Crosby stresses the price of conformance and the price of non-conformance classifications. These translate into the prevention/appraisal and internal/external failure categories, respectively, and the 'cost of quality' information is used as a managerial tool in assessing cost-related expenditures, for tracking costs, and to augment other forms of information on the operational and strategic decision-making process (Shank and Govindarajan 1994). Crosby's approach begins with discrediting the assumption that there is a correlation between quality and cost. He maintains that doing a job right the first time is more cost-effective than making mistakes, tracking them, and correcting them. Companies without the benefit of this wisdom probably spend more doing inferior work than they would if they adopted a clear, uncompromising and high-quality standard of zero defects (Burati et al. 1991). While there are some differences in the approaches and philosophies pertaining to cost of quality, there is much consensus regarding the broader nature of what constitutes a continuous improvement process and how customer satisfaction may be achieved.

The concept of cost of quality has been developed and used in manufacturing industries (Feigenbaum 1983). As for the application of the concept in construction industry, some research work has been reported in the United States. Construction Industry Institute's (CII) preliminary analysis of nine industrial-type projects indicates that the cost of 'failing to meet the quality standards' is in the range of 12 per cent to 15 per cent of the total project cost (Needs and Ledbetter 1991). At a national conference on quality assurance in the building community (USA), it was suggested that the cost of poor quality was at least 7.5 per cent of the value of new non-residential work (Shilstone 1983). In another study, the causes of quality deviations in the design and construction were investigated in nine fast-track industrial construction projects. Analysis of the data showed that deviations on the projects accounted for an average of 12.4 per cent of the total project cost (Arditi and Gunaydin 1998).

Use of quality concepts did not start in Japanese construction companies until the mid-Seventies. People in construction industry were sceptical about 'quality control' success in construction due to the above-named differences between the construction and manufacturing industries. The 1973 oil embargo and the steep increase in oil prices adversely affected the prospects of future construction contracts in Japan. Construction companies started thinking about methods for reducing the cost of operations. In other words, the decrease of potential work quantity stimulated a drive to decrease the cost while keeping the quality levels as high as possible (Gilly et al. 1988).

In India, a large number of workshops have been organized by Consultancy Development Centre, Indian Oil Corporation and other private and multinational construction organizations to educate their staffs on the concept of ISO 9000 certification and type of documents. However, very little research work has been reported in the area of evaluating cost of quality. Rao (2000) has emphasised the achievement of quality and performance through training of workmen, proper contract clauses and certification.

13.7 ISO STANDARDS

The growing need for common quality standards throughout the world in manufacturing, inspection and test specification, and the need for standardization led to the formation of an international committee with the objective of producing an international quality standard. The committee considered many national inputs, especially the stringent quality requirements for defence contractors, and in 1987 produced a series of standards. These standards were called ISO 9000. These are a set of guidelines to effectively manage the important activities in an organization which affect quality. These standards only specify generic guidelines—applicable to any industry/service organization. These are system standards and not product standards, and are generally considered as a milestone on the path to total quality management (TQM). These establish a metric to evaluate continuous improvement. International Organization for Standardization (ISO) located at Geneva, Switzerland, is the approved body for issue and guidance of international standards today.

13.7.1 Benefits of ISO 9000

The ISO standards are advisory in nature and have worldwide acceptance under two-party (for example, client–contractor) contractual situations. There are agencies that certify that a particular company is following ISO standards. ISO 9000 is more than a certificate hanging on the wall of an organization. Some of the benefits of implementing ISO standards are listed below:

- It is increasingly becoming the requirement for global export/tender and increases access to global supply to large indigenous companies.
- It is increasingly becoming an effective marketing strategy and provides decisive edge over competition. It also helps in acquiring new customers.

- It ensures consistently dependable processes, less field failures, less wasted time, materials, and efforts, and reduction in scrap and rework. Thus, it helps in improving overall productivity and profit.

Some documented case histories of ISO 9000 benefits (http://www.isocenter.com/9000/benefits.html) for different categories of business include:

- Ten per cent sales increase directly attributable to ISO and a reduction in costs yielding $300,000 per year in case of a wholesale distributor
- Return on investment being achieved in less than two years for a manufacturing-assembly shop
- Savings of $250,000 in the first year following registration in case of a service and repair shop
- Eighteen per cent reduction in customer focus and 25 per cent increase in production backlog in case of hardware manufacturers
- In case of process-control systems and instrumentation manufacturer, inventory reduction of 50 per cent; product-cost reduction of 5 per cent; decrease in lost workdays of 80 per cent; increase in on-time deliveries of 12 per cent; reduction in credit memos of 70 per cent; increase in market share of 15 per cent
- Two per cent increase in overall margin in case of a keypad manufacturer

A company following the ISO standards provision may wish to get certified by an assessment agency and thereby get ISO registration. Of course, these registrations are audited at certain specified intervals by both external and internal auditors.

13.7.2 Principles of Quality Management Systems

The ISO 9000 series of standards, being generic in nature, can easily be tailored to fit the needs of a construction organization. The crux of these standards is to say what it is doing to ensure quality, to do what it says, and to document what it has done is in accordance with what it has said. The quality management systems adopted by ISO 9000 (http://www.iso.org/iso/iso9000-14000/iso9000/qmp.html) is based on the following broad principles:

1. Customer focus: This involves understanding current and future customer needs and meeting or exceeding the same.
2. Leadership: Organizations rely on leaders and, therefore, the leaders should create and maintain an internal environment wherein people are involved completely, trying to achieve the organization's objectives in a collective manner.
3. Involvement of people: Organizations must encourage people's full involvement so that their abilities are used for the benefit of the organization.
4. Process approach: This involves managing activities and related resources as a process to achieve the desired results efficiently. Organizations are most efficient when they use a process approach.
5. Systems approach to management: Organizations must achieve effectiveness and efficiency by identifying interrelated processes and treating them as a system. Organizations must use a systems approach to manage the interrelated processes.

6. Continual improvement: Organizations must improve their overall performance on a continual basis and this should be one of the permanent objectives.
7. Factual approach to decision-making: In order to make an effective decision, organizations must make their decisions based on an analysis of factual data and information.
8. Mutually beneficial supplier relationship: Organizations and their suppliers are interdependent and, thus, a cordial relation benefits both parties.

13.7.3 ISO 9001–2000 Family of Standards

ISO 9001 standard has eight sections. Out of the eight sections, the first three are provided for information and the remaining five lay down the requirements to be followed by the organization that is implementing the standard. The International Organization for Standardization developed a series of international standards for quality systems (ISO 9000, 9001 and 9004).

1. ISO 9000 describes the fundamentals of quality management systems and specifies the terminology for quality management system and that used in the other two standards.
2. ISO 9001 specifies requirements for a QMS wherein an organization needs to demonstrate its ability to provide products that fulfil customer and applicable regulatory requirements, and aim to enhance customer satisfaction.
3. ISO 9004 provides guidelines that consider both the effectiveness and efficiency of the QMS.
4. ISO 19011 provides guidance on auditing quality and environmental management system.

The **outline of different sections of ISO 9001–2000** is given in Table 13.1. The main sections deal with scope, normative reference, terms and definitions, quality management system, management responsibility, resource management, product realization, and measurement analysis and improvement.

Table 13.1 Sections and subsections—ISO 9001–2000

1. Scope		Scope specifies the purpose of the standard. For example, for a construction company the contents of the scope could be—procurement, construction, inspection, testing and commissioning.
2. Normative reference		The section provides the reference to all the concepts, definitions and applicable documents.
3. Terms and definitions		The section deals with the definition of key terms such as client, supplier, third-party inspection, authorized inspection agency, quality assurance and control, quality plan, inspection, calibration, test certificates, and so on. The section also explains the meaning of abbreviations used anywhere in the document.
4. Quality management system	4.1 General requirements	It deals with identification of processes for product realization, determination of sequence involved with the processes, ensuring resources required for the processes, monitoring, measuring and analysing the processes, and finally, the implementation part.

(Continued)

Table 13.1 Continued

	4.2 Documentation requirements	It deals with establishing and maintaining quality manual, and control of documents and records.
5. Management responsibility	5.1 Management commitment	It deals with the commitment of management to the development, implementation and continual improvement of the quality management system.
	5.2 Customer focus	The section seeks to ensure that customer requirements are determined and met to the complete satisfaction of the customer.
	5.3 Quality policy	It deals with the quality policy of the company in line with the company's mission.
	5.4 Planning	The section deals with the objectives, planning for quality management system, project quality plan and the details of procedures for maintaining and controlling the quality records.
	5.5 Responsibility, authority and communication	It contains the organization chart specifying the responsibility and authority for the key positions marked for implementing/managing the quality system at project level.
	5.6 Management review	This relates to review of the quality system at specified intervals for the stated objectives, with reference to quality, cost, timely completion, safety, etc.
6. Resource management	6.1 Provision of resources	The objective and scope of resource management are specified. For example, for a construction company the objective could be to establish and maintain a procedure for resources in construction activities at site, and the scope could be describing the methods adopted for the administration of resources to the satisfaction of the customers.
	6.2 Human resources	It deals with personnel executing the work, their competence, their awareness towards achieving quality objectives, and the training details of personnel.
	6.3 Infrastructure	It deals with the requirement and maintenance of infrastructure to achieve the desired quality level.
	6.4 Work environment	It deals with the provision of a suitable work environment that can influence positively the motivation and performance of personnel involved with the processes.
7. Product realization	7.1 Planning of product realization	In the context of a construction organization, it could be establishing and maintaining the planning and control manual for construction activities, establishing work methods, and the contract administration procedures. The details of activities and the persons responsible for executing these are also defined.

(Continued)

	7.2 Customer-related process	It deals with the requirements specified by the product, and these could even be the ones not stated by the customer but essential for proper functioning of the product. It also deals with the communication arrangement to be followed by the organization for communicating with the customer.
	7.3 Design and development	This section deals with the planning aspect for product design and development, the determination of inputs for design and development, the description of outputs resulting from the design and development, and the review process for the design and development.
	7.4 Purchasing	This section deals with the purchasing processes such as defining who is responsible for purchasing, what process is adopted for the purchasing, the inspection process of the purchased product, and so on.
	7.5 Production and service provision	This section deals with establishing, implementing, maintaining and validating the procedures for control of production and service provision. The process of treating and storing the client-supplied materials (customer property) is also provided. The process of informing the customer of any loss, damage, incompleteness, or other discrepancy regarding any products supplied by them is also mentioned.
	7.6 Control of monitoring and measuring devices	The section deals with establishing and maintaining the procedure to calibrate and maintain the inspection and test equipment, to ensure that the equipment are capable of performing to required accuracy.
8. Measurement analysis and improvement	8.1 General	It deals with establishment of procedures for implementing, monitoring, measurement, analysis and improvement process in all activities to achieve customer satisfaction, product quality and QMS effectiveness.
	8.2 Monitoring and measurement	It deals with monitoring the information relating to customer's perception as to whether the organization has met customer requirement. The details on internal audit process are also specified.
	8.3 Control of non-conforming products	It describes the responsibilities and methods used by the management for control of non-conforming product/system/process and to ensure that the defective processes and products are prevented from use.
	8.4 Analysis of data	It deals with establishing and maintaining a system for application of statistical techniques that will enable decision-making and demonstrate the suitability and effectiveness of QMS.
	8.5 Improvement	It deals with establishing, implementing and maintaining a procedure for continual improvement and for elimination of potential non-conformities.

13.8 CONQUAS—CONSTRUCTION QUALITY ASSESSMENT SYSTEM

Construction Quality Assessment System, also known as CONQUAS, is a standard quality assessment system introduced by Building and Construction Authority (BCA) of Singapore. The system objectively measures constructed works against workmanship standards and specifications. In order to measure the project quality, the system uses a sampling approach to represent the whole project. The samples are distributed as uniformly as possible throughout the project, and the number of samples is dependent on the size of the building. The emphasis in this system is on 'doing it right the first time'. Once a project has been evaluated and a score assigned, there is no re-scoring in the CONQUAS—that is, rectification and correction made after the assessment is not taken into consideration. Over the years, the CONQUAS system has gained acceptability as a benchmarking tool across several countries including India. Some Indian companies have also got their projects evaluated under CONQUAS.

BCA conducts evaluation of all types of buildings, viz. commercial and industrial, institutional, public housing and landed housing. The assessment is conducted at the invitation of developer/contractor for private sector projects, while it is a must for all public sector projects. For the scoring, the project is divided in three major components—structural, architectural, and mechanical and electrical (M&E) works. These components are further divided into different subcomponents. For example, under structural components the different subcomponents are formwork, rebar, finished concrete, concrete quality, steel reinforcement quality, NDT-UPV test for concrete uniformity, and NDT-Electro-cover meter test for concrete cover. Under architectural components, inspection of internal finishes, roofs and external walls, etc., are carried out, while under M&E, inspection of air conditioning, mechanical ventilation, electrical works, fire-protecting works, sanitary and plumbing works, etc., are covered. The assessment is done primarily through on-site testing/inspection prior to installation/construction, during the installation/construction, and after the installation/construction.

The CONQUAS score of a building is obtained by summing the scores obtained in each of the three main components mentioned above. The maximum score in each of the three components varies according to the type of building being assessed. For example, in case of commercial, industrial and institutional buildings referred to as CAT A, the structural components have a weightage of 25 per cent, architectural works have a weightage of 50 per cent, and M&E works have a weightage of 20 per cent. The individual subcomponents under each of the three components are also given weightage. For example, under structural components, formwork, rebar, finished concrete, concrete quality, steel reinforcement quality, NDT-UPV test for concrete uniformity, and NDT-Electro-cover meter test for concrete cover are given a weightage of 15 per cent, 20 per cent, 25 per cent, 5 per cent, 5 per cent, 15 per cent and 15 per cent, respectively. The quality of these subcomponents is assessed against standards. If the subcomponent complies with the requirements laid out in the standards, 'S' is given against that requirement; else, an 'X' is recorded. The number of 'S' obtained against a subcomponent determines the quality score. This process of awarding an 'S' or an 'X' is carried out for all the subcomponents, and the final CONQUAS score is obtained. The companies implementing CONQUAS have an opportunity to benchmark their workmanship quality on an international basis. Besides, the companies with a consistently high CONQUAS score gain competitive advantages and their reputation in the international and domestic market also gets better.

The advantage with this system is that there is no subjectivity involved in the measurement of workmanship. Thus, product quality can be easily measured by gathering data for different projects spread across the country. The data gathered can be analysed to see whether the product meets the conformity requirement. The data can also be used for establishing characteristics and trends of product including opportunities for preventive action. The data collected can also be used for measuring supplier's workmanship.

13.9 AUDIT

Audit is a systematic and independent examination to determine (1) whether quality activities and related results comply with planned arrangements; (2) whether these arrangements are implemented effectively and are suitable to achieve objectives; and (3) whether quality policy is understood and implemented properly.

Auditing builds confidence in management. It also points out to system deficiencies, if any, as well as highlights system weaknesses before a potential problem occurs. It has been found to be a convenient framework for investigating problems in particular areas. It also allows personnel from other departments to know how their work affects others. Auditing creates opportunity for interchange of ideas and thereby results in improvement of process in a cost-effective manner. It also results in increased motivation for improving performance.

Audit Types

First-party audit: This is conducted by, or on behalf of, the organization itself for internal purposes.

Second-party audit: This is conducted by customers of the organization or by other persons on behalf of the customer.

Third-party audit: This is conducted by external independent organizations, usually accredited, and provides certification or registration of conformity with requirements such as ISO 9001.

Why to Audit?

This is a mandatory requirement laid down in ISO 9000. It helps in determining system conformity against a quality system standard/procedure. It also helps to determine the system effectiveness to meet the objectives as well as provide the auditee with information to use in improving the system.

ISO 9001–2000 requirements for internal audit

1. Audits are conducted at planned intervals to determine quality management system conformity with planned arrangements and laid-down policies and objectives. It needs to see whether the quality management systems are effectively implemented and maintained.
2. Plan a program of audits, covering processes/areas to be audited. This is considered based on the status and importance, and the audit history.
3. Define audit criteria, scope and frequency, as well as methods of auditing.
4. Choose an objective, impartial and trained auditor.
5. Documented procedure defining: This involves planning for conducting audits, reporting process and records-keeping process.

6. Management responsibility: Taking actions (without delay) and eliminating non-conformity and their causes are part of management responsibility.
7. Follow-up activities shall include verifying action taken and reporting results.

Guidance documents	
Audit guidelines	
ISO 19011	Guidance on auditing quality and environmental management system
Audit standards	
ISO 10011-1	Auditing
ISO 10011-2	Qualification criteria for auditors
ISO 10011-3	Managing an audit programme

Non-conformance

Non-conformance is the non-fulfilment of requirement. It can be there for quality management system, or ISO 9001: 2000, or customer satisfaction, or legislation, or regulatory body. Non-conformance can be either major or minor. It is graded as a major non-conformance if no evidence of adherence to a procedure/system element is found or if there is a major risk to final product or service quality. The non-conformance is a minor one if there is limited evidence of compliance with the procedure or there is no appreciable risk to final product or service quality. The non-conformance report should contain the relevant clause of the audit standard, the reference of procedure, the location, the mention of particular activity where non-conformance has been observed, the nature of problem, the evidence, and the scale of problem mentioning whether non-conformance is to be graded as major or minor.

13.10 CONSTRUCTION PRODUCTIVITY

Productivity is defined as the quantum of production of any work within the estimated cost, with an acceptable quality standard under the defined duration with respect to nature of work. Increased productivity means within the defined time frame the production increases without any cost variation per unit. This implies more sales in the same period with the same overheads.

With respect to the construction industry, the following factors govern productivity.

1. Well-planned work: This is work done through the thought process covering all aspects of planning.
2. Skilled manpower: If suitable and skilled manpower is deployed, we stand to save substantially in labour and material. Material wastage, rework and rectification can be avoided. Proper screening of labour should be done before they are actually pressed into service.
3. Good and suitable equipment: We should use equipment of good condition. Also, periodic maintenance of the equipment enhances productivity.
4. Defined methodology: Prior to starting any new activity, it is preferable to work out a step-by-step method of working and foresee the possible pitfalls in the process. This will enable trouble-free accomplishment of any task.
5. Right type of hand tools: The workers must be provided with the right type of tools to carry out any work.

6. Neat and tidy workplace: A good housekeeping habit can make this happen. If the workplace is easily accessible, the worker will not have any problem in carrying out his task. A site that is scattered with materials will result in accidents, material wastage, etc.
7. Staff productivity: The staff and supervisory personnel must be proactive. The usage of modern methods and equipment leads to better quality and productivity.

- Optimum usage of inputs, effective utilization of construction materials, and recycling the shuttering items lead to economy in expenditure resulting in boosting up the normal productivity norms.
- Awareness of, and in-house training programmes in, the latest versions in construction industry can prove to be the best method to increase productivity.
- Motivation and moral support to the needy at workplace can boost the capability of an individual, resulting in higher productivity.
- Coordination with interrelated disciplines can help to execute a task without delay.
- Advanced communication networks aid in speedy transfer of requisite inputs/documents required to complete a target well in time, resulting in better productivity.
- Curtailing unnecessary overheads can bring down production cost against sales price, resulting in profit.

13.10.1 Typical Causes of Low Labour Productivity

Worker's Low Morale

- Non-fulfilment of employment terms and conditions by the management
- Insecurity of employment
- Substandard working conditions
- Frequent transfers
- Frequent changes in the scope of work and work methodology
- Conflicts between supervisors and workers

Poor Pre-work Preparation by Supervisors

- Excess workers employed for the task
- Insufficient instructions for the execution of work
- Incorrect sequencing of work activities
- Shortage of tools and materials at the site
- Wastage resulting from frequent shifting of materials and poor-quality/defective work

Directional Failures of the Project Management

- Failure to set performance targets
- Failure to make provision for timely resources support

- Failure to provide feedback
- Failure to motivate workers

The type of manpower required at a typical project site and the typical productivity values of civil construction activities are given in Appendices 11 and 12 respectively of this text. These are provided for information and preliminary planning.

REFERENCES

1. Arditi, D. and Gunaydin, H.M., 1998, 'Factors that affect process quality in the life cycle of building projects', *ASCE Journal of Construction Engineering and Management*, 124(3), pp. 194–203.
2. Bajpai, A.K. and Willey, P.C.T., 1989, 'Questions about quality costs', *The International Journal of Quality and Reliability Management*, 6(6), pp. 9–17.
3. Battikha, M.G., 2002, 'QUALICON: Computer-based system for construction quality management', *Journal of Construction Engineering and Management*, 128(2), pp. 164–173.
4. Besterfield, D.H., Michna, C.B., Besterfield, G.H. and Sacre, M.B., 2005, *Total Quality Management*, 3rd ed., New Delhi: Prentice Hall.
5. Blank, L. and Solorzano, J., 1978, 'Using quality cost analysis for management improvement', *Industrial Engineering*, 10(2), pp. 46–51.
6. Brown, F.X. and Kane, R.W., 1984, 'Quality cost and profit performance', *Quality Cost: Ideas and Applications*, American Society for Quality Control, Milwaukee, WI.
7. Burati, J.L., Farrington, J.J. and Ledbetter, W.B., 1992, 'Causes of quality deviations in design construction', *Journal of Construction Engineering and Management*, 118(1), pp. 34–49.
8. Burati, L.B., Michael, F.M. and Kalidindi, S.N., 1991, 'Quality management in construction industry', *Journal of Construction Engineering and Management*, ASCE 117(2), pp. 341–359.
9. Burati, Jr, J.L., Matthews, M.F. and Kalidindi, S.N., 1991, 'Quality Management in Construction Industry', *ASCE Journal of Construction Engineering and Management*, 117(2), pp. 341–359.
10. Campanella, J. (ed.), 1990, *Principles of Quality Costs*, 2nd ed., ASQC Quality Press, Milwaukee.
11. Chitkara, K.K., 2006, *Construction Project Management. Planning, Scheduling and Controlling*, 10th reprint, New Delhi: Tata McGraw-Hill.
12. Collins, Jr, F.C., 1996, *Quality: The Ball in Your Court*, New Delhi: Tata McGraw-Hill.
13. Crosby, P., 1979, *Quality is Free*, New York: McGraw-Hill.
14. Davis, K., Ledbetter, W.B. and Burati, J.L., 1989, 'Measuring design and construction quality costs', *Journal of Construction Engineering and Management*, ASCE, 115(3), pp. 385–400.
15. Deffenbaugh, R.L., 1993, 'Total Quality Management at Construction Jobsites', *ASCE Journal of Management in Engineering*, 9(4), pp. 382–389.
16. Deming, W.E., 1986, *Out of the Crisis*, Centre for Advanced Engineering Study, MIT, Cambridge, Mass.
17. Eldin, N. and Hikle, V., 2003, 'Pilot study of quality function deployment in construction projects', *Journal of Construction Engineering and Management*, 129(3), pp. 314–329.

18. Feigenbaum, A.V., 1983, *Total Quality Control*, 3rd ed., New York: McGraw-Hill.
19. Feigenbaum, A.V., 1961, *Total Quality Control*, New York: Mc-Graw Hill.
20. Gilly, B.A., Touran, A. and Asai, T., 1988, 'Quality control circles in construction', *Journal of Construction Engineering and Management*, ASCE 113(3), pp. 427–439.
21. International Organization for Standardization (ISO), 1987, Family of quality management standards, ISO 9000, Geneva, Switzerland, http://www.iso.org/iso/iso9000-14000/iso9000/qmp.html
22. Juran, J., 1951, *Quality Control Handbook*, 1st ed., New York: McGraw-Hill.
23. Juran, J.M. and Gryna, F.M. (ed.), 1988, *Juran's Quality Control Handbook*, 4th ed., New York: McGraw-Hill.
24. Logothetis, N., 1992, *Managing for total quality: From Deming to Taguchi and SPC*, 1st ed., New Delhi: Prentice Hall.
25. Love, C.E., Guo, R. and Irwin, K.H., 1995, 'Acceptable quality level versus zero-defects: Some empirical evidence', *Computers and operations research*, 22(4), pp. 403–417.
26. Needs, T.A. and Ledbetter, W.B., 1991, *Quality performance and management in engineering/construction*, AACE Transactions, New York.
27. Pheng, L.S. and Teo, J.A., 2004, 'Implementing Total Quality Management in Construction Firms', *ASCE Journal of Management in Engineering*, 20(1), pp. 8–15.
28. Rao, S.K., 2000, 'Aspects of quality and performance in construction industry', Proceedings of the Third National Conference in Construction, February 10–11, 2000, Construction Industry Development Council, Vol. 3, pp. 49–54.
29. Schexnayder, C.J. and Mayo, R.E., 2004, *Construction Management Fundamentals*, Singapore: McGraw-Hill.
30. Shank, J.K. and Govindarajan, V., 1994, 'Measuring the "cost of quality": A strategic cost management perspective', *Journal of Cost Management*, 8(2), pp. 5–17.
31. Shilestone, J.M., 1983, 'Welcome, background and program objectives', Proceedings of the National conference on Quality Assurance in the Building Community, Shilestone and Associates, Inc., Washington, D.C.
32. The Construction Quality Assessment System (CONQUAS), details available at http://www.bca.gov.sg/professionals/IQUAS/others/CON21.pdf
33. www.isocenter.com/9000/benefits.html.

REVIEW QUESTIONS

1. State whether True or False:
    a. Lower bidder system is one of the major setbacks for achieving construction quality.
    b. Quality can be defined as:
        1. Fitness for purpose
        2. Conformance with specification

3. Meeting or exceeding the needs of customer

4. Value for money

5. Customer satisfaction

6. Just in time

7. Reduction in variability

   c. Construction quality is about finishing the project safely, on time, within budget, and without claims and litigation.

   d. 'Zero defect' and 'total quality control' are two major catchwords for which construction project should strive.

   e. A major cause of controversy in quality control is delegation of responsibility and authority pertaining to quality assurance.

   f. Linear responsibility chart (LRC) describes all the persons within the quality control programme, their responsibilities, authority and interrelationships relative to quality control task.

   g. Quality control is essentially the activities and techniques employed to achieve and maintain the quality of a product, a process and a service.

   h. Three major quality control methods commonly used on construction projects are inspection, testing and sampling.

   i. Quality assurance is broadly the prevention of quality problems through planned and systematic activities to fulfil desired purposes efficiently, effectively and economically.

   j. TQM is a way of planning, organizing and understanding each activity that depends on each individual at each level.

   k. Juran's quality trilogy involves planning, control and improvement.

   l. State whether the following are true or false:

      1. Quality costs = prevention costs + appraisal costs

      2. Quality control costs = internal failure cost + external failure cost

      3. Failure costs = quality control costs + failure costs

2. Discuss the concept of 'quality' for construction industry. How do you define construction quality?

3. Discuss inspection, quality control and quality assurance in a construction project.

4. Explain the ISO 9000 family structure and its benefits. How will you develop a quality system in your organization if you wish to be accredited with ISO 9000?

5. Discuss Juran's suggested steps for quality improvement.

6. What is Juran's quality trilogy?

7. What is total quality management? To what extent is it different from quality assurance system?

8. What are the various types of checklists and inspection reports that have to be designed to ensure proper monitoring and control of quality assurance on the job?

9. Explain the concept and meaning of productivity in general and also in particular with respect to construction. What are the factors affecting productivity? Also, enumerate the main hurdles that are often encountered in construction projects which tend to keep productivity down.
10. What do you mean by total quality management?
11. Discuss the contributions of Deming, Juran and Crosby in the field of quality management.
12. Estimate the quality cost under different heads.
13. Discuss in brief different ISO standards for quality.
14. Discuss the importance of audit in the context of quality conformance.

14

Risk and Insurance in Construction

> Introduction, risk, risk identification process, risk analysis and evaluation process, response management process (risk treatment strategies), insurance in construction industry, common examples of business and project risk, risks faced by Indian construction companies assessing international projects

14.1 INTRODUCTION

Risk management is at the core of any business or organization, and construction industry and construction companies are no exception to this. This is central to any business regardless of size, activity, or sector. Construction companies can lose substantial sums of money as a result of failure to identify and evaluate risk in time. Companies may even forego their opportunity to take advantage of potentially beneficial opportunities arising in the course of their activities if risks are not recognized in good time. Risk management is, therefore, as much about looking ahead to identify further opportunities as it is about avoiding or mitigating losses.

It is often claimed that formal risk management does not begin until the first actual risk assessment has taken place. Risks are rarely ignored when initial plans are made; however, it is very rare to identify all the risks systematically during the initial stages of planning projects. It is well known that managers and their teams generally know what could go wrong and what worthwhile opportunities might occur. Without the benefit of systematic risk analysis, however, it is not always possible for them to exploit their knowledge to the full. Even when an analysis is undertaken, a team will not always maintain and update it equally; sometimes, when risks are foreseen, they are dismissed on the grounds that 'it couldn't happen here'. Thus, through all the phases of a construction project, risk assessment must be adopted as part of a continuous review process. By doing so, the many risks to the business originating out of construction projects can be identified and managed.

The benefits of systematic risk identification and risk management include:

- more realistic business and project planning;
- actions being implemented in time to be effective;
- greater certainty of achieving business goals and project objectives;
- appreciation of, and readiness to exploit, all beneficial opportunities;

```
┌─────────────────────────────┐
│          Context            │
│         Business            │
│         Projects            │
│  What is at risk and why?   │
└─────────────┬───────────────┘
              ▼
┌─────────────────────────────┐
│    Risk identification      │
│ What (and where) are the risks? │
└─────────────┬───────────────┘
              ▼
┌─────────────────────────────┐
│       Risk analysis         │
│  What is known about them?  │
└─────────────┬───────────────┘
              ▼
┌─────────────────────────────┐
│      Risk evaluation        │
│   How important are they?   │
└─────────────┬───────────────┘
              ▼
┌─────────────────────────────┐
│      Risk treatment         │
│ What should be done about them? │
└─────────────────────────────┘
```

Figure 14.1 A simplified risk management process

- improved loss control;
- improved control of project and business costs;
- increased flexibility as a result of understanding all options and their associated risks;
- greater control over innovation and business development; and
- fewer costly surprises through effective and transparent contingency planning.

In this chapter, we discuss the process for identifying, assessing and controlling risk within a broad framework. The main features of this process in a simplified form are illustrated in Figure 14.1. The risk management process described here can be applied to each aspect of business activity and at each level of decision-making.

For a construction organization, projects are the principal means by which a business moves forward. In order to manage risk effectively, the business, project, or sub-project goals need to be dearly identified. This is because it is only in relation to an organization's or an individual's specified goals that risk arises. Confusion over project objectives is itself a major cause of project failure.

Identification of stakeholders is also an important part in clarifying goals and assessing risk. Unless the stakeholders are identified and understood at an early stage, the true extent of the management task and the source of much risk can go unrecognized. Identifying stakeholders also helps define the relationship between the business and its environment, and the context in which its projects will be carried out. Though it may seem easy, not all stakeholders are easily recognizable. There can be many organizations with influence and vested interests than is readily acknowledged. In order to achieve 'viewpoint-oriented' planning, different stakeholders must be identified and taken into account.

14.2 RISK

In literature, the word 'risk' is used to convey different meanings and is synonymously used with terms such as 'hazard' and 'uncertainty'. The researchers in this area do not follow a uniform and consistent definition. Although the term 'risk' has downsides (losses or damages) as well as upsides (profits or gains), more often it is assumed to convey the negative notion only.

Risk in this book follows the definition adopted by Al-Bahar and Crandall (1990), who define it as the exposure to the chance of occurrence of uncertainty. Uncertainty here represents the probability that an event will occur. Accordingly, the risk is assumed to be a function of uncertainty of an event and the likely loss or gain (note) from the event.

As pointed out earlier, modern construction projects are increasingly becoming complex, thereby giving rise to uncertainty at every stage of the project. The risk management process is not systematically performed in construction projects by most of the project stakeholders. Application of certain 'rules of thumb' based on the experience and judgement of stakeholders is most common in the context of construction projects. Unfortunately, risk management in construction projects is assumed to be no more than insurance management where the objective is to find the optimal insurance coverage for insurable risks. As we shall see shortly, risk management is much more than just 'insurances'.

Risk management according to Al-Bahar and Crandall (1990) is a formal, orderly process for systematically identifying, analysing and responding to risk events throughout the life of a project to obtain the optimum or acceptable degree of risk elimination or control. The different processes of risk management are discussed briefly in the following paragraphs.

14.3 RISK IDENTIFICATION PROCESS

Risk identification is the process of identifying all potential sources of project risks and their likely consequences, besides finding out the causes of those risks. Al-Bahar and Crandall (1990) define risk identification process as 'the process of systematically and continuously identifying, categorizing, and assessing the initial significance of risks associated with a construction project.'

As mentioned earlier, risks arise on account of uncertainties and, thus, risk identification aims to identify where the uncertainties exist. Some of the methods used for risk identification are—(1) brainstorming, (2) interviews, (3) questionnaires, (4) availing services of specialists, and (5) past experience.

The identification of risks should be done with a positive approach, and here, the objectives should be to identify not only the risks that present threat but also the presence of opportunities coupled with such threats. The identification of risk should begin early to have maximum impact on the project.

As shown in Figure 14.2, there are six steps involved in the risk identification process. The following sections will discuss each step separately.

```
┌─────────────────────────────────────┐
│         Preliminary checklist        │
└─────────────────────────────────────┘
                  ↓
┌─────────────────────────────────────┐
│ Defining consequences for each risk in checklist │
└─────────────────────────────────────┘
                  ↓
┌─────────────────────────────────────┐
│            Risk mapping              │
└─────────────────────────────────────┘
                  ↓
┌─────────────────────────────────────┐
│          Risk classification         │
└─────────────────────────────────────┘
                  ↓
┌─────────────────────────────────────┐
│ Risk category summary-sheet preparation │
└─────────────────────────────────────┘
```

Figure 14.2 Risk identification process

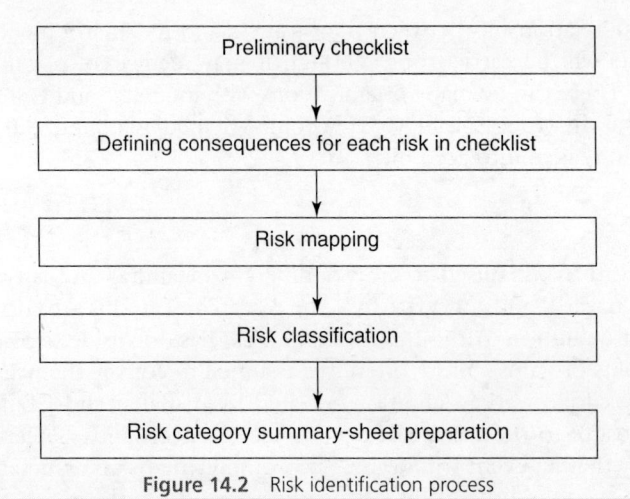

14.3.1 Preliminary Checklist

Preparation of a preliminary checklist is the first and the most important step towards identification of risks. The checklist includes all the risks that may affect project quality, performance, productivity and economy of construction. Brainstorming and questionnaire survey along with past experience help in preparing the preliminary checklist. Some companies have a standard template to prepare the checklist.

14.3.2 Risk Events Consequences Scenario

After the preparation of checklist, the consequences for each of the risks identified are defined. The consequences could be in the form of economic gains or losses, injury to the personnel involved with the work, physical damage, and time- and cost-related savings/overrun. These consequences are tried to be brought to a common scale, preferably in monetary terms.

14.3.3 Risk Mapping

Risk mapping is performed next. Risk map is a two-dimensional plot between probability of occurrence of the uncertainty and its potential severity. Risk map shows Iso-risk curves that are useful for a manager to understand the relative importance of each potential risk. Iso-risk curves contain points of equivalent risk, though there may be differences in probability values and potential severity (see Figure 14.3).

14.3.4 Risk Classification

After the risk mapping, risks are classified under categories. Though several researchers tried different means of classifying risk, the classification proposed by Al-Bahar and Crandall (1990) is reported here. The classification is based on the nature and the potential impact of the risk, and has six broad categories:

- Acts of God such as flood and earthquake
- Physical damage such as damage to structure, worker, equipment, etc.

Figure 14.3 Risk mapping

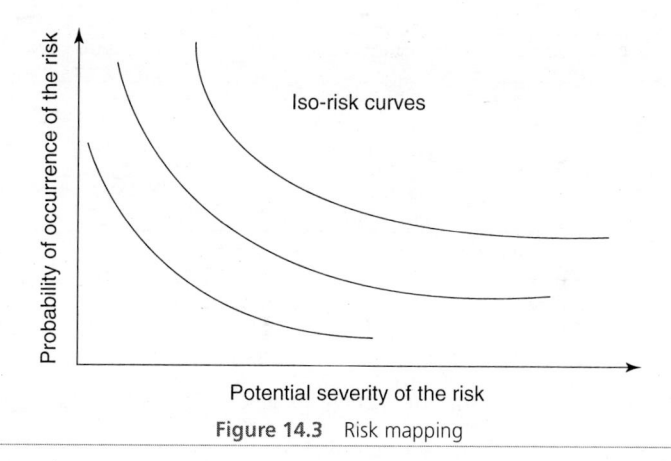

- Financial- and economic-related such as inflation and exchange-rate fluctuation
- Political- and environment-related such as changes in laws and regulations, and war and civil disorder
- Design-related such as incomplete design scope and defective design
- Construction-related such as labour productivity and different site conditions

14.3.5 Risk Category Summary Sheet

Preparation of risk category summary sheet is the last step in risk identification process. The objective of preparing the summary sheet is to involve all the participants in the project management team. The involvement of all the participants is important, as it would not be prudent to delegate the responsibility of judging the risk to a single person. In the risk category summary sheet, each risk event is listed and described, and their interaction is studied collectively by the project management team.

14.4 RISK ANALYSIS AND EVALUATION PROCESS

Risk analysis is the systematic use of available information to characterise the risks, determine how often the specified events could occur, and judge the magnitude of their likely sequence (BS 31100:2008). On the other hand, risk evaluation is the process to decide risk management priorities by evaluating and comparing the level of risk against predetermined standards, target risk levels, or other criteria.

It is not enough to identify risk. From the risk-mapping exercise, some of the risks identified are considered by project management to be more significant and selected for further analysis. What is needed now is to determine their significance quantitatively, through probabilistic analysis, before the response management stage.

The risk analysis and evaluation process is the vital link between systematic identification of risks and rational management of the significant ones. It forms the foundation for decision-making between different management strategies. With respect to this text, risk analysis and evaluation is defined as 'a process which incorporates uncertainty in a quantitative manner, using probability theory, to evaluate the potential impact of risk.' Figure 14.4 is a schematic

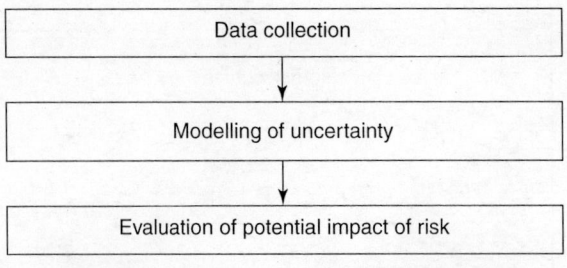

Figure 14.4 Risk analysis and evaluation process

representation of the various components of the process. The following sections will discuss each component separately.

There could be a number of risks associated with a project. There is no point attempting to concentrate on each one of them as the time and effort spent on them may not be commensurate with the returns. Risk mapping establishes the relative importance of different risks in qualitative terms. Now, the important and significant risks are analysed further to decide on the appropriate mitigation strategy.

14.4.1 Data Collection

Collecting data relevant to a given type of risk is an important and intricate step in the risk analysis and evaluation process. Although some contractors do keep data as the work progresses, most of the time it may not be in a structured form and, hence, difficult to put in use. One has to then rely on assessment based on experts' opinion. The data so gathered are put in a proper format to draw appropriate conclusions.

14.4.2 Modelling Uncertainty

This step is aimed to quantify the likelihood of occurrence and potential consequences based on all available information about the risk under consideration (Al-Bahar and Crandall 1990). The likelihood of occurrence of a risk event is measured in terms of probability values based on past data, intuition and expert opinion. The potential consequences of the risk are measured in monetary terms.

14.4.3 Evaluation of Potential Impact of Risk

This is the next step after the uncertainties have been modelled quantitatively. The evaluation of the potential impact of a risk is important in order to get an overall picture of different risks associated with the project. Some of the tools used for such evaluation are—expected value, Monte Carlo simulation, and influence diagram.

14.5 RESPONSE MANAGEMENT PROCESS (RISK TREATMENT STRATEGIES)

After evaluating the risks based on their financial impact, appropriate risk treatment strategies are formulated. The objectives of risk management strategies are to remove the potential impact as far as possible and to increase the control of risk. Some of the risk treatment strate-

```
┌─────────────────────────────────┐
│        Risk avoidance           │
└─────────────────────────────────┘
              ↓
┌─────────────────────────────────┐
│ Loss reduction and risk prevention │
└─────────────────────────────────┘
              ↓
┌─────────────────────────────────┐
│   Risk retention and assumption │
└─────────────────────────────────┘
              ↓
┌─────────────────────────────────┐
│         Risk transfer           │
└─────────────────────────────────┘
              ↓
┌─────────────────────────────────┐
│           Insurance             │
└─────────────────────────────────┘
```

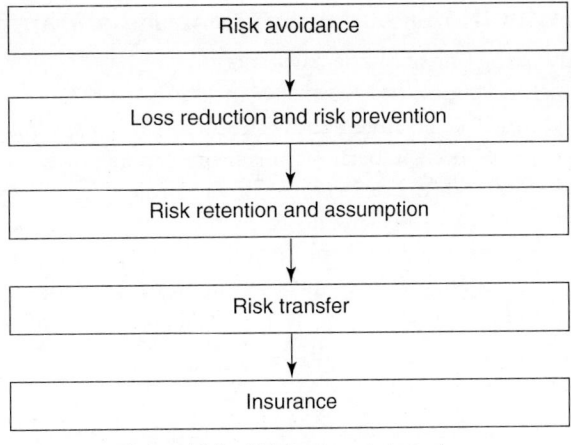

Figure 14.5 Risk treatment strategies

gies within the framework of risk management are shown in Figure 14.5 and discussed in subsequent sections.

14.5.1 Risk Avoidance

Risk avoidance is a fairly common strategy. By avoiding risks, construction companies know that the loss or gain associated with a given risk will not be encountered by them. For example, if a construction company perceives that tunnelling projects are associated with a lot of risks, they would not venture into such projects. Obviously, this also means that the company will not be able to reap the large gains associated with such projects.

14.5.2 Loss Reduction and Risk Prevention

The objective of loss reduction and risk prevention is to reduce the probability of a risk and to reduce the financial severity of the risk should the risk occur. For example, if it were known that theft is a problem in a given locality, the construction company would try to build in security-related costs in their bid. The loss reduction and risk prevention also has a bearing on the insurance premium that the insurer charges from the insured. For example, a company with a low level of loss reduction and a risk prevention programme will be charged a higher premium by the insurer, compared to a company with a high level of loss reduction and a risk prevention programme in place.

14.5.3 Risk Retention and Assumption

Risk retention and assumption is the assumed value of the financial impact of risk. It could be of planned or unplanned type. While planned risk retention and assumption means that a conscious effort has been made to estimate the financial impact of the risk involved in the project, in unplanned risk retention and assumption the construction company may not recognize the existence of risk. Thus, in the latter case, the construction company unconsciously assumes the loss that could occur (Al-Bahar and Crandall 1990). In the unplanned category, one can also envisage a situation wherein the construction company has underestimated the magnitude of the financial impact of a given risk.

14.5.4 Risk Transfer (Non-insurance or Contractual Transfer)

Risk transfer is the transferring of risk by the construction company to the other project participants such as subcontractor, vendor, or specialist contractors. In risk transfer, the construction company enters into a contractual arrangement with subcontractors, vendors, or specialist contractors. The basic principle is that the party who is more capable or better placed to maintain control should be able to take risk. For example, if the project involves construction of pile foundation, a specialist contractor specializing in piling works would be in a better position to bear the risk. Risk transfer is different from insurance in the sense that the risk is not being transferred to an insurance agency; rather, it is being transferred to the specialist who has adequate historical data and is in a better position to evaluate and undertake the risk.

14.5.5 Insurance

Insurance is a device by means of which the risks of two or more persons or firms are combined through actual or promised contributions to a fund out of which claimants are paid. Insurance is a contractual relationship that exists when one party, for a consideration, agrees to reimburse another for loss caused by designated contingencies.

The first party is called the insurer or underwriter; the second, the insured or policyholder; the contract is the insurance policy; the legal consideration is the premium; the loss of life or property in question is the exposure; and the contingency is the happening of the insured event.

Considering the importance that construction organizations attach to insurance as a means to manage risk, it is discussed separately in the following sections.

14.6 INSURANCE IN CONSTRUCTION INDUSTRY

In the construction industry, insurance is one of the most important ways to tackle risk. In fact, insurance is considered as a synonym for risk management in the industry. The majority of construction companies rely on insurance policies for different risk scenarios. They purchase a number of insurance policies depending on the project and contractual requirement. While selecting a given type of policy, a company considers the severity of potential risk, the probability of occurrence of the risk, and the available risk mitigation measures it has under its disposal.

14.6.1 Fundamental Principles of Insurance

The fundamental features of the principles of insurance are given in Table 14.1.

14.6.2 Insurance Policies for a Typical Construction Organization

Some commonly used insurance policies in the context of a large construction organization are given in Table 14.2.

14.6.3 Project Insurance

An element of risk is inherent in all types of engineering projects. It is ever-present in the world of commerce and industry, be it during construction or during operational stage. The insurance policies may be taken by the owner as well as the contractor. An owner organization may go in for policies such as marine-cum-erection (including third-party liability), delayed start-up (advance loss of profit), and so on. The contracting organization may go in for policies such as:

Table 14.1 Fundamental features of the principles of insurance

1.	Insurable interest	(a) There must be life, property, or financial interest capable of being covered. b) Such life, property, etc. must be the subject matter of insurance. c) The insured must be in a legally recognized relationship with the subject matter of insurance, in consequence of which the insured may benefit from its continued safety or the absence of liability, or may be prejudiced by its damage or destruction or the creation of liability.
2.	Utmost good faith	Insurance contracts are contracts of 'utmost good faith'. They are based upon mutual trust and confidence between the insured and the insurer. The proposer has a legal duty to voluntarily disclose all material facts that he knows and also what he ought to know.
3.	Indemnity	Indemnity means 'compensation for loss or injury sustained'. It means to make good the loss or damage subject to sum insured, or more simply, 'an exact financial compensation.'
4	Subrogation	This is the transfer of right and remedies of the insured to the insurer, who has indemnified the loss.
5.	Proximate cause	The active efficient cause that sets in motion a train of events that bring about a result, without the intervention of any force started and working actively from a new and independent source.
6.	Consideration	In insurance contract, payment of premium is the consideration from the insured, and promise to indemnify is the consideration from the insurer. The contract does not come into existence unless the premium is paid.
7.	Average	Average means under-insurance, liability being limited to that proportion of the loss or sum insured bears to the value at risk.
8.	Contribution	Contribution is the 'right of an insurer who has paid a loss under a policy to recover a proportionate amount from other insurers liable for loss.' If an insured has two or more policies on the same subject matter, he cannot recover his loss from all, as this would be more than his actual loss.

Table 14.2 Insurance policies commonly taken by a typical construction organization

S. No.	Policy name	Brief explanation
1.	Fire insurance	Policy 'A' or 'B' covering offices, main depots, transit houses, etc. Fire policy 'C' covering job sites.
2.	Workmen's compensation insurance	To take care of the legal liability of the employer towards the employees under W.C. Act due to accident.
3.	Group personal accident policy	For providing compensation to the employees who meet with accident resulting in fatal or non-fatal injuries.
4.	Group mediclaim policies	To take care of medical expenses necessitated due to hospitalization and domiciliary hospitalization on account of accidents/sickness/disease.
5.	Overseas mediclaim policy	To take care of medical expenses incurred due to sickness/ accident necessitating treatment abroad.

(Continued)

Table 14.2 Continued

S. No.	Policy name	Brief explanation
6.	Cash insurance (cash-in-transit/safe)	For covering loss of cash while it is in transit or in safe.
7.	Fidelity guarantee	For loss caused to the employer due to infidelity of the employees who are responsible for dealing with cash or stores.
8.	Motor insurance	For covering the vehicles (viz., motorcycles/scooters, private cars/jeeps, commercial vehicles) against accidental damage and third-party liability. Goods-carrying vehicles against own damage and liability.
9.	Machinery breakdown policy	For covering loss or damage to plant and machinery due to accidental failure caused by electrical or mechanical breakdown.
10.	Marine insurance	For covering loss or damage to materials, stores and spares, plant, machinery, etc. during transit.
11.	Contractor's plant and machinery insurance (CPM)	For plant and machinery that are essentially used at project sites.
12.	Contractor's all-risk insurance (CAR)	For covering mainly civil jobs.
13.	Erection all risk/storage-cum-erection insurance	For covering mechanical and electrical jobs.
14.	Marine-cum-erection insurance	This is an all-risk policy covering civil, mechanical and electrical jobs with extension of marine/transit risk cover.
15.	Electronic equipment insurance	To have all-risk protection for loss or damage to computers.
16.	Public liability insurance	For legal liability of the owner in respect of fatal/non-fatal injury caused to third-party personnel or damage to third-party property arising out of accident for which insured is held liable under law.
17.	Burglary/theft insurance	The policy covers materials such as temporary structures at site, stores material, spares, tools, etc. located at different project sites in the country.
18.	Group savings-linked insurance scheme	Covering all permanent employees.
19.	Baggage insurance	Covers accompanied baggage of the insured during travel against the risks of loss, damage by accident, fire, theft, etc.

- Construction plant and equipment insurance
- Employer's liability/workmen's compensation
- Marine vessels policy
- Motor vehicles policy
- Cash insurance and temporary properties of employers as well as of contractor

Some important and commonly applicable insurance policies are discussed briefly in the following paragraphs.

14.6.4 Marine-cum-Erection Insurance

The policy covers the contract works and equipment while in transit and during construction/erection at site, during testing and commissioning, and also during the defects liability period. Besides, the policy also covers damage to the surrounding property and liability to third parties.

For goods in transit, the policy covers all risks of physical loss or damage, including war and strike, riots and civil commotion. However, the policy does not cover loss or damage due to—insufficient or inadequate packing; inherent defects; unseaworthiness of vessels; financial default of vessel owners; radioactive contamination; and consequential losses caused by delay.

During construction and erection, the policy covers all risks of physical loss or damage to the project works, including environment perils, location perils, handling perils and negligent acts. Some examples of each of these perils are given below:

Environment perils—storm, flood, earthquake, etc.
Location perils—fire, lightning, theft and burglary, etc.
Handling perils—collision, impact, etc.
Negligent acts—carelessness, negligence, faults in erection, malicious act, etc.

The policy during construction and erection excludes war risks; normal wear and tear; rust, corrosion and erosion; cessation of work; wilful acts or wilful negligence; and consequential (financial) losses.

During the defects liability period, the policy covers contractor's liability for the damage caused by contractor on site during regular maintenance, as well as the damage caused by faults in erection on site. The exclusions during defects liability period are in respect of gradual pollution; damage to project works; the insured's own employees; the vehicles licensed for road use; and marine vessels or aircraft.

14.6.5 Contractor's All-risk Insurance (CAR Insurance)

This engineering policy offers protection against loss or damage to the contract work. There is also a provision to extend cover against any third-party claims while executing the project. All civil engineering works, from a small residential building to a huge bridge, are susceptible to damage from a wide range of causes including fire, explosion, flood, storm, impact and internal defects. Exposure to such damage commences at the time of first delivery of the materials to the contract site and continues to exist till completion of the work. Even after completion of the work, there is exposure especially during the maintenance period, after the civil engineering work is handed over to the owner/client. The contract usually describes the responsibilities of the contractor for loss or damage during the period of the contract and the subsequent maintenance period. The CAR policy can provide the contractor with a comprehensive insurance coverage.

This insurance cover is useful for (a) all civil engineering works including massive dams, bridges, tunnels and docks, (b) residential and office buildings,(c) water treatment plants, canals and roads, and (d) airports, factories, etc.

The policy covers loss or damage to the subject matter from any unforeseen or accidental cause that is not specifically excluded under the policy. Some of the more important causes of loss indemnified under a CAR policy are:

a. Fire, lightning, explosion, impact, aircraft damage
b. Flood, inundation, storm, cyclone, hurricane, etc.
c. Earthquake, subsidence, rockslide and landslide
d. Theft, burglary, riot and strike damage

Some general exclusions in this policy are nuclear perils; war group perils; and wilful act or gross negligence on the part of the insured. Specific exclusions of the policy include deductible excess; faulty design; inventory losses; defective material; bad workmanship; wear and tear, deterioration, normal atmospheric conditions, rusts, and scratching of painted or polished surfaces; cost of rectification or errors unless resulting in physical damage; and loss or damage to vehicles used on the road or waterborne or airborne craft/vehicles.

The period of insurance commences from the first unloading of the property at the contractor's site and expires on the date specified. The cover also ceases for that part of the insured contract work taken over by the principal prior to the expiry date.

The sum insured will be the total of the estimates of possible outlay or outgo under the following heads—contract price; materials or items supplied by the principal; any additional items not included in contract price and materials supplied by the principal; landed cost of imported items as at construction site; construction plant and machinery (restricted to five per cent of the contract value); clearance and removal of debris; insured's own surrounding property; extra charges for overtime, express freight, etc.; increased replacement value for contract price and materials supplied by the principal; and third-party liability.

The policy is also applicable during the maintenance period, usually 12 months after the completed work is handed over. However, the liability under this is restricted to loss or damage caused by the insured in the course of obligatory maintenance under the contract.

The additional cover such as clearance and removal of debris, construction plant and machinery, surrounding property, third-party liability and escalation may be provided in the policy depending on the insured's requirement.

14.6.6 Marine/Transit Insurance

Marine insurance may be defined as 'an agreement whereby the insurer undertakes to indemnify the insured against marine losses, incidental to marine adventure.' The common modes of marine transit are sea, rail/road, air and registered post. The standard types of perils include—inland transit (rail and road); fire; lightning; collision; overturning; derailment; theft; non-delivery; and breakage.

The various types of policies listed under marine/transit insurance are:

1. Specific policies: These are policies issued for a particular shipment. As and when a shipment is made, the insured approach insurers and take out a policy insuring that particular shipment only.
2. Open covers: These are agreements between the insured and the insurers to cover all shipments/dispatches during the agreed period of insurance, which is normally one year. Open policy is normally issued for inland transit and open cover for overseas shipments.

3. Cover note: These are issued by insurers when a policy cannot be issued for want of important details like name of the vessel, bill of lading number and date, RR/LR/AWB number and date, etc. These are valid for a particular period only and should be replaced by policies when dispatch particulars are obtained.
4. Certificate of insurance: This is normally issued when specific policies are not required to be issued under open policies, mainly to comply with bank stipulations.

14.6.7 Fire Policy

This policy covers factories, depots and stores. The risks covered under the policy are—fire/lightning; explosion/implosion; aircraft damages; riot/strike/malicious and terrorism damages; impact damages; storm/cyclone/typhoon/hurricane/tornado/flood and inundation; landslide/subsidence including rockslide; bursting and/or overflowing of water tanks apparatus and pipes; missile-testing operation; leakage from automatic sprinkler; and bushfire. The policy may have an additional cover for spontaneous combustion, earthquake, debris removal, etc.

14.6.8 Plant and Machinery Insurance

This is a comprehensive policy covering unforeseen and sudden physical damage to the property deployed/used at various locations/projects anywhere in the country. The types of cover are applicable for the following:

1. Material damage: This covers equipment of all types, including earth-moving equipment, all types of cranes, pumps (including concrete pumps), concrete mixers, welding machines, compressor, DG sets, batching plants, transformers, drilling rigs, piling hammer, hot mix plant, concrete pavers, truck-mounted boom placers, formwork materials and scaffolding materials.
2. Third-party liability: Under this, the insurance company indemnifies the insured against the legal liability for accidental loss or damage caused to the property of other persons and against the legal liability for fatal or non-fatal injury to any person other than employees.
3. Increased cost of working: The insurance company indemnifies additional costs incurred by the insured, to ensure continued operation with substitute equipment, subject to limit specified.
4. Inland transit: This insurance covers all risk of loss or damage to the subject matter. Transit insurance starts from the time the goods leave the warehouse/store until delivery to the final destination.

The indemnity limit varies depending on the type of cover.

14.6.9 Liquidity Damages Insurance

In general, under the provisions of construction agreements, the contractor is responsible to the owner for delay and/or under-performance of the project caused by technological failure or fault on the part of the contractor (including his subcontractors and suppliers). In such circumstances, the contractor is obliged to pay liquidated damages to the owner in amounts that should equate to the financial obligations of the owner to the project lenders. Cover for such damages is available to the contractor by way of liquidated damages insurance, which is designed to protect the contractor for liability assumed under contract for the payment of liquidated damages to the owner for late completion and/or performance shortfall, following errors and omissions on the part of the contractor, subcontractors and/or suppliers in connection with the work to be performed under the terms of the construction agreement. Such work could cover the engineering, design, procurement, construction and commissioning of the project.

14.6.10 Professional Indemnity Policy

The need for this policy is felt by construction companies that offer design and consultancy services. Besides, some of the contract conditions these days stipulate insurance policy to cover legal liability arising out of design defects. The policy offers an indemnity limit of ₹ 100 crore covering various locations in the country.

14.7 COMMON EXAMPLES OF BUSINESS AND PROJECT RISK

Some of the common project risks identified in BS 31100:2008 are given in Box 14.1. The list though is not comprehensive.

14.8 RISKS FACED BY INDIAN CONSTRUCTION COMPANIES ASSESSING INTERNATIONAL PROJECTS

The construction industry is a high-risk and low profit-margin industry. In the area of international construction, the risks are much more due to the political, legal, financial and cultural complexities involved. The Indian construction industry has only two companies that figure in the list of 'Top 225 International Contractors 2006' published by *Engineering News Record* (ENR) in August 2006, and India's market share of US$2.15 billion (Exim Bank 2006) is just 0.05 per cent of the estimated US$3–4 trillion international construction market, according to Economic Times Intelligence Group (ETIG 2005). In India, there is a lack of international experience but the opportunities for growth for companies willing to enter the international arena are enormous. The domestic construction industry is growing at a rapid pace, due to very high demand for housing and increase in spending on infrastructure. The Indian construction industry contributes substantially to the GDP (about 6 per cent), while its contribution to exports is small (2.13 per cent). Indian companies are expanding rapidly and will soon need to look to the international market for further growth, and also for acquiring the latest technology. The Indian industry is labour-intensive and the application of latest technologies for construction is seen mainly in projects of national importance. These factors would have an influence on the Indian companies' attitude towards addressing the risks in the international construction market. In this section, we view the risks from the Indian perspective.

14.8.1 Risks in International Construction

International construction projects are more complex due to conditions such as multiple ownership, elaborate financial provisions and different political ideologies, and thereby contain higher risks as compared to domestic market (Han et al. 2005). Wang (2004) has carried out a detailed analysis of international construction risks. Twenty-eight critical risks associated with international construction projects in developing countries have been

Box 14.1 Common examples of business and project risk

Business and Common Project Tasks

Human resource related
- management competence
- corporate policies
- management practices
- poor leadership
- inadequate authority
- poor stag-selection procedures
- lack of clarity over roles and responsibilities
- vested interests
- perceptual errors regarding risks
- individual or group interests
- personality clashes

Political/Societal-related business risks
- unexpected regulatory controls or licensing requirements
- changes in tax or tariff structure
- nationalization
- change of government
- war and disorder
- failure to obtain appropriate approval
- higher than anticipated compensation costs

Environmental risks
- natural disasters (floods, landsliders, etc.)
- Storms/tempests, pollution incidents
- aircraft/ship

Legal
- unforseen inclusion of contingent liabilities
- loss of intellectual property rights
- failure to achieve satisfactory contractual arangement

Economic/Financial
- exchange-rate fluctuation
- interest-rate instability
- inflation
- shortage of working capital
- failure to meet revenue targets

Commercial
- under-performance to specification
- management under performance
- collapse of contractors
- insolvency of promoter
- failure of suppliers to meet contracts (quality or quantity or timescale)
- cost and time overruns
- failure of plant and machinery
- insufficient capital revenues
- market fluctuations
- fraud

Technical/Operational
- inadequate design
- professional negligence
- human error/incompetence
- structural failure
- operational lifetime lower than expected
- dismantaling/decommissioning costs
- safety
- performance failure
- residual maintenance problems

identified, categorized into three (country, market and project) hierarchy levels, and their criticality evaluated and ranked. The top ten critical risks are—*approval and permit, change in law, justice reinforcement, credit worthiness of local entities, political instability, cost overrun, corruption, inflation and interest rates, government policies, and government influence on disputes.* The influence relationship among the risks in the three risk hierarchy levels has also been identified and confirmed. The risks at country level are more critical than those at market level, while the latter are more critical than those at project level. In addition, the risks at higher hierarchy level have dominating impacts on the risks at lower level. For the identified risks, the researcher has also provided and evaluated some practical mitigation measures.

Following case study approach, Bing and Tiong (1999) have proposed a risk management model for international construction joint ventures (JVs). The risk management process consists of three typical phases—(1) identification, (2) analysis and (3) treatment. The researchers have further identified a set of 25 risk factors applicable to international construction joint ventures. Bing et al. (1999) have grouped these risk factors into three main groups—(1) internal; (2) project specific; and (3) external. They examine the most effective mitigating measures adopted by construction professionals in managing these risks for their construction projects in East Asia. Based on an international survey of contractors, it has been found that the most critical risk factors exist in the *financial aspects of JVs, government policies, economic conditions* and *project relationship.* Turnbaugh (2005) provides guidelines to identify and quantify possible risk elements to the project, and then outline potential risk mitigation and control measures. Ten major areas of risk and a summary of potential risk elements and indicators as well as preferred risk responses have been identified.

In addition to identifying and evaluating the risk factors in general, some researchers have also tried to explore a particular factor in international construction risk scenario. For example, Kapila (2001) has stressed the financial risk factors associated with international construction ventures. He further examines the most effective mitigation measures adopted by construction professionals in managing these risks for their construction projects, and suggests strategies to minimize foreign-exchange risk and to better manage foreign-exchange dealings. Stallworthy and Kharbanda (1983) have emphasised project financing and rated it above technological excellence in export project development. Ashley and Bonner (1987) have studied the political risks in international construction.

Han et al. (2005) describe findings from experiments done to investigate the risk attitude and bid-decision behaviour in the selection of international projects. The participants demonstrated either weak risk seeking in profit situations or strong risk seeking towards loss situations when choosing between conflicting options of risky opportunities and sure payoffs. On the other hand, another experimental test attempting to investigate bid behaviour when making a realistic bid or no-bid situation in a complicated international construction project reveals the prevailing risk aversion. Further, they find the experimental supports for some of the errors and biases due to risk attitude that commonly exists in bid decisions in this area.

Hastak and Shaked (2000) have recommended an international construction risk assessment model (ICRAM-1) that assists the user in evaluating the potential risk involved in expanding operations in an international market by analysing risk at the macro (or country environment), market and project levels. They have discussed some of the existing models for country risk assessment, and further presented potential risk indicators at the macro, market and project levels. ICRAM-1 provides a structured approach for evaluating the risk indicators involved in an international construction operation and is designed to examine a specific project in a foreign country. It can be used as a tool to quantify the risk involved in an international construction investment as one of the preliminary steps in project evaluation.

It may be concluded that the researchers investigating the risk aspects in international construction have worked primarily in the areas of risk identification, classification, analysis, evaluation through risk assessment models, and developing the strategy for risk mitigation. Researchers have also focused their attention on studying the bid/no-bid situation, the attitude of contractors in selection of international project, and the entry strategy for foreign construction markets.

Based on the literature review and the views of experts, a list of risk factors applicable in international construction, referred to in the study as R1, R2,… R14, have been identified. They are summarized in Table 14.3.

Table 14.3 Summary of risk factors

S. No.	Risk	Id	Brief description
1.	Poor government responsiveness	R_1	Delay in approvals; inconsistent approach towards tax laws, foreign firms, environmental laws, expatriate laws, finance laws, etc.; corruption levels
2.	Weak legal system	R_2	Not universally understood; not effective and efficient; weak protection of intellectual property
3.	Political instability	R_3	Unstable government; inconsistency in approach of central and state/provincial governments; probability of nationalization of projects
4.	Cultural differences	R_4	Inability to reconcile differences in work culture and language values, racial prejudices between foreign and local partners, attitude of the public towards foreign firms
5.	Force majeure	R_5	Natural and manmade disasters that are beyond the firm's control—e.g., flood, earthquake, war, etc.
6.	Poor financial capability of the local partner	R_6	Financial soundness of the local partner
7.	Foreign-exchange risk	R_7	Exchange rate and interest rate fluctuations; unexpected inflation
8.	Inaccurate assessment of market demand	R_8	Inaccurate assessment of market demand made by owner or local partner
9.	Low project team cohesion	R_9	Poor interpersonal relations among multinational team members

(*Continued*)

Table 14.3 Continued

S. No.	Risk	Id	Brief description
10.	Ambiguous project-scope definition	R_{10}	Ambiguous scope definition due to different systems and standards in foreign countries, and unfamiliar contract conditions; inadequate design detailing
11.	Poor cost management and control	R_{11}	Delay or default in payments by owner; inadequate cash flow
12.	Poor project management	R_{12}	Inadequate or poor planning and control due to lack of organization structure or incompetence of project team, and due to difficulty in assessing capabilities in foreign countries
13.	Poor productivity and quality	R_{13}	Low productivity and quality standards of the local workforce due to outdated technology and inadequate training and supervision
14.	Weak safety ethos	R_{14}	Inadequate emphasis on safety leading to high accident rate

REFERENCES

1. Al-Bahar, J.F. and Crandall, K.C., 1990, 'Systematic risk management in construction projects', *ASCE Journal of Construction Engineering and Management*, 116(3), pp. 533–546.

2. Ashley, D. B. and Bonner, J.J., 1987, 'Political risks in international construction', *Journal of Construction Engineering and Management*, ASCE, 113(3), pp. 447–467.

3. Bing, L. and Tiong, R.L.K., 1999, 'Risk management model for international construction joint ventures', *Journal of Construction Engineering and Management*, ASCE, 125(5), pp. 377–384.

4. BS 31100:2008 British Standard on Risk management: Code of practice.

5. Devaya, M.N., 2007, 'Study of construction export potential of Indian construction industry', Masters thesis, Department of Civil Engineering, Indian Institute of Technology, Delhi.

6. Han, S.H., Diekmann, J.E. and Ock, J.H., 2005, 'Contractors risk attitudes in the selection of international construction projects', *Journal of Construction Engineering and Management*, ASCE, 27(4), pp. 283–292.

7. Hastak, M. and Shaked, A., 2000, 'ICRAM-1: Model for International Construction Risk Assessment', *Journal of Management in Engineering*, ASCE, 16(1), pp. 59–69.

8. Kapila, P. and Hendrickson, C., 2001, 'Exchange rate management in international construction ventures', *ASCE Journal of Management in Engineering*, 17(4), pp. 186–191.

9. Stallworthy, E.A. and Kharbanda, O.P., 1983, *International Construction*, Gower Publishing Company Ltd, England.

10. Turnbaugh, L., 2005, 'Risk management in large capital projects', *Journal of Professional Issues in Engineering Education and Practice*, ASCE, October 2005, pp. 275–280.

11. Wang, S. Q., 2004, 'Risk management framework for construction projects in developing countries', *Construction Management and Economics*, 22(3), pp. 237–252.

REVIEW QUESTIONS

1. State whether True or False:
   a. Realistic business and project planning, and cost saving are some of the benefits of systematic risk management.
   b. Risk management is a formal and orderly process for systematically identifying, analysing and responding to risk events throughout the life of a project.
   c. Risk identification process is the process of systematically and continuously identifying, categorizing and assessing the initial significance of risks associated with a construction project.
   d. For a construction organization, projects are the principal means by which a business moves forward.
   e. Identification of stakeholders is not an important part in clarifying goals and assessing risk.
   f. Iso-risk curves are used for risk identification and mapping.
   g. Risk analysis and evaluation process includes data collection, modelling of uncertainty, and evaluation of potential impact of risk.
   h. Risk treatment strategies are risk avoidance, loss reduction and risk prevention, risk retention and assumption, risk transfer, and insurance.
   i. The three broad types of risk factors are internal, project-specific and external.
   j. Brainstorming and interviews are some of the methods for risk identification.

2. Match the following:

(1) What is at risk and why?	(a) risk analysis
(2) What (and where) are the risks?	(b) risk treatment
(3) What is known about them?	(c) risk identification
(4) How important are they?	(d) context (business/projects)
(5) What should be done about them?	(e) risk evaluation

3. Arrange the following risk identification process in sequence—(a) risk classification, (b) defining consequences for each risk in checklist, (c) risk category summary-sheet preparation, (d) preliminary checklist, (e) risk mapping.
4. What are the benefits of systematic risk identification and risk management?
5. What are the different steps involved in risk management process?
6. What is meant by risk?
7. Discuss the importance of insurance in risk management.
8. Discuss different risk treatment strategies.

9. Discuss some important insurance policies prevalent in construction industry.
10. Discuss the importance of risk analysis and risk evaluation.
11. What is the right method of responding to any risk in a construction project? Give examples of (a) risk avoidance, (b) risk reduction, (c) risk transfer and (d) risk sharing.
12. Can you share and transfer the risk simultaneously?
13. What are the techniques used to identify the risk in a new project.
14. Large infrastructure projects such as airports and highways face many known and unknown risks at various stages of project execution. Using an example of a highway project, explain these risks for all the phases.
15. How do you handle unknown types of risks that may arise during the execution of a construction project?
16. What are the areas in construction project risks where insurance is possible?
17. Can manmade risks be always covered by insurance?
18. Comment on a project insurance policy by referring to general provisions, specific provisions and exclusions in a CAR policy.
19. Write short notes on the risks in international contracts.

15

Construction Safety Management

Introduction, evolution of safety, accident causation theories, foundation of a major injury, health and safety act and regulations, cost of accidents, roles of safety personnel, causes of accidents, principles of safety, safety and health management system, research results in safety management

15.1 INTRODUCTION

Construction industry is an integral part of infrastructure development in a nation, leading to economic growth and development. Economic growth is not possible without contribution from construction. However, construction is inherently hazardous. This is evident from the comparative statistics of fatal and non-fatal injuries that take place year after year in different industries (Figure 15.1) including construction. According to one study, construction accounts for most of the injuries next to mining industry. Hinze and Applegate (1991) cite the accident facts (1990) and report that the frequency of occurrence of disabling injuries incurred by construction workers is roughly twice the frequency rate of other industries and the death rate is roughly three times that of other industries.

The reasons behind the statistics are manifold. As pointed out earlier, construction is mostly a one-off activity and situations encountered in construction projects are unique. Construction workers, especially in our country, are mostly unskilled and uneducated (according to CIDC country report 2005–06, the unskilled workforce constitutes 73.1 per cent of the total construction workforce of about 33 million people). Further, they have to work under stiff outdoor climatic conditions. The workers are mostly employed on temporary basis and they often change their employers. Even on the same site, their services may be required for different construction activities under different charge hands and foremen. Today, construction is practised under the framework of multi-layered subcontracting system, with workers of different skill sets and age groups having varied training needs.

Despite the unique situations faced by the construction industry, statistics show that there is a significant reduction in the number of fatal and non-fatal injuries over the years (see Table 15.1 and Figure 15.2). Human life is precious and it should be the constant endeavour of all stakeholders to make the construction site a safe place to work. In addition to the human aspect, construction injuries also have legal and economic aspects associated with them, requiring all the more attention. Construction safety is of paramount importance in any construction.

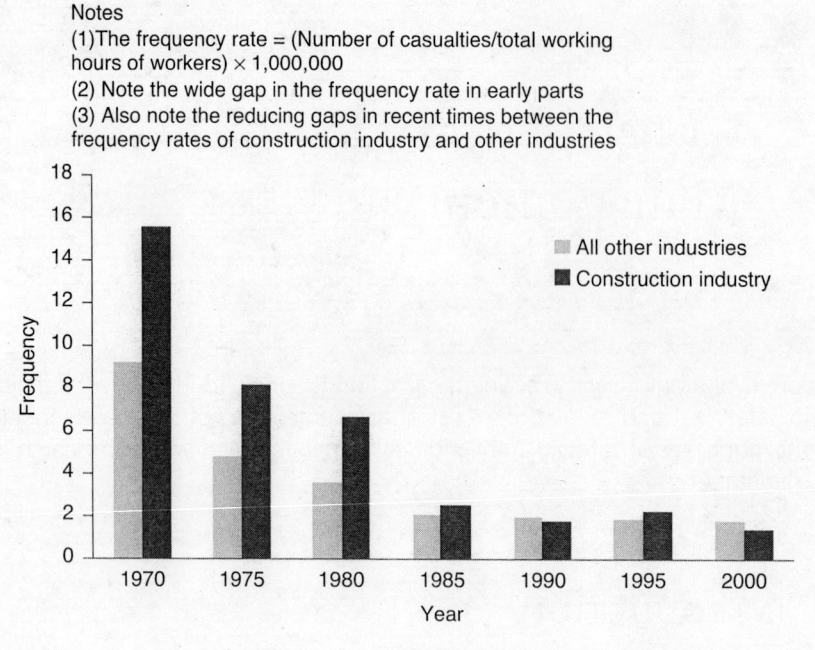

Notes
(1) The frequency rate = (Number of casualties/total working hours of workers) × 1,000,000
(2) Note the wide gap in the frequency rate in early parts
(3) Also note the reducing gaps in recent times between the frequency rates of construction industry and other industries

Figure 15.1 Comparison of the frequency rates for construction and all other industries in Japan

Table 15.1 Statistics of man–hours worked and frequency rate of accidents over the years for a large Indian construction company

Year	Man-hours worked (in millions)	Frequency rate
1995–96	225.1	3.79
1996–97	267.2	3.02
1997–98	306.6	2.04
1998–99	375.5	1.53
1999–2000	320	1.4
2000–01	278.8	1.28
2001–02	283.56	1.04
2002–03	332.57	0.8
2003–04	377.9	0.59
2004–05	447.5	0.4
2005–06	454.8	0.34
2006–07	499.3	0.33
2007–08		0.24

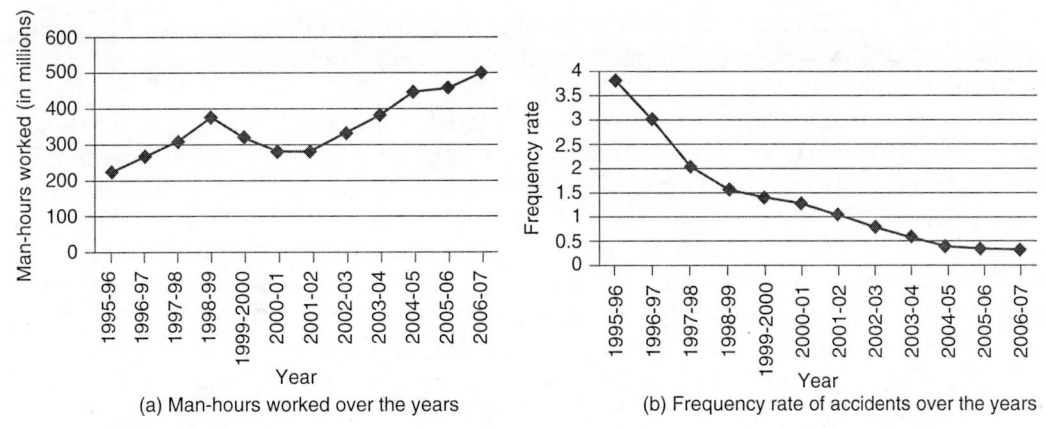

Figure 15.2 Details of man-hours worked and decline of frequency rate[1] of accidents over the years

(a) Man-hours worked over the years

(b) Frequency rate of accidents over the years

The objective of this chapter is to introduce to the readers the various façades of construction safety. In the initial part of this chapter, we discuss the evolution of safety and accident causation theories. Later, some key terms associated with construction safety are defined and discussed. In the middle part, we discuss some of the commonly observed accidents at construction sites, health and safety regulations, etc. Finally, we close the chapter by discussing the safety principles, the elements of safety management, and the results of some researches carried out in construction safety.

15.2 EVOLUTION OF SAFETY

It was in the year 1867 that workers' compensation laws were drafted for the first time to provide some measure of protection to industrial workers. This small but significant step can be marked as the birth of the modern safety movement. In the initial days, the scope of 'safety' was restricted to accident prevention and to analyse the cause of accident. These included removal of physical hazards and improving the industrial environment. Heinrich did considerable scientific work in the area of industrial accident prevention during this era. Later, the scope of 'safety' was enlarged to protect workers' health.

Organizations such as International Labour Organization (ILO) and World Health Organization (WHO) promoted occupational health of industrial workers globally. The word 'industrial' was replaced with 'occupational' to embrace all types of employment, instead of restricting it to factories and mines. The International Labour Organization identified the need for action for reducing the risks of accidents and adopted a Convention (No. 62) concerning minimum safety standards in the building industry, as far back as in June 1937. Thereafter, the organization adopted a comprehensive Convention (No. 167) and Recommendation (No. 175) on safety and health on the construction industry in June 1988.

Propagation and promotion of safety and health in industries was not an easy task. It was seen as a costly and unproductive investment, and profit-oriented industries were reluctant to

---

[1] The frequency rate shown is that of a large construction company and should not be generalized for the whole construction industry. There are many accidents that go unreported and employers still go scot-free by paying minimal compensation to workers. The employers in general are, thus, really not worried and do not consider it worthy to invest in safety, though the situation is bound to improve due to increasing awareness.

pay attention to it. Therefore, to fulfil the social obligation of protecting workers engaged in diverse occupations, various legislations were enacted and regulatory authorities appointed in different countries.

The 'safety' movement in India formally started when the National Safety Council was set up in 1966. The council recognized that safety is the responsibility of management and in this endeavour, the active support of workers is needed. In order to generate awareness, March 4 of every year is celebrated as National Safety Day. Lok Sabha passed a construction workers' bill in August 1996 to regulate the work conditions on all construction sites. The bill is applicable to all employers who employ 50 or more workers on any day at one go or in relay. The bill proposed the appointment of a director general of inspection with various subordinate inspectors to check adherence to the provisions of the bill, including workers' wages, working hours, temporary accommodation and other welfare measures. Subsequently, in 1998, the central rules on Building and Other Construction Workers (Regulation of Employment and Conditions of Service) also came in existence, and these are applicable to central establishments.

Various state governments are in advanced stages of framing the rules to be applicable in their respective states. Where rules are already in place, the strictness in enforcing them needs a lot of improvement. Although awareness about safety is increasing, a lot more is still needed from different stakeholders to reach up to global standards in occupational health and safety.

15.3   ACCIDENT CAUSATION THEORIES

An accident is defined as an event that is unplanned, undesired, unexpected and uncontrolled, and one that may or may not result in damage to property or injury to person, or both, in the course of employment. ILO defines occupational accident as an unexpected and unplanned occurrence, including acts of violence, arising out of or in connection with work, which results in one or more workers incurring a personal injury, disease, or death. In this section we discuss why accidents occur. A number of theories have been developed over the years to understand and explain accident causation. We briefly discuss a few major theories.

Heinrich, one of the early researchers (his book *Industrial Accident Prevention* was first published in 1931), propagated the domino theory to explain the causation of accidents. According to Heinrich, accidents are the result of a chain of sequential events. He compared these events with dominoes. When one of the dominoes falls, it triggers the collapse of the next domino, and the next, and so on. The five dominoes used by Heinrich in his theory are—(1) social environment and ancestry, (2) fault of person, (3) unsafe act and/or unsafe condition, (4) accident, and (5) injury. He suggested that removing a key factor such as an unsafe act or an unsafe condition would prevent the start of the chain reaction. This five-domino model suggested that through inherited or acquired undesirable traits such as stubbornness, greed and recklessness, people become bad-tempered, inconsiderate, greedy and reckless, which may lead them to commit unsafe acts or cause the existence of mechanical or physical hazards, which in turn cause injurious accidents. The easiest domino to control, according to Heinrich, is the unsafe act and/or unsafe condition, which is the central factor of his domino theory. Heinrich studied a large number of accident cases from the insurance records and found that 88 per cent of accidents were attributable to unsafe act of persons involved, 10 per cent were due to unsafe mechanical/physical condition, and the remaining 2 per cent were a result of natural calamities. Heinrich believed that people are the fundamental reason behind any accident. He was also of the opinion that management is responsible for prevention of accidents. He advocated strict supervision, remedial training and discipline to eliminate unsafe conditions.

The domino theory presented above was quite simple in explaining accident causation and it was easy to implement as well. Using this theory, it was possible to pinpoint the person responsible for the lapses leading to accident. Although his work faced criticism because of oversimplifying the human behaviour and the statistics on the contribution of unsafe acts and unsafe conditions, Heinrich's work still remains the foundation for many other researches undertaken subsequently. Over the years, the domino theory has been updated by many researchers. These models are known as management models or updated domino models in which the emphasis of accident causation is placed on faulty management system.

One updated model was given by Vincoli (his book *Basic Guide to Accident Investigation and Loss Control* was first published in 1994), in which he replaced the original five dominoes of Heinrich with the following:

1. Management: Loss of control
2. Origins: Basic causes
3. Immediate causes: Symptoms
4. Contact: Incident
5. Loss: People–property

According to Vincoli, lack of control by management initiates the process that eventually results in incidents. Vincoli distinguishes between incident and accident. He defines incident as any event that has the possibility of creating a loss, and a loss event as an 'accident'. Failure of management to fulfil its responsibility leads to basic causes from which incidents arise. The basic causes belong to personnel factors and job factors. Some examples of personnel factors are—lack of understanding or ability; improper motivation (bad attitude); and illness and mental or personal (non-work-related) problems. Examples of job factors are—inadequate work; bad design or maintenance; low-quality equipment; and normal or abnormal wear and tear. The unsafe acts and conditions are symptoms of root causes that the first and second dominoes represent. Management allowing these factors to continue unchecked leads to incidents. Incident has the possibility of creating a loss (minor, serious, or catastrophic), and a loss event is an 'accident'.

Petersen (1971) believed that root causes of accidents often relate to the management system. He developed a non-domino-based management model in 1971, known as multiple causation model. According to Petersen, a number of causes and sub-causes combined together in random fashion are responsible for an accident, and it is not possible to attribute the causation of an accident to a single cause as suggested simplistically in the domino theory.

In order to support his argument, Petersen gave an example of a worker falling off a defective stepladder. The domino theory would attribute the cause of this accident to unsafe act/unsafe condition. The climbing of a defective ladder is an unsafe act, while the presence of a defective ladder would constitute the unsafe condition. The remedial measure would be to remove the defective ladder. However, this is treating the problem only superficially. The root cause is not yet identified and there could be a similar instance of a defective ladder present at other work sites, and similar accidents can happen in the future as well.

Using the multiple causation theory, the investigator would address issues such as—why the defective ladder was not found in normal inspections; why the supervisor allowed its use; whether the injured employee knew that he/she should not use the ladder; whether the employee was properly trained; whether the employee was reminded that the ladder was defective; and whether the supervisor examined the job first. Once all these causes and sub-causes are analysed, it would lead to an improved situation. Thus, using the multiple causation model the investigator would recommend improved inspection procedures, improved training, better definition of responsibilities, and pre-job planning by supervisors.

Accident causation was also tried to be explained using human error theories. Notable among them are the behavioural model and the human factor model. Proponents of the behavioural model believe that accidents are caused mainly due to the fault of workers. The behavioural models are based on the accident proneness of a person. It says that accidents are not randomly distributed and certain characteristics inherent in a person make him/her accident-prone. Under a similar set of circumstances, an accident-prone person is more likely to be involved in the accident than a person who is not accident-prone. The theory emanates from the analysis of accidents of a large population in which the majority of people have no accidents, small percentages have one accident, and a very small percentage (these are also referred to as 'accident repeaters') has multiple accidents. The explanation of the accidents of this small percentage of population is given by their innate propensity for accidents. Nevertheless, the theory could not explain the causation of a number of accidents, and some researchers believe the theory could explain accident causation in 10 per cent–15 per cent of accident cases at best. A number of theories were developed to explain the mystery of accident repeaters. Notable among these is 'goals-freedom-alertness theory'. The theory states that safe work performance is the result of a psychologically rewarding work environment. Under the goals-freedom-alertness theory, accidents are viewed as low-quality work behaviour occurring in an unrewarding psychological climate, which does not contribute to a high level of alertness. The essence of the theory is that management should let a worker have a well-defined goal and give the worker the freedom to pursue that goal. The result will be that the worker focuses on the task that leads to that goal. The worker's attentiveness to the job will reduce the probability of being involved in an injury. In other words, a worker who knows what to do on a job will be well focused on the task to be performed and, therefore, will be safe.

Ferrel's theory (cited in Heinrich et al. 1980) is an important accident causation model under the 'human factor model' category. Ferrel developed the accident causation model based on a causal chain of human factors. He believed that accidents are caused due to human errors that result from (1) overload beyond the capacity of a human being, (2) incorrect response by the person due to incompatibility with which the human being is subjected to, and (3) performing an improper activity due to either lack of awareness or deliberately taking the risk. He emphasised the overload and incompatibility factors.

The 'accident root-cause tracing model' (ARCTM) developed by Abdelhamid and Everett (2000) helps in identifying the root cause of an accident in an easy manner. According to this model, accidents occur due to one or more of the three root causes—(1) failing to identify an unsafe condition that existed before an activity was started or that developed after an activity was started, (2) deciding to proceed with a work activity after the worker identifies an existing unsafe condition, and (3) deciding to act unsafe regardless of initial conditions of the work environment. In this model, a series of questions are asked, such as—how did the unsafe condition exist or develop; why did the worker decide to proceed with the work despite identifying the existing unsafe condition; does the worker know the correct procedure of doing the work; has the worker always/occasionally proceeded with the work despite identifying unsafe condition; and so on. Such systematic questioning guides the investigator to find out why the accident occurred, how the root cause of the accident developed, and how it could be eliminated.

Hinze developed the 'distractions theory' to explain accident causation in situations where there is an existence of (1) a recognized safety hazard or a mental distraction, and (2) a well-defined work task. He defined hazard as—(1) a physical condition with an inherent quality that can cause harm, and (2) the preoccupation with work-related or non-related issues such as approaching deadline, anticipated parties, or death in the family. The lower the distractions from a known hazard, the greater is the probability of completing a task safely. On the other hand, the higher the level of focus on the distractions posed by the hazard, the lower is the probability of achieving the task safely. Further, the theory claims that under similar hazardous and well-defined situations, the worker with the more heavy mental baggage (mental distraction) has the maximum chance not to complete the task in a safe manner.

15.4 FOUNDATION OF A MAJOR INJURY

The recognition that 'any unintended occurrence' is an accident is the first requirement of hazard control. It is altogether a different matter whether such accidents result in an injury or not. It can be observed that on many occasions the accidents do not cause any injury due to a number of reasons. In construction, for example, objects falling from a height may miss a person by a whisker. Such instances are also known as near misses and they are as important as accidents involving injuries.

According to a study conducted by Heinrich (1959), the ratio of 'no injury' to 'minor injury' to 'major injury' is **300:29:1** (refer to Figure 15.3). Underlying these minor injuries are numerous unsafe practices and unsafe conditions that, fortunately, may not result in any incident. Bird and Loftus (1982) have updated this ratio with further information on property damage accidents. According to this study, the ratio of 'near-miss accidents' to 'property damaging accidents' to 'accidents involving minor injuries' to 'accidents involving serious or disabling injuries' is **600:30:10:1** (see Figure 15.4). The moral of these ratio studies is that accident prevention must start with prevention of unsafe practices and unsafe conditions as well as of minor injuries.

15.4.1 Unsafe Conditions

An unsafe condition is one in which the physical layout of the workplace or work location, and the status of tools, equipment and/or material are in violation of contemporary safety standards (Abdelhamid and Everett 2000). A few examples of unsafe conditions are:

1. Defects of agencies such as rough, sharp, or slippery work, defective equipment, overloaded tools or equipment, defective ladders at site, and improperly constructed scaffolds
2. Dress or apparel hazards such as lack of protective equipment and improper clothing
3. Environmental hazards such as inadequate aisle space, insufficient work space, inadequate ventilation and improper illumination

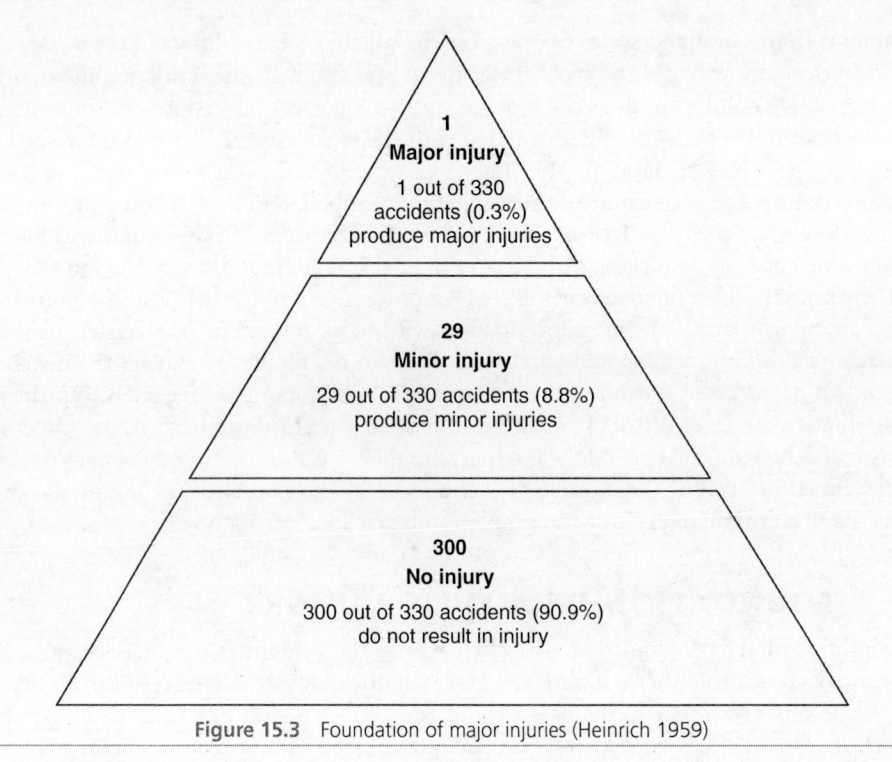

1
Major injury
1 out of 330 accidents (0.3%) produce major injuries

29
Minor injury
29 out of 330 accidents (8.8%) produce minor injuries

300
No injury
300 out of 330 accidents (90.9%) do not result in injury

Figure 15.3 Foundation of major injuries (Heinrich 1959)

1
Serious or disabling injuries

10
Minor injury, i.e., any reported injury less than serious

30
Property-damage accidents of all types

600
No visible injury or damage

Figure 15.4 Foundation of major accidents/injuries (Bird and Loftus 1982, cited in Mining Safety Handbook)

4. Placement hazards such as inadequately guarded, unguarded, unshielded, or protruding ends of reinforcing rods, protruding nails and wire ties, and unshored trenches

Some photographs showing unsafe conditions are given in Figure 15.5 and Figure 15.6. Toole (2002) identified some of the root causes of construction accidents and classified these under unsafe conditions. These conditions include lack of proper planning, deficient enforcement of safety, absence of safety equipment, unsafe methods or sequencing, unsafe site conditions such as poor housekeeping, broken ladder and structurally deficient work platform. Abdelhamid and Everett (2000) distinguish unsafe conditions in two categories depending on their occurrence in the work sequence and the person responsible to trigger them. Accordingly, we have an unsafe condition that exists before the work starts and an unsafe condition that develops after an activity has started. These two types of unsafe conditions are due to one of the following causes:

1. **Management actions/inactions:** For example, management may fail to provide proper or adequate personal protective equipment; fail to maintain or safeguard tools and equipment; and/or violate workplace standards by allowing slippery floors, insufficient ventilation, poor housekeeping, etc.
2. **Worker's or co-worker's unsafe acts:** A worker or co-worker may be inexperienced or new on site, or may choose to act unsafe, all of which may lead to unsafe conditions for other workers. Examples of unsafe acts leading to unsafe conditions include removing machine safeguards, working while intoxicated, working with insufficient sleep, sabotaging

Figure 15.5 Unsafe conditions in labour hutment

Figure 15.6 Unsafe conditions—deep excavation without barricade and heavy vehicles plying very near the edge of deep excavation

equipment, disregarding housekeeping rules, unauthorized operation of equipment, and horseplay.
3. **Nonhuman-related event(s):** Non-human-related events that may lead to unsafe conditions include systems, equipment, or tool failures, earthquakes, storms, etc.
4. **Unsafe condition is a natural part of the initial construction site conditions:** Examples of unsafe conditions that are a natural part of the initial construction site conditions include uneven terrain, concealed ditches, and scattered metallic or non-metallic materials, etc. These unsafe conditions are usually removed during initial site preparations.

15.4.2 Unsafe Acts

An unsafe act may be an act of commission (doing something that is unsafe) or an act of omission (failing to do something that should have been done). Not every unsafe act produces an injury or a loss, but by definition it has the potential for producing an accident. A worker may commit unsafe acts regardless of the initial conditions of the work (i.e., whether the condition was safe or unsafe). A few examples of unsafe acts are the decision to proceed with work in unsafe conditions, disregarding standard safety procedures such as not wearing a hard hat or safety glasses, working while intoxicated, and working with insufficient sleep. Some more examples of unsafe acts are:

- cleaning, oiling, adjusting, or repairing of moving electrically energized or pressurized equipment

- failure to use available personal protective equipment (life-saving devices) such as safety helmet or hard hat, safety belt, fall arresters and safety net
- improper use of equipment (overloading, etc.)
- improper use of hands or body parts (gripping objects insecurely, using hands instead of hand tools, etc.)
- taking unsafe positions or postures (under suspended loads, riding on forks of lift trucks, etc.)
- unsafe placing, mixing, combining, etc.

Some photographs showing unsafe acts are given in Figures 15.7 to 15.10. Some of the root causes of construction accidents identified by Toole (2002) under unsafe acts include:

1. Not using safety equipment
2. Poor attitude towards safety—for example, a tradesperson who has been trained on the proper use of ladders refuses to face the ladder when walking down it
3. Isolated, sudden deviation from prescribed behaviour

The causes that lead to unsafe acts may include lack of knowledge or skill (unawareness of safe practices, unskilled, improper attitude such as disregard of instruction, etc.), and physical or mental deficiency (defective eyesight or hearing, fatigue, etc.). The detection and correction of unsafe acts must be given priority. It is preferable to be more careful on jobs with a history of accidents. There should not be any leniency in observing the safety rules.

Figure 15.7 Unsafe act—workers taking shelter under heavy vehicles

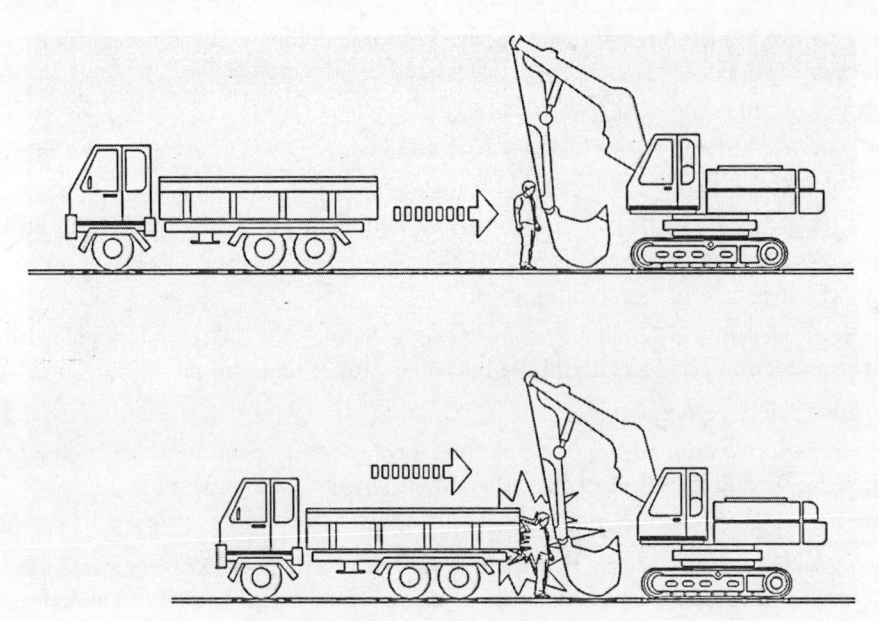

Figure 15.8 Unsafe act—worker in-between the reversing truck (without reverse horn) and the excavator

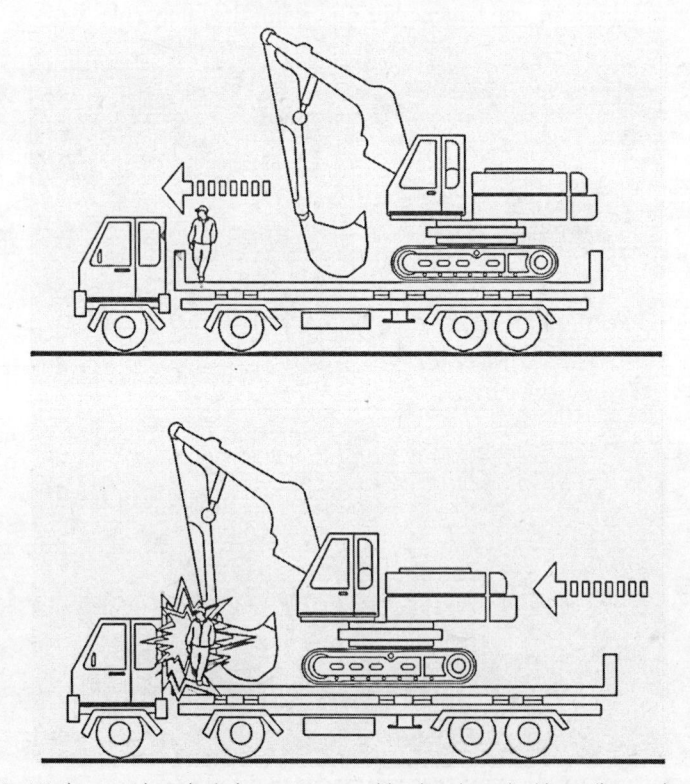

Figure 15.9 Worker near loosely tied excavator—sudden break applied by trailer crushed the worker

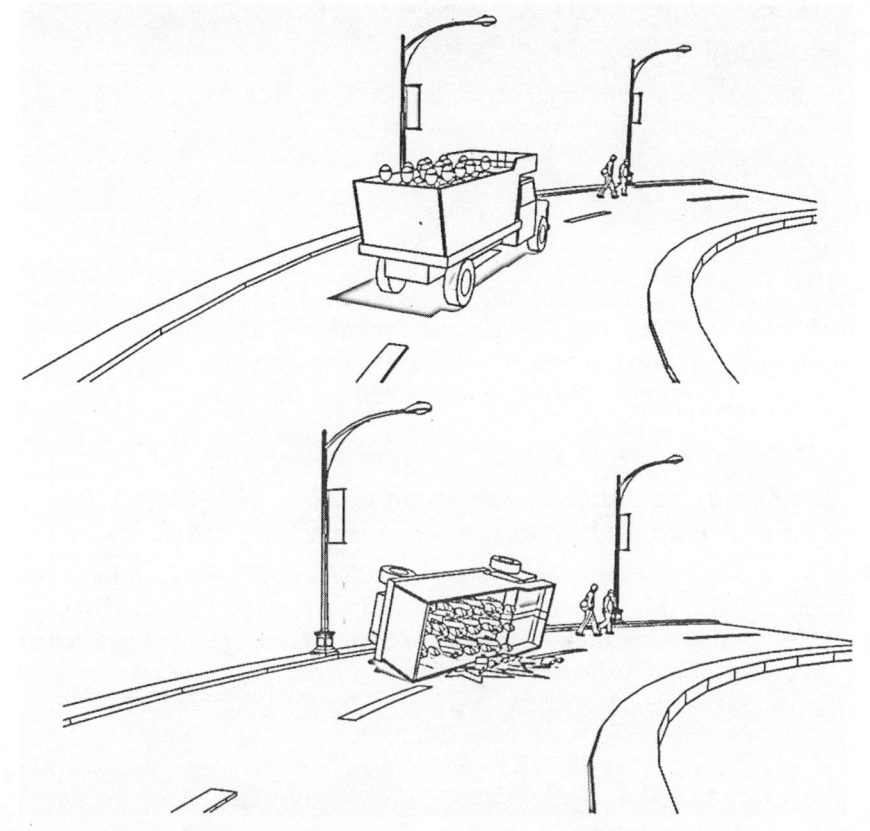

Figure 15.10 Unsafe act—workers being transported in open dumper

Some of the methods through which unsafe acts can be eliminated are:
- Initial job instructions
- Priority to engineering
- Stressing the after-effects of an accident
- Appeal to the worker's love for his family
- Showing disapproval of unsafe acts
- Education

15.5 HEALTH AND SAFETY ACT AND REGULATIONS

There are two major pieces of legislation governing health and safety law. These are:
- Building and Other Construction Workers (Regulation of Employment and Condition of Services) Act 1996
- Building and Other Construction Workers (Regulation of Employment and Condition of Services) Central Rules 1998

Allied to these are several statutory instruments governing safety. These are:
- Factories Act 1948
- The Delhi Building and Other Constructions Workers (Regulation of Employment and Condition of Services) Rules 2002
- Indian Electricity Act 1948
- Indian Electricity Regulations 1956
- Motor Vehicle Act 1998

Although the Central Rules 1998 came into existence over 10 years back, the formulation and implementation of these rules at state level is yet to materialize. In order to make these rules operational, the states need to (1) establish a state advisory committee, (2) establish an expert committee, (3) establish a welfare board, (4) appoint inspectors, and (5) formulate rules applicable in the specific state.

15.5.1 Building and Other Construction Workers (Regulation of Employment and Condition of Services) Act 1996

The Building and Other Construction Workers (Regulation of Employment and Conditions of Services) Act, 1996, is divided into 11 chapters and contains 64 sections. The objectives of the safety and health provisions therein are—(1) to regulate the employment and conditions of service; (2) to provide for safety, health and welfare measures; and (3) to extend social security to the building and construction workers.

The provisions are applicable to even small establishments employing 10 workers in any building and other construction work, on any day of the year. An individual employing workers in relation to construction of his residence of value not more than ₹ 10 lakh is not an establishment according to this act.

The term 'building and other construction work' has been defined to bring all such other activities under the definition of the term by notification. The act also puts the government departments and the contractors on the same footing as the employer. Any department of government employing construction workers directly without any contractor is liable for consequences of non-observance of law.

The act is applicable to every establishment that employs or had employed ten or more workers directly or through a contractor/subcontractor (undertaking to produce a given result or supplying building workers for any work of the establishment, excluding supply of goods or articles of manufacture). However, any building or construction work to which Factories Act, 1948, or Mines Act, 1952, apply is exempted from the BOCW provisions.

Chapter 7 of the act deals with safety and health measures. The important sections in this chapter are sections 38 and 39.

Section 38 deals with the safety committee and safety officers. It states that in every establishment wherein five hundred or more building workers are ordinarily employed, the employer shall constitute a safety committee consisting of such number of representatives of the employer and the building workers as may be prescribed by the state government. It is to be ensured that the number of persons representing the workers shall in no case be less than the persons representing the employer. Moreover, in every establishment the employer shall appoint a safety officer who shall possess such qualifications and perform such duties as may be prescribed.

Section 39 deals with the requirement of notice for certain accidents. For example, death and disability for 48 hours or more have to be informed by the employer to the authority. Enquiry by the authority within a month is a must in case of death to five or more persons.

The responsibility of employers is defined in Section 44 of Chapter 9. It states that an employer shall be responsible for providing constant and adequate supervision of any building or other construction work in his establishment so as to ensure compliance with the provisions of the act relating to safety, and for taking all practical steps necessary to prevent accidents.

Chapter 10 of the act deals in penalties and procedures. Section 47 is about penalty for contravention of provisions regarding safety measures. Section 48 discusses penalty for failure to give notice of the commencement of the building or other construction work, and Section 48 deals with penalty for obstructions.

15.5.2 Building and Other Construction Workers (Regulation of Employment and Conditions of Service) Central Rules, 1998

The Central Rules are divided into five parts, and contains 30 chapters, 12 schedules and 26 forms. Part III comprises 20 chapters in which chapters VI to XXV deal with safety and health. Chapter II, Rule Number 5 deals with responsibilities and duties of employers, architects, project engineers and designers, building workers, etc. The employers have detailed responsibilities including:

- to comply with requirements of rules
- not to permit an employee to do anything not in accordance with the generally accepted principles of safe operating practices connected with building and other construction work
- not to allow lifting appliances, lifting gear, lifting devices, transport equipment, vehicles, or any other device or equipment to be used by the building workers which do not comply with provision of rules
- to maintain the latrines, urinals, washing facilities and canteen in a clean and hygienic condition
- to ensure that adequate measures are taken to protect building workers against the harmful effects of excessive noise or vibration (Rule 34, Chapter VI)
- to ensure that the site is provided with fire-extinguishing equipment sufficient to extinguish any probable fire (Rule 35, Chapter VI). The employer is to ensure that a sufficient number of trained persons are available to operate the fire-extinguishing equipment. The employer is also responsible to ensure that fire extinguishers are properly maintained and inspected at regular intervals of not less than one year
- to ensure that emergency action plan is in place and is approved by the director general (Rule 36)
- to ensure that all mechanical equipment are provided with required safety features such as guarded moving parts (Rule 37)
- to ensure that workers do not lift or carry weight beyond prescribed limit, either by his hand or overhead (Rule 38)
- to ensure that health and safety policy specifying the steps to be taken to ensure health and safety of workers is in place, and is duly approved by the director general (Rule 39)

Table 15.2 Cost associated with an injury

Direct cost for workers	Direct cost for employer	Indirect cost to workers	Indirect cost for employer
• The loss of income • Healthcare cost	• Cost of ambulance service • Cost of medical and ancillary treatment • Cost of medication • Cost of hospitalization • Cost of disability benefits to be given to the injured • Cost of compensation payments • Payment of work not performed • Repair or replacement of damaged m/c and equipment • Cost incurred in legal proceedings • Cost of investigation	• The pain and suffering of the injury or illness • Family sufferings • The possible loss of a job • Negative impact on morale of other family members	• Cost of lost time of injured worker, crewmember, foreman, supervisor, first-aid attendant, safety officer and other executives • Cost due to damage to equipment, tools, property and materials • Cost due to loss of productivity of crewmembers and cost of idling of equipment • Cost of replacement for the injured/ill worker • Cost to train the replacement worker and cost of low productivity of new worker in the initial periods • Cost of loss of reputation and public relations • Cost for continuing wages and other benefits due to injured worker • Cost of overhead such as utilities and telephone

- to ensure the control of dust, gases, or fumes, and other harmful substances that may cause injuries if exposed beyond permissible concentration (Rule 40)

15.6 COST OF ACCIDENTS

It is estimated that global losses on account of accidents every year are almost more than the total loss due to World Wars I and II combined. According to NSC-US estimates of 1996, occupational death and injuries cost the nation about US$121 billion. The total cost of accidents can be broken up into direct cost and indirect cost. Direct cost could be further broken up into direct cost to workers and direct cost to employers. Similarly, indirect cost can also be broken up into indirect cost to workers and indirect cost to employers.

Although there are no hard and fast rules to classify direct and indirect costs, some examples of direct cost and indirect cost are shown in Table 15.2 purely for illustration purposes. While classifying, we have assumed the definition of direct costs as those that are directly attributed to or associated with injuries. These costs are easier to quantify and can be recorded quite accurately. On the other hand, indirect costs are assumed to be those costs that are hidden and quantifying them would be very difficult as no historical records exist for them. Some people consider direct costs as only those costs that are recorded in the account books of a firm.

Researches have been continuously carried out on the cost aspect of accidents to prove that accidents are unwanted and avoidable expenses, and that their cost is much more than is usually visible to the eye. Everett and Frank Jr (1996) found that the total cost of accidents amounted to between 7.9 per cent and 15 per cent of the total costs of construction, even without considering some intangible costs such as decreased employee morale, loss of image and greater turnover of employees.

Some researchers in the past have attempted to compute the ratio of indirect cost to direct cost. The values of these computed ratios vary. For example, Heinrich (1941) computed it to be 4:1, while Bird and Loftus (1979) computed it to be 10:1. Construction Safety Association of Ontario (CSAO) has found that the average ratio of indirect to direct costs in Ontario construction is 5:1. On the basis of nearly 500 construction worker injuries, Hinze and Applegate (1991) found that the ratio of indirect to direct costs varied from 4.2 to 20.3 depending on the severity of the injury. The analysis also demonstrated that even the costs of minor injury could be considerable. As mentioned, there is no agreement in classifying indirect and direct costs, and hence, the ratios of indirect cost to direct cost often vary from one study to another. Nonetheless, the ratios point out that the indirect costs are quite high when compared to direct costs associated with an injury.

For contractors reluctant to invest in health and safety, the study conducted by Japan Industrial Safety and Health Association (JISHA) could be quite interesting. In the study, JISHA (2000) tried to find the quantitative relationship between expenses incurred on safety and their effects. The study revealed that investment in safety pays, yielding 2.7 times as much returns as the expenses incurred on safety—in other words, a benefit-to-cost ratio of 2.7. CSAO is also of the opinion that although the investment in safety may cost money in the short term, it definitely pays in the long run. Joseph (1999) in his study on projects involving steel erection showed that actual profit can increase by 5 per cent to 10 per cent for an investment of 1 per cent to 3 per cent in safety, and concluded that safety costs money and can save money. Empirical researches have also shown that the higher the safety investment in a project, the lower the injury rates and the higher the profit.

15.7 ROLES OF SAFETY PERSONNEL

According to OSHA, contractors are responsible for providing a safe place of employment to their workers. If the contractor employs subcontractors, then he has the additional responsibility of providing a hazard-free workplace to the workers of subcontractors. The owner, designers and engineers also play an important role in the safety practices prevailing at site. Some of the design decisions go a long way in ensuring the overall safety of a project. Toole (2002), based on a survey, finds that there is no uniform agreement on the site safety responsibilities that should be assumed by design engineers, contractors and subcontractors.

Section 209 of the Delhi Building and Other Construction Workers (Regulation of Employment and Conditions of Service) Rules, 2002, states the requirement of a safety officer. It says that appointment of safety officer is a must for works employing 500 or more workers. For strength up to 1,000, one safety officer is a must, while two safety officers are required for workers up to 2,000. Likewise, depending on the strength of workers present at construction site, the number of safety officers required to be present is specified.

Some employers believe that by appointing the safety officer they have met the statutory requirement, and that the safety officer would be accountable to ensure the safety of workers at site. However, this is not the case entirely. The various duties of a safety officer as stipulated in the law will help clarify the confusion, and these are mentioned below:

- To advise the building workers for effective control of injuries
- To advise on safety aspects to carry out detailed safety studies
- To check and evaluate effectiveness of action
- To advise on purchasing and ensuring quality of personal protective equipment
- To carry out safety inspections of building
- To investigate all fatal and other selected accidents
- To investigate the cases of occupational disease
- To advise on maintenance of records related to accidents, etc.
- To promote the working of the safety committee
- To organize campaigns, contests and other such activities
- To design and conduct training and educational programme
- To frame safe rules and safe working practices
- To supervise and guide safety precautions

It can be seen that the role of a safety officer is that of an advisor, similar to any other support function in an organization. Thus, it is not appropriate to hold the safety officer accountable for accidents occurring at the workplace. The onus of ensuring safety lies with the line personnel, to build in the required safety measures in all the work executed under their supervision.

Employers need to ensure that safety officers are provided with all such facilities, equipment and information that are necessary to enable them to perform their duties effectively. The law also specifies that the safety officer should not be required or permitted to do any work that is unconnected to, inconsistent with, or detrimental to the performance of the duties prescribed above.

15.8 CAUSES OF ACCIDENTS

The Times of India (July 12, 2004) citing ILO reported that worldwide some 6,000 workers die everyday from occupational accidents and diseases. The report further said that of the almost 270 million accidents recorded each year, about 350,000 result in fatalities—in other words, one in every 771 accidents results in death. About 2.2 million work-related fatalities and 60 million work-related illnesses occur annually.

Some of the causes leading to accidents, such as unsafe acts of workers and co-workers, unsafe conditions, management action/inaction, workers' attitude and the stresses to which they are subjected, were cursorily mentioned while discussing accident causation models. Here, we summarize some commonly observed causes of accidents without using any classification. Accidents can occur at construction sites on account of:

- Lack of planning and organization
- Defects in technical planning

- Fixing unsuitable time limits and targets too difficult to achieve
- Assignment of work to incompetent contractors
- Insufficient or defective supervision
- Lack of cooperation among different crafts
- Inadequate examination of equipment
- Inadequate preparation for work
- Inadequate instructions concerning work
- Employment of unskilled or insufficiently trained workers
- Inadequate supervision of work
- Workers' irresponsible behaviour, unauthorized action and carelessness
- Faulty construction such as collapse of walls and other building parts
- Lack of necessary equipment and use of unsuitable equipment
- Structural or other defects in equipment being used
- Lack of safety devices or measures
- Unsuitable building materials and
- defective processing of building materials
- Collapse of stacks, masses of earth, etc.
- Collapse and overturning of ladders, scaffolds, stairs, beams, etc.
- Fall of objects, tools, pieces of work, etc.
- Fall of persons from ladders, stairs, roofs, scaffolds, buildings, through some openings, etc.

Analysis results of about 500 accidents in a large construction company are shown in Table 15.3. It shows that falls from elevation/height constitute about 32 per cent of all fatalities. This is followed by mechanical impact (caught in/between machinery), fall of objects, vehicles and electrocutions, in that order.

Further analysis of falls from the source data of Table 15.3 reveals that falls from height are most common (close to 40 per cent of total 'fall accident' cases; see Figure 15.11) in projects involving height of less than 10 metres. This can be attributed to negligence, overconfidence and failure to use protective equipment at lesser heights.

Causes of accident mentioned in Table 15.3 were categorized into the following categories:

- unsafe act by victim
- unsafe act by co-worker
- unsafe condition
- victim at fault as well as unsafe condition
- others

Table 15.3 Causes of fatalities (based on about 500 accidents in a large construction company)

Type of accident	Number of accidents	Percentage (%)	Break-up (%) of major accidents	
Fall from height	125	32		
Mechanical impact (caught in/between machinery)	39	10	Taking shelter under machinery	59%
			Maintenance during operation	14%
			Failure of machines	13%
			Absence of helper	8%
			Inexperienced operator	6%
Fall/hit by objects	39	10	Area not cordoned	41%
			Vicinity of hazardous substances	34%
			Improper material stacking	21%
			Not wearing helmets	4%
Accident involving vehicles	35	9	Unauthorized travel	60%
			Sloping ground	24%
			Rash driving	8%
			Vehicle in poor condition	8%
Electrocution	31	8	No ELCB/unearthed body	37%
			Punctured cable	29%
			Violation of work permit	24%
			Not authorized electrician	10%
Collapse of structures	31	8		
Others	27	7		
Fall from height (through opening)	27	7		
Involving passenger lift	8	2		
Collapse of excavation	4	1		
Drowning	4	1		
Fire	4	1		
Combined (more than one of above causes)	16	4		

The analysis results are shown in Figure 15.12. It was found that in the majority of cases (about 53 per cent), the victims themselves were at fault. The analysis further showed that victims were at fault mainly due to their incompatibility with the work they were performing, non-adherence to personnel protective equipment (PPE), and their carelessness and overconfidence (see Figures 15.13 to 15.15 as well).

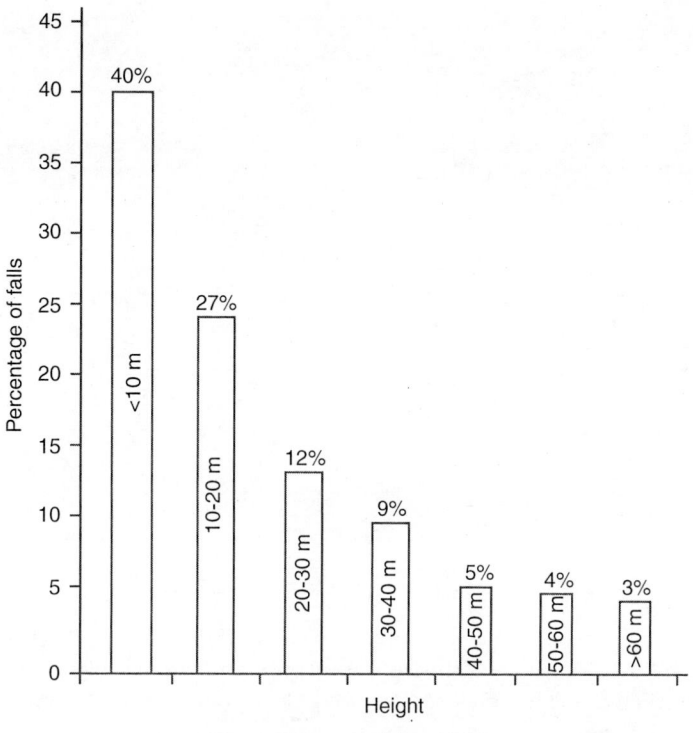

Figure 15.11 Analysis of falls

Improper earthing of the body of machinery, poorly maintained cables and violation of work permits during shutdowns were the major causes of electrocution. Accidents involving machinery were mainly caused due to unauthorized movement or resting behind the machinery, and also due to carrying out of maintenance when the machine was in operation, or vice versa.

Absence of guardrails and working platforms along with improper housekeeping and lack of supervision were the major unsafe conditions causing most construction falls. Since non-adherence to personal protective equipment caused nearly 90 per cent of 'falls from height' to be fatalities, it was further analysed. It emerged that both improper tying of safety belt and failure to use safety belt were almost equally responsible for the 'fall from height' accidents.

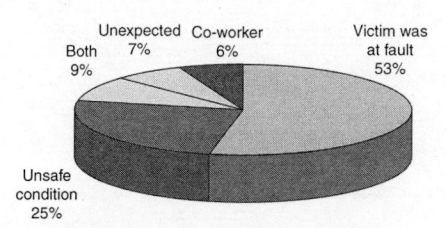

Figure 15.12 Causes of accident

Figure 15.13 Victims at fault
- Carelessness/overconfidence of the victim 33%
- Wrong person for the job 34%
- Non-adherence to PPE 33%

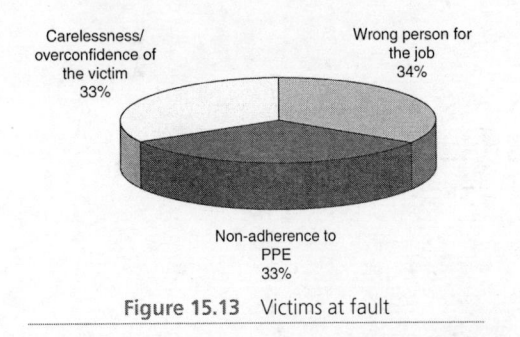

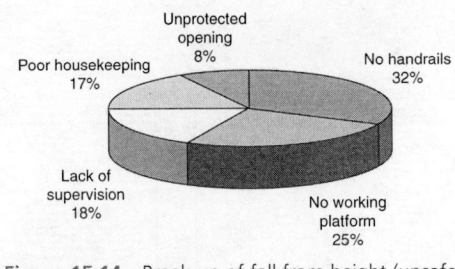

Figure 15.14 Break-up of fall from height (unsafe conditions)
- Unprotected opening 8%
- No handrails 32%
- No working platform 25%
- Lack of supervision 18%
- Poor housekeeping 17%

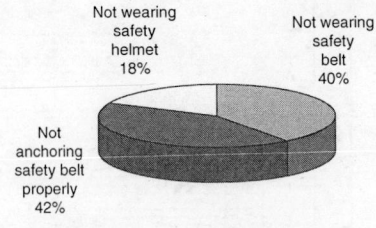

Figue 15.15 Non-adherence to PPE
- Not wearing safety helmet 18%
- Not wearing safety belt 40%
- Not anchoring safety belt properly 42%

Travelling in vehicles/trailers meant for transporting of materials was the major reason for accidents involving vehicles. The other reasons were sloppy ground conditions, rash driving and poor condition of vehicles.

Accidents due to falling of objects resulted due to hazardous areas left uncordoned, improper stacking of materials, and carrying out of secondary activities in the vicinity of hazardous activity.

Based on the assumption that identifying the causes of fatalities could be helpful in developing a prevention strategy, a number of similar studies were carried out in the United States, the United Kingdom and other countries. The analysis results pertaining to the United States and the United Kingdom are shown in Table 15.4 and Table 15.5, respectively. It may be mentioned that by showing the statistics here, we are not trying to compare the results for different countries. This is not possible anyway, as there is lack of common classification and the periods in which the data were collected are not uniform either. However, one thing that is common is that falls from height or elevation are one of the major causes of accidents in construction industry. The majority of accidents can be averted through common sense, some awareness and slight investment in safety gadgets.

Table 15.4 Distribution of construction accidents (USA—1985–89)

Falls from elevation	33%
Shocks electrical	17%
Struck by incidents	22%
Caught in-between incidents	18%
Other	10%

Table 15.5 Distribution of construction accidents (UK—1998–99)

Falls from height	35%
Slips, trips, or falls	21%
Struck by moving or falling objects	18%
Injured while lifting, handling and carrying	10%
Others	16%

15.9 PRINCIPLES OF SAFETY

Dan Petersen (1989) mentions ten basic principles of safety. These are:

1. An unsafe act, an unsafe condition, and an accident are all symptoms of something wrong in the management system.
2. We can predict that certain sets of circumstances will produce severe injuries. These circumstances can be identified and controlled.
3. Safety should be managed like any other company function. Management should direct the safety effort by setting achievable goals and by planning, organizing and controlling to achieve them.
4. The key to effective line-safety performance is management procedures that fix accountability.
5. The function of safety is to locate and define the operational errors that allow accidents to occur. This function can be carried out in two ways:
   - by asking why accidents happen—searching for their root causes
   - by asking whether certain known effective controls are being utilized
6. The causes of unsafe behaviour can be identified and classified. Some of the classifications are—overload (the improper matching of a person's capacity with the load); traps; and the worker's decision to error. Each cause is one that can be controlled.
7. In most cases, unsafe behaviour is normal human behaviour; it is the result of normal people reacting to their environment. Management's job is to change the environment that leads to unsafe behaviour.
8. There are three major subsystems that must be dealt with in building an effective safety system:
   a. The physical
   b. The managerial
   c. The behavioural
9. The safety system should fit the culture of the organization.
10. There is no one right way to achieve safety in an organization; however, for a safety system to be effective, it must meet certain criterions. The system must:

- force supervisory performance
- involve middle management
- have top management visibly showing their commitment
- have employee participation
- be flexible
- be perceived as positive

Many of these principles are just principles without tested techniques available, while others have well-tested methods available (Petersen 1989).

15.10 SAFETY AND HEALTH MANAGEMENT SYSTEM

Petersen in his principles of safety management mentions that safety should be managed like any other company function. Thus, in order to manage the project schedule, a construction company plans the manpower and other resources; these plans are monitored and, depending on the deviation, control measures are applied. In a similar manner, the management should direct the safety effort by having a proper safety and health management system in place. The key functions of the safety management systems are—planning for safety; organizing for safety; issuing directions for safety; and coordinating and controlling various safety issues. Planning for safety may include developing a safety and health policy, evaluating the policy from time to time, setting goals for safety and creating a budget for safety-related expenditures. Organizing for safety includes activities such as development of a safety organization structure, defining the roles and responsibilities, delegating authorities, and education and training for safety. Developing a proper communication system, standard operating procedures and a safety manual are parts of the directing function. Constitution of a safety committee can be considered under the coordinating function, while under the controlling function we define the mechanism for accident reporting, investigation, record keeping, and so on. Some of the important components of a safety management system are discussed next.

15.10.1 Safety Policy and Organization

It may be recalled that the third principle of Dan Petersen mentions that *'safety should be managed like any other company function'*. This implies that safety is a line responsibility. For the line management to accept this responsibility, the top management should clearly explain and stress the same by issuing a document in the form of guidelines, i.e., *a safety policy*. In the other part of the same principle, it is said that *'management should direct achievable goals by planning, organizing and controlling'*.

A safety policy is management's first step in implementing the above-said principle. A written safety policy removes any confusion regarding the objectives, directives and distribution of responsibilities in this regard. Construction companies formulate safety policy to show management's commitment to provide a safe and healthy work environment to all its employees. These are prominently displayed in different languages so that every worker is familiar about the existing policy.

As per Rule 39 of the Building and Other Construction Workers (Regulation of Employment and Conditions of Service) Central Rules, 1998, it is mandatory to have a written statement of policy in respect of safety and health of building workers if the number of employees is fifty or more. As per this rule, the safety policy should contain:

i. the intentions and commitments of the establishment regarding health, safety and environmental protection of building workers;
ii. the arrangements made to carry out the policy referred to in clause (a) specifying the responsibility at different levels of hierarchy;
iii. the responsibilities of the principal employer, contractor, subcontractor, transporter, or other agencies involved in the building or other construction work;
iv. the techniques and methods for assessment of risk to safety, health and environmental and remedial measures therefore; and
v. the arrangements for training of building workers, trainers, supervisors, or other persons engaged in the construction work.

The rule also mentions that the establishment shall revise the policy in case there is an expansion or modification in the scope of work carried out by the organization, or if a new building or other construction work, substances, articles, or techniques are introduced having implication on the health and safety of building workers.

Thus, it can be seen that safety policy is a document that, if properly prepared and implemented, can play a vital role in enhancing construction safety in an organization.

15.10.2 Safety Budget

In order to generate safety awareness and to provide different safety gadgets to workers and employees, construction companies plan for a safety budget. Some organizations have a separate safety budget for each of the project sites. From this budget, the purchase of PPE and other safety gadgets are made. However, investment in safety is yet to be realized as a necessary expenditure. It is commonly observed that project managers of sites with low margin tend to cut down on their safety budget. In order to safeguard against such practices adopted by some project managers, experts suggest delinking of safety budget from site margin. They suggest that expenditure in safety be taken under general overhead, and the safety budget be made at head office. The head office must supply the required safety gadgets to different project sites irrespective of the margins being realized at these sites.

15.10.3 Safety Organization

The organization of safety on the construction site is determined by the size of worksite, the system of employment, and the way in which the project is being organized. Safety and health records should be kept which facilitate the identification and resolution of safety and health problems on the site. In construction projects where subcontractors are used, the contract should clarify the responsibilities, duties and safety measures that are expected of the subcontractors' workforce. These measures may include the provision and use of specific safety equipment, the methods of carrying out specific tasks safely, and the inspection and appropriate use of tools. The person in-charge at the site should also ensure that materials, equipment and tools meet minimum safety standards. Training should be conducted at all levels including for managers, supervisors and workers; subcontractors and their workers may also need to be trained in site-safety procedures because teams of specialist workers may mutually affect each other's safety.

There should also be a system so that site management has prompt information about unsafe practices and defective equipment. Safety and health duties should be specifically assigned to certain persons. Some examples of duties are listed below:

- Provision, construction and maintenance of safety facilities such as access roadways, pedestrian route, barricade and overhead protection
- Construction and installation of safety signs
- Safety provisions peculiar to each trade
- Testing of lifting machinery such as cranes, goods hoists and lifting gears
- Inspection and rectification of access facilities such as scaffolds and battens
- Inspection and cleaning of welfare facilities such as toilets, clothing, accommodation and canteen
- Transmission of relevant parts of the safety plan to each work group
- Emergency and evacuation plans

One of the important functions of safety management system is safety organization. Section 40-B of Factories Act, 1948, also requires that every factory wherein 1,000 or more workers are ordinarily employed should have qualified safety officer(s).

Obviously, since the line has primary safety responsibility, the safety officer is, and must be, a staff function and is not directly responsible for safety results. His job responsibilities involve activities that help the line achieve safety standards. The safety officer has no authority over the line. He may have a great deal of influence but this is quite different from authority. In order to make the safety officer influential, it is important that he reports directly to the top executive and is given proper status. The safety professional in any organization obtains results by using either of two methods—1) makes recommendations to line executive, who issues an order, or 2) obtain acceptance of his suggestion voluntarily from line supervisors without taking the chain of command route. More often than not, the purposes of the safety officer are accomplished by the second method.

15.10.4 Education and Training

Unsafe action is one of the main contributing factors for most of the industrial accidents. While the action of any individual at any given time is controlled by various factors, the individual's knowledge and skill also play an important role. It is important to expose the workers, supervisors and line managers to various educational and training programmes that impart sufficient knowledge and help them to improve their skills in their work. Such training programmes should not be a one-time affair and the experienced employees should also be given refresher training. Change of jobs, whether it is to a different department or to a different employer, means adjustment for the individual, and he or she should be provided with a well-structured appreciation programme. In all such training programmes, pertinent safety information should be given adequate coverage and particular emphasis should be given on the company's policy towards safety.

Induction Programmes

Induction programmes for new recruits focus on basic safety requirements of the project and significant features of the construction work relating to safety. New recruits are given relevant safety literature, booklets, etc., that lay emphasis on the importance of safety at the workplace. How to avoid accidents, how to contribute to project goals related to safety performance, illustration of safe work methods for relevant construction activities, how to keep the workplace tidy and injury-free, how to use personnel protective equipment, and so on, are some of the topics covered in an induction programme. Besides, induction programmes inform new workers about the main hazards and risks involved in the construction activity of the project, the emergency procedures, the safety rules and regulations to be followed at work site, etc. Companies make it mandatory that all workers and employees attend the safety induction course prior to starting their work.

Toolbox Talks

Considered to be one of the most effective ways to eliminate accidents, toolbox talks (TBT) are a short meeting of the individual work groups assigned for a particular task before its physical commencement. The meeting is held each morning very near to the workplace and is attended by the staff of general contractor and subcontractor (if the work is sublet by the general contractor), supervisor, foreman and workers. Some of the issues discussed at a TBT are the work plan and procedures for the day, the identification of possible hazards, the PPE, the safety tools and equipment to be used, and the visual checks on the health of each worker.

15.10.5 Safety Plan

Modern construction requires a safety plan to be prepared before the actual commencement of construction activity. The safety plan gives the method statement in detail to be adopted for the particular construction activity. It includes the description of the activity to be accomplished, the time available for the activity, the details of hazards, if any, present to the workforce, and the allocation of resources for managing safety aspects. The contractor is supposed to prepare the safety plan and submit it to the owner, who approves it if it is found appropriate.

15.10.6 Safety Manual

Most construction companies these days establish and maintain a comprehensive set of safety rules and regulations in order to achieve better safety performance of the projects executed by them. Safety manuals contain a wide range of details such as—safe ways to operate and maintain the construction equipment, safe ways to store and handle construction materials, procedures to report hazards in construction activities to the concerned person, and cleanliness and housekeeping to be observed at workplace. The new recruits (workers and subcontractors) in the companies are encouraged to be familiar with the provisions of the safety manual.

15.10.7 Safety Committee

The safety committee is made with the primary objective of bringing together people with particular responsibilities for safety, so that they can formally address issues and take appropriate actions with regard to the work-site safety management objectives. It is desirable to conduct safety meetings frequently. Needless to say, such safety committees can be effective only when the duties and responsibilities of committee members are clearly spelt out. The committee is entrusted with the following responsibilities:

- To confirm if the management of safety and health is being properly carried out by all the parties concerned
- To ensure that the construction work is being performed safety and smoothly, complying with safety rules and regulations
- To conduct safety inspection of the entire site prior to the safety committee meeting
- To coordinate and control congested or hazardous working conditions of the subcontractors
- To resolve safety issues submitted by any subcontractor
- To increase subcontractors' safety knowledge and safety awareness
- To enforce safety training programmes
- To participate and organize safety promotional activities
- To promote and maintain housekeeping and waste disposal at the highest standards
- To review safety statistics of previous month and to review safety practices laid down by management

As mentioned under health and safety regulations under BOCW (RE & CS) Central Rules, 1998, the requirement of safety committee is laid out for building and construction works wherein 500 or more workers are ordinarily employed. According to Section 208 of the Delhi Building and Other Construction Workers (Regulation of Employment and Conditions of Service) Rules, 2002, safety committee is a must for building construction works employing more than 500 workers. The committee must have an equal number of representatives of employer and building workers. The main functions of such committees are given elsewhere in this text.

15.10.8 Incentive Programmes

Construction companies develop different procedures to recognize and acknowledge good safety performance by individual workers, teams and subcontractors. Companies have different means to show appreciation of the good work done by individuals, teams and subcontractors.

15.10.9 Accident Reporting, Investigation and Record Keeping

It is one of the important parts of a safety management system to report and investigate accidents. It is also one of the requirements under Section 88 of Factories Act, 1948, that the occurrence of reportable accidents and certain dangerous occurrences be notified to the inspector of factories.

Accident reporting and investigation should invariably start from the foreman/supervisor and line managers concerned. The safety officer should be involved in investigation of serious and fatal accidents. In all cases, the accident investigation should aim at tracing contributing factors to determine their underlining causes, instead of only being a fault-finding exercise.

Apart from reporting accidents to the inspector of factories, which is a legal requirement, it is equally or rather more important that the chief executive of the organization is kept well-informed about the accident records and their investigation reports. This can be done by sending him a monthly report, which should necessarily include:

1. frequency rate of accidents;
2. severity rate of accidents;
3. total time lost due to accidents;
4. cost of accidents;
5. findings; and
6. actions recommended to avoid recurrence.

These details should be provided not only for the company as a whole, but also for each individual plant or department. Such presentations are more likely to convince the supervisors or line managers of the desirability of devoting attention more actively to the prevention of accidents than had been done in the past.

Apart from reporting and recording major accidents, it is extremely important that accidents of minor nature be recorded as well. The accident that results in a minor injury (or non-injury) today, may cause serious injury or even death of a worker tomorrow.

15.10.10 Incident Investigation and Analysis

Accident investigation is conducted primarily to determine direct cause(s) for the accident and to determine what other causes contributed to the direct cause. It is also aimed at preventing similar accidents from occurring in the future by properly documenting different facts pertaining to the accident, including information on cost aspects associated with the accident. Thus, in essence, it is aimed at fact-finding and not fault-finding. The investigation should be carried out for every accident, incident and near-misses.

More often than not, accidents are caused due to multiple causes such as operator error, faulty design of the system, mechanical failure, procedures, inadequate training, and environment. In many cases, the tendency of the investigator is to pinpoint a single cause such as 'operator error' and so on as the cause of an accident. This may not help the purpose of conducting an accident investigation.

Some other reasons cited for not meeting the objectives of accident investigations are—investigation is done by an inexperienced and uninformed investigator; reluctance of an investigator to accept the facts; interpreting the facts wrongly; and forming biased opinion. Sometimes, delay in investigating accidents and poor interviewing techniques may also lead to failure. Also, if adequate time is not spent in investigation, and facts are not segregated from opinion expressed by witnesses, the accident investigation may fail to achieve its objective.

The investigation may be performed by the immediate supervisor, the head of department, the safety officer, or even the top management people individually or in a team. An accident can also be investigated by any expert third-party or a cross-functional team formed from within the organization.

In order to assist the investigation process, it is desirable that the investigator keeps himself equipped with a tool kit consisting of—camera(s), including films; lighting/torch; tape recorder/personal Dictaphone; mobile telephone/communication; thermometer; tape measure; clipboard/pens; coloured chalk; tape black/yellow/quarantine marked; PPE; protective gloves/latex; absorbent cloths/tissues; and resealable plastic bags. The kit helps in gathering and recording information.

The responsibilities of the investigating team include—assuming control of the accident scene; preventing further injury/damage/loss; preserving evidence; recording facts; preventing interference with environmental evidence; isolating and interviewing genuine witnesses/staff/persons in the vicinity of accident; checking relevant documentation; communicating and liaisoning with other team members and management; coordinating ongoing investigation information; preparing complete accident investigation report; determining causes of accidents and suggesting preventive measures; and dealing with the media.

The three important tasks to be performed in the accident investigation process are:

Gathering Information

The facts could be in the form of physical evidence such as reports, equipment details, documents and procedural manuals, or in the form of eyewitness account. The background information related to the project site is also useful. It is advisable to secure the accident site. The interviews of people connected with the accident are also a useful source of information and it is required that the investigating team be good at interviewing.

For physical evidence, some of the points that must be noted are—position of injured workers; equipment being used; materials being used; safety devices in use; position of appropriate guards; position of controls of machinery; damage to equipment; housekeeping of area; weather conditions; and lighting and noise levels.

Eyewitness accounts are primary sources of firsthand information related to the accident and it is desirable to interview any eyewitness as early as possible. It is to be noted that the investigator should not get carried away by perceptions and opinions of eyewitnesses, and it is essential to filter out the facts from the perceptions and opinions of eyewitnesses. The statements, if any, taken from the eye witnesses should preferably be in the written form and in a language in which the witness is comfortable.

The interviews of eyewitnesses reveal useful information and it is desirable to ask open-ended questions. The witnesses are kept at ease at the time of interviewing. Some of the do's suggested for interviewing are—explaining the purpose of the interview; letting the witness talk and listen; confirming that the statement obtained is correct; maintaining eye contact with the interviewee; and making short notes during the interview. Some don'ts are—intimidating the witness; interrupting; prompting; asking directional questions; showing one's own emotions; and making notes while witness is talking.

There are certain typical questions asked during the interview. Some examples are—Where were you at the time of the accident? What were you doing at the time? What did you see and hear? What were the environmental conditions such as weather, lights and noise? What was the injured doing at the time of accident? In your opinion, what caused the accident? How can such accidents be prevented in future?

The evidences are to be recorded in a structured form, such as an investigation form developed by the construction organization. It is imperative that relevant portions of this form be filled up. For example, some sections of the form may be required to be filled up by medical personnel, some by supervisor(s), some by manager(s), and some by the safety officer.

While recording evidence related to equipment, points such as model/make, serial number and age of the equipment are noted. The intended and actual use of the equipment is noted.

The maintenance record of the equipment and any modification made in the equipment since its first installation are also noted.

The evidences on documents and procedures dealing with risk assessment; technical instructions; permits to work; method statements; records of training; and records of previous problems/accidents are collected. All documentation must be made accessible to the investigating team. The evidences used must be recorded carefully, cross-referenced and signed by their originators.

The evidence of surroundings of the accident is also recorded. For this, the area/deal with any injured person(s) is isolated first and the position of injured person(s) is recorded based on witness account, after which some schematic sketch is prepared. The prevailing environment condition is noted, as are the time and date of the accident. The details of the injured person such as his/her name, date of birth, employment and training details, and number of years of experience are recorded as well.

The four key areas for which evidence must be recorded are—witness of the accident, equipment, documents and procedural manuals, and the injured person himself. The five-factor model for investigation suggests recording the details of (1) task; (2) manual; (3) environment; (4) personnel; and (5) management.

Some of the questions that must be addressed for the task in which the accident took place are—was a safe work procedure followed; were conditions changed to unsafe during the course of execution of the task; were appropriate tools used for performing the task; were safety devices working properly; and so on. If the answer to any of these questions is 'no', the causes of their absence are further explored.

Questions related to materials include—was there an equipment failure; what caused it to fail; was the machine properly designed; what hazardous substances were involved; and was the required PPE used. Environment-related questions that need attention are—what was the weather condition; is housekeeping a problem; was it too hot or too cold; was noise a problem; was there adequate light; were gases, fumes and dust present; and so on.

Some questions relating to personnel involved with the injury are—were the workers experienced; were they trained; were they physically fit; and were they tired. Some issues related to management that needs to be addressed are—were the safety rules communicated; were the work procedures available and enforced; was the supervision adequate; was the safety plan in place; and were the workers trained to do the work.

Analysis of Data

This involves developing the sequence of events and identifying the causes that led to the accident. What happened; why it happened; how it happened; and how can it be prevented are some of the issues that are considered. Essentially, this part of the accident investigation is concerned with finding the answers to the sequence of events leading to the accident, the identification of causes and failures that led to the accident, and the corrective measures that need to be taken to avoid recurrence of such accidents. The identification of causes leading to the accident must not only show the direct cause, but also point out the indirect causes such as system weaknesses that produced the surface causes for the accident.

Recommending Corrective Action

This involves suggesting and recommending improvements to avoid the recurrence of such accidents in future, as well as preparing a comprehensive accident investigation report.

In a typical situation, the corrective measures suggested at the end of an investigation process relate to strengthening the engineering control measures (such as elimination, substitution, isolation and physical protection) or the management control measures (such as development of safe procedures, supervision and training), and to the strict usage of personal protective equipment. Depending on the situation and the lapses, the measures suggested could either be in isolation or in combination of the above-suggested measures.

The investigation should provide a coherent and structured description of the events leading up to the accident. While preparing the report, the objectives of the report and their intended user should be kept in mind. It must be simple in language and avoid ambiguous sentences. Adjectives should be used minimally. The conclusions should be brief and adhere to the point. Finally, the report should contain a rational view of 'what should be done' as remedial measures to avoid the recurrence of such accidents.

In brief, a successful accident investigation process should incorporate—a causal model representing a 'system-based approach'; inputs of key, relevant individuals; procedures/protocols to structure and support the investigation; identification of both immediate and underlying causes; recommendations that address the latter; and development of an accessible database.

15.10.11 Accident Statistics and Indices

Accident statistics are maintained primarily for—(1) designing preventive measures and making people safety-conscious; (2) enabling inter- and intra-company comparisons in different time periods; and (3) understanding whether a particular preventive measure adopted by management has resulted in improvement. Based on the principle of 'what gets measured, gets improved', a number of safety indices/indicators have been proposed. Given below are the definition and explanation of some of the indicators taken from IS:3786.

Frequency Rate

The accident frequency rate is expressed in terms of the number of deaths and injuries in occupational accidents per one million work-hours in the aggregate. The frequency rate is obtained by dividing the number of lost-time injuries and the number of reportable lost-time injuries (multiplied by one million according to IS:3786; OSHA suggests multiplication by 200,000) in occupational accidents that occurred during the survey period, by the total number of man-hours for all workers who were exposed to risks in the same period. Thus, we have:

$$F_A = \frac{\text{Number of lost-time injury}}{\text{Man-hours worked}} \times 1,000,000 \quad (15.1)$$

$$F_B = \frac{\text{Number of reportable lost-time injury}}{\text{Man-hours worked}} \times 1,000,000 \quad (15.2)$$

It may be noted that Indian Standards distinguish between lost-time injury and reportable lost-time injury. Lost-time injury is the one in which the injury require only a little bit of medical attention and the worker returns to his work quickly. A reportable lost-time injury is one that may result in worker absenteeism for more than 48 hours (a clear two days after accident has occurred, leaving the day of accident) and is supposed to be reported by statute to the appropriate authority. The reportable injury may be fatal or non-fatal, with different degrees of injuries. In order to get the frequency rate of fatal injuries alone, the above formula can be modified in the following manner:

$$F_C = \frac{\text{Number of fatal injuries}}{\text{Man-hours worked}} \times 1{,}000{,}000 \quad (15.3)$$

F_C is also referred to as fatality rate. ILO suggests computation of frequency rate for fatal and non-fatal injuries separately. It further suggests that the denominator should ideally be the number of hours actually worked by workers in the reference group. If this is not possible, it may be calculated on the basis of normal hours of work, taking into account entitlements to periods of paid absence from work, such as paid vacations, paid sick leaves and public holidays. IS Code suggests that man-hours should include managerial, supervisory, professional, technical, clerical and other workers including contractors' labour.

Severity Rate

The severity rate of new cases of occupational injury is defined in the following manner:

$$S_A = \frac{\text{Man-days lost due to lost-time injury}}{\text{Man-hours worked}} \times 1{,}000{,}000 \quad (15.4)$$

$$S_B = \frac{\text{Man-days lost due to reportable lost-time injury}}{\text{Man-hours worked}} \times 1{,}000{,}000 \quad (15.5)$$

In all the above formulae, the reference period such as the month, the quarter, or the year must be clearly identified.

In order to simplify the above computations, IS:3786 has characterized the man-days lost due to fatality and different types of disablement. For example, if the accident has resulted in the death of a worker, the corresponding lost time is equivalent to 6,000 man-days. Code specifies that lost time corresponding to loss of one thumb is considered equivalent to 1,800 man-days; loss of one eye is considered equivalent to lost time of 2,400 man–days; and so on. The lost time prescribed corresponding to fatality and other types of disablement varies across countries. For example, Japanese code considers lost time in fatality as equivalent to 7,500 man-days.

Incidence Rate

It is defined as the ratio of the number of injuries to the number of persons during the period under review. It is expressed as the number of injuries per 1,000 persons employed. Like the frequency rate, the incidence rate may be calculated both for lost-time injuries and reportable lost-time injuries as given below:

$$\text{Lost-time injury incidence rate} = \frac{\text{Number of lost-time injuries}}{\text{Average number of persons employed}} \times 1{,}000 \quad (15.6)$$

$$\text{Reportable lost-time injury incidence rate} = \frac{\text{Number of reportable lost-time injuries}}{\text{Average number of persons employed}} \times 1{,}000 \quad (15.7)$$

According to ILO guidelines, incidence rate may be calculated separately for fatal and non-fatal injuries. The number of workers in the reference group should be the average for the reference period. In calculating the average, account should be taken of the hours normally worked by those persons.

Other Indices

Recent researches in the safety field has indicated that the parameters mentioned above cannot in isolation accurately capture the safety perormance of a contractor and, accordingly, researchers have suggested other measures such as EMR (experience modification rate), loss ratio, and workmen's compensation claim frequency.

EMR is employer-specific and takes into account the claim history of the employer, and the frequency and the severity of injuries associated with the employer. It gives proper weightage to minor injuries and major injuries in such a way that excessive occurrence of minor injuries penalizes the employer as well as not more severe alteration of EMR with occurrence of major injuries.

Loss ratio is the ratio of cost of claims to premium paid to the insuring agency. Loss ratio of more than one indicates that the cost of claims is more than the premiums paid to the insuring agency. Insurance company paid more in claims than it received as premiums from the contracting company.

$$\text{Workmen's compensation claim frequency index (WCCFI)} = \frac{\text{Number of WC claims}}{\text{Total man-hours}} \times 200{,}000 \quad (15.8)$$

15.10.12 Safety Inspection

Periodic inspections are extremely important to ensure that the activities in a project are performed according to management's intentions. The inspection should be carried out by line supervisors and managers, usually in consultation with the safety officer. The inspection should be aimed to control potential hazards if noticed during inspection. Normally, the personnel carrying out inspection have a checklist and the compliance or non-compliance corresponding to each point in the checklist is noted. Findings of the inspection should be made known to those responsible for taking corrective action.

15.10.13 Safety Audit

A safety audit is a systematic measurement and evaluation of the way in which an organization manages its health and safety programme against a series of specific and attainable standards. Fundamentally, a safety audit subjects each aspect of an organization's activities to a critical examination with the objective of minimizing injury and loss. The entire programme should be audited internally and externally (at some specified period by a third party) to determine its overall strengths and weaknesses. This step should be followed by recommendations of achievable targets to be met in subsequent audits. Safety auditing is, therefore, an ongoing process underpinning effective health and safety management (Ahmed 2000, *Journal of Management in Engineering*, November–December 2000). The management system ensures that such audits are carried out at agreed frequency and that corrective actions are taken.

15.10.14 Workers' Health and First-aid Facilities

Healthy workers are the biggest asset for any industrial unit. Management system should take due care of workers' health. At all such places in a factory where there is a risk of toxic chemical exposure or poisoning or any other occupational disease, the workers should be subjected to periodical medical examination. This will help management to take timely corrective measures and will save them from the botherations of protracted litigations, payment of compensations and other penalties. This will also induce a sense of confidence in workers and make them feel that their health is being really taken care of. This is also one of the requirements under sections 87 and 41(c) of the Factories Act, 1948. The management system should also ensure that an adequate number of first-aid boxes are maintained in good condition and replenished periodically to ensure that these always contain all the necessary items.

Effective safety management system is most essential to ensure safety in any organization. The best technical safety systems such as machine guarding, fire safety, chemical safety and waste disposal may be of little or no yield in the absence of a well-established safety management system.

15.11 RESEARCH RESULTS IN SAFETY MANAGEMENT

Analysis of accidents conducted over a long period of time has suggested certain crucial aspects, as explained here.

New workers are more vulnerable to injuries at project sites. There is evidence that frequency of injuries comes down if the workers are given formal orientation before they are inducted into construction activities at a project site. In the orientation, workers should be introduced to the key persons such as project manager, safety manager, concerned foreman, charge hand and fellow crewmembers. They should also be briefed about the job layout, the hazardous areas, the objectives of the project, the key dates in the project, and the do's and don'ts at the project site. There must be a regular feedback mechanism to check whether the worker needs orientation again.

Also, a high rate of employee turnover is found to have an adverse impact on the frequency of injuries. That is, if a company is able to retain its workforce for project after project, it has better safety records. Workers working with familiar colleagues have fewer chances to be involved in an accident.

The chances of accidents are reduced if the top management involves itself in safety-related reviews of project sites. The involvement could be in the form of site visits or it could even be telephonic contacts. If workers perceive that top management is concerned about the safety performance on a project, it helps in reduction of accidents.

If the evaluation criteria include overall performance including safety, and does not concentrate on meeting cost and schedule objectives alone, the frequency of accidents reduces. If the top management is involved in reviewing the accident statistics, it has a favourable impact on the frequency of injuries.

It is noteworthy that if the margin on a project is less and the project has been won amidst stiff competition, chances are that the accident frequency will go up. It has been found that projects in which there were four or less bidders have shown less frequency of injuries when compared to projects where there were more than four bidders. The reason is that companies having won the bid on low margin try to cut corners to maximize their gains, and the first casualty in such cases is the safety budget for the project. One possible correction as suggested by Hinze is to consider safety in the category of general overheads and to delink it from the project cost. By doing so, the project manager would not be compromising on safety appliances such as safety helmet, safety belt and other basic safety accessories even for projects having low margin.

It has been found that the accident rates for contractors that keep records of accidents by projects are substantially lower than those for companies that do not keep these records.

REFERENCES

1. Abdelhamid, T.S. and Everett, J.G., 2000, 'Identifying Root Causes of Construction Accidents', *Journal of Construction Engineering and Management*, ASCE, 126 (1), pp. 52–60.
2. Abudayyeh, O. et al., 2003, 'Analysis of Occupational Injuries and Fatalities in Electrical Contracting Industry', *Journal of Construction Engineering and Management*, 129(2), pp. 152–158.
3. Accident Facts, 1990, National Safety Council, Chicago authority, Chicago.
4. *Accident Prevention Manual for Industrial Operations*, N.S.C., Chicago, 1982.
5. Ahmed, S.M., Kwan, J.C., Ming, F.Y.W. and Ho, D.C.P., 2000, 'Site-Safety Management in Hong Kong', *Journal of Management in Engineering*, ASCE, 16(6), pp. 34–42.
6. Bird, F. and Loftus, R., 1976, *Loss Control Management*, Loganville, Ga: Institute Press.
7. Blake, R.B., 1973, *Industrial Safety*, New Jersey: Prentice Hall, Inc.
8. Construction Industry Development Council (CIDC), Country Report 2005–06, available at http://www.cidc.in.
9. Construction Safety Association of Ontario (CSAO), http://www.csao.org.
10. Davies, V.J. and Thomasin, K., 1990, 'Construction Safety Handbook', London: Thomas Telford Ltd.
11. Delhi Building and Other Construction Workers (Regulation of Employment and Conditions of Service) Rules, 2002.
12. Everett, J.G. and Frank, P.B., Jr, 1996, 'Costs of accidents and injuries to the construction industry', *Journal of Construction Engineering and Management*, ASCE 122(2), pp. 158–164.
13. Everett, J.G. and Yang, I.T., 1997, 'Workers' comp. premiums: Disparities in penalties for identical losses', *Journal of Construction Engineering and Management*, 123(3), pp. 312–317.
14. Fredericks, T. et al., 2002, 'Mechanical Contracting Safety Issues', *Journal of Construction Engineering and Management*, 128(2), pp. 186–193.
15. Garza De la, J.M. et al., 1998, 'Analysis of Safety Indicators in Construction', *Journal of Construction Engineering and Management*, 124(4), pp. 312–314.
16. Hancher, D.E. et al., 1997, 'Improving workers compensation management in construction', *Journal of Construction Engineering and Management*, 123(3), pp. 285–291.
17. Heinrich, H.W., 1959, *Industrial Accident Prevention: A Scientific Approach*, McGraw-Hill, USA.
18. Heinrich H.W., 1980, *Industrial Accident Prevention*, New York: McGraw-Hill.

19. Heinrich, H.W., Petersen, D. and Roos, N.R., 1980, *Industrial Accident Prevention*, New York: McGraw-Hill.
20. Hinze, J.W., 2008, *Construction Safety*, Prentice Hall.
21. Hinze, J. and Applegate, L.L., 1991, 'Costs of construction injuries', *Journal of Construction Engineering and Management*, ASCE, 117 (3), pp. 537–550.
22. Hinze, J. and Wilson, G., 2000, 'Moving Toward A Zero Injury Objective', *Journal of Construction Engineering and Management*, 126(5), pp. 399–403.
23. Hinze, J. et al., 1998, 'Identifying Root Causes of Construction Injuries', *Journal of Construction Engineering and Management*, 124(1), pp. 67–71.
24. Huang, X. and Hinze, J., 2003, 'Analysis of Construction Worker Fall Accidents', *Journal of Construction Engineering and Management*, 129(3), pp. 262–271.
25. Hudson, R., 1985, *Construction Hazard and Safety Handbook*, Butterworth.
26. International Conference on Safety in the Built Environment (BUSI-88) held in Portsmouth, England, July 13–15, 1988, edited by Jonathan, D.S., E. & F.N. Spon, London.
27. International Labour Organization, 'Occupation safety and health (ILO-OSH), 2001', Guidelines on occupational safety and health management systems, Geneva, Switzerland.
28. IS 18001:2000, Indian Standard on Occupational Health and Safety Management systems—Specification with Guidance for Use, Bureau of Indian Standards, New Delhi.
29. IS 3786:1983, Methods for computation of frequency and severity rates for industrial injuries and classification of industrial accidents.
30. Japan Industrial Safety and Health Association (JISHA), Cost-Benefit Analysis of Safety Measures (September 2000, pp. 1-32), available at http://www.jisha.or.jp/english/corner/benefits.pdf.
31. Joseph, A.J., 1999, 'Safety Costs Money and Can Save Money', Proceedings of the Second International Conference of CIB Working Commission W99/Honolulu/Hawaii/24–27 March 1999, pp. 223–228.
32. Journal by insurance company surveyors and loss assessors – Mumbai, published by insurance companies, available at http://www.elcosh.org/en/document/520/d000505/improving-safety-can-save-you-money.html.
33. Krishnan, N.V., 1997, *Safety Management in Industry*, Mumbai: Jaico Publishing House.
34. Lee, S. and Halpin, D.W., 2003, 'Predictive Tool for Estimating Accident Risk', *Journal of Construction Engineering and Management*, 129(4), pp. 431–436.
35. Lees, F.P., 1990, *Loss Prevention in Process Industries*, 2nd ed., London: Butterworth Publications.
36. Mohamed, S., 2002, 'Safety climate in construction site environments', *Journal of Construction Engineering and Management*, 128(5), pp. 375–384.
37. Occupational safety and health administration (OSHA), 2005, 29 CFR, Washington, D.C., (http://www.osha.gov), October 2005.
38. Petersen, D., 1971, *Techniques of Safety Management*, New York: McGraw-Hill, Inc.
39. Petersen, D., 1981, *Techniques of Safety Management*, Tokyo: McGraw-Hill.

40. Petersen, D., 1989, *Safe behavior reinforcement*, Goshen, NY: Aloray.
41. Ridley, J., 1983, *Safety at Work*, London: Butterworth & Co.
42. *Safety and Good Housekeeping*, N.P.C., New Delhi, 1985.
43. Suraji, A. et al., 2001, 'Development of Causal Model of Construction Accident Causation', *Journal of Construction Engineering and Management*, 127(4), pp. 337–344.
44. The Building and Other Construction Workers (Regulation of Employment and Conditions of Service) Act, 1996 (BOCW Act), Commercial Law Publishers (India) Pvt. Ltd, 2007, New Delhi.
45. The Factories Act, 1948, Universal Law Publishing Co. Pvt. Ltd, 2007, New Delhi.
46. *The Times of India*, July 12, 2004.
47. Toole, T.M., 2002, 'Construction Site Safety Roles', *Journal of Construction Engineering and Management*, 128(3), pp. 203–210.
48. Vincoli, J.W., 1994, *Basic Guide to Accident Investigation and Loss Control*, John Wiley & Sons.
49. Wilson, Jr, J.M. and Koehn, E., 2000, 'Safety Management: Problems Encountered and Recommended Solutions', *Journal of Construction Engineering and Management*, 126(1), pp. 77–79.
50. World Health Organization (WHO), http://www.who.int/en/.

REVIEW QUESTIONS

1. State whether True or False:
    a. Accidents are defined as an event that is planned, desired, expected and controlled, and without any loss in property, time, money and other assets.
    b. Five dominoes used by Heinrich are—social environment and ancestry, fault of person, unsafe act/condition, accident and injury.
    c. 'Goals freedom alertness theory' states that safe work performance is the result of a psychologically rewarding work environment.
    d. Accident root cause tracing model' proposes three root causes of accidents—failure to identify an unsafe condition, deciding to proceed with a work activity after worker identifies an existing unsafe condition, and deciding to act unsafe regardless of initial work conditions.
    e. An unsafe act may be an act of commission (doing something that is unsafe) or an act of omission (failing to do something that should be done).
    f. Costs of accidents involve—direct costs for worker and employer and indirect costs for employer and worker.
2. Match the following according to Vincoli's modified dominoes:
    a. Management   (i) Basic causes
    b. Origins   (ii) Symptoms
    c. Immediate causes   (iii) Incident

d. Contact (iv) People–Property
   e. Loss (v) Loss of control
3. Match the following:

   A Frequency rate   1 [No. of lost-time injuries × 1,000]/avg. no. of persons employed

   B Severity rate   2 [No. of lost-time injury × 1,000,000]/man-hours worked

   C Incidence rate   3 [Man-days lost due to lost-time injury × 1,000,000]/man-hours worked

4. Arrange the following causes of fatalities according to their statistical occurrences (in increasing order)
   a. Fall from height
   b. Accident involving vehicles
   c. Fire
   d. Drowning
   e. Collapse of excavation
   f. Electrocution
   g. Mechanical impact
   h. Fall/Hit by object
5. Unsafe conditions are mainly because of:
   a. Management actions/inactions
   b. Worker's or co-worker's unsafe acts
   c. Nonhuman-related events
   d. Unsafe condition
   e. All of above
6. What are the safety philosophy and principles of accident prevention?
7. Give a detailed account of accident investigation and its prevention.
8. What are safety-related problems in the Indian construction industry? What are the various hazards associated with the construction industry as regards safety?
9. What is safety? How do you ensure safety at construction sites?
10. What are the elements of a safety policy? Discuss some safety policies studied by you.
11. Define the roles and responsibilities of a safety officer deputed on a tunnelling site in a remote location. What measures can be taken by the safety officer to ensure safety at site? Give examples you know of or are familiar with.
12. Bring out the difference between an accident, an incident and a near-miss, and give examples of the same. How are these issues correlated?

13. What are the legal requirements regarding safety as applicable to construction sites and the construction industry?
14. How are safety provisions covered in contract document? Who is finally responsible for safety at project site?
15. What is a permit? Why is it required? Does having permit ensure adherence to all safety rules and regulations? Elaborate.
16. What are safety audits? Why is safety audit necessary? Explain the types of audits that can be done on site as well as on the organization.
17. With a flowchart, explain Heinrich's injury causation model.
18. Give a detailed account on accident investigation and its prevention.
19. Discuss the occupational health issues in construction industry.
20. Why is safety gaining such importance in the present-day construction scenario?
21. Accidents can be said to be caused through a combination of unsafe acts and conditions. Carefully read the following situations and identify the principal cause of the accident. Give your answer as given below:

    a. In cases the accident is caused by unsafe act(s)

    b. In cases the accident is caused by unsafe condition

    c. In cases the accident is caused by a combination of (a) and (b)

    d. In cases where the accident can be only attributed to reasons beyond the control of the site

1. A trench 1 m wide and 3 m deep is excavated and the walls are secured with appropriately driven steel sheets. Workers walking on the heap of earth removed from the trench slip and fall into the trench, causing one of them to have a fractured hand.
2. A worker is carrying out a painting job on a 10m high wall using a ladder. A person assigned to wash the floor of the room, without realizing that another worker was on the job, threw water on the floor, causing the ladder to slip, and causing injury to the painter.
3. A long trailer carrying reinforcement bars was standing near the stores waiting for instruction from the stores officer regarding the place of unloading. The stores officer directed the trailer driver to unload the reinforcement in the reinforcement yard. While reversing the trailer, it hit a stores worker who was standing behind the trailer and in the process received major head injury. (He was not wearing safety helmet thinking nothing would fall in the stores area.)
4. A worker is carrying reinforcing bars of diameter 20 mm, measuring between 3 m and 4 m, on his bicycle. The bars were projecting about 1.5 m–2 m on both sides longitudinally (pointed in the direction of bicycle wheels). While taking a turn at roundabout, it hit two pedestrians causing scratches to both of them.
5. An RCC slab had an opening of size 600 mm x 600 mm to carry fire-fighting pipelines from one floor to another floor. A worker going for lunch at about 1.00 p.m. fell through the opening from the second floor to the basement. The result was a fatal accident.
6. Lift shaft openings at each floor were left out in an eight-storey building. Lift people were busy installing the lift in it. At closing hours of day shift (about 8.30 p.m.), lift workers forgot to put the

barriers at the shaft location of fifth floor. During the night, a worker who had joined just two days back fell from this opening and lost his life.

7. A halogen lamp was kept on the floor (workers tend to work near halogen lamp during acute winter due to the warmth in its vicinity) near the deshuttered materials that had just been removed from the slab of the pre-heater building of a cement plant. The deshuttering material mostly consisted of plywood soaked with diesel and grease (used as deshuttering agents), and timber. At around 4.00 a.m., a major fire broke out damaging about 20 running metres of the pre-heater building.

8. A contractor while planning for a major RCC slab had procured a large quantity of cement. Cement bags were stacked one over the other and at some places 20–25 bags were stacked at one place. During the concreting operation, these cement bags were being shifted. At one point of time, cement bags started falling from the top and two workers got trapped underneath. It resulted in major injury to one of the workers.

9. Contractor has just finished concreting of a major RCC slab (it was cast just three days back). Another major RCC slab was planned in the next four days in its vicinity. Carpentry gang was working on this new slab formwork. This gang found some shortage of props of 4.1 m height. One of the carpenters of this gang sent his helpers to bring these props. The helpers went and started removing the props from underneath the newly cast RCC slab, resulting in the collapse of the entire slab progressively and causing major injuries to the helpers.

10. Construction work on a silo was going on at 70 m height. Reinforcement tying was in progress at this height during the night. A security guard while doing routine safety rounds at ground level on the site passed through the construction area. A piece of reinforcement bar fell from top causing serious head injury to the security guard.

11. A construction worker wearing a dhoti and a shawl was working at a sand-washing plant (it removes silt from sand mechanically and has a number of moving parts) during the night. During the morning hours (about 3.00 a.m.), he was resting under the conveyor belt. Somehow his dhoti and shawl got trapped in the conveyor belt and he was pushed to the main screening area. The worker lost his life.

12. Work on construction of a natural draught cooling tower (its diameter at base is about 100 m and height is about 120 m) was at full swing. The site had a skyrack (used for vertical transportation of construction workers and supervisors) for going to height. One day, eight workers were travelling (well within the limit) in it going to start the work in the night shift. Suddenly, the skyrack developed some fault at 15 m height and had a free fall. Four workers sustained minor injuries, three received major injuries, and one died on way to hospital.

13. There was leakage from the undercarriage. To identify the location of leakage, the operator opened the engine cover and started to look for the leakage point in the radiator. The machine was in running condition. The operator inserted his right hand and got struck with the moving fan blade. He received cut injury on his right-hand finger requiring specialist medical treatment.

16

Project Monitoring and Control System

Introduction, updating, project control, schedule/time/progress control, cost control, control of schedule, cost and technical performance—earned value method, illustrations of cost control system, management information system

16.1 INTRODUCTION

In the previous chapters, we discussed the concepts of planning and different types of scheduling. As is known, a construction project works in a dynamic environment and situations, constraints, etc., keep on changing. Hence, it is absolutely necessary to monitor the project at regular intervals and adopt suitable controlling measures in order to keep the project on track. For this purpose, construction organizations have project monitoring and control systems in place.

Traditionally, the schedule, cost and quality have been used as the control parameters for construction projects. The purpose of a control system is to see that the progress, cost and quality obtained in any project is within the agreed time schedule, the agreed cost, and the agreed specification, respectively. While there are many other parameters, such as disputes and accidents, which one would like to control, the focus of this chapter is to discuss aspects of time and cost control. It is believed that controlling these parameters would take care of other parameters.

Monitoring and control system also helps in providing feedbacks to management on the different schedules prepared earlier, in order to ensure that the project is progressing as per schedule. In case of slips/deviations from schedules, the system helps in taking corrective and timely action. The basic objective of any monitoring system is to monitor projects by measuring physical progress, costs and profits against targets, and to help in taking corrective action. It also provides data for preparing reports for management information system (MIS).

In this chapter, we discuss schedule control, cost control, and monitoring and control of different performance parameters. The chapter ends with a brief discussion on management information system applicable in construction.

16.2 UPDATING

Updating can be defined as planning and programming of the remaining portion of an activity job by introducing the latest information available. At the end of any day of work, the activities of the project must either be completed, in progress, or they may not have started yet.

Further, the actual progress may not be according to the originally envisaged schedule. Also, some of the activities that must have been completed as on the day of updating may not have been completed; worse still, some may not have started. Some of the activities may not have achieved the required percentage of progress that was planned on the day of updating. It is also possible that some new activities that were not in the original plan might have to be taken up. This may at times bring about a change in the network logic (sequencing of events). All these situations require updating of a project plan at an appropriate interval or frequency.

The updating frequency (how often the updating should be undertaken) depends upon the nature and type of the project, and the contractual provisions. It could be weekly, monthly, or even quarterly. The updating frequency may be lower during the start of the project, while more frequent updating would be needed towards completion of the project. Further, if there are any major changes in the objectives and scope of the project, or if the status of a stakeholder in the project has changed drastically, the project requires an immediate updating.

16.2.1 Updating Using Bar Chart

In addition to being an effective technique for overall project planning, the work progress can very clearly and easily be superimposed on the bar chart by plotting actual-progress bars alongside the scheduled-progress bars, as shown in Figure 16.1. As can be seen, the progress has been monitored at the end of seventh month (shown by a thick line). The top bar shows the planned duration and the bottom one shows the actual progress. The excavation is complete though it has taken one extra month to get completed. PCC, though it started late, is completed in time.

Unfortunately, bar charts have limited application for detailed work because many interrelationships of activities are not defined. They become very difficult to construct as the number of activities increases—a situation inherent in repetitive construction. Bar charts are difficult to update, do not show interdependencies of activities, and do not integrate costs or resources with the schedule. They only relate the listed activities to the timescale. They do not indicate interdependence of activities, and specifically for linear-type projects cannot indicate variations in rate of progress.

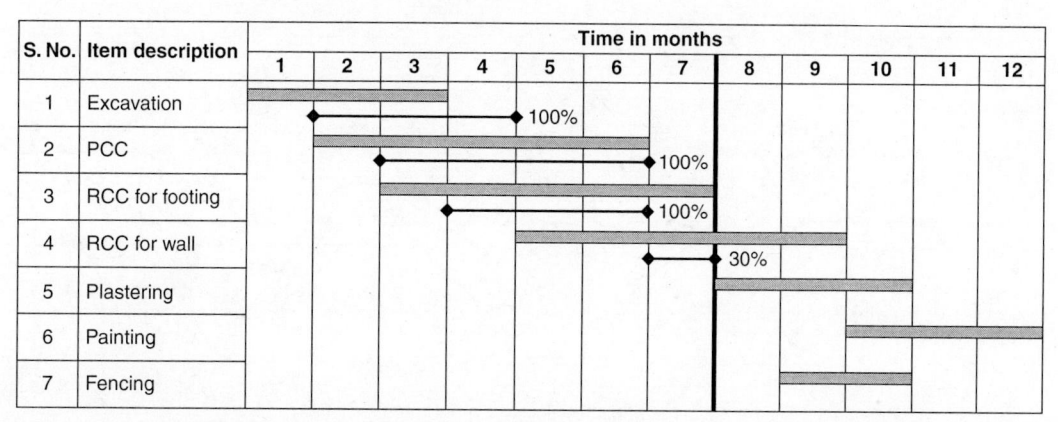

S. No.	Item description	\multicolumn{12}{c}{Time in months}
		1	2	3	4	5	6	7	8	9	10	11	12
1	Excavation					100%							
2	PCC							100%					
3	RCC for footing							100%					
4	RCC for wall								30%				
5	Plastering												
6	Painting												
7	Fencing												

Figure 16.1 Bar chart for showing progress of the construction of boundary wall

16.2.2 Updating Using PERT/CPM

The usefulness of CPM is preserved throughout the course of a project because of its updating capabilities. Starting with the status of activities as at the instant of updating, the CPM can be drawn for the remainder of the activities; any new critical paths and floats are identified, and selective management controls/directions can be affected.

The following steps are suggested:

1. All completed activities are given zero duration.
2. All activities in progress are given duration equal to the number of days remaining for completion.
3. All activities not yet started are given duration as reported in the latest estimate.
4. Any correlation change, i.e., changed network logic between activities or parts thereof, yet to be completed are brought forth and necessary changes in the arrow network diagram are duly envisaged, discussed and communicated/recorded.
5. To keep up the calendar scheduling of the time baseline unchanged and also keep the facts as they are, the earliest occurrence time of the first activity of the network being updated is marked as, say, X. Where X is the day on which the network is being updated. For example, if the updating is done at the end of the 15th day (i.e. if the updated activities are to begin on the morning of the 16th day) the early occurrence time of the first activity would be marked as 15.
6. New critical paths are then computed for further control/monitoring.

Example: Updating PERT/CPM network

We take the example represented in Figure 16.2 to illustrate the updating process. There are a total of seven activities and the durations taken to complete each of them are shown in the network. The critical path is 1-2-4-5-6 and the project duration is 16 days.

Let us assume that monitoring is being done after 10 days and the following observations are made on the progress of the activities:

- Activity *A* is complete.
- Activity *B* is complete.

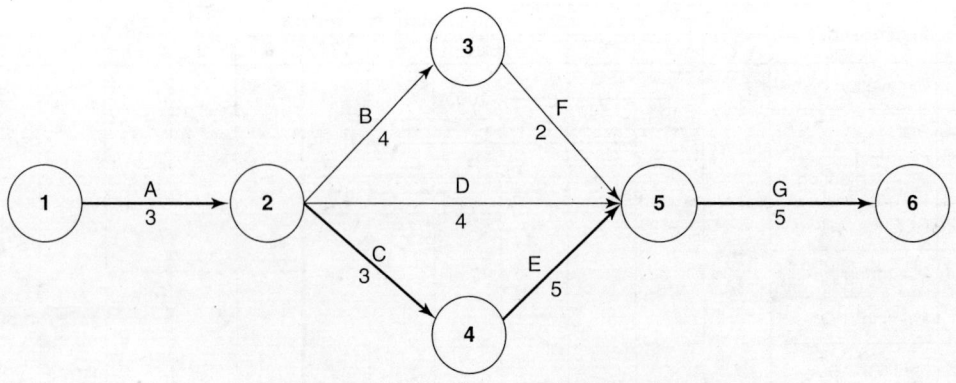

Figure 16.2 Example for illustrating updating of network

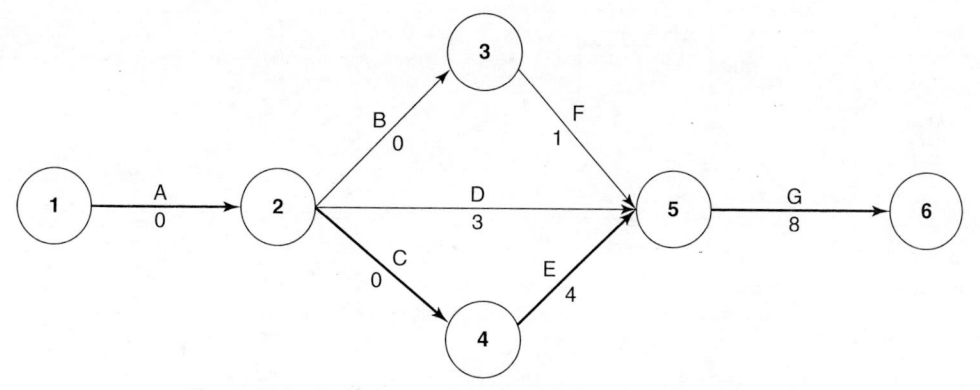

Figure 16.3 Revised network at the end of 10 days of project start

- Activity C is complete.
- Activity D will take 3 more days to finish.
- Activity E still needs 4 days to complete.
- Activity F is underway and will take another day to complete.
- Duration of activity G has been revised to 8 days.

For updating the network, a revised network is plotted. The durations of completed activities such as A, B and C are reduced to 0. Activity D is expected to take three more days, which is indicated in the revised network. Similarly, activity E is expected to take four more days to complete and, accordingly, this is shown in the network. Activity F needs another day to complete and this is shown in the network as well. Finally, the revised duration of activity G, which is estimated to take eight days now, is entered in the network. The revised updated network is shown in Figure 16.3.

In order to calculate the estimated project duration at the end of 10 days, the early occurrence time of event 1 is made equal to 10, and forward pass and backward pass is carried out in the manner explained previously. The early occurrence and late occurrence times of all the events are shown in Figure 16.4. The critical path of the revised network has activities (4–5) and (5–6), and thus, these are the critical activities. The revised project duration is 22 days, which means that there is a delay of 6 days (22 − 16) from the original programme.

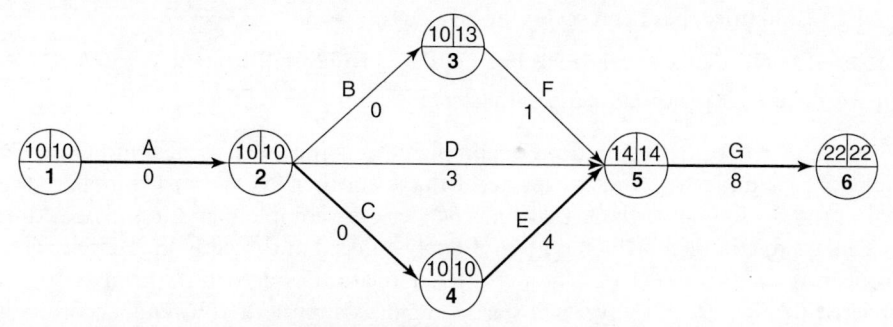

Figure 16.4 Revised updated network

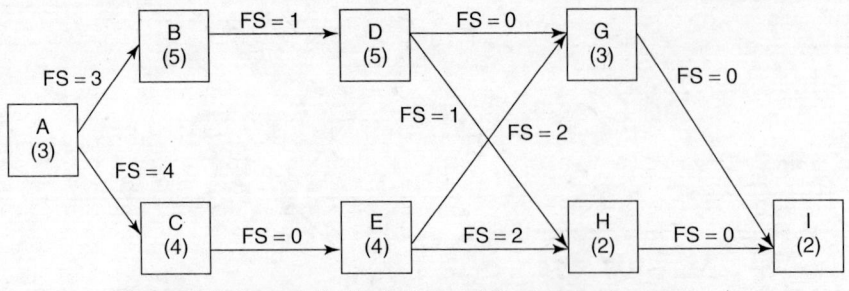

Figure 16.5 Example for illustrating updating in precedence network

16.2.3 Updating Using Precedence Network

For updating the precedence network a new activity with duration equal to the time interval between the start of the programme and the date of updating is introduced. Next, the activities that are completed as on the date of updating are identified. Duration of completed activities is changed to zero. Next, the revised durations of 'in progress' activities and 'not yet started' activities are changed depending on any new constraint and change in the requirement. Forward pass and backward pass as explained earlier are performed, and the duration of the project along with critical activities are identified in a similar manner.

Example: Updating precedence network

There are a total of eight activities and the durations taken to complete each of them are shown in the network (See Figure 16.5). The type of relationship along with the leads and lags are shown on the arrows connecting two activities. The project duration is 22 days and incidentally all activities are critical.

Monitoring is being done at the end of 14 days and the following observations are made:

1. Activity A is complete.
2. Activity B is complete.
3. Activity C is complete.
4. Activity D will take another five days to complete.
5. Activity E is 50 per cent complete and will take another six days to complete.
6. Activity G's duration has been revised to five days.
7. Activity H's precedence relation has been updated to FS = 2 from activity D.
8. Activity I will start three days after activity G.

For updating the network, the durations of completed activities such as A, B and C are reduced to 0. Activity D is underway and is expected to take another five days to complete, which is indicated in the revised network. Activity E is 50 per cent complete and it is expected to take another six days to complete, which is appropriately entered in the network. Activity G's duration has been revised to five days. The precedence relation of activity H from activity D has been updated to FS = 2. Activity I will start three days after activity G and, accordingly, the

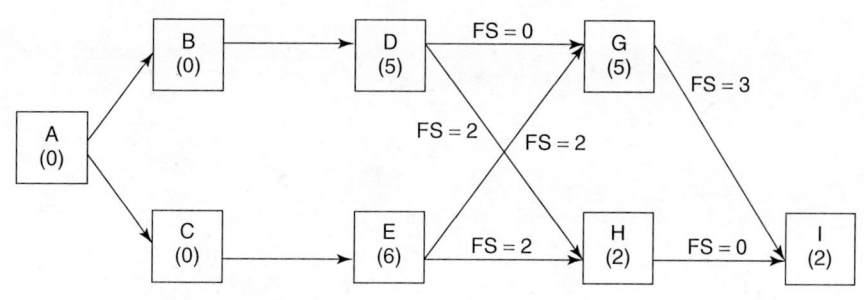

Figure 16.6 Updated network at the end of 14 days from the project start

relationship has been shown as FS = 3. The revised updated network incorporating all the changes is shown in Figure 16.6.

In order to calculate the estimated project duration after 14 days, the early start time of activity A is made equal to 14, and forward pass and backward pass are carried out in the manner explained previously. The early start, early finish, late start and late finish times of each activity are shown in the revised network given in Figure 16.7. The critical path of the revised network consists of activities E-G-I. The revised project duration is 32 days, which means that there is a delay of 10 days from the original programme.

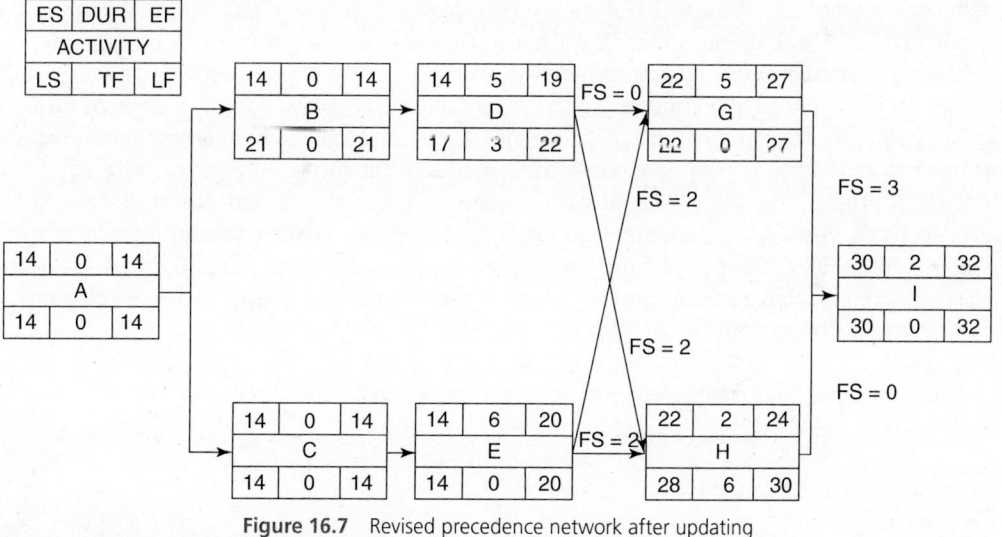

16.3 PROJECT CONTROL

If projects are left to run on their own, they may end up spending more money than initially planned and may also take longer to complete, and probably also be of quality lower than the

Figure 16.7 Revised precedence network after updating

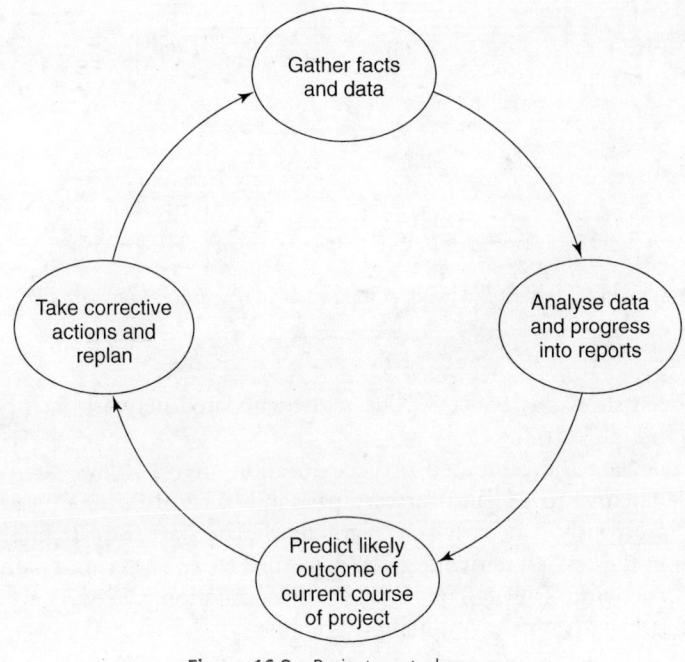

Figure 16.8 Project control process

agreed standard. Thus, every project requires some degree of control. The control is an integral part of the project management process, and it aims at regular monitoring of planned versus actual achievement. In case deviations are noticed, there has to be revision in the plan. The timeliness of application of control process is very important, and any control mechanism should attempt to discover the deviation at the earliest so that corrective measures can be applied.

It is very rare for a construction project to proceed exactly as per the plan. Thus, there is justification for applying the control process on a continuous basis. A typical control process involves gathering facts and data, analysing them, predicting the likely outcome based on the current data and taking appropriate corrective action. This is shown in Figure 16.8.

Since a project control system provides a basis for management decision, it should be simple to work with and at the same time be able to draw immediate attention to significant deviations (difference between planned and achieved accomplishment). It should also indicate the area where corrective actions are needed to overcome the deviation. Also, the key control measures should be identified carefully so that the results of control are worth the time and effort spent.

There are three important elements to be controlled in a project:

- Schedule/time/progress control
- Cost control
- Quality control

These are explained in subsequent sections.

16.4 SCHEDULE/TIME/PROGRESS CONTROL

The tools for progress control are bar charts or critical path networks. Whichever technique is used, the project manager should abide by the following steps:

- Establish 'targets' or 'milestones'—times by which identifiable complete sections of work must be completed
- As each target event occurs, compare actual against targeted performance
- Assess the effect of performance to date on future progress
- If necessary, re-plan so as to achieve original targets or to reach as close as possible in achieving them
- Request appropriate action from those directly responsible for the various activities

Planning and control techniques achieve nothing unless they are translated into action, and it is the responsibility of the project manager to see that this happens. Depending on the duration and the type of the project as well as the contractual provisions, monitoring could be done on weekly, fortnightly, monthly, or bimonthly basis. The most common monitoring period is on a monthly basis.

16.4.1 Monthly Progress Report

This report helps in reviewing monthly progress with respect to quantities of work done and invoicing made vis-à-vis the plans indicated in schedule of invoicing. This report also sets the targets for the following month (on the basis of actual performance till then) or for the next two to three months. Cumulative progress is reported by marking progress of each item on a bar chart. This report is not a true indicator of activity-wise progress in relation to construction and schedule of milestone events. If an activity-wise review is needed, the report will have to be used along with the progress of milestones. Where close monitoring is required, job status for the project (or for critical structures in the project) may be reported to the higher management by marking progress on a copy of the construction schedule. This copy can be updated continuously for regular review.

16.4.2 Measuring Progress at Site

The physical progress at the time of monitoring can be judged by any one of the following methods depending on the situation.

Quantities installed: In this method, the number of an item installed is physically counted and compared to the planned total quantity.

Percent complete: In this method, an estimate by the actual persons responsible for completing the activity or group of activities is made in terms of percentage.

Effort by support services: This method is employed when the tasks are more a function of time and cannot be easily quantified like logistics and the work by project planners. These are measured by time lapsed compared to planned total time allowed.

By milestones: In this method, the percentage completion value is assigned based on pre-defined estimated work for a predefined milestone. For example: Completion of substructure for a multi-storeyed building = 25 per cent; superstructure completion = 40 per cent; finishes completion = 15 per cent; services completion = 20 per cent.

Tied to contract terms of payment: In this method, the progress and contract terms of payments are related.

Two-point measurement: This method of measurement of progress is limited to two predefined measurement points—for example, 50 per cent when all the materials required for the work are on site, and 50 per cent on completion of the activity. Another example could be 60 per cent on supply part of the activity and 40 per cent on the erection part.

Single-point measurement: This method is employed when there are numerous small-value activities in a project. Here, the measurement point could be 100 per cent on completion only.

16.4.3 Typical Reports to Aid the Progress Review

Some of the typical reports used for aiding progress control are:

- Actual versus planned bar-chart schedules and time analysed network reports
- Activity status reports with early and scheduled starting and completion dates, for current and balance activities and floats
- Activity reports generated by departments, resources, costs and sales
- About two to three months' look-ahead plan
- Report on critical activities
- Report on slipping tasks
- Report on achievement of milestone activities
- A format of three bars showing original planned, revised planned and actual achieved till date.

16.5 COST CONTROL

There are different ways in which cost control can be exercised for a construction project. As mentioned earlier, the important points that must be emphasised while devising cost control for a construction project are the simplicity of the system and the response time. There is no point in having a cost control system that is complicated and involves a number of staff to man it, while the response time is very high.

Harris and McCaffer (2005) quote the example of a thermostat while illustrating the cost control process. A thermostat observes the outside temperature, compares it with some predefined standard, exercises control majors and adjusts the temperature. As described previously, the project cost control also can be thought of as a three-step process:

1. Observe the cost expended for an item, an activity, or a group of activities
2. Compare it with available standards. The standard could be a predefined accepted cost estimate (ACE) or it could be the tender estimate
3. Compute the variance between the observed and the standard, signalling a warning sign immediately to the concerned people so that timely and possible corrective measure can be taken

This is where the simplicity of the system comes into the picture. The more complicated the system, the more number of people it requires to be involved and possibly the more the response time needed to pinpoint the variation between the observed and the standard. The response time becomes very important since there is no point in finding the variance at a time when the activity showing variance is already completed, although this data can be useful for other projects.

The collection of cost for a project needs involvement of a number of people including planning engineer, billing engineer, plant and equipment engineer, timekeeper, storekeeper, accountant and head-office staff. The process of recording cost data against different cost codes and analysing them against these codes, descriptions, quantities, rates, etc., and then comparing them with some standard such as accepted cost estimate, detecting variances, if any, and taking timely corrective measures may all look simple in the beginning. However, the process is not as easy in real-world projects, as we shall see shortly. There are different control systems that can be adopted depending on the requirement. These are explained below.

16.5.1 Profit/loss at the Completion of Contract

In this system, total cost for the contract is calculated at the conclusion of contract, and profit or loss generated by the contract is calculated. Needless to say, though the system is very simple, it cannot actually be termed as a 'control system' since the information gathered by this process can hardly be used for the project for which calculations have been made. At best, the information may be useful for future projects by way of refinement of the existing databank. Moreover, this method can at best be used for very small projects.

16.5.2 Stage-wise Completion of Cost

In this system, cost calculation is done for different stages or milestones of a construction project. At the completion of each of these stages, cost is compared with standard cost of each package. The disadvantage of the system is more or less similar to the system described earlier.

16.5.3 Standard Costing

The system is used mostly in the manufacturing industries.

16.5.4 S-Curve

S-curve can be used to monitor the cost of a construction project. The S-curve very nearly depicts the progress profile of a construction project, which is characterized by slow progress at the beginning and rapid progress towards the middle, followed by slow progress again towards the end. The amounts planned to be invoiced (billed) and the actual bill amount raised can also be monitored, through a graph (S-curve). A similar graph can also be plotted for the amount planned to be spent vis-à-vis the amount actually spent.

Once the contracts have been awarded, it is relatively easier to estimate the 'bill value' for each activity on a detailed bar chart. By analysing the nature of each activity, one can estimate the way in which this value will be distributed on, say, a monthly basis. These amounts will then be summed to give the total estimated monthly value, and the figure plotted on a graphic form as shown in Figure 16.9. Now, a second curve (shown as dotted line) can be drawn based on the work that is actually performed.

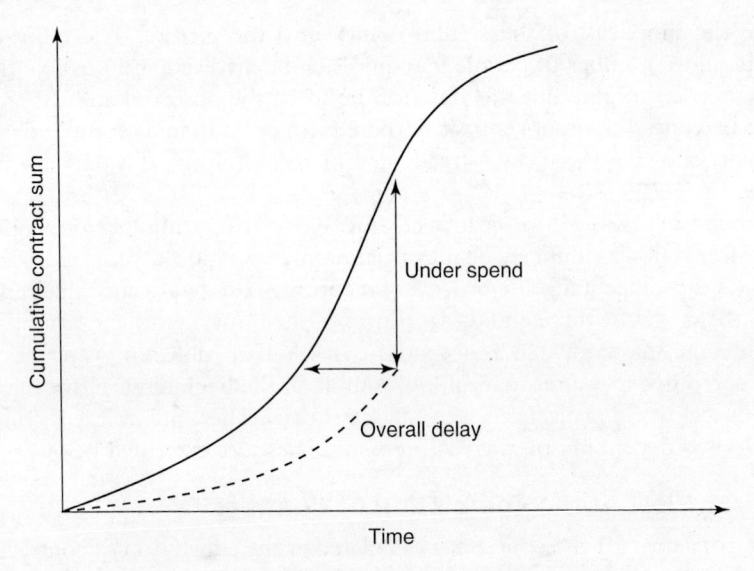

Figure 16.9 S-curve showing planned and actual bill value

It is clear from Figure 16.9 that the amount that should have been billed as on the day of monitoring has not been done, and that is the reason the second curve is running below the first curve. This is an 'under spend' situation. Similarly the second curve is also showing a shift towards the right. This shows an 'overall delay' indicating that the project is running behind schedule. Thus, the financial value may be used as a single general measure of the progress of a project.

The continuous curves shown here can, of course, be drawn only if it is possible to estimate the value of the work done in detail as in case of contracts based on bill of quantities. The project manager's ability to obtain a single and reasonable objective measure of overall progress is diminished where the contract payments are based on some other method.

After the bar chart has been derived from the project network, the estimated cost for different activities is calculated on either a monthly or a weekly basis. This could be based on either early start of activities or late start of activities. Accordingly, we will have two sets of cost. At this point of time, let us leave the discussion of cost based on early start and late start. The planned cost month-wise for a project is given in Column 3 of Table 16.1 and the cumulative value is given in Column 4 of the same table. As can be seen, the project is planned for 12 months duration, from January 2007 to December 2007. These planned or estimated costs have been arrived at by estimating labour, material, plant and equipment, and subcontractor costs—also known as direct cost for all the activities. Certain proportions of total indirect cost for the project are added as well.

In a very unlikely situation, the work will be executed exactly as per the plan. In most of the situations, though, the work will consume either more than the estimated duration or less than the estimated duration. For our example, consider the situation in which the project is estimated to take more than the originally estimated duration. The completion is now sched-

Table 16.1 Cost details of example project

S. No.	Month	Estimated cost of activities on monthly basis (₹ lakh)	Cumulative cost at the end of month (₹ lakh)	Revised estimated cost of activities on monthly basis (₹ lakh)	Cumulative cost at the end of month (₹ lakh)	Actual cost of activities on monthly basis (₹ lakh)	Cumulative actual cost at the end of month (₹ lakh)
(1)	(2)	(3)	(4)	(5)	(6)	(7)	(8)
1	January 2007	5	5	5	5	5	5
2	February 2007	10	15	10	15	10	15
3	March 2007	20	35	20	35	25	40
4	April 2007	25	60	20	55	25	65
5	May 2007	35	95	25	80	25	90
6	June 2007	50	145	30	110	30	120
7	July 2007	50	195	40	150	45	165
8	August 2007	40	235	40	190	50	215
9	September 2007	30	265	30	220	30	245
10	October 2007	20	285	25	245	30	275
11	November 2007	10	295	20	265	20	295
12	December 2007	5	300	15	280	15	310
13	January 2008			10	290	10	320
14	February 2008			5	295	10	330
15	March 2008			5	300	5	335

uled at the end of March 2008. The revised monthly cost is shown in Column 5 of Table 16.1. Notice that total estimated cost is still the same, which is equal to the earlier value (refer to Column 6).

It is interesting to note that the actual cost incurred for a particular activity will have very less likelihood of exactly matching the estimated cost. For this example, consider the actual cost month-wise as given in Column 7 of Table 16.1. The cumulative cost now stands at ₹ 335 lakh as given in Column 8 of the same table.

The cumulative cost figures given by columns 4, 6 and 8 have been drawn against time and presented in Figure 16.10. It can be observed that the project is behind schedule and over budget.

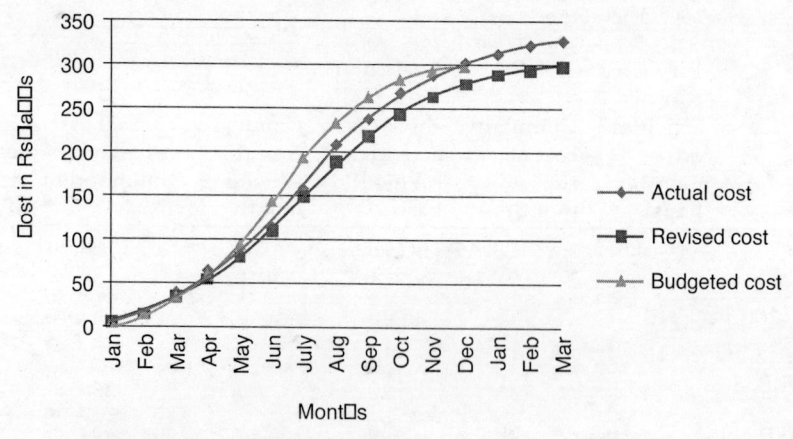

Figure 16.10 S-curves for budgeted cost, revised cost and actual cost

16.5.5 Unit Costing

In this system, project cost report is prepared usually on a monthly basis. This report assists in reviewing the cost every month, comparing actual cost vis-à-vis ACE[1] of major items, and initiating corrective actions in case of overruns in cost. This also helps in regularly assessing the projected total cost and the expected profit at site level, which is also known as site contribution.

The ACE of an item could be exactly matching the bid rate or sales prices (estimated 'no profit, no loss' scenario), lower than the bid rate (estimated profit scenario), or higher than the bid rate (estimated loss scenario). The ACE becomes the standard of comparison in future monitoring endeavours—any excessive expenditure over the ACE is treated as a warning sign and the control measures are promptly applied.

The basic input for preparing the project cost report is the cost statement. While the planning engineer prepares the project cost report, the site accountant provides the input cost statement. It is clear that close coordination between the accountant and the engineer is required in the preparation of this report. In this system of cost control, the steps involved are described below.

Step 1 Setting up the cost code
This is similar to that explained for the S-curve system.

Step 2 Preparation of cost statement
Cost statement is the report containing the cost code-wise details of expenses incurred during the month and till that month. The statement consists of the cost code-wise details of labour

---

[1] As soon as the project is awarded to the contracting organization and the letter of intent issued by the client, the contractor appoints a project manager and a planning engineer for the project. The two then go through the entire contract document and prepare the estimates of executing each bid item. They also prepare the estimate for the overheads of the project.
Cost codes are established and the budgeted estimates for each of the cost codes are prepared. These budgeted estimates are scrutinized by the top management of the contracting organization and a mutually agreeable budgeted estimate is agreed to. This is called the accepted cost estimate (ACE) or zero cost by the contracting organization.

cost, material cost, plant and equipment cost, subcontractor cost and overhead cost. The details are worked for the month and till the time of the monitoring. Further, the total cost for each of the cost codes is worked out using the following relation:

$$\text{Total cost} = (\text{cost of labour} + \text{material} + \text{plant} + \text{subcontractor} + \text{overhead}) - \text{deferred expense} + \text{provision for expenses} \quad (16.1)$$

The way to extract cost of labour, material, plant and subcontractor has already been discussed. Provision for expenses and deferred expenses are explained below.

Provision for Expenses
This includes foreseeable loss and expenses to be incurred in closing down of the projects. The delayed recoveries by clients for materials issued to the contractor, electricity and water supplied, etc., are also to be taken into account. Similarly, provision should also be made for the expenses that have been incurred though their exact cost is not known (for example, customs duty for imported items, commission to clearing agency, etc.) at the time of preparing the report.

Deferred Expenses
For some activities, the initial investment required would be very high. The investment is made for completing the entire activity. For instance, if the activity is only 10 per cent complete, it would be unfair to book the total investment for the activity in the cost statement. In such cases, only a part of the expenses, preferably on a pro rata basis, is shown in the cost statement and the remaining amount is taken as deferred expense, which is adjusted afterwards in proportion to the progress of activity.

Adding up the total cost for each of the cost codes gives the total project cost for the month and up to the month. This information is passed on to the planning engineer for further processing.

Step 3 Preparation of project cost report
Planning engineer prepares this report normally on a monthly basis after getting the cost statement from the accountant. For preparing the report, the planning engineer works out the variance in cost for each of the cost codes. The steps involved in preparation of project cost report are given below.

- **Estimate to date**
  This is found as follows:
  Calculate the quantity done to date. Let it be Q_d. Take cost to date data for this cost code from the cost statement. Let it be C_d. Compute the unit cost to date. This is denoted by E_d. Hence,

$$E_d = \frac{C_d}{Q_d} \quad (16.2)$$

- **Estimate for balance quantity**

  Calculate remaining quantity to be done (Q_b). Estimate the cost likely to be incurred (C_b). Compute the estimated unit cost for the balance quantity (E_b). This is given as

  $$E_b = \frac{C_b}{Q_b} \qquad (16.3)$$

- **Revised estimates**

  The revised estimate for a cost code is computed by first adding the quantities Q_d and Q_b. The revised total cost is the sum of C_d and C_b. The revised unit cost (E_r) is calculated by dividing the revised total cost, C_r, by the revised total quantities, Q_r. Expressed mathematically,

  $$E_r = \frac{C_r}{Q_r} = \frac{(C_d + C_b)}{(Q_d + Q_b)} \qquad (16.4)$$

- **Original estimate**

  Original estimate, E_o, is the estimated cost for a particular cost code. This is computed by dividing the original estimated cost (C_o) by original quantities (Q_o). This is expressed mathematically as,

  $$E_o = \frac{C_o}{Q_o} \qquad (16.5)$$

The variance (V) for each of the cost code is the difference of E_o and E_r. That is, variance

$$V = E_o - E_r \qquad (16.6)$$

The process is repeated for all the cost codes of the project to arrive at the total project cost at the time of monitoring. This information tells us whether the total project is over-running the budgeted cost or under-running the budgeted cost.

The typical format of a project cost report is given in Figure 16.11 and a small example is worked out to illustrate the approach. We take up the example of boundary wall construction discussed earlier in Chapter 6. Let us suppose that the contractor has been awarded the contract of constructing a boundary wall at a price of ₹ 4,979,375, and that he has agreed to complete the contract in about two months (62 days to be precise). The unit rate payable to the contractor for different activities is as given in Column 5 of Table 16.2.

Step 1
For illustrating the method, let there be seven cost codes for our example project corresponding to earthwork, formwork, reinforcement, concrete, structural steel, barbed wire and overhead. The cost codes of these are given in Table 16.3.

Step 2
Prepare the budgetary cost estimate for all activities of the project according to the cost codes. As mentioned earlier, this is the time to prepare the ACE or zero cost. This is shown in Table 16.4 for the mentioned activities.

PROJECT MONITORING AND CONTROL SYSTEM | 585

Original Contract Value: _____ Original Contract Duration: _____
Revised Contract Value: _____ Revised Contract Duration: _____

Cost code	Description	UOM	$E_d = \dfrac{C_d}{Q_d}$			$E_b = \dfrac{C_b}{Q_b}$			$E_r = \dfrac{C_r}{Q_r}$			$E_o = \dfrac{C_o}{Q_o}$			$V = E_o - E_r$	Remarks
			Q_d	C_d	E_d	Q_b	C_b	E_b	Q_r	C_r	E_r	Q_o	C_o	E_o		

Figure 16.11 Sample project cost report

Table 16.2 Extract from bill of quantity for the boundary wall construction project

Activity ID	Activity description	Quantity	Unit	Unit rate (₹)	Amount
1	Earthwork	500	m³	100	50,000
2	Cement concrete	40	m³	3,750	150,000
3	Concrete raft	120	m³	8,500	1,020,000
4	Concrete wall	312.5	m³	10,750	3,359,375
5	Barbed wire fencing	500	m	800	400,000
				Total	4,979,375

It may be noted that the estimate to complete could be the same as the unit cost to date, or it could be more or it could be less than the unit cost to date. This depends on a number of factors such as the nature of balance work the likely changes in cost component, and the escalation. Also, the revised quantity should be used in case there is a change in quantity against a particular cost code from the tendered quantity.

Total cost is arrived at by adding up the different cost components as mentioned above, and is entered in the project cost report (see Figure 16.12) under the column 'project status to

Table 16.3 Cost-code details for the example project

S. No.	Description	Cost code
1	Earthwork	100
2	Formwork	200
3	Reinforcement	300
4	Concrete	400
5	Structural steel	500
6	Barbed wire	600
7	Overhead	700

Table 16.4 Budgeted expenditure

Activity	Cost code		100	200	300	400	500	600	700	Total
	Accepted cost estimate									
	Rate	Amount								
EW	80	40,000	36,500						3,500	40,000
CC	3,000	120,000		8,000		100,000			12,000	120,000
CR	6,833	819,960		70,000	450,000	217,964			81,996	819,960
CW	8,576	2,680,000		625,000	990,000	797,000			268,000	2,680,000
FE	850	425,000					180,000	202,500	42,500	425,000
Total		4,084,960	36,500	703,000	1,440,000	1,114,964	180,000	202,500	407,996	4,084,960

date', against the various cost codes. The quantities of work done are then posted against each cost code to derive the unit cost. The unit costs derived should be compared with the ACE. The derived unit costs assist in identifying areas where there is an overrun, and further review will highlight the necessity of cost control measures. These unit costs are consulted to arrive at unit rates for 'estimates to complete' for the balance portion of work.

Thereafter, 'estimates to complete' unit costs are determined keeping in mind the unit cost to date, the nature of balance work, the quantities involved and the likely changes in components of cost, escalations, etc., for carrying out works at future dates.

The revised estimate is arrived at by adding 'project status to date' and 'estimates to complete'. The original estimate will comprise of revised quantities as reflected in revised estimates columns.

The revised contract price has to be arrived at based on revised quantities and agreed rates as per the contract. All escalations payable by the client and other claims as and when raised have to be included in the revised contract value. The revised contract value for the revised quantities minus revised cost estimates gives the estimated site contribution (profit) at the end of the project as on date.

16.6 CONTROL OF SCHEDULE, COST AND TECHNICAL PERFORMANCE—EARNED VALUE METHOD

The methods discussed earlier are not amenable to integrating both schedule and cost control aspects. On the other hand, earned value method can successfully integrate the schedule and cost control aspects.

Earned value is a methodology for determining the cost, schedule and technical performance of a project by comparing it with the planned or budgeted performance. It is assumed that once the budget has been made, it is more or less fixed.

PROJECT MONITORING AND CONTROL SYSTEM | 587 |

NAME OF THE PROJECT:		CONSTRUCTION OF BOUNDARY WALL													MONTH: July 2008		
CONTRACTUAL DATE OF COMMENCEMENT		01.06.2008				ORIGINAL CONTRACT VALUE			4979375						BOUNDARY WALL		
CONTRACTUAL DATE OF COMPLETION		02.08.2008				REVISED CONTRACT VALUE			4979375								
EXPECTED DATE OF COMPLETION																	
			PROJECT STATUS TO DATE			ESTIMATE TO COMPLETE			REVISED ESTIMATE FOR REV. QTY			ORIGINAL ESTIMATE FOR REV. QTY					
Sl. No.	COST CODE	DESCRIPTION	UOM	QTY	TOTAL COST	UNIT COST	QTY	TOTAL COST	UNIT COST	QTY	TOTAL COST	UNIT COST	QTY	TOTAL COST	UNIT COST	VARIANCE
1	100	Earthwork	m^3	500	30,000	60	0	0	0	500	30,000	60	500	40,000	80	10,000
2	200	Formwork	m^2	1,060	185,500	175	1,890	425,250	225	2,950	610,750	207	2,950	908,600	308	297,850
3	300	Reinforcement	t	12.60	478,800	38,000	19.40	814,800	42,000	32	1,293,600	40,425	32	1,280,000	40,000	−13,600
4	400	Concrete	m^3	205.75	617,250	3,000	266.75	800,250	3,000	472.50	1,417,500	3,000	472.50	1,431,675	3,030	14,175
5	500	Structural steel	t	0	0	0	3	165,000	55,000	3	165,000	55,000	3	135,000	45,000	−30,000
6	600	Barbed wire	m	0	0	0	500	450,000	900	500	450,000	900	500	425,000	850	−25,000
7	700	Overhead			150,000			200,000			350,000		0	407,996		57,996
		Total cost			1,461,550			2,855,300			4,316,850			4,628,271		311,421
		Gross margin	662,525.00								662,525			351,104		
		Site contribution (%)									13.31			7.05		

Figure 16.12 Cost report for the example project

The performance measurement parameter could be cost, schedule or any other technical parameter such as man-hour. The comparison is done in terms of a common monetary term (such as rupees or dollar) or in terms of man-hour, equipment hour, 'quantity of material', etc., assigned to the work.

Earned value is based on the idea that the value of the product of the project increases as tasks are completed. Therefore, the earned value is a measure of the real progress of the project.

The earned value method involves the concepts of work package and earned value to analyse the performance of a project. It not only tells us about the progress to be achieved to be on schedule, but also clarifies whether the resources being spent are commensurate with the budget. This aspect will be clearer with the help of an example.

The earned value method is also used for forecasting the likely course of a project by extrapolating from the amount of work already put into a project. It is possible to forecast the project cost at completion and also the probable completion date. The forecasts made are also used to measure variances and define trends. The project manager can take appropriate actions should there be unwanted variances and unfavourable trends.

The implementation of earned value method involves the following steps:

Defining the scope of the works The scope of the works is frozen at an early stage in the project life cycle and changes, if any, in due course are suitably addressed.

Setting up a work breakdown structure (WBS) The project is broken down in logical groupings and levels, and these are shown in a hierarchy. The lowest level is usually made up of work packages that are the smallest self-contained grouping of work tasks considered necessary for the level of control needed. The details on WBS can be read in Chapter 6.

Developing a project master schedule The project master schedule contains details of scheduled start and completion of the entire work package.

Allocating costs to each work package The principal cost elements are labour, materials, and plant and equipment. These cost elements have been described in detail in previous chapters. The cost components are summed up for each work package.

Establishing a practical way of measuring the actual work completed The different ways in which the actual work is measured were described earlier. The selection of a particular method depends on the type of contract, the terms of payment, etc. However, 'simplicity' of the adopted system is the guiding philosophy most of the times.

Setting the performance measurement baseline The baseline indicates the plan to spend in different time periods for all the work packages, over the planned duration of the project. The period could be weeks or months.

16.6.1 Terminologies of Earned Value Method

Some of the commonly used terminologies in 'earned value' are discussed below with the help of Figure 16.13.

Budgeted Cost for Work Scheduled (BCWS)

This is defined as the budget or plan for all work packages planned to be completed. The BCWS curve is derived from the work breakdown structure (WBS), the project budget and the project master schedule. The cost of each work package is calculated period to period, and the cumulative cost of work packages is shown based on the planned completion dates shown in the master schedule.

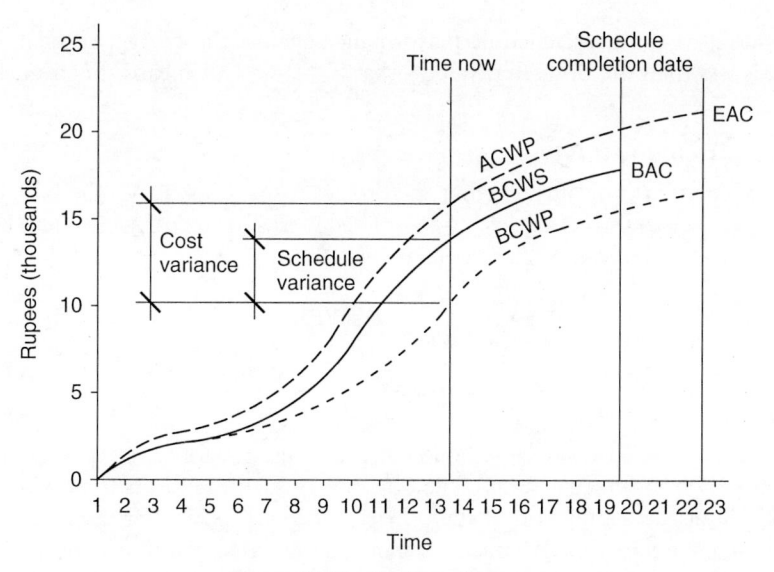

Figure 16.13 The three 'S' curves

Budgeted Cost of Work Performed (BCWP)

The planned costs of the work allocated to the completed activities are the earned value. The BCWP is calculated from the measured work complete and the budgeted costs for that work.

$$Earned\ value\ (BCWP) = Percentage\ completion\ of\ project \times Project\ budget \qquad (16.7)$$

Actual Cost of Work Performed (ACWP)

This is the real cost of the work charged against the completed activities. The ACWP curve is found by actual measurement of the work completed. Actual costs are recorded from invoices and workmen's time sheets. This appears a daunting task but it can be very simple with sufficient planning and organizing.

Variances

Schedule and cost variances can both be calculated in monetary terms (or in man-hour terms) from the data needed to produce the S-curves. Schedule variance (SV) is the difference between the earned value and the planned budget, and is calculated from the following expression:

$$SV = BCWP - BCWS \qquad (16.8)$$

Cost variance (CV) is the difference between the earned value and the actual cost of the works. This is calculated from the following expression:

$$CV = BCWP - ACWP \qquad (16.9)$$

Schedule Performance Index (SPI)

Schedule performance index indicates whether the project is on time. SPI is a ratio of earned value and the budgeted value of completed works, and is computed by the following expression:

$$SPI = \frac{BCWP}{BCWS} \qquad (16.10)$$

The index is used to compare schedule performance against the budget or plan. In case the earned value is less than the budgeted value—that is, SPI < 1, it means the project is 'behind schedule'.

Cost Performance Index (CPI)

Cost performance index indicates whether the project is spending as per the budget. CPI is a ratio of earned value and the actual cost of completed works. If the earned value is less than the actual cost, CPI < 1, it indicates a 'cost overrun' situation.

$$CPI = \frac{BCWP}{ACWP} \tag{16.11}$$

Estimate at Completion (EAC)

As discussed earlier, the earned value methodology can be used to forecast the likely course a project will take based on the current status. The objective of such forecast is to provide an accurate projection of cost at the completion of the project. One such parameter used for forecasting is the 'estimate at completion' (EAC), which is defined as the actual cost to date plus an appropriate estimate of costs for remaining work. The EAC can be calculated in multiple ways and some of these are given in Table 16.5. The formulas given for EAC use different parameters such as ACWP, ETC, BCWP, CPI and BAC, all of which except for 'budget at completion' (BAC) have been defined earlier. BAC is the sum of all budgets allocated to a project and is invariably equal to the total BCWS for the project.

Variance at Completion (VAC)

The expression given in Table 16.5 for computation of EAC can also be used to determine the variance at completion (VAC) for the project. The VAC is the difference between BAC and EAC, and is given by the following expression:

$$VAC = BAC - EAC \tag{16.12}$$

Table 16.5 Expressions for computing EAC

S. No.	Expression for computing EAC	Remarks
1	EAC = actual cost (ACWP) + estimate to complete (ETC)	It assumes that all remaining work is independent of the burn-rate incurred to date.
2	EAC = ACWP/BCWP × BAC	It assumes that the 'burn-rate' will be the same for the remainder of the project.
3	EAC = BAC/CPI	The formula is an extension of the previous formula. It assumes that the 'burn-rate' remains constant for the remainder of the project.
4	EAC = ACWP/BCWP × [work completed and in progress] + [cost of work not yet begun]	It assumes that the work not yet begun will be completed as planned.
5	EAC = ACWP + [(BAC − BCWP)/CPI]	This formula is a very subjective way of calculating EAC.

A positive value of VAC means the forecast is for an 'under-run', while a negative VAC is understood to have an 'overrun' projection for the project.

Example

Suppose that a 5 km length of road is to be constructed in 5 months at a budgeted cost of ₹ 500 lakh. It is proposed that 1 km of road length will be completed every month and, thus, ₹ 100 lakh is planned or budgeted for each month. At the end of 2 months, suppose that 1.5 km of road length is completed at an actual cost of ₹ 180 lakh. Is the project under-spending or overspending? Is the project behind schedule or ahead of schedule?

The project is clearly behind schedule since only 1.5 km road length has been completed against a planned progress of 2 km road length. The project appears to be 'under budget' because the budgeted cost at the end of month 2 was supposed to be ₹ 200 lakh (₹ 100 lakh for month 1 and ₹ 100 lakh for month 2) and the actual cost during this period has been ₹ 180 lakh. However, this seemingly 'under budget' schedule scenario is due to the project being 'behind schedule', which is why less money has been spent. In fact, according to the earned value method, there is a 'cost overrun' of ₹ 30 lakh at this point of time and a schedule overrun of ₹ 50 lakh. It may be noticed that earned value methodology has measured both schedule and cost in a common unit of measurement, which in this case is rupees, although conventionally the schedule overrun is measured in unit of time.

Let us take another example in which a house is to be constructed and is scheduled to take 12 months. Its budgeted cost of construction is ₹ 3,000,000. The month-wise plan for costs to be incurred in the different activities of house construction is given in Table 16.6. Also, let us assume that the construction progress is monitored at the end of September. The budgeted cost of work performed and the actual cost of work performed have been found, and the values are given in Table 16.6.

Table 16.6 BCWS, BCWP and ACWP values (₹ '000) for house construction example

	January	February	March	April	May	June	July	August	September	October	November	December
Budgeted cost of work scheduled (BCWS)	50	100	200	250	350	500	500	400	300	200	100	50
Cumulative value of BCWS	50	150	350	600	950	1,450	1,950	2,350	2,650	2,850	2,950	3,000
Budgeted cost of work performed (BCWP)	50	100	200	200	250	300	400	400	300			
Cumulative value of BCWP	50	150	350	550	800	1,100	1,500	1,900	2,200			
Actual cost of work performed (ACWP)	50	100	250	250	250	300	450	500	300			
Cumulative value of ACWP	50	150	400	650	900	1,200	1,650	2,150	2,450			

Figure 16.14 The BCWS, BCWP and ACWP curves for house construction example

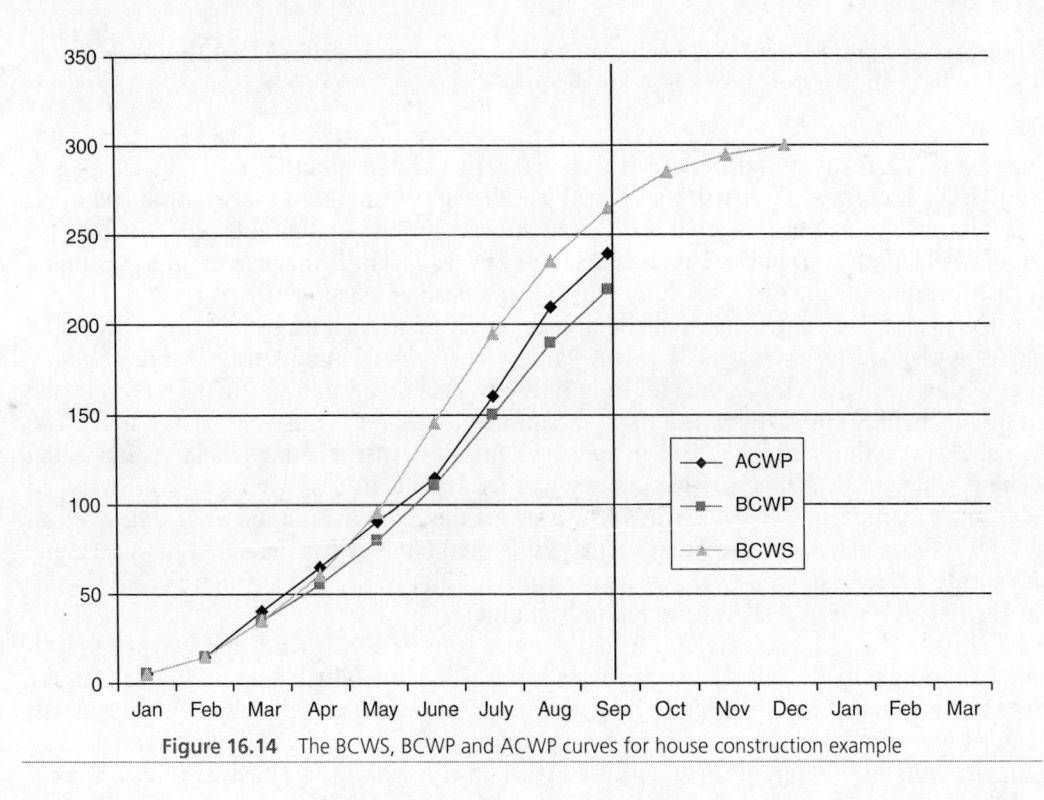

The three S-curves drawn corresponding to the cumulative values of BCWS, BCWP and ACWP given in Table 16.6 are shown in Figure 16.14.

In Figure 16.14, the BCWS curve shows that house construction is planned for accomplishment over a 12-month time frame, with a total budget amount of ₹ 3,000,000. The 'time now' line (end of September) shows that ₹ 2,650,000 of the project resources was planned to be completed at this point in the project. Another way to look at this is that the house construction was planned to be 88.3 per cent complete (₹ 2,650,000/₹ 3,000,000) at this point in time. In this example, the house construction was planned to have accomplished ₹ 2,650,000 worth of work in nine months, but the real accomplishment was only ₹ 2,200,000, which is the earned value in this period as shown in Figure 16.14. Comparing the earned value with the budgeted value measures the rupees value of work accomplished versus the rupees value of work planned. This difference is 'schedule variance'. Thus, schedule variance in this case is ₹ 2,200,000 − ₹ 2,650,000 = −₹ 450,000, a 'behind schedule' shown appropriately in Figure 16.14.

As mentioned earlier, the **earned value** for the work performed compared with the **actual cost** incurred for the work performed gives the cost variance and is a measure of cost efficiency. The actual cost of work has been determined (Table 16.6) and it is found that the actual costs for performing the ₹ 2,200,000 work was ₹ 2,450,000. Thus, there is a variance of ₹ 2,200,000 − ₹ 2,450,000 = −₹ 250,000, which is a 'cost overrun' condition.

The above variances can also be explained in terms of SPI and CPI. Thus, SPI and CPI for the example problem are:

$$\text{SPI} = \frac{\text{BCWP}}{\text{BCWS}}$$
$$= \frac{2{,}200{,}000}{2{,}650{,}000} = 0.83$$

Since the SPI is < 1, the project is behind schedule.

$$\text{CPI} = \frac{\text{BCWP}}{\text{ACWP}} = \frac{2{,}200{,}000}{2{,}450{,}000} = 0.898$$

The CPI being < 1, the project is over budget.

It may be noted that the earned value approach has reduced both cost and schedule variance to a common denominator, i.e., Rupees or a ratio. Hence, we can compare the two and come to a conclusion that the project is behind schedule by ₹ 450,000 and behind budget by ₹ 250,000—i.e., it is more behind schedule than it is behind budget.

The estimate at completion using the expression [EAC = ACWP + ((BAC − BCWP)/CPI) = 2,450,000 + ((3,000,000 − 2,200,000)/0.898)] is found to be ₹ 3,340,000, which means that if the project continues to progress as it has done so far, it will be completed at a cost of ₹ 3,340,000, i.e., ₹ 340,000 more than the original budget. The EAC can also be calculated using the other formulae given in Table 16.5. The result will be different than what we have using this formulae. The most appropriate method for a given situation must be chosen by the project manager. Also, it is up to the project manager to assess the reasons for the project being behind schedule and over budget, and take corrective actions.

The above calculations have been performed for a single work package. A project consists of a number of work packages. The status of the overall project (whether it is on time and within budget) can be found out using the same set of formulae illustrated for the performance evaluation of a work package. For this, the BCWS, BCWP and ACWP values for all the work packages are summed up and used in the formulas. An example will explain this.

Suppose a project consists of 13 work packages—1, 2, 3, _____, 13. The schedule start and completion of each work package is shown in the bar chart given in Figure 16.15. The budgeted cost of each work package is also shown on the bar on a monthly basis. Thus, from Figure 16.15 it is clear that work package 1 (earthwork) is scheduled to start from month 1 and get completed in month 6—that is, the scheduled duration is 6 months. In month 1, it is budgeted to cost ₹ 0.25 lakh, while in months 2 to 6, the budgeted costs are ₹ 0.75 lakh, ₹ 1.0 lakh, ₹ 1.25 lakh, ₹ 1.25 lakh and ₹ 0.5 lakh, respectively. The budgeted cost of each work package is shown above the bar of the corresponding work package.

We can now draw a graph (see Figure 16.16) of time versus cumulative cost/price, which is like the S-curve discussed earlier. Further, let us assume that the progress of the project is monitored at the end of month 3, and relevant information related to each work package is captured as in Table 16.7.

In Table 16.7, the project BCWS, BCWP and ACWP have been calculated by adding up the BCWS, BCWP and ACWP of the thirteen work packages. The project has a schedule variance of 70.1 − 76.9 = −6.8 lakh and a cost variance of 70.1 − 74.1 = −4 lakh. Thus, at the end

Work Package	1	2	3	4	5	6	7	8	9	10
Earthwork	0.25	0.75	1	1.25	1.25	0.5				
Concrete work		8	16	16	20	12	8			
Formwork			4.2	8.4	8.4	10.5	6.3	4.2		
Reinforcement work			11.25	22.5	22.5	28.125	16.875	11.25		
Brickwork				3.4	5.1	6.8	6.8	6.8	3.4	1.7
Plastering					1.5	2.25	3	3	3	1.5	0.75
Painting					0.5	0.75	1	1	1	0.5	0.5
Flooring						11	16.5	22	22	22	11	5.5
Waterproofing								6	6		
Aluminium work						33.75	33.75	33.75	33.75		
Electrical work								8.75	8.75	8.75	5.25	3.5
Sanitary and plumbing work				1	2	3	4	3	3	2	2
Road work							8	8		
Value in each month	0.25	24.2	52.3	68.25	92.55	92.6	85.375	50.525	21.95	12
Cumulative value at end of month	0.25	24.45	76.75	145	237.55	330.15	415.53	465.05	488	500

All values in ₹ lakhs. Time (month).

Figure 16.15 Bar chart showing the budgeted schedule and cost of all work packages

of month 3, the overall project is running behind schedule and is also over budget. The schedule and cost performance of individual work packages as described earlier is monitored and the appropriate corrective action taken. It is also possible to forecast the budget at completion and the estimated cost at completion from the total BCWS, BCWP and ACWP values as described earlier.

16.7 ILLUSTRATIONS OF COST CONTROL SYSTEM

Example 1
The cost control methods described above can be best explained by illustrating a short example project that consists of five activities. It is illustrated in a step-by-step manner.

Step 1
Freeze the cost codes. Suppose that the construction company has made the following cost codes:

Step 2
Prepare the budgetary cost estimate for all the activities of the project according to the cost codes. This is shown below for the mentioned activities.

PROJECT MONITORING AND CONTROL SYSTEM | 595

Cumulative cost

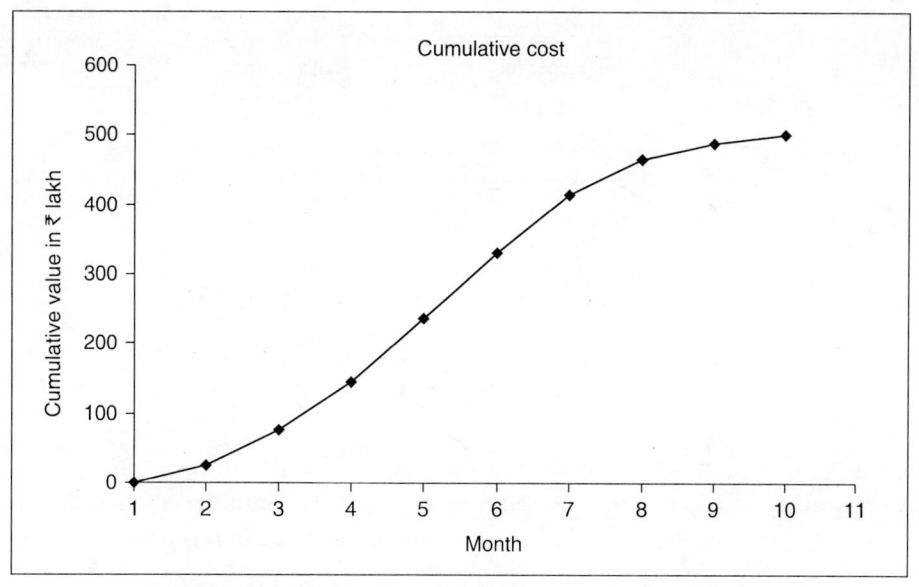

Figure 16.16 Time versus cumulative value for the example problem

Table 16.7 Information related to each work package (BCWS, BCWP and ACWP values in ₹ lakh)

Work package number	Scheduled progress	Actual progress	BCWS[1] (₹ lakh)	BCWP[2] (₹ lakh)	ACWP[3] (₹ lakh)
1	40%	50%	2.0	2.5	3.0
2	30%	20%	24	16	18
3	30%	30%	12.6	12.6	13.6
4	30%	30%	33.9	33.9	35.0
5	10%	15%	3.4	5.1	4.5
6	0%	0%	0	0	0
7	0%	0%	0	0	0
8	0%	0%	0	0	0
9	0%	0%	0	0	0
10	0%	0%	0	0	0
11	0%	0%	0	0	0
12	5%	0%	1.0	0	0
13	0%	0%	0	0	0
Total			76.9	70.1	74.1

[1]BCWS has been derived from the bar chart given in Fig 16.15
[2]BCWP has been derived from (BCWS × actual progress ÷ scheduled progress)
[3]ACWP is based on the record of actual cost incurred in a work package

Table 16.8 Cost code details

S. No.	Description	Cost code
01	Earthwork	100
02	Formwork	200
03	Reinforcement	300
04	Concrete	400
05	Structural steel	500
06	Barbed wire	600
07	Overhead	700

Step 3

Prepare the construction schedule. For the sample project, the initial construction schedule is given in Figure 16.17.

Step 4

Collect the cost data for each of the cost codes at the time of monitoring. Let's suppose that the company is monitoring the project at the end of day 30. The collected cost data is shown in Table 16.10.

Step 5

Update the progress of the schedule at the end of day 30. The updated schedule is given in Figure 16.18.

Actual costs at the end of every day up to the end of day 30 are given in Table 16.11, and the cost-wise expenditure at the end of day 30 are given in Table 16.12.

Table 16.9 Budgetary cost estimate for all the activities of the example project

Activity	Cost code ACE amount	100	200	300	400	500	600	700	Total
EW	40,000	36,500						3,500	40,000
CC	120,000		8,000		100,000			12,000	120,000
CR	819,960		70,000	450,000	217,964			81,996	819,960
CW	2,680,000		625,000	990,000	797,000			268,000	2,680,000
FE	425,000					180,000	202,500	42,500	425,000
Total	4,084,960	36,500	703,000	1,440,000	1,114,964	180,000	202,500	407,996	4,084,960

Figure 16.17

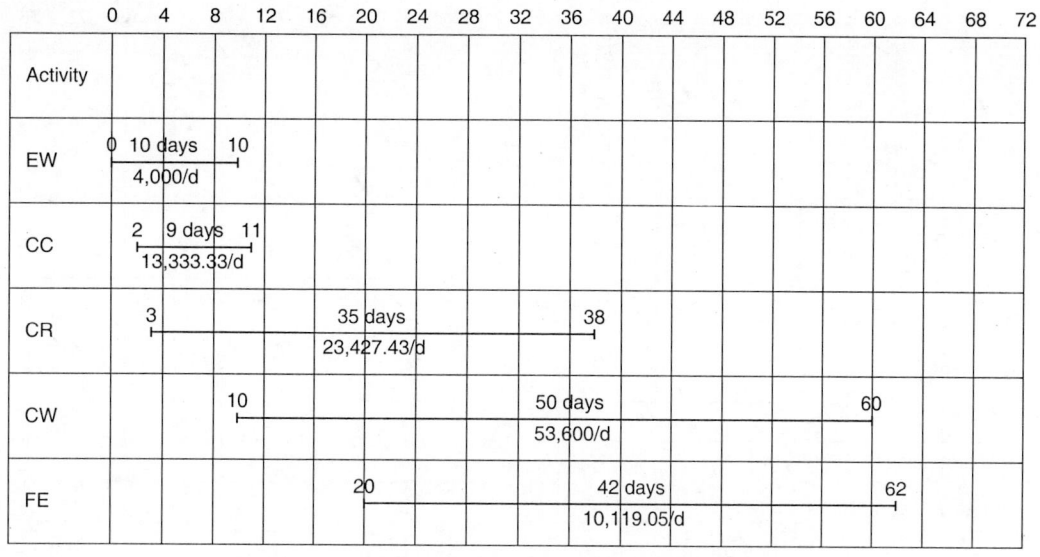

	0	4	8	12	16	20	24	28	32	36	40	44	48	52	56	60	64	68	72
Activity																			
EW	0	10 days 4,000/d	10																
CC			2	9 days 13,333.33/d	11														
CR		3					35 days 23,427.43/d			38									
CW				10							50 days 53,600/d					60			
FE						20					42 days 10,119.05/d						62		

Figure 16.17 Initial construction schedule for the example project

Step 6

Now calculate the variance cost code-wise by subtracting the value for each of the cost codes at the end of day 30 and the actual cost for each of the cost codes at the end of day 30. The variances with the negative sign (refer to Table 16.13) indicate that the actual cost incurred for the particular cost code is in excess of the budgeted cost and are the areas of concern that need proper attention.

16.8 MANAGEMENT INFORMATION SYSTEM

It is a system to convert data from internal and external sources into information, and communicate the same in an appropriate form to managers at all levels to enable them to make timely and appropriate decisions for planning, directing and controlling the activities for which

Table 16.10 Details of cost data at the time of monitoring

Activity	ACE amount	% completion	100	200	300	400	500	600	700	Total
EW	40,000	100%	36,500						3,500	40,000
CC	120,000	100%		8,000		100,000			12,000	120,000
CR	819,960	60%		42,000	270,000	130,778			49,198	491,976
CW	2,680,000	30%		187,500	297,000	239,100			80,400	804,000
FE	425,000	0					0	0	0	0
Total	4,084,960		36,500	237,500	567,000	469,878	0	0	145,098	1,455,976

598 | CONSTRUCTION PROJECT MANAGEMENT

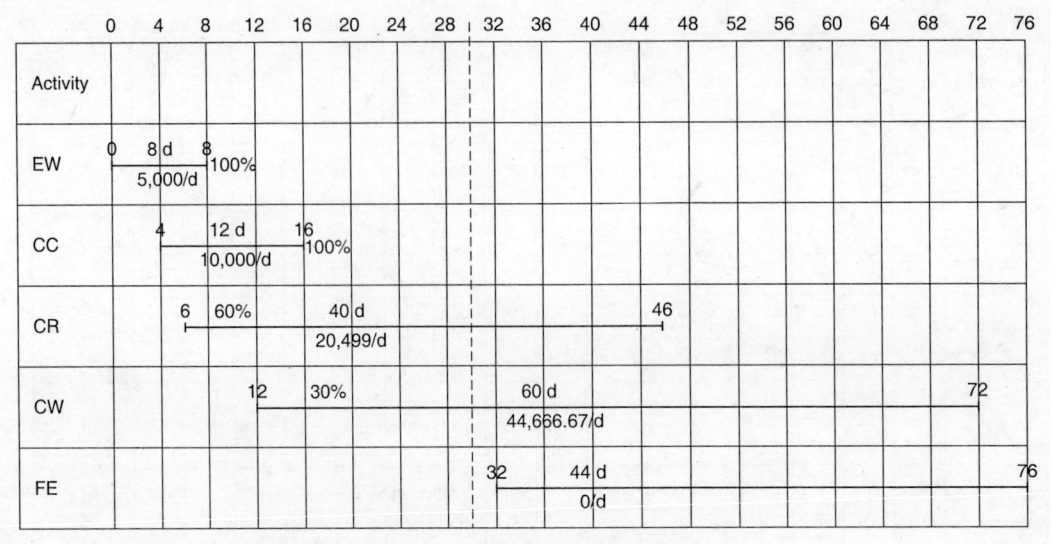

Figure 16.18 Updated schedule at the end of day 30

Table 16.11 Details of actual costs incurred

Days	1 2 3 4	5 6 7 8	9 10 11 12	13 14 15 16	17 18 19 20	21 22 23 24	25 26 27 28	29 30
Cost/day	20,000	31,250	42,500	43,250	70,500	73,305	77,070	72,525

Table 16.12 Actual costs incurred cost code-wise

Activity	Cost code ACE amount	100	200	300	400	500	600	700	Total
EW	40,000	36,500						3,500	40,000
PCC	120,000		12,000		100,000			12,000	124,000
CR	819,960		42,000	286,000	130,778			49,198	507,976
CW	2,680,000		208,313	329,967	262,640			103,654	904,575
FE	425,000					0	0	0	0
Total	4,084,960	36,500	262,313	615,967	493,419	0	0	168,352	1,576,551

Table 16.13 Cost code-wise variance details

Cost code	100	200	300	400	500	600	700	Total
Variance	0	−24,813	−48,967	−23,540	0	0	−23,254	−120,575

they are responsible. Thus, the objectives of the management information system (MIS) are to provide information relating to actual performance against budgets for the projects, strategic business units, and the corporate as a whole, and to highlight areas of shortfall in achievement so that decisions regarding the required remedial actions may be taken.

The comparison of the performance is always made with respect to the budget, which is normally prepared at a suitable interval at all levels. The budget so prepared at different levels is basically a sort of commitment by the lower level of hierarchy to its immediate upper level of hierarchy. Any revision in the budget is very carefully examined and changes are accepted only if they are well-justified.

The decisions may be required to be taken at strategic, tactical and operational levels of management, and accordingly, may require different information at these levels. For example, at strategic level, the information needed would be pertaining to trends in industry, investment pattern, government regulation and policies, competition strategies, technological changes, etc. The information at tactical levels of management would be pertaining to short-term planning, cash-flow projection, financial results, sales analysis, budgetary control, variance statements, capacity utilization, etc. Finally, the information at operational levels would be day-to-day progress at project sites, item-wise stock, pending purchases, subcontractor's performance, etc.

Depending on the type of business, the information type would also vary. A construction organization executes projects that are characterized by their geographical spread and difficult terrain. It is a challenge to share and access information vital for the decision-making.

A typical MIS report used in construction industry would consist of monthly progress report; S-curve for budget invoice, actual invoice, budgeted cost and actual cost; project cost report; and project performance report giving details of status of invoicing, contribution and collection. Some of the common reports associated with construction project MIS are monthly progress report, project cost report and project performance report.

These reports are prepared by the planning engineer in consultation with the project manager and the project accountant. Normally, the above reports are prepared on a monthly basis and sent to higher offices such as regional, divisional, or head office, as the case may be. After reviewing these reports, the higher offices may suggest corrective actions as and when required.

At least once in three months (more often in the case of large projects or projects requiring critical managerial attention), the site manager and the site planning engineer meet the higher-ups (such as regional manager, divisional manager, or general manager, as the case may be) who are based at higher offices, and discuss the support required by the sites in terms of major resources to meet the physical targets, resourced deployment, etc. For the subsequent months till the next review meeting, the actual performance is reported against these revised schedules.

The regional manager will have the responsibility for taking follow-up actions in close coordination with, and under the direction of, his superior, such as the general manager, on the action plan decided during the review meetings.

At the sites, the site planning and project manager should hold weekly meetings with the site engineer/supervisor in charge of the various sections. At these meetings, the physical targets for the various sections for the next week will be set and the resource requirements in terms of materials, manpower equipment, etc., identified. The site planning engineer is normally responsible for ensuring that these are provided to the various sections on time. The week's performance against targets is also reviewed in these meetings and follow-up actions are taken.

The MIS system could broadly consist of flash reports and detailed reports. Since timely action is the essence of any MIS, the flash reports are generated from profit centres as on a particular date. The major data required from sites for preparing the flash reports are—estimated invoicing, material and labour costs, and actual site overheads. These reports could be approximate in nature and serve the purpose of promptly reaching out to the top management professionals of the organization. Construction organizations normally follow a cut-off date for reporting the data and, accordingly, the date of preparation of flash report or detailed report is also kept fixed for considering and estimating the quantum of work. The evaluation on these cut-off dates need not always be based on client certification (since this could lead to delays in obtaining data), but could be estimated based on the quantum of work actually carried out. The variation, if any, between the client certification and the estimation has to be corrected in the subsequent reports when the detailed MIS report is prepared.

Although reports related to schedule progress are not difficult to prepare on a regular basis, there could be problems in regard to reports related to costs, especially material cost. As was mentioned earlier, this is because a number of persons from different departments are involved in some way or the other in providing the cost data. Coordinating with all of them is a time-consuming process and most of the times, this is the reason for delay in such reports.

Considering the importance of MIS for decision-making, a large number of organizations are using information technology tools for gathering data and generating suitable information in a tailormade manner. In fact, organizations have started maintaining a separate MIS cell to provide appropriate and accurate information at the right time and at an optimum cost to managers to monitor the business activities, to avoid information overload, and to avoid generating redundant information. The personnel in MIS cell collect data, extract relevant information from the data collected, and disseminate the same at different levels of management to assist in their decision-making process.

REFERENCES

1. Austen, A.D. and Neale, R.H., 1995, *Managing Construction Projects: A Guide to Process and Procedures*, International Labour Organization.
2. Chitkara, K.K., 2006, *Construction Project Management: Planning, Scheduling and Controlling*, 10th reprint, New Delhi: Tata McGraw-Hill.
3. Harris, F. and McCaffer, R., 2005, *Modern Construction Management*, 5th ed., Blackwell Publishing.

REVIEW QUESTIONS

1. State whether True or False:
    a. The traditional control parameters of construction projects are schedule, cost and quality.
    b. Updating can be defined as planning and programming of the remaining portion of an activity by introducing the latest information available.

c. Interrelationships of activities can be shown on bar charts.

d. Which one is the most effective way of updating the project schedule—(a) PDM, (b) CPM, (c) PERT, or (d) Gantt Chart?

e. S-curve can be used to monitor the cost of a construction project.

f. Earned value is based on the idea that the value of the project increases as tasks are completed, and therefore, the earned value is a measure of the real progress of a project.

g. If SPI is greater than 1, the project is behind schedule.

h. If CPI is less than 1, the project has cost overrun.

i. The objective of the MIS is to provide information on actual performance against budgets for the project and to highlight areas of shortfall in achievement so that any decision regarding the required remedial actions may be taken.

j. The three common reports in MIS are monthly progress report, project cost report and project performance report.

2. Match the following:

(1) Cost performance index (CPI)	(a) % project completion × project budget
(2) Earned value	(b) BCWP - BCWS
(3) Schedule variance	(c) BCWP - ACWP
(4) Cost variance	(d) BCWP/BCWS
(5) Schedule performance index (SPI)	(e) BCWP/ACWP

3. How do you compare actual progress versus expected progress using an S-curve? Illustrate with a suitable example.

4. Table Q4.1 gives the budgeted monthly cash-flow requirements for a construction project. At the end of October 2006, the total actual project expenditure is reported as ₹ 114 lakh and the project progress is reported as 50 per cent.

Draw an S-curve for cumulative cash flows for the project and calculate cost and schedule performance indices as at the end of October 2006. What is the delay in the project as on date in number of months?

Table Q4.1 Data for Question 4

Month	May 2006	June 2006	July 2006	August 2006	September 2006	October 2006	November 2006	December 2006	January 2007	February 2007	March 2007
Cash flow	8.5	11.5	13.5	15.5	30.5	27.0	22.5	20.0	20.0	12.5	2.5

5. Calculate the cost variance (CV), the cost performance index (CPI), the schedule variance (SV), and the schedule performance index (SPI), and analyse the cost and schedule performance for each of the projects given in Table Q5.1.

Table Q5.1 Data for Question 5

Project (1)	BCWS (2)	BCWP (3)	ACWP (4)	CV (5) = (3) − (4)	CPI (6) = (3)/(4)	Cost analysis (7)	SV (8) = (3) − (2)	SPI (9) = (3)/(2)	Schedule analysis (10)
1	50,000	50,000	50,000	0	1	On target	0	1	On target
2	50,000	40,000	60,000	−20,000	0.67	Over budget	−10,000	0.8	Behind schedule
3	40,000	40,000	50,000						
4	50,000	50,000	40,000						
5	40,000	50,000	50,000						
6	60,000	40,000	60,000						

17

Construction Claims, Disputes, and Project Closure

Claim, dispute, correspondence, project closure

17.1 CLAIM

During the execution of a project, several issues arise that cannot be resolved among project participants. Such issues typically involve contractor requesting for either time extension or reimbursement of an additional cost, or sometimes both. Such requests by the contractor are referred to as 'claim'. If the owner accedes to the claim of contractor and grants him extension of time or reimbursement of additional cost, or both, the issue is sorted out.

However, if the owner does not agree to the claim put out by contractor and there are differences in the interpretations, the issue takes the form of a dispute. Claims are becoming an inevitable and unavoidable burden in modern projects involving new technologies, specifications and high expectations from the owner.

The claim mentioned above can also be put up by the owner. It is, therefore, imperative for all the parties to be fully acquainted with the procedures and systems, including recourse to certain preventive actions as found necessary and required.

17.1.1 Sources of Claims

As mentioned, the claim may arise due to the owner or the contractor. The claim may be on account of any one of the following causes:

- There may be defects and loopholes in the contract document. For example, the contract document may not be clear, may have dual meanings at different places, or may not have sufficient details. Also, an unresponsive contract administration may lead to contractor raising the claim.
- There may be delay in release of areas as per contract. Besides, site conditions differ to a large extent from those described in the contract document.
- The owner may desire to get the work done at a faster pace than is required by the contract document.
- There may be delay in supply of power, water and other materials from the owner.
- There may be hold on works due to delay in release of drawings and other inputs.

- There may be delay in release of payments to the contractor.
- The scope of work may be substantially modified by the owner.
- There may be levy of liquidated damages (LD) on the contractor. Other recoveries from bills may also lead to contractor raising the claim.
- There may be delay on the part of contractor in completion of works due to inadequate mobilization of labour, material and plant.
- There may be loss of profit and investment to the owner due to delays caused by the contractor.
- Construction claims can also arise on account of inclement weather.

17.1.2 Claim Management

The major issues in claims and disputes are identification of issues and the party responsible for the claim, and ascertaining the time and cost impact of the claim. The party raising the claim has to notify the claims once they have been identified. Further, it is the responsibility of the party raising the claim to substantiate the facts. Depending on the decision of the other party against which the claim is made, the claim may be settled amicably or it may take the form of a dispute.

In the following paragraphs, the claim management process has been explained from the perspective of a contractor:

Claim Identification

The contractor studies the instructions in the form of drawings as well as oral or written instructions provided by the owner/engineer. If it contains extra works, the same is read against the provisions of the contract.

Claim Notification

After it is established by the contractor that it is an extra work, the contractor is required to inform the engineer within the time frame stipulated and clarify his intention to claim extra rates for the same. This is very important because failure on contractor's part regarding this shall entail its rejection by the engineer.

Claim Substantiation

The contractor has to fully establish the claim including his entitlement under the contract, giving reference to the relevant clauses. The claim is supported by necessary backup calculations. Backup documents like letters, vouchers and drawings are also enclosed. For period-related claims such as extended stay costs and interest on delayed payments, it is required to revalidate the claim at periodic intervals and submit the same to the engineer until the end of the relevant period.

Decision of Engineer/Owner

The owner/engineer is supposed to convey his decision on the claim to the contractor within a time frame specified in the contract. If the claim is not allowed, the same needs to be stated along with reasons. The value of claim allowed shall also be stated.

Further Action by Contractor

The contractor has to refer the claim for adjudication if provided, within a specific time frame after receiving the decision from the engineer, if the same is being disallowed. The adjudication process is carried out as per the provisions set out in the contract.

17.1.3 Some Guidelines to Prepare the Claims

- It is always preferable to link the claim to contractual provisions.
- Prepare the base for its establishment.
- Indicate intention and submit it within the time frame provided in the contract.
- Submit with all backup documents, calculations, etc.
- Be fair in projecting the figures and do not inflate.

17.2 DISPUTE

Given the uncertainties involved in a construction project and the magnitude of funds involved, it is only natural to have disagreements between parties, but these need to be resolved. While most of such day-to-day differences are resolved in an amicable manner, without having to resort to a more formal mechanism, the parties at times agree to disagree and seek redressal through independent intervention. Although, in principle, the discussion falls under the purview of construction law, effort has been made to discuss some of the aspects related to disputes and dispute resolution with as little legalese as possible.

Technically, a dispute implies assertion of a claim by one party and repudiation thereof by another. Thus, neither a mere claim without repudiation, nor a pair of claim and counterclaim, can be called a dispute.

Example

In a letter dated June 30, 1998, a contractor makes a claim for ₹ 5 lakh of additional work in dewatering of foundations, resulting from unforeseen rains in the month of March and the item not being provided for in the contract. The details of the rainfall and the costs incurred in terms of manpower, equipment, power consumption, overtime payments made, etc., are annexed to the claim along with photographs. In a letter dated July 13, 1998, the owners acknowledge the receipt of the aforesaid letter from the contractor, and inform them (the contractor) of the owners' decision to (a) impose LD for delay of work, and (b) recover ₹ 2 lakh (towards LD) in the next RA.

Does the matter define a dispute?

From the information given, it is not at all clear whether it is the delay that is attributable to the rains or it is the liquidated damages. It should also be noted that though the contractor has asked for an extra payment, he has not asked for an extension of time to complete the project. The above can be, therefore, considered as basically an example of a claim and counterclaim, and does not constitute a dispute.

17.2.1 Causes of Disputes

The geneses of many disputes often lie in the contract document itself—it is often observed that tenders are hastily made and sufficient attention is not paid to ensure that (a) all the required information and details are appropriately incorporated in the tender document, (b) the documents are internally consistent, i.e., there is no contradiction in the provisions of general conditions, special conditions and drawings, and (c) specifications, where required, are available. Of course, incompleteness, inaccuracy and inconsistency of information are only part of the reasons for disputes in a construction project. The following paragraphs briefly discuss some of the common causes of litigation.

Incorrect Ground Data

Such data includes information about ground conditions, depth of groundwater table, rainfall and temperature data, availability of power and water, etc. The estimates of a contractor are based on the ground data provided with the tender documents, though depending upon the size of the project and the means of a contractor, the latter also at times carries out an independent assessment of the data provided. Obviously, any difference between the ground reality during execution and the conditions provided in the contract could easily be the reason for disputes.

Use of Faulty and Ambiguous Provisions and/or Language in Contracts

The language of the contract should be clear and such that it is not open to different interpretations. Use of ambiguous language and/or provisions could open a floodgate of avoidable litigation. It is also important that the contract clearly lays down specific procedures that are to be adopted in the event of contingencies. A well-defined hierarchy of documents that will prevail in the event of a discrepancy, often goes a long way in determining the appropriate course of action without having to resort to arbitration. Also, at times, absence of appropriate provisions to handle technical inspections by the client or owner, or third parties, could become a source for litigation, as such inspections themselves require money and at times result in observations that need appropriate rectification action, which may have financial implications or cause avoidable delay. An ill-defined or a vaguely defined hierarchy of documents that will be deemed to prevail in the event of a dispute could be a cause for dispute, as the following example clearly brings out.

Some drawings for the material-handling plant showed handrails along only a central walkway, whereas some showed similar railings along all other areas where covering plates fabricated from MS flats were provided. The confusion was compounded by a statement in the special conditions of the contract, which stated that railings were to be provided along the central walkways as per drawings. The owners insisted that railings should be provided for all covers, which was provided by the contractor, who, however, made a case that such railings were 'extra' and not a part of the original lump-sum contract. The arbitrator citing confusion in drawings, and precedence in documents, insisted that the provisions of the SCC prevail. In this case, the 'natural meaning' of the provisions of the SCC was taken as follows—the document implied the existence of other covered areas where railings need not be provided, and therefore upheld the contractor's claim.

Deviations

The contract should be so designed that there are as few extra items and/or deviations as possible. In other words, the scope of work in any contract should be unambiguously defined, and this obviously calls for thorough preparation on the part of the client/owner before actually floating an enquiry.

Unreasonable Attitudes

It should be borne in mind that in order to complete the work professionally, it is important that the parties involved resort to unilateral action to preserve an environment of mutual trust. Thus, both the client and the contractor need to have a professional approach to the project, including areas where there could be disagreement on interpretation, etc. Measures such as suspension of the contract or invoking of clauses related to imposition of liquidated damages should be resorted to only in the most extreme cases, as they vitiate the atmosphere of the project, and also affect the work on other contracts. Delays in payment of bills should also be avoided to ensure that the contractor does not get cash-strapped, which will obviously affect his ability to perform.

Contractor Being of Poor Means

It is important that the contractor identified to do a job possesses the required human, financial and technical resources. In the absence of any of these, it is very likely that the contractor will look for an escape route for leaving the project, and may try to force a suspension or determination (termination) of the contract, or take the matter into arbitration/litigation to cut his losses.

Unfair Distribution of Risk

This could be a major reason for not only avoidable litigation but also increase in the cost of the project. Indian contracts typically are heavily loaded against the contractor, who obviously tries to cover the risks he is 'forced' to take by either hiking the rates, or taking an approach of 'crossing the bridge when we come to it', and the latter is almost a certain prescription for litigation if adverse conditions are encountered.

17.2.2 Dispute Avoidance Vs Dispute Resolution

In view of the above, an appropriate strategy for dispute avoidance and dispute resolution needs to be drafted, and put in place even before the onset of works in a project. Given that the client or the owner usually takes the lead in drafting contracts, the onus is largely on him to ensure that both dispute avoidance and dispute resolution are adequately addressed in the contract. To that end, it may be noted that once the important reasons as listed above are appropriately taken care of, the possibility of disputes can be largely minimized.

Contracts need to be drawn up with a professional mindset, and a fair distribution of risk between the contractor and the owner can go a long way in laying the foundation of a healthy relationship, which in turn creates a conducive work environment at sites.

Special care needs to be taken in drafting dispute-prone clauses and identifying potential neutral agencies, which could be called upon to mediate any dispute.

Also, it may be considered advisable to provide for a binding mechanism for alternate dispute resolution (ADR), which could include constitution of dispute review board (DRB), comprising persons drawn from the contractor's and the owner's sides, which should meet periodically to review any pending or potential dispute. The persons identified to serve on such a board could preferably be such that they are not directly involved in the day-to-day decision-making and contract operation. This gives them a certain amount of objectivity in approach, which is very important to one's ability to be able to work towards a solution of a problem, which almost invariably involves a certain amount of give-and-take. Also, it makes it easier for people involved in the project to accept their verdict without loss of face, and so on.

It should be borne in mind that in a project the possibility of dispute arising all of a sudden is reasonably bleak, and if an independent institution looks at potential problems, there is always the likelihood of some solution being found before the work in the project begins to suffer. It may be pointed out that a discussion on areas of differences in opinion should be part of any review meeting that discusses the progress in a project.

17.2.3 Mechanisms of Dispute Resolution

Apart from the normal legal process, emphasis here is on the alternative dispute resolution mechanisms generally available in construction contracts. Such mechanisms could include negotiation, mediation, conciliation and arbitration. While the first three mechanisms are briefly touched upon in the following paragraphs, the subject of arbitration has been dealt with in greater detail in view of its importance.

Negotiation

This could refer to a focused discussion on the dispute among the engineers from all interested parties, with the intention of resolving differences without the involvement of third parties, as happens in the case of mediation and arbitration. Indeed, this is an informal process in the legal sense, but if an agreement is reached through the process, it may have the usual legal significance. The negotiation process is fast and does not involve additional expenses. The discussions are held between the parties across the table in a cordial and peaceful atmosphere.

Mediation and Conciliation

Mediation and conciliation are essentially an informal process in which the parties are assisted by one or more neutral third parties in their efforts towards settlement. These mediators do not sit in judgement but try to advise and consult impartially with the parties with the object of assisting in bringing about a mutually agreeable solution to the problem. Naturally, under the conditions, they have no power to impose an outcome on disputing parties.

Mediation and conciliation are voluntary in the sense that the parties participate of their own free will, and a neutral third party simply assists them in reaching a settlement. The process is private, confidential and conducted without prejudice to any legal proceedings. The process is non-binding unless an agreement is reached. Of course, once an agreement is reached, and the parties have signed it, the document (or the understanding) is as binding as any other agreement would be.

Although the process is largely informal, the following could be identified as parts or stages in a mediation process. In the pre-mediation stage, there has to be a basic agreement among the parties to the mediation process, including the identification of a mediator. Mediation could be direct or indirect, and could involve meeting(s) with parties, presentation(s) being made by them, putting together of facts, negotiations and a settlement. Finally, a mediator may also like to be involved in the process of compliance with the settlement reached.

Given that mediation is an informal process, it has certain inherent advantages over the more formal and legal process. For example, it could be a lot less time-consuming and even involve lesser costs, and the outcome could be more satisfying to the parties, while also minimizing further disputes. It also opens channels of communication, and could contribute greatly to preserving or enhancing a professional relationship. Further, the exercise may be said to empower the parties and give them greater confidence in their ability to handle disputes.

Arbitration

Arbitration is perhaps the most commonly used mechanism for settlement of technical disputes in a construction project. It is a quasi-judicial process to the extent that legal protocol is largely observed, and it is important that the arbitrator, who basically acts as a judge, understands legal procedures. In India, the Arbitration and Conciliation Act, 1996, provides the legal framework for the arbitration process. In principle, collection and interpretation of evidence, examination and cross-examination of witnesses, etc., are some examples of essentially legal matters, which an arbitrator needs to have a sound understanding of. However, a basic belief in principles of natural justice and a practical approach are the hallmark of a successful arbitrator. He should be able to guide and provide a direction to the proceedings, which could be quite tough, especially when the parties to the dispute are represented by professional lawyers. In fact, the law has now added a new dimension to the arbitration process by empowering the arbitrators to conciliate and help the parties in arriving at a fair compromise or an equitable settlement of the case before him.

As far as the number of arbitrators is concerned, much like the judicial system, technical disputes can also be resolved by single arbitrators, or a panel of several arbitrators, and though the parties are free to determine the number of arbitrators, it should be ensured that the number is odd, so that a situation of a 'tie' in an award is preempted. Often, one arbitrator each is nominated by the contractor and the owner, and these individuals together choose a third colleague (brother) arbitrator, to complete the constitution of a bench of arbitrators.

17.2.4 Causes Leading to Arbitration

Some of the causes leading to arbitration as identified by Iyer (1996) are:

- Incorrect ground data
- Contracts containing faulty and ambiguous provisions
- Faulty administration of contract
- Deviations
- Suspension of works
- Contractor being of poor means
- Default by contractor
- No publicity involved
- Unreasonable attitude adopted by contractor
- Overpayment
- Levy of compensation for delay
- Delay in payment of bills
- Observation arising out of technical examination of works

17.2.5 Advantages of ADR over Legal Proceedings in a Court

Alternative dispute resolution (ADR) has clear merits over formal legal proceedings in a court of law, and is often preferred over the latter. Though the award has legal sanction and can be imposed, the process is less formal and quasi-judicial in nature, which allows a certain degree of flexibility and ease to the parties. Of course, an arbitrator can always seek expert legal advice on matters of law. The process is ideally suited for technical disputes—for example, the arbitrator can be appropriately selected and a visit to the site made as may be required.

Since the arbitrator works on a lesser number of cases at any given time, the settlement of cases is quicker and less expensive. Also, given the fact that the parties may have their offices at places different from the site of the project, it becomes much more convenient if the time and place of a hearing are fixed based on the mutual convenience of parties. Since the hearings are not open to the public, the overall relationships are less affected. This aspect is important considering the fact that the parties often want to avoid needless publicity as it adversely affects their professional standing and relationships.

17.2.6 Some Do's and Dont's to Avoid Dispute

The following factors can be kept in view in respect of disputes:
1. When contractors are faced with lack of work and idle overheads, they tend to under-quote and take up jobs at cutthroat rates, which often land them in a soup. Thus, it may be said that 'contractors are so afraid of dying that they commit suicide.'
2. Good claims-management practice means:
    i. eliminating risks to the extent possible before entering into contract, and
    ii. during the course of work, to have variation orders settled without elevating them to the status of claims.
3. 'Be careful how the law of the land interprets 'no damages clause' in favour of the owner.'
4. 'Don't throw good money after bad money in pursuing bad claims.'
5. 'Don't set up ego barriers in settling disputes during the course of work.'
6. Negotiated contracts have fewer claims than lump-sum contracts. Most owners' engineers tend to assume that the contractors have covered all risks while quoting. Highlight the areas not covered in the quotation.
7. Study contract conditions and local laws thoroughly.
8. Educate and train staff to act early at the lower level so that disputes do not escalate and rise to the highest decision-making level.
9. Do not pile up claims to the end, which result in the owner's engineer also getting hemmed in. Settle them early, before the amount looks big.
10. Provide analysis and documentation early, and not at the end.
11. Relate every claim to the project schedule drawn up in the beginning. The base plan should be prepared right at the beginning of the contract, and this should not be lost sight of either mentally or physically. Keep a copy of the original schedule in your cash box or bank locker so that it is not lost.
12. Concurrent delays cannot be seen from bar chart. It is better to use CPM network so that the floats are known. Do the CPM network right in the beginning of the project. 'You can build a project without CPM but you cannot build delay-claim without it.' Update the CPM network through periodical monitoring so that you can prove delays, as the onus of proof is on the contractor.
13. Claim is a three-legged table:
    i. Liability, which means contractual facts
    ii. Causation, which means connection
    iii. Damages, which means claims presented

14. Submission of claims should be subject to IRAC test as follows:
    i. Issue—Are you entitled to recovery?
    ii. Review—Contractual and factual right to recover
    iii. Analysis—If necessary, get expert opinion and judgements available
    iv. Chronology—Perspective to be drawn up
15. Submissions should be understandable by the arbitrator. If available, enclose periodical progress photographs, which can speak volumes.
16. Do not lose credibility by submitting untenable or exorbitant claims.
17. Have the submissions examined by your own people, or experts. 'You never have a second chance to make the first impression.'
18. Have your claims settled during the course of execution when you have leverage.
19. While negotiating international contracts, suggest a formula from outside the country, preferably a neutral country, to settle the disputes.
20. Suggest a mediator from a neutral country to advise during the course of contract.

17.3 CORRESPONDENCE

Correspondence is defined as communication between parties concerned, either in writing or in oral form. A communication is said to be complete only when

- Originator is able to make receiver understand his text
- Receiver in turn replies him back

It is imperative that all correspondence with client or engineer must serve the purpose they are intended for.

Some do's and don'ts while corresponding with the client/engineer are given below.

Do's

- Be specific on point to be conveyed
- Inform client/engineer on all matters concerning contract without any delay
- Where required, correlate matter to relevant contractual clause, and explain the case
- Send reply to all letters from client or engineer, however trivial the matter may be
- Put message in simple language
- Make letters easily understandable by the other party
- Must serve building up of records
- Should not consider correspondence as a weapon for warfare
- Matters like extra claims and rate analysis for extra items of work should always be supported with necessary calculations
- Nature of correspondence should be in a manner paving the way for resolving points/disputes between parties

Don'ts
- Tenor of correspondence should not be aggressive
- Should not be overzealous
- Facts should not be misrepresented
- Facts and figures should not be exaggerated
- Be reflective of facts of the case
- Correspondence should not to be made for the sake of record creation

17.4 PROJECT CLOSURE

This is the last phase of a construction project and is as important as any other phase in the project. This is a process of completing and documenting all the construction tasks required to complete the project. A poor project closure (or close-out, as it is referred to sometimes) leaves the client unsatisfied and may prove to be a cause for not getting repeat business. Thus, project closure should be meticulously planned. Considering the importance of project closure, some companies have developed certain templates/checklists to assist them in the process of project close-out. It is helpful for both contractors and owners to have short project-closure time. While an owner faces least interference in moving in and acquiring the constructed facilities, the contractor also can quickly move out to other project locations if the project closure does not become lengthy. A poorly planned project closure may take more than a year to complete. The outputs from the project closure phase help to execute the next project with much more efficiency and control, as we will see subsequently.

Project closure consists of a number of tasks. These are given in Figure 17.1.

As can be seen from Figure 17.1, the project closure phase can be divided into the following broad headings:

1. Construction closure
2. Financial closure
3. Contract closure
4. Project manager's closure
5. Lessons learned from the project

These are briefly discussed in the following sections.

17.4.1 Construction Closure

This involves preparation of the project punch list, which is a list of deficiencies identified during the combined inspection of constructed facilities by the representatives of client, contractor, consultant and architect. During the regular inspection also, deficiencies are reported to the contractor by the architect, the consultant, and the client's representative. The punch list is prepared usually towards the end of the project when all major construction activities are completed. The punch list is formally handed over to the contractor, who takes steps to rectify the deficiencies thus pointed out. There may be a situation in which some of the deficiencies pointed out in the punch list may not be part of the contract, and the contractor in such cases usually asks the owner for extra payment.

CONSTRUCTION CLAIMS, DISPUTES, AND PROJECT CLOSURE | 613

```
Construction closure
  • Preparation of punch list
  • Certificate of substantial completion
  • Certificate of occupancy
  • Demobilization of resources and utilities
```

```
Financial closure
  • Final payment
  • Release of bank guarantees
```

```
Contract closure
  • Preparation of as-built drawings
  • Submission of operation and
    maintenance (O&M) manual
  • Submission of warranties
  • Submission of test reports
```

```
Project manager's closure
  • Preparation of as-built estimate
  • Analysis of cost and time overrun
  • Obtaining feedback from external
    agencies
  • Obtaining feedback from internal people
```

```
Lessons learned from the project
  • Were the project objectives met?
  • Was the project executed with client's
    satisfaction?
  • Could it have been done in a better way?
```

Figure 17.1 List of tasks to be performed in project closure phase

Certificate of Substantial Completion

For a contractor, obtaining the certificate of substantial completion is an important milestone event as it ends the contractor's liability for liquidated damages (LD). Substantial completion refers to a situation in which the project is sufficiently completed. In other words, even though some minor deficiencies may be present (all the deficiencies pointed out in the punch list may not have been attended to), the constructed facility can now be used for its intended function. Usually, the unattended or yet-to-be-rectified deficiencies are attached with the certificate of substantial completion.

Certificate of Occupancy

This is usually issued by the municipality under whose jurisdiction the project location falls. It indicates that the constructed facility complies with the entire codal requirement and is safe to be occupied. Fire- and elevator-related inspection by municipal authorities is required before the certificate of occupancy is issued.

Demobilization or Release of Resources
This consists of demobilization (release) of resources such as staff and workers, and is as important as their mobilization. The closure of office, removal of unused materials lying in store, and disconnecting water, electricity and sewerage lines are all part of the demobilization process.

17.4.2 Financial Closure
Financial closure consists of writing applications for final payment, release of various bank guarantees, and settlement of any change order issued by the client.

Final Payment
The contractor has to apply for the release of final payment after he has attended to all the deficiencies pointed out in the punch list. The request for release of retention money is also made.

Release of Various Bank Guarantees
During the course of execution of project, the contractor submits a number of bank guarantees to the owner. A written request is made to the client to release the bank guarantees.

17.4.3 Contract Closure
Construction contract usually specifies the requirement of contract closure and, thus, the contractor should prepare a list of requirements for contract closure as per the contract between him and the owner. Some commonly mentioned requirements are:

Submission of As-built Drawings
During the project execution process, due to site constraint there might be some changes in the as-built facility from that as specified in the contract drawings. Thus, it is very important to prepare the as-built drawings by estimating the actual dimension and condition of the constructed facility. The as-built drawings of all the trades such as civil, electrical and mechanical disciplines should be compiled and submitted to the owner. As-built drawings are a great help during the operation and maintenance (O&M) stage of a constructed facility, and in most modern construction projects, it is mandatory for the contractor to provide as-built drawings.

Submission of Operation and Maintenance Manual
Modern projects involve a number of mechanical and electrical appliances—for example, elevator, cooling tower, air-handling unit and diesel-generator set. The manufacturers of these appliances provide operation and maintenance manual associated with these appliances. The O&M manual outlines the operational and maintenance procedure and also specifies maintenance interval for the appliances. Thus, it is very important for the owner to possess these documents.

Submission of Warranties
It is the duty of the contractor to collect all the warranties and guarantees from vendors, subcontractors and suppliers, and submit these to the owner.

Submission of Test Reports

During the execution of project, a number of tests are conducted on materials, appliances and systems that are installed in the project. The test records need to be compiled and submitted to the owner for future reference.

17.4.4 Project Manager's Closure

This includes tasks such as preparation of an as-built estimate, analysis of actual cost versus estimated cost, analysis of items where cost overrun was high, and conduct of meetings with external agencies such as client, architect and consultants for understanding their feedback on various project management aspects. Meetings with own staff and subcontractor should also be held to get their feedback on various issues.

17.4.5 Lessons Learned from the Project

This involves collection and compilation of all records associated with the project, and preparing archives of important project records. It also involves documenting the important issues faced in the project and their resolution. This helps in planning for such type of issues in the early stages of other projects. The following questions should be addressed for betterment of future projects:

- Did the project meet its requirements and objectives?
- Was the customer satisfied?
- Was the project schedule met? Was the project completed within the stipulated cost?
- Was the level of achieved quality acceptable?
- Were the risks identified and appropriately mitigated?
- What better ways can be employed to improve project execution and its management?

Compiling records of lessons learned may help in productivity improvement of the team in future projects.

REFERENCES

1. Iyer, K.C., 1996, Identification and evaluation of dispute-prone clauses in Indian construction contracts, Doctoral thesis submitted at Indian Institute of Technology Madras, Chennai, India.

REVIEW QUESTIONS

1. State whether True or False:
    a. Claims involve both owner and contractor, and are unavoidable in modern projects.
    b. Loopholes in the contract document are one of the sources of claim.
    c. Modification in the scope of work without notification by the client can be a cause of claim.
    d. Claim management involves claim identification and notification to the concerned party.

e. Linking the claim to contractual provisions is imperative in making claim of any kind.
f. A dispute implies assertion of a claim by one party and repudiation thereof by another.
g. Ambiguity in contractual provisions is one of the major causes of disputes.
h. Mechanisms of dispute resolution are—negotiation, mediation and conciliation, and arbitration.
i. Alternative dispute resolution (ADR) is not a preferable method of dispute handling over formal legal proceedings in a court of law.
j. Financial closure and contract closure do not belong to project closure phase.

2. Match the following in the context of 'claim is a three-legged table':

(i) Liability	(1) means claim presented
(ii) Causation	(2) means contractual facts
(iii) Damages	(3) means connections

3. Match the following in the context of 'submissions of claims… subject to IRAC':

(i) Issue	(1) Contractual and factual right to recover
(ii) Review	(2) Are you entitled to recovery?
(iii) Analysis	(3) Perspective to be drawn
(iv) Chronology	(4) If necessary, get expert opinion and judgements available

4. How will you differentiate the claims that occur in the construction business? Which claims can be considered as contractual in nature?

5. Write short notes on (a) penalty and bonus, (b) dispute resolution by conciliation, and (c) utility of arbitration in construction contracts.

6. Describe the various dispute resolution mechanisms adopted by construction industry.

7. Claims are inevitable in construction contracts. Discuss the different aspects of successful claims management systems for a contractor.

8. In a flyover project, which is a lump-sum quoted job including design, the contract provides for a specific length of the flyover including the name of the locations to be covered by it. During execution, the length was found to be more than that quoted for. Discuss the action to be taken by the contractor.

9. In a building project, delays take place, and the contractor applies for 'extension of time' by citing various reasons as attributable to the owner, before the expiry of the contractual completion date. No action is taken on the above by the owner. Meanwhile, the owner levies liquidated damages on the contractor and recovers the same from the contractor's bills. Is the owner correct in his action? Comment.

10. During heavy monsoon, the site of works got completely submerged in water, including plant and machinery, materials and infrastructure. How will you attribute the delay in works due to the above? What actions would you take as a contractor to salvage the situation including the losses?

11. A project (fixed-rate contract) was scheduled to commence on a particular date as per the letter of intent issued by the owner. Owing to certain reasons attributable to the owner, the handing-over of the site got delayed. Ultimately, it was handed over to the contractor after a delay of six months. As the contractor, what action would you like to take?

12. In a civil project, the lead for stacking the excavated earth was mentioned as 100 m. Due to site conditions, it could not be stacked within the above distance. As per the approval given by the owner, it was stacked at a distance of 500 m from the excavated area. What action would you contemplate as the contractor?

13. The contract stipulated the usage of a particular material from a particular location. Due to unforeseen circumstances, the said material could not be obtained from the location specified. It was procured from a different location with the owner's approval, at a different cost. What is the action required from the contractor?

18

Computer Applications in Scheduling, Resource Levelling, Monitoring, and Reporting

Introduction, popular project management software, functions of project management software, illustration of MS project, illustration of Primavera

18.1 INTRODUCTION

Modern construction management has fast evolved into an independent field of study to include a gamut of issues including financing of large infrastructure projects, construction equipment, planning and scheduling, and tender and contract management. The emergence of computers and their widespread use in the construction industry for data storage, retrieval and analysis have on the one hand made the job of the construction manager easier, and on the other hand made it more complex. Off-the-shelf software tailor-made for use in the construction industry is now commercially available. While it is not the intention of this chapter to present an exhaustive list of such software, or the nuances of using them to their full capability, it is essential to highlight some of the features of the more popular software.

The following sections briefly discuss some of the programmes that are readily available to the manager for planning, scheduling, levelling, monitoring and reporting functions. Once the basic data has been entered into the software, it is possible to schedule, reschedule and monitor at frequent intervals without much effort.

Although the advantages of the project management software may not be remarkably visible for a small project, it really pays off when used in management of large projects. In fact, in most of the large projects these days, it is contractually mandatory to use software for project scheduling and monitoring. This is perhaps the area that has caught the imagination of programmers to the largest extent, and a wide range of project management software is available in the market, varying in the degree of sophistication and features, and in the price.

Software is helpful at both planning and execution phases of the project. In the former, software helps to create a well-thought-out plan and, at times, also a more creative solution, based on logical sequencing of events, besides helping a planner to foresee problems. During the execution of a project, software is helpful for promoting effective coordination and better resource allocation.

Some of the advantages associated with project management software are:

- Speed and accuracy have greatly improved.
- It is not a very costly proposition and even medium and small companies can afford these software.

- Most of the software are user-friendly and one can use them quite easily.
- The software can handle complex problems involving multiple stakeholders as well as a number of constraints.
- It is easy to modify the inputs, and maintaining the records has also become easier.
- Decision-making has become easier. It is possible to try different possible alternatives and select the best alternative with the assistance of 'what if' analysis in project management software.

Nevertheless, it should be borne in mind that the use of (any) software does not make a project manager more effective—it is only a tool to make the manager more efficient. It is only the time required for carrying out calculations that can be reduced, and the basic models, etc., have still to be developed by a manager. Simply put, project management software does not help define the scope of work, or communicate to the client, or to carry out resource allocation. It only facilitates these processes and eliminates the possibility of computational errors, though judgemental errors can still be made.

The initial few steps are common to both manual planning and software-based planning. For example, in software-based management of projects, we need to think about the objectives of the project, the constraints, and the assumption on which the overall plan is based. There has to be clear and measurable objectives, and the plan should include a list of deliverables and milestones or dates of delivery. Due to the presence of unknown factors in the beginning of a project, the software-based plan also utilizes the planner's previous knowledge and experience to come up with 'educated guesses' for many of the input parameters.

At the initial stage, often there are many unknown factors—for example, the resource requirement of key activities and the overall available resources for different stages of the project. Experienced planners can rely on their previous knowledge and come up with 'educated guesses'. The accuracy of these assumptions will determine the quality of the plan. Constraints on a project are factors that are likely to limit the project manager's options. Typically, the three major constraints are schedule, resources and scope. Any change in one of these constraints usually affects the other two, and also affects the quality of the overall plan.

Given the unknown factors and potential changes in constraints, the planner needs to define a scope management plan to deal with the same during execution of a project.

18.2 POPULAR PROJECT MANAGEMENT SOFTWARE

The history of project management software is perhaps as old as the history of the computer. While the early project management software were limited in features, their modern counterparts are very powerful. Artemis, Can Plan, Hard Hat Manager, Microsoft Project, Primavera Project Planner, Primavera Suretrak Project Kick Start and Scitor's Business Solutions—PC Suite are some of the project management software available in the market, besides Microsoft Excel, which is also used by managers. Based on a survey of the actual user base of software, it was found that MS Project was by far the most popular (about 48 per cent), followed by Primavera Project Planner (about 14 per cent).

18.2.1 Primavera

Primavera Systems, Inc. is the world's leading provider of project, program and portfolio management software solutions. It provides the software foundation that enables all types of businesses to excel in managing their portfolios, programs, projects and resources. Primavera helps companies make better portfolio investment decisions, improve governance, prioritize their project investments and resources, and deliver tangible results back to the business. Primavera has product solutions specific to certain industries like construction, aerospace, manufacturing and power.

18.2.2 Milestone Professional

It is a fast and easy software to schedule, manage and report projects. It has Gantt chart software for creating presentation-ready project management charts and also works with Microsoft Project. It can function as an add-on tool for Microsoft Project. It can create presentation reports, combine cost and schedule, manage large projects, create reports from Microsoft Office Project, calculate earned value, and distribute schedules via print, email and Internet.

18.2.3 'Candy'—Construction Project Modelling and Project Control

'Candy'—Construction Project Modelling and Project Control is a single-package, project control system designed by construction professionals specifically for the construction industry. Estimations, valuations, planning, cash flow and forecasting components can be integrated . The operation in candy is similar to manual methods. One of Candy's most powerful features is the unique facility to dynamically link money and time, i.e., linking the bill of quantities to the programme of work.

18.2.4 AMS Realtime Projects

This is a powerful, easy-to-use tool that provides integrated project, resource scheduling and cost management. It supports the needs of individual project managers and provides consolidation, aggregation, analysis and management through powerful multi-project facilities.

18.2.5 Project KickStart

Project KickStart helps save time and money by organizing thoughts quickly with easy user interface. It has straightforward task management and progress tracking, both of which are essential for efficient, successful projects. Its other features include basic budgeting and cost tracking. It also supports Gantt chart to keep the project on time—or better still, ahead of schedule—and helps in keeping the project organized with centralized document repository.

18.2.6 MS Project

MS Project is a popular software offering a number of functions such as scheduling, resource levelling, tracking and reporting, in a user-friendly manner. In appearance, it is almost like a spreadsheet. Preparation of schedule and identification of critical path are easily achievable in MS Project. MS Project distinguishes between work resources and material resources. Tracking is possible by entering the information on percent completion of task. The newer versions are equipped with work breakdown-structure tools, risk analysis tools and multiple project-planning tools.

18.3 FUNCTIONS OF PROJECT MANAGEMENT SOFTWARE

While the features of different software may differ, all project management software make an effort to address the following key areas:

1. Scheduling function
2. Resource management including labour, equipment and materials management
3. Monitoring of a project during execution to assess compliance with schedules and estimates prepared prior to commencement
4. Generation of appropriate progress reports, which could be different depending upon the target reader (audience)—reports for internal circulation within a contractor's organization could be quite different from those submitted to the client. Similarly, the issues of interest for a financer could be quite different from those for an end user.

The following paragraphs discuss the above functions in slightly greater detail.

18.3.1 Scheduling Function

This function focuses on creating a project schedule using software with available information on tasks, resources, etc. As has been mentioned elsewhere, once a project concept has been identified, the most important phase of a successfully managed project is a definition or planning phase wherein the project is detailed and planned. During this phase, a plan is created based on the defined requirements of a project.

As mentioned already, the first step in drawing up a plan is to establish the project's requirements in terms of goals and objectives, and use this information to establish a work breakdown structure (WBS), which has been described elsewhere. In a large or complex project, it may be necessary to develop a high-level work breakdown structure. In order to be able to come up with a schedule output, we need to feed in the WBS, the duration of the activity, their dependence, and an initial commitment of resources to different tasks. This output serves as a draft initial plan, which becomes the basis for further refinement. Once a tentative schedule is available, effort is directed to improve it, and the following could be some of the strategies adopted for the purpose.

- Applying minimum late start time
- Applying minimum late finish time
- Looking for highest resource demand
- Redistributing to achieve best resource usage
- Looking for opportunities for parallel activities to provide more float
- Cascading activities calling for similar skills to improve efficiency and avoid 'downtime'

18.3.2 Resource Management Function

It is very common to face resource constraints when a project is undertaken. It is never felt that there are enough resources for the project. Under the circumstance, a look at resource scheduling may help.

As was discussed earlier, resource scheduling is a way of determining schedule dates on which activities should be performed to rationalize demand for resources or avoid exceeding given limits or availability. Depending on the arrangement of activities on the project, you may be able to do this without jeopardizing the end date. More often, it means a prolonging of the project.

On the other hand, resource levelling aims to examine resource requirements during specific periods of the project. The primary objective here is to minimize the variations in resource demand to improve efficiency or reflect reality by modifying activities within available float. In other words, modify resource loading for each unit of time—that is, day, week, or month.

By now, it must have been realized that resource scheduling is not that simple. For example, one can only redistribute the same resource but cannot substitute a different resource unless it has the same skill set. The project's new end date may exceed the mandated end date, requiring more juggling. The resource levelling may change the route of the critical path requiring a recalculation of the schedule network.

As mentioned elsewhere, construction projects involve consumption of different resources and a planner needs to know the timing and quantum of a resource required. Software often address this question on the basis of the information provided for the scheduling function.

It needs to be borne in mind that the scheduling function itself cannot be carried out unless some initial assumptions about the resources required and committed, as well as the productivity are known. To that extent, the resource management function provides a basis for refinement of these initial estimates or assumptions. It may also be borne in mind that this function of software can be easily exploited during the planning phase, even before physical commencement of work at site.

In Chapter 7, we presented two class of scheduling problems—resource levelling and resource allocation. A number of heuristic-type computer programs are available which address the two types of problems. The decision rules employed for such computer programs vary and, accordingly, the final resource-loading chart may also vary. Essentially, these programs tend to smooth out the resource requirement and they also prioritize the activities competing for the same resources on a particular day, based on some 'decision rules'. Commercial software employs a number of decision rules and sometimes the rules are not even known to the user due to proprietary concerns. Some commonly used decision rules employ early start–total float, late start–total float, etc., for prioritizing activities competing for the same resources.

We had utilized visual inspection and jugglery for resource-levelling problem, while for resource allocation a typical decision rule was used. A number of heuristics exist utilizing a number of 'decision rules'. Some of these have been patented and are not in public domain. An illustration of resource levelling using software has been given in latter sections.

18.3.3 Tracking or Monitoring Function

Once a project gets underway, it needs to be ensured that relevant information about resources and tasks underway or completed is updated and mid-course corrections appropriately made. In fact, a manager needs to constantly update data such as the likely date of completion or the likely cost at the time of completion, on the basis of the latest available information.

Given the complexity of this function in a large project, there are software with inbuilt modules to address these issues, and provide the manager with appropriate information to be able to compare current progress status with (original or a modified) schedule, sort and arrange data in a useful manner, identify specific portions of a project in terms of problem areas, and monitor resources to make sure they are being used efficiently. The function is also useful in foreseeing any likely conflicts in resource utilization by different agencies (within the organization), and taking measures to resolve them by readjusting the schedule of activities within a certain framework. While the matter of 'what if' analysis is dealt with in greater length elsewhere, it may be noted here that this feature in certain software helps a planner to explore different possibilities before making a decision.

18.3.4 Reporting Function

Using good project management software, different kinds of reports can be generated to address specific concerns of managers at site or head office, corporate heads of companies, clients, or bankers. Such reports can be prepared with the intention of review, feedback and updating progress. It may be borne in mind, however, that all these reports are based on the input data provided to the software and follows the simple dictum of 'garbage-in-garbage-out'. Therefore, utmost care needs to be taken at the time of updating information.

Also, these reports hardly ever add any new information, and are essentially formatted to present one set or the other of raw data, appropriately packaged. Finally, it is up to a project manager to communicate this data (or report) in a manner that the target audience gets the intended message.

The reports could be in graphic or tabular form. Some of the reports generated by various software are given elsewhere in this chapter.

18.3.5 Additional Functions

In addition to the above-mentioned functions, some of the available software incorporates the following as well.

Calendars

Modern projects involve multiple participants from different countries and time zones, abiding by different workweeks and different sets of holidays. Further, some activities may require different workweeks relative to other activities. For example, curing of concrete works may be required to be done on all days up to a particular number of days, while one need not work on all days for other activities. Also, in some projects some specialists may visit the site only for a few days in a month. Under such constraints, the calendar feature of software becomes very handy. One can create multiple calendars not only for different activities but also for different resources.

'What If' Analysis

Most of the software can be used for performing 'what if' analysis. For example, the user may be interested in knowing the effect of increase in a particular resource on the project schedule and project cost. It may also be desired to know the impact of reduction in duration of project on the project cost. Since software can perform the analysis faster, the user would be better equipped to know the impact of changes made in a particular variable on project schedule and project cost, and thus would be in a better position to take decisions.

Budgeting and Cost Control

It is possible to associate cost information with each activity and each resource in a project in modern project management software. The normal cost and overtime cost pertaining to human and equipment resources are entered, while one-time cost or per-use cost is entered for material resources. The accounting and budgeting codes associated with each of the resources can be set up. The cost data and code entered are used to calculate the budgeted cost of the project. Further, at the time of monitoring when the data on actual cost is provided, the software can calculate the variance in budgeted cost versus the actual cost and forecast the cost at completion.

Sorting and Filtering

Sorting feature allows the user to view information in a desired order, such as the list of activities in the order of their early start date or the total float, and so on. Most programs allow multiple levels of sorting (for example, first filter early start date, second filter total float, and so on). Using the filtering feature, the user selects only certain data that meet certain specified criteria. For example, if the user wants information only on those tasks that use tower crane, it can be done through filtering option.

In addition to the above-mentioned features, some of the project management software can handle multiple projects and, thus, can be quite useful for corporate applications. Compatibility of the software with import and export data from other applications is also an important feature. Some of the software come with Internet capabilities which facilitate easy communication of project data among different users. The software have inbuilt security features and they can provide password access either to the project file or to some specific data which the user does not want to disclose to other users.

While some of the software may have only a few of the above features, some others may have all. The frequent newer releases provided by software developers go on adding new features and improving the existing ones. As mentioned earlier, there are large numbers of vendors selling project management software with different degrees of sophistication. It may become very confusing while taking the software-purchase decision.

While choosing software, the user needs to evaluate the function for which he/she is intending to purchase the software and, accordingly, look for the software that closely meets its requirement. Some other criteria for evaluating software are—its user-friendliness; its integration capability with other applications; the requirement of installation; and vendor support.

In the following sections, we illustrate two of the most popular project management software—MS Project and Primavera Project Planner.

18.4 ILLUSTRATION OF MS PROJECT

We explain certain terminologies specific to this software and then discuss some of the primary features.

18.4.1 Definitions of Some Terminologies

Although technical terms have been defined in individual chapters related to planning and scheduling, inventory control, and accounts, it may be appropriate to focus on the definition of some of them in light of the software that are available for professional construction project management. The following terminologies are frequently used in the context of MS Project application:

Task

A task is a work package with clear deliverables, a well-defined starting date and duration. The duration for any task may be determined by dividing the total work involved by the resources committed to the work. For example, if it is stated that three painters work 8 hours each day for 2 days to complete a task, the total work involved can be stated to be 48 man-hours (3 men × 2 day × 8 hours/day); and the resources committed to the task total 24 man-hours per day (3 men × 8 hours per day); thus, the duration for completing the task would be 2 days. Naturally, if the resources committed to the job (task) were increased (or decreased), the duration would appropriately decrease (or increase). It should be pointed out that this is a rather simplistic, though illustrative, way of defining a task. In practice, different types of tasks may be encountered as suggested below:

Fixed work   A task where the work is fixed and changes in the duration and units do not affect the work to be done

Fixed duration   Total span of active working time (from start to finish of a task) required for completing a task

Fixed units   Task in which the assigned units (or resources) are a fixed value and any changes to the amount of work or the task's duration do not impact the task's units.

Task Hierarchy

In any large project, there could be hundreds of tasks associated with the different phases. In MS Project terminology, summary task is equivalent to a phase and the tasks within a phase are called subtasks. The start date of summary task is the start date of the first subtask, while the finish date of summary task is the finish date of its last subtask. The subtasks, in turn, can have their own lower-level subtasks. Thus, the whole project schedule becomes a task hierarchy with several levels.

Milestones

Milestones are intermediate goals that need to be achieved in order to realize the overall objective of a project. They are also review points where project progress can be assessed. In MS Project, a milestone can be set by entering a task with zero duration.

Resources

MS Project distinguishes resources into work resources and material resources.

Work resources refer to people and equipment that directly participate in, and are obviously required for, completion of a task. These resources actually expend time or work on the task, and are always referred to with the context of time. Usually, they are expressed in terms of how much time or maximum units the resource is available for the project. In other words, if a resource is simply dedicated to a project, the value assigned is 100 per cent. Similarly, if a resource is available to a project for one day in a five-day week, the value assigned is 20 per cent. On the other hand, if five electricians as a consolidated group work only on a certain project, a value of 500 per cent is assigned for the resource.

Material resources do not participate in a task in an active manner but are required for its completion. Supplies, stock and consumable items fall under the larger umbrella of material resources. Some examples of material resources are cement, concrete, timber and steel. It may be noted that the concept of maximum units (discussed in the context of work resources) is not applicable in the case of material resources. Further, unlike calendars for work resources, material resource calendars are not prepared and these resources cannot be 'levelled' either. Material resource[s] can be created for the project, or can be shared from other project[s] or from a standard resource pool.

Baseline Plan

A set of original start and finish dates, durations, work, and cost estimates constitute the baseline. Although some adjustment can be done before commencing the actual project execution (for example, during the fine-tuning of the overall plan), the final baseline serves as a benchmark (or reference point) for monitoring and tracking as the work progresses. Any changes are measured with reference to the baseline.

Interim Plan

As the project progresses, data from activities completed, resources used, etc., start becoming available and there is a need to generate 'interim plans' to facilitate the task of a project manager. The software has an inbuilt ability to save information related to current start and finish dates of tasks, and generate several interim plans.

18.4.2 Working with MS Project

As explained earlier, MS Project is the most widely used project management software. It is very easy to understand and is user-friendly. It can carry out the major functions required of project management software, namely scheduling, resource levelling, tracking, and generating reports. Further, new features are added every now and then, and the versatility of the software is improving with every new version. Familiarity with any version will make the user understand and adapt to the newer version quite easily. We cover the basic functions of MS Project 2002 in a step-by-step manner.

Providing Input on Project and Task

Input on project After completing the initial planning (scope, objectives, assumptions and constraints of the project are determined), the planner starts a new file in the MS Project program. At this point of time, the window screen looks like the one shown in Figure 18.1. The default view in MS Project program is the Gantt Chart view, which can be changed to other views such as network diagram view, task usage view, and so on.

The project information such as start date and finish date if applicable, the planning process such as schedule from start date or schedule from finish date, project priority on a scale of 0 to 1000, and some other general information are entered by clicking on 'project information'. Entering all these information is not compulsory and if nothing is specified, the program considers the default option and proceeds. The information on different tasks of the project is entered next.

Input on working times Working calendar is defined in which the working days in a week and the working hours are specified. By default, MS Project considers 8 working hours per day, 5 working days in a week, and 20 working days in a month—thus, it considers 40 working hours in a week and 800 working hours in a month. The user can make changes to these values if required. Different resources of the project do not follow the same calendar. For example, some specialist may be required on the project only for a few days in a week or on some particular days in a month. In such cases, separate calendars can be made and attached to the specialist resource. Similarly, separate calendars can be made for different activities. For example, for the activity 'curing of concrete' we can have a calendar that has seven working days in a week.

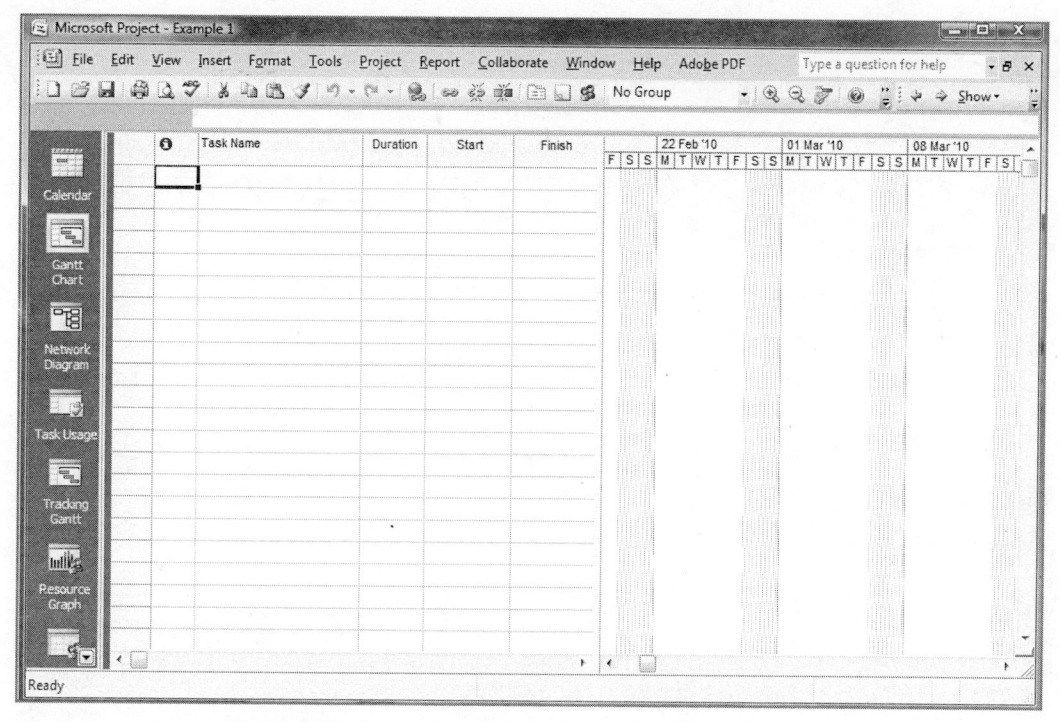

COMPUTER APPLICATIONS IN SCHEDULING, RESOURCE LEVELLING, MONITORING, AND REPORTING

Figure 18.1 Screen view after the File—New command is invoked

Input on task list The task is either individually entered or it can be copied and pasted in MS Project window as shown in Figure 18.1. The tasks are normally entered in the order that they occur. For each of the tasks, the duration and its predecessor(s) is (are) defined. The task duration could be in terms of minutes, hours, days, weeks and months.

Tasks should be entered in the order they occur. For each task, enter its task name, duration and the relationship. It is advisable not to enter the start date and finish date of tasks manually, and to let the program calculate these dates on their own.

Information on milestones, if any, in the project is also entered at this point of time. For creating a milestone, the user just needs to assign zero under duration column against the milestone. The milestones are important review points in a project.

Hierarchy among different tasks can be created by invoking the indent and outdent options of MS Project. For example, let us assume that a project consists of the tasks plain cement concrete (PCC) and reinforced cement concrete (RCC). Further, let us assume that RCC has three subtasks—formwork, reinforcement and concreting. Thus, RCC is known as summary task in the MS Project terminology. Clearly, the start date of a summary task is the start date of its first subtask and the finish date of the summary task is the finish date of its last subtask. For making RCC the summary task, the tasks formwork, reinforcement and concreting are selected and indent option is pressed. The final result is as shown in Figure 18.2.

Additional information on any task can be provided by selecting the task, clicking the right button of mouse, and selecting the 'task information' option. In addition, the information per-

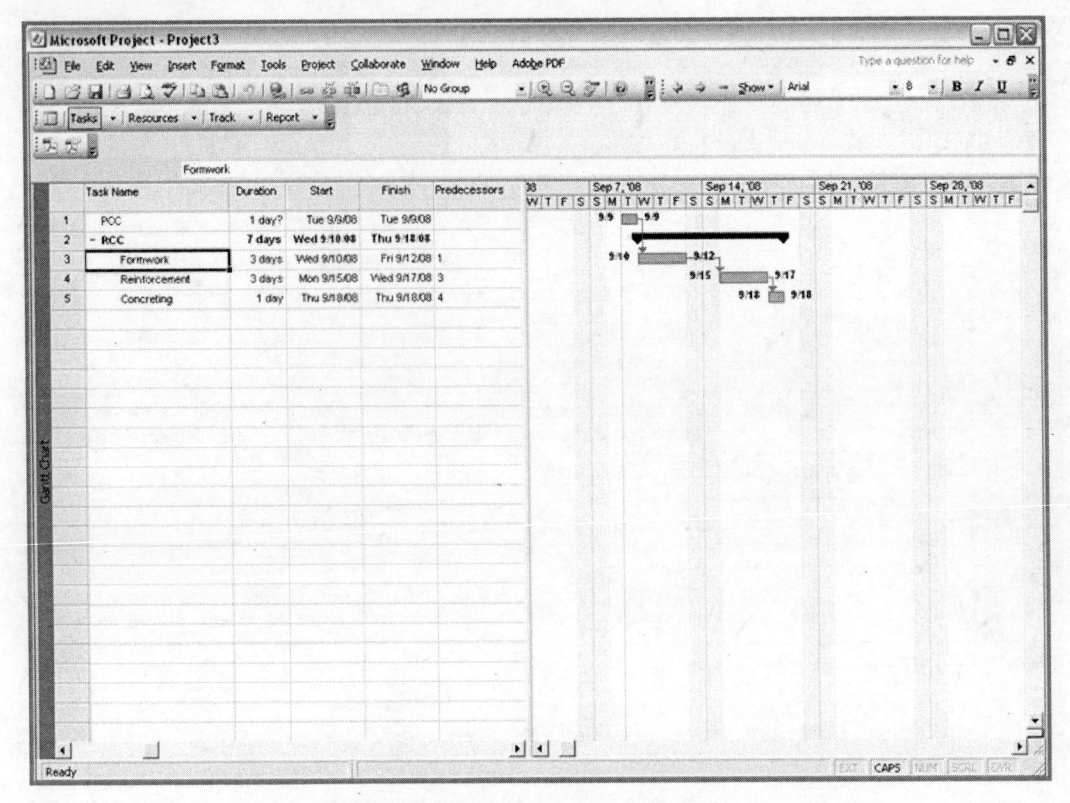

Figure 18.2 Illustration of summary task

taining to a task can be given by inserting a column and selecting from the default options. If the default option does not have the column that the user desires, the user can create his own column wherein he can enter the information pertaining to that task.

MS Project offers the option of specifying the deadline and constraints associated with a task. Deadline can be directly entered in the form of a date. For example, the completion deadline of activity xyz could be assigned as 09/09/2008, or it can be selected by clicking on a date on the calendar. An indicator is displayed if the deadline for a task is set. Examples of constraints given in MS Project program are—as soon as possible, as late as possible, finish no earlier than, and so on. Constraints and deadlines options are used when the user wants to set a specific start or finish date for a task.

The user can also split a task as many times as necessary, if work on the task is interrupted and then resumes later in the schedule.

In addition to the above features, there are many other features associated with task in MS Project program, such as identifying risks to the project, publishing project information on the Web, adding documents to project, and so on.

Schedule tasks The aim of task scheduling is to establish relationships between tasks and define task dependencies.

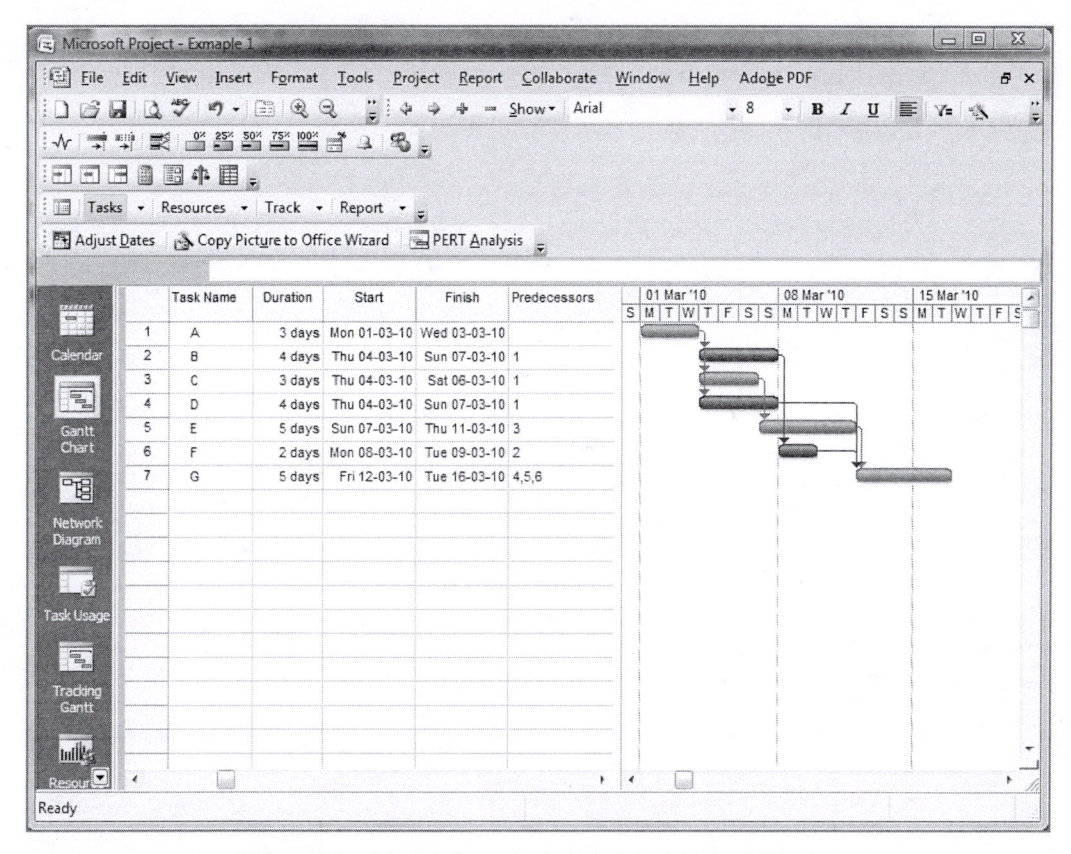

COMPUTER APPLICATIONS IN SCHEDULING, RESOURCE LEVELLING, MONITORING, AND REPORTING | 629 |

Figure 18.3 Schedule for example project drawn using MS Project

After all the tasks are entered along with their respective duration, the information on task dependencies is specified by means of specifying the predecessor and/or the successor of each of the tasks. A task whose start or finish depends on another task is the successor. The task that the successor is dependent on is the predecessor. The information is entered using predecessor and successor columns of the program. The overlap in tasks can be created by specifying lead or lag time between the two tasks. For creating the schedule, the easiest option is to use Gantt Chart Wizard. Using this, MS Project automatically calculates the start and finish dates for each of the tasks, and also identifies the critical tasks and critical path(s) of the project. It is possible to use all the types of relationship (SS, SF, FS and FF with leads and lags) in MS Project program. The default relationship type is FS with zero lead and lag.

The schedule for the example project (discussed in section 7.2-see Figure 7.1) is shown in Figure 18.3. It can be clearly seen that duration of project is 16 days, starting on 1st Mar 2010 and ending on 16th Mar 2010. The critical activities identified by the software are A, C, E, and G, which are the same as obtained through manual computations.

If the user is not happy with the schedule generated, he can adjust the schedule by utilizing the constraints option of the software. Further, the user can increase or decrease the resource

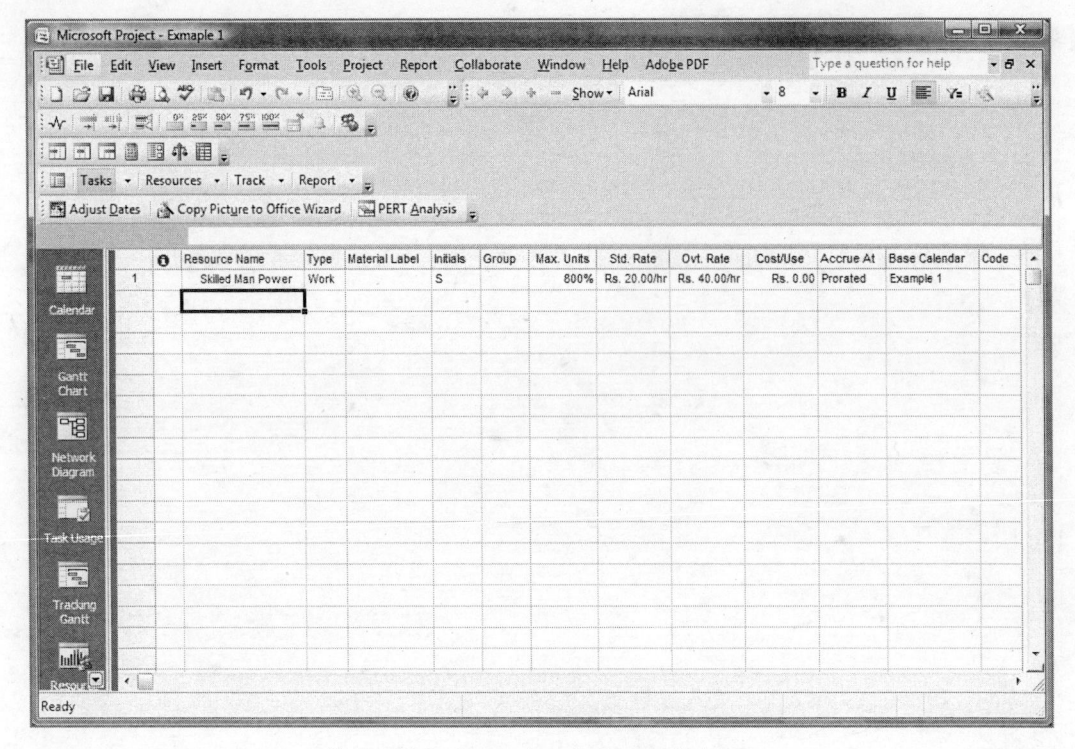

Figure 18.4 Screen view of resource dictionary

assigned for the project. All such adjustments can be carried out quickly and desired project schedules can be prepared.

Providing Input on Resources

Resource management function is carried out in MS Project in two steps. In the first step, the user identifies the different work and material resources proposed and committed for the project. In soft ware parlance, this is called 'creating resource dictionary'. In the resource dictionary (See Figure 18.4), the working times of work resources are also to be specified. It may so happen that different resources have different working times and, hence, different calendars. Further, in order to perform the cost analysis, the user has to specify the rates (options of standard rates, overtime rates, and per-use rates exist) for different work and material resources. The fixed cost associated with work resources can also be entered if it is known. The costs of resources are accrued on 'prorated' basis by default in MS Project. The other options available for distribution of cost are 'at start' or 'at end' of task.

In the next step, the resources are assigned to the different tasks of the project (see Figure 18.5). The user can specify one or more resources to a task and also specify whether a work resource works full-time or part-time on a task. MS Project automatically decreases the duration of a task in case more resources are assigned to a task. For example, a task with two days duration and one assigned resource has 16 hours of work. With effort-driven scheduling, if a second resource is assigned to it, the task still has 16 hours of work but its duration

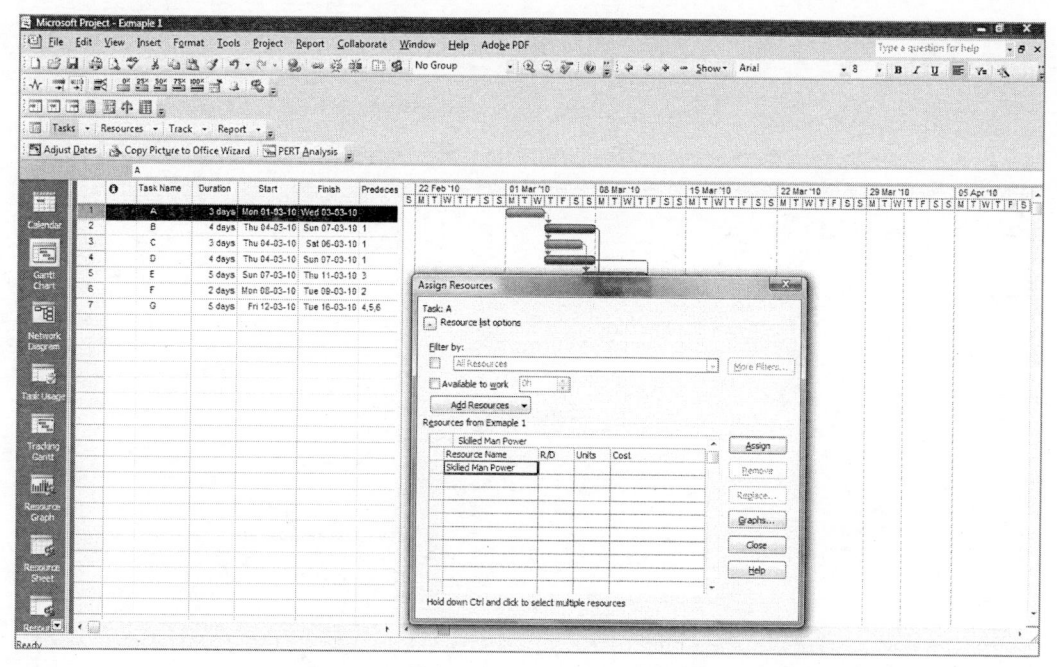

COMPUTER APPLICATIONS IN SCHEDULING, RESOURCE LEVELLING, MONITORING, AND REPORTING

Figure 18.5 Resource assignment view for activity A

is reduced to one day. MS Project has the option to switch off effort-driven scheduling before assigning another resource. The task will then have 32 hours of work and still have duration of two days.

Using the resource usage view, the user can view the assignment of different resources to different tasks.

For example, the user can find out how many hours each resource is scheduled to work on specific tasks. Besides this, one can also see which resource is over-allocated (shown in Figure 18.6). It may be noticed that Figure 18.6 is the software result of the example problem taken in section 7.2. Compare Figure 18.6 with Figure 7.2 (Resource–loading chart based on early start). Figure 18.6 can be obtained by clicking on **View → Resource Graph**.

The user can also exercise the option of levelling in MS Project by invoking the command **Tools → Level**.

The resulting levelled resource profile is shown in Figure 18.7a and 18.7b. Compare Figure 7.4 with Figure 18.7 a -Resource assignment profile of Men (time constrained leveling) - The results obtained by software and manual computations match. The Figure is obtained by clicking **Tools → Level** and selecting the leveling order as Standard and choosing the option 'Level only within available slack'. This is MS Project Output corresponding to the time constrained leveling. In Figure 18.7 a, it may be noted that the project duration is still 16 days and the peak resource requirement has come down from 11 to 10 units.

Compare Figure 7.15 with Figure 18.7b -Resource assignment profile of Men (Non time constrained leveling). The Figure is obtained by clicking **Tools → Level** and selecting the

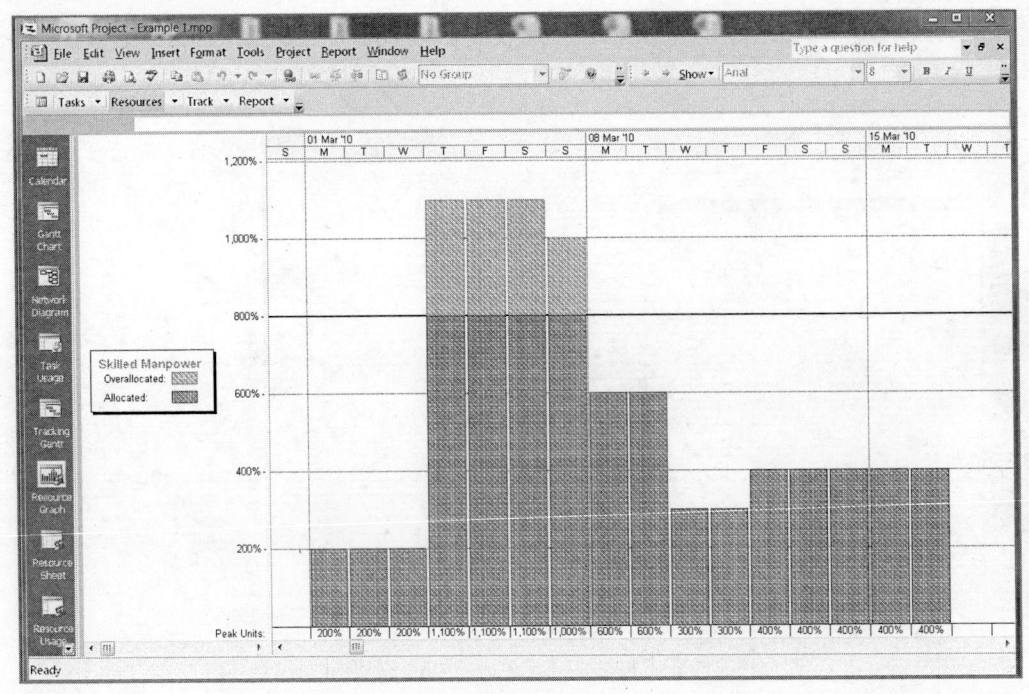

Figure 18.6 Resource–loading chart based on early start

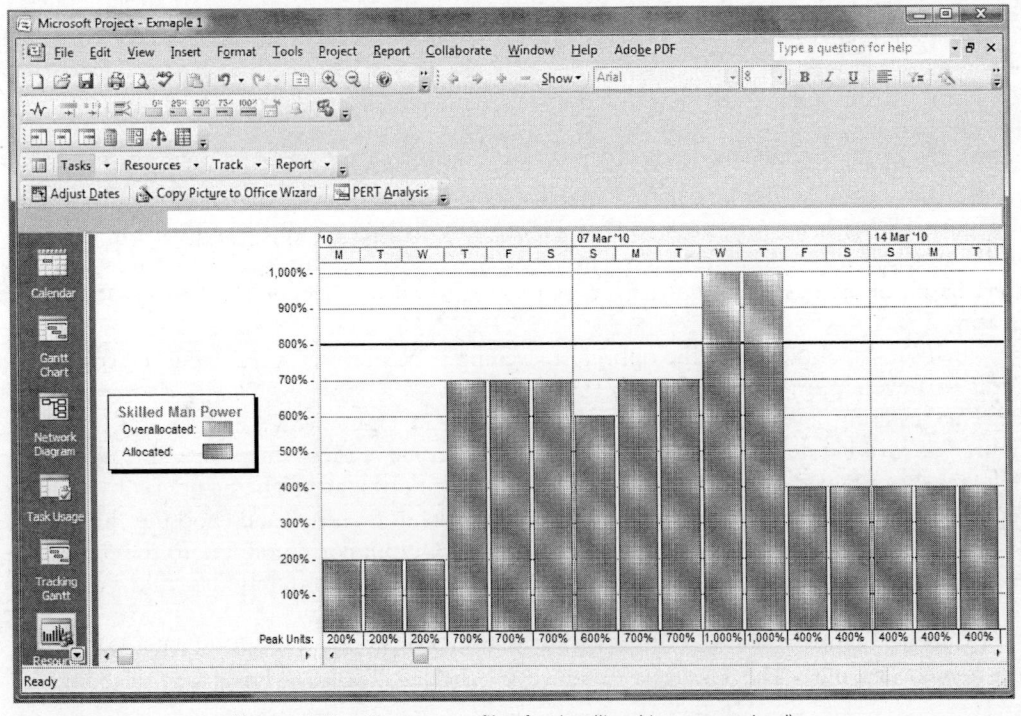

Figure 18.7a Resource profile after levelling (time-constrained)

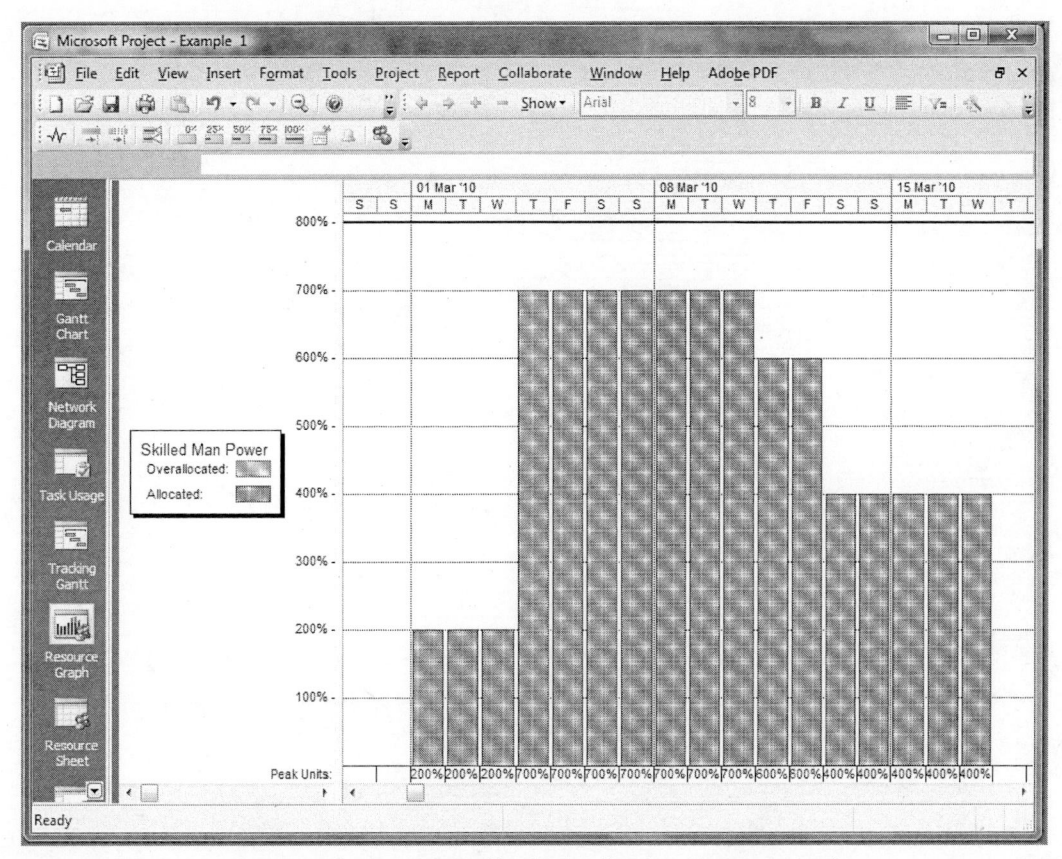

Figure 18.7b Resource profile after levelling (non-time constrained)

leveling order as ID and choosing the option 'Levelling can adjust individual assignments on a task'. This is MS Project Output corresponding to the non-time constrained leveling. In Figure 18.7b, it may be noted that the project duration is now 17 days and the peak resource requirement is within the maximum available 8 units.

The user can also see the current cost of the project corresponding to the current assignment of resources and revise the allocation if the current cost does not suit the requirement of the user. The cost gets updated each time MS Project recalculates the project.

Providing Input for Tracking

Using MS project, the user can keep project information up-to-date and identify problems early. For this, the user needs to track project information such as time, money and scope.

On the day of monitoring, the user enters information on the actual progress achieved for a particular activity and its actual start date. The process is repeated for all the affected activities as on the date of monitoring. If there is any change in other parameters of the activity—for example, cost, scope and planned dates—they are also updated. Once the information is provided, the user can get the updated project schedule by invoking the command—**View → Tracking Gantt**. The user gets a report as shown in Figure 18.8. The Figure is the software solution of

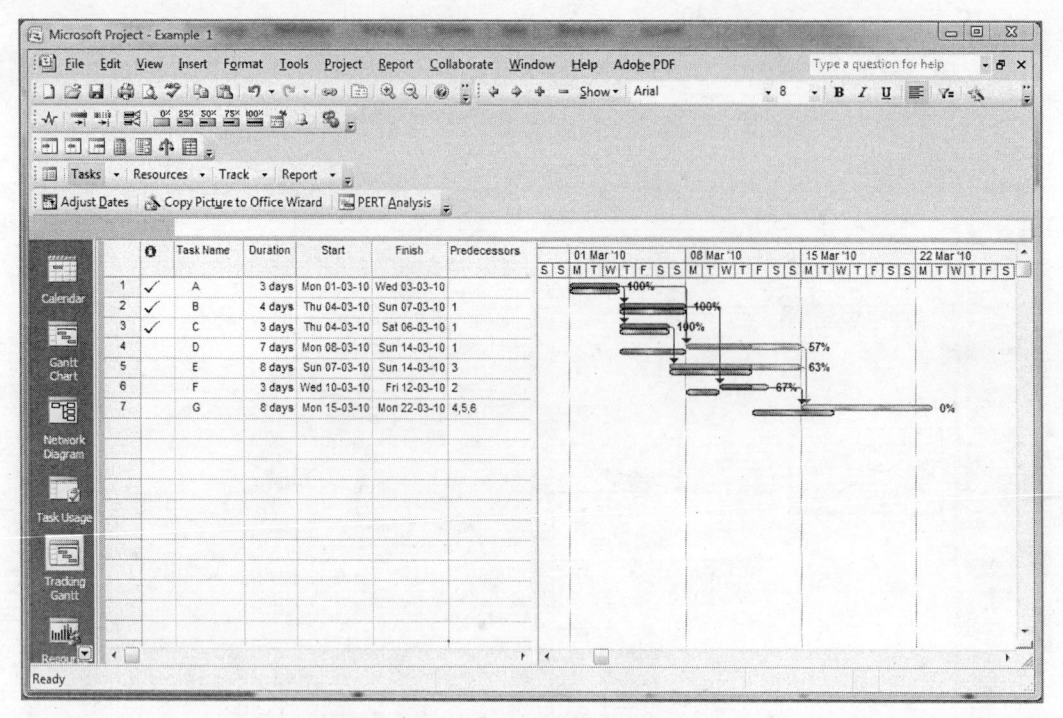

Figure 18.8 Tracking Gantt Chart for the example project

example problem illustrated in Section 16.2.2 of Chapter 16. It may be noted that the revised project duration is now 22 days, a delay of 6 days from the original project duration. The results match with the manual computations performed earlier in Chapter 16 (Compare Figure 16.4 with Figure 18.8).

The tracking Gantt Chart shows the current schedule with the original schedule, along with the progress achieved as on the date of monitoring for each task. The chart helps the user in identifying the trouble spots—for example, it identifies the tasks that varies from the initial plan (baseline plan). The variance can be good as well as bad depending on the nature and severity of variance. A negative variance may caution the user to take urgent measures immediately. The user controls the variations by adjusting task dependencies, reassigning resources, or deleting some tasks to meet deadlines. MS Project uses the actual values entered by the user to reschedule the remaining portions of a project. Thus, it may so happen that the end date gets changed or there is a new set of critical activities on the date of monitoring.

MS Project can also track the cost of the project by individually comparing each task for its budgeted cost and actual cost. For this, the user has to enter the actual cost associated with each task. MS Project then compares the actual cost with the baseline cost and calculates the variance in cost for each task. By doing so, it is possible to identify cost overruns in a task early and adjust either the schedule or the budget accordingly. It is also possible to view the project's current, baseline, actual and remaining costs to see whether the budget is being adhered to. MS Project uses earned value analyses to compare baseline schedule with the actual schedule for each task in

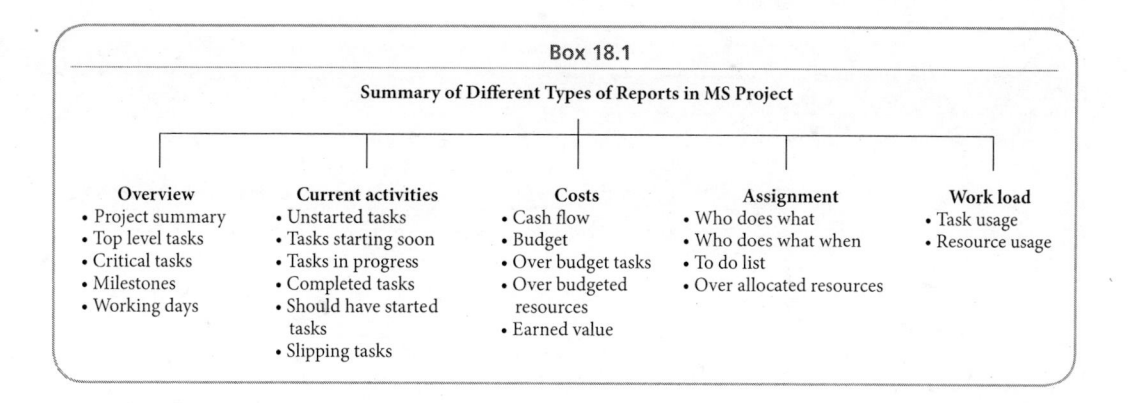

COMPUTER APPLICATIONS IN SCHEDULING, RESOURCE LEVELLING, MONITORING, AND REPORTING | 635 |

Box 18.1

Summary of Different Types of Reports in MS Project

Overview	Current activities	Costs	Assignment	Work load
• Project summary	• Unstarted tasks	• Cash flow	• Who does what	• Task usage
• Top level tasks	• Tasks starting soon	• Budget	• Who does what when	• Resource usage
• Critical tasks	• Tasks in progress	• Over budget tasks	• To do list	
• Milestones	• Completed tasks	• Over budgeted resources	• Over allocated resources	
• Working days	• Should have started tasks	• Earned value		
	• Slipping tasks			

---

terms of cost. Besides the variance in cost and schedule, MS Project can also predict the cost at completion.

Reports

MS Project offers the option of reports mostly in tabular form and with limited customization.

In order to generate reports, the user clicks on view and then on reports. The user has the option to choose the reports from five categories—(1) overview, (2) current activities, (3) costs, (4) assignments, and (5) work load. The custom icon shown on the screen (see Figure 18.9) is used to customize the look of the report obtained from each of the five categories. The list of reports corresponding to each of these categories of reports can be seen by clicking on the desired report type. The screen appearance shown in Figure 18.9 is obtained by clicking on the overview icon. It can be observed that five different reports are available under this category. Similarly, other types of reports can be obtained by clicking on the relevant icon. The different reports corresponding to each of the five categories of reports are listed in Box 18.1.

18.5 ILLUSTRATION OF PRIMAVERA

The terminologies frequently used in the context of Primavera application have been explained here.

18.5.1 Adding a New Project

This is done by invoking the command File → New. At this point, the screen would look like Figure 18.10. The software accepts the project name in four characters only. Providing the number/version, the project title (up to 36 characters), and the company's name (up to 36 characters) is optional. The planning unit could be hours, days, weeks, or months, and needs to be specified. The default planning unit is 'day' in the software, and the default start day of a week is Monday. Also, by default the project start date is taken as the current date on which the schedule is created. Project end date is also optional.

When all the above information has been entered, one simply clicks 'add' and the screen shows the activity form. In this form, different activities of the projects are entered. Although there are a number of options for entering the activities in the software, we will discuss the

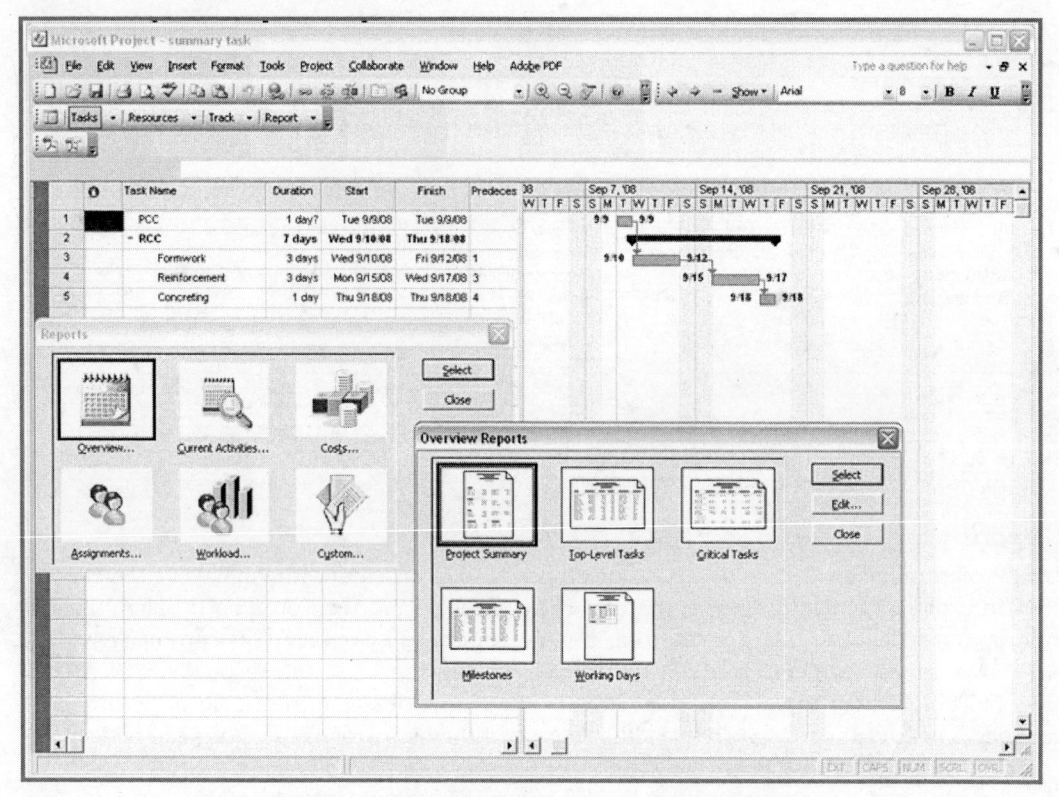

Figure 18.9 Screen view after invoking tools-reports-overview

inserting of an activity directly in the activity table. For this, we click + in edit bar to add an activity or press + on the keyboard. If an activity is highlighted, the new activity is inserted below the highlighted activity. An activity is characterized by activity ID, activity description, activity duration and activity type. The software accepts activity ID up to 10 characters, activity description up to 48 characters, and activity original duration up to 4 characters. Primavera has an option of dealing with nine types of activities—task, start flag, independent, finish flag, meeting, hammock, start milestone, finish milestone, and WBS. If the user has not assigned any type of task out of the nine mentioned, the software treats the activity as 'Task' by default.

18.5.2 Preparing Schedule

For preparing the schedule using this software, the four minimum inputs required to be given are—(1) activity identity, (2) activity description, (3) activity duration, and (4) the relationship among different activities in the form of assigning relationship in the predecessor and successor columns.

Activity identity is an important feature and the software can accept up to 10 characters. Although for some small projects the benefits of activity identity cannot be realized, for larger projects involving a number of activities of different types, the advantages are quite visible. For example, we can create activity identity such as CIVILB1241, representing civil engineering-

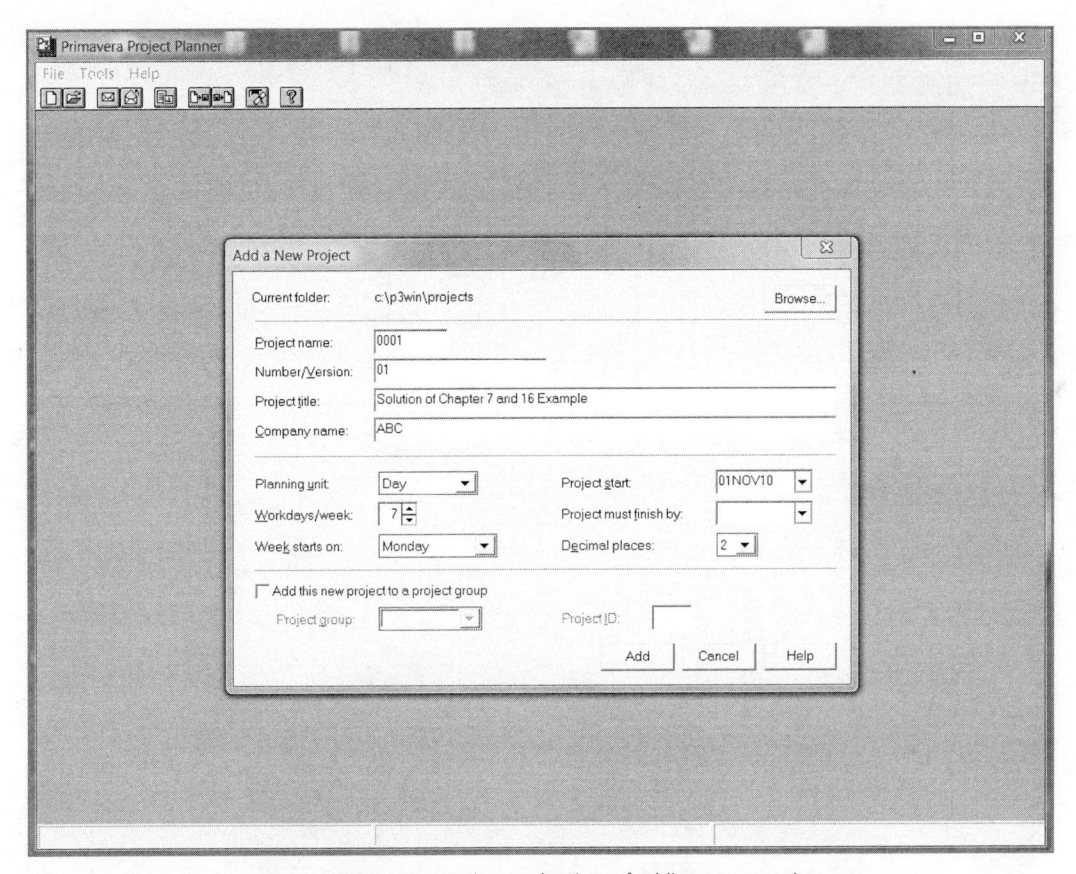

COMPUTER APPLICATIONS IN SCHEDULING, RESOURCE LEVELLING, MONITORING, AND REPORTING | 637 |

Figure 18.10 Screen view at the time of adding a new project

related activity number 241 of Building 1. Similarly, ELECTB2003 could represent electrical engineering-related activity number 003 of Building 2. Now, when the activities are sorted out based on activity identities, it would be easier to visualize the different activities under different works such as civil, electrical, and so on. The screenshot has been captured in Figure 18.11 showing the activity ID of two activities in project BCDC.

The relationship among different activities is assigned by first activating F7 command and then choosing predecessor or successor as the case may be. Primavera P3 supports four types of activity relationships—finish-to-start with lag; start-to-start with lag; finish-to-finish with lag; and start-to-finish with lag.

For illustrating the preparation of schedule using P3, the example project discussed in section 7.2 (see Figure 7.1) is taken. The four inputs mentioned above for each of the 7 activities are entered. In order to perform forward pass and back pass, and thus to view bar chart and network, choose tools and select schedule, or press F9. Choose a data date and press OK. Upon invoking these commands, Primavera generates the schedule report in Primavera Look. If the user finds deviations, it is advisable to quit Primavera Look and make necessary changes

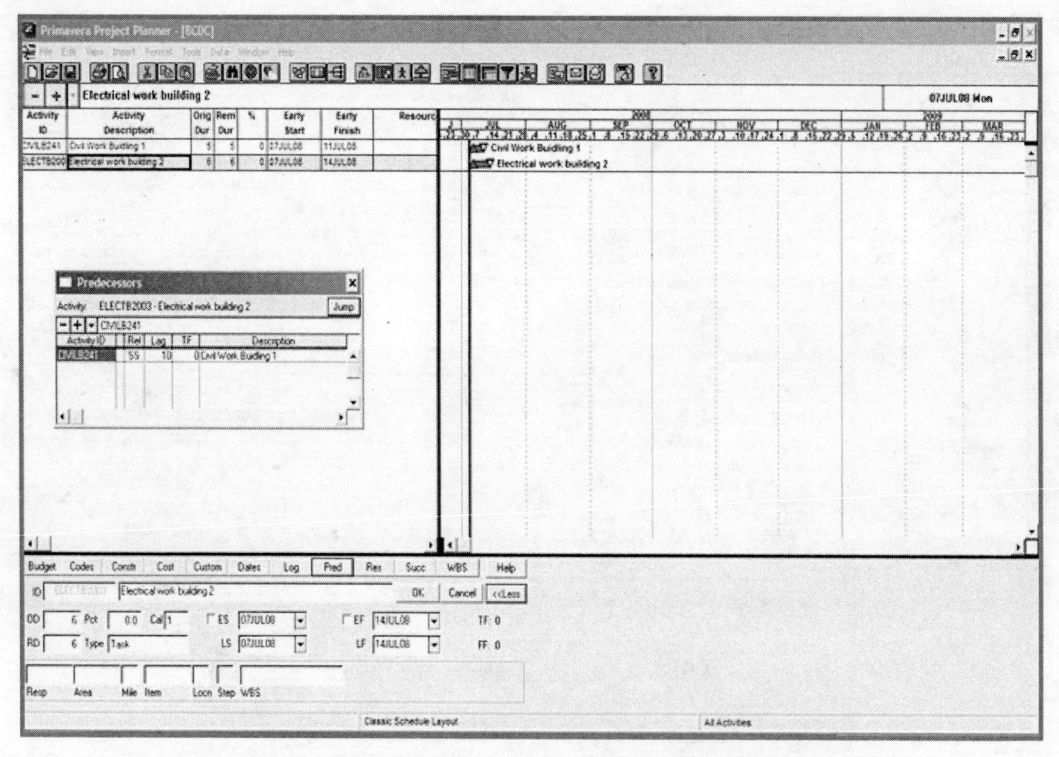

Figure 18.11 Screen view of the activity ID of two activities in project BCDC

in the plan and reschedule the network to generate a fresh report. One such report generated for a project is shown in Figure 18.12a and its corresponding network in bar chart view is given in Figure 18.12b. It can be clearly seen that duration of project is 16 days, starting on 1st Nov 2010 and ending on 16th Nov 2010. The critical activities identified by the software are A, C, E, and G, which are the same as obtained through manual computations."

A common mistake while preparing schedule using this software is that in addition to providing the first three inputs, the user also inputs the start and finish dates by manually working out these dates. This obviously is not a correct approach since the capacity of the software to calculate these dates on its own is not utilized.

18.5.3 Resource Levelling

Resource levelling is achieved in three steps in the soft ware. As a first step, one creates a resource dictionary by clicking on 'data' followed by 'resources'. Here, one can enter all types of resources—namely human, material and equipment. For each of the resources, one has to enter its (a) eight-digit code, (b) unit of measurement, (c) information on whether the resource is driving or not, and (d) the description of the resource. Further, as can be seen from the lower parts of Figure 18.13, one can record the normal and maximum limits for each of the resources along with their prices. The software has the option to take care of different normal and maximum limits besides price, through different periods of time in a project. The resource

COMPUTER APPLICATIONS IN SCHEDULING, RESOURCE LEVELLING, MONITORING, AND REPORTING | 639 |

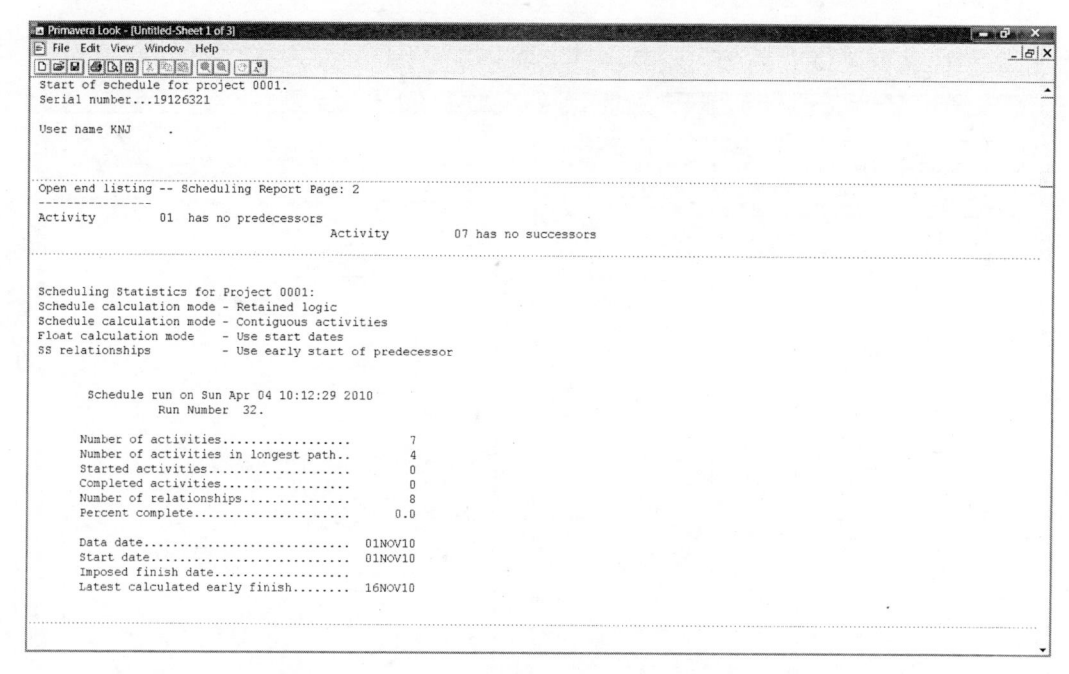

```
Start of schedule for project 0001.
Serial number...19126321

User name KNJ


Open end listing -- Scheduling Report Page: 2
---------------
Activity       01  has no predecessors
                                  Activity        07 has no successors


Scheduling Statistics for Project 0001:
Schedule calculation mode - Retained logic
Schedule calculation mode - Contiguous activities
Float calculation mode    - Use start dates
SS relationships          - Use early start of predecessor

       Schedule run on Sun Apr 04 10:12:29 2010
            Run Number  32.

      Number of activities.................   7
      Number of activities in longest path..  4
      Started activities...................   0
      Completed activities.................   0
      Number of relationships..............   8
      Percent complete.....................  0.0

      Data date............................  01NOV10
      Start date...........................  01NOV10
      Imposed finish date..................
      Latest calculated early finish.......  16NOV10
```

Figure 18.12a Screen view of schedule report generated by P3

dictionary need not be created every time afresh and it can be imported from other files and then modified accordingly.

In the next step, the resources are assigned to different activities as the case may be (see Figure 18.14). For this, 'assign' followed by 'resources' command is invoked, and the required units of different resources are assigned to different activities. The screen shot of resource profile (resource assignment on a periodic basis in a graphic form) after the assignment of resources to different activities are shown in Figures 18.15a and 18.15b. The resource profile can be viewed after rescheduling the project by invoking Tools → Schedule. The attentive readers might have noticed that the two Figures 18.15a and 18.15b are the resource profile based on early start and late start respectively (before leveling) for the example taken for illustration in section 7.2 of Chapter 7. Figures 7.2 and 7.3 should compare with Figures 18.15a and 18.15b. If the resource profile is not up to the liking of the planner and there are too many peaks and troughs besides the situation of over-allocation (exceeding the maximum limit) of resources, one needs to run the 'level' command.

The last step in this exercise is to run the command 'tools' followed by 'level'. The screen view after invoking the Tools → Level command is shown in Figure 18.16. It may be noticed that P3 has a number of options for leveling the resources. The option shown in Figure 18.16 is corresponding to the non-time constrained leveling of resources. The resource profile after leveling is shown in Figures 18.17a and 18.17b for non-time constrained and time-constrained cases respectively. Figures 18.17a should compare with Figure 7.15. In Figure 18.17a, it may be noted that the project duration is now 17 days and the peak resource requirement is within the maximum available 8 units.

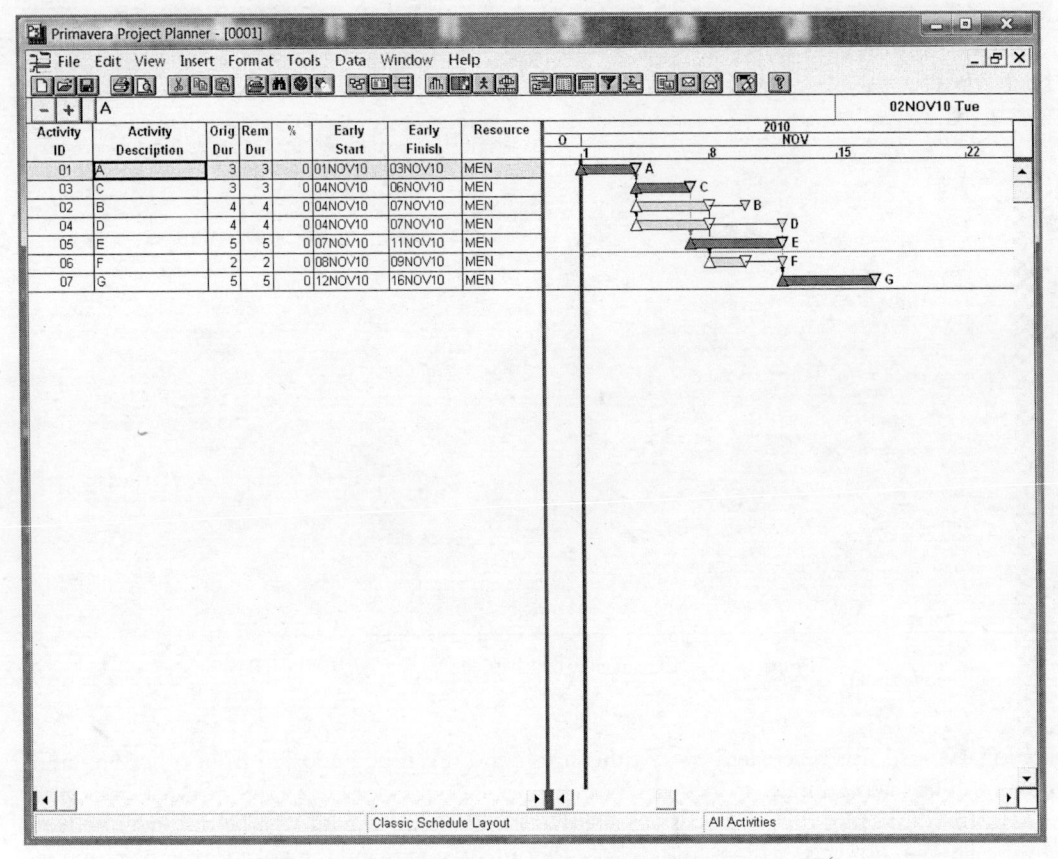

Activity ID	Activity Description	Orig Dur	Rem Dur	%	Early Start	Early Finish	Resource
01	A	3	3	0	01NOV10	03NOV10	MEN
03	C	3	3	0	04NOV10	06NOV10	MEN
02	B	4	4	0	04NOV10	07NOV10	MEN
04	D	4	4	0	04NOV10	07NOV10	MEN
05	E	5	5	0	07NOV10	11NOV10	MEN
06	F	2	2	0	08NOV10	09NOV10	MEN
07	G	5	5	0	12NOV10	16NOV10	MEN

Figure 18.12b Schedule for example project drawn using P3

In Figure 18.17b, it may be noted that, on days 8 and 9, the resource requirement is 10 while on days 10 and 11, the resource requirement is 7. This is in contrast with Figure 7.4, wherein the resource requirement for days 8 and 9 were 7 while on days 10 and 11, the resource requirement was 10. In Figure 18.17 b. Thus there is slight variation in the resource profile (Figures 18.17b and 7.4) in the case of time-constrained leveling. This is because of different algorithm used for leveling. However, in both the figures, it may be noted that the project duration is still 16 days and the peak resource requirement has come down from 11 to 10 units.

Resource levelling uses the resource limits and then applies the resource for the most efficient use.

Resource levelling in Primavera allows levelling of up to 500 resources at one time, within the maximum and normal limits defined in the resource dictionary. For levelling, it delays the activities to ensure that the maximum limit is not exceeded. It tries to schedule activities within the available float of an activity and allows activities to be prioritized. It can perform either forward or backward levelling. The details on this can be found by referring to the 'help' command available with the soft ware.

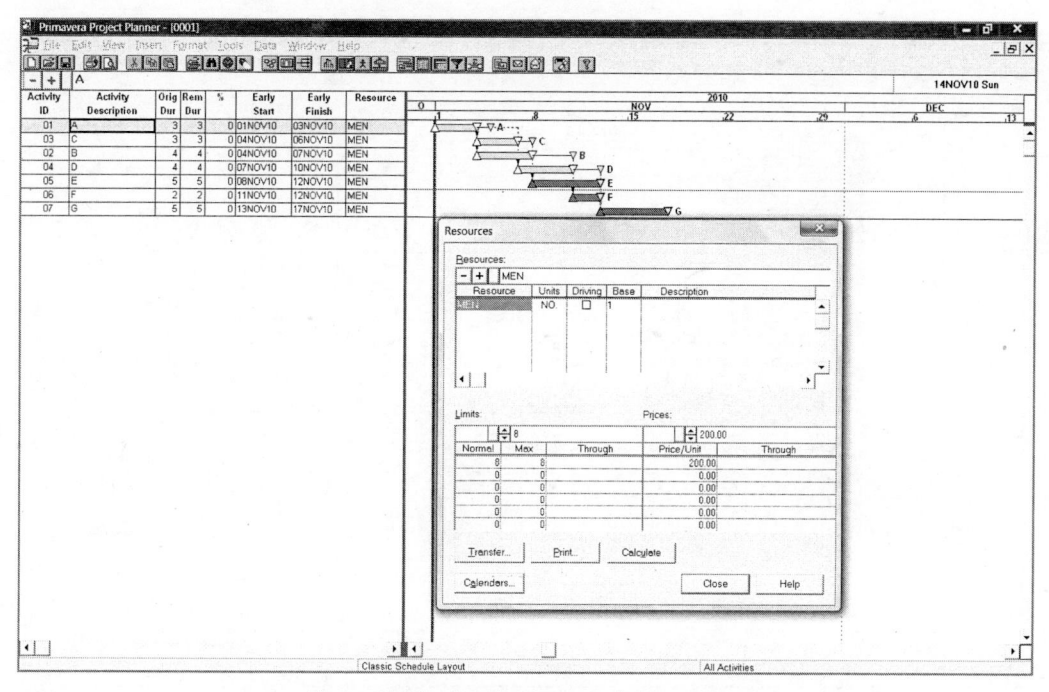

COMPUTER APPLICATIONS IN SCHEDULING, RESOURCE LEVELLING, MONITORING, AND REPORTING | 641 |

Figure 18.13 Creating resource dictionary in P3

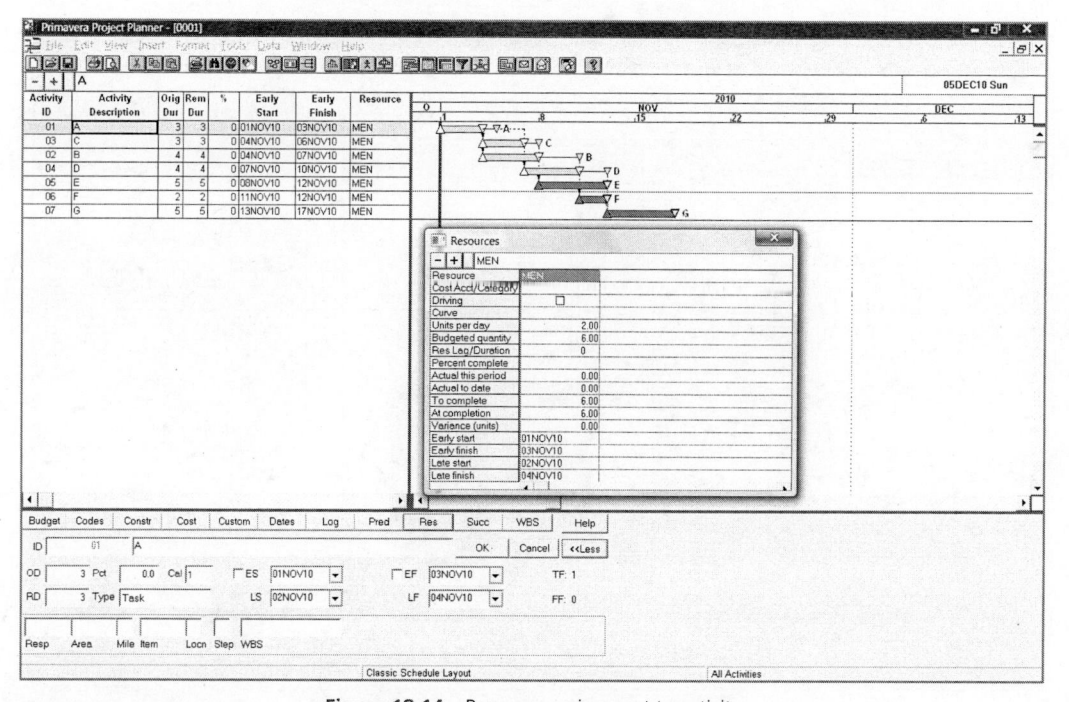

Figure 18.14 Resource assignment to activity

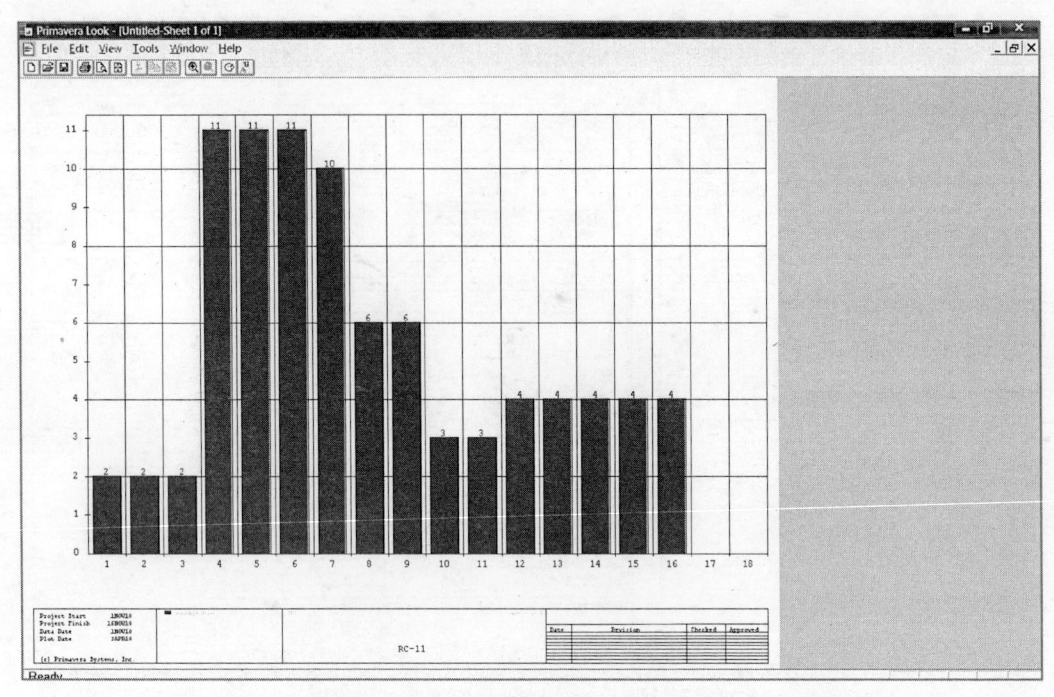

Figure 18.15a Resource profile based on early start (before levelling)

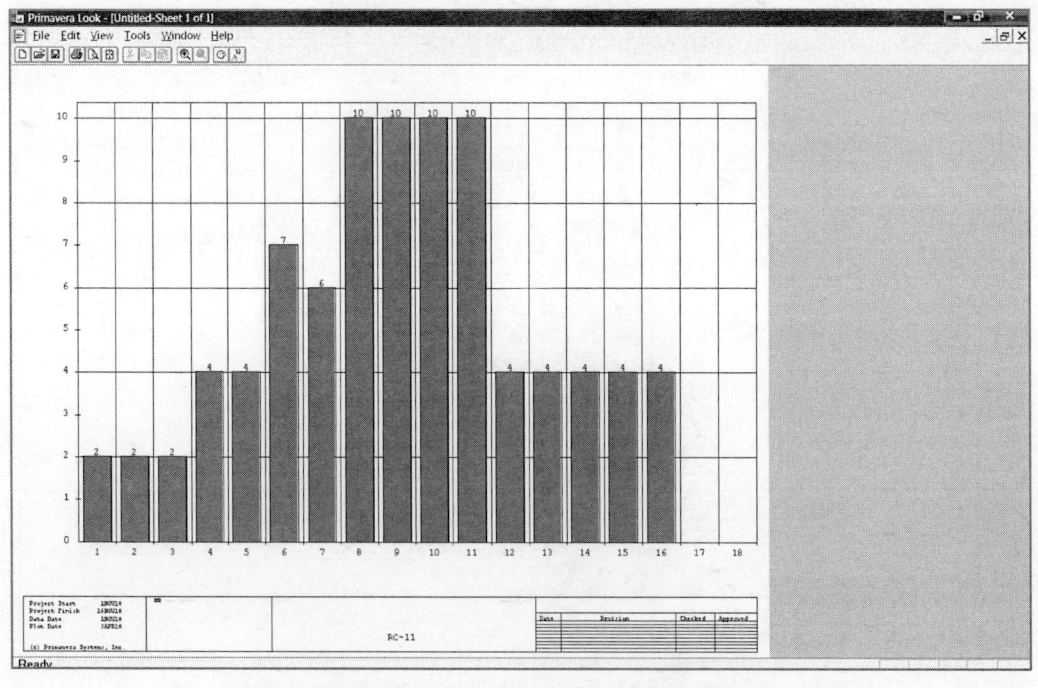

Figure 18.15b Resource profile based on late start (before levelling)

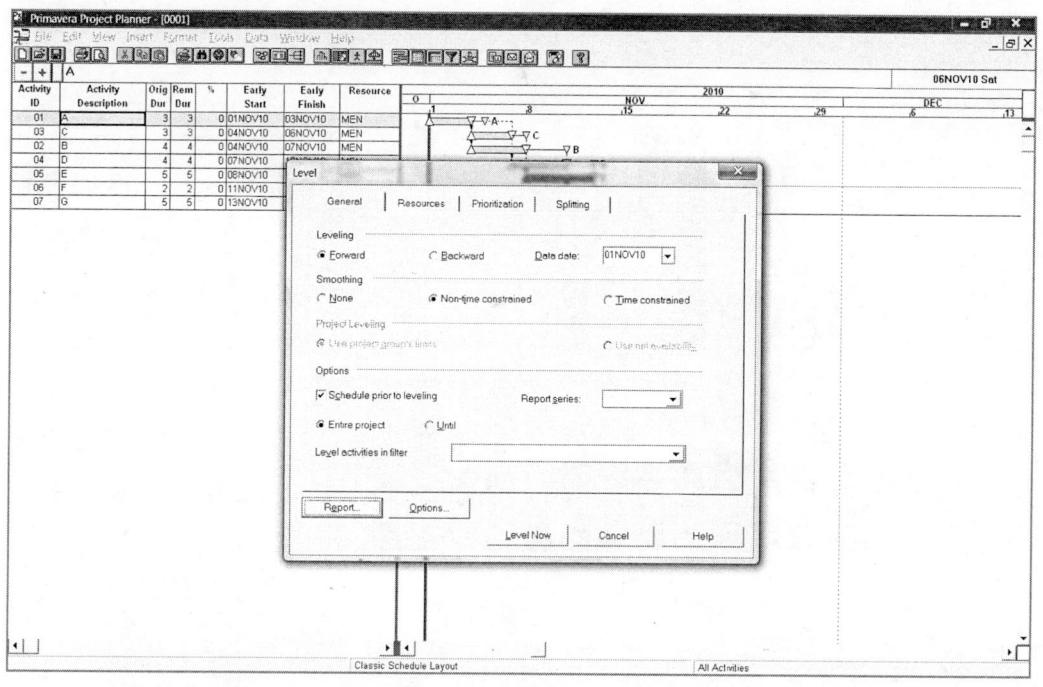

COMPUTER APPLICATIONS IN SCHEDULING, RESOURCE LEVELLING, MONITORING, AND REPORTING | 643 |

Figure 18.16 Screen view after invoking Tools → Level (note the options selected above)

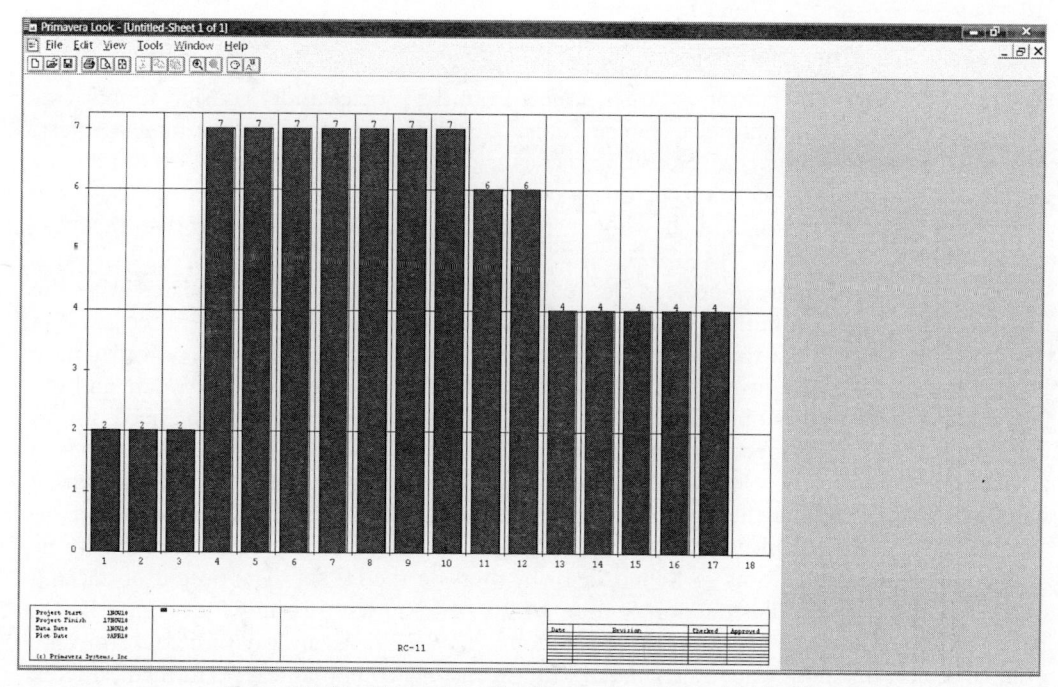

Figure 18.17a Resource profile after non-time constrained levelling

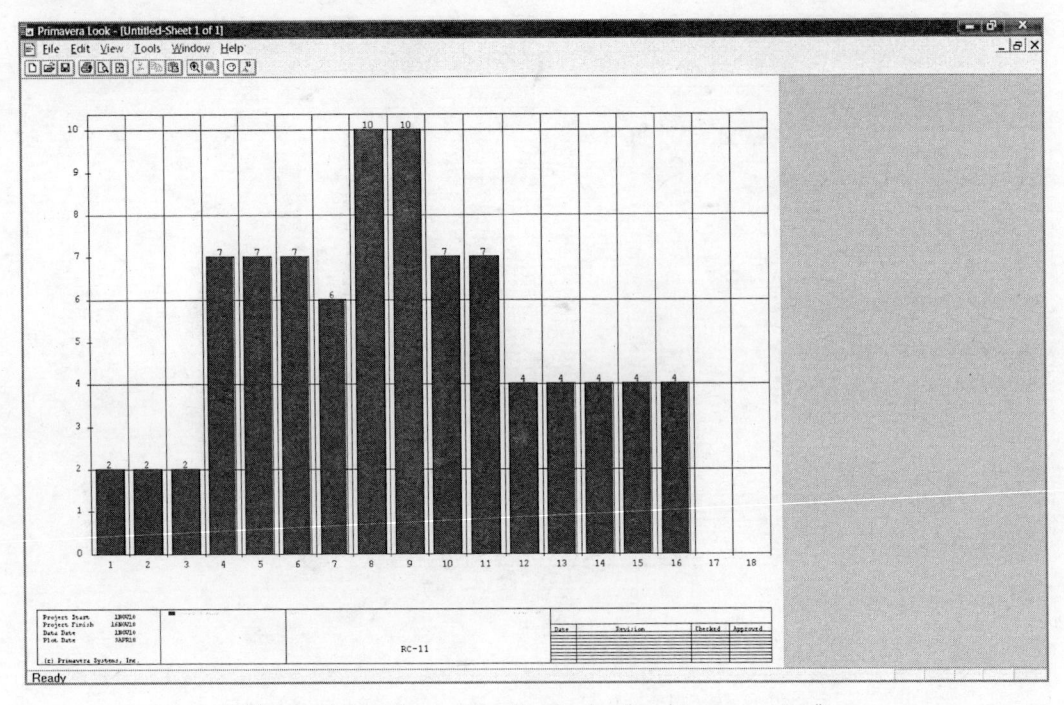

Figure 18.17b Resource profile after levelling (time-constrained)

18.5.4 Tracking the Project

As a first step in project tracking, a target project from the project under consideration is created. For creating a target project, choose Tools → Project Utilities and select target project. The target project created is given a four-character name such as TRG1, and other relevant details are filled up. The logic of creating a target project is to make an exact replica of the project under consideration, and allowing changes in the target project only so that the original project is not disturbed.

After creating the target project file, it is opened. The activities are updated with the information available as on the day of tracking. For any activity, there could be three cases on the day of tracking. It may have been started or it may not have been started, or it may have been completed. For the first case, the planner needs to specify the actual start date and the percentage completion as on the day of tracking, while for the second case one need not do anything. One can also specify the remaining duration for an activity. For the third case, the actual finish date is to be specified. This exercise is carried out for all the activities that have witnessed changes during the day of planning and the day of tracking, or between two consecutive tracking days. Once this is over, the project is rescheduled on a given data date by invoking the command **Tools → Schedule** or by pressing the F9 key. The output of tracking of the example problem illustrated in Section 16.2.2 of Chapter 16 is shown in Figure 18.18. It may be noted that the revised project duration is now 22 days, a delay of 6 days from the original project duration. The results match with the manual computations performed earlier in Chapter 16 (Compare Figure 16.4 with Figure 18.8).

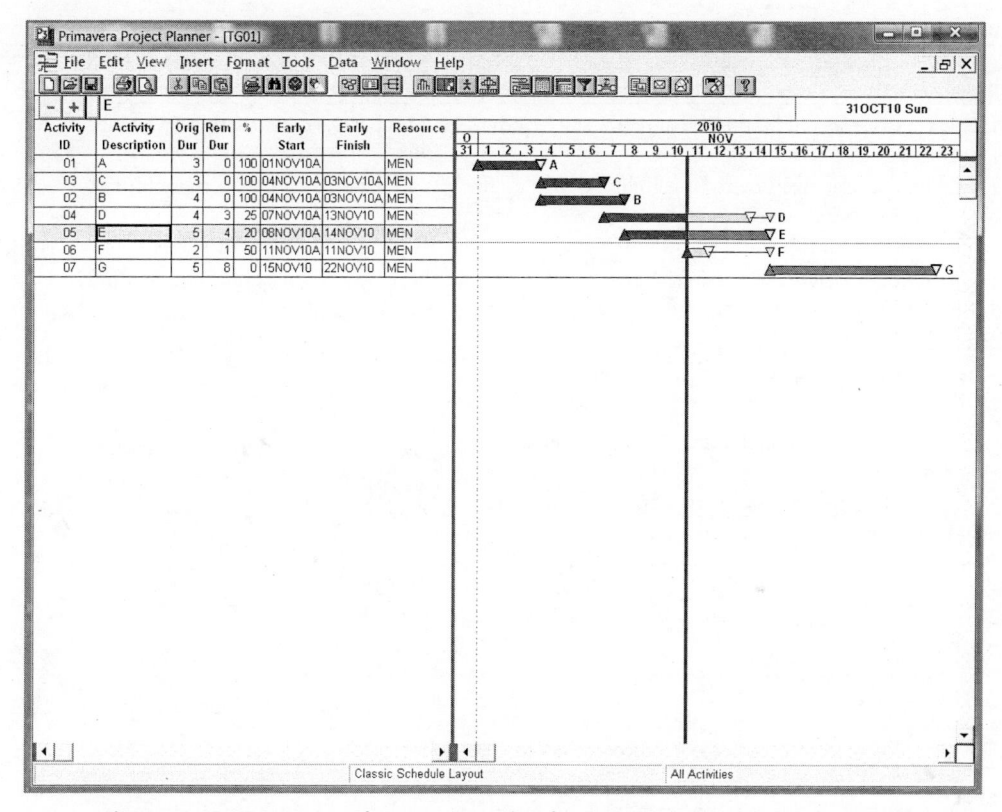

COMPUTER APPLICATIONS IN SCHEDULING, RESOURCE LEVELLING, MONITORING, AND REPORTING | 645 |

Figure 18.18 Screen view after progress update for the example illustrated in Chapter 16

18.5.5 Reporting

A large number of reports can be generated in Primavera. The reports could be in tabular as well as graphical forms. Some of the reports that can be generated are given in Table 18.1.

For generating a particular type of report, choose Tools → Tabular Reports or Graphic Reports as per the requirement and select the report type. The steps are shown in Figures 18.19 to 18.21.

Thus, in order to get one of the schedule-related tabular reports, we choose Tools → Tabular Reports → Schedule and select an existing template from the list and click 'modify', or click 'add' to add a template. Select the required report and click 'run'.

18.5.6 Some Additional Features

Adding Constraint

Suppose there is an activity that can start no earlier than or finish no earlier than a particular date. In such cases, we will not be able to describe the relationship using the four types of relationships (FS, FF, SS, SF) with the provision of lead–lag factor mentioned earlier. Under such situations—and many other situations such as 'start not later than' or 'finish no

Table 18.1 Some typical tabular and graphical reports generated by P3

Tabular report	Graphical report
Schedule report—sorted by total float, early start, activity ID, and so on	Bar charts—construction summary by phase, cost, and so on
Schedule report organized by responsibility, department, late start, and so on	Bar charts by phase, activity logs, and so on
Schedule report—60 days look ahead	Highlighting responsibilities
Schedule report—detailed precedence analysis, summary with budgets, comparison to target	Activity schedule/resource loading
	Two-month early date window
Schedule report with resource usage	Responsibility summary with costs
Schedule report comparison to target	Target showing current early finish
Resource control—detailed by activity, resource, and so on	Current vs target comparison
Earned value report, resource use—monthly/weekly report	Logic by phase, pure logic diagram
	Stacked bars for resource and cost
Cost control—summary by activity, detailed by resource, detailed by cost category, detailed by cost account, and so on	Total usage for all resources
	Cumulative and monthly costs
Cost, price and rates report, earned value reports; tabular cost—monthly project cash flow; current/target early start and variance	Area profile—current vs target
	Cumulative cash flow
	Resource profiles

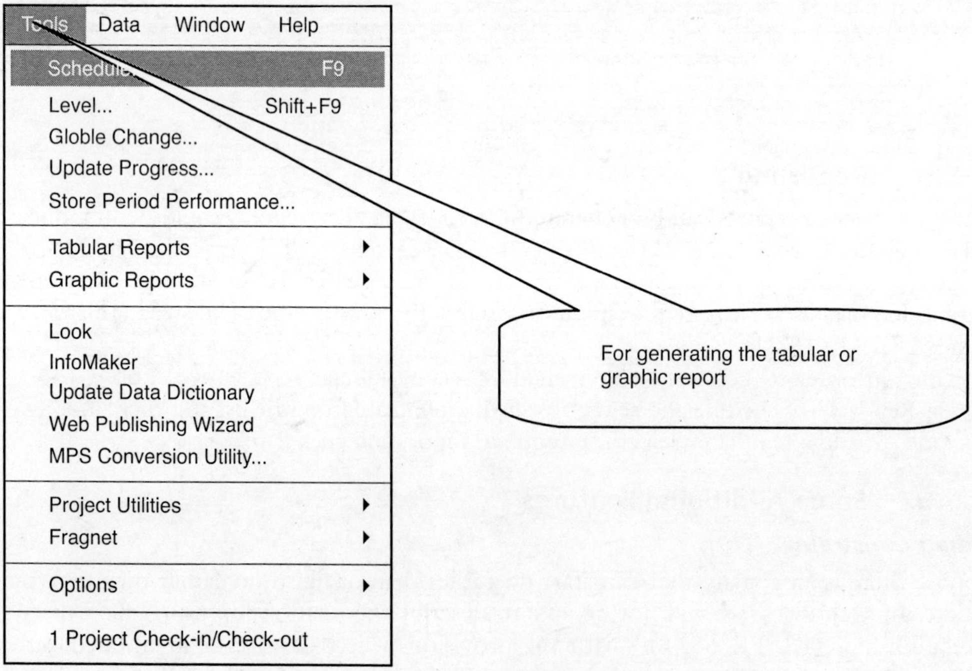

Figure 18.19 Screen view to generate tabular and graphic reports

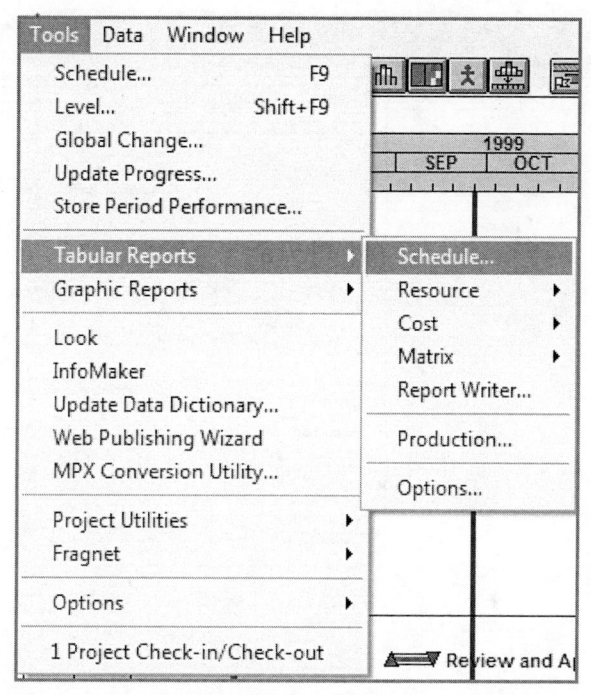

COMPUTER APPLICATIONS IN SCHEDULING, RESOURCE LEVELLING, MONITORING, AND REPORTING | 647 |

Figure 18.20 Screen view after invoking the command Tools → Tabular Reports → Schedule

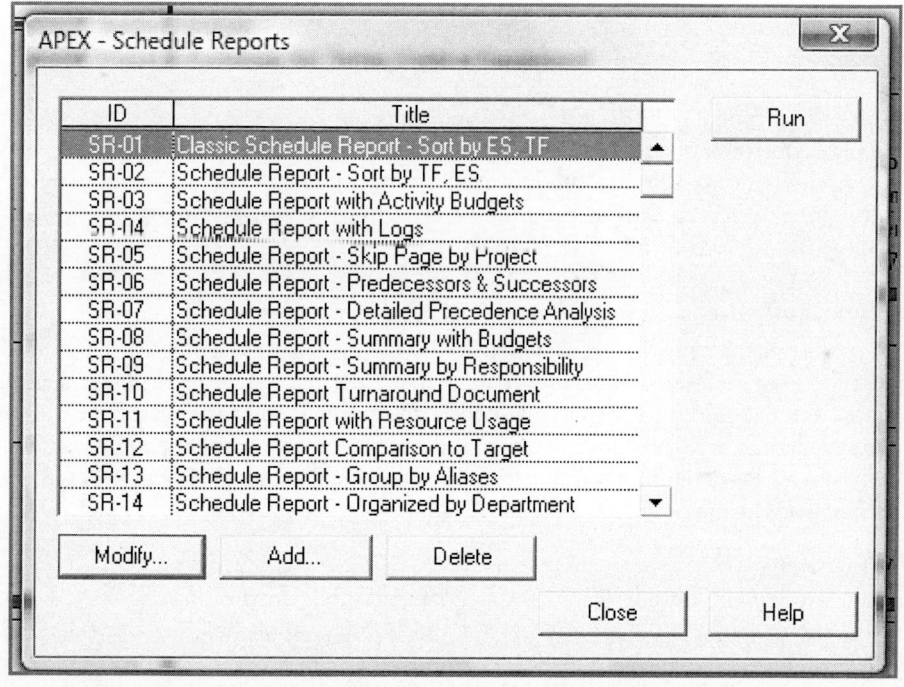

Figure 18.21 Different options of tabular reports under schedule-related report

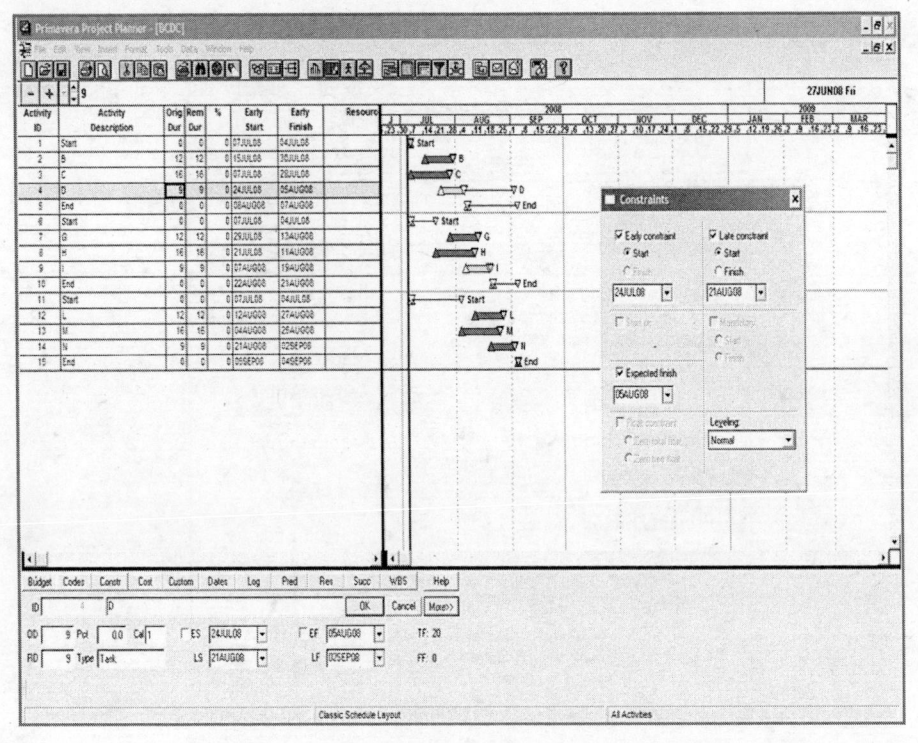

Figure 18.22 Screen view after invoking the constraints option

earlier than'—one can use the 'constraints' option given in the software. The assignment of a particular constraint type is achieved by clicking on the constraint option appearing in the lower window and clicking on the required constraints from the default constraints list of the software. The screen shot at the time of invoking the constraints option is shown in Figure 18.22.

Activity Box Templates

Primavera project planner provides default activity box templates that can be used and modified depending on the user's needs. For making changes in the default activity box templates, in PERT mode, choose Format → Activity box configuration and select Modify Template. The cells of the default activity box can be changed according to the user's needs, and the content of a cell can also be selected from a host of options available from the pull-down menu. The change in activity box configuration procedure is captured in Figure 18.23.

Developing a Calendar

For creating a calendar, choose Data → Calendar and click on the option → Add. Specify the name of the calendar and assign workdays and holidays applicable for this calendar (see Figure 18.24). Once the calendar is made, assign it to the relevant activities or to a resource or

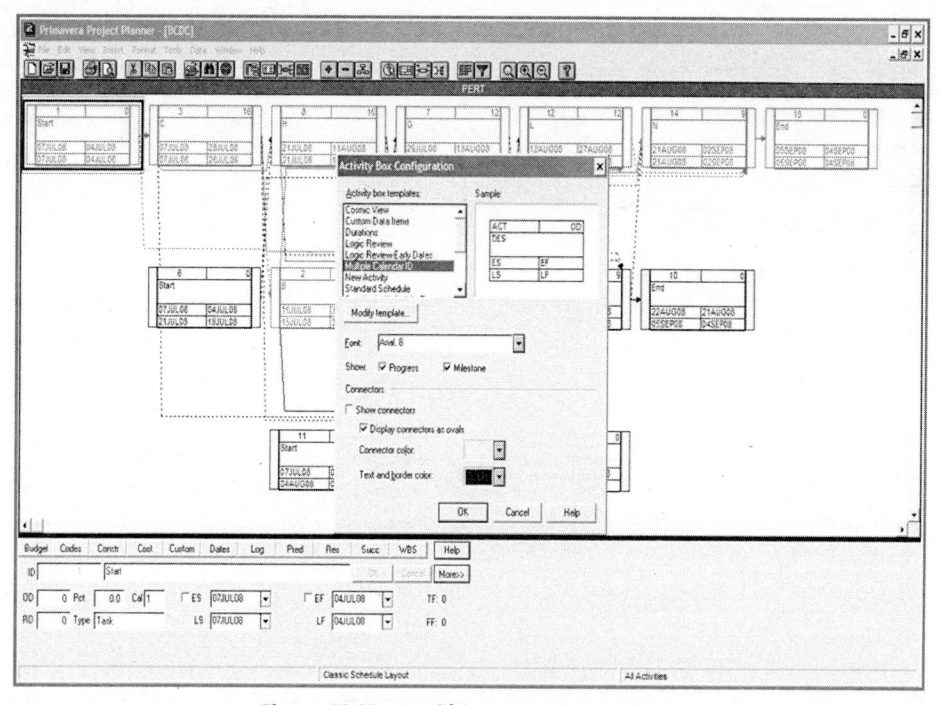

COMPUTER APPLICATIONS IN SCHEDULING, RESOURCE LEVELLING, MONITORING, AND REPORTING

Figure 18.23 Modifying activity box template

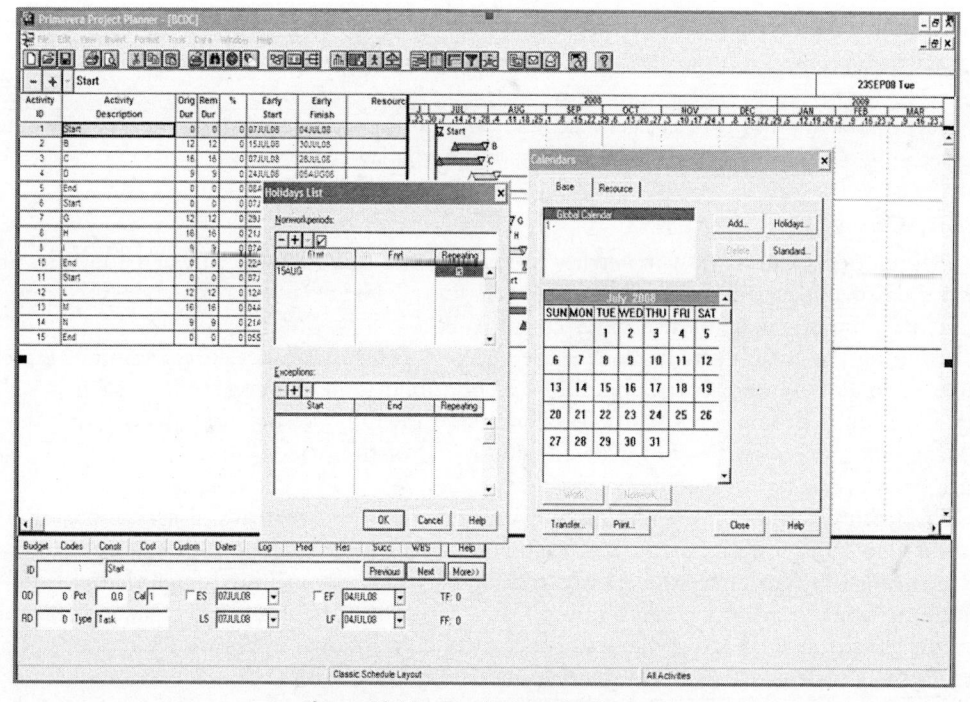

Figure 18.24 Developing calendar in P3

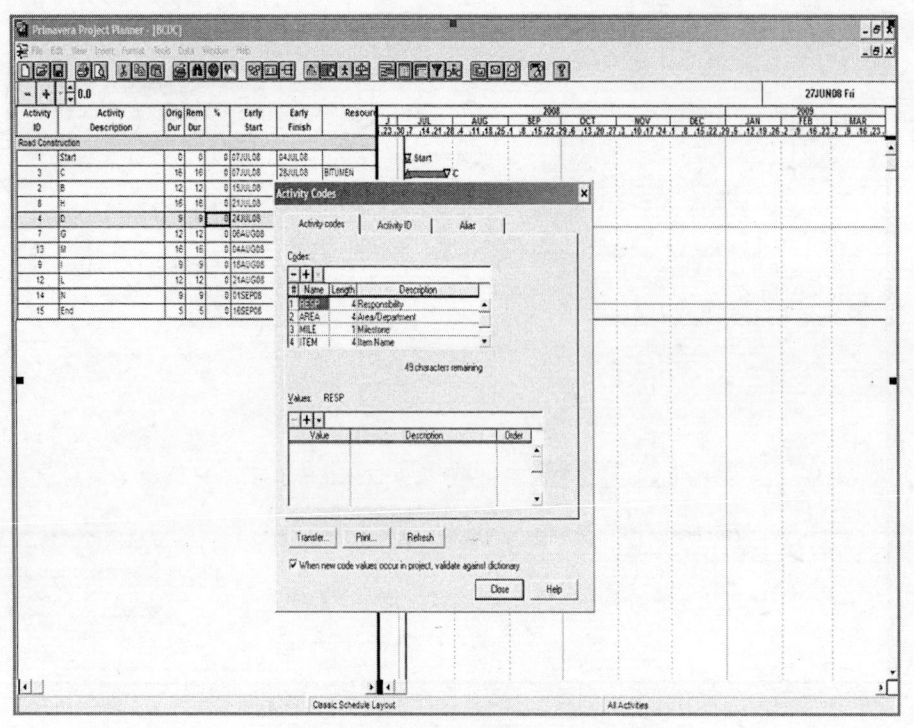

Figure 18.25 Developing activity codes

resources that would follow this calendar. If a holiday is repeating every year on the same date, the repeating option is 'double-clicked' to mark the holiday every year.

Activity Codes

For a large project, the option of activity codes becomes quite handy. The activity codes are used for classifying an activity based on a set of attributes. The default activity codes are responsibility, area, milestone, item, location, step and WBS.

For using the default activity codes, one needs to enter the values corresponding to each activity code. For example, corresponding to responsibility, one can create values such as KNJ, DKS and PKO, standing for KN Jha, DK Singh and PK Ojha, respectively. Notice that three characters have been used here for coding the name of the person and, accordingly, in the length column one has to specify 3.

This feature is useful for grouping activities into similar categories. For example, at any point of time if a user wants to know the activities in which KNJ, DKS and PKO are involved, he/she needs to go to FORMAT → organize and select 'responsibility' from the pull-down menu. For defining activity codes, choose Data → Activity Codes and select an activity code from the already created activity codes. The activity codes shown in Figure 18.25 correspond to the activity code RESP (Responsibility) and its associated values.

One can also create activity codes other than the default codes given in the software. For this, one has to choose Data → Activity Codes and create the new code by clicking on + key. The code can take up to four characters and there can be a maximum of 20 codes per project. The desired length (up to a maximum of 10 characters) of the activity code is specified, and the values corresponding to each activity code along with its description and the order in which it has to appear in the lower window are specified.

REFERENCES

1. AMS REALTIME Projects, http://www.amsrealtime.com/products/projects.htm.
2. Chitkara, K.K., 2006, *Construction Project Management: Planning, Scheduling and Controlling*, 10th reprint, New Delhi: Tata McGraw-Hill.
3. Construction Computer System Planning System, http://www.ccssa.com.
4. Issues and Considerations in *Project Management*, available at http://www.maxwideman.com/issacons.
5. Lock, D., 2003, *Project Management*, 8th edition, Gower Publishing Limited.
6. Microsoft Project, 1998, Reference Manual, Project 98 Windows, Microsoft Corporation, One Microsoft Way, Redmond, WA.
7. Milestone Professional, http://www.kidasa.com.
8. MS Project, http://www.office.microsoft.com/project.
9. Primavera Reference Manual, Ver. 3.0 for Windows, Primavera, Two Bala Plaza, Bala Cynwyd, PA 19004.
10. Primavera, http://www.primavera.com.
11. Project KickStart, http://www.projectkickstart.com.

REVIEW QUESTIONS

1. State whether True or False:
    a. Project management software may give wrong results because of the speed with which it performs the forward and backward pass calculations.
    b. Resource levelling function can easily be done using project management software.
    c. More customized reports are possible in MS Project when compared to Primavera.
    d. Software may not be much useful during planning phase of projects.
    e. Project monitoring is not possible using MS Project.

f. 'What if' analysis cannot be performed using Primavera.

g. MS Project does not have budgeting and cost control functions of project management.

h. MS Project uses 'baseline plan' terminology.

i. Primavera does not provide facility for PDM network but MS Project does.

j. Primavera has more features compared to MS Project.

k. MS Project is more user-friendly compared to Primavera.

2. Activity network exercise: In a project there are nine activities (A, B, C, D, E, F, G, H and I). In the accompanying table, the duration to complete these activities, their dependence (predecessor), and the relationship among the activities are given. Also given is the total float of some activities.

   a. Sketch the activity network and calculate ES and EF for every activity.

   b. Identify the critical path.

   c. Calculate the LS and LF for every activity.

   d. Complete TF and FF for every activity.

   e. How did you find the duration of activity I?

   f. Can you shorten the project by shortening activity D? Explain.

   g. What will be the project duration if activity D takes 18 weeks to complete instead of 14 weeks?

   h. Will the critical path change?

Activity code	Duration (weeks)	Predecessor	Type of relationship	Total float
A	7	–		
B	9	–		
C	12	–		
D	14	A	FF = 7	
		B	FS = 2	
		C	SS = 8	
E	11	D	FS = 0	
F	17	D	SS = 11	
G	13	E	FS = 0	0
H	6	E	FS = 3	
		F	FS = 0	
I	?	F	SS = 12	0

COMPUTER APPLICATIONS IN SCHEDULING, RESOURCE LEVELLING, MONITORING, AND REPORTING

CASE STUDIES

Case 1: Office Construction - Computer Application in Network Preparation and Resource Levelling

Description

You have to construct a building to set up your office. You own a plot of 100 sq yard (30 ft × 30 ft) in a prime location. For the time being, you are in a position to construct ground floor alone due to financial constraints. So, you have decided to construct a simple RC structure. It consists of individual footing for 16 columns (columns are spaced every 10 ft in both directions) and a plinth beam at about 1 m from the existing road level. All the columns are going up to roof level and you plan to have an RC roof slab and beam. In-between space is constructed by using locally available bricks. There is earth filling up to plinth level and after proper compaction of filled-up earth, flooring is laid. Other activities include plastering, painting, doors and windows, plumbing and electrical work. It is estimated that the entire building will be completed in about three months. For your ready reference, the estimated duration for individual activities and their dependence as well as resources consumed by them are listed in the following table.

Table Activity details

Act. no.	Activity description	Duration (days)	Depends on	Resources
1	Site clearance	2		2 USK
2	Excavation in foundations	6	1	6 USK
3	Foundations including plinth beam	10	2	6 USK, 2 SK
4	Column casting up to roof level	10	3	4 USK, 3 SK
5	Slab and beam of roof	21	4	6 USK, 3 SK
6	Earth filling and flooring	10	5FS + 10d	4 USK, 1 SK
7	Brickwork	10	5FS + 7d	4 USK, 2 SK
8	Doors/windows	3	7FF	2 USK, 1 SK
9	Plastering	10	7SS + 7d	2 USK, 1 SK
10	Painting	7	9SS + 7d	1 USK, 1 SK
11	Plumbing	7	7FF + 3d	1 USK, 1 SK
12	Electrical works	7	11SS	1 USK, 1 SK
13	Handover	1	10, 11, 12	1 USK, 1 SK

In the above table, USK represents unskilled worker and SK represents skilled worker.
FS represents 'finish to start' relationship.
SS represents 'start to start' relationship.
FF represents 'finish to finish' relationship.
5FS + 10d means the activity in question can start 10 days after activity no. 5 has been finished.
You may assume any other missing data.
Tasks

1. Draw the network for the above project.
2. Identify the critical activities.
3. Draw the resource histogram.
4. Level the resources when you have no limitation on their availability.

Case 2: Cross-Country Pipeline Project - Computer Application in Network Preparation

Description

In a cross-country pipeline project, a 600 mm diameter line having a thickness of 6 mm is to be laid over a distance of 100 km. Pipes are available in 5 m lengths, and welding is to be carried out at all joints. It has been found that in 50 km of the length referred to as ZONE A, the pipes are laid 2 m below ground level on a 100 mm thick PCC bed. In the remaining portion referred to as ZONE B, the pipeline is supported at 5 m c/c on precast RCC supports, resting on 100 mm PCC levelling course. The project is required to be completed within three months.

In the project, assume the following:

1. Precast supports are available through a supplier @ ₹ 3,000 per piece (including transportation to place of installation)
2. Cost of PCC is ₹ 1,500/m³
3. Cost of pipe is ₹ 60/kg
4. There is no bottleneck in procurement of materials, but there is a 10-day lead time for material to reach site (after placing the order)
5. The depth of excavation for buried pipeline length is 2 m (top of PCC) from ground level
6. Wage of a welder is ₹ 200/day of 8 hours
7. In case of overtime work, the rate of payment is double (that of normal working hours)

Tasks

1. List the activities, their estimated quantities, precedence table, manpower schedule, and total estimated cost of the project.
2. Draw the network for this project.

Following are the possible activities for this project:

Act. no.	Activity	Qty	Remarks
1	Survey and layout	100 km	$50,000 \times 1.5 \times 2.1$
2	Excavation of trench in Zone A	157,500 m³	$50,000 \times 1.5 \times 2.1$
3	Laying of PCC in trench in Zone A	7,500 m³	$50,000 \times 1.5 \times 0.1$
4	Laying of pipes in position in Zone A	50 km	
5	Welding of joints in Zone A	18,840 m	$(50,000/5) \times 3.14 \times 0.6$
6	Excavation of footing in Zone B	3,240 m³	$(50,000/5) \times 0.9 \times 0.6 \times 0.6$
7	Laying of PCC in footing in Zone B	360 m³	$(50,000/5) \times 0.6 \times 0.6 \times 0.1$
8	Laying of RCC footing in Zone B	10,000 nos	
9	Erecting precast column in Zone B	10,000 nos	
10	Erecting pipeline in Zone B	50 km	
11	Welding of joints in Zone B	18,840 m	$(50,000/5) \times 3.14 \times 0.6$
12	Testing and commissioning	100 km	
13	Backfilling in Zone A	157,500 m³	
14	Backfilling in Zone B	3,240 m³	

19
Factors Behind the Success of a Construction Project

General, project performance measurement, criteria for project performance evaluation, project performance attributes, effect of other elements on project performance, the theory of 3Cs and the iron triangle

19.1 GENERAL

In any construction project, a number of scarce resources are at stake. If a project is executed well—under the stipulated time and cost, and with the desired quality—it gives immense satisfaction to the participants. The project is then termed successful. Researchers from the area of project management have tried to find the attributes or factors that make a project either a success or a failure. The inherent objectives behind these researches have been to emulate the success factors and eliminate the factors responsible for failure. A number of studies have been carried out since the 1960s to identify the performance-affecting factors of projects in different countries, and the findings are recorded in several international literatures. Besides, the professionals have also shared their experiences in these literatures.

The term 'success' itself has undergone a sea change in the complex project environment with so many stakeholders. Success for one participant may be failure for another participant. The Denver airport project in the USA reveals that what is viewed as failure today may be treated as success in future (Griffith et al. 1999). Besides, the construction projects today are no longer confined to a single discipline but are generally multidisciplinary. Modern projects involve multiple players such as designers, contractors, subcontractors, construction managers, consultants and specialists from different disciplines. In a multi-agency environment, it is natural to have a clash of objectives among different participants. The objective of project management is to ensure success of the project, which is not just managing the schedule, cost and quality, generally known as 'the iron triangle'. Apart from 'the iron triangle', a number of performance-measuring parameters/criteria are cited to call a project successful, such as satisfaction of project participants, technical performance of the project, and number of disputes at the completion of project. Thus, the measurement of performance also depends to a great extent on the criteria employed to measure it (PMBOK 2000).

19.2 PROJECT PERFORMANCE MEASUREMENT

In every walk of life, we need to measure performance either in order to draw valuable conclusions about the individual or to compare the competing individuals.

Measuring the performance of a project in terms of success or failure is, in fact, a difficult task. This is because success/failure has different meanings for different participants. Further, defining the success of a project has always been full of ambiguity and there is no universal definition of success. McCoy (1986) observes that a standardized definition of project success does not exist; nor is there an accepted methodology to measure it.

Success is viewed from the different perspectives of individuals and the goals related to a variety of elements including technical, financial, educational, social and professional issues. Each industry, project team, or individual has a definition of success. Failures and successes are relative terms and they are highly subjective (Parfitt and Sanvido 1993).

The definition of success or failure can even change from project to project. As stated, success for one participant may be failure for another (Iyer and Jha 2004b, de Wit 1988). When we say this, consider a situation in which the participants are an architect, an engineer, an accountant, a human resources manager, and a chief executive officer. Observe the disparity in points of view of these participants.

The architect may consider success in terms of aesthetic appearance; the engineer in terms of technical competence; the accountant in terms of rupees spent under budget; the human resources manager in terms of employee satisfaction; and the chief executive officer in terms of the stock market (Freeman and Beale 1992). For the same project, notice the different yardsticks employed by different participants to measure the success.

The perception of success or failure is also time-dependent. The aforementioned Denver airport project reveals that what was viewed as a failure during construction phase is now treated as a success due to high inflow of revenue and improved lifestyle of local inhabitants (Griffith et al. 1999). The development phases of most North Sea oil projects suffered from budget and schedule overruns, but the projects were saved from disaster by the substantial increases in the price of oil in 1973 and 1979. The production phase also did not live up to expectation and on an average the production rates had been lower and the reserve estimates reduced, but again, this was compensated by the price increases. The slump in oil prices in 1986 again changed success into failure, at least temporarily, indicating how a project appearing to be a high failure or a success can turn out to be, respectively, a success or a failure with time (de Wit 1988).

In spite of the difficulties involved, some researchers have tried to define the success of a project. Parfitt and Sanvido (1993) quote the definition of overall success of a project given by de Wit, which is as follows:

The project is considered an overall success if the project meets the technical performance specifications and/or the mission to be performed, and if there is a high level of satisfaction concerning the project outcome among: key people in the parent organization, key people in the project team, and key users or clientele of the project effort.

Traditionally, success is defined as the degree to which project goals and expectations are met and the project requirements are fulfilled. However, modern projects involving multiple designers, contractors, subcontractors, construction managers, consultants and specialists from different disciplines, and increasing domain of project requirements have compounded the problem further, and understanding the success of a project has become all the more complicated.

From the above discussion, we can realize the difficulty in measuring the performance of a project in terms of success and failure. Nonetheless, there are certain criteria that are in vogue for measuring the performance of a project and are described below.

19.3 CRITERIA FOR PROJECT PERFORMANCE EVALUATION

Criteria are the set of principles or standards by which judgement is made (Lim and Mohamed 1999) and are considered to be the rule of the game. Traditionally, project performance is evaluated using schedule, cost and quality performances, also known as the 'iron triangle' (Atkinson 1999). Subsequently, different researchers have proposed different sets of success evaluation criteria in addition to the iron triangle. You can make an analogy of these criteria with the performance measurement of a student in a particular course—marks obtained, attendance, discipline, and so on. Some of the criteria used by Baker et al. (1983), Ashley et al. (1987), Freeman and Beale (1992), Maloney (1990), Norris (1990), Parfitt and Sanvido (1993), Songer and Molenar (1997), and Lipovetsky et al. (1997) for evaluating the performance of a project are given in the following paragraphs.

Budget or Cost Performance

The project if completed at or under the contracted cost is characterized as a successful project. The cost-success criterion could be measured in terms of cost over/under-run as a percentage of initial budgets.

Schedule or Time Performance

The project if completed on or before the contractual finish time is a successful project. The time-success criterion could be measured in terms of over/under-runs as a percentage of the initial plan.

Quality Performance—Whether Specifications Have Been Met

If the completed project meets or exceeds the accepted standards of workmanship in all areas and conforms to user's expectations, it is regarded as a successful project. In other words, 'the project must produce what it said it would produce' (PMI 1996). Quality typically includes such measures as the amount of rework required.

Safety

If the project honours health and safety rights of the people involved with the project by ensuring safe working conditions, it is regarded as a successful project. According to Crane et al. (1999), safety performance can be measured by compiling safety statistics such as lost-time incidents.

Dispute

If the project is completed with the least number of litigations resulting from disagreements among participants, it indicates that the project is a successful one. Dispute could be measured in terms of either the number of disputes or the monetary value involved in the dispute, or through qualitative measures such as whether there are minor or major disputes, and so on.

Stakeholder Satisfaction

If the completed project meets or exceeds the stakeholders' goals, it is termed as a successful project. The key stakeholders include architect, client, contractor, engineer, project manager and subcontractor/vendor. The project is also considered successful by employees if it has enabled personal growth for the employees of contractor and owner organizations.

Other Criteria

Some other criteria used to measure the performance of a project are:

- Whether the performance of constructed facility is as per the expectation and specifications, and whether the design goals have been met
- Whether the project has met the functionality requirement as expected
- Whether any technical innovation has been achieved during the implementation of project
- Whether the efficiency of project execution was as per requirement
- Whether managerial and organizational expectations have been met during the project implementation
- Whether the financial performance/profitability is as per the expectations of the stakeholders
- Whether the project has made any substantial and positive impact on the customer
- Whether the project implementation has influenced the involved organization's business favourably and whether it has helped the organization better prepare for the future
- Whether the project has benefited the customer, and whether it has led to favourable influence on the national infrastructure

A close look at the success criteria suggests that these can be kept under two broad categories—objective and subjective. Making an analogy with the student example, marks obtained in a course can be objectively defined, while discipline and other criteria can be taken as subjective. In the case of project evaluation, the objective criteria are those that are tangible and measurable, such as schedule, cost, quality, safety and dispute, while the subjective or intangible criteria may include client satisfaction, contractor satisfaction and project management team satisfaction.

Based on the above discussion, it can be concluded that distinguishing a project along broad terms such as success or failure is always going to be contradictory. Similarly, there is hardly any coherence in the opinion on how to measure the performance of a project—i.e., what set of criteria to employ.

Construction projects are vital for the growth of a nation and so is the need to make all-out efforts in ensuring the successful outcome of a project. In the next section, we discuss some of the attributes that are considered key to ensure the success of a project.

19.4 PROJECT PERFORMANCE ATTRIBUTES

Project attributes are the variables that influence the outcome of a project. The attributes can be people (project participants and their traits), resources, technology, working environment and system, or tasks. 'Project success is repeatable and it is possible to find certain success attributes,' has been the genesis of many research works in this area (Ashley et al. 1987). Also, there are certain attributes termed as failure attributes, which when present lead to failure of the project. Finding the success attributes and maximizing them is as important as finding the failure attributes and minimizing them. Accordingly, researchers have put their energy into identifying success attributes and failure attributes, with the common objective of enhancing the chances of project success. The success and failure attributes are discussed separately in the next two sections.

19.4.1 Success Attributes/Factors

Some of the success attributes identified from literatures (Sayles and Chandler 1971, Martin 1976, Baker et al. 1983, Cleland and King 1983, Locke 1984, Morris and Hough 1987, Schultz et al. 1987, Chan et al. 2001, Ashley et al. 1987, Chan et al. 2001a, Mansfield and Odeh 1991, McNeil and Hartley 1986, Thompson 1991) are listed in Box 19.1. They are classified under some major heads purely for ease in reading. The attributes are also referred to as critical success factors (CSF). CSFs are those key areas of activity in which favourable results are absolutely necessary for a particular manager to reach his or her own goals ... those limited number of areas where 'things must go right' (Rockart 1982).

19.4.2 Failure Attributes

Avots (1969) concludes that *choice of wrong project manager, unplanned project termination* and *unsupportive top management* are the main reasons for project failure. Hughes (1986) in another study identifies that projects fail because of improper basic managerial principles such as *improper focus of the management system, rewarding the wrong actions and lack of communication of goals*. Chitkara (1998) points out *inadequate project formulation and improper management of projects* as the primary reasons for project failures. According to Bonnal et al. (2002), the skill sets needed to manage the pre-project phase and the project phase are quite different. Inability to recognize this may lead to wrong selection of personnel for the pre-project and project phases, and ultimately lead to the failure of a project. Researchers have identified the factors responsible for causing schedule overrun, cost overrun and so on, in different parts of the world. These factors have been taken from the studies conducted by Mansfield et al. (1994), Chan and Kumaraswamy (1997), Dumont et al. (1997), and Jha (2004), and are summarized thus:

- Poor contract management—this may be due to lack of adequate experience and training at the senior management level
- Inadequate technical manpower
- Very low level of productivity
- Inadequate finances for short- and long-term purposes, and absence of specialization
- Unforeseen ground conditions and changes in site conditions
- Shortage of materials and plant items
- Design changes
- Price fluctuations
- Inaccurate estimates prepared by contractor

Box 19.1

Success Attributes/Factors

Project formulation-related factors
- Properly defined goals, and organizational philosophy
- Accurate initial cost estimates
- The clarity of understanding of project goals by team members
- Participation of key stake holders in developing the project plan
- Clarity of scope and work definition
- Proper risk and liability assessment
- Clear identification of end-users needs, and constraints imposed by end users
- Preparation of project plans, and determination of deliverables and important milestones

Project management related factors
- Scheduling
- Planning and review
- Control systems and responsibilities
- Control and information mechanics
- Planning and control techniques

Top management related factors
- Organize and delegate authority
- Select project team
- Absence of bureaucracy
- Top management support
- Logistic requirements
- Financial support
- Facility support
- Executive development and training
- Appoint competent project manager
- Make project commitments unknown

Project manager-related factors
- Onsite project manager
- Project manager's previous experience has minimal impact on the project's performance and size of the previously managed project does affect the manager's performance
- Technical knowledge of project manager
- Competent project manager
- Characteristics of project manager

Resource related factors
- Allocate sufficient resources
- Better management of human resources on construction projects by making use of the motivation factor
- Adequate funding to completion
- Manpower and organization
- Adequate Project team capability

Commitment of project team members
- Continuing involvement in the project
- Goal commitment of project team
- Project team motivation
- Project team commitment

Client-related factors
- Involvement of client
- Competence of client
- Client consultation and personnel selection and training

Monitoring and feedback-related factors
- Acquisition, information and communication channels
- Project review through regular progress meetings
- Set up control mechanisms

Stakeholders related
- Contractor's competencies

Communication

Coordination

Other miscellaneous factors
- Power and politics within the organization
- Environmental events
- Urgency of the project
- Politics
- Community involvement

- Too many change orders (variations of works)
- Fraudulent practices and kickbacks
- Poor site management and supervision
- Slow speed of decision-making involving all project teams
- Poor scope definition

19.5 EFFECT OF OTHER ELEMENTS ON PROJECT PERFORMANCE

Another area in literature explores the effect of certain attributes/factors on the project success. McNeil and Hartley (1986) emphasise the role of project planning on project performance and suggest that skilled people need to be intimately involved in the planning process. Lim and Ling (2002) find the client's role to be an important ingredient in achieving project success, and Chan et al. (2001a) assert inter-organizational teamwork as a major factor in ensuring project success.

Bower et al. (2002) through case studies find that *incentives* play an important role in construction contracts, though these alone cannot ensure project success. The incentives may be in the form of cost, performance, and schedule and delivery. They further suggest that incentives if applied correctly can focus both the client and the contractor on the appropriate business objectives that will lead to successful project results.

Dvir et al. (2003) examine the relationship between project planning efforts and project success. The planning efforts were measured in terms of 'requirements' definition, development of technical specifications, and project management processes and procedures. These findings suggest that project success is insensitive to the level of implementation of management processes and procedures. On the other hand, the project success is positively correlated with the investment in 'requirements' definition and the development of technical specifications. The criteria for success measurement were meeting planning goals, end-user benefits and contractor benefits.

Thomas et al. (1998) establish a positive and quantitative link between communications effectiveness and project success. Their analysis has revealed that 41 per cent of the variation in perceptions of success can be attributable to variation in communications effectiveness.

Study by Sadeh et al. (2000) has revealed that the contract type has an effect on the success of defence projects. Based on data collected on 110 defence development projects implemented in Israel, they have found that when technological uncertainty at the start of projects is high, cost-plus contracts result in better performance. Fixed-price contracts are better suited for projects with lower levels of technological uncertainty. A combination of cost-plus and fixed-price contracts is suggested for use in projects suffering from high technological uncertainty, starting with a cost-plus contract for the early stages of the project and switching to a fixed-price contract when the uncertainty is reduced.

The results of a study of 280 construction projects indicate that partnered projects achieve superior results in controlling costs, in technical performance, and in satisfying customers, compared with those projects managed in an adversarial, guarded adversarial, and even informal partnering manner (Larson 1995).

19.6 THE THEORY OF 3CS AND THE IRON TRIANGLE

Jha (2004) conducted a study to (a) identify critical factors responsible for success or failure of projects, and (b) evaluate the relative impact of critical success factors on the performance of construction projects. The performance of the project was measured on the following four performance criteria in the study—(1) adherence to schedule, (2) adherence to cost, (3) adherence to quality performance, and (4) 'no dispute'.

A two-stage questionnaire-survey approach was considered appropriate to address the objectives of the study. A total of 114 responses were received for the first-stage questionnaire and 90 responses for the second stage, out of 450 and 300 questionnaires mailed, respectively. While the first stage of the questionnaire helped in achieving the first objective (a), the second stage questionnaire helped in addressing the second objective (b).

A total of 55 attributes affecting the project performance objectives of schedule, cost, quality and no-dispute compliances were identified through case studies, project management literatures and personal interviews of experts in construction industry. Responses on the extent of effects of these attributes on the four performance criteria were sought on a five-point ordinal scale. For example, in the scale for schedule performance criteria, 1 refers to 'adverse delay', 2 to 'significant delay', 3 to 'marginal delay', 4 to 'no effect', and 5 to 'helps speeding up progress'. From the analyses of responses, the mean values (μ) for each of the attributes were calculated for all the four performance criteria. These attributes were then segregated in three groups based on the mean values. The first group of attributes ($\mu \geq 4.5$) was considered to contribute positively in achieving the stated performance objectives; the second group of attributes ($3.5 < \mu < 4.5$) was considered neutral, causing neither positive effect nor negative effect in achieving the performance objectives; and the third group of attributes ($\mu \leq 3.5$) affected adversely. They were accordingly referred to as success, neutral and failure attributes, respectively. The number of attributes appearing in the first group was 31; in the second group, it was 2; and the remaining 22 attributes were in the third group. Only two groups of attributes, success and failure attributes, were taken for further study.

Depending on the performance-measuring criteria, the analysis resulted in different sets of success attributes ($\mu \geq 4.5$). For example, when schedule is the criterion, a total of 31 attributes have emerged in this group; and when cost is the criterion, a total of 30 attributes have emerged. Similarly, for quality performance a total of 29 attributes, and for no-dispute performance a total of 28 attributes have emerged. Mean values of all the success attributes and their rank orders under the four performance criteria are summarized criterion-wise in Table 19.1.

The rank orders of the success attributes in different evaluation criteria suggest that 'positive attitude of PM and project participants' is the most important attribute when schedule, quality and no-dispute criteria are of prime importance in gauging the project performance, while 'coordinating ability and rapport of PM with top management' takes supreme importance when cost criterion is considered. Some of the top-ranking success attributes across the four performance criteria are observed to be—positive attitude of PM and project participants; proper rapport and coordination among PM, top management, owner's representatives, team members and sub-contractor; selection of PM with proven track record at an early stage by top management; effective monitoring and feedback by the project team members and the PM himself; and leadership quality and technical ability of the PM. The top five success attributes in the four project performance criteria are reproduced in Figure 19.1 for ready reference.

The attributes falling under group 3 (for $\mu \leq 3.5$) in the four performance criteria are given in Table 19.2.

Table 19.1 Rank of success attributes (µ ≥ 4.5) based on performance criteria

Act no.	Coordination activities	Schedule Mean	Schedule Rank	Cost Mean	Cost Rank	Quality Mean	Quality Rank	No-dispute Mean	No-dispute Rank
1	Positive attitude of project manager (PM) and project participants	4.84	1	4.74	4	4.75	1	4.81	1
2	Selection of PM with proven track record at an early stage by top management	4.82	2	4.63	8	4.56	14	4.68	7
3	Effective monitoring and feedback by the project team members	4.82	3	4.78	2	4.73	5	4.67	8
4	Authority to take day-to-day decisions by the PM's team at site	4.81	4	4.61	10	4.69	6	4.59	17
5	Coordinating ability and rapport of PM with his team members and subcontractor	4.80	5	4.59	15	4.54	18	4.68	6
6	Project manager's authority to take financial decision, selecting key team members, etc.	4.75	6	4.54	16	4.58	12	4.60	16
7	Leadership quality of PM	4.75	7	4.59	13	4.74	4	4.63	14
8	Understanding of responsibilities by various project participants	4.74	8	4.59	12	4.64	8	4.65	10
9	Coordinating ability and rapport of PM with other contractors at site	4.74	9	4.52	18	4.51	19	4.68	5
10	Top management's backing up the plans and identifying critical activities	4.73	10	4.50	20	4.44	23	4.56	20
11	Commitment of all parties to the project	4.73	11	4.54	17	4.59	11	4.64	11
12	Top management's enthusiastic support to the project manager and project team at site	4.72	12	4.59	14	4.60	9	4.62	15
13	Effective monitoring and feedback by PM	4.70	13	4.75	3	4.75	2	4.70	3
14	Coordinating ability and rapport of PM with top management	4.70	14	4.78	1	4.69	7	4.75	2
15	Understanding operational difficulties by the owner engineer, thereby taking appropriate decisions	4.70	15	4.68	6	4.50	20	4.59	18
16	Project manager's technical capability	4.69	16	4.70	5	4.74	3	4.58	19
17	Coordinating ability and rapport of PM with owner representatives	4.69	17	4.51	19	4.47	22	4.69	4

(Continued)

Table 19.1 Continued

18	Monitoring and feedback by top management	4.68	18	4.61	9	4.55	16	4.52	23
19	Monitoring and feedback by client	4.63	19	4.33	29	4.40	24	4.34	28
20	Construction control meetings	4.62	20	4.42	23	4.58	13	4.67	9
21	Training the human resources in the skill demanded by the project	4.61	21	4.38	26	4.60	10	4.52	22
22	Delegating authority to project manager by top management	4.57	22	4.49	22	4.47	21	4.63	13
23	Availability of resources (funds, machinery, material) as planned throughout the project duration	4.50	25	4.35	28	4.55	15	4.54	21
24	Regular budget update	4.49	26	4.50	20	4.30	28	4.51	24
25	Favourable social environment	4.46	27	4.12	N/A	4.06	N/A	4.21	N/A
26	Scope and nature of work well-defined in the tender	4.46	28	4.60	11	4.40	25	4.40	26
27	Favourable political and economic environment	4.45	29	4.39	24	4.09	N/A	4.48	25
28	Ability to delegate authority to various members of his team by PM	4.35	30	4.38	25	4.29	29	4.26	N/A
29	Developing and maintaining a short and informal line of communication among project team	4.33	1	4.22	30	4.39	26	4.35	27

Schedule performance	Cost performance	Quality performance	Dispute performance
• Positive attitude of PM and project participants • Effective monitoring and feedback by the project team members • Selection of PM with proven track record at an early stage by top management • Authority to take day-to-day decisions by the PM's team at site • Coordinating ability and rapport of PM with his team members and subcontractor	• Effective monitoring and feedback by PM • Coordinating ability and rapport of PM with top management • Effective monitoring and feedback by the project team members • Positive attitude of PM and project participants • Project manager's technical capability	• Positive attitude of PM and project participants • Effective monitoring and feedback by PM • Project manager's technical capability • Leadership quality of PM • Effective monitoring and feedback by the project team members	• Positive attitude of PM and project participants • Coordinating ability and rapport of PM with top management • Effective monitoring and feedback by PM • Coordinating ability and rapport of PM with owner's representatives • Coordinating ability and rapport of PM with other contractors at site

Figure 19.1 Five most important success attributes

Table 19.2 Rank of failure attributes (μ ≤ 3.5) based on performance criteria

Act no.	Coordination activities[a]	Schedule Mean	Schedule Rank	Cost Mean	Cost Rank	Quality Mean	Quality Rank	No-dispute Mean	No-dispute Rank
1	Conflicts between PM and other outside agency such as owner, subcontractor, or other contractors	1.54	1	1.83	6	2.54	14	1.87	2
2	Poor human resource management and labour strike	1.54	1	1.58	1	2.01	2	2.01	6
3	Conflicts between PM and top management	1.55	3	1.76	4	2.16	6	2.06	8
4	Inadequate project formulation in the beginning	1.56	4	1.72	2	2.35	10	1.90	3
5	Negative attitude of PM and project participants	1.56	4	1.74	3	1.71	1	1.65	1
6	**Reluctance in timely decision by PM**	1.65	6	1.88	7	2.37	12	2.10	10
7	Vested interest of client representative in not getting the project completed in time	1.66	7	1.78	5	2.36	11	2.01	6
8	Conflicts among team members	1.77	8	2.00	11	2.12	5	2.12	11
9	Ignorance of appropriate planning tools and techniques by PM	1.81	9	2.00	11	2.21	8	2.23	13
10	**Hostile political and economic environment**	1.84	10	2.16	15	2.76	16	2.54	16
11	Tendency to pass on the blame to others	1.86	11	2.19	16	2.12	4	2.10	9
12	Lack of understanding of operating procedure by the PM	1.87	12	1.99	10	2.19	7	2.12	11
13	**Reluctance in timely decision by top management**	1.88	13	2.07	13	2.75	15	2.49	15
14	Mismatch in capabilities of client and architect	1.88	13	1.88	8	2.09	3	1.97	5
15	Holding key decisions in abeyance	1.91	15	1.98	9	2.30	9	1.94	4
16	**Harsh climatic condition at the site**	1.96	16	2.09	14	2.39	13	2.96	18
17	**Hostile social environment**	2.01	17	2.25	17	2.90	17	2.55	17
18	**Project completion date specified but not yet planned by the owner**	2.19	18	2.42	18	3.05	20	2.37	14
19	**Uniqueness of the project activities requiring high technical know-how**	2.75	19	2.77	19	3.29	21	3.18	21
20	**Size and value of the project being large**	3.32	20	3.32	21	3.58	22	3.00	20
21	**Aggressive competition at tender stage**	3.32	21	3.64	22	2.99	19	2.97	19
22	**Crisis-management skill of PM**	3.84	22	3.81	23	3.84	N/A	4.10	N/A
23	**Urgency emphasised by the owner while issuing tender**	3.85	N/A	2.91	20	2.91	18	3.31	22
24	The capability of project participants to market the end product to the intended users	4.00	N/A	4.03	N/A	4.05	N/A	4.09	N/A

[a] Activities shown in bold face have varied intensity of adverse effects on the four performance criteria.

Schedule performance	Cost performance	Quality performance	Dispute performance
• Poor human resource management and labour strike • Negative attitude of PM and project participants • Conflict between PM and other outside agency such as owner, subcontractor, or other contractors • Conflicts between PM and top management • Inadequate project formulation in the beginning	• Poor human resource management and labour strike • Negative attitude of PM and project participants • Inadequate project formulation in the beginning • Vested interest of client representative in not getting the project completed in time • Conflicts between PM and top management	• Negative attitude of PM and project participants • Poor human resource management and labour strike • Mismatch in capabilities of client and architect • Conflicts among team members • Tendency to pass on the blame to others	• Negative attitude of PM and project participants • Conflict between PM and other outside agency such as owner, subcontractor, or other contractors • Inadequate project formulation in the beginning • Holding key decisions in abeyance • Mismatch in capabilities of client and architect

Figure 19.2 Five most important failure attributes

It is seen that depending on the performance-measuring criterion, the analysis has resulted in different rank orders of failure attributes. Also, when schedule is the performance-measuring criterion, a total of 22 attributes have emerged in this group, and when cost is the criterion, a total of 23 attributes are found. Similarly, for quality and no-dispute performances a total of 22 failure attributes have emerged.

While 'conflicts between PM and other outside agency such as owner, subcontractor, or other contractors' has the most adverse effect on achieving the schedule performance, 'poor human resource management and labour strike' emerges as the root cause for cost escalation at project sites. 'Negative attitude of PM and project participants' is the prime reason for underperformance on quality account and increase in dispute. Some of the other top-ranking critical attributes having adverse performance on the four performance criteria are—inadequate project formulation in the beginning; conflicts between PM and top management; mismatch in capabilities of client and architect; tendency to pass on the blame to others; holding key decisions in abeyance; conflicts among team members; and vested interest of client representative in not getting the project completed in time. The importance of these attributes revealed in this study is consistent with the findings of other researchers. For easy reference, the top five attributes under each criterion are summarized in Figure 19.2.

In order to understand the success and failure attributes in a better way, and to reduce their numbers from the existing 55, success and failure groups of attributes were subjected to factor analyses separately.

The factor analyses results for the four performance criteria are presented in Figures 19.3a to 19.6b. Figures 19.3a and 19.3b illustrate the extent of variance explained by various factors towards project success and failure in schedule performance criterion. Similarly other pairs of figures like 19.4a and 19.4b, 19.5a and 19.5b and 19.6a and 19.6b illustrate variances explained by various factors in cost, quality and no-dispute performance criteria.

It can be observed that there are a number of common factors across all four performance criteria, while a few other factors emerged predominantly in only select performance criteria. Taking the union of all common and uncommon factors across the four performance criteria, there were 11 success factors and 9 failure factors. These factors are summarized in Table 19.3.

FACTORS BEHIND THE SUCCESS OF A CONSTRUCTION PROJECT | 667 |

(a) Success factors

- Owners competence (F_6): 5.36%
- Commitment of all project participants (F_5): 6.75%
- Favorable working condition (F_4): 8.72%
- Monitoring and feedback & Good coordination among project participants (F_3): 15.24%
- Top management support (F_2): 15.86%
- Project manager's competence (F_1): 27.95%

(b) Failure factors

- Aggressive competition during tender stage (F_{18}): 7.50%
- Harsh climatic condition at site (F_{17}): 7.50%
- Indecisiveness of project participants (F_{16}): 9.98%
- Owner's incompetence (F_{15}): 10.05%
- Hostile socio economic environment (F_{14}): 11.13%
- Project manager's ignorance and lack of knowledge (F_{13}): 11.34%
- Conflict among project participant (F_{12}): 15.34%

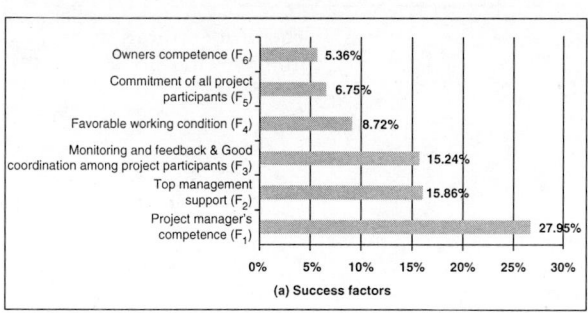

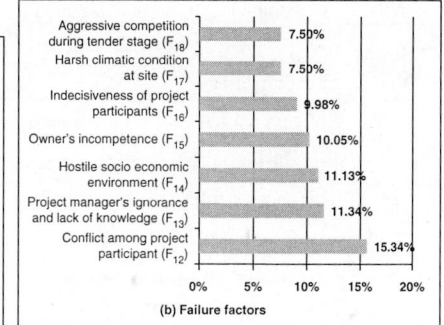

Figure 19.3 Factor profile for schedule criterion

(a) Success factors

- Favorable working condition & Owners competence (F_4 and F_6): 6.16%
- Commitment of all project participants (F_5): 8.13%
- Good coordination among project particapants (F_9): 8.26%
- Monitoring and feedback (F_3): 9.28%
- Top management support (F_2): 11.41%
- Project manager's competence (F_1): 32.43%

(b) Failure factors

- Aggressive competition during tender stage (F_{18}): 6.67%
- Indecisiveness of project participants (F_{16}): 8.17%
- Hostile socio economic environment & Harsh climatic condition at site (F_{14} and F_{17}): 8.25%
- Project manager's ignorance and lack of knowledge (F_{13}): 10.79%
- Faulty project conceptualization (F_{20}): 15.20%
- Conflict among project participant (F_{12}): 19.95%

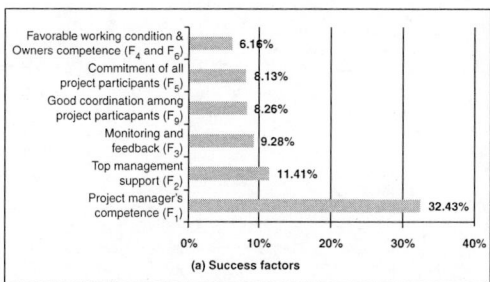

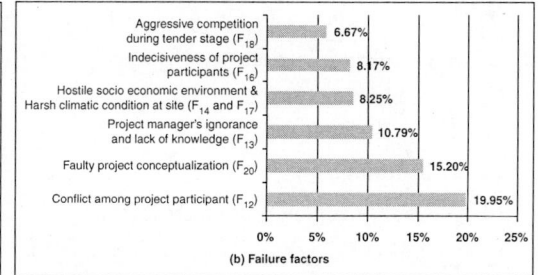

Figure 19.4 Factor profile for cost criterion

(a) Success factors

- Interaction between project participants-external (F_8): 4.56%
- Monitoring and feedback (F_3): 8.16%
- Owners competence (F_6): 9.32%
- Interaction between project participants-internal (F_7): 9.62%
- Top management support (F_2): 25.85%
- Project manager's competence (F_1): 19.52%

(b) Failure factors

- Aggressive competition during tender stage (F_{18}): 6.80%
- Project manager's ignorance and lack of knowledge (F_{13}): 11.66%
- Hostile socio economic environment & Harsh climatic condition at site (F_{14} and F_{17}): 12.57%
- Faulty project conceptualization (F_{20}): 15.76%
- Conflict among project participant (F_{12}): 23.53%

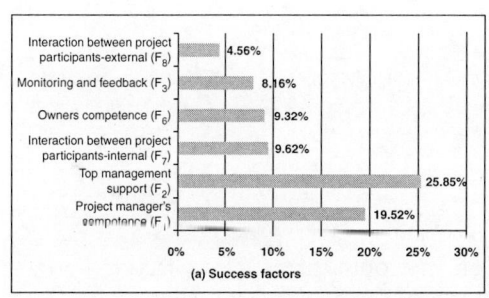

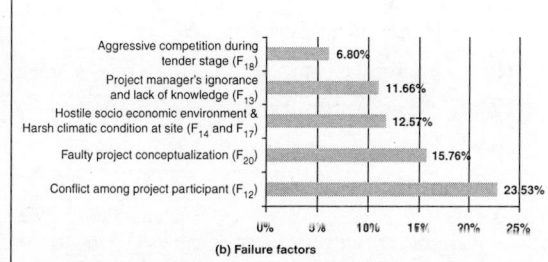

Figure 19.5 Factor profile for quality criterion

(a) Success factors

- Regular budget updtae (F_{11}): 3.86%
- Availability of trained resources (F_{10}): 6.80%
- Favorable working condition (F_4): 7.71%
- Owners competence (F_6): 9.23%
- Top management support (F_2): 20.10%
- Project manager's competence (F_1): 28.38%

(b) Failure factors

- Harsh climatic condition at site (F_{17}): 7.69%
- Faulty project conceptualization (F_{20}): 8.39%
- Indecisiveness of project participants & Nagetive attitude of project participants (F_{16}): 8.58%
- Hostile socio economic environment (F_{14}): 10.94%
- Project manager's ignorance and lack of knowledge (F_{13}): 14.90%
- 'Conflict among project participant (F_{12}): 17.54%

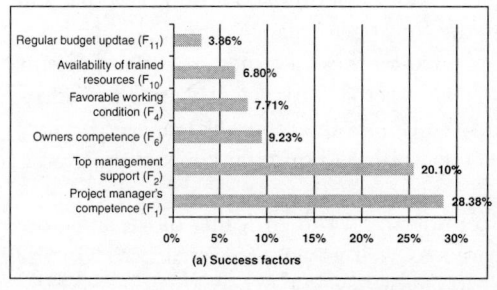

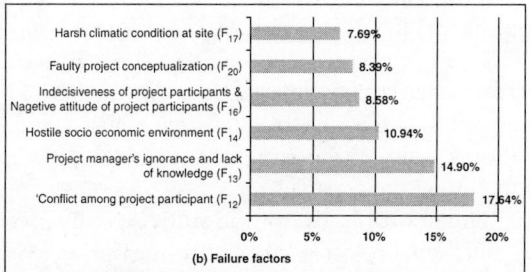

Figure 19.6 Factor profile for no-dispute criterion

Table 19.3 Pooled list of factors identified under various performance evaluation parameters

S. No.	Factor names	Factor identification number after pooling
	Success factors	
1	Project manager's competence	F_1
2	Top management support	F_2
3	Monitoring and feedback	F_3
4	Favourable working condition	F_4
5	Commitment of all project participants	F_5
6	Owner's competence	F_6
7	Interaction between project participants—internal	F_7
8	Interaction between project participants—external	F_8
9	Good coordination among project participants	F_9
10	Availability of trained resources	F_{10}
11	Regular budget update	F_{11}
	Failure factors	
12	Conflict among project participants	F_{12}
13	Project manager's ignorance and lack of knowledge	F_{13}
14	Hostile socio-economic environment	F_{14}
15	Owner's incompetence	F_{15}
16	Indecisiveness of project participants	F_{16}
17	Harsh climatic condition at site	F_{17}
18	Aggressive competition during tender stage	F_{18}
19	Negative attitude of project participants	F_{19}
20	Faulty project conceptualization	F_{20}

It can be observed from figures 19.3a to 19.6b that only three success factors—project manager's competence (F_1), top management support (F_2), and owner's competence (F_6) (Figures 19.3a, 19.4a, 19.5a and 19.6a)—are common across all the four performance criteria. From the failure factors in figures 19.3b, 19.4b, 19.5b and 19.6b, it can be observed that the four factors F_{12}, F_{13}, F_{14} and F_{17} are common across the four performance criteria.

It can be pointed out here that the variance explained by these factors across the different performance criteria differed. It is also established in statistics that variance explained by factors may not be a measure of intensity or importance of the factors in any performance criterion, but only indicates the grouping of variables in a given factor based on their concomitant variation among the variables.

The next objective of the study was thus set to understand the criticality of these factors on different project performance-measuring criterions as well as on an overall basis. In the second stage, questionnaire is developed using the 20 factors as explanatory variables and the contribution of these factors (variables) to actual performance of the choice project as response variable, and responses are sought as explained below. It may be pertinent here to tell the reader about the choice project.

Respondents were asked to select a project of their own choice, which they had executed or with which they were associated. This was the so-called choice project. Respondents were asked to judge the extent of contribution the identified success and failure factors made to the choice project in the context of performance. In order to understand the criticality of these factors on the project performance-measuring criterion as well as on an overall basis, the second-stage questionnaire survey was undertaken.

In the questionnaire, responses on the extent of contribution of the 20 factors towards the performance of the choice project are sought on an 11-point scale (−5 to +5 through 0, with −5 indicating high negative contribution, 0 being no effect, and +5 indicating high positive contribution). The performance rating of the choice project is obtained in a 10-point scale (1 to 10, with 1 being very poor performance and 10 being very good performance).

With the responses on performance rating of the choice project as values of response variable and the extent of contribution of various factors as values of explanatory variables, multiple regression is applied as given in Equation (19.1).

$$\text{Performance rating} = f(F_1, F_2, F_3, \ldots, F_{20}) \quad (19.1)$$

Since the response variable has been a discrete variable, the multinomial logistic regression was considered more appropriate.

The conclusions drawn from the multinomial logistic regression analyses are summarized below.

- Extent of contribution of various success or failure factors varies with current-level performance ratings of the project. However, six factors (monitoring and feedback; availability of trained resources; regular budget update; owner's incompetence; negative attitude of project participants; and faulty project conceptualization) have not been found to cause significant influence on the project outcome.

- None of the factors has been found to have significant influence on all four performance criteria. Among the 11 success factors, only two, top management support (F_2) and owner's competence (F_6), are found to significantly influence at least three performance criteria. While F_2 contribute in improvement in cost, quality and no-dispute performance criteria, F_6 contribute in schedule, quality and no-dispute performances. Similarly, among the failure factors, project manager's ignorance and lack of knowledge, and indecisiveness of project participants significantly influence three project performance criteria—schedule, cost and no dispute.

- When schedule compliance is the prime objective, seven factors (see Figure 19.7) are observed to have significant influence on the schedule outcome. Three factors—commitment of the project participants; owner's competence; and conflict among project participants—have been found to possess the capability to enhance performance level, while the remaining four factors—coordination among project participants; project manager's ignorance and lack of knowledge; hostile socio-economic environment; and indecisiveness of project participants—tend to retain the schedule performance at its existing level.

- Factors F_4, F_9 and F_{12}—respectively, coordination among project participants; favourable working condition; and conflict among project participants—are found to be important in

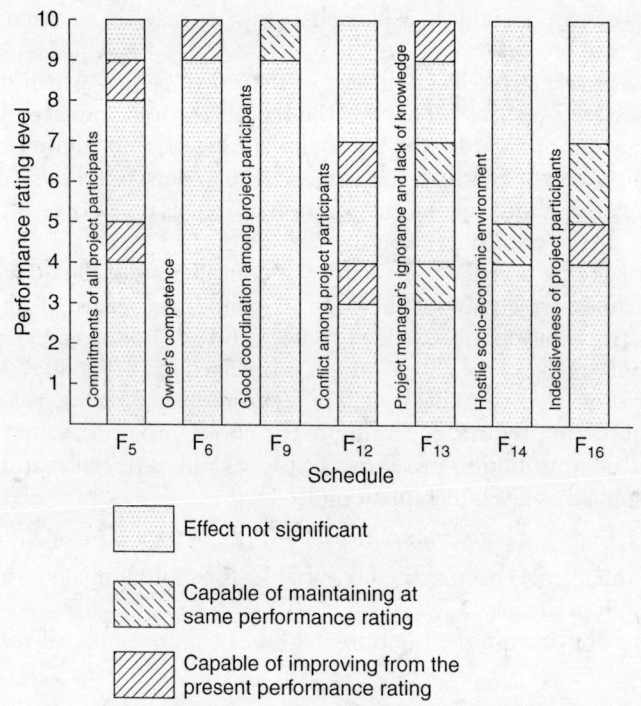

Figure 19.7 Project success and failure factors and their impact at different performance-rating levels in the schedule performance criteria

enhancing cost performance of the project (see Figure 19.8). On the other hand, important factors like top management support; commitment of project participants; project manager's ignorance; and indecisiveness of project participants tend to keep the cost performance of the project at the same level.

- While no failure factor has emerged to be significantly affecting the quality performance of the project (see Figure 19.9), five success factors have significant influence on the quality performance. Three factors—project manager's competence; top management support; and interaction between project participants (external)—contribute significantly in enhancing the project quality performance from its existing level, while the remaining two factors, owner's competence and interaction between project participants (internal), tend to retain the quality performance at the existing level itself. The emergence of project manager's competence and top management support as positive contributor to improving quality reestablishes the findings of quality gurus that management is more responsible to achieve the desired quality in any system.

- A total of six factors have emerged to be significant in the context of no-dispute criteria (see Figure 19.10). Out of these, the factors of top management support; favourable working condition; and owner's competence contribute significantly in avoiding disputes. However, these factors enhance the probability of avoiding dispute only when performance on no-dispute rating is of average nature. None of the success factors considered in the present study seems to have dispute-avoiding potential at either low or high dispute ratings.

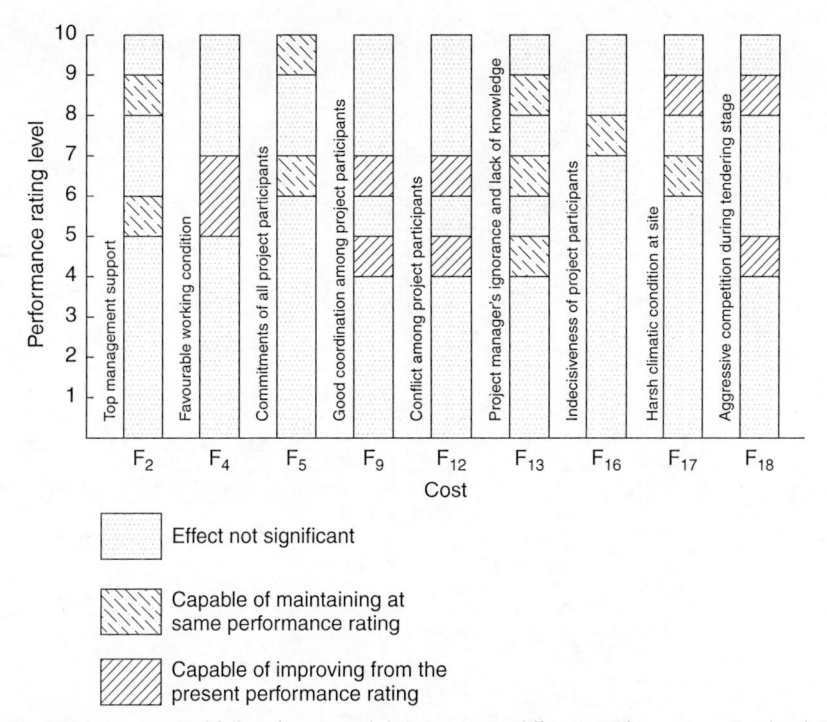

Figure 19.8 Project success and failure factors and their impact at different performance-rating levels in the cost performance criteria

To sum up, four factors—project manager's competence; commitment of project participants; owner's competence; and coordination among project participants—have been predominantly contributing towards enhancing the performance level across the three parameters of schedule, cost and quality, or at least sustaining these at the same level when the performance level is already high. In a practical situation, while project manager's competence is considered to be of utmost importance, hardly any importance is given to owner's competence. However, as the results reveal, owner's competence also contributes in improving the performance level. Hence, competency of one should be considered to be complementing the competency of the other to further enhance the performance level, instead of viewing them in isolation. Hence, a common name, 'competence', is suggested that includes competence of both project manager and owner, and thus, the three Cs—'competence', 'commitment' and 'coordination'—become the key factors for the success of the project. This is clearly evident from Figure 19.11. The Figure can prove to be a guideline for project professionals.

Among the three Cs, it is generally difficult to assign relative importance to one over another. As seen in the literature on total quality management, where commitment is given heavy importance, in project performance also, this factor can be treated as a driving force. Commitment will keep the team motivated towards successful completion of the project on time. However, this alone will not lead to fruitful results without proper direction as to how the work is to be achieved. Direction to carry out the work within the budgeted time, cost and quality can be obtained only through the competence of project officials at various levels. Competency enables the project participants to take appropriate decisions including

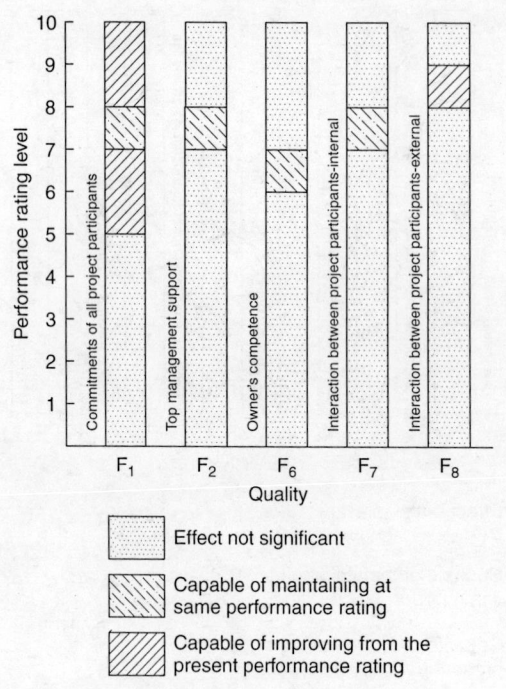

Figure 19.9 Project success and failure factors and their impact at different performance-rating levels in the quality performance criteria

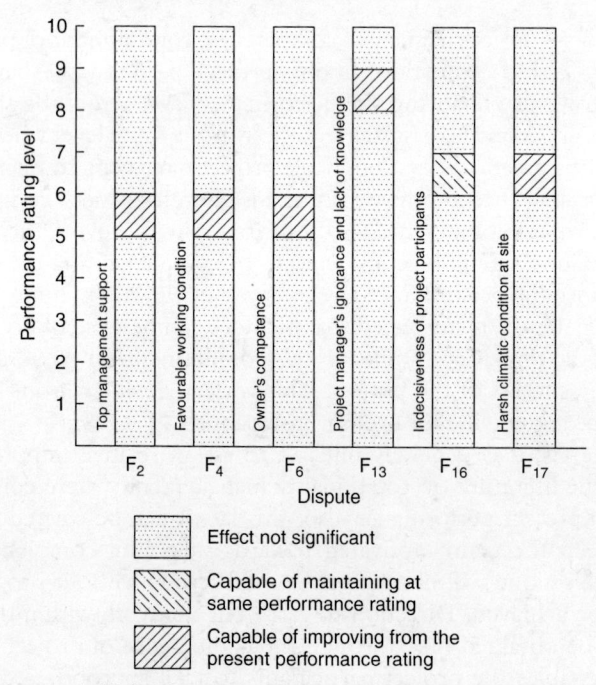

Figure 19.10 Project success and failure factors and their impact at different performance-rating levels in the dispute performance criteria

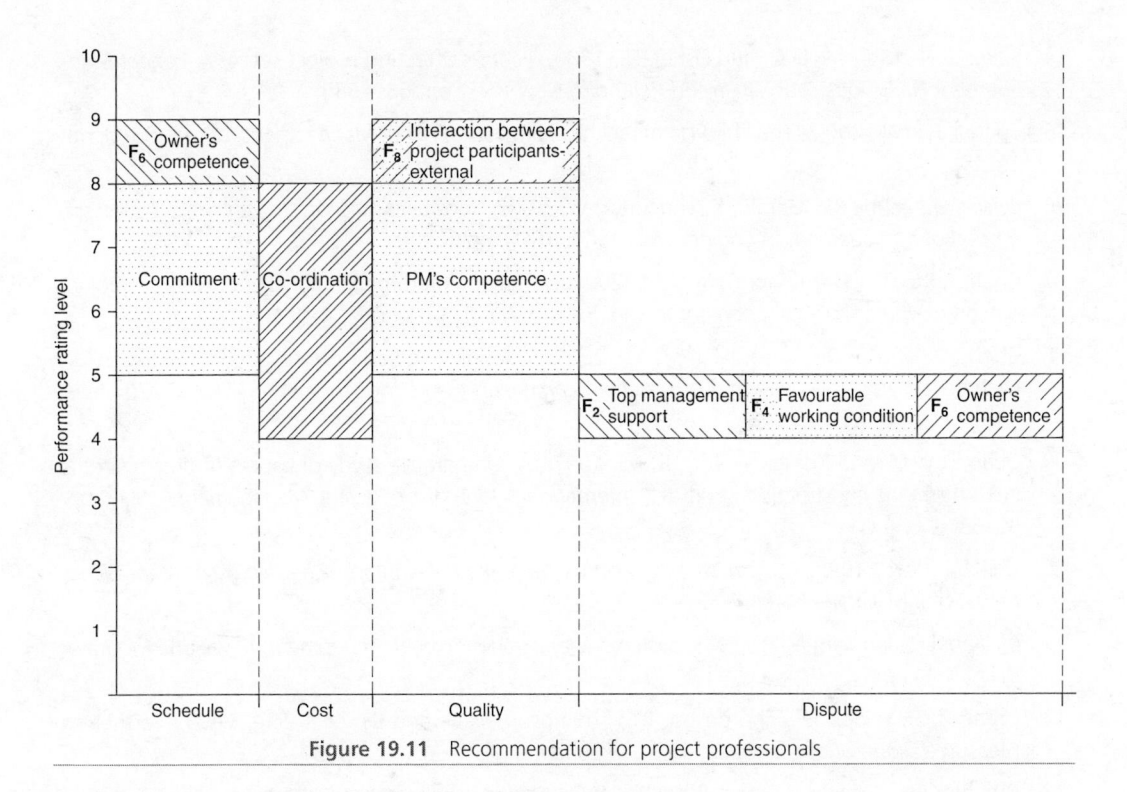

Figure 19.11 Recommendation for project professionals

corrective actions, if necessary, on time. Timely decision by the competent person keeps the project within the stipulated time and cost, and this gives a level of satisfaction to the project participants, which eventually boosts the motivational level and thereby also the commitment of project participants. Thus, competency becomes another enabler for the success of the project. In fact, the two factors of commitment and competency could suffice for any project of a single discipline and when it is not dependent on any external agency. However, most project activities are multidisciplinary in nature and with involvement of several other agencies, flow of resources or information among various participants on time is of utmost importance to keep the project objectives to a satisfactory level. Flow of resources or information among various participants is possible only with good coordination. Hence, all three factors—commitment, competency and coordination—qualify as the three pillars for a successful project.

REFERENCES

1. Ashley, D., Jaselskis, E. and Lurie, C.B., 1987, 'The determinants of construction project success', *Project Management Journal*, 18(2), pp. 69–79.

2. Atkinson, R., 1999, 'Project management: Cost, time and quality, two best guesses and a phenomenon, it's time to accept other success criteria', *International Journal of Project Management*, 17(6), pp. 337–342.

3. Avots, I., 1969, 'Why does project management fail?', *California Management Review*, 12(1), pp. 77–82.

4. Baker, B.N., Murphy, D.C. and Fisher, D., 1983, 'Factors affecting project success', *Project Management Handbook, Van Nostrand Reinhold*, New York, pp. 669–685.
5. Bonnal, P. et al., 2002, 'The life cycle of technical projects', *Project Management Journal*, 33(1), pp. 12–19.
6. Bower, D., Ashby, G., Gerald, K. and Smyk, W., 2002, 'Incentive mechanisms for project success', *Journal of Management in Engineering*, ASCE, 18(1), pp. 37–43.
7. Chan, A.P.C., Ho, D.C.K. and Tam, C.M., 2001a, 'Effect of interorganizational teamwork on project outcome', *Journal of Management in Engineering*, ASCE, 17(1), pp. 34–40.
8. Chan, A.P.C., Ho, D.C.K. and Tam, C.M., 2001b, 'Design and build project success factors: Multivariate analysis', *Journal of Construction Engineering and Management*, ASCE, 127(2), pp. 93–100.
9. Chan, D.W.M. and Kumaraswamy, M.M., 1997, 'A comparative study of causes of time overruns in Hong Kong construction projects', *International Journal of Project Management*, 15(1), pp. 55–63.
10. Chitkara, K.K., 1998, *Construction project management: Planning, scheduling and controlling*, New Delhi: McGraw-Hill.
11. Cleland, D.I. and King, W.R., 1983, *Systems Analysis and Project Management*, New York: McGraw-Hill.
12. Crane, T.G., Felder, J.P., Thompson, P.J., Thompson, M.G. and Sanders, S.R., 1999, 'Partnering Measures', *Journal of Management in Engineering*, ASCE, 15(2), pp. 37–42.
13. Dumont, P.R., Gibson, G.E. and Fish, J.R., 1997, 'Scope management using project definition rating index', *Journal of Management in Engineering*, ASCE, 13(5), pp. 54–60.
14. Dvir, D., Raz, T. and Shenhar, A.J., 2003, 'An empirical analysis of the relationship between project planning and project success', *International Journal of Project Management*, 21(2), pp. 89–95.
15. Freeman, M. and Beale, P., 1992, 'Measuring project success', *Project Management Journal*, 23(1), pp. 8–17.
16. Griffith, A.F., Gibson, G.E., Hamilton, M.R., Tortora, A.L. and Wilson, C.T., 1999, 'Project success index for capital facility construction projects', *Journal of Performance of Constructed Facilities*, ASCE, 13(1), pp. 39–45.
17. Hughes, M.W., 1986, 'Why projects fail: The effects of ignoring the obvious', *Industrial Engineering*, 18, pp. 14–18.
18. Jha, K.N. and Iyer, K.C., 2004b, 'Critical factors affecting schedule performance of Indian construction projects', *Proceedings of 4th International Conference on Construction Project Management (ICCPM 2004)*, Nanyang Technological University, Singapore, 4–5 March, pp. 71–80.
19. Jha, K.N., 2004, 'Factors for the success of a construction project: An empirical study', Doctoral thesis submitted at Indian Institute of Technology, Delhi.
20. Lim, C.S. and Mohamed, M.Z., 1999, 'Criteria for project success: An exploratory reexamination', *International Journal of Project Management*, 17(4), pp. 243–248.
21. Lim, E.H. and Ling, F.Y.Y., 2002, 'Model for predicting client's contribution to project success', *Engineering Construction and Architectural Management*, 9(5/6), pp. 388–395.

22. Lipovetsky, S., Tishler, A., Dvir, D. and Shenhar, A., 1997, 'The relative importance of project dimensions', *R&D Management*, 27(2), pp. 97–106.
23. Locke, D., 1976, *Project management*, St Martins Press, New York.
24. Maloney, W.F. and McFillen, J.M., 1995, 'Job characteristics: Union–nonunion differences', *Journal of Construction Engineering and Management*, ASCE, 121(1), pp. 43–54.
25. Mansfield, N.R. and Odeh, N.S., 1991, 'Issues affecting motivation on construction projects', *International Journal of Project Management*, 9(2), pp. 93–98.
26. Mansfield, N.R., Ugwu, O.O. and Doran, T., 1994, 'Causes of delay and cost overruns in Nigerian construction projects', *International Journal of Project Management*, 12(4), pp. 254–260.
27. Martin, C.C., 1976, *Project Management*, Amaco, New York.
28. McCoy, F.A., 1986, 'Measuring success: Establishing and maintaining a baseline', Project Management Institute Seminar/Symposium, Montreal, Canada, pp. 47–52.
29. McNeil, H.J. and Hartley, K.O., 1986, 'Project planning and performance', *Project Management Journal*, 17(1), pp. 36–44.
30. Morris, P.W. and Hough, G.H., 1987, *The Anatomy of Major Projects*, New York: John Wiley and Sons.
31. Norris, W.E., 1990, 'Margin of profit: Teamwork', *Journal of Management in Engineering*, ASCE, 6(1), pp. 20–28.
32. Parfitt, M.K. and Sanvido, V.E., 1993, 'Checklist of critical success factors for building projects', *Journal of Management in Engineering*, ASCE, 9(3), pp. 243–249.
33. Project Management Institute (PMI), 1996, *A guide to the project management body of knowledge*, Upper Darby.
34. Rockart, 1982, 'The changing role of the information systems executive: A critical success factors perspective', *Sloan Management Review*, Fall, pp. 3–13.
35. Rubin, I.M. and Seeling, W., 1967, 'Experience as a factor in the selection and performance of project managers', *IEEE Transactions on Engineering Management*, 14(3), pp. 131–134.
36. Sadeh, A., Dvir, D. and Shenhar, A., 2000, 'The role of contract type in the success of R&D defense projects under increasing uncertainty', *Project Management Journal*, 31(3), pp. 14–22.
37. Sayles, L.R. and Chandler, M.K., 1971, *Managing Large Systems*, New York: Harper and Row.
38. Schultz, R.L., Slevin, D.P. and Pinto, J.K., 1987, 'Strategy and tactics in a process model of project implementation', *Interfaces*, 17(3), pp. 34–46.
39. Songer, A.D. and Molenaar, K.R., 1997, 'Project characteristics for successful public–sector design–build', *Journal of Construction Engineering and Management*, ASCE, 123(1), pp. 34–40.
40. Thomas, S.R., Tucker, R.L. and Kelly, W.R., 1998, 'Critical communications variables', *Journal of Construction Engineering and Management*, ASCE, 124(1), pp. 58–66.
41. Thompson, P., 1991, 'The client role in project management', *International Journal of Project Management*, 9(2), pp. 90–92.

REVIEW QUESTIONS

1. State whether True or False:
   a. A project is considered successful when the project meets or exceeds the expectations of the stakeholders.
   b. A project is considered successful when the project sponsor announces the completion of the project.
   c. Iron triangle includes three parameters, namely schedule, cost and quality.
   d. The 3 Cs are commitment, coordination and competency.
   e. Failures and successes of a project are independent of each other.
   f. Iron triangle provides a platform for the evaluation of project performance.
   g. Safety and stakeholders' satisfaction are other two vital criteria for project performance evaluation apart from the iron triangle.
   h. Maximizing the success attributes is as important as minimizing the failure attributes.
   i. In the context of project evaluation, objective evaluation criteria are those that are intangible and non-measurable, such as client satisfaction and contractor satisfaction.
   j. In the context of project evaluation, subjective evaluation criteria are those that are tangible and measurable, such as schedule and cost.
   k. In general, a project fails because of competent project manager, planned project termination and supportive top management.
   l. The five process groups according to PMBOK are terminating, static, stagnating, leveraging and endless.
   m. Project scope management is one of the nine knowledge areas according to PMBOK.
2. What are the traditional parameters on which project performance is measured?
3. What is meant by 'iron triangle'? Discuss the different views held by researchers in the context of project performance measurement.
4. Can you suggest some methodology to measure project success?
5. What are the typical project success and failure attributes?
6. What are the major success attributes for ensuring timely completion of a project?
7. Name three success factors that are common across all the four project-performance parameters.
8. Name some failure factors that are common across all the four project-performance parameters.

20
Linear Programming in Construction Management

Introduction, linear programming, problems in construction, formulation, graphical solution, simplex method, dual problem, sensitivity analysis and their application to civil engineering

20.1 INTRODUCTION

Linear programming was developed by George Dantzig in year 1947 while working on research projects for the US Air Force. Early application of linear programming was mainly in military operations. Subsequently its application spread in solving a number of business problems including construction.

In this chapter, our attempt is to learn the fundamentals of linear programming in the context of construction. Here the term 'construction' has been defined as an act or a process of constructing. It consists of a series of actions to produce either a new set of buildings and infra-structure or may involve alterations in the existing buildings and infrastructure (Radosavljevic and Bennett 2012). A construction project is a part of construction work that is being attempted or undertaken. It is very clear that for any firm including construction, resources are never enough. This is where a manager has to play a role. He or she has to ensure that his objectives are achieved within the limited resources, and various other environmental and social constraints. Linear programming tries to address this problem mathematically by converting the objective and constraints in the form of linear equations and inequalities. It may be pointed here that linear programming assumes 'linear relationship' among various variables considered in a problem although in real life situation many times it would not be justifiable to assume a linear relationship. In addition, although a construction firm may have multiple objectives, we shall be treating single objective problems in this chapter. Readers should refer to advanced texts on 'operations research' to deal with multiple objective situations.

In this chapter, general formulation of a linear programming is explained. Subsequently, graphical and the simplex method is explained to solve these problems. The conversion of a primal problem into its dual is explained in the subsequent section. Further the process of conducting sensitivity analysis is explained with the help of an example.

20.2 LINEAR PROGRAMMING

Let us take example of a small time manufacturer with one multipurpose brick making machine which can run for eight hours and can produce six different bricks B_1, B_2, B_3, B_4, B_5, and B_6. The time consumed in making these bricks is 10, 15, 20, 5, 25, and 30 minutes respectively. The profits per unit from these bricks are ₹ 2, 2.5, 3, 1.5, 4, and 5. The brick manufacturer desires to maximize his profit. Assume that manufacturer can sell all the bricks produced irrespective of their types at the given per unit profit.

The solution can be understood with the help of Table 20.1.

It can be clearly seen from the Table 20.1 that it is possible to produce a maximum of 48, 32, 24, 96, 20, and 16 bricks of B_1, B_2, B_3, B_4, B_5, and B_6 respectively. The maximum possible profit from each of these bricks is ₹ 96, 80, 72, 144, 80, and 80 respectively. Thus we find that if the manufacturer concentrates on manufacturing B_4 bricks, he would obtain the maximum profit of ₹ 144.

The manufacturer would definitely be interested in knowing: (a) is this the maximum profit? (b) is there any other alternative production plan of bricks that can still maximize his profit?

The answer to such problems can be obtained by formulating and solving the above problem as a linear programming model.

Let's take another example in which a small time subcontractor is into manufacturing wooden windows and doors for government buildings. For every window and door that he sells, he earns ₹ 500 and ₹ 600 respectively. Each of the windows and doors are processed in two departments—carpentry and painting. In the carpentry department, the subcontractor has employed eight carpenters who work for 8 hours daily (thus total available time in this department = 8 × 8 = 64 hours) and in the painting department he has employed 8 full time painters and one part time painter. A full time painter works for eight hours daily while the part time painter works only for four hours a day (thus total available time in this department = 8 × 8 + 4 = 68 hours).

Fabricating one window requires 3 hours in the carpentry and 1 hour in painting department. On the other hand, one door requires 2 hours in the carpentry and 4 hours in painting department. The subcontractor has good contacts with the government officials and whatever quantity of windows and doors he produces are easily sold. The subcontractor desires to

Table 20.1 Brick manufacturing example

Brick type	Time consumed per brick (minute)	Profit obtained per unit (₹)	Maximum possible number of bricks	Maximum possible profit (₹)
B_1	10	2	480/10 = 48	48 × 2 = 96
B_2	15	2.5	480/15 = 32	32 × 2.5 = 80
B_3	20	3	480/20 = 24	24 × 3 = 72
B_4	5	1.5	480/5 = 96	96 × 1.5 = 144
B_5	24	4	480/24 = 20	20 × 4 = 80
B_6	30	5	480/30 = 16	16 × 5 = 80

Table 20.2 Window and door manufacturing example

Product (1)	Consumption per unit (Hours)		Maximum possible production		Profit/Unit (₹)	Total Profit (₹)
	Carpentry (2)	Painting (3)	64 hours (4)	68 hours (5)	(6)	(7)
Window	3	1	64/3 = 21.33	68/1 = 68	500	64/3 × 500 = 10,666.66
Door	2	4	64/2 = 32	68/4 = 17	600	17 × 600 = 10,200

maximize his profit under the given constraints. The solution can be understood with the help of Table 20.2.

For solving the above problem, it has been assumed that a window or a door is complete only when they have been processed in both the carpentry and painting departments. Thus for finding the maximum possible production, we have taken the minimum values out of the two columns (4) and (5) corresponding to window and door rows of Table 20.2. It may be found that the maximum profit corresponds to a situation where the manufacture makes 64/3 windows and the corresponding profit is ₹ 10,666.67.

As in first example, here also we are not sure whether the solution obtained above is the optimum one leading to the maximum profit for the manufacturer. We would also like to know if the profit can be increased for different production combinations of windows and doors.

Linear programming comes to rescue in solving such problems. For the brick manufacturing example, the objectives and constraints can be easily found out. The objective function here is to maximize and the constraint is the eight hours working of the brick making machine.

20.3 FORMULATION OF LINEAR PROGRAMMING PROBLEMS

In general, a linear programming model can be represented as shown below

Maximize or minimize

$$Z = \sum_{j=1}^{n} c_j \times x_j \tag{20.1}$$

Subject to

$$\sum_{j=1}^{n} a_{ij} \times x_j \leq b_i, \quad i = 1, 2, 3, \ldots, m$$

$$x_j \geq 0, \quad j = 1, 2, 3, \ldots, n$$

The generalized linear programming problem formulation is explained in Table 20.3 as well. In the model, m represents resources, n represents number of activities, x_j represents activity

Table 20.3 Generalized linear programming problem formulation

Resource	Resource consumption per unit activity				Available quantities of resources
1	a_{11}	a_{12}	...	a_{1n}	b_1
2	a_{21}	a_{22}	...	a_{2n}	b_2
...	...	...	...	...	...b_i
m	a_{m1}	a_{m2}	...	a_{mn}	b_m
per unit contribution	c_1	c_2	...	c_n	

and is also referred to as 'decision variables', c_j refers to per unit contribution also known as cost coefficient or profit coefficient as the case may be, and b_i represents resource availability. Z represents objective (goal) function which is an overall measure of performance.

For the brick manufacturer example, the various variable values are shown in Table 20.4.

Note that in this problem, the only resource is the multipurpose brick making machine which is performing six activities—manufacturing bricks of types B_1, B_2, B_3, B_4, B_5, and B_6. The values of per unit contribution c_j for different activities are given in the last row of Table 20.4. Let x_1, x_2, x_3, x_4, x_5 and x_6 represent the quantities of bricks of types B_1, B_2, B_3, B_4, B_5, and B_6 produced respectively.

The objective function then would be:

Maximize

$$Z = 2x_1 + 2.5x_2 + 3x_3 + 1.5x_4 + 4x_5 + 5x_6$$

Subject to:

$$10x_1 + 15x_2 + 20x_3 + 5x_4 + 24x_5 + 30x_6 \leq 480$$

$$x_1 \geq 0$$
$$x_2 \geq 0$$
$$x_3 \geq 0$$
$$x_4 \geq 0$$
$$x_5 \geq 0$$
$$x_6 \geq 0$$

For the window and door manufacturer problem, the various variable values are given in Table 20.5.

Note that in this problem, there are two resources—carpentry department and painting department. These two resources are involved in two activities—window and door manufacturing. The values of per unit contribution c_j for different activities are given in the

Table 20.4 Variable values in the generalized linear programming problem for brick manufacturing example

Resource	Resource consumption per unit activity (Activity here is manufacturing bricks of types $B_1, B_2, B_3, B_4, B_5,$ and B_6)						Available quantities of resources
1 = Multipurpose brick making machine	$a_{11} = 10$	$a_{12} = 15$	$a_{13} = 20$	$a_{14} = 5$	$a_{15} = 24$	$a_{16} = 30$	$b_1 = 8$ hours = 480 minutes
per unit contribution (₹)	$c_1 = 2$	$c_2 = 2.5$	$c_3 = 3$	$c_4 = 1.5$	$c_5 = 4$	$c_6 = 5$	

last row of Table 20.5. Let x_1 and x_2 represent the quantities of windows and doors produced respectively.

The objective function then would be:

Maximize

$$Z = 500x_1 + 600x_2$$

Subject to:

$$3x_1 + 2x_2 \leq 64$$

$$x_1 + 4x_2 \leq 68$$

$$x_1 \geq 0$$

$$x_2 \geq 0$$

20.4 GRAPHICAL SOLUTION OF LINEAR PROGRAMMING PROBLEMS

Graphical method of solving linear programming problems is a simple to use method for two variables (although it can be used for three variables also with some difficulties). Since the real life problems involve more than three variables, graphical methods are normally not used for solving real world LP problems. However to a beginner, graphical method gives a lot of insight in visualizing a problem.

Table 20.5 Variable values in the generalized linear programming problem for window and door manufacturing example

Resource	Resource consumption per unit activity (Activity here is manufacturing window and door)		Available quantities of resources
	Window	Door	
1 = Carpentry department	$a_{11} = 3$	$a_{12} = 2$	$b_1 = 64$ hours
2 = Painting department	$a_{21} = 1$	$a_{22} = 4$	$b_2 = 68$ hours
per unit contribution (₹)	$c_1 = 500$	$c_2 = 600$	

In order to solve a linear programming problem graphically after expressing the problem mathematically, we first graph all the constraints, mark the feasibility region which satisfies all the constraints, and finally test the corner points of the feasible solution.

The mathematical model for the window and door manufacturing problem is reproduced below:

Maximize

$$Z = 500x_1 + 600x_2$$

Subject to:

$$3x_1 + 2x_2 \leq 64 \tag{20.2}$$

$$x_1 + 4x_2 \leq 68 \tag{20.3}$$

$$x_1 \geq 0 \tag{20.4}$$

$$x_2 \geq 0 \tag{20.5}$$

It is clear that the non-negativity constraints, $x_1 \geq 0$ and $x_2 \geq 0$ restrict the solution space area into the first quadrant.

Now, constraint represented by (20.2) is graphed in Figure 20.1. It can be seen that A (0, 32) and B (64/3, 0) clearly satisfy the equation $3x_1 + 2x_2 = 64$. In order to graph the inequality $3x_1 + 2x_2 \leq 64$, we choose the appropriate side which satisfies the inequality. To identify the side correctly, we can consider origin (0, 0) as a reference point in all cases except when the line does not pass through the origin itself. For example, when we put $x_1 = 0$, and $x_2 = 0$ in $3x_1 + 2x_2 \leq 64$, we find that the inequality is satisfied ($3 \times 0 + 2 \times 0 = 0$ which is less than 64). Thus, the given inequality would contain the origin (0,0). The appropriate side for this inequality is shaded in Figure 20.1.

With reference to Figure 20.1, it can be clearly seen that any combination of windows and doors on line AB will consume all the 64 hours available in the carpentry department, for example, 16 windows and 8 doors. On the other hand, any point on left of line say $X(2, 6)$ will result in unused capacity of the carpentry department. It may also be noted that any point on the right of line AB, say point $Y(15, 20)$ exceeds the available carpentry hours. Thus, any combination of windows and doors which lie on right of AB is not possible without violating the constraint.

In Figure 20.2 constraint represented by (20.3) is plotted. It can be clearly seen that ($x_1 = 0$, $x_2 = 17$) and ($x_1 = 68$, $x_2 = 0$) clearly satisfy the equation $x_1 + 4x_2 = 68$. Inequality $x_1 + 4x_2 \leq 68$ is shown in Figure 20.2.

With reference to Figure 20.2, it can be clearly seen that any combination of windows and doors on line CD will consume all the 68 hours available in the painting department, for example, $P(16,13)$ that is 16 windows and 13 doors. On the other hand, any point on left of line say point $Q(20,8)$ will result in unused capacity of the painting department. It can also be noted that any point on right of line say R(40,10) exceed the number of hours. In other

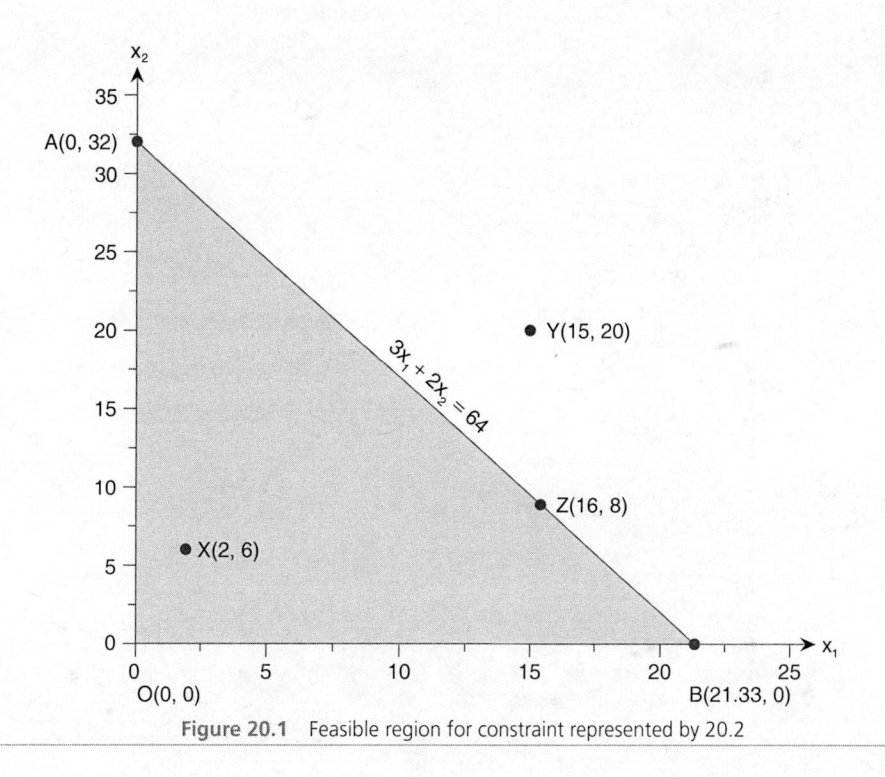

Figure 20.1 Feasible region for constraint represented by 20.2

Figure 20.2 Feasible region for constraint represented by 20.3

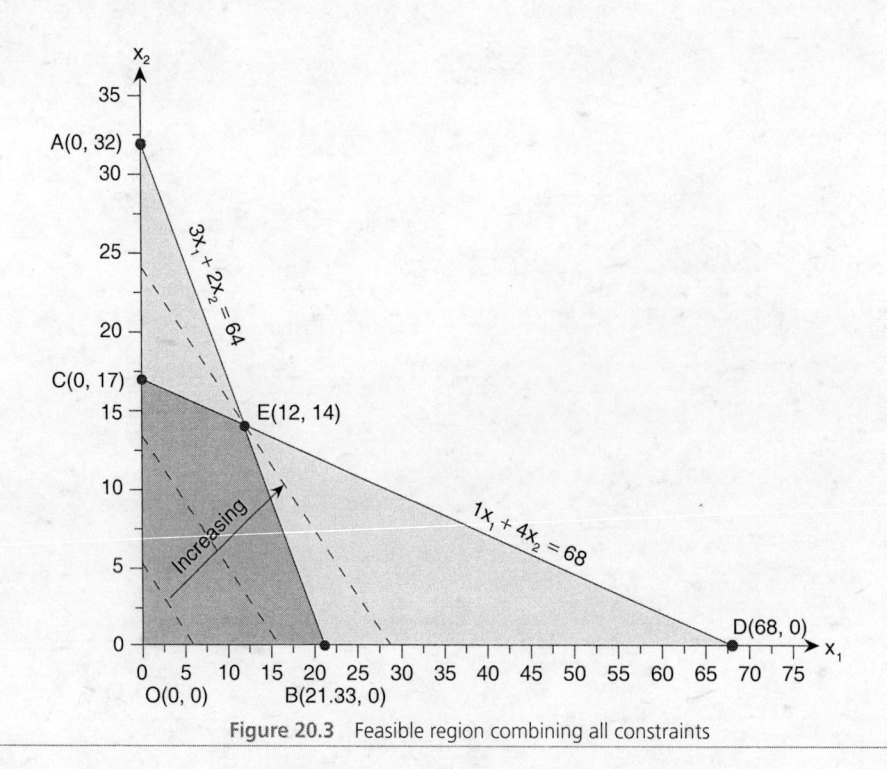

Figure 20.3 Feasible region combining all constraints

words, any combination of windows and doors which lie on right side of line CD is not possible without violating the constraint.

Figures 20.1 and 20.2 are combined and redrawn now as shown in Figure 20.3. Feasible solution, optimum feasible solution, corner point solutions—feasible or infeasible etc., are explained with the help of Figure 20.3 in the context of window and door manufacturing example. Any value of x_1 and x_2 satisfying the constraints (in this case—two constraints) is referred to as 'feasible solution'.

In Figure 20.3, points O, A, B, C, D, and E are referred to as corner points, and OBEC is termed as the feasible region. The corner points such as O, B, E, and C (those on the boundaries of feasible region) are corner point feasible solutions. In Figure 20.3, OBEC is the collection of all feasible solutions. By feasible solution, it is understood that x_1 and x_2 values at these points would satisfy all the constraints. For example, the coordinates of O(0,0), B(64/3,0), E(12,14), and C(0,17) would satisfy constraints (20.2) and (20.3) both. In contrast, points D (68,0) and A (0,32) represent corner point infeasible solution as they violate at least one of the given constraints.

Corresponding to each corner point feasible solution the objective function values can be determined. For example, corresponding to corner points O, B, E, and C, the objective function values are 0; 10,666.67; 14,400; and 10,200 respectively.

An optimal solution is a feasible solution corresponding to the largest value in case of a maximization objective and the smallest value corresponding to a minimization objective.

For the present maximization problem, the optimal solution is given by point E(12,14) as it results in the largest profit of objective function $Z = 12 \times 500 + 14 \times 600 = 14,400$.

However, in order to find the optimal solution using graphical method, we draw the objective function or profit line considering a positive random value say 3,000 in this case. Thus, we plot $Z = 500x_1 + 600x_2 = 3,000$. This line is moved towards right by increasing the value of Z arbitrarily to values such as 4,000, 5,000 and so on. The optimal solution occurs at E(12,14), which is one of the points in the feasible region OBEC. Beyond this, any increase in Z will make the solution infeasible. The optimum production combination is E(12, 14), *i.e.*, 12 windows and 14 doors giving a profit of $12 \times 500 + 14 \times 600 = ₹14,400$.

Sensitivity Analysis Using Graphical Method

In the above section, we have found that under the given two constraints the optimal profit is obtained when the manufacturer produces 12 windows and 14 doors. Suppose the manufacturer wants to analyze following situations.

The impact on the current profit if he/she makes changes in the profit coefficients of windows and doors.

The impact on objective function if there is a change in carpentry and painting man-hours availabilities. The manufacturer would also like to know which is the more productive resources he has out of the carpentry and painting resources. In other words, the manufacturer is interested to know the worth of each of the resources employed.

The above changes can be studied by reformulating the problem addressing the change the manufacturer decides to make and then solving the problem as explained earlier. However, reformulating the problem again and again and solving them is a cumbersome process. Rather, we conduct a sensitivity analysis to address the impact of changes made. We consider the above three cases one by one under sensitivity analysis using graphical method.

1. Changes in the Profit/cost Coefficient

Here the manufacturer would like to find the changes in objective function for changes made in the objective function coefficients. We can find out the range of an objective function coefficient given other coefficients so that the optimal point remains the same. Looking at Figure 20.4, we can conclude that as long as the slope of objective function (line shown in broken dashes), *i.e.*, ratio c_1/c_2 occupies a value in between the slopes of the two lines (lines 1 and 2), the optimal point won't change. In extreme cases, the c_1/c_2 value could be the slopes of either line. This situation would correspond to multiple solutions for the problem. We can also state the above condition in terms of c_2/c_1. Thus to assure that the optimal point is still at E (12,14) for the window-door problem, we can write the following conditions.

Slope of line $x_1 + 4x_2 = 68 \leq c_1/c_2 \leq$ slope of line $3x_1 + 2x_2 = 64$
That is

$$\frac{1}{4} \leq \frac{c_1}{c_2} \leq \frac{3}{2}$$

In terms of c_2/c_1
Slope of line $3x_1 + 2x_2 = 64 \leq c_2/c_1 \leq$ slope of line $x_1 + 4x_2 = 68$
That is $\frac{2}{3} \leq \frac{c_2}{c_1} \leq \frac{4}{1}$

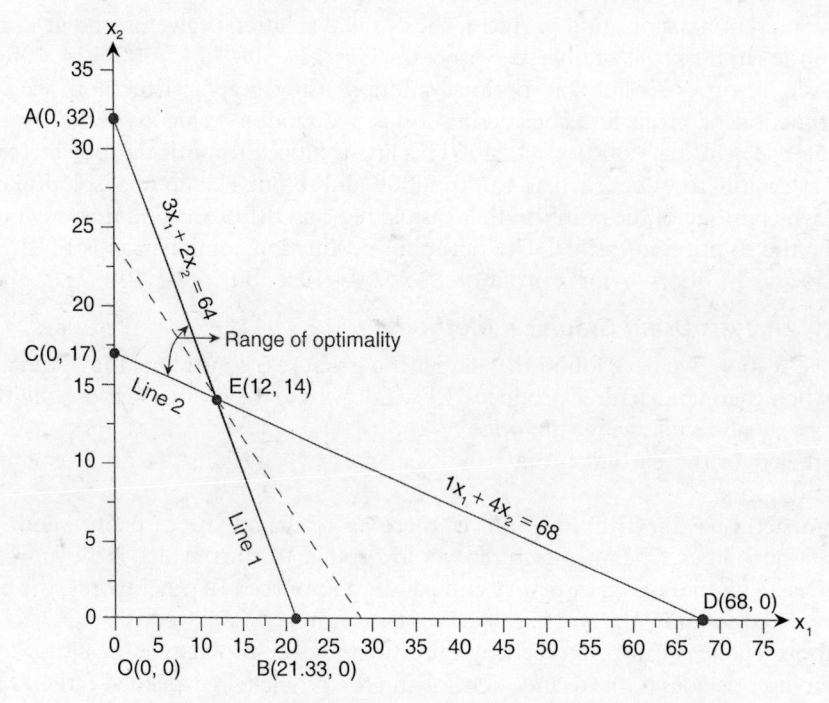

Figure 20.4 Finding range of optimality for window-door problem

Now given the value of c_2, we can find out the range of values for c_1 which would still assure that the optimal point is still $E(12,14)$. Knowing the objective value coefficients, we can easily find the value of objective function. For example for $c_2 = 600$, the range of values for c_1 would be : $\frac{1}{4} \times 600 \leq c_1 \leq \frac{3}{2} \times 600$, in other words, for any value of c_1 between 150 and 900 (both values included) the optimal point would still be $E(12,14)$.

Similarly given the value of c_1, we can find out the range of values for c_2 which would still assure that the optimal point is still $E(12,14)$. Knowing the objective value coefficients, we can easily find the value of objective function. For example for $c_1 = 500$, the range of values for c_2 would be: $\frac{2}{3} \times 500 \leq c_2 \leq 4 \times 500$, in other words for any value of c_2 between 333.33 and 2,000 (both values included) the optimal point would still be $E(12,14)$.

2. Changes in the Right Hand Sides of the Constraint Equations

We would like to study the impact of changes in the right-hand sides of the constraint equations. It may be recalled that the right-hand sides represent resource availability. For example, if the manufacture would like to increase or decrease the carpentry hours or painting hours. What will be the impact? Does my optimal point change? Do I need to reframe the problem and solve it again. In the sensitivity analysis, we try to find out the range of R.H.S values in such a manner that the optimal point E is still at the intersection of the two existing constraints (Lines 1 and

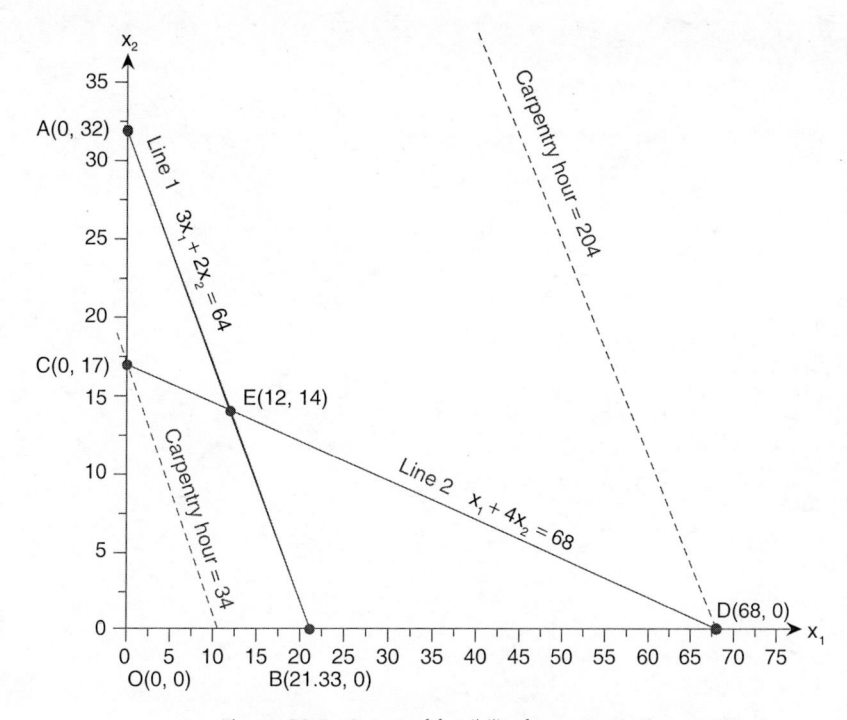

Figure 20.5 Range of feasibility for carpentry time

2 in this case). In some problems, there might be more than 2 constraints say 3 constraints. Assume that optimal point is obtained at the intersection of lines 2 and 3. Thus, sensitivity analysis is aimed at finding the range of values of R.H.S such that the optimal point is still obtained at the intersection of lines 2 and 3 and not at the intersection of say lines 1 and 2 etc.

For the window and door problem looking at the Figure 20.5, in order to find the range of values of R.H.S for line 1, we mark the two extreme points C and D. Now at C and D, we draw lines parallel to other constraint (line 2 in this case). Now we put the coordinates of points C(0,17) and D(68,0) in the equation of line 2 to find out the range of R.H.S of line 1. Thus, we find that as long as the R.H.S value of line 1 is between 34 and 204 (both values inclusive), we are assured of the optimal point resulting from the intersection of lines 1 and 2.

Likewise, we find the range of R.H.S values for line 2. It can be seen from Figure 20.6 that as long as the R.H.S value of line 2 is between 21.33 and 128 (both values inclusive), we are assured of the optimal point resulting from the intersection of lines 1 and 2.

3. Finding the Worth of Resources

Worth of resources has been defined here as the contribution in objective function per unit change in the resource value. For example, we are interested in finding out what would be the impact in objective function if we change the carpentry hour from 64 to 65 hours. By how much amount the profit is going to increase? Likewise for every unit increase/decrease in painting hours, how much does the profit increase/decrease?

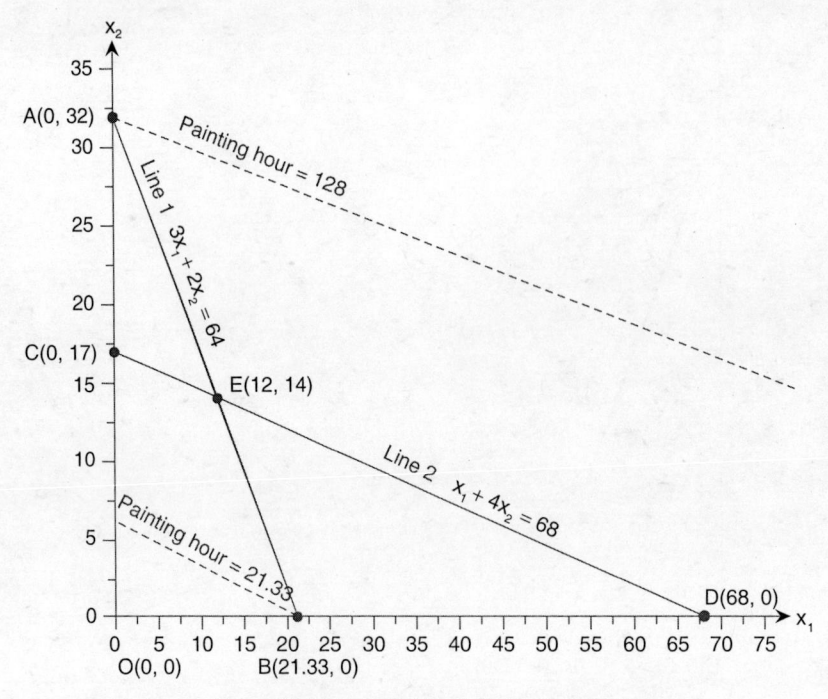

Figure 20.6 Range of feasibility for painting time

This is calculated using the following expression:

$$\text{Worth of carpentry resource} = \frac{\begin{pmatrix}\text{Objective function value at the highest permissible value of}\\\text{resource} - \text{Objective function value at the lowest permissible}\\\text{value of resource}\end{pmatrix}}{\text{Range of permissible resource value}}$$

The lowest and highest permissible values are obtained based on the condition that optimal point is still obtained by the intersection of the existing constraints. Now, objective function value Z at D (68, 0)

$$= 500 \times 68 + 600 \times 0$$
$$= 34{,}000$$

and objective function value Z at C (0, 17)

$$= 500 \times 0 + 600 \times 17$$
$$= 10{,}200$$

Hence, worth of carpentry resource

$$= \frac{34{,}000 - 10{,}200}{204 - 34}$$
$$= 140$$

The above result shows that for every 1 hour increase in carpentry time there will be an increase of ₹140 in the range of 34 and 204, in the optimum value of Z.

In a similar manner, worth of painting resource can be found out to be

$$\frac{19{,}200 - 10{,}666.66}{(128 - 21.33)} = 80$$

In other words, the result shows that for every 1 hour increase in painting time in the range of 21.33 and 128, there will be an increase of ₹ 80 in the optimum value of Z.

20.5 SIMPLEX METHOD

Simplex method provides a systematic method to solve linear programming problems. The various steps involved in simplex method are given below:

Step 1: As in the case of graphical method, a mathematical model identifying the decision variables, objective function and various constraints are formulated for the given problem.

Step 2: Converting the constraints into equation. Most of the time constraints are expressed in terms of inequalities such as $\leq$ and $\geq$. Sometimes the constraints may also be in the form of equation. Simplex method requires all the constraints to be converted into equation form using slack and surplus variables which have been explained below with examples.

Step 2(a): *Converting $\leq$ Inequalities into Equations*

Let's assume one of the constraints in a problem is

$$c_1 x_1 + c_2 x_2 \leq b_1$$

In the above inequality, c_1 and c_2 are the rate at which decision variables x_1, and x_2 consume a particular resource which has a maximum availability limit b_1. In other words, left-hand side is the usage of the resource by the activities (variables) whereas the right-hand side is the maximum available resource.

We introduce another variable say s_1 (also known as slack variable) which is defined as the difference of right-hand side (R.H.S) and left-hand side (LHS). In other words s_1 is the difference between availability and usage of a given resource. The difference would then represent the unutilized part of the maximum available resource limit.

Thus the given inequality can be written as:

$$c_1 x_1 + c_2 x_2 + s_1 = b_1$$

$$s_1 \geq 0$$

In a similar manner all inequalities with $\leq$ are converted into equation form.

Step 2(b): *Converting ≥ Inequalities into Equations*

Let's assume one of the constraints in a problem is

$$c_1 x_1 + c_2 x_2 \geq b_1$$

For such cases, we introduce another variable say S_1 (also known as surplus variable) which is defined as the difference of left-hand side and the right-hand side. In other words, S_1 represents the amount by which 'usage' exceeds the available resources. It is also sometimes called as 'negative slack'. The given inequality can be written as:

$$c_1 x_1 + c_2 x_2 - S_1 = b_1$$
$$S_1 \geq 0$$

It must be noted that the right-hand side (R.H.S) of the equation obtained from inequalities must always be non-negative. In case the R.H.S is negative, multiply each side with (–1) to get non-negative R.H.S It may be recalled that in such cases the inequalities sign will also get reversed. That is ≥ will become ≤ and ≤ will become ≥.

Step 3: *Setting up an Initial Simplex Solution*

Simplex method is an iterative process. In order to start the process, we need to identify an initial solution. Whenever possible, the simplex method starts with the origin as the initial corner point feasible solution.

Step 3(a): *≤ Inequalities*

For constraints such as $6x_1 + 7x_2 \leq 25$, as explained earlier we convert it into equation form as below:

$$6x_1 + 7x_2 + s_1 = 25$$

The initial solution for the simplex method is provided by, $x_1 = 0$ and $x_2 = 0$, thus getting $s_1 = 25$.

Step 3(b): *≥ Inequalities*

For constraints such as $6x_1 + 7x_2 \geq 25$, as explained earlier, we convert it into equation form as below:

$$6x_1 + 7x_2 - S_1 = 25$$

As before, for $x_1 = 0$ and $x_2 = 0$, we get $S_1 = -25$, which has no meaning in practical situation. In order to avoid negative values as the initial solution, another variable (known as artificial variable) is added to above equation as shown below:

$6x_1 + 7x_2 - S_1 + A_1 = 25$, which provides an initial non-negative solution $A_1 = 25$ for $x_1 = 0$, $x_2 = 0$, and $S_1 = 0$. It can be noticed that the artificial variable now acts as a slack variable.

One of the requirements in LP problems is that all variables whether they are real, slack, or surplus must be non-negative all the time.

Step 3(c): = Constraints

For constraints such as $6x_1 + 7x_2 = 25$, it is easy to guess that $x_1 = 4.25$, $x_2 = 0$ would yield an initial feasible solution. However for equations containing large number of constraints and variables, this way of providing an initial solution would not work (Try 8 constraints with 5 variables—can you solve them easily to get an initial solution?).

In order to provide an easily obtainable starting solution the constraints already in the equation form must also be added with an artificial variable to act like a slack variable.

Thus the constraint such as:
$6x_1 + 7x_2 = 25$ is converted into the following form.

$$6x_1 + 7x_2 + A_1 = 25$$

This provides a starting solution $A_1 = 25$, for $x_1 = 0, x_2 = 0$

The artificial variables mentioned earlier for the two cases have no meaning in physical sense. They simply act as a tool for computation in solving linear programming problems. Indeed in a feasible solution, artificial variables would turn out to be zero indicating that every constraint of the problem has been satisfied. In fact in a feasible solution, artificial variables would not be there in the solution space.

Step 4: Generalizing the Objective Function and the Constraints

Before we start setting up the simplex table for beginning the iteration process, we rewrite the objective function and various constraints of the problem in a manner so that all variables (real or artificial) must appear in all equations.

Step 5: Preparing Initial Simplex Table

The general form of the simplex table is shown in Table 20.6. In the table, the extreme left column shows the iteration number followed by the list of basic variables, thereafter, coefficients of real variables are written. The coefficients of slack, surplus, and artificial variables depending on the case are written subsequently. In the second last column, we write the solutions to the system of equations which are the values of basic variables. In the last column, computed ratio is written.

Table 20.6 General form of a simplex table

		\multicolumn{11}{c}{Coefficients of}		
		Real Variables			Slack Variables			Surplus Variables			Artificial Variables				
Iteration	Basic	x_1 x_2 ... x_n	s_1 s_2 ... s_n	S_1 S_2 ... S_n	A_1 A_2 ... A_n	Solution	Ratio
	Z						
	s_1						
	S_1						
	A_1						

Step 6: Selecting an entering and leaving variable using the optimality and feasibility conditions respectively. These are explained in the next section.

Step 7: Determining the new basic solution by using Gauss Jordan computations. This is also explained in the next section.

Step 8: Repeating steps 6 and 7 till optimal solution has been achieved.

20.5.1 Understanding Simplex Iterations

The iterative nature of simplex method can best be understood with the help of graphical solution as obtained in Figure 20.3.

As can be recollected, there are two constraint equations $3x_1 + 2x_2 + s_1 = 64$ and $x_1 + 4x_2 + s_2 = 68$ and the four variables $x_1, x_2, s_1,$ and s_2 associated with this problem. It is common that for solving these two equations at any time, two out of the four variables must be zero.

In the simplex method, iteration is started from the origin O(0,0) where the value of variables x_1 and x_2 are equal to 0. Thus this iteration provides initial solution in which $x_1 = 0$, $x_2 = 0$, $s_1 = 64$, and $s_2 = 68$ leading to $Z = 0$. As mentioned earlier, the variables x_1 and x_2 are termed as non-basic while the variables s_1 and s_2 are termed as basic variables.

Since the problem is to maximize Z, we try to find means through which Z is maximized quickly. The simplex iteration tries to identify variable which should be increased to increase Z; however, the search is limited to increasing one variable at a time.

The value of Z can be increased by moving along x-axis towards point B or y-axis towards C. Remember that iteration won't allow us to go to point D or A as both these points fall beyond the feasible region marked by OBEC in Figure 20.3.

The simplex iteration will direct us to move along y-axis to the point C(0,17), as it would result in more increase in Z (at a rate of ₹ 600/door unit) compared to moving along x-axis where the increase in Z would be at a rate of ₹ 500/window unit. At the point C, $x_1 = 0$, $s_2 = 0$, $x_2 = 17$, and $s_1 = 30$, leading to an improved $Z = 10,200$. Remember at this stage, x_1 and s_2 are the non-basic variables while x_2 and s_1 are basic variables. It is also said that s_2 has left the solution space and x_2 has entered the solution space. In other words, s_2 is the leaving variable and x_2 is the entering variable.

After C, the simplex iteration will further direct us to reach the point E(12,14), where there is further improvement in Z. New and optimum value of Z would be 14,400. The simplex iteration will terminate here as the optimum value has been reached.

Since graphical representation will not be possible for all the cases, the simplex method prescribes certain rules to systematically perform all the iteration and reach the optimal point.

The rules are helpful in deciding the selection of entering and leaving variable, and the corresponding rules are referred to as optimality and feasibility conditions. The rules are mentioned for the maximization and minimization objectives separately.

For maximization objective:

(a) **Optimality condition:** The selection of entering variable is done on the basis of the most negative Z-row coefficient (or the largest positive objective function coefficient). If ties are observed, they are broken arbitrarily. The optimum is reached if all the Z-row coefficients of non-basic variables have become non-negative.

(b) **Feasibility condition:** Selection of the leaving variable is based on the smallest non-negative ratio (with strictly positive denominator). Here also, ties are broken arbitrarily.

For minimization objective:

(a) **Optimality condition:** The selection of entering variable is done on the basis of most positive Z-row coefficient. If ties are observed, they are broken arbitrarily. The optimum is reached if all the Z-row coefficients of non-basic variables have become non-positive.
(b) **Feasibility condition:** This is same as in case of maximization objective.

Once the entering and exiting variables corresponding to an iteration are obtained, new basic feasible solution is obtained by Gauss – Jordan row operations. Before understanding the Gauss–Jordan row operations, we define pivot column, pivot row and pivot element first. A pivot column is the column with the maximum negative Z-row coefficient in the simplex table for maximization objective while it is the column with the maximum positive z-row coefficient in case of a minimization objective. A pivot row, on the other hand, is associated with the basic variable having the smallest value of ratio or intercept. This is true for both maximization and minimization objectives. The coefficient at the intersection of pivot row and pivot column is called pivot element.

Now after selecting the pivot column; pivot row; and pivot element, the next set of values for the new set of variables are obtained using the following two expressions.

$$New\ Pivot\ Row = \frac{Current\ Pivot\ Row}{Pivot\ Element} \qquad (20.6)$$

Other rows of the simplex tableau are obtained from the following expression:

$$Other\ Rows = Current\ Row - (Pivot\ Column\ Coefficient\ Corresponding\ to\ that\ Row) \times New\ Pivot\ Row \qquad (20.7)$$

The above steps of the simplex method are illustrated below with an example.

20.5.2 Illustration of Simplex Method—A Maximization Problem

We redo the example of maximizing $Z = 500x_1 + 600x_2$
Subject to

$$3x_1 + 2x_2 \leq 64 \qquad (20.8)$$

$$1x_1 + 4x_2 \leq 68 \qquad (20.9)$$

$$x_1 \geq 0 \qquad (20.10)$$

$$x_2 \geq 0 \qquad (20.11)$$

In generalized form, we write the objective function and constraints as shown below:

Maximize

$$Z = 500x_1 + 600x_2 + 0s_1 + 0s_2$$

Table 20.7a Initial simplex table for the window and door problem

Basic	Coefficients				Solution	Ratio
	x_1	x_2	s_1	s_2		
Z	−500	−600	0	0	0	
s_1	3	2	1	0	64	
s_2	1	4	0	1	68	

In equation form:

$$Z - 500x_1 - 600x_2 - 0s_1 - 0s_2 = 0$$

Subject to:

$$3x_1 + 2x_2 + s_1 = 64 \Rightarrow 3x_1 + 2x_2 + 1s_1 + 0s_2 = 64 \quad (20.12)$$

$$1x_1 + 4x_2 + s_2 = 68 \Rightarrow 1x_1 + 4x_2 + 0s_1 + 1s_2 = 68 \quad (20.13)$$

$$x_1 \geq 0,\ x_2 \geq 0,\ s_1 \geq 0,\ s_2 \geq 0 \quad (20.14)$$

Initial simplex table is shown is Table 20.7a.

As discussed earlier the initial table shows s_1 and s_2 as basic variables and $Z = 0$. Decision variables x_1 and x_2 are non-basic at this stage and are equal to zero. At this stage, $s_1 = 64$, and $s_2 = 68$. Now the simplex iteration will increase the value of the non-basic variable one at a time and check whether an optimum solution has been obtained. Since this is a maximization problem, selection of entering basic variable is done on the basis of most negative Z-row coefficient (or the largest positive objective function coefficient). For the given maximization problem, the entering variable happens to be x_2 with the most negative Z-row coefficient = −600.

Selection of the leaving basic variable is based on the least positive ratio. In this case, s_2 is the leaving variable with the least ratio of 68/4 = 17. Thus, the pivot column and pivot row have been identified. These are shown shaded in Table 20.7b. The pivot element 4 is also shown in this table.

After the entering and leaving variables are obtained, new basic feasible solution is obtained by Gauss – Jordan row operations.

New pivot row is obtained using the Equation (20.6) and the computation is shown below. Please note that the pivot element is 4. Also note that in this step the basic variable x_2 has entered in place of the variable s_2.

Basic	x_1	x_2	s_1	s_2	Solution
x_2	$=\dfrac{1}{4}=0.25$	$=\dfrac{4}{4}=1$	$=\dfrac{0}{4}=0$	$=\dfrac{1}{4}=0.25$	$=\dfrac{68}{4}=17$

LINEAR PROGRAMMING IN CONSTRUCTION MANAGEMENT | 695

Table 20.7b Iteration 1 of the window and door problem

Iteration	Basic	x_1	x_2	s_1	s_2	Solution	Ratio
	Z	−500	−600	0	0	0	
1	s_1	3	2	1	0	64	64/2 = 32
	s_2	1	4	0	1	68	68/4 = 17

Other rows (Z-row and s_1 row) of the simplex tableau are obtained from Equation (20.7). For example, for computing Z-row for iteration 2, we identify the pivot coefficient of this row which is −600. The current row element is −500 for x_1 column corresponding to Z-row. Similarly we identify the new pivot row element for x_1 column, which is 0.25 (See Table 20.7c).
Thus,

$$\text{New } Z\text{-row} = \text{Current } Z\text{-row} - (-600) \times \text{New Pivot Row}$$

$$\text{New } s_1 - \text{row} = \text{Current } s_1 \text{ row} - (2) \times \text{New Pivot Row}$$

Computation of Z-row according to the above expression is shown below:

Basic	x_1	x_2	s_1	s_2	Solution
Z	= −500 − (−600) $\times \frac{1}{4}$ = −350	= −600 − (−600) $\times 1 = 0$	= 0 − (−600) $\times 0 = 0$	0 − (−600) $\times \frac{1}{4} = 150$	= 0 − (−600) $\times 17 = 10,200$

Similarly the computation of s_1 row is as shown below:

Basic	x_1	x_2	s_1	s_2	Solution
s_1	= 3 − (2) $\times \frac{1}{4} = 2.5$	= 2 − (2) $\times 1 = 0$	= 1 − (2) $\times 0 = 0$	0 − (2) $\times \frac{1}{4}$ = −0.5	= 64 − (2) $\times 17 = 30$

In Table 20.7c, the above computations have been compiled.

Table 20.7c Iteration 2 of the window and door problem

Iteration	Basic	x_1	x_2	s_1	s_2	Solution	Ratio
	Z	−350	0	0	150	10,200	
2	s_1	2.5	0	1	−0.5	30	12
	x_2	0.25	1	0	0.25	17	68

Table 20.7d Iteration 3 of the window and door problem

Iteration	Basic	Coefficients x_1	x_2	s_1	s_2	Solution	Ratio
3 (Optimal)	Z	0	0	140	40	14,400	
	x_1	1	0	0.4	−0.2	12	
	x_2	0	1	−0.1	0.3	14	0

For the next iteration, pivot column and pivot row as shown in Table 20.7c are identified. The pivot element is 2.5. Using the expressions for new pivot row and other rows, all the rows have been computed. The values are shown in Table 20.7d.

In Table 20.7d we notice that there is no negative value in Z-row coefficient which indicates that the optimality condition has been achieved. The optimal solution as obtained through the simplex method matches with the solution obtained graphically. The optimum production level is (12,14) and objective function $Z = 14,400$.

20.6 PRIMAL DUAL

The linear programming example presented earlier (also referred to as Primal Problem) can also be stated in an alternative manner known as Dual Problem. The solutions of Primal and Dual Problems are equivalent and alternative procedures are used for their solution. It will be noted that the dual of a primal contains a lot of economic information useful to a manager. Besides in some situations solving a dual problem would be computationally much easier and this is one of the major advantages of Primal Dual conversion, although with the advent of software this advantage may not be that evident. Software can solve even complicated LP problems swiftly and there would be no apparent need for this conversion.

Conversion of window and door manufacturing example problem into its dual is illustrated first. It may be remembered that the objective in this problem was to produce such a combination of windows and doors that will maximize the profit for the contractor within the given set of constraints. The objective function and the set of constraints for this problem are reproduced below.

Maximize

$$Z = 500x_1 + 600x_2$$

Subject to:

$$3x_1 + 2x_2 \leq 64$$
$$x_1 + 4x_2 \leq 68$$
$$x_1 \geq 0$$
$$x_2 \geq 0$$

For the dual of the above problem, the objective function now would be to minimize the opportunity cost of not using the two resources (carpentry and painting) in an optimal manner.

Assume that the variable y_1 represents hourly contribution or worth of carpentry department's time and y_2 represents hourly contribution or worth of painting department's time. These variables are also referred as the dual value of one hour of carpentry and painting department's time respectively.

The objective function of the dual problem for window and door example can then be written as below:

Minimize opportunity cost

$$W = 64y_1 + 68y_2$$

Subject to:

$$3y_1 + y_2 \geq 500 \quad (20.15)$$
$$2y_1 + 4y_2 \geq 600 \quad (20.16)$$

The right-hand side constant 500 and 600 are the earnings from one window and one door respectively. The coefficients of y_1 and y_2 are the amounts of each scarce resource (3 hours of carpentry time and 1 hour of painting time) that are required to produce a window.

The first inequality states that the total imputed value or potential worth of the scarce resources needed to produce a window must be at least equal to the earnings derived from the window. In a similar manner the second inequality can also be interpreted.

20.6.1 Construction of the Dual from the Primal

In this section a generalized method is presented to convert a primal problem into dual problem and vice versa.

The primal in equation form as explained in section 20.3 is reproduced below:

Maximize or minimize

$$Z = \sum_{j=1}^{n} c_j \times x_j$$

Subject to:

$$\sum_{j=1}^{n} a_{ij} \times x_j \leq b_i, i = 1, 2, 3, \ldots\ldots\ldots\ldots m$$

$x_j \geq 0, j = 1, 2, 3, \ldots\ldots\ldots\ldots n$ (It includes the surplus, slack, and artificial variable, if any)

For the conversion, a dual variable is defined for each primal (constraint) equation. Also a dual constraint is defined from each primal variable (See Table 20.8). The constraint coefficients of a primal variable define the left-hand side coefficients of the dual constraints and its objective coefficient defines the right-hand side. The objective coefficients of the dual equal the right-hand side of the primal constraint equation whereas the right-hand side of the primal constraints define the coefficients of dual objective function.

Table 20.8 Illustration of rules for constructing the dual from primal

Dual variables	Primal variables							
	x_1	x_2	x_3	...	x_j	...	x_n	
	c_1	c_2	c_3	...	c_j		c_n	
y_1	a_{11}	a_{12}	a_{13}	...	a_{1j}		a_{1n}	b_1
y_2	a_{21}	a_{22}	a_{23}	...	a_{2j}		a_{2n}	b_2
...	...	...	...	...	...		...	...
y_m	a_{m1}	a_{m2}	a_{m3}	...	a_{mj}		a_{mn}	b_m
					j^{th} dual constraint			Dual objective coefficients

20.6.2 Rules for Constructing the Dual Problem

(a) For maximization objective of primal

Step 1: Make sure all primal constraints have non- negative right-hand side and all variables are non-negative.

Step 2: The dual problem will have minimization objective

Step 3: All constraints type for the dual problem would be of ≥ types irrespective of inequality types of primal problem.

Step 4: Variable signs for the dual problem would be unrestricted.

(b) For minimization objective of primal

Step 1: Make sure all primal constraints have non- negative right-hand side and all variables are non-negative.

Step 2: The dual problem will have maximization objective

Step 3: All constraints type for the dual problem would be of ≤ types irrespective of inequality types of primal problem

Step 4: Variable signs for the dual problem would be unrestricted.

The above procedure for dual conversion from primal is illustrated below:

Example 20.6.1
Primal

Maximize
$$Z = 12x_1 + 5x_2 + 4x_3$$

Subject to:
$$2x_1 + x_2 + x_3 \leq 10$$
$$x_1 - 2x_2 + 3x_3 = 12$$
$$x_1 \geq 0, x_2 \geq 0, x_3 \geq 0$$

Primal in equation form

Maximize
$$Z = 12x_1 + 5x_2 + 4x_3 + 0s_1$$

Subject to:
$$2x_1 + x_2 + x_3 + 1s_1 = 10$$
$$x_1 - 2x_2 + 3x_3 + 0s_1 = 12$$
$$x_1 \geq 0, x_2 \geq 0, x_3 \geq 0, s_1 \geq 0$$

It may be noted that for converting a primal into its dual, there is no need of introducing an artificial variable in 'equal to' constraints.

Let's assume that y_1 and y_2 are dual variables.

Thus, dual problem would be stated as:

Minimize
$$W = 10y_1 + 12y_2$$

Subject to:
$$2y_1 + 1y_2 \geq 12$$
$$1y_1 - 2y_2 \geq 5$$
$$1y_1 + 3y_2 \geq 4$$

$$\left.\begin{array}{l} 1y_1 + 0y_2 \geq 0 \\ y_1 \text{ unrestricted} \\ y_2 \text{ unrestricted} \end{array}\right\} \Rightarrow y_1 \geq 0 \text{ and } y_2 \text{ unrestricted}$$

Readers can further make dual of this dual and convince themselves that dual of dual is the primal problem itself.

Example 20.6.2
Primal

Minimize
$$Z = 4x_1 + 5x_2$$

Subject to:
$$2x_1 + x_2 \leq 3$$
$$x_1 + 2x_2 \geq 5$$
$$x_1 \geq 0, x_2 \geq 0$$

Primal in equation form
Minimize

$$Z = 4x_1 + 5x_2 + 0s_1 + 0S_2$$

Subject to:

$$2x_1 + x_2 + 1s_1 + 0S_2 = 3$$
$$x_1 + 2x_2 + 0s_1 - 1S_2 = 5$$
$$x_1 \geq 0, x_2 \geq 0, s_1 \geq 0, S_2 \geq 0$$

Let's assume that y_1 and y_2 are dual variables.
   Thus, dual problem would be stated as:

Maximize

$$W = 3y_1 + 5y_2$$

Subject to:

$$\left.\begin{array}{l}2y_1 + 1y_2 \leq 4 \\ 1y_1 + 2y_2 \leq 5 \\ 1y_1 + 0y_2 \leq 0 \\ 0y_1 - 1y_2 \leq 0 \\ y_1 \text{ unrestricted} \\ y_2 \text{ unrestricted}\end{array}\right\} \Rightarrow y_1 \leq 0, y_2 \geq 0$$

Example 20.6.3
Primal
Maximize

$$Z = 6x_1 + 5x_2$$

Subject to:

$$2x_1 + x_2 = 5$$
$$6x_1 - x_2 \geq 6$$
$$x_1 + 4x_2 \leq 4$$

x_1 is unrestricted, $x_2 \geq 0$

Primal in equation form by substituting $x_1 = x_1^+ - x_1^-$
Maximize

$$Z = 6x_1^+ - 6x_1^- + 5x_2 + 0S_1 + 0s_2$$

Subject to:

$$2x_1^+ - 2x_1^- + x_2 + 0S_1 + 0s_2 = 5$$
$$6x_1^+ - 6x_1^- - x_2 - 1S_1 + 0s_2 = 6$$
$$x_1^+ - x_1^- + 4x_2 + 0S_1 + 1s_2 = 4$$
$$x_1^+, x_1^-, S_1, s_2, x_2 \geq 0$$

Let's assume that y_1 and y_2 are dual variables.
Thus, dual problem would be stated as:
Minimize

$$W = 5y_1 + 6y_2 + 4y_3$$

Subject to:

$$\left.\begin{array}{l}2y_1 + 6y_2 + y_3 \geq 6 \\ -2y_1 - 6y_2 - y_3 \geq -6 \Rightarrow 2y_1 + 6y_2 + y_3 \leq 6\end{array}\right\} \Rightarrow 2y_1 + 6y_2 + y_3 = 6$$

$$\left.\begin{array}{l}y_1 - y_2 + 4y_3 \geq 5 \\ 0y_1 - y_2 + 0y_3 \geq 0 \\ 0y_1 + 0y_2 + 1y_3 \geq 0 \\ y_1 \text{ unrestricted} \\ y_2 \text{ unrestricted} \\ y_3 \text{ unrestricted}\end{array}\right\} \Rightarrow y_1 \text{ unrestricted}, y_2 \leq 0, y_3 \geq 0$$

20.7 BIG-M METHOD

Big-M method is also known as the Charne's Penalty Method. This method starts with the linear program in the equation form. If an equation does not have a slack, an artificial variable A, is added to form a starting solution similar to convenient all slack basic solution. However since the artificial variables are not the part of the original LP model, they are usually assigned a higher penalty in the objective function in order to minimize the effect of that variable and eventually turning it out to zero in the optimal solution. This will be always the case when the problem has the feasible solution.

Steps involved in the Big-M method

- Convert LP problem into a set of linear equations.
- Depending on constraints, inequalities and equalities are converted into equations using slack, surplus and artificial variables.
- In case of a maximization objective, a higher penalty is subtracted from the objective function while in case of a minimization objective a higher penalty is added to the objective function.
- Rests are similar to the steps followed in all slack simplex method.

Example 20.7.1

Maximize

$$Z = 30x_1 - 10x_2$$

Subject to:

$$20x_1 + 10x_2 \geq 20$$
$$10x_1 + 30x_2 \leq 30$$
$$x_2 \leq 40$$
$$x_1, x_2 \geq 0$$

Solution

Convert all the constraint equations into equations using surplus, artificial, and slack variables.

$$20x_1 + 10x_2 - S_1 + A_1 = 20$$
$$10x_1 + 30x_2 + s_2 = 30$$
$$x_2 + s_3 = 40$$
$$x_1, x_2, S_1, s_2, s_3 \geq 0$$

New modified Z equation is

$$Z = 30x_1 - 10x_2 - MA_1 + 0S_1 + 0s_2 + 0s_3$$

Please note the subtraction of penalty MA_1 in the above objective function in case of maximizations problem.

In equation form,

$$Z - 30x_1 + 10x_2 + MA_1 - 0S_1 - 0s_2 - 0s_3 = 0$$

The constraint equations are reproduced below:

$$20x_1 + 10x_2 - S_1 + A_1 = 20$$
$$10x_1 + 30x_2 + s_2 = 30$$
$$x_2 + s_3 = 40$$
$$x_1, x_2, S_1, s_2, s_3 \geq 0$$

There are three constraint equations and six variables: x_1, x_2, S_1, A_1, s_2, and s_3. Thus at any time three variables out of the six must be zero. As part of initial solution, s_2, s_3, and A_1 are basic variables and the remaining three variables x_1, x_2, and S_1 are zero (non-basic). The initial simplex tableau is generated using the equation form of the problems as shown in Table 20.9a.

It can be seen from the table that the initial solution consists of $A_1 = 20$, $s_2 = 30$, and $s_3 = 40$, while $x_1 = 0$, $x_2 = 0$, and $S_1 = 0$. The Z value shown in Table 20.9a is 0 whereas for the initial variable values, Z should be equal to $-20M$ ($30 \times 0 - 10 \times 0 - M \times 20 + 0 \times S_1 + 0$

LINEAR PROGRAMMING IN CONSTRUCTION MANAGEMENT | 703 |

Table 20.9a Initial Simplex table for the example problem 20.7.1

Iteration	Basic	x_1	x_2	A_1	S_1	s_2	s_3	Solution	Ratio
	Z	−30	10	M	0	0	0	0	
Initial tableau	A_1	20	10	1	−1	0	0	20	
	s_2	10	30	0	0	1	0	30	
	s_3	0	1	0	0	0	1	40	

$\times 30 + 0 \times 40 = -20M$). We say that Z is not consistent for the initial variable values and thus it should be made consistent first before proceeding further. The inconsistency is removed by manipulating Z-row as below.

$$\text{New } Z\text{-row} = \text{Current } Z\text{-row} - M \times A_1 \text{ row}.$$

The computation for new Z-row using above expression is shown below:

Basic	x_1	x_2	A_1	S_1	s_2	s_3	Solution
Z	$= -30 - (20)$ $\times M =$ $-30 - 20M$	$= 10 - (10)$ $\times M = 10$ $-10M$	$= M - (1)$ $\times M = 0$	$= 0 - (-1)$ $\times M = M$	$= 0 - (0)$ $\times M = 0$	$= 0 - (0)$ $\times M = 0$	$= 0 - (20)$ $\times M =$ $-20M$

The revised initial simplex table after the above manipulation is shown in Table 20.9b. The Z-row coefficient −30−20M is the most negative hence x_1 is entering and A_1 is the leaving variable as shown in iteration 1.

Once the simplex table is made consistent, usual process of identifying the entering and leaving variables are done followed by the Gauss-Jordan row operations to find the next basic feasible solution and iterations are carried out until the optimal feasible solution is attained. Iterations 2 and 3 are shown in Table 20.9c and 20.9d.

Hence the optimal solution is obtained in the 3rd iteration giving

$$Z = 90, x_1 = 3, x_2 = 0, A_1 = 0, S_1 = 40, s_2 = 0 \text{ and } s_3 = 40.$$

Table 20.9b Iteration 1 of the example problem 20.7.1

Iteration	Basic	x_1	x_2	A_1	S_1	s_2	s_3	Solution	Ratio
		−30−20M	10−10M	0	M	0	0	−20M	
1	A_1	20	10	1	−1	0	0	20	1
	s_2	10	30	0	0	1	0	30	3
	s_3	0	1	0	0	0	1	40	

Table 20.9c Iteration 2 of the example problem 20.7.1

Iteration	Basic				Coefficients				Solution	Ratio
		x_1	x_2	A_1	S_1	s_2	s_3			
2	Z	0	25	1.5 + M	−1.5	0	0	30	
	x_1	1	0.5	0.05	−0.05	0	0	1	−20
	s_2	0	25	−0.5	0.5	1	0	20	40
	s_3	0	1	0	0	0	1	40	∞

Table 20.9d Iteration 3 of the example problem 20.7.1

Iteration	Basic				Coefficients				Solution	Ratio
		x_1	x_2	A_1	S_1	s_2	s_3			
3 Optimal	Z	0	100	M	0	3	0	90	
	x_1	1	3	0	0	0.1	0	3	
	S_1	0	50	−1	1	2	0	40	
	s_3	0	1	0	0	0	1	40	

Example 20.7.2:

Minimize

$$Z = 3x_1 + 2x_2$$

Subject to:

$$x_1 + 3x_2 = 6$$
$$5x_1 + 4x_2 \geq 20$$
$$7x_1 + 5x_2 \leq 35$$
$$x_1, x_2 \geq 0$$

Solution:
Convert all the constraint equations into equations using surplus, artificial, and slack variables.

$$x_1 + 3x_2 + A_1 = 6$$
$$5x_1 + 4x_2 - S_1 + A_2 = 20$$
$$7x_1 + 5x_2 + s_2 = 35$$
$$x_1, x_2, A_1, A_2, S_1, s_2 \geq 0$$

New modified Z equation is

$$Z = 3x_1 + 2x_2 + MA_1 + MA_2 + 0S_1 + 0s_2$$

LINEAR PROGRAMMING IN CONSTRUCTION MANAGEMENT | 705

Please note the penalties MA_1 and MA_2 have been added in minimization problem instead of subtracting them as previously done in maximization problem. .

In equation form

$$Z - 3x_1 - 2x_2 - MA_1 - MA_2 - 0S_1 - 0s_2 = 0$$

The constraint equations are reproduced below:

$$x_1 + 3x_2 + A_1 = 6$$
$$5x_1 + 4x_2 - S_1 + A_2 = 20$$
$$7x_1 + 5x_2 + s_2 = 35$$

There are three constraint equations and six variables: $x_1, x_2, S_1, A_1, A_2,$ and s_2. Thus at any time three variables out of the six must be zero. As part of initial solution, A_1, A_2, and s_2 are basic variables and the remaining three variables $x_1, x_2,$ and S_1 are zero (non-basic). The initial simplex tableau is generated using the equation form of the problems as shown in Table 20.10a.

It can be seen from the table that the initial solution consists of $A_1 = 6$, $A_2 = 20$, and $s_2 = 35$, while $x_1 = 0$, $x_2 = 0$, and $S_1 = 0$. The Z value shown in Table 20.10a is 0 whereas for the initial variable values, Z should be equal to $+26M$ ($3 \times 0 + 2 \times 0 + M \times 6 + M \times 20 + 0 \times 0 + 0 \times 35 = 26M$). We say that Z is not consistent for the initial variable values and thus it should be made consistent first before proceeding further. The inconsistency is removed by manipulating Z-row as below.

$$\text{New Z-row} = \text{Current Z-row} + M \times A_1 \text{ row} + M \times A_2 \text{ row}.$$

Basic	x_1	x_2	s_1	s_2	A_1	A_2	Solution
Z	$= -3 + (1)$ $\times M + (5)$ $\times M =$ $-3 + 6M$	$= -2 + (3)$ $\times M + (4)$ $\times M = -2$ $+ 7M$	$= 0 + (0)$ $\times M$ $+ (1)$ $\times M$ $= -M$	$= 0 + (0)$ $\times M + (0)$ $\times M = 0$	$= -M + (1)$ $\times M + (0)$ $\times M = 0$	$= -M + (0)$ $\times M + (1)$ $\times M = 0$	$= 0 + (6)$ $\times M$ $+ (20)$ $\times M$ $= 26M$

Table 20.10a Initial Simplex table for the example problem 20.7.2

Iteration	Basic	x_1	x_2	S_1	s_2	A_1	A_2	Solution	Ratio
Initial	Z	-3	-2	0	0	-M	-M	0	
	A_1	1	3	0	0	1	0	6	
	A_2	5	4	-1	0	0	1	20	
	s_2	7	5	0	1	0	0	35	

Table 20.10b Iteration 1 of the example problem 20.7.2

Iteration	Basic	x_1	x_2	S_1	S_2	A_1	A_2	Solution	Ratio
	Z	$-3+6M$	$-2+7M$	$-M$	0	0	0	$26M$	
1	A_1	1	3	0	0	1	0	6	2
	A_2	5	4	-1	0	0	1	20	5
	S_2	7	5	0	1	0	0	35	7

It may be noticed that the Z-row has been made consistent by adding M times A_1 and A_2 rows. The revised initial simplex table is shown in Table 20.10b.

The Z-row coefficient $-2+7M$ is the most positive hence is entering and A_1 is the leaving variable (least ratio) as shown in iteration 1.

The usual process of Gauss-Jordan row operations is performed to find the next basic feasible solution and iterations are carried out until the optimal feasible solution is attained. Iterations 2 and 3 are shown in Table 20.10c and 20.10d.

Hence the optimal solution is obtained in the 3rd iteration giving

$$Z = \frac{128}{11}, x_1 = \frac{36}{11}, x_2 = \frac{10}{11}, A_1 = 0, A_2 = 0, S_1 = 0 \text{ and } s_2 = \frac{83}{11}.$$

The above solution is also obtained graphically as shown in Figure 20.7

Table 20.10c Iteration 2 of the example problem 20.7.2

Iteration	Basic	x_1	x_2	S_1	S_2	A_1	A_2	Solution	Ratio
	Z	$\frac{-7}{3}+\frac{11M}{3}$	0	$-M$	0	$\frac{2}{3}+\frac{7M}{3}$	0	$4+12M$	
2	x_2	$\frac{1}{3}$	1	0	0	$\frac{1}{3}$	0	2	6
	A_2	$\frac{11}{3}$	0	-1	0	$-\frac{4}{3}$	1	12	$\frac{36}{11}$
	S_2	$\frac{16}{3}$	0	0	1	$-\frac{5}{3}$	0	25	$\frac{75}{16}$

Table 20.10d Iteration 3 of the example problem 20.7.2

Iteration	Basic	x_1	x_2	S_1	s_2	A_1	A_2	Solution	Ratio
3	Z	0	0	$\dfrac{-7}{11}$	0	$-M-\dfrac{2}{11}$	$-M+\dfrac{7}{11}$	$\dfrac{128}{11}$	
	x_2	0	1	$\dfrac{1}{11}$	0	$\dfrac{5}{11}$	$-\dfrac{1}{11}$	$\dfrac{10}{11}$	
	x_1	1	0	$-\dfrac{3}{11}$	0	$-\dfrac{4}{11}$	$\dfrac{3}{11}$	$\dfrac{36}{11}$	
	s_2	0	0	$\dfrac{16}{11}$	1	$\dfrac{3}{11}$	$-\dfrac{16}{11}$	$\dfrac{83}{11}$	

Figure 20.7 Graphical solution of example problem 20.7.2

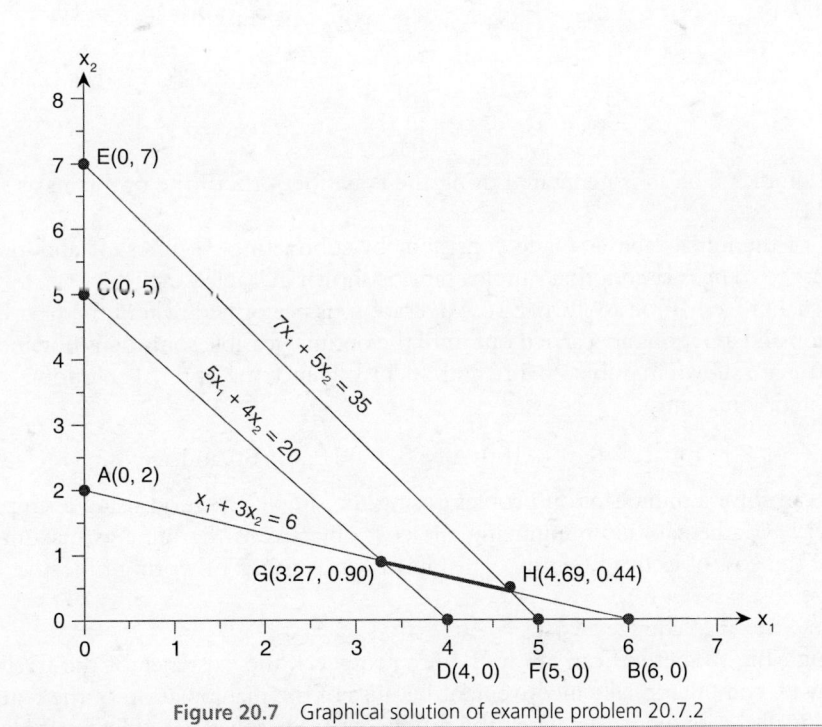

Example 20.7.3

Maximize
$$Z = 3x_1 + 4x_2$$

Subject to:
$$x_1 + x_2 = 100$$
$$x_1 \leq 30$$
$$x_2 \geq 15$$
$$x_1, x_2 \geq 0$$

Solution:
Convert all the constraint equations into equations using surplus, artificial and slack variables.

$$x_1 + x_2 + A_1 = 100$$
$$x_1 + s_1 = 30$$
$$x_2 - S_2 + A_2 = 15$$
$$x_1, x_2, A_1, A_2, s_1, S_2 \geq 0$$

New modified Z equation is

$$Z = 3x_1 + 4x_2 - MA_1 - MA_2 + 0s_1 + 0S_2$$

Or in equation form

$$Z - 3x_1 - 4x_2 + MA_1 + MA_2 - 0s_1 - 0S_2 = 0$$

The initial simplex tableau is generated using the equation form of the problems as shown in Table 20.11a.

Z-row of the initial table is made consistent by subtracting M times A_1 and A_2 rows as discussed earlier. The revised initial simplex table is shown in Table 20.11b

The usual process of Gauss-Jordan row operations is performed to find the next basic feasible solution and iterations are carried out until the optimal feasible solution is obtained. Iterations 2 and 3 are shown in Table 20.11c and 20.11d. Hence the optimal solution is obtained in the 3rd iteration giving

$$Z = 400, x_1 = 0, x_2 = 100, A_1 = 0, A_2 = 0, s_1 = 30, \text{and } S_2 = 85.$$

Note: We can solve a minimization problem using the simplex method using a simple mathematical trick. Mathematically minimizing the cost objective is the same as maximizing the negative of the cost objective function. This means that instead of writing objective function as Minimize cost = $5x_1 + 6x_2$

We can write as Maximize (-cost) = $-5x_1 - 6x_2$

Although Big-M method can always be used to check the existence of the feasible solution, it may be computationally inconvenient because of the manipulation of the constant M. Another difficulty arises especially when the problem is to be solved on a digital computer. To use the digital computer, M must be assigned some value which must be larger than the values $c_1, c_2,...$, in the objective function.

Table 20.11a Initial Simplex table for the example problem 20.7.3

Iteration	Basic	x_1	x_2	s_1	S_2	A_1	A_2	Solution	Ratio
Initial	Z	−3	−4	0	0	M	M	0	
	A_1	1	1	0	0	1	0	100	
	s_1	1	0	1	0	0	0	30	
	A_2	0	1	0	−1	0	1	15	

Table 20.11b Iteration 1 of the example problem 20.7.3

Iteration	Basic	x_1	x_2	s_1	S_2	A_1	A_2	Solution	Ratio
1	Z	−3−M	−4−2M	0	M	0	0	−115M	
	A_1	1	1	0	0	1	0	100	100
	s_1	1	0	1	0	0	0	30	
	A_2	0	1	0	−1	0	1	15	15

Table 20.11c Iteration 2 of the example problem 20.7.3

Iteration	Basic	x_1	x_2	s_1	S_2	A_1	A_2	Solution	Ratio
2	Z	−3−M	0	0	−4−M	0	4+2M	60−85M	
	A_1	1	0	0	1	1	−1	85	85
	s_1	1	0	1	0	0	0	30	
	x_2	0	1	0	−1	0	1	15	

Table 20.11d Iteration 3 of the example problem 20.7.3

Iteration	Basic	x_1	x_2	s_1	S_2	A_1	A_2	Solution	Ratio
3 (Optimal)	Z	1	0	0	0	4+M	M	400	
	S_2	1	0	0	1	1	−1	85	
	s_1	1	0	1	0	0	0	30	
	x_2	1	1	0	0	1	0	100	

20.8 TWO-PHASE METHOD

In order to overcome the limitations of Big-M, two-phase method is followed. In Big-M method, the use of the penalty M, which by definition must be large relative to the actual objective function coefficients of the model, can result in round off error that may affect the accuracy of the simplex calculations. The two-phase method solves this difficulty by eliminating the constant M altogether. As the name indicates the method solves the linear program in two phases: (1) in the first phase, an attempt is made to find the starting basic feasible solution; which is done by eliminating the artificial variables, (2) in the second phase, the optimal solution is determined for the original problem.

Example 20.8.1

Minimize

$$Z = 7.5x_1 + 3x_2$$

Subject to:

$$3x_1 + x_2 \geq 3$$
$$x_1 + x_2 \geq 2$$
$$x_1, x_2 \geq 0$$

Solution:

Convert all the constraint equations into equations using surplus & artificial variables.

$$3x_1 + x_2 - S_1 + A_1 = 3$$
$$x_1 + x_2 - S_2 + A_2 = 2$$
$$x_1, x_2, S_1, S_2, A_1, A_2 \geq 0$$

In Phase I
Minimize, $a = A_1 + A_2$
In equation form, $a - A_1 - A_2 = 0$
The constraint equations are:

$$3x_1 + x_2 - S_1 + A_1 = 3$$
$$x_1 + x_2 - S_2 + A_2 = 2$$
$$x_1, x_2, S_1, S_2, A_1, A_2 \geq 0$$

Table 20.12a Initial Simplex table for the example problem 20.8.1 (Phase 1)

Iteration Phase 1	Basic	x_1	x_2	S_1	S_2	A_1	A_2	Solution	Ratio
	a	0	0	0	0	−1	−1	0	
Initial	A_1	3	1	−1	0	1	0	3	
	A_2	1	1	0	−1	0	1	2	

LINEAR PROGRAMMING IN CONSTRUCTION MANAGEMENT | 711 |

Table 20.12b   Iteration 1 of the example problem 20.8.1 (Phase 1)

Iteration Phase 1	Basic	x_1	x_2	S_1	S_2	A_1	A_2	Solution	Ratio
	a	4	2	-1	-1	0	0	5	
1	A_1	3	1	-1	0	1	0	3	1
	A_2	1	1	0	-1	0	1	2	2

Table 20.12c   Iteration 2 of the example problem 20.8.1 (Phase 1)

Iteration Phase 1	Basic	x_1	x_2	S_1	S_2	A_1	A_2	Solution	Ratio
	a	0	2/3	1/3	-1	-4/3	0	1	
2	x_1	1	1/3	-1/3	0	1/3	0	1	3
	A_2	0	2/3	1/3	-1	-1/3	1	1	3/2

Table 20.12d   Iteration 3 of the example problem 20.8.1 (Phase 1)

Iteration Phase 1	Basic	x_1	x_2	S_1	S_2	A_1	A_2	Solution	Ratio
	a	0	0	0	0	-1	-1	0	
3	x_1	1	0	-1/2	1/2	1/2	-1/2	1/2	
	x_2	0	1	1/2	-3/2	-1/2	3/2	3/2	

Phase 1 produces $x_1 = 1/2$, $x_2 = 3/2$ and a = 0. It may also be noted that there is no artificial variable in the optimal solution.

The initial simplex tableau is generated using the equation form of the problems as shown in Table 20.12a. This simplex tableau is made consistant by adding A_1 and A_2 rows to the a-row as shown in Table 20.12b. The usual process of Grauss-Jordan row operations is performed to find the next basic feasible solution and iterations are carried out until the optimal feasible solution is obtained. Iterations 2 and 3 are shown in Tables 20.12c and 20.12d.

PHASE 2
Now Minimize $Z = 7.5x_1 + 3x_2 - 0S_1 - 0S_2$
    Subject to

$$x_1 - \frac{1}{2}S_1 + \frac{1}{2}S_2 = \frac{1}{2}$$

$$x_2 + \frac{1}{2}S_1 - \frac{3}{2}S_2 = \frac{3}{2}$$

$$x_1, x_2, S_1, S_2 \geq 0$$

Table 20.12e Initial Simplex table for the example 20.8.1 (Phase 2)

Iteration (Phase 2)	Basic	x_1	x_2	S_1	S_2	Solution	Ratio
	Z	−7.5	−3	0	0	0	
Initial	x_1	1	0	−1/2	1/2	1/2	
	x_2	0	1	1/2	−3/2	3/2	

Initial simplex tableau of the phase 2 obtained from above equations and shown in Table 20.12e is inconsistent as the Z-row coefficients corresponding to basic variables are non-zero numbers. Thus to make it consistent, Z-row manipulation is done as below:

$$\text{New Z-row} = \text{Current Z-row} + 7.5 \times x_1 \text{ row} + 3 \times x_2 \text{ row}$$

The results after above manipulation is shown in Table 20.12f.

Hence the optimal solution is obtained in the 1st iteration of phase 2 and may also be checked graphically as $Z = 8.25, x_1 = 0.5, x_2 = 1.5, A_1 = 0, A_2 = 0, S_1 = 0$ and $S_2 = 0$ (See Figure 20.8).

Table 20.12f Iteration 1 of the example 20.8.1 (Phase 2)

Iteration (Phase 2)	Basic	x_1	x_2	S_1	S_2	Solution	Ratio
	Z	0	0	−2.25	−0.75	8.25	
1	x_1	1	0	−1/2	1/2	1/2	
	x_2	0	1	1/2	−3/2	3/2	

20.9 DUAL SIMPLEX METHOD

This algorithm is required for carrying out sensitivity analysis (discussed subsequently). In this method, the LP starts optimum and infeasible. Successive iterations are carried out to move toward feasibility without violating optimality. The algorithm ends when the feasibility is restored. Contrast this method with the earlier one where we start with feasible and non-optimal and go on till optimality is achieved. In this method, the starting table must have an optimum objective row with at least one infeasible (<0) basic variable. *The leaving variable is the basic variable having the most negative value with ties broken arbitrarily.* If all the basic variables are non-negative, algorithm ends. *The entering variable is determined from among the non-basic variables as the one corresponding to the minimum of* $\left| \dfrac{Z_j - C_j}{\alpha_{rj}} \right|$ for $\alpha_{rj} < 0$. The term $Z_j - c_j$ is the objective coefficient of the Z-row in the simplex table. The term α_{rj} is the negative constraint coefficient of the simplex table corresponding to the row of the leaving variable

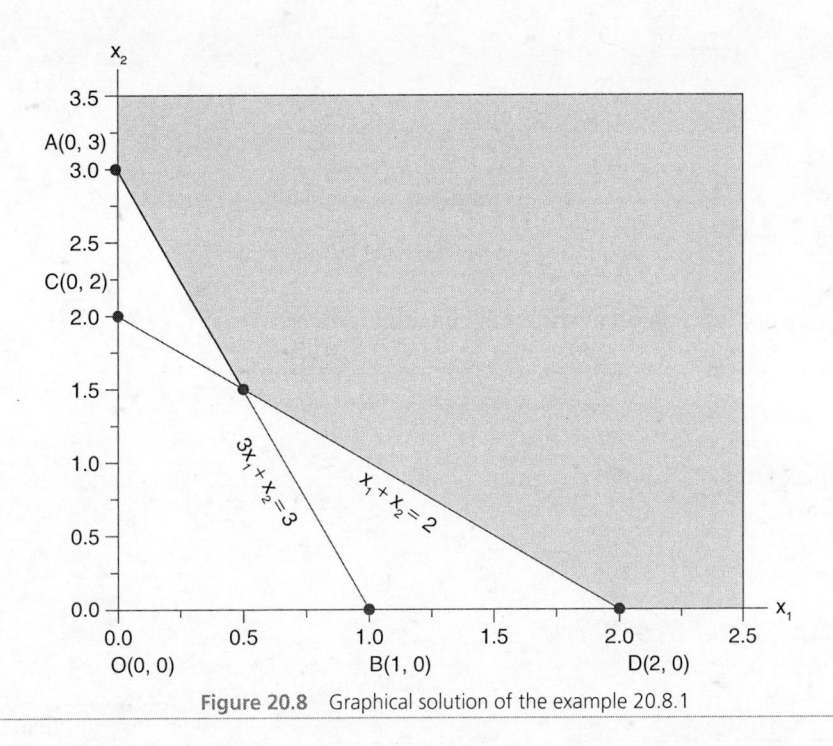

Figure 20.8 Graphical solution of the example 20.8.1

and column of the non-basic variable. The dual simplex method is explained with the help of following examples.

Example 20.9.1
Minimize $Z = 8x_1 + 3x_2$
  Subject to:

$$3x_1 + x_2 \geq 3$$
$$x_1 + x_2 \geq 2$$
$$4x_1 + 3x_2 \leq 12$$
$$x_1, x_2 \geq 0$$

All the $\geq$ constraints are first converted to $\leq$ constraints by multiplying with −1 and hence changing the inequality sign.

$$-3x_1 - x_2 \leq -3$$
$$-x_1 - x_2 \leq -2$$
$$4x_1 + 3x_2 \leq 12$$
$$x_1, x_2 \geq 0$$

Then these $\leq$ constraints are converted to equation form as in case of all slack simplex method.

Thus,

$$-3x_1 - x_2 + s_1 = -3$$
$$-x_1 - x_2 + s_2 = -2$$
$$4x_1 + 3x_2 + s_3 = 12$$
$$x_1, x_2, s_1, s_2, s_3 \geq 0$$

Objective function may then be written in equation form as :

$$Z - 8x_1 - 3x_2 - 0s_1 - 0s_2 - 0s_3 = 0$$

Initial simplex table may then be populated with coefficients of above equations as shown in Table 20.13a.

Table 20.13a Initial simplex table for example 20.9.1

Iteration	Basic	Coefficients					Solution
		x_1	x_2	s_1	s_2	s_3	
	Z	-8	-3	0	0	0	0
	s_1	-3	-1	1	0	0	-3
	s_2	-1	-1	0	1	0	-2
Initial	s_3	4	3	0	0	1	12
	$\left|\dfrac{Z_j - C_j}{\alpha_{rj}}\right|$	8/3	3/1				

It may be noted that none of the Z-row coefficients are positive thus the initial solution is optimal as the objective is to minimize. Also, initial solution is represented by $s_1 = -3$ and $s_2 = -2$ and thus infeasible.

According to dual optimality condition, s_1 (−3) is the leaving variable and the entering variable is x_1 (corresponding to the minimum $\left|\dfrac{Z_j - C_j}{\alpha_{rj}}\right| = 8/3$). This is shown in Table 20.13b. Table 20.13c shows the second iteration.

The optimal solution is obtained in the 3rd iteration giving

$$Z = 8.5, x_1 = 0.5, x_2 = 1.5, s_1 = 0, s_2 = 0, s_3 = 5.5 \text{ (see Table 20.13d)}$$

Table 20.13b Iteration 1 of the example 20.9.1

Iteration	Basic	x_1	x_2	s_1	s_2	s_3	Solution
	Z	−8	−3	0	0	0	0
	x_1	−3	−1	1	0	0	−3
1	s_2	−1	−1	0	1	0	−2
	s_3	4	3	0	0	1	12
	$\left\|\dfrac{Z_j - C_j}{\alpha_{rj}}\right\|$	8/3	3/1				

Table 20.13c Iteration 2 of the example 20.9.1

Iteration	Basic	x_1	x_2	s_1	s_2	s_3	Solution
	Z	0	−1/3	−8/3	0	0	8
	x_1	1	1/3	−1/3	0	0	1
2	s_2	0	−2/3	−1/3	1	0	−1
	s_3	0	5/3	4/3	0	1	8
	$\left\|\dfrac{Z_j - C_j}{\alpha_{rj}}\right\|$		1/2	8			

Table 20.13d Iteration 3 of the example 20.9.1

Iteration	Basic	x_1	x_2	s_1	s_2	s_3	Solution
	Z	0	0	−5/2	−1/2	0	8.5
3	x_1	1	0	−1/2	1/2	0	1/2
(Optimal)	x_2	0	1	1/2	−3/2	0	3/2
	s_3	0	0	1/2	5/2	1	11/2

Example 20.9.2

Minimize

$$Z = 5x_1 + 4x_2$$

Subject to:

$$x_1 + 3x_2 \geq 9$$
$$x_1 + x_2 = 6$$
$$x_1, x_2 \geq 0$$

All the ≥ and equality constraints are first converted to ≤ constraints by multiplying with −1 and hence changing the inequality sign. We also write $x_1 + x_2 = 6$ as $x_1 + x_2 \geq 6$ and

$x_1 + x_2 \leq 6$. and thereafter $x_1 + x_2 \geq 6$ is multiplied with –1. Thus, constraints are now written as below:

$$-x_1 - 3x_2 \leq -9$$
$$\left.\begin{array}{r}-x_1 - x_2 \leq -6 \\ x_1 + x_2 \leq 6\end{array}\right\}$$
$$x_1, x_2 \geq 0$$

Then these $\leq$ constraints are converted to equation form as

$$-x_1 - 3x_2 + s_1 = -9$$
$$-x_1 - x_2 + s_2 = -6$$
$$x_1 + x_2 + s_3 = 6$$
$$x_1, x_2, s_1, s_2, s_3 \geq 0$$

Objective function may then be written in equation form as

$$Z - 5x_1 - 4x_2 - 0s_1 - 0s_2 - 0s_3 = 0$$

Initial simplex table may then be populated with coefficients of above equations as shown in Table 20.14a. Intermediate and final iterations are shown in Tables 20.14b and 20.14c.

The optimal solution is obtained in the 2nd iteration giving

$$Z = 24, x_1 = 0, x_2 = 6, s_1 = 9, s_2 = 0, s_3 = 0.$$

Table 20.14a Initial simplex table for the example 20.9.2

Iteration	Basic	\multicolumn{5}{c	}{Coefficients}	Solution
		x_1	x_2	s_1	s_2	s_3	
	Z	–5	–4	0	0	0	0
	s_1	–1	–3	1	0	0	–9
	s_2	–1	–1	0	1	0	–6
Initial	s_3	1	1	0	0	1	6
	$\left|\dfrac{Z_j - C_j}{\alpha_{rj}}\right|$	5	4/3				

Table 20.14b Iteration 1 of the example 20.9.2

Iteration	Basic	\multicolumn{5}{c	}{Coefficients}	Solution
		x_1	x_2	s_1	s_2	s_3	
	Z	–11/3	0	–4/3	0	0	12
	x_2	1/3	1	–1/3	0	0	3
	s_2	–2/3	0	–1/3	1	0	–3
1	s_3	2/3	0	1/3	0	1	3
	$\left|\dfrac{Z_j - C_j}{\alpha_{rj}}\right|$	11/2		4			

Table 20.14c Iteration 2 of the example 20.9.2

Iteration	Basic	\multicolumn{5}{c	}{Coefficients}	Solution
		x_1	x_2	s_1	s_2	s_3	
	Z	−1	0	0	−4	0	24
2	x_2	1	1	0	−1	0	6
(Optimal)	s_1	2	0	1	−3	0	9
	s_3	0	0	0	1	1	0

Example 20.9.3
Dual simplex with artificial constraint

Maximize

$$Z = 2x_1 - x_2 + x_3$$

Subject to:

$$2x_1 + 3x_2 - 5x_3 \geq 4$$
$$-x_1 + 9x_2 - x_3 \geq 3$$
$$4x_1 + 6x_2 + 3x_3 \leq 8$$
$$x_1, x_2, x_3 \geq 0$$

All the constraints are first converted to equation form using slack or surplus variables as

$$2x_1 + 3x_2 - 5x_3 - S_1 = 4$$
$$-x_1 + 9x_2 - x_3 - S_2 = 3$$
$$4x_1 + 6x_2 + 3x_3 + s_3 = 8$$
$$x_1, x_2, S_1, S_2, s_3 \geq 0$$

Objective function may then be written in equation form as

$$Z - 2x_1 + x_2 - x_3 - 0S_1 - 0S_2 - 0s_3 = 0$$

Initial simplex table may then be populated with coefficients of above equations as given in Table 20.15a

The initial table shows that the current solution $S_1 = -4, S_2 = -3, s_3 = 8$ is neither feasible nor optimum. Thus, the condition of dual simplex that of optimum and infeasible is not met.

Hence in order to apply the dual simplex algorithm additional artificial constraint equation $x_1 + x_3 \leq M$ is used and all rows are manipulated to make it an optimal infeasible solution table.

Further dual simplex method is applied to move the solution towards the feasible solution. Because the Z-row coefficients of x_1 and x_3 are negative hence they have been chosen

Table 20.15a Initial simplex table for the example 20.9.3

Iteration	Basic	\	\	\	Coefficients	\	\	Solution
		x_1	x_2	x_3	S_1	S_2	S_3	
	Z	−2	1	−1	0	0	0	0
Initial	S_1	−2	−3	5	1	0	0	−4
	S_2	1	−9	1	0	1	0	−3
	S_3	4	6	3	0	0	1	8

Table 20.15b Initial simplex table with an artificial constraint equation for the example 20.9.3

Iteration	Basic				Coefficients				Solution
		x_1	x_2	x_3	S_1	S_2	S_3	S_4	
	Z	−2	1	−1	0	0	0	0	0
Initial with	S_1	−2	−3	5	1	0	0	0	−4
an artificial	S_2	1	−9	1	0	1	0	0	−3
constraint	S_3	4	6	3	0	0	1	0	8
	S_4	1	0	1	0	0	0	1	M

as variables in the artificial constraint equation. The artificial constraint equation is written in equation form and appended to the simplex tableau as

$$x_1 + x_3 + s_4 = M \quad \text{(see Table 20.15b)}$$

Now all the rows are manipulated using s_4 row as pivot row. For illustration Z-row is manipulated using the artificial constraint equation as

$$\text{New Z-row} = \text{Old Z-row} - (-2) \times s_4 \text{ row}$$

Simplex tableau in iteration 1 is giving an optimal infeasible solution which is the necessary condition for applying dual simplex algorithm. Table 20.15c also shows that s_3 is exiting variable and s_4 is the entering basic variable. Table 20.15d shows the results of second iteration.

The optimal solution is obtained in the 3rd iteration (see Table 20.15e) giving

$$Z = 2.095, x_1 = 1.286, x_2 = 0.476, x_3 = 0$$

Table 20.15c Iteration 1 of the example 20.9.3

Iteration	Basic	\	\	Coefficients					Solution
		x_1	x_2	x_3	S_1	S_2	S_3	S_4	
1	Z	0	1	1	0	0	0	2	$2M$
	S_1	0	−3	7	1	0	0	2	$-4+2M$
	S_2	0	−9	0	0	1	0	−1	$-3-M$
	s_3	0	6	−1	0	0	1	−4	$8-4M$
	x_1	1	0	1	0	0	0	1	M
	$\dfrac{Z_j - C_j}{\alpha_{rj}}$			1				1/2	

Table 20.15d Iteration 2 of the example 20.9.3

Iteration	Basic			Coefficients					Solution
		x_1	x_2	x_3	S_1	S_2	S_3	S_4	
2	Z	0	4	0.5	0	0	0.5	0	4
	S_1	0	0	6.5	1	0	0.5	0	0
	S_2	0	−10.5	0.25	0	1	−0.25	0	−5
	S_4	0	−1.5	0.25	0	0	−0.25	1	$M-2$
	x_1	1	1.5	0.75	0	0	0.25	0	2
	$\dfrac{Z_j - C_j}{\alpha_{rj}}$		4/10.5 = 0.38				0.5/0.25 = 2		

Table 20.15e Iteration 3 of the example 20.9.3

Iteration	Basic			Coefficients					Solution
		x_1	x_2	x_3	S_1	S_2	S_3	S_4	
3 (Optimal)	Z	0	0	0.5950	0	0.381	0.405	0	2.095
	S_1	0	0	6.5	1	0	0.5	0	0
	x_2	0	1	−0.024	0	−0.095	0.024	0	0.476
	S_4	0	0	0.214	0	−0.143	−0.214	1	$M-1.286$
	x_1	1	0	0.786	0	0.143	0.214	0	1.286

20.10 SPECIAL CASES IN SIMPLEX

In this section some special situations encountered in simplex method are illustrated. Solution procedures for such situations are also explained. The following situations may be encountered while solving a linear programming problem using simplex method.

1. There is a tie in the entering basic variable.
2. There is a tie for the basic leaving variable.
3. There is no leaving basic variable.
4. There are multiple optimal solutions.

Special cases in simplex method has been discussed with examples for following cases.

- Infeasibility
- Unboundedness
- Degeneracy
- Multiple optimal solution

20.10.1 Infeasibility

Infeasibility comes when there is no solution that satisfies all of the problem's constraints. In simplex method, an infeasible solution is indicated by looking at the final table. This situation will never occur if all the constraints are of type ≤ because the slacks provide a feasible solution. For other type of constraints if there is no feasible solution at least one artificial variable will show positive value in the optimum iteration

Draw the graph and see physically what is happening when artificial variable is positive.

Physically the infeasible solution indicates incorrect problem formulation. Alternatively the infeasibility may also be checked graphically.

Infeasibility of solution is explained both graphically and using simplex algorithm using an example.

Example 20.10.1

Maximize

$$Z = 2x_1 + 3x_2$$

Subject to:

$$3x_1 + 4x_2 \leq 12$$
$$x_1 + x_2 \geq 5$$
$$x_1, x_2 \geq 0$$

The problem is written in equation form as

$$Z - 2x_1 - 3x_2 - 0s_1 - 0S_2 + MA_1 = 0$$
$$3x_1 + 4x_2 + s_1 = 12$$
$$x_1 + x_2 - S_2 + A_1 = 5$$
$$x_1, x_2, s_1, S_2, A_1 \geq 0$$

The initial simplex table corresponding to above equations is shown in Table 20.16a. However, Table 20.16a is inconsistent because of non-zero coefficient in the Z-row corresponding to the current basic variable. Before applying simplex algorithm it is made consistent as

Table 20.16a Initial simplex table for the example 20.10.1

Iteration	Basic	x_1	x_2	s_1	S_2	A_1	Solution	Ratio
	Z	−2	−3	0	0	M	0	
	s_1	3	4	1	0	0	12	
	A_1	1	1	0	−1	1	5	

Table 20.16b Revised initial simplex table for the example 20.10.1

Iteration	Basic	x_1	x_2	s_1	S_2	A_1	Solution	Ratio
	Z	−2−M	−3−M	0	M	0	−5M	
	s_1	3	4	1	0	0	12	3
	A_1	1	1	0	−1	1	5	5

Table 20.16c Iteration 1 of the example 20.10.1

Iteration	Basic	x_1	x_2	s_1	S_2	A_1	Solution	Ratio
	Z	0.25(1−M)	0	0.25(3+M)	M	0	9−2M	
1	x_2	3/4	1	1/4	0	0	3	4
	A_1	1/4	0	−1/4	−1	1	2	8

shown in Table 20.16b. The intermediate and final iterations are shown in Tables 20.16c and 20.16d.

The presence of artificial variable A_1 having positive value in the optimum tableau suggest that the solution to above problem is infeasible.

Table 20.16d Iteration 2 of the example 20.10.1

Iteration	Basic	x_1	x_2	s_1	S_2	A_1	Solution	Ratio
	Z	0	(M−1)/3	(M+2)/3	M	0	8−M	
2	x_1	1	4/3	1/3	0	0	4	
	A_1	0	−1/3	−1/3	−1	1	1	

Figure 20.9 Graphical illustration of infeasibility of the example 20.10.1

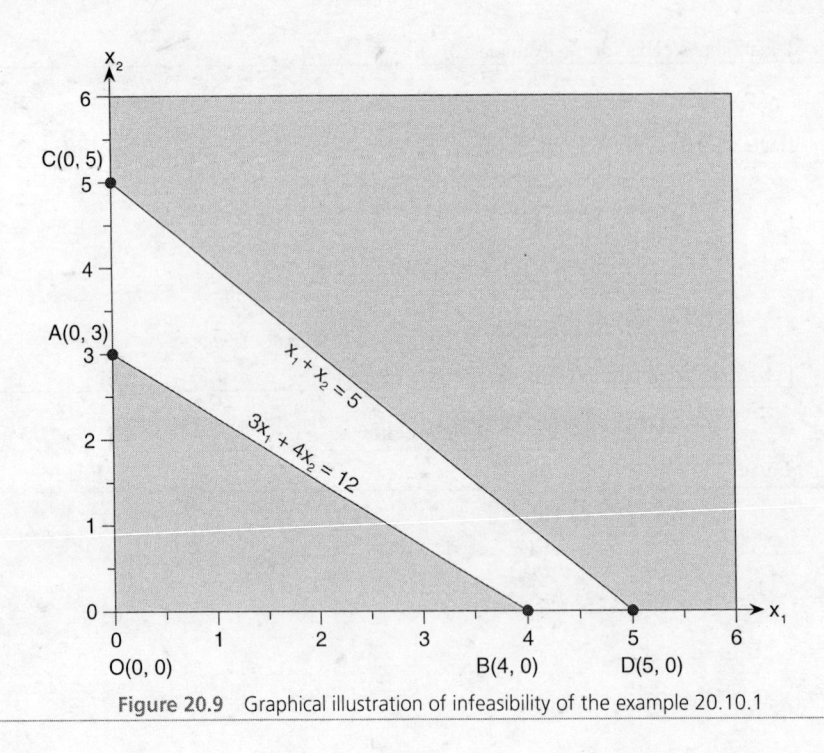

Alternatively the infeasibility may also be checked graphically as observed in the Figure 20.9.

The two constraint equations in this case when plotted have no common feasible region suggesting that both the constraints will never be satisfied

20.10.2 Unboundedness

Unboundedness describes linear programs that do not have finite solutions. It occurs in maximization problems, for example, when a solution variable can be made infinitely large without violating a constraint. Unboundedness can be identified prior to reaching the final table. If all the ratios turn out to be negative or undefined, it indicates that the problem is unbounded. This situation again indicates that the model is poorly constructed.

Example 20.10.2

Maximize

$$Z = x_1 + 2x_2$$

Subject to:

$$-x_1 + x_2 \leq 10$$
$$x_2 \leq 20$$
$$x_1, x_2 \geq 0$$

Table 20.17a Initial simplex table for the example 20.10.2

Iteration	Basic	Coefficients x_1	x_2	s_1	s_2	Solution	Ratio
	Z	−1	−2	0	0		
Initial	s_1	−1	1	1	0	10	
	s_2	0	1	0	1	20	

Table 20.17b Iteration 1 of the example 20.10.2

Iteration	Basic	Coefficients x_1	x_2	s_1	s_2	Solution	Ratio
	Z	−1	−2	0	0		
1	s_1	−1	1	1	0	10	10
	s_2	0	1	0	1	20	20

Table 20.17c Iteration 2 of the example 20.10.2

Iteration	Basic	Coefficients x_1	x_2	s_1	s_2	Solution	Ratio
	Z	−3	0	2	0	20	
2	x_2	−1	1	1	0	10	
	s_2	1	0	−1	1	10	10

The problem is written in equation form as

$$Z - x_1 - 2x_2 - 0s_1 - 0s_2 = 0$$
$$-x_1 + x_2 + s_1 = 10$$
$$x_2 + s_2 = 20$$
$$x_1, x_2, s_1, s_2 \geq 0$$

The initial simplex table corresponding to above equations is shown in Table 20.17a. The intermediate iterations are shown in Tables 20.17b to 20.17d.

Table 20.17d Iteration 3 of the example 20.10.2

Iteration	Basic	Coefficients x_1	x_2	s_1	s_2	Solution	Ratio
	Z	0	0	−1	3	50	
3	x_2	0	1	0	1	20	∞
	x_1	1	0	−1	1	10	−10

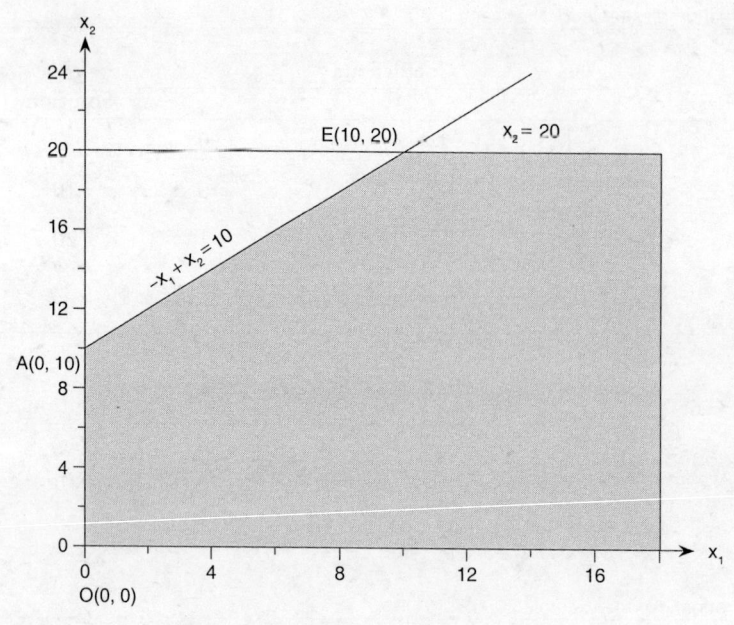

Figure 20.10 Graphical illustration of unboundedness for the example 20.10.2

The last iteration (see Table 20.17d) shows none of the ratio is positive indicating the problem is unbounded. The unboundedness of the problem may be graphically verified from the Figure 20.10.

20.10.3 Degeneracy

It may develop when a problem contains a redundant constraint; that is presence of one or more of the constraints in the formulation make another unnecessary. For example out of the following three constraints, the third constraint is unnecessary.

$$x_1 \leq 10$$
$$x_2 \leq 10$$
$$x_1 + x_2 \leq 20$$

Degeneracy arises when ratio calculation is made. If there is a tie for the smallest ratio, this is a signal that degeneracy exists. As a result of this when next table is developed, one of the variables in the solution will have a value of zero. Degeneracy could lead to cycling or circling in which the simplex algorithm alternates back and forth between the same non-optimal solution. Degeneracy may also be temporarily observed. In case the selection of one of the variables having tied ratio results into degeneracy, select the other variable with same ratio and proceed.

Example 20.10.3

Maximize

$$Z = 2x_1 + 4x_2$$

Subject to:

$$2x_1 + 3x_2 \leq 12$$
$$7x_1 + 6x_2 \leq 24$$
$$x_1, x_2 \geq 0$$

The initial table corresponding to above equations is shown in Table 20.18a. Iteration 1 is shown in Table 20.18b. The Table 20.18b shows same intercept or ratio for both s_1 and s_2 hence, this indicates degeneracy and in next iteration one of the basic varable will have value equal to zero. Breaking the ties arbitrarily and assuming that s_1 is the exiting variable we get the optimal solution having one of the basic variable $s_2 = 0$ indicating that the problem is a case of degeneracy *i.e.* one of the constraint is redundant (see Table 20.18c).

Table 20.18a   Initial simplex table for the example 20.10.3

Iteration	Basic	x_1	x_2	s_1	s_2	Solution	Ratio
	Z	−2	−4	0	0	0	
Initial	s_1	2	3	1	0	12	
	s_2	7	6	0	1	24	

Table 20.18b   Iteration 1 of the example 20.10.3

Iteration	Basic	x_1	x_2	s_1	s_2	Solution	Ratio
	Z	−2	−4	0	0	0	
1	s_1	2	3	1	0	12	4
	s_2	7	6	0	1	24	4

Table 20.18c   Iteration 2 of the example 20.10.3

Iteration	Basic	x_1	x_2	s_1	s_2	Solution	Ratio
	Z	2/3	0	4/3	0	16	
2	x_2	2/3	1	1/3	0	4	
	s_2	3	0	−2	1	0	

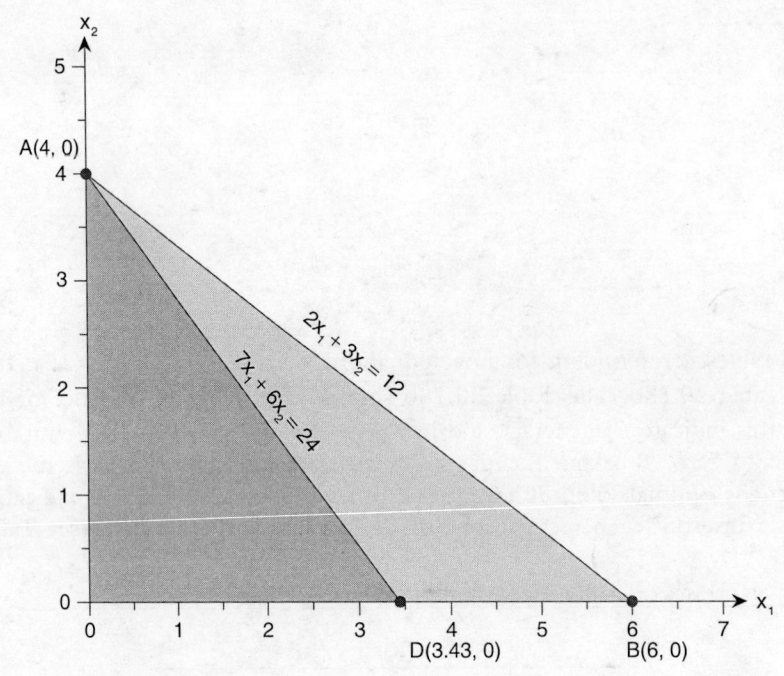

Figure 20.11 Graphical representation of degeneracy for the example 20.10.3

The degeneracy may be observed from the plot of the constraints also as shown in Figure 20.11. In this case the constraint $7x_1 + 6x_2 \leq 24$ is redundant.

20.10.4 Multiple or Alternative Optima

Multiple, or alternate, optimal solutions can be spotted when the simplex method is being used by looking at the final table. Look at the coefficients of non-basic variables in z-row. In case any non-basic variable has 0 value in z-row, it indicates alternate optima. Letting the non-basic variable with 0 in z-row coefficient will not alter the optimum value. This can be verified by observing the solution using graphical method also as shown in Figure 20.12. Multiple optimal values are useful in practice because it offers number of alternatives to a decision maker.

Example 20.10.4

Maximize

$$Z = 3x_1 + 6x_2$$

Subject to:

$$5x_1 + 3x_2 \leq 15$$
$$x_1 + 2x_2 \leq 4$$
$$x_1, x_2 \geq 0$$

Initial simplex tableau for the above problem is shown in Table 20.19a. Optimal solution is shown in Table 20.19b.

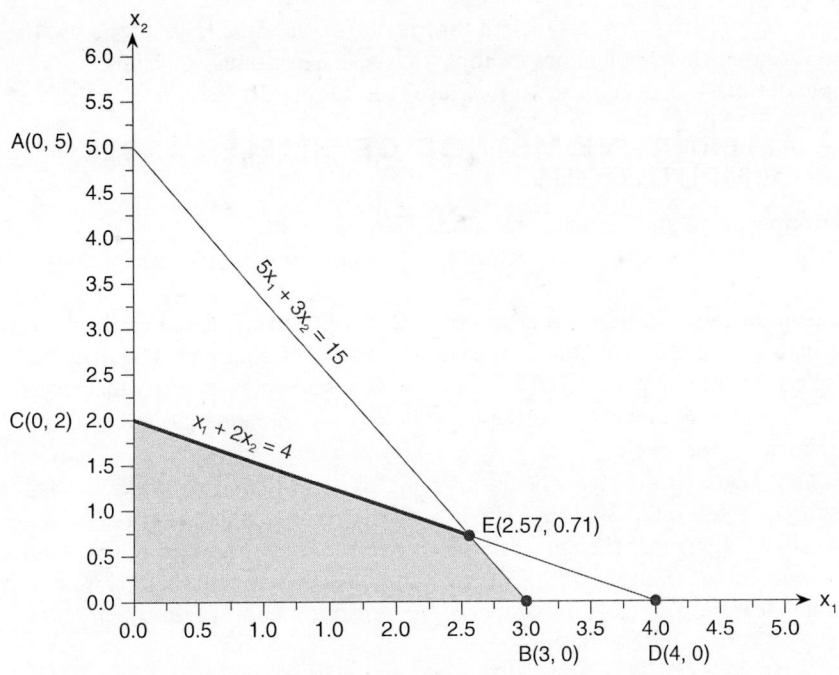

LINEAR PROGRAMMING IN CONSTRUCTION MANAGEMENT | 727 |

Figure 20.12 Graphical representations of multiple optima for the example 20.10.4

Table 20.19a Initial simplex table for the example 20.10.4

Iteration	Basic	\multicolumn{4}{c}{Coefficients}	Solution	Ratio
		x_1	x_2	s_1	s_2		
	Z	−3	−6	0	0	0	
Initial	s_1	5	3	1	0	15	5
	s_2	1	2	0	1	4	2

Table 20.19b Iteration 1 of the example 20.10.4

Iteration	Basic	x_1	x_2	s_1	s_2	Solution	Ratio
	Z	0	0	0	3	12	
1 (Optimal)	s_1	3.5	0	1	−3/2	9	
	x_2	1/2	1	0	1/2	2	

The presence of zero Z-row coefficient for the non-basic variable x_1 in the optimal tableau 20.19b indicate that the problem has multiple or alternate optimal solutions.

Multiple optima situation is shown graphically in Figure 20.12.

20.11 ALTERNATIVE METHOD OF SIMPLEX TABLE COMPUTATIONS

In this section we learn an alternative method of simplex table computations. We learn about identity and inverse matrices. We also learn how to derive the dual variable values in different iterations.

Initial simplex table for a primal is shown in Table 20.20a. It is imperative here to have a close look at this table especially under the starting variables columns. The diagonal elements under these columns are equal to 1. They form what is known as 'identity matrix'.

P_1, P_2, P_m, and P_n represent constraint coefficient vectors corresponding to real variables x_1, x_2, x_m and x_n respectively.

Subsequent Gauss–Jordan row operations on this identity matrix produces what is known as 'inverse matrix' (see Table 20.20b). The inverse matrix in any iteration can be quite useful.

It is possible to generate the entire simplex table for an iteration using the original primal data (as given in Table 20.20a) and the inverse matrix associated with that iteration. For convenience, we divide a simplex table in two parts: (a) constraint columns both left and right-hand sides and (b) objective row coefficients.

(a) Constraint column computations
This is performed using the following expression:

$$\begin{pmatrix} \text{Constraint column in} \\ \text{iteration } i \end{pmatrix} = \begin{pmatrix} \text{Inverse in} \\ \text{iteration } i \end{pmatrix} \times (\text{Original constraint column}) \quad (20.17)$$

In fact it is also possible to generate the dual coefficients at any iteration using the following expression:

$$\begin{pmatrix} \text{Value of} \\ \text{dual variables in iteration } i \end{pmatrix} = \begin{pmatrix} \text{Row vector of original} \\ \text{objective coefficients of} \\ \text{primal basic variables} \\ \text{in the same order in} \\ \text{iteration } i \end{pmatrix} \times \begin{pmatrix} \text{Inverse matrix at} \\ \text{iteration } i \end{pmatrix} \quad (20.18)$$

Table 20.20a General form of a simplex starting table

	Real variables				Starting variables					
	x_1	x_2	...	x_n	s_1	s_2	...	s_n		
Objective Z-row →	c_1	c_2	...	c_n	0	0	0	0	=	R.H.S
	a_{11}	a_{12}	a_{13}	...	1	0	...	0	=	b_1
	a_{21}	a_{22}	a_{23}	...	0	1	...	0	=	b_2
	...	...	...	...	...	...	...	0	=	...
	a_{m1}	a_{m2}	a_{m3}	...	0	0	...	1	=	b_m
	P_1	P_2	P_m	P_n	Location of Identity matrix					

Table 20.20b General form of a simplex table at iteration i

	Real variables				Starting variables				
	x_1	x_2	...	x_n	s_1	s_2	...	s_n	
Objective Z-row →	$Z_1 - c_1$	$Z_2 - c_2$	...	$Z_n - c_n$	y_1	y_2	y_{n-1}	y_n	= R.H.S
					Location of inverse matrix in iteration i resulting from Gauss-Jordan row operations				= = = =

Further in any simplex iteration, the objective function coefficients of a variable may be computed either using Equation (20.19a) or (20.19b) as given below

$$\begin{pmatrix} \text{Primal Objective} \\ \text{function coefficient} \\ \text{of variable } x_j \end{pmatrix} = \begin{pmatrix} \text{Left Hand Side of} \\ j_{th} \\ \text{Dual constraint} \end{pmatrix} - \begin{pmatrix} \text{Right Hand Side of} \\ j_{th} \\ \text{Dual constraint} \end{pmatrix} \quad (20.19a)$$

$$\begin{pmatrix} \text{Primal Objective} \\ \text{function coefficient} \\ \text{of variable } x_j (Z_j - C_j) \end{pmatrix} = \begin{pmatrix} \text{Row vector of} \\ \text{Dual} \\ \text{variable}(\Upsilon)^T \end{pmatrix} \times \begin{pmatrix} \text{Constraint column} \\ \text{vector} \\ \text{corresponding} \\ \text{to } x_j (P_j) \end{pmatrix} - \begin{pmatrix} \text{Objective function} \\ \text{coefficient of } x_j (C_j) \end{pmatrix}$$

(20.19b)

In case the optimal inverse matrix as shown in Table 20.20c is known beforehand, dual values and Z-row coefficients may be computed directly without intermediate steps using following expressions.

$$\begin{pmatrix} \text{Constraint column in} \\ \text{optimal iteration} \end{pmatrix} = \begin{pmatrix} \text{Optimum Inverse} \\ \text{Matrix} \end{pmatrix} \times (\text{Original constraint column}) \quad (20.20)$$

$$\begin{pmatrix} \text{Optimal Value of} \\ \text{Dual Variables} \end{pmatrix} = \begin{pmatrix} \text{Row vector of the Objective} \\ \text{Coefficients} \\ \text{corresponding to Optimal} \\ \text{Primal Basic Variables} \end{pmatrix} \times (\text{Optimum Inverse Matrix}) \quad (20.21)$$

The computation details are illustrated with the help of the following example.

Example 20.11.1
A prefabricated building element company manufactures three standard structural elements: beam, wall panels and roof decks for low-cost housing. The production process requires three types of skilled labour, for formwork, reinforcement, and concreting activities. Table 20.21a shows availabilities of these resources, their usage in producing respective elements and profits per unit.

Table 20.20c General form of a simplex table at optimal iteration

	Real variables				Starting variables				
	x_1	x_2	...	x_n	s_1	s_2	...	s_n	
Objective Z-row →	$Z_1 - c_1$	$Z_2 - c_2$	...	$Z_n - c_n$	$y_1 o$	$y_2 o$	$y_{n-1} o$	$y_n o$	= R.H.S
					Location of optimal inverse matrix				=
									=
									=
									=

It is desired to maximize the profit for the company. Determine the structural element combination that the manufacturer would produce for optimum profit.

Solution

From the concepts of simplex all slack solution method, it is not difficult to solve the above problem; however, we want to revisit the solution steps to understand the terms such as identity and inverse matrix.

The profit function for the given primal problem can be written as:
Maximize

$$Z = 600x_1 + 450x_2 + 200x_3$$

Subject to:

$$3x_1 + 2x_2 + x_3 \leq 445$$
$$2x_1 + x_2 + x_3 \leq 300$$
$$2x_1 + 2x_2 + x_3 \leq 400$$
$$x_1, x_2, x_3 \geq 0$$

Where x_1 = Number of beam elements produced;
x_1 = Number of wall panel elements produced; and
x_3 = Number of roof deck elements produced.

The dual of the above problem can be written in the following form:
Minimize

$$Z' = 445y_1 + 300y_2 + 400y_3$$

Subject to

$$3y_1 + 2y_2 + 2y_3 \geq 600$$
$$2y_1 + y_2 + 2y_3 \geq 450$$
$$y_1 + y_2 + y_3 \geq 200$$
$$y_1 \geq 0$$
$$y_2 \geq 0$$
$$y_3 \geq 0$$

Table 20.21a Data for the example 20.11.1

Resource for Activity	Beam (x_1)	Wall Panel (x_2)	Roof Deck (x_3)	Daily Availability
Formwork labor	3	2	1	445
Reinforcement labor	2	1	1	300
Concreting labor	2	2	1	400
Profit per unit (₹)	600	450	200	

In equation form, the primal objective function and the constraints can be written as:

$$Z - 600x_1 - 450x_2 - 200x_3 - 0s_1 - 0s_2 - 0s_3 = 0$$

$$3x_1 + 2x_2 + x_3 + s_1 = 445 \quad (20.22)$$

$$2x_1 + x_2 + x_3 + s_2 = 300 \quad (20.23)$$

$$2x_1 + 2x_2 + x_3 + s_3 = 400 \quad (20.24)$$

$$x_1 \geq 0, x_2 \geq 0, x_3 \geq 0, s_1 \geq 0, s_2 \geq 0, \text{ and } s_3 \geq 0$$

An initial simplex table, for the primal is shown in Table 20.21b. Please note, the 'identity matrix' under starting variables s_1, s_2, and s_3 in Table 20.21b. This matrix is highlighted in Table 20.21b. Subsequent Gauss–Jordan row operations on this identity matrix produces what is known as inverse matrix. As mentioned the inverse matrix in any iteration can be quite useful.

We now illustrate, the generation of an entire simplex table using the original primal data (as given in Table 20.21b) and the inverse associated with the iteration. For illustration purpose the Example 20.11.1 has been solved by commonly used simplex algorithms and corresponding inverse matrix has been used for regenerating entire table. Table 20.21c shows iteration 1.

We compute (a) constraint columns on both left and right-hand sides; and (b) objective row coefficients.

(b) Constraint column computations
Constraint column coefficient vectors may be computed using Equation 20.17. The calculation steps are illustrated for second iteration.

Table 20.21b Initial simplex table for the example 20.11.1

Iteration	Basic	x_1	x_2	x_3	s_1	s_2	s_3	Solution	Ratio
	Z	−600	−450	−200	0	0	0	0	
1	s_1	3	2	1	1	0	0	445	
	s_2	2	1	1	0	1	0	300	
	s_3	2	2	1	0	0	1	400	

Inverse matrix for second iteration and original constraints columns have been taken from Table 20.21d and Table 20.21b, respectively, as

$$\begin{pmatrix} \text{Inverse in} \\ \text{iteration 2} \end{pmatrix} = \begin{pmatrix} \dfrac{1}{3} & 0 & 0 \\ -\dfrac{2}{3} & 1 & 0 \\ -\dfrac{2}{3} & 0 & 1 \end{pmatrix}$$

$$\text{Original constraint column of } x_1 = \begin{pmatrix} 3 \\ 2 \\ 2 \end{pmatrix}$$

$$\text{Original constraint column of } x_2 = \begin{pmatrix} 2 \\ 1 \\ 2 \end{pmatrix}$$

$$\text{Original constraint column of } x_3 = \begin{pmatrix} 1 \\ 1 \\ 1 \end{pmatrix}$$

$$\text{Original constraint column of R.H.S or solution} = \begin{pmatrix} 445 \\ 300 \\ 400 \end{pmatrix}$$

Substituting above values in Equation (20.17):

$$(x_1 \text{ column in iteration 2}) = \begin{pmatrix} \dfrac{1}{3} & 0 & 0 \\ -\dfrac{2}{3} & 1 & 0 \\ -\dfrac{2}{3} & 0 & 1 \end{pmatrix} \times \begin{pmatrix} 3 \\ 2 \\ 2 \end{pmatrix} = \begin{pmatrix} 1 \\ 0 \\ 0 \end{pmatrix}$$

$$(x_2 \text{ column in iteration 2}) = \begin{pmatrix} \dfrac{1}{3} & 0 & 0 \\ -\dfrac{2}{3} & 1 & 0 \\ -\dfrac{2}{3} & 0 & 1 \end{pmatrix} \times \begin{pmatrix} 2 \\ 1 \\ 2 \end{pmatrix} = \begin{pmatrix} \dfrac{2}{3} \\ -\dfrac{1}{3} \\ \dfrac{2}{3} \end{pmatrix}$$

$$(x_3 \text{ column in iteration 2}) = \begin{pmatrix} \dfrac{1}{3} & 0 & 0 \\ -\dfrac{2}{3} & 1 & 0 \\ -\dfrac{2}{3} & 0 & 1 \end{pmatrix} \times \begin{pmatrix} 1 \\ 1 \\ 1 \end{pmatrix} = \begin{pmatrix} \dfrac{1}{3} \\ \dfrac{1}{3} \\ \dfrac{1}{3} \end{pmatrix}$$

LINEAR PROGRAMMING IN CONSTRUCTION MANAGEMENT | 733 |

Table 20.21c Iteration 1 of the example 20.11.1

Iteration	Basic	\multicolumn{6}{c	}{Coefficients}	Solution	Ratio
		x_1	x_2	x_3	s_1	s_2	s_3		
	Z	−600	−450	−200	0	0	0	0	
1	s_1	3	2	1	1	0	0	445	148.3
	s_2	2	1	1	0	1	0	300	150
	s_3	2	2	1	0	0	1	400	200

$$(RHS \text{ or solution column in iteration 2}) = \begin{pmatrix} \frac{1}{3} & 0 & 0 \\ -\frac{2}{3} & 1 & 0 \\ -\frac{2}{3} & 0 & 1 \end{pmatrix} \times \begin{pmatrix} 445 \\ 300 \\ 400 \end{pmatrix} = \begin{pmatrix} \frac{445}{3} \\ \frac{10}{3} \\ \frac{310}{3} \end{pmatrix}$$

The dual coefficients can be computed using Equation (20.18) as shown below:

$$(y_1, y_2, y_3) = (600, 0, 0) \times \begin{pmatrix} \frac{1}{3} & 0 & 0 \\ -\frac{2}{3} & 1 & 0 \\ -\frac{2}{3} & 0 & 1 \end{pmatrix} = (200, 0, 0)$$

Please note, the order of objective function coefficients (600 0 0) corresponding to the basic variables (x_1, s_2, s_3), respectively in iteration 2.

(c) Objective row coefficients computations

At any simplex iteration, the objective function coefficients of a variable x_j may be computed using Equations (20.19a or 20.19b).

Thus, Z-row entries for the second iteration may be calculated using Equation (20.19b) as follows. It may be recollected that dual vector $Y = (200\ 0\ 0)$ from the previous section.

$$Z_1 - c_1 = Y \times P_1 - c_1 = (200\ 0\ 0) \times \begin{pmatrix} 3 \\ 2 \\ 2 \end{pmatrix} - 600 = 0$$

$$Z_2 - c_2 = Y \times P_2 - c_2 = (200\ 0\ 0) \times \begin{pmatrix} 2 \\ 1 \\ 2 \end{pmatrix} - 450 = -50$$

$$Z_3 - c_3 = Y \times P_3 - c_3 = (200\ 0\ 0) \times \begin{pmatrix} 1 \\ 1 \\ 1 \end{pmatrix} - 200 = 0$$

Table 20.21d Iteration 2 of the example 20.11.1

Iteration	Basic	Coefficients						Solution	Ratio
		x_1	x_2	x_3	s_1	s_2	s_3		
2	Z	0	50	0	200	0	0	89,000	
	x_1	1	$\frac{2}{3}$	$\frac{1}{3}$	$\frac{1}{3}$	0	0	$\frac{445}{3}$	222.5
	s_2	0	$-\frac{1}{3}$	$\frac{1}{3}$	$-\frac{2}{3}$	1	0	$\frac{10}{3}$	
	s_3	0	$\frac{2}{3}$	$\frac{1}{3}$	$-\frac{2}{3}$	0	1	$\frac{310}{3}$	155

$$Z_4 - c_4 = Y \times P_4 - c_4 = (200\ 0\ 0) \times \begin{pmatrix} 1 \\ 0 \\ 0 \end{pmatrix} - 0 = 200$$

$$Z_5 - c_5 = Y \times P_5 - c_5 = (200\ 0\ 0) \times \begin{pmatrix} 0 \\ 1 \\ 0 \end{pmatrix} - 0 = 0$$

$$Z_6 - c_6 = Y \times P_6 - c_6 = (200\ 0\ 0) \times \begin{pmatrix} 0 \\ 0 \\ 1 \end{pmatrix} - 0 = 0$$

$$Z_7 - c_7 = Y \times P_7 - c_7 = (200\ 0\ 0) \times \begin{pmatrix} 445 \\ 300 \\ 400 \end{pmatrix} - 0 = 89,000$$

The above calculations are also shown by conventional Gauss–Jordan row reduction as part of iteration 2 in Table 20.21d.

The third iteration which happens to generate optimal profit can also be generated in a similar manner as shown below.

Dual variables and z-row coefficients in iteration 3 can be computed using Equations 20.18 and 20.19b respectively as shown below:

$$(y_1, y_2, y_3) = (600, 0, 450) \times \begin{pmatrix} 1 & 0 & -1 \\ -1 & 1 & 0.5 \\ -1 & 0 & 1.5 \end{pmatrix} = (150\ 0\ 75)$$

$$Z_1 - c_1 = Y \times P_1 - c_1 = (150\ 0\ 75) \times \begin{pmatrix} 3 \\ 2 \\ 2 \end{pmatrix} - 600 = 0$$

$$Z_2 - c_2 = Y \times P_2 - c_2 = (150\ 0\ 75) \times \begin{pmatrix} 2 \\ 1 \\ 2 \end{pmatrix} - 450 = 0$$

$$Z_3 - c_3 = Y \times P_3 - c_3 = (150\ 0\ 75) \times \begin{pmatrix} 1 \\ 1 \\ 1 \end{pmatrix} - 200 = 25$$

$$Z_4 - c_4 = Y \times P_4 - c_4 = (150\ 0\ 75) \times \begin{pmatrix} 1 \\ 0 \\ 0 \end{pmatrix} - 0 = 150$$

$$Z_5 - c_5 = Y \times P_5 - c_5 = (150\ 0\ 75) \times \begin{pmatrix} 0 \\ 1 \\ 0 \end{pmatrix} - 0 = 0$$

$$Z_6 - c_6 = Y \times P_6 - c_6 = (150\ 0\ 75) \times \begin{pmatrix} 0 \\ 0 \\ 1 \end{pmatrix} - 0 = 75$$

$$Z_7 - c_7 = Y \times P_7 - c_7 = (150\ 0\ 75) \times \begin{pmatrix} 445 \\ 300 \\ 400 \end{pmatrix} - 0 = 96{,}750$$

$$(R.H.S \text{ or solution column in iteration 3}) = \begin{pmatrix} 1 & 0 & -1 \\ -1 & 1 & 0.5 \\ -1 & 0 & 1.5 \end{pmatrix} \times \begin{pmatrix} 445 \\ 300 \\ 400 \end{pmatrix} = \begin{pmatrix} 45 \\ 55 \\ 155 \end{pmatrix}$$

The above calculations are also shown by conventional Gauss–Jordan row reduction as part of iteration 3 as shown in Table 20.21e.

The optimal profit is ₹ 96,750 and $x_1 = 45$ and $x_2 = 155$ for the example problem.

Please note, the **'optimal inverse matrix'** under the starting variables $s_1, s_2,$ and s_3 **columns** in Table 20.21e. This matrix is highlighted in the table. The optimal inverse matrix is very important for all calculations in the context of changes made in the R.H.S, objective function, and addition of a new activity.

Table 20.21e Iteration 3 of the example 20.11.1

Iteration	Basic	x_1	x_2	x_3	s_1	s_2	s_3	Solution	Ratio
	Z	0	0	25	150	0	75	96,750	
3	x_1	1	0	0	1	0	−1	45	
(Optimal)	s_2	0	0	0.5	−1	1	0.5	55	
	x_2	0	1	0.5	−1	0	1.5	155	

20.11 SENSITIVITY ANALYSIS

Construction projects are dynamic in nature. In order to cater to the changing circumstances, plans need to be revised, resources need to be rearranged for the changed circumstances, and so on. For example, in a given linear programming problem, the following situations may arise:

1. There may be changes in the right-hand side of the constraint equations. For example, there may be increase or decrease in available resources in one or more operation stages of production of precast element.
2. There may be changes in the objective function coefficients. For example, in case of bulk orders, supplier may offer a discounted price on one or more products.
3. There may be an additional constraint. For example, due to specific client requirement the product may have to be sent to another operation during production phase requiring a constraint stating the relationship between resource usage and availability for the new operation.

It would indeed be cumbersome to solve the problem again and again to suit the above changing circumstances. Sensitivity analysis is, thus, aimed to investigate the changes in the optimal solution resulting out of the changes made in the various input parameters of the linear programming problem as mentioned in the above three points.

We consider the Example 20.11.1 to illustrate the above mentioned three situations. All the situations discussed in subsequent section are shown in a schematic diagram in Figure 20.13.

20.11.1 Changes in the Constraints

In this section, we discuss the changes made in the right-hand side of constraints. We also discuss the range of R.H.S values so that the solution still remains feasible, and finally, we

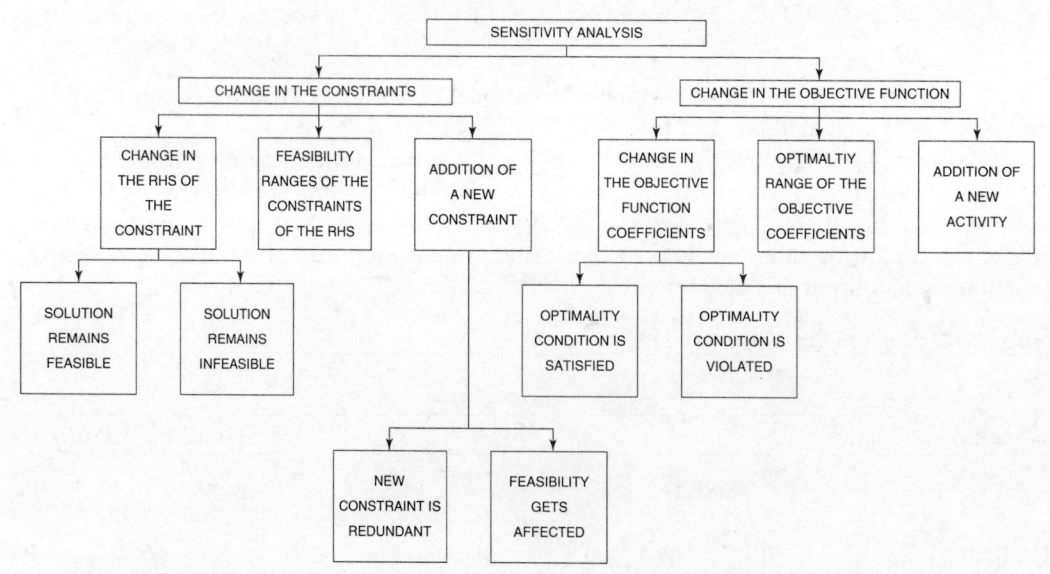

Figure 20.13 Schematic representations of different cases considered in sensitivity analysis

see the effect of adding a new constraint. All these cases are explained with the help of Example 20.11.1.

Case 1: Changes made in the R.H.S of a constraint

On account of changes made in the right-hand side of the constraints, there might be two possibilities. In one situation, the current basic solution remains feasible and, in the other situation, the current basic solution becomes infeasible.

(a) The solution remains feasible

Suppose, the precast structural element manufacturer wants to increase the resources for various activities by 20% to 534, 360 and 480, respectively. Then the changed value of basic variables may be calculated using the formula:

$$\begin{pmatrix} \text{New right-hand side} \\ \text{of tableau in iteration } i \end{pmatrix} = \begin{pmatrix} \text{Inverse in} \\ \text{iteration } i \end{pmatrix} \times \begin{pmatrix} \text{New right-hand side} \\ \text{of the constraints} \end{pmatrix}$$

Using the above formula, the revised optimum production quantity can be calculated as:

$$\begin{pmatrix} x_1 \\ s_2 \\ x_2 \end{pmatrix} = \begin{pmatrix} 1 & 0 & -1 \\ -1 & 1 & 0.5 \\ -1 & 0 & 1.5 \end{pmatrix} \times \begin{pmatrix} 534 \\ 360 \\ 480 \end{pmatrix} = \begin{pmatrix} 54 \\ 66 \\ 186 \end{pmatrix}$$

Please note, the order of the new right-hand side $(x_1, s_2, x_2)^T$, which is in the same order where the basic variables appeared in the last iteration as given by Table 20.21e.

Hence, the manufacturer's profit after increasing resources by 20% will reach to

$$Z = 54 \times 600 + 186 \times 450 = 116,100.$$

In this case, the increase in resources has not affected the feasibility of the solution as non-negativity constraints are satisfied as the value of $x_1 = 54$, $s_2 = 66$, and $x_2 = 186$, all feasible values.

(b) The solution becomes infeasible

Suppose, the surplus of 55 man-days in reinforcement labour (in the last iteration as given by Table 20.21e is planned to be utilized in formwork. In other words, let's assume that the new resource allocation for formwork, reinforcement, and concreting is 500 (445 + 55), 245 (300 – 55), and 400, respectively. Please note, it is assumed here that there is a possibility of interchangeability of labour among different activities.

As before, the changed value of basic variables may be calculated as follows:

$$\begin{pmatrix} x_1 \\ s_2 \\ x_2 \end{pmatrix} = \begin{pmatrix} 1 & 0 & -1 \\ -1 & 1 & 0.5 \\ -1 & 0 & 1.5 \end{pmatrix} \times \begin{pmatrix} 500 \\ 245 \\ 400 \end{pmatrix} = \begin{pmatrix} 100 \\ -55 \\ 100 \end{pmatrix}$$

Hence, $Z = 100 \times 600 + 100 \times 450 = 105,000.$

However, the resulting solution is infeasible, because $s_2 = -55$ in the solution mix is violating the non-negativity constraint $s_2 \geq 0$.

Table 20.22a Changes in R.H.S making infeasible, but optimal solution

Iteration	Basic	x_1	x_2	x_3	s_1	s_2	s_3	Solution
	Z	0	0	25	150	0	75	105,000
	x_1	1	0	0	1	0	−1	100
4	s_2	0	0	0.5	−1	1	0.5	−55
(Optimal, but infeasible)	x_2	0	1	0.5	−1	0	1.5	100
	$\left|\dfrac{Z_j - C_j}{\alpha_{rj}}\right|$				150			

In such cases of the new solution becoming infeasible the dual simplex algorithm is used to retrieve the feasibility of the solution as shown in Tables 20.22a and 20.22b.

Hence, the solution is $(x_1, s_1, x_2) = (45, 55, 155) \, Z = 96{,}750$.

Reader may check the feasibility of the solution with another resource constraint vector (495, 250, 450) in which case 50 resources are added to formwork and concreting team, whereas 50 resources are withdrawn from the reinforcement team

$$\begin{pmatrix} x_1 \\ s_2 \\ x_2 \end{pmatrix} = \begin{pmatrix} 1 & 0 & -1 \\ -1 & 1 & 0.5 \\ -1 & 0 & 1.5 \end{pmatrix} \times \begin{pmatrix} 495 \\ 250 \\ 450 \end{pmatrix} = \begin{pmatrix} 45 \\ -20 \\ 180 \end{pmatrix}$$

Hence, the solution is $(x_1, s_1, x_2) = (25, 20, 200), Z = 105{,}000$. The calculations are shown in Tables 20.22c and 20.22d.

Case 2: Feasibility range of the elements of the right-hand side

We can also find the range of values of R.H.S, so that the current solution remains feasible.

Suppose, we need to find the feasible range of resources for formwork.

In case, the manufacturer wants to ascertain the range of a particular resource for which current solution will remain feasible, the same can be found with the help of feasibility

Table 20.22b Applying dual simplex for retrieving feasibility of the solution

Iteration	Basic	x_1	x_2	x_3	s_1	s_2	s_3	Solution	Ratio
	Z	0	0	100	0	150	150	96,750	
5	x_1	1	0	0.5	0	1	−0.5	45	
(Feasible and optimal)	s_1	0	0	−0.5	1	−1	−0.5	55	
	x_2	0	1	0	0	−1	1	155	

Table 20.22c Changes in R.H.S making infeasible but optimal solution

Iteration	Basic	x_1	x_2	x_3	s_1	s_2	s_3	Solution
	Z	0	0	25	150	0	75	108,000
	x_1	1	0	0	1	0	−1	45
4 (Optimal, but infeasible)	s_2	0	0	0.5	−1	1	0.5	−20
	x_2	0	1	0.5	−1	0	1.5	180
	$\left\lvert\dfrac{Z_j - C_j}{\alpha_{rj}}\right\rvert$				150			

(non-negativity) condition using the optimal inverse matrix. The following example illustrates the determination of the range of formwork resource for the current solution to remain feasible.

Let us assume that the right-hand side is now changed to

$$\begin{pmatrix} 445 + D_1 \\ 300 \\ 400 \end{pmatrix},$$

where D_1 represents the change in the capacity of resource for formwork which may be positive or negative in practical situations. The current basic solutions remain feasible if all the current basic variables are positive, that is

$$\begin{pmatrix} x_1 \\ s_2 \\ x_2 \end{pmatrix} = \begin{pmatrix} 1 & 0 & -1 \\ -1 & 1 & 0.5 \\ -1 & 0 & 1.5 \end{pmatrix} \times \begin{pmatrix} 445 + D_1 \\ 300 \\ 400 \end{pmatrix} \geq \begin{pmatrix} 0 \\ 0 \\ 0 \end{pmatrix}$$

$$\begin{pmatrix} x_1 \\ s_2 \\ x_2 \end{pmatrix} = \begin{pmatrix} 45 + D_1 \\ 55 - D_1 \\ 155 - D_1 \end{pmatrix} \geq \begin{pmatrix} 0 \\ 0 \\ 0 \end{pmatrix}$$

Table 20.22d Applying dual simplex for retrieving feasibility of the solution

Iteration	Basic	x_1	x_2	x_3	s_1	s_2	s_3	Solution	Ratio
	Z	0	0	100	0	150	150	105,000	
5 (Feasible and Optimal)	x_1	1	0	0.5	0	1	−0.5	25	
	s_1	0	0	−0.5	1	−1	−0.5	20	
	x_2	0	1	0	0	−1	1	200	

The above condition gives the following limits to the value of D_1

$$(x_1 \geq 0) \Rightarrow (45 + D_1) \geq 0 \text{ or } D_1 \geq -45$$
$$(s_2 \geq 0) \Rightarrow (55 - D_1) \geq 0 \text{ or } -D_1 \geq -55 \text{ or } D_1 \leq 55$$
$$(x_2 \geq 0) \Rightarrow (155 - D_1) \geq 0 \text{ or } D_1 \leq 155$$

Thus, the current basic solution remains feasible if $-45 \leq D_1 \leq 55$ or $400 \leq$ formwork resources ≤ 500. In other words, the current set of basic feasible variables x_1, x_2 and s_2 will remain feasible when the available resource for formwork lies in the range of 400 to 500, provided that other resources remain unchanged. The procedures illustrated here is applicable when the resources are considered individually only.

Case 3: Addition of a new constraint

The addition of a new constraint to an existing linear programing problem may lead to one of the following two cases:

(a) The new constraint is redundant

If the solution remains unchanged even after solving the problem with new constraint this indicates that the new constraint is redundant and, hence, it may be removed from the problem statement. Alternatively, if the current solution satisfies the additional constraint, then the solution remains unchanged.

Assume, a finishing activity is added which require one man-day for each element and the resource limit for finishing activity is 210, hence the constraint equation may be written as

$$x_1 + x_2 + x_3 \leq 210$$

The current solution $x_1 = 45, x_2 = 155$ and $x_3 = 0$ satisfies the additional constraint equation, hence the solution will remain unchanged and constraint is redundant.

Alternatively, if it is desired to check the effect of addition of the above constraint, one need not start afresh. Rather, one can start with the last optimum feasible solution table (see Table 20.21e) to obtain the revised optimum solution. For this, another row and column are added for the new constraint $x_1 + x_2 + x_3 + s_4 = 210$ as shown in Table 20.22e.

Table 20.22e Addition of a new constraint in the example 20.11.1

Iteration	Basic	x_1	x_2	x_3	s_1	s_2	s_3	s_4	Solution
	Z	0	0	25	150	0	75	0	96,750
	x_1	1	0	0	1	0	−1	0	45
Initial	s_2	0	0	0.5	−1	1	0.5	0	55
	x_2	0	1	0.5	−1	0	1.5	0	155
	s_4	1	1	1	0	0	0	1	210

Table 20.22f Addition of a new constraint in the example 20.11.1

Iteration	Basic	x_1	x_2	x_3	s_1	s_2	s_3	s_4	Solution
	Z	0	0	25	150	0	75	0	96,750
	x_1	1	0	0	1	0	−1	0	45
Initial	s_2	0	0	0.5	−1	1	0.5	0	55
	x_2	0	1	0.5	−1	0	1.5	0	155
	s_4	0	0	0.5	0	0	−0.5	1	10

The initial simplex tableau as shown in Table 20.22e after addition of the fourth constraint is inconsistent, because of the presence of non-zero coefficients (1 and 1) in the basic variables column of x_1 and x_2, respectively. Hence, it is made consistent by manipulating s_4 row as shown below:

$$\text{New } s_4 \text{ row} = \text{Current } s_4 \text{ row} - 1 \times x_1 \text{ row} - 1 \times x_2 \text{ row}$$

The above simplex tableau (see Table 20.22f) satisfies the optimality and feasibility conditions, hence no further iterations are to be carried out. It may further be noticed that the solution remained unchanged even after addition of fourth constraint hence the additional constraint is redundant.

(b) The current solution may violate the new constraint

In such cases, the feasibility is first retrieved using dual simplex method and then optimality is achieved by further iterations.

Let us consider a situation in which the client requires the above mentioned structural elements with textured paint finish, hence the manufacturer will add another activity in the production sequence and require resources for that activity. Let's also assume that the daily resource available for the textured painting is 250 man-days. Consider that the beam, wall panels and roof decks require 3, 1, and 1 man-days painting time, respectively. Thus, the new additional constraint for the original problem takes the following form.

$$3x_1 + x_2 + x_3 \leq 250$$

The current solution $x_1 = 45, x_2 = 155$ and $x_3 = 0$ does not satisfy the additional constraint equation, hence the constraint is not redundant as was in the previous case.

Alternatively, if it is desired to check the effect of addition of the above constraint, one need not to start afresh. Rather, one can start with the last optimum feasible solution table (see Table 20.21e) to obtain the revised optimum solution. For this, another row and column are added for the new constraint $3x_1 + x_2 + x_3 + s_4 = 210$ as shown in Table 20.22g.

The initial simplex tableau (Table 20.22g) after addition of the fourth constraint is inconsistent because of the presence of non-zero coefficients (3 and 1) in the basic variables column of x_1 and x_2, respectively. Hence, it is made consistent by manipulating s_4 row as shown below:

$$\text{New } s_4 \text{ row} = \text{Current } s_4 \text{ row} - 3 \times x_1 \text{ row} - 1 \times x_2 \text{ row}$$

Table 20.22g Addition of a new constraint in the example 20.11.1

Iteration	Basic	Coefficients							Solution
		x_1	x_2	x_3	s_1	s_2	s_3	s_4	
	Z	0	0	25	150	0	75	0	96,750
	x_1	1	0	0	1	0	−1	0	45
Initial	s_2	0	0	0.5	−1	1	0.5	0	55
	x_2	0	1	0.5	−1	0	1.5	0	155
	s_4	3	1	1	0	0	0	1	250

The simplex tableau in the 1st iteration is giving infeasible solution, hence Dual Simplex Algorithm is applied (see Table 20.22h) in order to retrieve the feasibility and then usual row manipulations are done to get an optimal feasible solution (see Table 20.22i).

The 2nd iteration gives optimal solution for the above problem with additional constraint. In this case, addition of the new constraint has reduced the optimum profit, hence it is not redundant. Note that the addition of additional constraint will never improve the profit function.

20.11.2 Changes in the Objective Function Coefficients

In this section, we discuss the changes made in the initial objective coefficients. Such addition may affect the optimality of the current solution. We would also like to find the range of objective function coefficients which would still keep the current solution optimal. We also discuss the impact of adding a new economic activity.

Case 1: Changes in the original objective coefficient

The changes in objective coefficients may result in either (a) optimality condition being satisfied, or (b) optimality condition not satisfied. These are discussed as follows:

Table 20.22h Iteration 1 of the example 20.11.1 after adding the new constraint

Iteration	Basic	Coefficients							Solution
		x_1	x_2	x_3	s_1	s_2	s_3	s_4	
	Z	0	0	25	150	0	75	0	96,750
	x_1	1	0	0	1	0	−1	0	45
	s_2	0	0	0.5	−1	1	0.5	0	55
Iteration 1	x_2	0	1	0.5	−1	0	1.5	0	155
	s_4	0	0	0.5	−2	0	1.5	1	−40
	$\left\| \dfrac{Z_j - C_j}{\alpha_{rj}} \right\|$				75				

Table 20.22i Iteration 2 of the example 20.11.1 after adding the new constraint

Iteration	Basic	x_1	x_2	x_3	s_1	s_2	s_3	s_4	Solution
	Z	0	0	62.5	0	0	187.5	75	93,750
	x_1	1	0	0.25	0	0	−0.25	0.5	25
2 (optimal)	s_2	0	0	0.25	0	1	−0.25	−0.5	75
	x_2	0	1	0.25	0	0	0.75	−0.5	175
	s_1	0	0	−0.25	1	0	−0.75	−0.5	20

(a) Optimality condition satisfied

In this case, the optimality condition is satisfied, hence the solution remains unchanged. However, the optimum value changes. This is illustrated with the previous example and reproduced as follows:

Maximize

$$Z = 600x_1 + 450x_2 + 200x_3$$

Subject to:

$$3x_1 + 2x_2 + x_3 \leq 445$$
$$2x_1 + x_2 + x_3 \leq 300$$
$$2x_1 + 2x_2 + x_3 \leq 400$$
$$x_1, x_2, x_3 \geq 0$$

Table 20.22j Dual of the primal problem 20.11.1

Primal	Dual
Max $Z = 550x_1 + 420x_2 + 175x_3 + 0s_1 + 0s_2 + 0s_3$ Subject to: $3x_1 + 2x_2 + x_3 + s_1 = 445$ $2x_1 + x_2 + x_3 + s_2 = 300$ $2x_1 + 2x_2 + x_3 + s_3 = 400$ $x_1, x_2, x_3, s_1, s_2, s_3 \geq 0$	Min $Z' = 445y_1 + 300y_2 + 400y_3$ Subject to: $3y_1 + 2y_2 + 2y_3 \geq 550$ $2y_1 + y_2 + 2y_3 \geq 420$ $y_1 + y_2 + y_3 \geq 175$ $y_1 \geq 0$ $y_2 \geq 0$ $y_3 \geq 0$

Let's assume that the objective coefficients have been revised to 550, 420, and 175 corresponding to x_1, x_2, and x_3, respectively. Thus, the revised primal and dual for the above example may be written as shown in Table 20.22j:

Step 1
Using the revised objective function coefficients corresponding to basic variables (x_1, s_2, x_2), i.e., (550, 0, 420) the new dual coefficients are calculated using the Equation (20.21) as

$$(y_1 y_2 y_3) = (\text{Revised Primal Coefficients of Basic Variables}) \times (\text{Optimal Inverse})$$

$$(y_1 y_2 y_3) = (550, 0, 420) \times \begin{pmatrix} 1 & 0 & -1 \\ -1 & 1 & 0.5 \\ -1 & 0 & 1.5 \end{pmatrix}$$

$$\Rightarrow (y_1 y_2 y_3) = (130, 0, 80)$$

Now, the revised primal Z-row coefficients may be computed for non-basic variables x_3, s_1, s_3, respectively by Equation (20.19b). Note that the Z-row coefficients for basic variables will still be zero, hence need not be calculated; however, for completeness new coefficients of the Z-row for all variables have been shown and reader may verify that the Z-row coefficients are zero for all currently basic variables; i.e., x_1, x_2, s_2.

$$(130, 0, 80) \times \begin{pmatrix} 3 \\ 2 \\ 2 \end{pmatrix} - 550 = 0 \quad \textit{corresponding to } x_1$$

$$(130, 0, 80) \times \begin{pmatrix} 2 \\ 1 \\ 2 \end{pmatrix} - 420 = 0 \quad \textit{corresponding to } x_2$$

$$(130, 0, 80) \times \begin{pmatrix} 1 \\ 1 \\ 1 \end{pmatrix} - 175 = 35 \quad \textit{corresponding to } x_3$$

$$(130, 0, 80) \times \begin{pmatrix} 1 \\ 0 \\ 0 \end{pmatrix} - 0 = 130 \quad \textit{corresponding to } s_1$$

$$(130, 0, 80) \times \begin{pmatrix} 0 \\ 1 \\ 0 \end{pmatrix} - 0 = 0 \quad \textit{corresponding to } s_2$$

$$(130, 0, 80) \times \begin{pmatrix} 0 \\ 0 \\ 1 \end{pmatrix} - 0 = 80 \quad \textit{corresponding to } s_3$$

Note that if the new Z-row calculations satisfy the optimality condition (non-negative coefficients, in case of maximization problem) the basic variable and their values remain unchanged; however, because of the revised cost coefficients, the objective function value changes.

The objective function may then be calculated with the revised cost coefficient as

$$Z = 550 \times 45 + 420 \times 155 = 89{,}850$$

(b) Optimality condition not satisfied

In this case, the optimality condition is violated and, hence, Primal Simplex Method is applied to regain optimality. To understand this, let us assume that the profit function for the above problem has changed to

Maximize

$$Z = 500x_1 + 400x_2 + 225x_3$$

Subject to:

$$3x_1 + 2x_2 + x_3 \leq 445$$
$$2x_1 + x_2 + x_3 \leq 300$$
$$2x_1 + 2x_2 + x_3 \leq 400$$
$$x_1, x_2, x_3 \geq 0$$

The primal and dual for the above example may be written as shown in Table 20.22k.

Note all the cost coefficients for the primal Z have changed from earlier coefficients of (600, 450, 200).

Step1
Using the revised objective function, coefficients corresponding to basic variables (x_1, s_2, x_2), i.e., (500, 0, 400), the new dual coefficients are calculated using Equation (20.21) as

Table 20.22k The primal and dual with revised objective function coefficients

Primal	Dual
Max $Z = 500x_1 + 400x_2 + 225x_3 + 0s_1 + 0s_2 + 0s_3$ Subject to: $3x_1 + 2x_2 + x_3 + s_1 = 445$ $2x_1 + x_2 + x_3 + s_2 = 300$ $2x_1 + 2x_2 + x_3 + s_3 = 400$ $x_1, x_2, x_3, s_1, s_2, s_3 \geq 0$	Min $Z' = 445y_1 + 300y_2 + 400y_3$ Subject to: $3y_1 + 2y_2 + 2y_3 \geq 500$ $2y_1 + y_2 + 2y_3 \geq 400$ $y_1 + y_2 + y_3 \geq 225$ $y_1 \geq 0$ $y_2 \geq 0$ $y_3 \geq 0$

$$(y_1 y_2 y_3) = (\text{Revised Primal Coefficients of Basic Variables}) \times (\text{Optimal Inverse})$$

$$(y_1 y_2 y_3) = (500, 0, 400) \times \begin{pmatrix} 1 & 0 & -1 \\ -1 & 1 & 0.5 \\ -1 & 0 & 1.5 \end{pmatrix}$$

$$\Rightarrow (y_1 y_2 y_3) = (100, 0, 100)$$

Now, the revised Primal Z-row coefficients may be computed for the non-basic variables x_3, s_1, s_3, respectively, by Equation (20.19b). Note that the Z-row coefficients for basic variables will still be zero, hence need not be calculated; however, for completeness, new coefficients of the Z-row for all variables have been shown and reader may verify that the Z-row coefficients are 0 for all the currently basic variables, i.e, x_1, x_2, s_2.

$$(100, 0, 100) \times \begin{pmatrix} 3 \\ 2 \\ 2 \end{pmatrix} - 500 = 0 \quad \text{corresponding to } x_1$$

$$(100, 0, 100) \times \begin{pmatrix} 2 \\ 1 \\ 2 \end{pmatrix} - 400 = 0 \quad \text{corresponding to } x_2$$

$$(100, 0, 100) \times \begin{pmatrix} 1 \\ 1 \\ 1 \end{pmatrix} - 225 = -25 \quad \text{corresponding to } x_3$$

$$(100, 0, 100) \times \begin{pmatrix} 1 \\ 0 \\ 0 \end{pmatrix} - 0 = 100 \quad \text{corresponding to } s_1$$

$$(100, 0, 100) \times \begin{pmatrix} 0 \\ 1 \\ 0 \end{pmatrix} - 0 = 0 \quad \text{corresponding to } s_2$$

$$(100, 0, 100) \times \begin{pmatrix} 0 \\ 0 \\ 1 \end{pmatrix} - 0 = 100 \quad \text{corresponding to } s_3$$

The objective function may then be calculated with the revised cost coefficient as

$$Z = 500 \times 45 + 400 \times 155 = 84{,}500$$

Note that in the present example, Z-row coefficient corresponding to x_3 is negative (−25 now as shown in Table 20.22l) which violates the condition for optimality. Hence, the variable with maximum negative Z-row coefficient, in this case, x_3 will be the entering variable in the next

Table 20.22l Simplex table after revision in z-row coefficients

Iteration	Basic	x_1	x_2	x_3	s_1	s_2	s_3	Solution	Ratio
3	Z	0	0	−25	100	0	100	84,500	
	x_1	1	0	0	1	0	−1	45	
	s_2	0	0	0.5	−1	1	0.5	55	110
	x_2	0	1	0.5	−1	0	1.5	155	310

iteration. Simplex tableau with revised calculation is shown in Table 20.22m, wherein primal simplex method is applied to attain the optimality in the 4th iteration.

Hence, the revised solution and objective function is $x_1 = 45, x_3 = 110, x_2 = 100$ and

$$Z = 45 \times 500 + 100 \times 400 + 110 \times 225 = 87{,}250$$

Case 2: Optimality range of the objective coefficients

Instead of decreasing/increasing the coefficient one by one or simultaneously, we may wish to find the range of cost coefficient in order to retain optimality.

For example, let us assume that cost coefficient for j^{th} variable is C_j. Now, we are interested in finding a range for C_j. The range of the cost coefficient may then be found by replacing the cost coefficient by $C_j + D_j$ where D_j may take positive or negative value.

To understand this, let us assume that the objective function for the precast element manufacturing problem is changed to

Maximize

$$Z = 600x_1 + 450x_2 + (200 + D_i)x_3$$

Table 20.22m Find iteration for the problem with revised objective function coefficients

Iteration	Basic	x_1	x_2	x_3	s_1	s_2	s_3	Solution	Ratio
4 (optimal)	Z	0	0	0	50	50	125	87,250	
	x_1	1	0	0	1	0	−1	45	
	x_3	0	0	1	−2	2	1	110	
	x_2	0	1	0	0	−1	1	100	

Subject to:

$$3x_1 + 2x_2 + x_3 \leq 445$$
$$2x_1 + x_2 + x_3 \leq 300$$
$$2x_1 + 2x_2 + x_3 \leq 400$$
$$x_1, x_2, x_3 \geq 0$$

Now, we need to find the optimality range for the D_3.

We follow the similar steps as done in the two examples shown above, *i.e.*, calculate the dual price and then using the dual price calculate the revised Z-row coefficients as follows:

$$(y_1 y_2 y_3) = (600, 0, 450) \times \begin{pmatrix} 1 & 0 & -1 \\ -1 & 1 & 0.5 \\ -1 & 0 & 1.5 \end{pmatrix}$$

$$(y_1 y_2 y_3) = (150, 0, 75)$$

Note that x_3 is a non-basic variable, hence the calculation of dual price is not affected by the variation in its cost coefficient.

Now, the revised primal Z-row coefficients may be computed for non-basic variables x_3, s_1, s_3, respectively, by the Equation (20.19b). Note that the Z-row coefficients for basic variables will still be zero, hence need not be calculated; however, for completeness new coefficients of the Z-row, for all variables, have been shown and reader may verify that the Z-row coefficients are zero for all current basic variables, *i.e.*, x_1, x_2, s_2.

$$(150, 0, 75) \times \begin{pmatrix} 3 \\ 2 \\ 2 \end{pmatrix} - 600 = 0 \quad \textit{corresponding to } x_1$$

$$(150, 0, 75) \times \begin{pmatrix} 2 \\ 1 \\ 2 \end{pmatrix} - 450 = 0 \quad \textit{corresponding to } x_2$$

$$(150, 0, 75) \times \begin{pmatrix} 1 \\ 1 \\ 1 \end{pmatrix} - (200 + D_3) = 25 - D_3 \quad \textit{corresponding to } x_3$$

$$(150, 0, 75) \times \begin{pmatrix} 1 \\ 0 \\ 0 \end{pmatrix} - 0 = 150 \quad \textit{corresponding to } s_1$$

$$(150, 0, 75) \times \begin{pmatrix} 0 \\ 1 \\ 0 \end{pmatrix} - 0 = 0 \quad \textit{corresponding to } s_2$$

$$(150, 0, 75) \times \begin{pmatrix} 0 \\ 0 \\ 1 \end{pmatrix} - 0 = 75 \quad \textit{corresponding to } s_3$$

Now, to satisfy the condition of optimality, all the Z-row coefficients must be positive. Hence, we may write

$$25 - D_3 \geq 0$$

or

$$D_3 \leq 25$$

In other words, the current solution will remain optimal as long as the objective coefficient C_3 for x_3 does not exceed $200 + 25 = 225$.

We now assume that the objective function for the precast element manufacturing problem is changed to

Maximize

$$Z = (600 + D_1)x_1 + 450x_2 + 200x_3$$

Subject to:

$$3x_1 + 2x_2 + x_3 \leq 445$$
$$2x_1 + x_2 + x_3 \leq 300$$
$$2x_1 + 2x_2 + x_3 \leq 400$$
$$x_1, x_2, x_3 \geq 0$$

Now, we need to find the optimality range for the D_1.

We follow the similar steps given in the two examples above, *i.e.*, calculating the dual price and then using the dual price to calculate the revised Z-row coefficients as follows:

$$(y_1 y_2 y_3) = \{(600 + D_1), 0, 450\} \times \begin{pmatrix} 1 & 0 & -1 \\ -1 & 1 & 0.5 \\ -1 & 0 & 1.5 \end{pmatrix}$$

$$(y_1 y_2 y_3) = \{(150 + D_1), 0, (75 - D_1)\}$$

Note that x_1 is basic variable. Hence, the calculation of dual price is affected by the variation in its cost coefficient.

Now, the revised primal Z-row coefficients may computed for non-basic variables x_3, s_1, s_3, respectively, by Equation (20.19b). Note that the Z-row coefficients for basic variables will still be zero, hence need not be calculated. However, for completeness new coefficients of the Z-row for all variables have been shown and reader may verify that the Z-row coefficients are zero for all current basic variables, *i.e.*, x_1, x_2, s_2.

$$\{(150 + D_1), 0, (75 - D_1)\} \times \begin{pmatrix} 3 \\ 2 \\ 2 \end{pmatrix} - (600 + D_1) = 0 \quad corresponding\ to\ x_1$$

$$\{(150 + D_1), 0, (75 - D_1)\} \times \begin{pmatrix} 2 \\ 1 \\ 2 \end{pmatrix} - 450 = 0 \quad corresponding\ to\ x_2$$

$$\{(150+D_1),0,(75-D_1)\} \times \begin{pmatrix} 1 \\ 1 \\ 1 \end{pmatrix} - (200) = 0 \quad \text{corresponding to } x_3$$

$$\{(150+D_1),0,(75-D_1)\} \times \begin{pmatrix} 1 \\ 0 \\ 0 \end{pmatrix} - 0 = (150+D_1) \quad \text{corresponding to } s_1$$

$$\{(150+D_1),0,(75-D_1)\} \times \begin{pmatrix} 0 \\ 1 \\ 0 \end{pmatrix} - 0 = 0 \quad \text{corresponding to } s_2$$

$$\{(150+D_1),0,(75-D_1)\} \times \begin{pmatrix} 0 \\ 0 \\ 1 \end{pmatrix} - 0 = (75-D_1) \quad \text{corresponding to } s_3$$

Now, to satisfy the condition of optimality all the Z-row coefficients must be positive. We may write

$$150 + D_1 \geq 0 \Rightarrow D_1 \geq -150$$
$$75 - D_1 \geq 0 \Rightarrow D_1 \leq 75$$

Or,
$$-150 \leq D_1 \leq 75$$

In other words, the current solution will remain optimal as long as the objective coefficient C_1 for real variable (beam element) lies in the range $450 \leq D_1 \leq 675$.

Case 3: Addition of a new activity

It is equivalent to adding a new variable. The addition of activity is desirable only when it makes any positive contribution in the profit, *i.e.*, it enhances the optimal value of the objective function.

The newly added activity can be thought of as a non-basic variable. This means that the dual values associated with the current solution remains unchanged. Also, in other words, if there is no change in optimality, it is not desirable to add the activity.

In the precast element manufacturing problem, the manufacturer decides to replace the non-profit making element, roof deck, by another roofing material, deck slab, for which the resource required in each new element is (1, 1, 2) and profit per deck slab is ₹ 325.

Let x_4 denote the new deck slab with the given resource requirement on each operation (1, 1, 2) for the production of each unit of x_4. Moreover, the new variable is initially considered non-basic and, hence, it does not affect the values of the current dual variable which remains unchanged as calculated in iteration 3 of Problem 20.11.1

$$(y_1 y_2 y_3) = (600, 0, 450) \times \begin{pmatrix} 1 & 0 & -1 \\ -1 & 1 & 0.5 \\ -1 & 0 & 1.5 \end{pmatrix}$$

$$\Rightarrow (y_1 y_2 y_3) = (150, 0, 75)$$

Table 20.22n Simplex table after addition of a new activity

Iteration	Basic	x_1	x_2	x_3	x_4	s_1	s_2	s_3	Solution	Ratio
3	Z	0	0	25	−25	150	0	75	96,750	
	x_1	1	0	0	−1	1	0	−1	45	
	s_2	0	0	0.5	1	−1	1	0.5	55	55
	x_2	0	1	0.5	2	−1	0	1.5	155	77.5

Now, knowing the dual variables, the Z-row coefficient corresponding to x_4, may be calculated using Equation (20.19b) as

$$1 \times y_1 + 1 \times y_2 + 2 \times y_3 - 325 = 1 \times 150 + 1 \times 0 + 2 \times 75 - 325 = -25$$

In order to achieve the optimality, the Gauss–Jordan row operation is done after selecting the entering and leaving variables. Before that, we need to calculate the coefficients of the x_4 column using the formula

$$(\text{Constraint column in iteration } i) = (\text{Inverse in iteration } i) \times (\text{Original constraint column})$$

$$= \begin{pmatrix} 1 & 0 & -1 \\ -1 & 1 & 0.5 \\ -1 & 0 & 1.5 \end{pmatrix} \times \begin{pmatrix} 1 \\ 1 \\ 2 \end{pmatrix}$$

$$= \begin{pmatrix} -1 \\ 1 \\ 2 \end{pmatrix}$$

The above computations are shown in Table 20.22n. Note that negative Z-row coefficient corresponding to x_4 suggest that the optimality condition has been violated by addition of activity x_4. New optimal solution is obtained by entering x_4 into basic column in place of s_2 as shown in Table 20.22o. Hence, the optimum profit has increased to Z = 98,125 by adding/replacing the new activity/product x_4 in place of non-profit making activity x_3.

Table 20.22o Final iteration for the case of addition of new activity

Iteration	Basic	x_1	x_2	x_3	x_4	s_1	s_2	s_3	Solution	Ratio
(Optimal)	Z	0	0	37.5	0	125	25	87.5	98,125	
	x_1	1	0	0.5	0	0	1	−0.5	100	
	x_4	0	0	0.5	1	−1	1	0.5	55	
	x_2	0	1	−0.5	0	1	−2	0.5	45	

REFERENCES

1. Hillier, F.S., Lieberman, G.J., 2009, *Operations Research*, 10th Reprint, Tata McGraw Hill, New Delhi, India.
2. Mohan, C., Deep, K., 2009, *Optimization Techniques*, First Edition, New Age International Publishers, New Delhi, India.
3. Radosavljevic, M. and Bennett, J. 2012, *Construction Management Strategies: A Theory of Construction Management*, John Wiley & Sons, West Sussex, U.K.
4. Render, B., Stair, R.M., and Hanna, M.E., 2000, *Quantitative Analysis for Management*, Eighth Edition, Pearson Education, New Delhi, India.
5. Sen, P.R., 2010, Operations Research-Algorithms and Applications, Prentice Hall India, New Delhi, India.
6. Taha, H.A., 2007, *Operations Research-An Introduction*, Seventh Edition, Prentice Hall India, New Delhi, India.
7. Wagner, H.M., 2005, *Principles of Operations Research*, 2nd Edition, Prentice Hall India, New Delhi, India.

REVIEW QUESTIONS

1. Maximize

   $Z = 2x_1 + x_2$

   Subject to

   $x_1 \leq 4$

   $x_2 \leq 4$

   $x_1 + x_2 \leq 5$

   $x_1 \geq 0, x_2 \geq 0$

   a. Plot the feasible region and circle all the Corner Point Feasible (CPF) solutions
   b. For each CPF solution, use this pair of constraint boundary equations to solve algebraically for the values of x_1 and x_2 at the corner point.

2. Maximize

   $Z = 2x_1 + x_2$

   Subject to,

   $3x_1 + x_2 \leq 8$

   $x_1 + x_2 \leq 4$

   $x_1, x_2 \geq 0$

   a. Use graphical analysis to identify all the corner point solutions for this model. Label each as either feasible or infeasible.

b. Calculate the value of the objective function for each of the CPF solutions. Use this information to identify an optimal solution.

3. Solve graphically the following linear optimization problem.

   Maximize

   $Z = 5x_1 + 6x_2$

   Subject to,

   $2x_1 + 3x_2 \leq 18$

   $2x_1 + x_2 \leq 12$

   $x_1 + x_2 \leq 8$

   $x_1 \geq 0, x_2 \geq 0$

4. Solve the following linear optimization problem.

   Maximize,

   $Z = 2x_1 + x_2$

   Subject to,

   $4x_1 + 3x_2 \leq 32$

   $x_1 \leq 5$

   $5x_1 + 6x_2 \leq 18$

   $x_1, x_2 \geq 0$

   Use graphical analysis to solve the above problem. Label all the corner point solutions, identify the feasible region and show clearly the profit line that runs through the optimal solution.

5. Solve the following linear program to determine the optimum values for x_1 and x_2:

   Minimize,

   $W = 1x_1 + 3x_2$

   Subject to,

   $x_1 + x_2 \geq 10$

   $x_1 + 2x_2 \geq 15$

   $6x_1 + x_2 \geq 18$

   $x_1 - x_2 \leq 12$

   $x_1, x_2 \geq 0$

6. A small construction company has 3 employees A, B, and C and it manufactures wooden doors and windows. It makes a profit of ₹ 600 and ₹ 450 respectively for every door and window that it

sells. Employee A makes 4 doors a day while employee B makes 6 windows per day. The employee C is a painter and can paint 18 m² per day. Each wooden door requires a paint of 4 m² and wooden window requires a paint of 2 m². How many doors and windows should the company produce each day to maximize total profit?

a. Formulate the linear programming model for this problem.

b. Use graphical method to solve this model.

c. In order to face the competition from another wooden door manufacturer, if the company decides to lower his profit from ₹ 600 per wooden door to ₹ 400 per wooden door, how would the optimum solution change (if it changes at all).

d. What will happen to the optimal solution of the original data if the employee 'A' decides to lower his working hours and instead of 4 doors, makes only 3 doors a day?

7. Apply simplex algorithm to solve the following:

Maximize

$15x_1 + 6x_2 + 9x_3 + 2x_4$

Subject to

$2x_1 + x_2 + 5x_3 + .6x_4 \leq 20$

$3x_1 + x_2 + 3x_3 + 0.25 x_4 \leq 24$

$7x_1 + x_4 \leq 70$

every $x_i \geq 0$

8. A cement company blends clinker, fly ash and granulated blast furnace slag (gbfs) in two types of blended cements. The relative profit associated with per kg of clinker used in cement 1 and cement 2 are ₹ 1 and 80 paise respectively. Similarly the same for use of flyash are ₹ 0.3 and Re 0.5 respectively and those for gbfs are ₹ 0.4, and ₹ 0.4 respectively. The total available quantity of clinker, flyash, and gbfs per day are 120 tons, 40 tons, and 100 tons respectively. To satisfy the customer demand the company must produce 100 tons of each cement. The maximum proportions of fly ash in any cement is restricted to 20% and minimum clinker content in cement 1 is 60% and that in cement 2 is 40%. Formulate the LP Problem and solve.

9. A construction project involves huge quantity of earth work and has 5 quarries to get the earth from and the details of the costs and times from each quarry are given below.

The owner of the first three quarries cannot give more than 3,000 m³ a day. Quantities of at least 2,000 m³ and 1,500 m³ have to be taken from Quarry 4 and 5 respectively.

Total no. of dump trucks available with the project is 60 and targeted amount of earthwork per day is 8,000 m³. Average volume of a dump truck can be taken as 8m³. Formulate the linear programming model for this problem.

Description	Unit	Quarry 1	Quarry 2	Quarry 3	Quarry 4	Quarry 5
Time taken for a dump truck to reach the site	minutes	30	45	35	25	20
Cost/m³ of earth work	₹.	80	100	85	75	70

10. Solve the following LP problem by Big-M method

    Maximize

    $Z = 4x_1 + x_2$

    Subject to:

    $3x_1 + x_2 = 3$

    $4x_1 + 3x_2 \geq 6$

    $x_1 + 2x_2 \leq 4$

    $x_1, x_2 \geq 0$

11. Find the dual of the linear program

    Max $Z = 3X_1 + X_2 + 6X_3$

    Subject to,

    $2X_1 + X_2 + 2X_3 \leq 5$

    $4X_1 + X_2 - 2X_3 \leq 3$

    $X_1 + 3X_2 + 2X_3 \leq 4$

    All variables ≥ 0

12. Find the dual of the linear program

    Max $Z = 5X_1 + 8X_2$

    Subject to,

    $4X_1 + 2X_2 \leq 12$

    $2X_1 + 3X_2 \leq 8$

    $X_1 + X_2 \leq 3$

    All variables ≥ 0

13. Find the dual of the linear program

    Max $Z = 4X_1 + 2X_2 + 7X_3$

    Subject to,

    $2X_1 + X_2 + X_3 \leq 3$

    $5X_1 + 2X_2 + 4X_3 \leq 2$

    All variables ≥ 0

14. Find the dual of the linear program

    Max $Z = 5X_1 + 3X_2 + 10X_3$

    Subject to,

    $5X_1 + X_2 + 3X_3 \leq 9$

    $3X_2 + X_3 \leq 3$

    $2X_1 + 2X_2 + 6X_3 \leq 7$

    All variables ≥ 0

15. Find the dual of the linear program

    Min $Z = 9X_1 + 3X_2 + 7X_3$

    Subject to,

    $5X_1 + 2X_3 \geq 5$

    $X_1 + 3X_2 + 2X_3 \geq 3$

    $3X_1 + X_2 + 6X_3 \geq 10$

    All variables ≥ 0

16. Solve the following linear programming problem by dual simplex method

    Minimize

    $x_1 + 2x_2 + 3x_3$

    Subject to:

    $x_1 + 3x_2 + 5x_3 \geq 7$

    $2x_1 + 5x_2 + 7x_3 \geq 10$

    x_1, x_2 and $x_3 \geq 0$

17. Each of the following table is associated with a maximization model. One or more of the following descriptions applies to each table:

    (i) optimal, (ii) shows problem is unbounded, (iii) shows problem is infeasible, (iv) there are alternate optima, (v) the problem is feasible, but the current solution is not optimal.

    Decide which of the above descriptions applies to each table. If there are alternate optima, identify at least two of the corner points that are optimal. Justify all your answers.

    (a)

Basic	z	x_1	x_2	s_1	s_2	Soln.
z	1	0	0	3	0	90
x_1	0	1	1/3	1/3	0	10
s_2	0	0	3	−1/2	1	12

(b)

Basic	z	x_1	x_2	x_3	s_1	s_2	s_3	Soln.
z	1	−1	0	0	1	0	2	1,400
x_2	0	1/2	1	0	−1/2	0	2/3	100
x_3	0	0	0	1	1/2	0	−1/4	100
s_2	0	0	0	0	−1/2	1	−1/3	200

(c)

Basic	x_1	x_2	s_1	s_2	Soln.
z	0	1	3	0	90
x_1	1	−1/3	1/3	0	10
s_2	0	−3	−1/2	1	12

(d)

Basic	x_1	x_2	s_1	s_2	Soln.
z	−9	−6	0	0	0
s_1	−1	1	1	0	11
s_2	−2	3	0	1	15

18. In converting a ≥ constraint into an initial simplex table
    (a) a slack variable is added.
    (b) a surplus variable is added.
    (c) a surplus variable is subtracted and an artificial variable is added.
    (d) an artificial variable is subtracted.
19. In the simplex method, slack variables are added to
    (a) all equality constraints.
    (b) all inequality constraints.
    (c) only ≤ constraints
    (d) only ≥ constraints

21

Transportation, Transshipment, and Assignment Problems

Network representation and LP formulation of transportation problems, introduction, transportation problems, formulation as LP problems, deriving the starting feasible solution using: Northwest corner rule, least or minimum cost method, and Vogel's approximation method (VAM), Phase II—Moving towards optimality, transshipment problem, assignment problem, unbalanced problems and restriction on assignment, some other applications

21.1 INTRODUCTION

Construction goods are transported from a number of sources to a number of destinations. Each of the sources has a defined supply capacity and each of the destinations has a defined demand. The goods may be transported either from the source directly to the destinations or they can be shifted to an intermediate destination before shifting them to their final destinations. Accordingly, these problems are classified as transportation problem and transshipment problem respectively. The objective of both the problems is to minimize the transportation cost. Similarly we often come across a situation wherein we have to assign certain jobs to certain professionals or vendors. The cost incurred definitely depends on the job and the person to whom it is assigned. Such problems known as assignment problems have an objective to minimize the total cost of assignment. It is possible to solve each of the mentioned problems by using methods described in previous chapter; however, in this chapter we will learn some special methods to solve such problems.

21.2 TRANSPORTATION PROBLEM

Consider a ready-mix concrete manufacturer who produces concretes of different grades and supplies to different locations. The manufacturer has manufacturing plants located at two different locations in the National Capital Region (NCR)—Gurgaon and Noida. The manufacturer is catering to four project sites currently with different daily demands. The manufacturer has worked out the transportation cost (See Table 21.1) to supply at each of the four sites from the two plant locations.

A transportation problem can be represented in a network form also. The data presented in Table 21.1 can also be shown in a network diagram as in Figure 21.1.

Table 21.1 Ready-mix concrete transportation problem

	To				
From	Project 1 (d_1)	Project 2 (d_2)	Project 3 (d_3)	Project 4 (d_4)	Supply (m^3)
Gurgaon RMC Plant (s_1)	₹ 80	₹ 60	₹ 100	₹ 90	100
Noida RMC Plant (s_2)	₹ 90	₹ 120	₹ 130	₹ 70	150
Demand (m^3)	75	50	60	65	

The two nodes s_1 and s_2 on the left represent the Gurgaon and Noida plants respectively, while the four nodes d_1, d_2, d_3, and d_4 on the right represent the four project sites requiring the ready-mix concrete.

c_{11}, c_{12}, c_{13}, and c_{14} are the per m^3 cost of transporting the ready-mix concrete from plant s_1 (Gurgaon plant) while c_{21}, c_{22}, c_{23}, and c_{24} are the per m^3 cost of transporting the ready-mix concrete from plant s_2 (Noida plant) to the four project sites d_1, d_2, d_3, and d_4 respectively.

It may be noted that for the above problem, the demand and supply both are equal to 250 m^3 (verify!!!). These problems are referred to as Balanced Transportation Problem (BTP).

21.2.1 Formulation as LP Problem

The above problem can be formulated in the following manner and can be solved using simplex method. However, as it can be observed that solution using simplex method would be little time consuming, more-so when the number of sources and destinations increase.

For the formulation
Let,

x_{11} = the ready-mix concrete quantity supplied from plant s_1 to d_1
x_{12} = the ready-mix concrete quantity supplied from plant s_1 to d_2
x_{13} = the ready-mix concrete quantity supplied from plant s_1 to d_3
x_{14} = the ready-mix concrete quantity supplied from plant s_1 to d_4
x_{21} = the ready-mix concrete quantity supplied from plant s_2 to d_1
x_{22} = the ready-mix concrete quantity supplied from plant s_2 to d_2
x_{23} = the ready-mix concrete quantity supplied from plant s_2 to d_3
x_{24} = the ready-mix concrete quantity supplied from plant s_2 to d_4

The objective function and constraints can be expressed as:

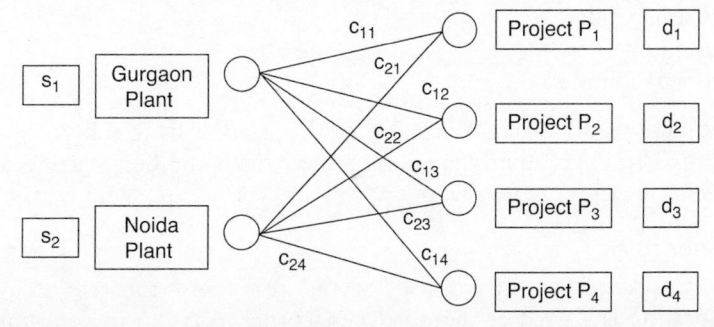

Figure 21.1 Network representation of ready-mix concrete transportation problem

Minimize transportation cost

$$Z = 80x_{11} + 60x_{12} + 100x_{13} + 90x_{14} + 90x_{21} + 120x_{22} + 130x_{23} + 70x_{24}$$

Subject to:

Supply constraints:

$$x_{11} + x_{12} + x_{13} + x_{14} = 100$$
$$x_{21} + x_{22} + x_{23} + x_{24} = 150$$

Demand constraints:

$$x_{11} + x_{21} = 75$$
$$x_{12} + x_{22} = 50$$
$$x_{13} + x_{23} = 60$$
$$x_{14} + x_{24} = 65$$

$$x_{11}, x_{12}, x_{13}, x_{14}, x_{21}, x_{22}, x_{23}, \text{ and } x_{24} \geq 0$$

Algorithms have been developed to solve such problems quickly. In this section, these algorithms have been illustrated with the help of small examples.

21.2.2 The Transportation Algorithm

We illustrate the transportation algorithm for m sources and n destination. It will be noticed that for a balance transportation problem, there would be one redundant constraint all the time. In other words, the transportation model will have $(m + n - 1)$ independent constraint equations and the starting basic solution consists of $(m + n - 1)$ basic variables. For example, for 2 sources and 4 destinations (ready-mix concrete problem), there would be $2 + 4 - 1 = 5$ independent constraint equations and 5 basic variables.

The transportation problem is solved in two phases:

- Phase I – Obtaining an initial feasible solution
- Phase II – Moving toward optimality

21.2.3 Phase I Obtaining an Initial Feasible Solution

Any one of the following methods can be used for obtaining the initial feasible solution. The methods are explained in the following sections:

1. Northwest Corner Rule
2. Least Cost Method
3. Vogel Approximation Method

It would be pertinent to note that the quality of the starting basic solution varies with the method employed. The Vogel Approximation method yields the best starting solution while Northwest corner rule yields the worst.

Northwest corner rule

This method is explained in the context of the ready-mix concrete example. In this method, the allocation is started at the northwest corner. As per this algorithm, maximum possible allocation is made to the selected cell. Once the allocation is made in a cell, the

Table 21.2a Allocation using northwest corner (step 1)

From	To Project 1 (d_1)	Project 2 (d_2)	Project 3 (d_3)	Project 4 (d_4)	Supply (m^3)
Gurgaon RMC Plant (s_1)	80 75	60	100	90	~~100~~; 25
Noida RMC Plant (s_2)	90	120	130	70	150
Demand (m^3)	~~75~~	50	60	65	

corresponding demand and supply are reduced by the allocated amount. For the problem, the northwest corner cell happens to be the cell s_1d_1 (See the shaded cell represented by s_1 row and d_1 column in Table 21.2a).

The demand for project 1 is 75 m^3 and the supply from plant s_1 is 100 m^3. Applying this method, the maximum possible allocation for this cell is 75 m^3 (minimum of 75 and 100 m^3). Thus, the total demand of project 1 is met and 75 m^3 in the bottom row is struck off. Also from the supply column of Gurgaon plant row, we subtract 75 m^3 from the 100 m^3 supply. The resultant 25 m^3 supply is now shown adjacent to struck off figure of 100 m^3.

It is observed that the allocation in cell s_1d_1 has exhausted the demand in project 1 and it cannot take any more. Thus the next cell, where the concrete quantity from Gurgaon RMC plant is allocated, is cell s_1d_2 (s_1 row d_2 column) which happens to be the next Northwest corner. The maximum supply that the Gurgaon RMC plant can provide is 100 m^3 out of which 75 m^3 is already given to project 1, thus project 2 is allocated 25 m^3 (as shown in Table 21.2b).

By doing so, supply from Gurgaon RMC plant is exhausted; however, the demand of project 2 is still not met out as is clear by observing the demand row corresponding to project 2 column. Out of 50 m^3 demand of project 2, 25 m^3 has been left out which has been shown by striking off 50 m^3 and writing 25 m^3 adjacent to it.

Next allocation is made in the cell s_2d_2 (see the shaded cell corresponding to s_2 row and d_2 column in Table 21.2c) which happens to be the next Northwest corner. The maximum quantity that can be allocated is 25 m^3 and this is what has been allocated in the shaded cell. By doing so, the demand requirement of project 2 is met; however, Noida RMC plant can still supply the remaining quantity 125 m^3. This has been shown by striking off 150 m^3 and writing 125 m^3 adjacent to it in the supply column of Table 21.2c.

Table 21.2b Allocation using northwest corner (step 2)

From	To Project 1 (d_1)	Project 2 (d_2)	Project 3 (d_3)	Project 4 (d_4)	Supply (m^3)
Gurgaon RMC Plant (s_1)	80 75	60 25	100	90	~~100~~; 25
Noida RMC Plant (s_2)	90	120	130	70	150
Demand (m^3)	~~75~~	~~50~~; 25	60	65	

Table 21.2c Allocation using northwest corner (step 3)

From	To Project 1 (d_1)	Project 2 (d_2)	Project 3 (d_3)	Project 4 (d_4)	Supply (m^3)
Gurgaon RMC Plant (s_1)	80 75	60 25	100	90	~~100~~, 25
Noida RMC Plant (s_2)	90	120 25	130	70	~~150~~, 125
Demand (m^3)	~~75~~	~~50~~, 25	60	65	

The next Northwest corner is cell $s_2 d_3$ (s_2 row and d_3 column) where the maximum allocation equal to the demand of project (60 m^3) has been done. By doing this, the demand of 60 m^3 of project 3 is met and the remaining supply quantity from Noida RMC plant is 65 m^3 (see demand row corresponding to project 3 and supply column in Table 21.2d corresponding to Noida RMC plant)

Table 21.2d Allocation using northwest corner (step 4)

From	To Project 1 (d_1)	Project 2 (d_2)	Project 3 (d_3)	Project 4 (d_4)	Supply (m^3)
Gurgaon RMC Plant (s_1)	80 75	60 25	100	90	~~100~~, 25
Noida RMC Plant (s_2)	90	120 25	130 60	70	~~150~~, ~~125~~, 65
Demand (m^3)	~~75~~	~~50~~, 25	~~60~~	65	

In the final step, the remaining supply of 65 m^3 is allocated to the cell $s_2 d_4$ (see the shaded cell in Table 21.2e) which satisfies the demand requirement of project 4. The supply and demand quantity of 65 m^3 is crossed out in the demand row and supply column.

The interpretation of initial solution is like this:

Table 21.2e Allocation using northwest corner (step 5)

From	To Project 1 (d_1)	Project 2 (d_2)	Project 3 (d_3)	Project 4 (d_4)	Supply (m^3)
Gurgaon RMC Plant (s_1)	80 75	60 25	100	90	~~100~~, 25
Noida RMC Plant (s_2)	90	120 25	130 60	70 65	~~150~~, ~~125~~, 65
Demand (m^3)	~~75~~	~~50~~, 25	~~60~~	~~65~~	

From the Gurgaon plant, transport 75 m^3 and 25 m^3 to projects 1 and 2 respectively and from the Noida plant, transport 25, 60, and 65 m^3 to projects 2, 3, and 4 respectively. The total transportation cost corresponding to this initial solution would be:

$$Z = 75 \times 80 + 25 \times 60 + 25 \times 120 + 60 \times 130 + 65 \times 70 = ₹\,22{,}850.$$

Least Cost Method

The above ready-mix problem is taken to illustrate the initial solution using the method of least cost. The problem statement is repeated in Table 21.3a. As will be seen soon, this method finds a better starting solution compared to the Northwest Corner Rule. The least cost method suggests allocation to the cheapest route.

In the first step, cell with the lowest unit cost is selected. According to this method, assign as much as possible to the cell with the smallest unit cost (ties are broken arbitrarily).

For this problem, the cell happens to be $s_1 d_2$ (see the shaded cell in the Table 21.3b) with the corresponding transportation cost of ₹ 60. The demand for project 2 is 50 m^3 as against possible supply of 100 m^3 by plant s_1. Thus a total of 50 m^3 is allocated through this route. Demand of 50 m^3 is crossed out in the bottom row corresponding to project 2 and 50 m^3 is subtracted from 100 m^3 in the s_1 row under supply column. This is as shown in Table 21.3b.

In the next step, the cell $s_2 d_4$ with the next least cost is chosen for allocation. The maximum quantity that can be transported through this route is 65 m^3 which is the lower of 65 m^3 demand and 150 m^3 supply. In the demand row 65 m^3 is crossed out corresponding to project 4 and 85 m^3 (150-65) is written adjacent to crossed out figure of 150 m^3. The computations corresponding to this step is shown in Table 21.3c.

The next cell ready for allocation is the cell $s s_1$ (see the shaded cell in the Table 21.3d). The maximum quantity that can be transported through this route is 50 m^3 (lower of remaining supply of 50 m^3 and project 1 demand of 75 m^3). In the demand row corresponding to project 1, 75 m^3 is crossed out and 25 m^3 is entered adjacent to 75 m^3. The total supply in the supply column corresponding to s_1 row is exhausted and thus 50 m^3 is crossed out.

In the next step, it is observed that for two cells ($s_1 d_4$ and $s_2 d_1$), there is a tie for the next least cost. The demand of project 4 is already met in previous steps; hence, the cell $s_1 d_4$ cannot be used for further transportation. Thus, the cell $s_2 d_1$ is used for transporting the lower of 25 m^3 and 85 m^3. By doing this, the demand of project 1 is met and supply remaining from s_2 is 60 m^3. The computations are shown in the Table 21.3e.

Table 21.3a Problem statement for illustration using least cost method

From	Project 1 (d_1)	Project 2 (d_2)	Project 3 (d_3)	Project 4 (d_4)	Supply (m^3)
Gurgaon RMC Plant (s_1)	₹ 80	₹ 60	₹ 100	₹ 90	100
Noida RMC Plant (s_2)	₹ 90	₹ 120	₹ 130	₹ 70	150
Demand (m^3)	75	50	60	65	

764 | CONSTRUCTION PROJECT MANAGEMENT

Table 21.3b Allocation using least cost method (step 1)

From	Project 1 (d_1)	Project 2 (d_2)	Project 3 (d_3)	Project 4 (d_4)	Supply (m^3)
Gurgaon RMC Plant (s_1)	80	60 50	100	90	~~100~~, 50
Noida RMC Plant (s_2)	90	120	130	70	150
Demand (m^3)	75	~~50~~	60	65	

Table 21.3c Allocation using least cost method (step 2)

From	Project 1 (d_1)	Project 2 (d_2)	Project 3 (d_3)	Project 4 (d_4)	Supply (m^3)
Gurgaon RMC Plant (s_1)	80	60 50	100	90	~~100~~, 50
Noida RMC Plant (s_2)	90	120	130	70 65	~~150~~, 85
Demand (m^3)	75	~~50~~	60	~~65~~	

In the final step, the remaining supply of 60 m^3 is allocated to the cell $s_2 d_3$ (see the shaded cell in Table 21.3f) which satisfies the demand requirement of project 3. The supply and demand quantity of 60 m^3 is crossed out in the demand row and supply column.

Thus, the initial starting solution yields following 5 basic variables:

$$X_{11} = 50, X_{12} = 50, X_{21} = 25, X_{23} = 60, \text{ and } X_{24} = 65$$

Table 21.3d Allocation using least cost method (step 3)

From	Project 1 (d_1)	Project 2 (d_2)	Project 3 (d_3)	Project 4 (d_4)	Supply (m^3)
Gurgaon RMC Plant (s_1)	80 50	60 50	100	90	~~100~~, ~~50~~
Noida RMC Plant (s_2)	90	120	130	70 65	~~150~~, 85
Demand (m^3)	~~75~~, 25	~~50~~	60	~~65~~	

Table 21.3e Allocation using least cost method (step 4)

From	To Project 1 (d_1)	Project 2 (d_2)	Project 3 (d_3)	Project 4 (d_4)	Supply (m^3)
Gurgaon RMC Plant (s_1)	80 50	60 50	100	90	~~100~~, 50
Noida RMC Plant (s_2)	90 25	120	130	70 65	~~150~~, ~~85~~, 60
Demand (m^3)	~~75~~, 25	~~50~~	60	~~65~~	

The total transportation cost (objective value) corresponding to this initial solution would be:

$$Z = 50 \times 80 + 50 \times 60 + 25 \times 90 + 60 \times 130 + 65 \times 70 = ₹\ 21,600.$$

Clearly the initial solution obtained using the least cost method yields better starting solution in comparison to the Northwest corner rule.

It may be noted that in both the methods above, if both, a row and column are satisfied simultaneously, only one is crossed out.

Vogel's Approximation Method

In this method, for each row and column, a penalty measure is determined first by subtracting the smallest unit cost from the next smallest unit cost element in the same row and column. In the next step, the row or column with the largest penalty is identified. Ties are broken arbitrarily. The maximum possible allocation is made to the variable in the selected row or column. Finally the supply and demand are adjusted. The process is repeated.

The ready-mix problem illustrated earlier is again taken to illustrate the initial solution using the Vogel approximation method. The problem statement is repeated in Table 21.4a. As will be seen soon, this method finds a better starting solution compared to the Northwest corner rule as well as the least cost method.

The penalties in each row and each column are computed first. For example, the penalty in s_1 row is the difference between the second smallest and the smallest cell values = 80 − 60

Table 21.3f Allocation using least cost method (step 5)

From	To Project 1 (d_1)	Project 2 (d_2)	Project 3 (d_3)	Project 4 (d_4)	Supply (m^3)
Gurgaon RMC Plant (s_1)	80 50	60 50	100	90	~~100~~, 50
Noida RMC Plant (s_2)	90 25	120	130 60	70 65	~~150~~, ~~85~~, ~~60~~
Demand (m^3)	~~75~~, 25	~~50~~	60	~~65~~	

Table 21.4a Problem statement for illustration using Vogel approximation method

From	To Project 1 (d_1)	Project 2 (d_2)	Project 3 (d_3)	Project 4 (d_4)	Supply (m^3)
Gurgaon RMC Plant (s_1)	₹ 80	₹ 60	₹ 100	₹ 90	100
Noida RMC Plant (s_2)	₹ 90	₹ 120	₹ 130	₹ 70	150
Demand (m^3)	75	50	60	65	

= 20 and for the s_2 row it is equal to 90 − 70 = 20. Similarly, the penalties in d_1, d_2, d_3, and d_4 columns are 10, 60, 30, and 20 respectively.

It is observed that the d_2 column has the maximum penalty value = 60, and thus, maximum possible allocation of 50 m^3 is made in the cell $s_1 d_2$ which is having minimum cost coefficient (minimum of (60,120)) for delivery to project 2. Demand of project 2 is fulfilled and the remaining supply from Gurgaon plant is 50 m^3. This is shown in Table 21.4b.

Row wise and column wise penalty calculation is performed again as shown in the Table 21.4c. It can be observed that the penalty for s_1 row has now been revised from 20 to 10. Remaining penalties remain unchanged. The penalty calculation for $d2$ column is not required as the demand has been fulfilled for this project. The maximum penalty is associated with d_3 column, and thus, the maximum possible quantity 50 m^3 is allocated to the cell $s_1 d_3$ (why not the cell $s_2 d_3$?). With this allocation, the supply from Gurgaon plant is exhausted and the remaining demand for project 3 is 10 m^3 as shown in Table 21.4c.

Row wise and column wise penalty calculation is performed again as shown in the Table 21.4d. As mentioned earlier, the penalty calculation for s_1 row is not required as the supply from source s_1 is already exhausted. It can be observed that the penalty for d_3 column has now been revised to 130 which is the maximum among all and thus the maximum possible allocation of 10 m^3 is made to the cell $s_2 d_3$. With this allocation, the demand of project 3 is fulfilled and the remaining supply capacity of Noida RMC plant is 140 m^3 as shown in Table 21.4d.

Row wise and column wise penalty calculation is performed again as shown in the Table 21.4e. It can be observed that the penalty for s_2 row is 20, while the penalties for d_1 and d_4 columns are 90 and 70 respectively. Maximum allocation of 75 m^3 is allocated to the cell

Table 21.4b Allocation using Vogel approximation method (step 1)

From	To Project 1 (d_1)	Project 2 (d_2)	Project 3 (d_3)	Project 4 (d_4)	Supply (m^3)	Penalty (row)
Gurgaon RMC Plant (s_1)	80	60 / 50	100	90	~~100~~, 50	80 − 60 = 20
Noida RMC Plant (s_2)	90	120	130	70	150	90 − 70 = 20
Demand (m^3)	75	~~50~~	60	65		
Penalty (column)	90 − 80 = 10	120 − 60 = 60	130 − 100 = 30	90 − 70 = 20		

Table 21.4c Penalty calculation and allocation using Vogel approximation method (step 2)

From	To Project 1 (d_1)	Project 2 (d_2)	Project 3 (d_3)	Project 4 (d_4)	Supply (m^3)	Penalty (row)
Gurgaon RMC Plant (s_1)	₹ 80	₹ 60 50	₹ 100 50	₹ 90	~~100~~,50	90 − 80 = 10
Noida RMC Plant (s_2)	₹ 90	₹ 120	₹ 130	₹ 70	150	90 − 70 = 20
Demand (m^3)	75	~~50~~	~~60~~, 10	65		
Penalty (column)	90 − 80 = 10	−	130 − 100 = 30	90 − 70 = 20		

$s_2 d_1$. With this allocation, the demand of project 1 is fulfilled and the remaining supply capacity of Noida RMC plant is 65 m^3 as shown in Table 21.4e.

Row wise and column wise penalty calculation is performed once again for the final time as shown in the Table 21.4f. It can be observed that the penalty for s_2 row is 70, and the penalty for d_4 column is also 70. Thus, maximum allocation of 65 m^3 is made to the cell $s_2 d_4$. With this allocation, the demand of project 4 is fulfilled and the remaining supply capacity of Noida RMC plant is also exhausted as shown in Table 21.4f.

Thus the initial starting solution yields following 5 basic variables;

$$x_{12} = 50, \ x_{13} = 50, \ x_{21} = 75, x_{23} = 10, \text{ and } x_{24} = 65$$

The total transportation cost (objective value) corresponding to this initial solution would be:

$$Z = 60 \times 50 + 100 \times 50 + 90 \times 75 + 130 \times 10 + 70 \times 65 = ₹\ 20,600$$

Clearly the initial solution obtained using the Vogel's approximation method yields better starting solution in comparison to the Northwest corner rule and the least cost method.

It may be noted that in this method also as in both the methods explained earlier, if both a row and column are satisfied simultaneously, only one is crossed out.

Table 21.4d Penalty calculation and allocation using Vogel approximation method (step 3)

From	To Project 1 (d_1)	Project 2 (d_2)	Project 3 (d_3)	Project 4 (d_4)	Supply (m^3)	Penalty (row)
Gurgaon RMC Plant (s_1)	₹ 80	₹ 60 50	₹ 100 50	₹ 90	~~100~~,50	−
Noida RMC Plant (s_2)	₹ 90	₹ 120	₹ 130 10	₹ 70	~~150~~,140	90 − 70 = 20
Demand (m^3)	75	~~50~~	~~60~~,10	65		
Penalty (column)	90	−	130	70		

Table 21.4e Penalty calculation and allocation using Vogel approximation method (step 4)

From	To Project 1 (d_1)	Project 2 (d_2)	Project 3 (d_3)	Project 4 (d_4)	Supply (m^3)	Penalty (row)
Gurgaon RMC Plant (s_1)	₹ 80	₹ 60 50	₹ 100 50	₹ 90	~~100~~,50	–
Noida RMC Plant (s_2)	₹ 90 75	₹ 120	₹ 130 10	₹ 70	~~150,140~~, 65	90 – 70 = 20
Demand (m^3)	~~75~~	~~50~~	~~60~~,10	65		
Penalty (column)	90	–	–	70		

21.2.4 Phase II—Moving Toward Optimality

After obtaining the starting solution using any of the three methods explained earlier, we use the simplex optimality condition to determine the entering variable as the current non-basic variables. If the optimality condition is satisfied, the process is stopped otherwise the leaving variable is determined using the simplex feasibility condition. Basis is changed now and the process is repeated till the optimality condition is satisfied.

Method of Multipliers

The method illustrated here for phase II computation is also known as the method of multipliers. The method has been explained for the ready-mix concrete transportation problem. Let's consider the starting solution provided by the Least Cost method. This is shown in Table 21.5.

As can be noted from the table, the starting solution is as below:

$x_{11} = 50$, $x_{12} = 50$, $x_{21} = 25$, $x_{23} = 60$, and $x_{24} = 65$ which yields $Z = ₹ 21,600$.

There are a total of 6 variables (5 basic and 1 non-basic variable) and 5 equations associated with this problem.

Table 21.4f Penalty calculation and allocation using Vogel approximation method (step 5)

From	To Project 1 (d_1)	Project 2 (d_2)	Project 3 (d_3)	Project 4 (d_4)	Supply (m^3)	Penalty (row)
Gurgaon RMC Plant (s_1)	₹ 80	₹ 60 50	₹ 100 50	₹ 90	~~100~~,50	–
Noida RMC Plant (s_2)	₹ 90 75	₹ 120	₹ 130 10	₹ 70 65	~~150,140, 65~~	70
Demand (m^3)	~~75~~	~~50~~	~~60~~,10	~~65~~		
Penalty (column)	–	–	–	70		

Table 21.5 Starting solution provided by the Least Cost method

From	To Project 1 (d_1)	Project 2 (d_2)	Project 3 (d_3)	Project 4 (d_4)	Supply (m^3)
Gurgaon RMC Plant (s_1)	80 50	60 50	100	90	100
Noida RMC Plant (s_2)	90 25	120	130 60	70 65	150
Demand (m^3)	75	50	60	65	

In the method of multipliers, we associate the multipliers u_i and v_j with row i and column j of the transportation table. For each current basic variable x_{ij}, these multipliers are having following relationships:

$$u_i + v_j = c_{ij}, \text{ for each basic } x_{ij} \quad (21.1)$$

The above equation will lead to the following set of equations for this problem:

$$u_1 + v_1 = 80 \quad (a)$$
$$u_1 + v_2 = 60 \quad (b)$$
$$u_2 + v_1 = 90 \quad (c)$$
$$u_2 + v_3 = 130 \quad (d)$$
$$u_2 + v_4 = 70 \quad (e)$$

In order to solve equations (a) to (e), we arbitrarily assign $u_1 = 0$, which leads to $v_1 = 80$, $v_2 = 60$, $u_2 = 10$, $v_3 = 120$, and $v_4 = 60$. This is also shown in Table 21.6

The values of u_i and v_j obtained by solving equations (a) to (e) are now used to evaluate the non-basic variables by computing –

$u_i + v_j - c_{ij}$, for each non-basic x_{ij} (see Table 21.7)

Table 21.6 Computations of u_i and v_j

	From	$v_1 = 80$ Project 1 (d_1)	$v_2 = 60$ Project 2 (d_2)	$v_3 = 120$ Project 3 (d_3)	$v_4 = 60$ Project 4 (d_4)	Supply (m^3)
$u_1 = 0$	Gurgaon RMC Plant (s_1)	₹80 50	₹60 50	₹100	₹90	100
$u_2 = 10$	Noida RMC Plant (s_2)	₹90 25	₹120	₹130 60	₹70 65	150
	Demand (m^3)	75	50	60	65	

Table 21.7 Computation for non-basic variables

Non-basic variable	$u_i + v_j - c_{ij}$
x_{13}	$u_1 + v_3 - c_{13} = 0 + 120 - 100 = 20$
x_{14}	$u_1 + v_4 - c_{14} = 0 + 60 - 90 = -30$
x_{22}	$u_2 + v_2 - c_{22} = 10 + 60 - 120 = -50$

This is equivalent to the z row computations in simplex method.

Basic	x_{11}	x_{12}	x_{13}	x_{14}	x_{21}	x_{22}	x_{23}	x_{24}
z	0	0	20	−30	0	−50	0	0

Since transportation cost is a minimization problem, the highest positive value x_{13} will be entering variable. Selection of x_{13} implies that we want to ship through this route. How much quantity can be transported through this route is another question that needs answer at this stage.

Let us assume a quantity θ that is shipped through this route. Now if θ quantity is to be shipped through the cell $s_1 d_3$, the cell $s_1 d_1$ has to offload θ quantity from the initial transported quantity 50, so that the total supply from source s_1 remains at 100. Now that the cell $s_1 d_1$ has been offloaded with θ quantity to a revised quantity 50 − θ, the cell $s_2 d_1$ has to be augmented with θ quantity making total shipment through this route to be 25 + θ, so that the demand of project d_1 remains satisfied. Having augmented the cell $s_2 d_1$ with θ quantity, we must offload θ quantity from the cell $s_2 d_3$ so that the supply from source s_2 remains at 150.

By making the above adjustments in shipment, we land up constructing a closed loop which starts and ends at the cell $s_1 d_3$. While constructing a loop, we can move horizontally across a row and vertically along a column. Any diagonal movement is not allowed. Table 21.8 shows this loop.

The maximum value of θ can be obtained in such a way that the following two conditions are satisfied.

1. The supply limits and the demand requirements remain satisfied, and
2. No negative shipments are allowed through any of the routes.

Table 21.8 First iteration after starting solution

		$v_1 = 80$	$v_2 = 60$	$v_3 = 120$	$v_4 = 60$	
		\multicolumn{4}{c}{To}	
	From	Project 1 (d_1)	Project 2 (d_2)	Project 3 (d_3)	Project 4 (d_4)	Supply (m^3)
$u_1 = 0$	Gurgaon RMC Plant (s_1)	₹ 80 50 − θ	₹ 60 50	₹ 100 θ	₹ 90	100
$u_2 = 10$	Noida RMC Plant (s_2)	₹ 90 25 + θ	₹ 120	₹ 130 60 − θ	₹ 70 65	150
	Demand (m^3)	75	50	60	65	

Table 21.9 Second iteration after starting solution

		$v_1 = 60$	$v_2 = 60$	$v_3 = 100$	$v_4 = 40$	
		\multicolumn{4}{c}{To}	
	From	Project 1 (d_1)	Project 2 (d_2)	Project 3 (d_3)	Project 4 (d_4)	Supply (m^3)
$u_1 = 0$	Gurgaon RMC Plant (s_1)	₹ 80	₹ 60 50	₹ 100 50	₹ 90	100
$u_2 = 30$	Noida RMC Plant (s_2)	₹ 90 75	₹ 120	₹ 130 10	₹ 70 65	150
	Demand (m^3)	75	50	60	65	

The application of above conditions leads to:

$$x_{11} = 50 - \theta \geq 0$$
$$x_{13} = 60 - \theta \geq 0$$

Thus, maximum value of $\theta = 50$

For $\theta = 50$, $x_{11} = 0$. Thus, x_{11} will leave the solution.
Hence x_{13} enters and x_{11} leaves.
Reduction in cost = $50 \times 20 = $ ₹ 1,000
New objective value = 21,600 – 1,000 = ₹ 20,600.
The revised transportation routes and the transported quantities are shown in Table 21.9.
As before, for each current basic variable x_{ij}, the multipliers and v_j will have following relationships:

$$u_i + v_j = c_{ij}$$

The above relationship leads to the following equations:

$$u_1 + v_2 = 60$$
$$u_1 + v_3 = 100$$
$$u_2 + v_1 = 90$$
$$u_2 + v_3 = 130$$
$$u_2 + v_4 = 70$$

For an arbitrarily assigned value of $u_1 = 0$, we get, $v_2 = 60$, and $v_3 = 100$

$$u_2 = 30, v_1 = 60, \text{ and } v_4 = 40.$$

As before, we use the computed values of u_i and v_j to evaluate the non-basic variables by computing –

$$u_i + v_j - c_{ij}, \text{ for each nonbasic } x_{ij} \text{ (see Table 21.10)}$$

Table 21.10 Computations for non-basic variables

Non-basic variable	$u_i + v_j - c_{ij}$
x_{11}	$u_1 + v_1 - c_{11} = 0 + 60 - 80 = -20$
x_{14}	$u_1 + v_4 - c_{14} = 0 + 40 - 90 = -50$
x_{22}	$u_2 + v_2 - c_{22} = 30 + 60 - 120 = -30$

This is equivalent to the z row computations in simplex method.

Basic	x_{11}	x_{12}	x_{13}	x_{14}	x_{21}	x_{22}	x_{23}	x_{24}
Z	−20	0	0	−50	0	−30	0	0

We find that z-row has all negative values for non-basic variables. Hence this is the optimum table.

∴ Objective value
$$= 50 \times 60 + 50 \times 100 + 75 \times 90 + 10 \times 130 + 65 \times 70 = 20{,}600$$

21.3 TRANSSHIPMENT PROBLEM

In construction, a contractor normally works at many project sites at any point of time. In this process, he has to mobilize a new project. He has to cater to different sites by pooling in resources. For example, for project site 1, he may mobilize material and equipment from sites 2, 3, and 4. In order to mobilize the site in an economical manner, it may so happen that the contractor first shifts material from sites 2, 3, and 4 to some central location and then they are transported to project site 1. In other words, materials are not directly shifted to the final destination. Rather they are shifted via an intermediate destination. It may be cheaper in practice to ship through intermediate or transient nodes before reaching the final destination.

Such problems, where a shipment may move to the final destination through an intermediate node or transshipment node are known as **transshipment problems** in the OR literature. As in the case of transportation problems, transshipment problems can also be represented by a network. For an illustration, the network representation for a transshipment problem with three sources, two intermediate nodes, and two destinations is shown in Figure 21.2.

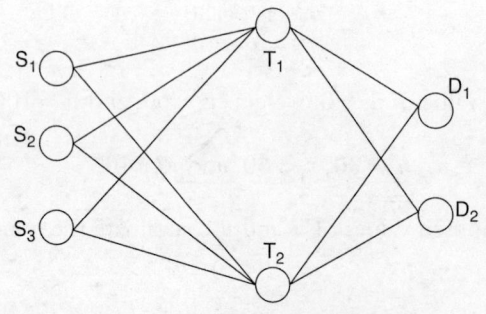

Figure 21.2 Network representation of a typical transshipment problem (3 sources, 2 intermediate nodes, and 2 destinations)

For solving such problems, there are special algorithms available in which the transshipment problem is converted into a larger transportation problem. In transportation problem, we saw how only direct shipments were allowed between a source and a destination. The use of buffer can be implemented while solving transshipment problems. The buffer amount should be sufficiently large to allow the entire original supply (or demand) units to pass through any of the transshipment nodes. Thus, the buffer amount B = Total supply (or demand)

An example is illustrated to explain the solution process of transshipment problems.

Example:

A small construction company is in the process of demobilizing two construction projects S_1 and S_2 and mobilizing two new projects D_1 and D_2. The materials available for demobilization at sites S_1 and S_2 are 12 tons and 18 tons respectively, while the demand of similar materials at sites D_1 and D_2 are 20 tons and 10 tons respectively. The company has two warehouses at different locations where loading and unloading facilities exist. The warehouses also have maintenance facilities and it is the policy of the company to repair and maintain all materials before sending it to the new sites. The transportation cost per ton of materials from old projects to warehouses and warehouses to new projects are given in the following network diagram. For example, from project S_1, materials can be dispatched at a cost of ₹ 100 per ton to warehouse T_1 and at ₹ 150 per ton to warehouse T_2. Likewise similar costs can be interpreted.

As can be seen from the network presentations, transshipment occurs in the network because the entire supply amount of 30 tons (12 + 18) from projects S_1 and S_2 could potentially pass through any warehouse T_1 and T_2 before reaching new projects D_1 and D_2.

From Figure 21.3, it can be observed that T_1 and T_2 act as both supply and demand source, while S_1 and S_2 are pure supply nodes. The nodes D_1 and D_2 act as pure demand nodes.

Using the above classification of pure node and pure supply, the problem can be converted into a regular transportation problem with 4 sources P_1, P_2, T_1, T_2 and 4 destinations T_1, T_2, D_1, D_2.

As mentioned earlier, the supply and demand at the transshipment node can be considered equal to the buffer amount which is the total supply or demand. Thus, buffer in this case = 18

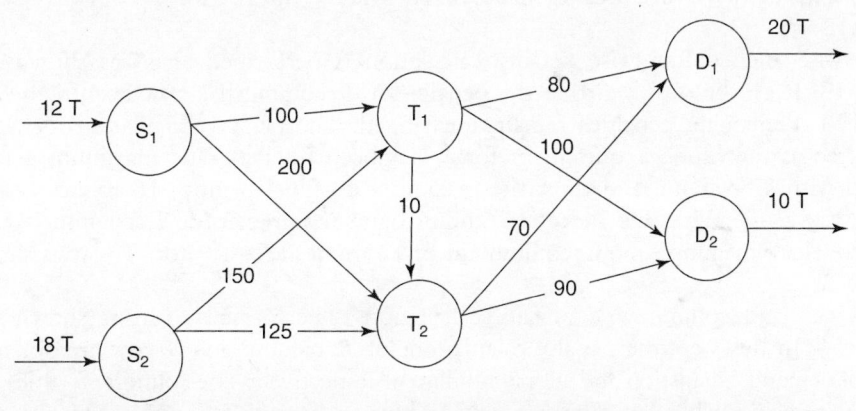

Figure 21.3 Network representation of example problem to illustrate transshipment

Table 21.11 Table showing the cost of transportation

	T_1	T_2	D_1	D_2	Supply
S_1	100	200	M	M	12
S_2	150	125	M	M	18
T_1	0	10	80	100	B
T_2	M	0	70	90	B
Demand	B	B	20	10	

+ 12 or 20 + 10 = 30. For routes where shipments are not possible, a prohibitive cost (very high cost M) is considered. For example, it is not possible to directly ship the materials from P_1 to D_1, thus cost corresponding to this route is shown as M. Likewise other cost values can be interpreted. The modified data is presented in Table 21.11 and is self-explanatory.

Buffer is either the sum of the total supply or demand and in this case equals to $B = 12 + 18 = 30$ or $20 + 10 = 30$.

Now initial basic solution to the balanced transportation problem can be obtained by one of the three methods discussed above. Here, Vogel's approximation method has been used to derive the starting solution; however, reader may choose any method for getting the initial basic solution followed by the optimality check using $u_i - v_j$ calculations.

For each row and column, a penalty measure is determined first by subtracting the smallest unit cost from the next smallest unit cost element in the same row and column. For example, the penalty in S_1 row is the difference between the second smallest and the smallest cell values = 200 – 100 = 100 and for the S_2 row it is equal to 150 – 125 = 25. Similarly, the penalties in d_1 and d_2 rows are 10 and 70 respectively. The penalties in T_1, T_2, D_1, and D_2 columns are 100, 10, 10, and 10 respectively. This has been shown in Table 21.12.

It is observed that the S_1 row and T_1 column have the maximum penalty value = 100, and thus, there is a tie. The maximum possible allocation of 12 tonnes is made in the cell S_1T_1 which is having minimum cost coefficient (100). Supply requirement of S_1 is thus fulfilled while there is a balanced demand of 18 tonnes (30–12) for T_1. This is shown in Table 21.12.

Row wise and column wise penalty calculation is performed again as shown in the Table 21.13. It can be observed that the penalty for T_1 column has now been revised from 100 to 150. Remaining penalties remain unchanged. The penalty calculation for S_1 row is not required as the supply requirement for S_1 has been fulfilled. The maximum penalty is associated with T_1 column, and thus, the maximum possible quantity 18 tonnes is allocated to the cell T_1T_1. With this allocation, the demand requirement of T_1 column is fulfilled while there is a remaining supply requirement of 12 tonnes (30–18) for T_1 row as shown in Table 21.13.

Row wise and column wise penalty calculation is performed again as shown in the Table 21.14. In this step, the penalty calculations of T_1 column and S_1 row are not needed because of demand exhaustion and supply fulfillment respectively. The column penalties for T_2, D_1, and D_2 remain unchanged. The row penalties for S_2, T_1, and T_2 are M – 125, 70, and 70

Table 21.12 Allocation using Vogel approximation method (step 1)

From	To T_1	T_2	D_1	D_2	Supply (T)	Penalty (row)
S_1	100 12	200	M	M	~~12~~	200 − 100 = 100
S_2	150	125	M	M	18	150 − 125 = 25
T_1	0	10	80	100	30	10 − 0 = 10
T_2	M	0	70	90	30	70 − 0 = 70
Demand (T)	~~30~~, 18	30	20	10		
Penalty (column)	100 − 0 = 100	10 − 0 = 10	80 − 70 = 10	100 − 90 = 10		

respectively. It can be observed that the penalty for S_2 row has now been revised to M − 125 which is the maximum among all, and thus, the maximum possible allocation of 18 tonnes is made to the cell $S_2 T_2$. With this allocation, the supply requirement of S_2 row is fulfilled and the remaining demand of T_2 column is 12 tonnes (30 − 18) as shown in Table 21.14.

Table 21.13 Allocation using Vogel approximation method (step 2)

From	To T_1	T_2	D_1	D_2	Supply (T)	Penalty (row)
S_1	100 12	200	M	M	~~12~~	−
S_2	150	125	M	M	18	150 − 125 = 25
T_1	0 18	10	80	100	~~30~~, 12	10 − 0 = 10
T_2	M	0	70	90	30	70 − 0 = 70
Demand (T)	~~30, 18~~	30	20	10		
Penalty (column)	150	10	10	10		

Table 21.14 Allocation using Vogel approximation method (step 3)

From	To T_1	T_2	D_1	D_2	Supply (T)	Penalty (row)
S_1	100 12	200	M	M	~~12~~	–
S_2	150	125 18	M	M	~~18~~	$M - 125$
T_1	0 18	10	80	100	~~30~~, 12	$80 - 10 = 70$
T_2	M	0	70	90	30	$70 - 0 = 70$
Demand (T)	~~30~~, 18	~~30~~, 12	20	10		
Penalty (column)	–	10	10	10		

Row wise and column wise penalty calculation is performed again as shown in the Table 21.15. In this step, the penalty calculations of T_1 column, besides S_1 and $S2$ rows are not needed for the reasons explained earlier. The column penalties for T_2, D_1, and D_2 remain unchanged and are equal to 10. The row penalties are maximum for T_1 and T_2 and incidentally there is a tie. An allocation of 12 tonnes is made to the lowest cost cell T_1T_2. By this allocation, the demand requirement of T_2 column is satisfied while the supply requirement of T_2 row has been reduced to 18 from 30 as shown in Table 21.15.

Row wise and column wise penalty calculation is performed again as shown in the Table 21.16. In this step, the penalty calculations of T1 and T2 columns besides S_1 and $S2$ rows are not needed. The column penalties for D_1 and D_2 are unchanged. The row penalties for T_1 and T_2 are tied at 20. Thus an allocation of 18 tonnes is made to the lowest cost cell T_2D_1. By this allocation, the supply requirement of T_2 row is satisfied while the demand requirement of D_1 row is now reduced to 2 as shown in Table 21.16.

Row wise and column wise penalty calculation is performed again as shown in the Table 21.17. In this step, the penalty calculations of T1 and T2 columns besides S_1, S_2, and T_2 rows are not needed. The column penalties for D_1 and D_2 are 80 and 100 respectively. The row penalty for T_1 is 20. Thus an allocation of 10 tonnes is made to the cell T_1D_2. By this allocation, the demand requirement of D_2 column is satisfied while the supply requirement of T_1 row is now reduced to 2 as shown in Table 21.17.

Row wise and column wise penalty calculation is performed for the final time again as shown in the Table 21.18. The row penalty for T_1 is 80 while column penalty for D_1 is also 80.

Table 21.15 Allocation using Vogel approximation method (step 4)

From	T_1	T_2	D_1	D_2	Supply (T)	Penalty (row)
S_1	100 12	200	M	M	~~12~~	–
S_2	150	125 18	M	M	~~18~~	–
T_1	0 18	10	80	100	~~30~~, 12	80 – 10 = 70
T_2	M	0 12	70	90	~~30~~, 18	70 – 0 = 70
Demand (T)	~~30, 18~~	~~30, 12~~	20	10		
Penalty (column)	–	10 – 0 = 10	10	10		

Thus an allocation of 2 tonnes is made to the cell $T_1 D_1$. By this allocation, the demand requirement of D_1 column and supply requirement of T_1 row is satisfied simultaneously as shown in Table 21.18.

Thus, the initial starting solution yields following 7 basic variables;

$$x_{11} = 12,\ x_{22} = 18,\ x_{31} = 18,\ x_{33} = 2,\ x_{34} = 10,\ x_{42} = 12,\ \text{and}\ x_{43} = 18$$

The total transportation cost (objective value) corresponding to this initial solution would be:

$$Z = 100 \times 12 + 125 \times 18 + 0 \times 18 + 80 \times 2 + 100 \times 10 + 0 \times 12 + 70 \times 18 = ₹\,5{,}870$$

Table 21.16 Allocation using Vogel approximation method (step 5)

From	T_1	T_2	D_1	D_2	Supply (T)	Penalty (row)
S_1	100 12	200	M	M	~~12~~	–
S_2	150	125 18	M	M	~~18~~	–
T_1	0 18	10	80	100	~~30~~, 12	100 – 80 = 20
T_2	M	0 12	70 18	90	~~30~~, ~~18~~	90 – 70 = 20
Demand (T)	~~30, 18~~	~~30, 12~~	20, 2	10		
Penalty (column)	–	–	80 – 70 = 10	100 – 90 = 10		

Table 21.17 Allocation using Vogel approximation method (step 6)

From	To T_1	T_2	D_1	D_2	Supply (T)	Penalty (row)
S_1	100 12	200	M	M	~~12~~	–
S_2	150	125 18	M	M	~~18~~	–
T_1	0 18	10	80 10	100	~~30,12~~,2	100 – 80 = 20
T_2	M	0 12	70 18	90	~~30~~, ~~18~~	–
Demand (T)	~~30,18~~	~~30,12~~	~~20~~,2	~~10~~		
Penalty (column)	–	–	80	100		

In the method of multipliers, we associate the multipliers u_i and v_j with row i and column j of the transportation table. For each current basic variable x_{ij}, these multipliers are having following relationships:

$$u_i + v_j = c_{ij}, \text{ for each basic } x_{ij}$$

Table 21.18 Allocation using Vogel approximation method (step 7)

From	To T_1	T_2	D_1	D_2	Supply (T)	Penalty (row)
S_1	100 12	200	M	M	~~12~~	–
S_2	150	125 18	M	M	~~18~~	–
T_1	0 18	10	80 2	100 10	~~30, 12, 2~~	**100**
T_2	M	0 12	70 18	90	~~30~~, ~~10~~	–
Demand (T)	~~30, 18~~	~~30, 12~~	~~20~~	~~10~~		
Penalty (column)	–	–	80	–		

Table 21.19 Computations of u_i and v_j

	From	T_1	T_2	D_1	D_2	Supply
		$v_1 = 100$	$v_2 = 110$	$v_3 = 180$	$v_4 = 200$	
			To			
$u_1 = 0$	S_1	100 / 12	200	M	M	12
$u_2 = 15$	S_2	150	125 / 18	M	M	18
$u_3 = -100$	T_1	0 / 18	10	80 / 2	100 / 10	30
$u_4 = -110$	T_2	M	0 / 12	70 / 18	90	30
	Demand	30	30	20	10	

The above equation will lead to the following set of equations for this problem:

$$u_1 + v_1 = 100 \quad \text{(a)}$$
$$u_2 + v_2 = 125 \quad \text{(b)}$$
$$u_3 + v_1 = 0 \quad \text{(c)}$$
$$u_3 + v_3 = 80 \quad \text{(d)}$$
$$u_3 + v_4 = 100 \quad \text{(e)}$$
$$u_4 + v_2 = 0 \quad \text{(f)}$$
$$u_4 + v_3 = 70 \quad \text{(g)}$$

In order to solve equations (a) to (g), we arbitrarily assign $u_1 = 0$, which leads to
$u_2 = 15$, $u_3 = -100$, $u_4 = -110$, $v_1 = 100$, $v_2 = 110$, $v_3 = 180$, and $v_4 = 200$. This is also shown in Table 21.19.

The values of u_i and v_j obtained by solving equations (a) to (g) are now used to evaluate the non-basic variables by computing –

$$u_i + v_j - c_{ij}, \text{ for each nonbasic } x_{ij} \text{ (see Table 21.20)}$$

This is equivalent to the z row computations in the simplex method.

Basic	x_{11}	x_{12}	x_{13}	x_{14}	x_{21}	x_{22}	x_{23}	x_{24}	x_{31}	x_{32}	x_{33}	x_{34}	x_{41}	x_{42}	x_{43}	x_{44}
z	0	−90	180−M	200−M	−35	0	195−M	215−M	0	0	0	0	−10−M	0	0	0

We find that z-row has all negative values for non-basic variables. Hence this is the optimum table.

$$\therefore \text{Objective value} = ₹ 5,870$$

Hence result may be summarized as shown in Table 21.21.

Table 21.20 Computations for non-basic variables

Non-basic variable	
x_{12}	$u_1 + v_2 - c_{12} = 0 + 110 - 200 = -90$
x_{13}	$u_1 + v_3 - c_{13} = 0 + 180 - M = 180 - M$
x_{14}	$u_1 + v_4 - c_{14} = 0 + 200 - M = 200 - M$
x_{21}	$u_2 + v_1 - c_{21} = 15 + 100 - 150 = -35$
x_{23}	$u_2 + v_3 - c_{23} = 15 + 180 - M = 195 - M$
x_{24}	$u_2 + v_4 - c_{24} = 15 + 200 - M = 215 - M$
x_{32}	$u_3 + v_2 - c_{32} = -100 + 110 - 10 = 0$
x_{41}	$u_4 + v_1 - c_{41} = -110 + 100 - M = -10 - M$
x_{44}	$u_4 + v_4 - c_{44} = -110 + 200 - 90 = 0$

21.4 ASSIGNMENT PROBLEM

In an assignment problem, the objective is to minimize the total cost of assigning n number of workers to n number of jobs. The cost of worker i performing a job j is denoted by C_{ij}. The inherent assumption made in this type of problem is: 'all workers are assigned a job and each job is performed'.

An assignment problem is illustrated with the help of an example. Let's assume that a ready-mixed concrete company has employed four marketing engineer to serve the four project sites.

It may be noted that an assignment problem can be viewed as a special type of transportation problem described earlier in which the supplies and demands are equal to 1.

Like a transportation problem, an assignment problem can also be represented in a network diagram. Figure 21.4 shows the representation of a problem involving four workers and four jobs. W_1, W_2, W_3, and W_4 are workers while J_1, J_2, J_3, and J_4 are the jobs to be assigned to these workers. The costs of performing jobs J_1, J_2, J_3, and J_4 by a worker W_1 are r

Table 21.21 Summary of Results

Transport from	Transport to	Quantity transported	Unit Cost of Transportation	Cost of transportation (₹)
S_1	T_1	12	100	1,200
S_2	T_2	18	125	2,250
T_1	T_1	18	0	0000
T_1	D_1	2	80	0160
T_1	D_2	10	100	1,000
T_2	T_2	12	0	0000
T_2	D_1	18	70	1,260
Total cost of transportation				5,870

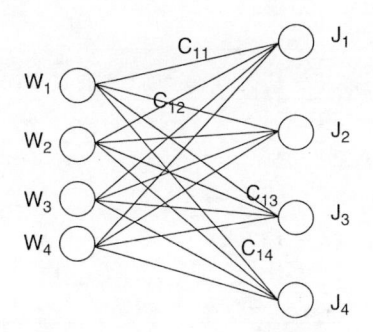

Figure 21.4 Network representation of an assignment problem

epresented by C_{11}, C_{12}, C_{13}, and C_{14}. Similarly other cost coefficients shown in the Figure 21.4 are interpreted.

An assignment problem is easily solved using the Hungarian method. For the illustration of this method, let's assume that a real estate builder has four upcoming projects—P_1, P_2, P_3, and P_4. He has invited bids from four known construction contractors with a promise that each one of them would be getting a project. The bid prices (₹ in million) submitted by the four contractors for the four projects are shown in the Table 21.22. It can be noted that the bid prices for contractor C_1 for the four projects are 100, 70, 200, and 145 million ₹ respectively. Likewise the bid prices for other contractors are also given in the table.

The various steps involved in the Hungarian method are given below. It may be noted that the Hungarian method is quite useful for hand computation.

Step 1

For the cost matrix of the problem, minimum of each row is identified (see Table 21.23) and it is subtracted from all the entries of the row (see Table 21.24). This is also referred to as row operation or row reduction.

Table 21.22 Data to illustrate assignment problem

	Project 1 – P_1	Project 2 – P_2	Project 3 – P_3	Project 4 – P_4
Contractor 1 – C_1	₹ 100 million	₹ 70 million	₹ 200 million	₹ 145 million
Contractor 2 – C_2	₹ 95 million	₹ 75 million	₹ 180 million	₹ 160 million
Contractor 3 – C_3	₹ 110 million	₹ 65 million	₹ 195 million	₹ 145 million
Contractor 4 – C_4	₹ 105 million	₹ 80 million	₹ 190 million	₹ 155 million

Table 21.23 Identifying row minimum for the example problem

	Project 1 – P_1	Project 2 – P_2	Project 3 – P_3	Project 4 – P_4	Row minimum
Contractor 1 – C_1	100	70	200	145	70
Contractor 2 – C_2	95	75	180	160	75
Contractor 3 – C_3	110	65	195	145	65
Contractor 4 – C_4	105	80	190	155	80

Table 21.24 Illustration of row operation

	Project 1 – P_1	Project 2 – P_2	Project 3 – P_3	Project 4 – P_4
Contractor 1 – C_1	30	0	130	75
Contractor 2 – C_2	20	0	105	85
Contractor 3 – C_3	45	0	130	80
Contractor 4 – C_4	25	0	110	75

Table 21.25 Identifying column minimum

	Project 1 – P_1	Project 2 – P_2	Project 3 – P_3	Project 4 – P_4
Contractor 1 – C_1	30	0	130	75
Contractor 2 – C_2	20	0	105	85
Contractor 3 – C_3	45	0	130	80
Contractor 4 – C_4	25	0	110	75
Column minimum	20	0	105	75

Step 2

Similar operation is followed for each column of the matrix. In other words, minimum of each column is identified (see Table 21.25) and subtracted from all column entries of the corresponding column (see Table 21.26). This is also known as column operation or column reduction.

Step 3

All 'zero' elements obtained after step 1 and step 2 are marked and shown in the Table 21.27. An optimal solution is identified as the feasible assignment associated with the zero elements of the matrix.

Table 21.26 Illustration of column operation

	Project 1 – P_1	Project 2 – P_2	Project 3 – P_3	Project 4 – P_4
Contractor 1 – C_1	10	0	25	0
Contractor 2 – C_2	0	0	0	10
Contractor 3 – C_3	25	0	25	5
Contractor 4 – C_4	5	0	5	0

Table 21.27 Marking of zero elements

	Project 1 – P_1	Project 2 – P_2	Project 3 – P_3	Project 4 – P_4
Contractor 1 – C_1	10	0	25	0
Contractor 2 – C_2	0	0	0	10
Contractor 3 – C_3	25	0	25	5
Contractor 4 – C_4	5	0	5	0

Table 21.28 Drawing of minimum number of lines that will cover all the zero entries

	Project 1 – P_1	Project 2 – P_2	Project 3 – P_3	Project 4 – P_4
Contractor 1 – C_1	10	0	25	0
Contractor 2 – C_2	0	0	0	10
Contractor 3 – C_3	25	0	25	5
Contractor 4 – C_4	5	0	5	0

Elements at the intersection of two lines

The location of zero entries does not allow one project per contractor. For example, if contractor 2 is assigned project 1, then row 2 will be eliminated, and we will be left with a situation in which there would be no taker for project 3 as there would be no zero entry in the project 3 column. This situation can be tackled by adding the following step in the steps mentioned earlier.

Step 4

In case there is no feasible solution assignment obtained with all zero entries in **step 3**, then we draw the minimum number of horizontal and vertical lines that will cover all the zero entries. This is shown again in the Table 21.28.

In the Table 21.28, the smallest uncovered element is selected and it is subtracted from every uncovered element. Besides, the smallest uncovered element is added to every element at the intersection of two lines.

The smallest uncovered element in the Table 21.28 is '5'. The subtraction and addition and the resulting matrix after the mentioned computations is shown in the Table 21.29

The process of drawing the minimum lines is repeated if no feasible assignment is found among the zero entries.

For allocation, we notice that row containing contractor 3 data has only one zero, and thus, contractor $C3$ is assigned project 2. This is marked with √ in the Table 21.30 in row 3. The remaining zeros in project 2 column are crossed out (x) as shown.

Table 21.29 Resulting matrix after subtraction and addition of the smallest uncovered element.

	Project 1 – P_1	Project 2 – P_2	Project 3 – P_3	Project 4 – P_4
Contractor 1 – C_1	10 – 5 = 5	0	25 – 5 = 20	0
Contractor 2 – C_2	0	0 + 5 = 5	0	10 + 5 = 15
Contractor 3 – C_3	25 – 5 = 20	0	25 – 5 = 20	5
Contractor 4 – C_4	5 – 5 = 0	0	5 – 5 = 0	0

Elements at the intersection of two lines

Table 21.30 Allocation of project – C_3 gets P_2

	Project 1 – P_1	Project 2 – P_2	Project 3 – P_3	Project 4 – P_4	
Contractor 1 – C_1	5	0×	20	0	
Contractor 2 – C_2	0	5	0	15	2
Contractor 3 – C_3	20	0√	20	5	
Contractor 4 – C_4	0	0×	0	0	1
		3		4	

Table 21.31 Allocation of project – C_1 gets P_4

	Project 1 – P_1	Project 2 – P_2	Project 3 – P_3	Project 4 – P_4	
Contractor 1 – C_1	5	0×	20	0√	
Contractor 2 – C_2	0	5	0	15	2
Contractor 3 – C_3	20	0√	20	5	
Contractor 4 – C_4	0	0×	0	0×	1
		3		4	

By crossing out the zeros, we notice that contractor 1 row now has only one uncrossed zero. Thus, project 4 is assigned to contractor C_1. Remaining zero in project 4 column is crossed out. This is shown in Table 21.31.

Now, it is noticed that there is no row or column with only one zero. In fact there are two uncrossed zeroes each in contractor C_2 and C_4 column. This indicates a multiple solution situation.

Arbitrarily in one case contractor C_2 is assigned project 1 and the corresponding cell is marked with √ as shown in Table 21.32. The other zero in the same row and same column is crossed out. This leaves only one zero uncrossed in project 3 column, and thus, contractor C_4 is assigned project P_3.

Thus corresponding to the above case, the assignment is as shown below:

Contractor C_1 – Project P_4, Cost – ₹ 145 million
Contractor C_2 – Project P_1 – Cost – ₹ 95 million

Table 21.32 Allocation of project – Alt. 1 – C_2 gets P_1 and C_4 gets P_3

	Project 1 – P_1	Project 2 – P_2	Project 3 – P_3	Project 4 – P_4	
Contractor 1 – C_1	5	0×	20	0√	
Contractor 2 – C_2	0√	5	0×	15	2
Contractor 3 – C_3	20	0√	20	5	
Contractor 4 – C_4	0×	0×	0√	0×	1
		3		4	

Table 21.33 Allocation of project – Alt. 2 – C_2 gets P_3 and C_4 gets P_1

	Project 1 – P_1	Project 2 – P_2	Project 3 – P_3	Project 4 – P_4	
Contractor 1 – C_1	5	0×	20	0√	
Contractor 2 – C_2	0×	5	0√	15	2
Contractor 3 – C_3	20	0√	20	5	
Contractor 4 – C_4	0√	0×	0×	0×	1
		3		4	

Contractor C_3 – Project P_2 – Cost – ₹ 65 million
Contractor C_4 – Project P_3 – Cost – ₹ 190 million

To get the alternate solution, project P_3 is assigned to contractor C_2. The corresponding cell is marked with √ in Table 21.33. The remaining zeroes in project 3 column and contractor C_2 row is crossed out. This leaves only one uncrossed zero in contractor C_4 row. Thus project 1 is awarded to contractor C_4.

Thus corresponding to the above case, the assignment is as shown below:

Contractor C_1 – Project P_4, Cost – ₹ 145 million
Contractor C_2 – Project P_3 – Cost – ₹ 180 million
Contractor C_3 – Project P_2 – Cost – ₹ 65 million
Contractor C_4 – Project P_1 – Cost – ₹ 105 million

It may be noted that resulting cost of the assignment in both the solutions is ₹ 495 million (verify!!!).

We take up another example to reinforce the above concepts. For this, we take the following minimal assignment problem in which the costs of performing four jobs $J_1, J_2, J_3,$ and J_4 by four persons A, B, C, and D are known. These are produced in Table 21.34.

For solving this problem using the Hungarian method, as a first step we identify the row minimum for each row. This is shown in Table 21.35. The row minimum is subtracted from each row element. The resulting matrix is shown in Table 21.36.

From the matrix represented by Table 21.36, each column's minimum is identified. This is shown in Table 21.37.

The column minimum is subtracted from each column element. The resultant matrix is shown in Table 21.38.

Table 21.34 Another data set for illustration of assignment problem

	J_1	J_2	J_3	J_4
A	40	60	30	70
B	40	30	20	10
C	50	80	50	40
D	70	90	40	70

Table 21.35 Identifying row minimum

	J_1	J_2	J_3	J_4	Row minimum
A	40	60	30	70	30
B	40	30	20	10	10
C	50	80	50	40	40
D	70	90	40	70	40

Table 21.36 Illustration of row operation

	J_1	J_2	J_3	J_4	Row minimum
A	10	30	0	40	30
B	30	20	10	0	10
C	10	40	10	0	40
D	30	50	0	30	40

Table 21.37 Identification of column minimum

	J_1	J_2	J_3	J_4	Row minimum
A	10	30	0	40	30
B	30	20	10	0	10
C	10	40	10	0	40
D	30	50	0	30	40
Column minimum	10	20	0	0	

A minimum number of vertical and horizontal straight lines necessary to cover zeroes in the matrix are now drawn. In this case, 4 lines can be drawn to cover all zeroes. If the number of lines needed to cover zeroes equal to the number of rows or columns, the optimal solution can be identified. It means a feasible assignment associated with the zero elements is possible.

For the assignment, we notice that row containing worker D cost data has only one zero, and thus, the job J_3 is assigned to worker D. This is marked with √ in the Table 21.38 in row 4. The remaining zero in J_3 column is crossed out (x) as shown in Table 21.38. By crossing out the zero in J_3 column, it is noticed that the A row now has only one uncrossed zero. Thus J_1

Table 21.38 Illustration of column operation and drawing of lines

	J_1	J_2	J_3	J_4	
A	0√	10	0×	40	1
B	20	0√	10	0×	2
C	0×	20	10	0√	3
D	20	30	0√	30	4

Table 21.39 Data reproduced from Table 21.34 to illustrate alternate solution method

	J_1	J_2	J_3	J_4
A	40	60	30	70
B	40	30	20	10
C	50	80	50	40
D	70	90	40	70

Table 21.40 Matrix after column reduction

	J_1	J_2	J_3	J_4
A	0	30	10	60
B	0	0	0	0
C	10	50	30	30
D	30	60	20	60

is assigned to worker A and the remaining zero in J_1 column is crossed out. By doing so, now there is only one uncrossed zero in row C. Thus, job J_4 is assigned to worker C. The remaining zero in J_4 column is crossed out which leaves row B with only one uncrossed zero. Thus, job J_2 is assigned to worker B.

The optimum cost of the above assignment = 40 + 30 + 40 + 40 = 150

In solving the above problem, first row operation or row reduction was performed. Subsequently, the column operation was performed. The problem could also have been solved by performing column operation or column reduction first followed by row operation. The problem is reproduced in Table 21.39.

The matrix resulting after column reduction is shown in Table 21.40. Row reduction in the Table 21.40 matrix results into the matrix shown in Table 21.41.

A minimum number of vertical and horizontal straight lines necessary to cover zeroes in the matrix are now drawn. In this case, 3 lines can be drawn to cover all zeroes (see Table 21.42). Thus, a feasible assignment cannot be made unless we have 4 lines which can cover all the zeroes.

Table 21.41 Matrix after row reduction

	J_1	J_2	J_3	J_4
A	0	30	10	60
B	0	0	0	0
C	0	40	20	20
D	10	40	0	40

Table 21.42 Drawing of minimum number of lines that will cover all the zero entries

	J_1	J_2	J_3	J_4
A	0	30	10	60
B	0	0	0	0
C	0	40	20	20
D	10	40	0	40

In the Table 21.42, the smallest uncovered element is selected and it is subtracted from every uncovered element. Besides, the smallest uncovered element is added to every element at the intersection of two lines. The result is shown in Table 21.43. It may be pointed out that the smallest uncovered element in the Table 21.42 is 20. With these operations, it is now possible to get the four lines as shown in Table 21.43.

For the assignment, we notice that row containing worker A cost data has only one zero, and thus, the job J_1 is assigned to worker A. This is marked with √ in the Table 21.43 in row A. The remaining zero in J_1 column is crossed out (x) as shown in Table 21.43. By crossing out the zero in J_1 column, it is noticed that the C row now has only one uncrossed zero. Thus J_4 is assigned to worker C and the remaining zero in J_4 column is crossed out. By doing so, now there is only one uncrossed zero in rows B and D. Thus, job J_2 is assigned to worker B and job J_3 is assigned to worker D.

The optimum cost of the above assignment = 40 + 30 + 40 + 40 = 150, which is same as obtained in the previous case.

21.5 UNBALANCED PROBLEM AND RESTRICTIONS ON ASSIGNMENT

The problems discussed in previous section had equal number of rows and columns. Such problems are also known as balanced assignment problems. However, in real life, situations may not warrant balanced problem formulation all the time. For example, there could be a situation in which there are more men available than the number of jobs to be assigned or vice versa. Such problems are referred to as unbalanced assignment problem. For solving unbalanced problems, a dummy row or column may be added depending on the case to convert to a balanced assignment problem. For example, if the number of rows is less than the number of columns, a dummy row is added. On the other hand if the number of columns is less than the

Table 21.43 Assignment of Jobs

	J_1	J_2	J_3	J_4
A	0√	10	10	40
B	20	0√	20	0x
C	0x	20	20	0√
D	10	20	0√	20

Table 21.44 Data for the example problem (Figures in ₹Lakhs)

Facilities	A	B	C	D	E
Office	10	12	11	15	9
Workshop	5	6	–	4.5	5.5
Stores	–	14	12	10	11
Laboratory	5.5	4	4.5	5	6

number of rows, a dummy column is added. The cell elements corresponding to a dummy row or a column are taken as zero, as in reality, the dummy row or column is non-existent.

In another variant of an assignment problem, there could be restriction on certain assignments. For example a particular job cannot be assigned to all the men available for assignment or a particular worker may not be able to do a certain job and so on. For solving such problems, a very high cost M is associated with the restricted assignment.

In this section, we will discuss these two aspects of an assignment problem—unbalanced and restriction on assignment. An example is solved to illustrate these two aspects.

Example:

A ready-mix concrete plant operator is planning to set up: (1) office, (2) plant and machinery workshop, (3) stores, and (4) laboratory. He has identified five possible locations A, B, C, D, and E for setting up these facilities. Because of limited space, plant and machinery workshop cannot be set up at location C. Also, stores cannot be set up at location A as there is likelihood of pilferage from this place. The costs of setting up of these facilities at different locations, in lakhs of rupees, are given in the Table 21.44. It is desired to find the suitable locations for setting up these facilities at a minimum cost.

This is an unbalanced problem as the number of facilities does not equal the number of set up locations. As mentioned earlier, a dummy facility row is created to make it a balanced assignment problem. Thus, a fictitious facility row is created and zero cost is assign corresponding to the different set up locations. Also where set up is restricted, a very high cost M is assigned to it. The revised matrix now is given in Table 21.45 which is clearly a balanced assignment problem and the solution steps mentioned in earlier sections are applicable.

The matrix resulting after row reduction is shown in Table 21.46.

Column reduction on the Table 21.46 matrix results into the matrix shown in Table 21.47.

Only four lines are possible to be drawn as against five required for optimal assignments. The least among uncovered elements is identified which happens to be 0.5. This is subtracted

Table 21.45 Conversion into balanced assignment problem

Facilities	A	B	C	D	E
Office	10	12	11	15	9
Workshop	5	6	M	4.5	5.5
Stores	M	14	12	10	11
Laboratory	5.5	4	4.5	5	6
Dummy	0	0	0	0	0

Table 21.46 Matrix after row reduction

Facilities	A	B	C	D	E	Row Minimum	
Office	1	3	2	6	0	9	4
Workshop	0.5	1.5	M – 4.5	0	1	4.5	
Stores	M – 10	4	2	0	1	10	
Laboratory	1.5	0	0.5	1	2	4	3
Dummy	0	0	0	0	0	0	1

[2]

from all uncovered elements and added to the elements which are at the junction of two lines. Hence the modified coefficients as obtained are shown in Table 21.48

The optimal assignment is as shown below:
Set up office at location E
Set up workshop at location A
Set up stores at location D, and
Set up laboratory at location B.
Setup dummy facility at location C, i.e., it has no assignment.
The resulting assignment cost is
= ₹ 9 lakhs + ₹ 5 Lakhs + ₹ 10 Lakhs + ₹ 4 Lakhs
= ₹ 28 Lakhs.

Table 21.47 Matrix after column reduction

Facilities	A	B	C	D	E	
Office	1	3	2	6	0	4
Workshop	0.5	1.5	M – 4.5	0	1	
Stores	M – 10	4	2	0	1	
Laboratory	1.5	0	0.5	1	2	3
Dummy	0	0	0	0	0	1
Minimum	0	0	0	0	0	

[2]

Table 21.48 Optimal assignment solution

Facilities	A	B	C	D	E	
Office	1	3	2	6.5	0√	5
Workshop	0√	1	M – 5	0×	0.5	
Stores	M – 10.5	3.5	1.5	0√	0.5	
Laboratory	1.5	0√	0.5	1.5	2	4
Dummy	0×	0×	0√	0.5	0×	1

[3] [2]

TRANSPORTATION, TRANSSHIPMENT, AND ASSIGNMENT PROBLEMS | **791** |

Table 21.49 Profit data for example 21.6.1

	D_1	D_2	D_3	D_4
A	16	10	14	11
B	14	11	15	15
C	15	15	13	12
D	13	12	14	15

21.6 SOME MORE APPLICATIONS

There are many possible applications of the Hungarian method illustrated in the last few sections. With little modifications in the solution steps mentioned earlier, following types of problems can also be addressed.

21.6.1 Maximization Problem

Such problems can be solved in two ways

Method I

In solving such problems, all cell elements C_{ij} of assignment matrix are changed to $-C_{ij}$ and exactly same steps as in minimization problems are repeated. An illustration is provided in Example 21.6.1.

Method II

In this method, all C_{ij} elements are subtracted from the highest cell element of the assignment matrix. Thereafter the usual process of row and column reduction, drawing of horizontal and vertical lines, and finally the appropriate assignment is carried out. An illustration is provided in Example 21.6.1.

Example 21.6.1. Maximization Problem

A company has a team of four salesmen and there are four districts where the company wants to start its business. After taking into account the capabilities of salesman and the nature of districts, the company estimates that the profit per day in rupees for each salesman in each district is as given in Table 21.49.

Find the assignment of salesman to various districts which will yield maximum profit.

Solution

Method 1 – Multiply all the elements with (−1) as shown in Table 21.50 and proceed as usual. Row operation on Table 21.50 leads to the Table 21.51.

Table 21.50 Multiplying all the cells with −1

	D_1	D_2	D_3	D_4
A	−16	−10	−14	−11
B	−14	−11	−15	−15
C	−15	−15	−13	−12
D	−13	−12	−14	−15

Table 21.51 Row operation

	D_1	D_2	D_3	D_4	Row Min
A	0	6	2	5	−16
B	1	4	0	0	−15
C	0	0	2	3	−15
D	2	3	1	0	−15

Table 21.52 Column operation

	D_1	D_2	D_3	D_4
A	0√	6	2	5
B	1	4	0√	0×
C	0×	0√	2	3
D	2	3	1	0√
Column minimum	0	0	0	0
	1	4	3	2

Column operation on Table 21.51 leads to the Table 21.52
The assignment is as below for the profit maximization.
A − D_1, B − D_3, C − D_2, D − D_4 and the maximum profit per day = 16 + 15 + 15 + 15 = 61

Method 2 – Subtract from the highest element (i.e., 16) among all the elements of the given profit data represented by Table 21.49. The resulting matrix is shown in Table 21.53.

Now apply Hungarian method to get the optimum solution. Row operation leads to Table 21.54 and subsequent column operation leads to Table 21.55.

The assignment is as below for the profit maximization.

Table 21.53 Subtracting all the elements from the highest element 16

	D_1	D_2	D_3	D_4
A	0	6	2	5
B	2	5	1	1
C	1	1	3	4
D	3	4	2	1

Table 21.54 Row operation

	D_1	D_2	D_3	D_4	Row Min
A	0	6	2	5	0
B	1	4	0	0	1
C	0	0	2	3	1
D	2	3	1	0	1

Table 21.55 Column operation

	D_1	D_2	D_3	D_4
A	0√	6	2	5
B	1	4	0√	0x
C	0x	0√	2	3
D	2	3	1	0√
Column minimum	0	0	0	0
	1	4	3	2

$A - D_1$, $B - D_3$, $C - D_2$, $D - D_4$ and the maximum profit per day $= 16 + 15 + 15 + 15 = 61$ which is same as obtained with Method 1.

21.6.2 Crew Assignment Problem

Using such types of problem formulation, an optimal assignment of crew members can be organized. In this section, a crew assignment problem is solved to illustrate the solution steps. This method can be used to plan the assignment of crew members in different locations by a transport company.

Example 21.6.2

A trip from Dehradun to Delhi takes six hours by bus. A typical time table of the bus service is given in Table 21.56.

The cost of providing this service by the transport company depends upon the time spent by the bus crew (driver and conductor) away from their places in addition to service time. There are five crews. There is a constraint that every crew should be provided with at least 4 hours of rest before the return trip again and should not wait for more than 24 hours for the return trip. The company has residential facilities for the crew at Dehradun as well as at Delhi. Find the optimal service line connections.

Solution:

Change over time = At least 4 hours.

Let the entire crew resides at Dehradun. Thus the waiting time in Delhi is as given in the Table 21.57. The computation of waiting time is illustrated for route 'A' crew assuming that it has to cater to routes 1, 2, 3, 4, and 5.

Table 21.56 Time table for bus service from Delhi–Dehradun–Delhi

Departure from Dehradun	Route no.	Arrival at Delhi	Arrival at Dehradun	Route no.	Departure from Delhi
5:00	A	11:00	11:00	1	5:00
7:00	B	13:00	14:00	2	8:00
11:00	C	17:00	20:00	3	14:00
18:00	D	00:00	00:00	4	18:00
00:30	E	06:30	06:00	5	00:00

794 | CONSTRUCTION PROJECT MANAGEMENT

Table 21.57 Waiting time of crew at Delhi assuming crew resides at Dehradun

	1	2	3	4	5
A	18	21	M	7	13
B	16	19	M	5	11
C	12	15	21	M	7
D	5	8	14	18	24
E	22.5	M	7.5	11.5	17.5

The route 'A' crew departs at 5:00 AM from Dehradun and arrives Delhi at 11:00 AM. Now if the same crew is assigned the route 1 (scheduled departure at 5:00 from Delhi), their waiting time in Delhi can be computed as below:

[Departure Time from Delhi (route 1) − Arrival Time at Delhi (route A)] = (5:00 − 11:00) = 18 hours.

If crew of route A is assigned the route 2, their waiting time in Delhi can be computed as below:

[Departure Time from Delhi (route 2) − Arrival Time at Delhi (route A)] = (8:00 − 11:00) = 21 hours.

If crew of route A is assigned the route 3, their waiting time in Delhi can be computed as below:

[Departure Time from Delhi (route 3) − Arrival Time at Delhi (route A)] = (14:00 − 11:00) = 3 hours.

Since the waiting time in this case is less than the minimum specified waiting time of 4 hours, the route 3 cannot be assigned to the crew of route A. Mathematically, this is achieved by assigning a penalty M in the cell A-3 as shown in Table 21.57.

If crew of route A is assigned the route 4, their waiting time in Delhi can be computed as below:

[Departure Time from Delhi (route 4) − Arrival Time at Delhi (route A)] = (18:00 − 11:00) = 7 hours.

If crew of route A is assigned the route 5, their waiting time in Delhi can be computed as below:

[Departure Time from Delhi (route 5) − Arrival Time at Delhi (route A)] = (00:00 − 11:00) = 13 hours.

Similar calculations may be done for finding waiting time of crew on route B being assigned route 1 to route 5. Likewise the waiting time of crews C, D, and E can be computed if they are assigned routes 1 to 5. The computed waiting times are entered in the appropriate cells of Table 21.57.

Now we assume that the entire crew resides at Delhi. Thus, the waiting time at Dehradun is as given in Table 21.58. The computation of waiting time is illustrated for route 5 crew assuming that it has to cater to routes A, B, C, D, and E.

Calculations for waiting time in Table 21.58 is done column wise as crew is assumed to be based in Delhi and crew of route 1–5 are assigned route A–E respectively on their return trip from Dehradun to Delhi.

Table 21.58 Waiting time of crew at Dehradun assuming crew resides at Delhi

	1	2	3	4	5
A	18	15	9	5	23
B	20	17	11	7	M
C	24	21	15	11	5
D	7	4	22	18	12
E	13.5	10.5	4.5	M	18.5

Calculation steps for Column 5 has been illustrated here:

The route 5 crew departs at 00:00 from Delhi and arrives at Dehradun at 06:00. Now if the same crew is assigned the route A (scheduled departure at 5:00 from Dehradun), their waiting time in Dehradun can be computed as below:

[Departure Time from Dehradun (route A) − Arrival Time at Dehradun (route 5)] = (05:00 − 06:00) = 23 hours.

If crew of route 5 is assigned the route B, their waiting time in Dehradun can be computed as below:

[Departure Time from Dehradun (route B) − Arrival Time at Dehradun (route 5)] = (07:00 − 06:00) = 1hour. Since the waiting time in this case is less than the minimum specified waiting time of 4 hours, the route B cannot be assigned to the crew of route 5. Mathematically, this is achieved by assigning a penalty M in the cell B-5 as shown in Table 21.58.

If crew of route 5 is assigned the route C, their waiting time in Dehradun can be computed as below:

[Departure Time from Dehradun (route C) − Arrival Time at Dehradun (route 5)] = (11:00 − 06:00) = 5 hours.

If crew of route 5 is assigned the route D, their waiting time in Dehradun can be computed as below:

[Departure Time from Dehradun (route D) − Arrival Time at Dehradun (route 5)] = (18:00 − 06:00) = 12 hours.

If crew of route 5 is assigned the route E, their waiting time in Dehradun can be computed as below:

[Departure Time from Dehradun (route E) − Arrival Time at Dehradun (route 5)] = (00:30 − 06:00) = 18.5 hours.

Similar calculations may be done for finding waiting time of crew on route 4 being assigned route A to route E. Likewise the waiting time of crews 3, 2, and 1 can be computed if they are assigned routes A to E. The computed waiting times are entered in the appropriate cells of Table 21.58.

As the crew can be asked to reside at Dehradun or at Delhi, minimum waiting time from the above operation can be computed for different route combinations by choosing the

Table 21.59 Waiting time of crew at Dehradun or Delhi assuming crew may reside either in Delhi or in Dehradun

	1	2	3	4	5
A	18	15	9	5	13
B	16	17	11	5	11
C	12	15	15	11	5
D	5	4	14	18	12
E	13.5	10.5	4.5	11.5	17.5

Table 21.60 Row reduction

	1	2	3	4	5	Row Minimum
A	13	10	4	0	8	5
B	11	12	6	0	6	5
C	7	10	10	6	0	5
D	1	0	10	14	8	4
E	9	6	0	7	13	4.5

minimum of two waiting times. For this, each cell in Table 21.57 is compared with corresponding cell of Table 21.58 and the minimum of the two is entered in the corresponding cell as shown in Table 21.59.

Now the above information can be used and assignment algorithm is applied for optimum solution using Hungarian Method. Row operation on Table 21.59 leads to Table 21.60 and subsequent column operation leads to Table 21.61.

Against the required numbers of five assignments only four assignments is possible at this stage reflected by minimum four lines needed to cover all zeros (see Table 21.61). Hence, we need to subtract the least cell value from all uncovered cells and add the same number at the intersections.

Least value cell among uncovered cells = 4. The resulting matrix is shown in Table 21.62.

Table 21.62 shows that minimum four lines are required to cover all zeros. Thus, all five assignments would not be possible even at this stage. Again the least of all uncovered element

Table 21.61 Column reduction

	1	2	3	4	5
A	12	10	4	0	8
B	10	12	6	0	6
C	6	10	10	6	0
D	0	0	10	14	8
E	8	6	0	7	13
Column Minimum	1	0	0	0	0

Table 21.62 Subtracting the least uncovered cell value from uncovered and adding the same to intersecting cells

	1	2	3	4	5	
A	8	6	0	0	8	
B	6	8	2	0	6	
C	2	6	6	6	0	4
D	0	0	10	18	12	3
E	8	6	0	11	17	
			1	2		

is subtracted from all uncovered and added to intersecting cell values and the table is checked for possible optimal assignment. The minimum of uncovered cell elements in Table 21.62 is 6. The revised matrix is shown in Table 21.63.

Table 21.63 shows that at least five lines are required to cover all zeros hence all five assignments would be possible at this stage. Hence assignments have been done in following order

First Assignment:

Third row (C.) has only one zero, hence assignment has been shown with (√) in the fifth column, i.e., (C – 5). Other zero of 5th column has been crossed out.

Second Assignment:

Fourth column has two zeros, hence, tie is broken randomly and assignment is made to B row, i.e., (B – 4). Other zeroes in 4th column and B rows are crossed.

Third Assignment:

Now 1st column has only one zero, hence, assignment can be in that cell as done in (D – 1). Other zero in the D row is crossed.

Fourth Assignment:

Second and third columns have two zeros and assignments could be done randomly to (E – 3) and (A – 2) respectively or (E – 2) or (A – 3) respectively.

Except assignment (C – 5), all other assignments have been done randomly by breaking ties, hence, it is a case of multiple optimal assignments.

Hence, one of the optimal assignments is: **A – 2, B – 4, C – 5, D – 1, and E – 3**

Table 21.63 Subtracting the least uncovered cell value from uncovered and adding the same to intersecting cells

	1	2	3	4	5
A	2	0√	0×	0×	2
B	0×	2	2	0√	0×
C	2	6	12	12	0√
D	0√	0×	16	24	12
E	2	0×	0√	11	11
	2	1	3	4	5

The waiting time of these assignments as well as the residence of crews are given below:

Crew	Residence at	Service no.	Waiting time hours
1	Delhi	(2 – A)	15 hours
2	Dehradun	(B – 4)	5 hours
3	Delhi	(5 – C)	5 hours
4	Dehradun	(D – 1)	5 hours
5	Delhi	(3 – E)	4.5 hours

Total minimum waiting time for the crew is <u>34.5 hours.</u>

Reader may check as many assignments as possible in this case as an alternative assignment. Note that the total waiting time for all optimal assignments must remain the same as 34.5 hours.

One such possible assignment could be A – 4, B – 1, C – 5, D – 2, and E – 3.

Crew	Residence at	Service no.	Waiting time hours
1	Delhi	(4 – A)	5 hours
2	Dehradun	(B – 1)	16 hours
3	Delhi	(5 – C)	5 hours
4	Dehli	(2 – D)	4 hours
5	Delhi	(3 – E)	4.5 hours

Total minimum waiting time for the crew is <u>34.5 hrs.</u>

21.6.3 Travelling Salesman Problem

In such problems an optimal route is found out for a travelling salesman on the condition that salesman starts his trip from the central office and after making trips to different destinations returns to the office. Parameters such as distances, time or cost between each pair of destinations is given (C_{ij}). The objective is to minimize the travel cost, travel time, or travel distance under the mentioned constraint. If it is assumed that there are n destinations and distances C_{ij} are known, then there are $(n-1)!$ possible ways for this tour. There could be two types—symmetrical and asymmetrical problems in this category. In the symmetrical problem, distances between each pair of destinations is independent of the direction of his journey while in an asymmetrical problem, the distance changes with the direction for one or more pair of destinations. Since, the salesman cannot go directly from destination i to i, all C_{ii} cells are assigned ∞. Example 21.6.3 illustrates the solution procedure for such problems.

Example 21.6.3

A salesman caters to the five regions A, B, C, D, and E. The salesman must start for his work every day in the morning from 'A' where his office is located. He must return to his office every day in the evening to prepare daily report. The travel cost from and to different regions are provided in Table 21.64. It is desired to choose a sequence for the salesman so that it results in the least travel cost.

TRANSPORTATION, TRANSSHIPMENT, AND ASSIGNMENT PROBLEMS | 799

Table 21.64 Travel cost data for the example 21.6.3

From	To				
	A	B	C	D	E
A	∞	4	7	3	4
B	4	∞	6	3	4
C	7	6	∞	7	5
D	3	3	7	∞	7
E	4	4	5	7	∞

Table 21.65 Row operation

	A	B	C	D	E	Row minimum
A	∞	1	4	0	1	3
B	1	∞	3	0	1	3
C	2	1	∞	2	0	5
D	0	0	4	∞	4	3
E	0	0	1	3	∞	4

Solution:

We carry out the row operation on the data provided in Table 21.64. The resultant matrix is shown in Table 21.65. Column operation on Table 21.65 leads to Table 21.66.

It can be seen that only four lines are needed to cover all the zeroes as against 5. Thus, subtract the least cell value, i.e., 1 from uncovered cells and add 1 at the intersections. The resulting matrix is shown in Table 21.67.

From Table 21.67, it is possible to get five lines to cover all zeroes. We can search the possible sequence from the zero locations of Table 21.67.

Table 21.66 Column operation

	A	B	C	D	E	Row minimum	
A	∞	1	3	0	1	3	
B	1	∞	2	0	1	3	
C	2	1	∞	2	0	5	
D	0	0	3	∞	4	3	2
E	0	0	0	3	∞	4	1
Column minimum	0	0	1	0	0		

3 4

Table 21.67 Subtracting the least uncovered cell value from uncovered and adding the same to intersecting cells.

	A	B	C	D	E	
A	∞	0̶	2	0̶	1	5
B	0̶	∞	1	0̶	1	4
C	1	0̶	∞	2	0̶	3
D	0̶	0̶	3	∞	5	2
E	0×	0×	0✓	4	∞	1

From the Table 21.67, a possible sequence: C→E→B→D→A is obtained. This could be the answer if the salesman has the liberty of starting from any city but it is not so as he has to start from 'A'.

Starting with 'A', we can also choose two sequences as below:

Sequence (1) A→D, D→B, and B→A. However this solution does not provide the solution of the traveling salesman as it gives A→D, D→B, B→A, and thus, it does not pass through C & E.

Sequence (2) A→B, B→D, and D→A. This sequence also has the same problem as it does not cover regions C and E.

Hence further calculation is needed. We try to find the next best solution which satisfies this restriction. The next minimum (non-zero) element in the cost matrix is 1. So, we bring 1 into the solution. But the element '1' occurs at three places. We consider all the cases separately until we get an optimal solution.

Start with making an assignment at the cell (1,5) instead of zero assignment at the cell (1,2) or (1,4). The resulting sequence: A→E, E→C, C→B, B→D, D→A has a cost = 21. Thus travel sequence is A→E→C→B→D→A and the corresponding cost is 21. There exists some alternative route as well. For example check A→D→B→C→E→A at the same cost.

REFERENCES

1. Hillier, F.S., Lieberman, G.J., 2009, *Operations Research*, 10th Reprint, Tata McGraw Hill, New Delhi, India.

2. Mohan, C., Deep, K., 2009, *Optimization Techniques*, First Edition, New Age International Publishers, New Delhi, India.

3. Render, B., Stair, R.M., and Hanna, M.E., 2000, *Quantitative Analysis for Management*, Eighth Edition, Pearson Education, New Delhi, India.

4. Sen, P.R., 2010, Operations Research-Algorithms and Applications, Prentice Hall India, New Delhi, India.

5. Taha, H.A., 2007, *Operations Research-An Introduction*, Seventh Edition, Prentice Hall India, New Delhi, India.

6. Wagner, H.M., 2005, *Principles of Operations Research*, 2nd Edition, Prentice Hall India, New Delhi, India.

REVIEW QUESTIONS

1. Find out the basic starting solution using (1) Northwest corner rule, (2) Least cost method, and (3) Vogel Approximation method

	Project 1	Project 2	Project 3	Supply (in 1,000)
Brick source 1	5	1	8	12
Brick source 2	2	4	0	14
Brick source 3	3	6	7	4
Demand (in 1,000)	9	10	11	

2. The assignment cost of assigning any operator to any machine is given in the following table. Solve the minimal assignment problem.

Machine \ Operators	A	B	C	D
I	10	5	13	15
II	3	9	18	13
III	10	7	3	2
IV	5	11	9	7

   Hint: I – B, II – D, III – C, IV – A and optimal assignment cost = 16

3. Solve the minimal assignment problem?

4	6	3	7
4	3	2	1
5	8	5	4
7	9	4	5

4. Solve the minimal assignment problem?

5	6	7	7
2	2	5	2
3	3	6	3
7	8	8	9

5. Solve the minimal assignment problem?

	A	B	C	D	E
A	11	7	13	9	7
B	6	10	9	9	10
C	9	12	18	13	6
D	4	6	11	8	8
E	9	11	11	7	8

6. Solve the minimal assignment problem?

11	7	13	9	7
6	10	9	9	10
9	12	18	13	6
4	6	11	8	8
9	11	11	7	8

7. Find the highest cost for the following matrix?

5	6	7	7
2	2	5	2
3	3	6	3
7	8	8	9

8. A machine operator processes 5 types of items on his machines each week and must choose a sequence for them. The setup cost per change depends on the item presently on the machine and the setup to be made according to the following table.

		(To change to Job)				
		A	B	C	D	E
	A	∞	4	7	3	4
(To change from Job)	B	4	∞	6	3	4
	C	7	6	∞	7	5
	D	3	3	7	∞	7
	E	4	4	5	7	∞

9. Given the following table of distances solve the traveling sales man problem?

	1	2	3	4	5
1	∞	2	14	8	6
2	4	∞	12	6	8
3	2	12	∞	4	2
4	2	10	8	∞	12
5	14	10	8	10	∞

10. A company has four territories open, and four salesman available for the assignment. The territories are not equally rich in their sales potential; it is estimated that a typical salesman operating in each territory would bring in the following annual sales:

Territory:	I	II	III	IV
Annual sales ₹	60,000	50,000	40,000	30,000

Four salesman are also considered to differ in their ability. It is estimated that working under the same conditions, their yearly sales would be proportional as follows:

Salesman	A	B	C	D
Proportion	7	5	5	4

If the criterion is to maximize expected total sales, then the intuitive answer is to assign the best sales man to the richest territory, the next best salesman to the second richest and so on. Verify this answer by the assignment technique.

11. A department has 5 employees with 5 jobs to be performed. The time (in hrs) each man will take to perform each job is given in the cost matrix below:

Jobs	I	II	III	IV	V
A	10	5	13	15	16
B	3	9	18	13	6
C	10	7	2	2	2
D	7	11	9	7	12
E	7	9	10	4	12

How should the jobs be allotted, one per employee, so as to minimize the total manhours?

Hint: A – II, B – I, C – V, D – III, E – IV and the minimum total time = 23 hrs.

12. Solve the following salesman problem given by the following data.

$C_{12} = 20$, $C_{13} = 4$, $C_{14} = 10$, $C_{23} = 5$, $C_{34} = 6$, $C_{25} = 10$, $C_{35} = 6$, $C_{45} = 20$

Where $C_{ij} = C_{ji}$ and there is no route between cities i and j if the value for C_{ij} is not shown.

Hint: The shortest route for the traveling salesman is 1→3→2→5→4→1

13. Solve the minimal assignment problem whose cost matrix is given below:

	1	2	3	4
A	2	3	4	5
B	4	5	6	7
C	7	8	9	8
D	3	5	8	4

Hint: The possible optimal solutions with each of cost ₹20 are:

1.) A – 2, B – 3, C – 4, D – 1
2.) A – 1, B – 3, C – 2, D – 4
3.) A – 3, B – 2, C – 4, D – 1
4.) A – 2, B – 3, C – 1, D – 4
5.) A – 3, B – 2, C – 4, D – 1

14. Four new machines M1, M2, M3, M4 are to be installed in a machine shop. There are 5 vacant places A, B, C, D and E that are available. Because of limited space, machine M2 cannot be placed at C and M3 cannot be placed at A. The cost matrix is shown below:

	A	B	C	D	E
M1	4	6	10	5	4
M2	7	4	-	5	4
M3	-	6	9	6	2
M4	9	3	7	2	3

15. Find the highest cost for the following matrix?

5	6	7	7
2	2	5	2
3	3	6	3
7	8	8	9

Hint: Maximum assignment value = 6 + 5 + 3 + 9 = 23

16. Solve the following traveling-sales man problem

	A	B	C	D
A	∞	46	16	40
B	41	∞	50	40
C	82	32	∞	60
D	40	40	36	∞

Hint: The traveling salesman route A→C→B→D→A with minimum cost = ₹128.

17. The best initial solution in a transportation problem is provided by:

   (a) Northwest corner rule.

   (b) Least cost method.

   (c) Vogel's approximation method.

18. When the total demand and supply are equal, the problem is referred to as:

   (a) unbalanced.

   (b) feasible.

   (c) balanced.

   (d) infeasible.

ANSWERS TO THE OBJECTIVE QUESTIONS

Chapter 1
Q.1
(a) True   (b) True   (c) False

Chapter 2
Q.1
(a) True   (b) False   (c) False   (d) True   (e) True
(f) True   (g) False   (h) False   (i) False   (j) True
(k) True   (l) True   (m) True   (n) True   (o) True
(p) False (q) True

Chapter 3
Q.1
(a) 129,700   (b) 8,135   (c) 50,000   (d) 129,687
(e) 19,275   (f) 8.00%

Chapter 4
Q.1
(a) True   (b) True   (c) True   (d) True   (e) True
(f) True   (g) True   (h) False (i) False

Q.2
(a) 5   (h) 4   (c) 3   (d) 2   (e) 1

Chapter 5
Q.1
(a) False (b) True   (c) True   (d) False (e) True
(f) True   (g) True   (h) True

Chapter 6
Q.1
(a) True   (b) True   (c) True   (d) True   (e) False
(f) False (g) True   (h) True   (i) True   (j) True
(k) True   (l) True   (m) True   (n) True   (o) True

(p) True	(q) True	(r) False	(s) False	(t) False
(u) False	(v) True	(w) True	(x) False	(y) False
(z) True	(aa) True	(bb) True	(cc) True	(dd) False

Chapter 7
Q.1

| (a) True | (b) True | (c) False | (d) True | (e) True |

Chapter 8
Q.1

(a) True	(b) True	(c) True	(d) True	(e) True
(f) False	(g) False	(h) True	(i) True	(j) True
(k) True				

Chapter 9
Q.1

(a) True	(b) True	(c) True	(d) False	(e) True
(f) True	(g) True	(h) True	(i) True	(j) True
(k) True	(l) False			

Chapter 10
Q.1

| (i) a | (ii) a | (iii) a | (iv) B | (v) b |
| (vi) True | (vii) False | (viii) False | (ix) True | (x) False |

Chapter 11
Q.1

| (a) True | (b) True | (c) True | (d) True | (e) True |
| (f) True | | | | |

Chapter 12
Q.1

| (a) True | (b) True | (c) True | (d) True | (e) True |
| (f) True | (g) True | (h) True | (i) True | |

Chapter 13
Q.1

| (a) True | (b) True | (c) True | (d) True | (e) True |
| (f) True | (g) True | (h) True | (i) True | (j) True |

(k) True   (l)
(1) False   (2) False   (3) False

Chapter 14
Q.1
(a) True   (b) True   (c) True   (d) True   (e) False
(f) True   (g) True   (h) True   (i) True   (j) True

Q.2
(1) (d)   (2) (c)   (3) (a)   (4) (e)   (5) (b)

Q.3
(d)---(b)---(e)---(a)---(c)

Chapter 15
Q.1
(a) False (b) True   (c) True   (d) True   (e) True
(f) True

Q.2
(a) (v)   (b) (i)   (c) (ii)   (d) (iii) (e) (iv)

Q.3
(A) (2)   (B) (3)   (C) (1)

Q.4
a---g =h---b---f---e =d =c

Q.5 (l)

Chapter 16
Q.1
(a) True   (b) True   (c) False (d) (a)   (e) True
(f) True   (g) False (h) True   (i) True   (j) True

Q.2
(1) (e)   (2) (a)   (3) (b)   (4) (c)   (5) (d)

Chapter 17
Q.1
(a) True   (b) True   (c) True   (d) True   (e) True
(f) True   (g) True   (h) True

Q.2 (F)

Q.3 (F)
Q.4
 (i) (2)   (ii) (3)   (iii) (1)

Q.5
 (i) (2)   (ii) (1)   (iii) (4)   (iv) (3)

Chapter 18
Q.1
 (a) False   (b) True   (c) False   (d) False   (e) False
 (f) False   (g) False   (h) True   (i) False   (j) True
 (k) True

Chapter 19
Q.1
 (a) True   (b) True   (c) True   (d) True   (e) False
 (f) True   (g) True   (h) True   (i) False   (j) False
 (k) False   (l) True   (m) True

Chapter 20
Q.18 (c)
Q.19 (c)

Chapter 21
Q.17 (c)
Q.18 (c)

APPENDICES

Appendix 1
0.5% interest factors for discrete compounding periods

n	F/P, I, n	P/F, I, n	F/A, I, n	A/F, I, n	P/A, I, n	A/P, I, n	A/G, I, n
1	1.005	0.9950	1.000	1.0000	0.9950	1.0050	0.0000
2	1.010	0.9901	2.005	0.4988	1.9851	0.5038	0.4988
3	1.015	0.9851	3.015	0.3317	2.9702	0.3367	0.9967
4	1.020	0.9802	4.030	0.2481	3.9505	0.2531	1.4938
5	1.025	0.9754	5.050	0.1980	4.9259	0.2030	1.9900
6	1.030	0.9705	6.076	0.1646	5.8964	0.1696	2.4855
7	1.036	0.9657	7.106	0.1407	6.8621	0.1457	2.9801
8	1.041	0.9609	8.141	0.1228	7.8230	0.1278	3.4738
9	1.046	0.9561	9.182	0.1089	8.7791	0.1139	3.9668
10	1.051	0.9513	10.228	0.0978	9.7304	0.1028	4.4589
11	1.056	0.9466	11.279	0.0887	10.6770	0.0937	4.9501
12	1.062	0.9419	12.336	0.0811	11.6189	0.0861	5.4406
13	1.067	0.9372	13.397	0.0746	12.5562	0.0796	5.9302
14	1.072	0.9326	14.464	0.0691	13.4887	0.0741	6.4190
15	1.078	0.9279	15.537	0.0644	14.4166	0.0694	6.9069
16	1.083	0.9233	16.614	0.0602	15.3399	0.0652	7.3940
17	1.088	0.9187	17.697	0.0565	16.2586	0.0615	7.8803
18	1.094	0.9141	18.786	0.0532	17.1728	0.0582	8.3658
19	1.099	0.9096	19.880	0.0503	18.0824	0.0553	8.8504
20	1.105	0.9051	20.979	0.0477	18.9874	0.0527	9.3342
21	1.110	0.9006	22.084	0.0453	19.8880	0.0503	9.8172
22	1.116	0.8961	23.194	0.0431	20.7841	0.0481	10.2993
23	1.122	0.8916	24.310	0.0411	21.6757	0.0461	10.7806
24	1.127	0.8872	25.432	0.0393	22.5629	0.0443	11.2611
25	1.133	0.8828	26.559	0.0377	23.4456	0.0427	11.7407
30	1.161	0.8610	32.280	0.0310	27.7941	0.0360	14.1265
35	1.191	0.8398	38.145	0.0262	32.0354	0.0312	16.4915
40	1.221	0.8191	44.159	0.0226	36.1722	0.0276	18.8359
45	1.252	0.7990	50.324	0.0199	40.2072	0.0249	21.1595
50	1.283	0.7793	56.645	0.0177	44.1428	0.0227	23.4624
55	1.316	0.7601	63.126	0.0158	47.9814	0.0208	25.7447
60	1.349	0.7414	69.770	0.0143	51.7256	0.0193	28.0064
65	1.383	0.7231	76.582	0.0131	55.3775	0.0181	30.2475
70	1.418	0.7053	83.566	0.0120	58.9394	0.0170	32.4680
75	1.454	0.6879	90.727	0.0110	62.4136	0.0160	34.6679
80	1.490	0.6710	98.068	0.0102	65.8023	0.0152	36.8474
85	1.528	0.6545	105.594	0.0095	69.1075	0.0145	39.0065
90	1.567	0.6383	113.311	0.0088	72.3313	0.0138	41.1451
95	1.606	0.6226	121.222	0.0082	75.4757	0.0132	43.2633
100	1.647	0.6073	129.334	0.0077	78.5426	0.0127	45.3613

1% interest factors for discrete compounding periods

n	F/P, I, n	P/F, I, n	F/A, I, n	A/F, I, n	P/A, I, n	A/P, I, n	A/G, I, n
1	1.010	0.9901	1.000	1.0000	0.9901	1.0100	0.0000
2	1.020	0.9803	2.010	0.4975	1.9704	0.5075	0.4975
3	1.030	0.9706	3.030	0.3300	2.9410	0.3400	0.9934
4	1.041	0.9610	4.060	0.2463	3.9020	0.2563	1.4876
5	1.051	0.9515	5.101	0.1960	4.8534	0.2060	1.9801
6	1.062	0.9420	6.152	0.1625	5.7955	0.1725	2.4710
7	1.072	0.9327	7.214	0.1386	6.7282	0.1486	2.9602
8	1.083	0.9235	8.286	0.1207	7.6517	0.1307	3.4478
9	1.094	0.9143	9.369	0.1067	8.5660	0.1167	3.9337
10	1.105	0.9053	10.462	0.0956	9.4713	0.1056	4.4179
11	1.116	0.8963	11.567	0.0865	10.3676	0.0965	4.9005
12	1.127	0.8874	12.683	0.0788	11.2551	0.0888	5.3815
13	1.138	0.8787	13.809	0.0724	12.1337	0.0824	5.8607
14	1.149	0.8700	14.947	0.0669	13.0037	0.0769	6.3384
15	1.161	0.8613	16.097	0.0621	13.8651	0.0721	6.8143
16	1.173	0.8528	17.258	0.0579	14.7179	0.0679	7.2886
17	1.184	0.8444	18.430	0.0543	15.5623	0.0643	7.7613
18	1.196	0.8360	19.615	0.0510	16.3983	0.0610	8.2323
19	1.208	0.8277	20.811	0.0481	17.2260	0.0581	8.7017
20	1.220	0.8195	22.019	0.0454	18.0456	0.0554	9.1694
21	1.232	0.8114	23.239	0.0430	18.8570	0.0530	9.6354
22	1.245	0.8034	24.472	0.0409	19.6604	0.0509	10.0998
23	1.257	0.7954	25.716	0.0389	20.4558	0.0489	10.5626
24	1.270	0.7876	26.973	0.0371	21.2434	0.0471	11.0237
25	1.282	0.7798	28.243	0.0354	22.0232	0.0454	11.4831
30	1.348	0.7419	34.785	0.0287	25.8077	0.0387	13.7557
35	1.417	0.7059	41.660	0.0240	29.4086	0.0340	15.9871
40	1.489	0.6717	48.886	0.0205	32.8347	0.0305	18.1776
45	1.565	0.6391	56.481	0.0177	36.0945	0.0277	20.3273
50	1.645	0.6080	64.463	0.0155	39.1961	0.0255	22.4363
55	1.729	0.5785	72.852	0.0137	42.1472	0.0237	24.5049
60	1.817	0.5504	81.670	0.0122	44.9550	0.0222	26.5333
65	1.909	0.5237	90.937	0.0110	47.6266	0.0210	28.5217
70	2.007	0.4983	100.676	0.0099	50.1685	0.0199	30.4703
75	2.109	0.4741	110.913	0.0090	52.5871	0.0190	32.3793
80	2.217	0.4511	121.672	0.0082	54.8882	0.0182	34.2492
85	2.330	0.4292	132.979	0.0075	57.0777	0.0175	36.0801
90	2.449	0.4084	144.863	0.0069	59.1609	0.0169	37.8724
95	2.574	0.3886	157.354	0.0064	61.1430	0.0164	39.6265
100	2.705	0.3697	170.481	0.0059	63.0289	0.0159	41.3426

2% interest factors for discrete compounding periods

n	F/P, I, n	P/F, I, n	F/A, I, n	A/F, I, n	P/A, I, n	A/P, I, n	A/G, I, n
1	1.020	0.9804	1.000	1.0000	0.9804	1.0200	0.0000
2	1.040	0.9612	2.020	0.4950	1.9416	0.5150	0.4950
3	1.061	0.9423	3.060	0.3268	2.8839	0.3468	0.9868
4	1.082	0.9238	4.122	0.2426	3.8077	0.2626	1.4752
5	1.104	0.9057	5.204	0.1922	4.7135	0.2122	1.9604
6	1.126	0.8880	6.308	0.1585	5.6014	0.1785	2.4423
7	1.149	0.8706	7.434	0.1345	6.4720	0.1545	2.9208
8	1.172	0.8535	8.583	0.1165	7.3255	0.1365	3.3961
9	1.195	0.8368	9.755	0.1025	8.1622	0.1225	3.8681
10	1.219	0.8203	10.950	0.0913	8.9826	0.1113	4.3367
11	1.243	0.8043	12.169	0.0822	9.7868	0.1022	4.8021
12	1.268	0.7885	13.412	0.0746	10.5753	0.0946	5.2642
13	1.294	0.7730	14.680	0.0681	11.3484	0.0881	5.7231
14	1.319	0.7579	15.974	0.0626	12.1062	0.0826	6.1786
15	1.346	0.7430	17.293	0.0578	12.8493	0.0778	6.6309
16	1.373	0.7284	18.639	0.0537	13.5777	0.0737	7.0799
17	1.400	0.7142	20.012	0.0500	14.2919	0.0700	7.5256
18	1.428	0.7002	21.412	0.0467	14.9920	0.0667	7.9681
19	1.457	0.6864	22.841	0.0438	15.6785	0.0638	8.4073
20	1.486	0.6730	24.297	0.0412	16.3514	0.0612	8.8433
21	1.516	0.6598	25.783	0.0388	17.0112	0.0588	9.2760
22	1.546	0.6468	27.299	0.0366	17.6580	0.0566	9.7055
23	1.577	0.6342	28.845	0.0347	18.2922	0.0547	10.1317
24	1.608	0.6217	30.422	0.0329	18.9139	0.0529	10.5547
25	1.641	0.6095	32.030	0.0312	19.5235	0.0512	10.9745
30	1.811	0.5521	40.568	0.0246	22.3965	0.0446	13.0251
35	2.000	0.5000	49.994	0.0200	24.9986	0.0400	14.9961
40	2.208	0.4529	60.402	0.0166	27.3555	0.0366	16.8885
45	2.438	0.4102	71.893	0.0139	29.4902	0.0339	18.7034
50	2.692	0.3715	84.579	0.0118	31.4236	0.0318	20.4420
55	2.972	0.3365	98.587	0.0101	33.1748	0.0301	22.1057
60	3.281	0.3048	114.052	0.0088	34.7609	0.0288	23.6961
65	3.623	0.2761	131.126	0.0076	36.1975	0.0276	25.2147
70	4.000	0.2500	149.978	0.0067	37.4986	0.0267	26.6632
75	4.416	0.2265	170.792	0.0059	38.6771	0.0259	28.0434
80	4.875	0.2051	193.772	0.0052	39.7445	0.0252	29.3572
85	5.383	0.1858	219.144	0.0046	40.7113	0.0246	30.6064
90	5.943	0.1683	247.157	0.0040	41.5869	0.0240	31.7929
95	6.562	0.1524	278.085	0.0036	42.3800	0.0236	32.9189
100	7.245	0.1380	312.232	0.0032	43.0984	0.0232	33.9863

3% interest factors for discrete compounding periods

n	F/P, I, n	P/F, I, n	F/A, I, n	A/F, I, n	P/A, I, n	A/P, I, n	A/G, I, n
1	1.030	0.9709	1.000	1.0000	0.9709	1.0300	0.0000
2	1.061	0.9426	2.030	0.4926	1.9135	0.5226	0.4926
3	1.093	0.9151	3.091	0.3235	2.8286	0.3535	0.9803
4	1.126	0.8885	4.184	0.2390	3.7171	0.2690	1.4631
5	1.159	0.8626	5.309	0.1884	4.5797	0.2184	1.9409
6	1.194	0.8375	6.468	0.1546	5.4172	0.1846	2.4138
7	1.230	0.8131	7.662	0.1305	6.2303	0.1605	2.8819
8	1.267	0.7894	8.892	0.1125	7.0197	0.1425	3.3450
9	1.305	0.7664	10.159	0.0984	7.7861	0.1284	3.8032
10	1.344	0.7441	11.464	0.0872	8.5302	0.1172	4.2565
11	1.384	0.7224	12.808	0.0781	9.2526	0.1081	4.7049
12	1.426	0.7014	14.192	0.0705	9.9540	0.1005	5.1485
13	1.469	0.6810	15.618	0.0640	10.6350	0.0940	5.5872
14	1.513	0.6611	17.086	0.0585	11.2961	0.0885	6.0210
15	1.558	0.6419	18.599	0.0538	11.9379	0.0838	6.4500
16	1.605	0.6232	20.157	0.0496	12.5611	0.0796	6.8742
17	1.653	0.6050	21.762	0.0460	13.1661	0.0760	7.2936
18	1.702	0.5874	23.414	0.0427	13.7535	0.0727	7.7081
19	1.754	0.5703	25.117	0.0398	14.3238	0.0698	8.1179
20	1.806	0.5537	26.870	0.0372	14.8775	0.0672	8.5229
21	1.860	0.5375	28.676	0.0349	15.4150	0.0649	8.9231
22	1.916	0.5219	30.537	0.0327	15.9369	0.0627	9.3186
23	1.974	0.5067	32.453	0.0308	16.4436	0.0608	9.7093
24	2.033	0.4919	34.426	0.0290	16.9355	0.0590	10.0954
25	2.094	0.4776	36.459	0.0274	17.4131	0.0574	10.4768
30	2.427	0.4120	47.575	0.0210	19.6004	0.0510	12.3141
35	2.814	0.3554	60.462	0.0165	21.4872	0.0465	14.0375
40	3.262	0.3066	75.401	0.0133	23.1148	0.0433	15.6502
45	3.782	0.2644	92.720	0.0108	24.5187	0.0408	17.1556
50	4.384	0.2281	112.797	0.0089	25.7298	0.0389	18.5575
55	5.082	0.1968	136.072	0.0073	26.7744	0.0373	19.8600
60	5.892	0.1697	163.053	0.0061	27.6756	0.0361	21.0674
65	6.830	0.1464	194.333	0.0051	28.4529	0.0351	22.1841
70	7.918	0.1263	230.594	0.0043	29.1234	0.0343	23.2145
75	9.179	0.1089	272.631	0.0037	29.7018	0.0337	24.1634
80	10.641	0.0940	321.363	0.0031	30.2008	0.0331	25.0353
85	12.336	0.0811	377.857	0.0026	30.6312	0.0326	25.8349
90	14.300	0.0699	443.349	0.0023	31.0024	0.0323	26.5667
95	16.578	0.0603	519.272	0.0019	31.3227	0.0319	27.2351
100	19.219	0.0520	607.288	0.0016	31.5989	0.0316	27.8444

4% interest factors for discrete compounding periods

n	F/P, I, n	P/F, I, n	F/A, I, n	A/F, I, n	P/A, I, n	A/P, I, n	A/G, I, n
1	1.040	0.9615	1.000	1.0000	0.9615	1.0400	0.0000
2	1.082	0.9246	2.040	0.4902	1.8861	0.5302	0.4902
3	1.125	0.8890	3.122	0.3203	2.7751	0.3603	0.9739
4	1.170	0.8548	4.246	0.2355	3.6299	0.2755	1.4510
5	1.217	0.8219	5.416	0.1846	4.4518	0.2246	1.9216
6	1.265	0.7903	6.633	0.1508	5.2421	0.1908	2.3857
7	1.316	0.7599	7.898	0.1266	6.0021	0.1666	2.8433
8	1.369	0.7307	9.214	0.1085	6.7327	0.1485	3.2944
9	1.423	0.7026	10.583	0.0945	7.4353	0.1345	3.7391
10	1.480	0.6756	12.006	0.0833	8.1109	0.1233	4.1773
11	1.539	0.6496	13.486	0.0741	8.7605	0.1141	4.6090
12	1.601	0.6246	15.026	0.0666	9.3851	0.1066	5.0343
13	1.665	0.6006	16.627	0.0601	9.9856	0.1001	5.4533
14	1.732	0.5775	18.292	0.0547	10.5631	0.0947	5.8659
15	1.801	0.5553	20.024	0.0499	11.1184	0.0899	6.2721
16	1.873	0.5339	21.825	0.0458	11.6523	0.0858	6.6720
17	1.948	0.5134	23.698	0.0422	12.1657	0.0822	7.0656
18	2.026	0.4936	25.645	0.0390	12.6593	0.0790	7.4530
19	2.107	0.4746	27.671	0.0361	13.1339	0.0761	7.8342
20	2.191	0.4564	29.778	0.0336	13.5903	0.0736	8.2091
21	2.279	0.4388	31.969	0.0313	14.0292	0.0713	8.5779
22	2.370	0.4220	34.248	0.0292	14.4511	0.0692	8.9407
23	2.465	0.4057	36.618	0.0273	14.8568	0.0673	9.2973
24	2.563	0.3901	39.083	0.0256	15.2470	0.0656	9.6479
25	2.666	0.3751	41.646	0.0240	15.6221	0.0640	9.9925
30	3.243	0.3083	56.085	0.0178	17.2920	0.0578	11.6274
35	3.946	0.2534	73.652	0.0136	18.6646	0.0536	13.1198
40	4.801	0.2083	95.026	0.0105	19.7928	0.0505	14.4765
45	5.841	0.1712	121.029	0.0083	20.7200	0.0483	15.7047
50	7.107	0.1407	152.667	0.0066	21.4822	0.0466	16.8122
55	8.646	0.1157	191.159	0.0052	22.1086	0.0452	17.8070
60	10.520	0.0951	237.991	0.0042	22.6235	0.0442	18.6972
65	12.799	0.0781	294.968	0.0034	23.0467	0.0434	19.4909
70	15.572	0.0642	364.290	0.0027	23.3945	0.0427	20.1961
75	18.945	0.0528	448.631	0.0022	23.6804	0.0422	20.8206
80	23.050	0.0434	551.245	0.0018	23.9154	0.0418	21.3718
85	28.044	0.0357	676.090	0.0015	24.1085	0.0415	21.8569
90	34.119	0.0293	827.983	0.0012	24.2673	0.0412	22.2826
95	41.511	0.0241	1012.785	0.0010	24.3978	0.0410	22.6550
100	50.505	0.0198	1237.624	0.0008	24.5050	0.0408	22.9800

5% interest factors for discrete compounding periods

n	F/P, I, n	P/F, I, n	F/A, I, n	A/F, I, n	P/A, I, n	A/P, I, n	A/G, I, n
1	1.050	0.9524	1.000	1.0000	0.9524	1.0500	0.0000
2	1.103	0.9070	2.050	0.4878	1.8594	0.5378	0.4878
3	1.158	0.8638	3.153	0.3172	2.7232	0.3672	0.9675
4	1.216	0.8227	4.310	0.2320	3.5460	0.2820	1.4391
5	1.276	0.7835	5.526	0.1810	4.3295	0.2310	1.9025
6	1.340	0.7462	6.802	0.1470	5.0757	0.1970	2.3579
7	1.407	0.7107	8.142	0.1228	5.7864	0.1728	2.8052
8	1.477	0.6768	9.549	0.1047	6.4632	0.1547	3.2445
9	1.551	0.6446	11.027	0.0907	7.1078	0.1407	3.6758
10	1.629	0.6139	12.578	0.0795	7.7217	0.1295	4.0991
11	1.710	0.5847	14.207	0.0704	8.3064	0.1204	4.5144
12	1.796	0.5568	15.917	0.0628	8.8633	0.1128	4.9219
13	1.886	0.5303	17.713	0.0565	9.3936	0.1065	5.3215
14	1.980	0.5051	19.599	0.0510	9.8986	0.1010	5.7133
15	2.079	0.4810	21.579	0.0463	10.3797	0.0963	6.0973
16	2.183	0.4581	23.657	0.0423	10.8378	0.0923	6.4736
17	2.292	0.4363	25.840	0.0387	11.2741	0.0887	6.8423
18	2.407	0.4155	28.132	0.0355	11.6896	0.0855	7.2034
19	2.527	0.3957	30.539	0.0327	12.0853	0.0827	7.5569
20	2.653	0.3769	33.066	0.0302	12.4622	0.0802	7.9030
21	2.786	0.3589	35.719	0.0280	12.8212	0.0780	8.2416
22	2.925	0.3418	38.505	0.0260	13.1630	0.0760	8.5730
23	3.072	0.3256	41.430	0.0241	13.4886	0.0741	8.8971
24	3.225	0.3101	44.502	0.0225	13.7986	0.0725	9.2140
25	3.386	0.2953	47.727	0.0210	14.0939	0.0710	9.5238
30	4.322	0.2314	66.439	0.0151	15.3725	0.0651	10.9691
35	5.516	0.1813	90.320	0.0111	16.3742	0.0611	12.2498
40	7.040	0.1420	120.800	0.0083	17.1591	0.0583	13.3775
45	8.985	0.1113	159.700	0.0063	17.7741	0.0563	14.3644
50	11.467	0.0872	209.348	0.0048	18.2559	0.0548	15.2233
55	14.636	0.0683	272.713	0.0037	18.6335	0.0537	15.9664
60	18.679	0.0535	353.584	0.0028	18.9293	0.0528	16.6062
65	23.840	0.0419	456.798	0.0022	19.1611	0.0522	17.1541
70	30.426	0.0329	588.529	0.0017	19.3427	0.0517	17.6212
75	38.833	0.0258	756.654	0.0013	19.4850	0.0513	18.0176
80	49.561	0.0202	971.229	0.0010	19.5965	0.0510	18.3526
85	63.254	0.0158	1245.087	0.0008	19.6838	0.0508	18.6346
90	80.730	0.0124	1594.607	0.0006	19.7523	0.0506	18.8712
95	103.035	0.0097	2040.694	0.0005	19.8059	0.0505	19.0689
100	131.501	0.0076	2610.025	0.0004	19.8479	0.0504	19.2337

6% interest factors for discrete compounding periods

n	F/P, I, n	P/F, I, n	F/A, I, n	A/F, I, n	P/A, I, n	A/P, I, n	A/G, I, n
1	1.060	0.9434	1.000	1.0000	0.9434	1.0600	0.0000
2	1.124	0.8900	2.060	0.4854	1.8334	0.5454	0.4854
3	1.191	0.8396	3.184	0.3141	2.6730	0.3741	0.9612
4	1.262	0.7921	4.375	0.2286	3.4651	0.2886	1.4272
5	1.338	0.7473	5.637	0.1774	4.2124	0.2374	1.8836
6	1.419	0.7050	6.975	0.1434	4.9173	0.2034	2.3304
7	1.504	0.6651	8.394	0.1191	5.5824	0.1791	2.7676
8	1.594	0.6274	9.897	0.1010	6.2098	0.1610	3.1952
9	1.689	0.5919	11.491	0.0870	6.8017	0.1470	3.6133
10	1.791	0.5584	13.181	0.0759	7.3601	0.1359	4.0220
11	1.898	0.5268	14.972	0.0668	7.8869	0.1268	4.4213
12	2.012	0.4970	16.870	0.0593	8.3838	0.1193	4.8113
13	2.133	0.4688	18.882	0.0530	8.8527	0.1130	5.1920
14	2.261	0.4423	21.015	0.0476	9.2950	0.1076	5.5635
15	2.397	0.4173	23.276	0.0430	9.7122	0.1030	5.9260
16	2.540	0.3936	25.673	0.0390	10.1059	0.0990	6.2794
17	2.693	0.3714	28.213	0.0354	10.4773	0.0954	6.6240
18	2.854	0.3503	30.906	0.0324	10.8276	0.0924	6.9597
19	3.026	0.3305	33.760	0.0296	11.1581	0.0896	7.2867
20	3.207	0.3118	36.786	0.0272	11.4699	0.0872	7.6051
21	3.400	0.2942	39.993	0.0250	11.7641	0.0850	7.9151
22	3.604	0.2775	43.392	0.0230	12.0416	0.0830	8.2166
23	3.820	0.2618	46.996	0.0213	12.3034	0.0813	8.5099
24	4.049	0.2470	50.816	0.0197	12.5504	0.0797	8.7951
25	4.292	0.2330	54.865	0.0182	12.7834	0.0782	9.0722
30	5.743	0.1741	79.058	0.0126	13.7648	0.0726	10.3422
35	7.686	0.1301	111.435	0.0090	14.4982	0.0690	11.4319
40	10.286	0.0972	154.762	0.0065	15.0463	0.0665	12.3590
45	13.765	0.0727	212.744	0.0047	15.4558	0.0647	13.1413
50	18.420	0.0543	290.336	0.0034	15.7619	0.0634	13.7964
55	24.650	0.0406	394.172	0.0025	15.9905	0.0625	14.3411
60	32.988	0.0303	533.128	0.0019	16.1614	0.0619	14.7909
65	44.145	0.0227	719.083	0.0014	16.2891	0.0614	15.1601
70	59.076	0.0169	967.932	0.0010	16.3845	0.0610	15.4613
75	79.057	0.0126	1300.949	0.0008	16.4558	0.0608	15.7058
80	105.796	0.0095	1746.600	0.0006	16.5091	0.0606	15.9033
85	141.579	0.0071	2342.982	0.0004	16.5489	0.0604	16.0620
90	189.465	0.0053	3141.075	0.0003	16.5787	0.0603	16.1891
95	253.546	0.0039	4209.104	0.0002	16.6009	0.0602	16.2905
100	339.302	0.0029	5638.368	0.0002	16.6175	0.0602	16.3711

7% interest factors for discrete compounding periods

n	F/P, I, n	P/F, I, n	F/A, I, n	A/F, I, n	P/A, I, n	A/P, I, n	A/G, I, n
1	1.070	0.9346	1.000	1.0000	0.9346	1.0700	0.0000
2	1.145	0.8734	2.070	0.4831	1.8080	0.5531	0.4831
3	1.225	0.8163	3.215	0.3111	2.6243	0.3811	0.9549
4	1.311	0.7629	4.440	0.2252	3.3872	0.2952	1.4155
5	1.403	0.7130	5.751	0.1739	4.1002	0.2439	1.8650
6	1.501	0.6663	7.153	0.1398	4.7665	0.2098	2.3032
7	1.606	0.6227	8.654	0.1156	5.3893	0.1856	2.7304
8	1.718	0.5820	10.260	0.0975	5.9713	0.1675	3.1465
9	1.838	0.5439	11.978	0.0835	6.5152	0.1535	3.5517
10	1.967	0.5083	13.816	0.0724	7.0236	0.1424	3.9461
11	2.105	0.4751	15.784	0.0634	7.4987	0.1334	4.3296
12	2.252	0.4440	17.888	0.0559	7.9427	0.1259	4.7025
13	2.410	0.4150	20.141	0.0497	8.3577	0.1197	5.0648
14	2.579	0.3878	22.550	0.0443	8.7455	0.1143	5.4167
15	2.759	0.3624	25.129	0.0398	9.1079	0.1098	5.7583
16	2.952	0.3387	27.888	0.0359	9.4466	0.1059	6.0897
17	3.159	0.3166	30.840	0.0324	9.7632	0.1024	6.4110
18	3.380	0.2959	33.999	0.0294	10.0591	0.0994	6.7225
19	3.617	0.2765	37.379	0.0268	10.3356	0.0968	7.0242
20	3.870	0.2584	40.995	0.0244	10.5940	0.0944	7.3163
21	4.141	0.2415	44.865	0.0223	10.8355	0.0923	7.5990
22	4.430	0.2257	49.006	0.0204	11.0612	0.0904	7.8725
23	4.741	0.2109	53.436	0.0187	11.2722	0.0887	8.1369
24	5.072	0.1971	58.177	0.0172	11.4693	0.0872	8.3923
25	5.427	0.1842	63.249	0.0158	11.6536	0.0858	8.6391
30	7.612	0.1314	94.461	0.0106	12.4090	0.0806	9.7487
35	10.677	0.0937	138.237	0.0072	12.9477	0.0772	10.6687
40	14.974	0.0668	199.635	0.0050	13.3317	0.0750	11.4233
45	21.002	0.0476	285.749	0.0035	13.6055	0.0735	12.0360
50	29.457	0.0339	406.529	0.0025	13.8007	0.0725	12.5287
55	41.315	0.0242	575.929	0.0017	13.9399	0.0717	12.9215
60	57.946	0.0173	813.520	0.0012	14.0392	0.0712	13.2321
65	81.273	0.0123	1146.755	0.0009	14.1099	0.0709	13.4760
70	113.989	0.0088	1614.134	0.0006	14.1604	0.0706	13.6662
75	159.876	0.0063	2269.657	0.0004	14.1964	0.0704	13.8136
80	224.234	0.0045	3189.063	0.0003	14.2220	0.0703	13.9273
85	314.500	0.0032	4478.576	0.0002	14.2403	0.0702	14.0146
90	441.103	0.0023	6287.185	0.0002	14.2533	0.0702	14.0812
95	618.670	0.0016	8823.854	0.0001	14.2626	0.0701	14.1319
100	867.716	0.0012	12381.662	0.0001	14.2693	0.0701	14.1703

8% interest factors for discrete compounding periods

n	F/P, I, n	P/F, I, n	F/A, I, n	A/F, I, n	P/A, I, n	A/P, I, n	A/G, I, n
1	1.080	0.9259	1.000	1.0000	0.9259	1.0800	0.0000
2	1.166	0.8573	2.080	0.4808	1.7833	0.5608	0.4808
3	1.260	0.7938	3.246	0.3080	2.5771	0.3880	0.9487
4	1.360	0.7350	4.506	0.2219	3.3121	0.3019	1.4040
5	1.469	0.6806	5.867	0.1705	3.9927	0.2505	1.8465
6	1.587	0.6302	7.336	0.1363	4.6229	0.2163	2.2763
7	1.714	0.5835	8.923	0.1121	5.2064	0.1921	2.6937
8	1.851	0.5403	10.637	0.0940	5.7466	0.1740	3.0985
9	1.999	0.5002	12.488	0.0801	6.2469	0.1601	3.4910
10	2.159	0.4632	14.487	0.0690	6.7101	0.1490	3.8713
11	2.332	0.4289	16.645	0.0601	7.1390	0.1401	4.2395
12	2.518	0.3971	18.977	0.0527	7.5361	0.1327	4.5957
13	2.720	0.3677	21.495	0.0465	7.9038	0.1265	4.9402
14	2.937	0.3405	24.215	0.0413	8.2442	0.1213	5.2731
15	3.172	0.3152	27.152	0.0368	8.5595	0.1168	5.5945
16	3.426	0.2919	30.324	0.0330	8.8514	0.1130	5.9046
17	3.700	0.2703	33.750	0.0296	9.1216	0.1096	6.2037
18	3.996	0.2502	37.450	0.0267	9.3719	0.1067	6.4920
19	4.316	0.2317	41.446	0.0241	9.6036	0.1041	6.7697
20	4.661	0.2145	45.762	0.0219	9.8181	0.1019	7.0369
21	5.034	0.1987	50.423	0.0198	10.0168	0.0998	7.2940
22	5.437	0.1839	55.457	0.0180	10.2007	0.0980	7.5412
23	5.871	0.1703	60.893	0.0164	10.3711	0.0964	7.7786
24	6.341	0.1577	66.765	0.0150	10.5288	0.0950	8.0066
25	6.848	0.1460	73.106	0.0137	10.6748	0.0937	8.2254
30	10.063	0.0994	113.283	0.0088	11.2578	0.0888	9.1897
35	14.785	0.0676	172.317	0.0058	11.6546	0.0858	9.9611
40	21.725	0.0460	259.057	0.0039	11.9246	0.0839	10.5699
45	31.920	0.0313	386.506	0.0026	12.1084	0.0826	11.0447
50	46.902	0.0213	573.770	0.0017	12.2335	0.0817	11.4107
55	68.914	0.0145	848.923	0.0012	12.3186	0.0812	11.6902
60	101.257	0.0099	1253.213	0.0008	12.3766	0.0808	11.9015
65	148.780	0.0067	1847.248	0.0005	12.4160	0.0805	12.0602
70	218.606	0.0046	2720.080	0.0004	12.4428	0.0804	12.1783
75	321.205	0.0031	4002.557	0.0002	12.4611	0.0802	12.2658
80	471.955	0.0021	5886.935	0.0002	12.4735	0.0802	12.3301
85	693.456	0.0014	8655.706	0.0001	12.4820	0.0801	12.3772
90	1018.915	0.0010	12723.939	0.0001	12.4877	0.0801	12.4116
95	1497.121	0.0007	18701.507	0.0001	12.4917	0.0801	12.4365
100	2199.761	0.0005	27484.516	0.0000	12.4943	0.0800	12.4545

9% interest factors for discrete compounding periods

n	F/P, I, n	P/F, I, n	F/A, I, n	A/F, I, n	P/A, I, n	A/P, I, n	A/G, I, n
1	1.090	0.9174	1.000	1.0000	0.9174	1.0900	0.0000
2	1.188	0.8417	2.090	0.4785	1.7591	0.5685	0.4785
3	1.295	0.7722	3.278	0.3051	2.5313	0.3951	0.9426
4	1.412	0.7084	4.573	0.2187	3.2397	0.3087	1.3925
5	1.539	0.6499	5.985	0.1671	3.8897	0.2571	1.8282
6	1.677	0.5963	7.523	0.1329	4.4859	0.2229	2.2498
7	1.828	0.5470	9.200	0.1087	5.0330	0.1987	2.6574
8	1.993	0.5019	11.028	0.0907	5.5348	0.1807	3.0512
9	2.172	0.4604	13.021	0.0768	5.9952	0.1668	3.4312
10	2.367	0.4224	15.193	0.0658	6.4177	0.1558	3.7978
11	2.580	0.3875	17.560	0.0569	6.8052	0.1469	4.1510
12	2.813	0.3555	20.141	0.0497	7.1607	0.1397	4.4910
13	3.066	0.3262	22.953	0.0436	7.4869	0.1336	4.8182
14	3.342	0.2992	26.019	0.0384	7.7862	0.1284	5.1326
15	3.642	0.2745	29.361	0.0341	8.0607	0.1241	5.4346
16	3.970	0.2519	33.003	0.0303	8.3126	0.1203	5.7245
17	4.328	0.2311	36.974	0.0270	8.5436	0.1170	6.0024
18	4.717	0.2120	41.301	0.0242	8.7556	0.1142	6.2687
19	5.142	0.1945	46.018	0.0217	8.9501	0.1117	6.5236
20	5.604	0.1784	51.160	0.0195	9.1285	0.1095	6.7674
21	6.109	0.1637	56.765	0.0176	9.2922	0.1076	7.0006
22	6.659	0.1502	62.873	0.0159	9.4424	0.1059	7.2232
23	7.258	0.1378	69.532	0.0144	9.5802	0.1044	7.4357
24	7.911	0.1264	76.790	0.0130	9.7066	0.1030	7.6384
25	8.623	0.1160	84.701	0.0118	9.8226	0.1018	7.8316
30	13.268	0.0754	136.308	0.0073	10.2737	0.0973	8.6657
35	20.414	0.0490	215.711	0.0046	10.5668	0.0946	9.3083
40	31.409	0.0318	337.882	0.0030	10.7574	0.0930	9.7957
45	48.327	0.0207	525.859	0.0019	10.8812	0.0919	10.1603
50	74.358	0.0134	815.084	0.0012	10.9617	0.0912	10.4295
55	114.408	0.0087	1260.092	0.0008	11.0140	0.0908	10.6261
60	176.031	0.0057	1944.792	0.0005	11.0480	0.0905	10.7683
65	270.846	0.0037	2998.288	0.0003	11.0701	0.0903	10.8702
70	416.730	0.0024	4619.223	0.0002	11.0844	0.0902	10.9427
75	641.191	0.0016	7113.232	0.0001	11.0938	0.0901	10.9940
80	986.552	0.0010	10950.574	0.0001	11.0998	0.0901	11.0299
85	1517.932	0.0007	16854.800	0.0001	11.1038	0.0901	11.0551
90	2335.527	0.0004	25939.184	0.0000	11.1064	0.0900	11.0726
95	3593.497	0.0003	39916.635	0.0000	11.1080	0.0900	11.0847
100	5529.041	0.0002	61422.675	0.0000	11.1091	0.0900	11.0930

10% interest factors for discrete compounding periods

n	F/P, I, n	P/F, I, n	F/A, I, n	A/F, I, n	P/A, I, n	A/P, I, n	A/G, I, n
1	1.100	0.9091	1.000	1.0000	0.9091	1.1000	0.0000
2	1.210	0.8264	2.100	0.4762	1.7355	0.5762	0.4762
3	1.331	0.7513	3.310	0.3021	2.4869	0.4021	0.9366
4	1.464	0.6830	4.641	0.2155	3.1699	0.3155	1.3812
5	1.611	0.6209	6.105	0.1638	3.7908	0.2638	1.8101
6	1.772	0.5645	7.716	0.1296	4.3553	0.2296	2.2236
7	1.949	0.5132	9.487	0.1054	4.8684	0.2054	2.6216
8	2.144	0.4665	11.436	0.0874	5.3349	0.1874	3.0045
9	2.358	0.4241	13.579	0.0736	5.7590	0.1736	3.3724
10	2.594	0.3855	15.937	0.0627	6.1446	0.1627	3.7255
11	2.853	0.3505	18.531	0.0540	6.4951	0.1540	4.0641
12	3.138	0.3186	21.384	0.0468	6.8137	0.1468	4.3884
13	3.452	0.2897	24.523	0.0408	7.1034	0.1408	4.6988
14	3.797	0.2633	27.975	0.0357	7.3667	0.1357	4.9955
15	4.177	0.2394	31.772	0.0315	7.6061	0.1315	5.2789
16	4.595	0.2176	35.950	0.0278	7.8237	0.1278	5.5493
17	5.054	0.1978	40.545	0.0247	8.0216	0.1247	5.8071
18	5.560	0.1799	45.599	0.0219	8.2014	0.1219	6.0526
19	6.116	0.1635	51.159	0.0195	8.3649	0.1195	6.2861
20	6.727	0.1486	57.275	0.0175	8.5136	0.1175	6.5081
21	7.400	0.1351	64.002	0.0156	8.6487	0.1156	6.7189
22	8.140	0.1228	71.403	0.0140	8.7715	0.1140	6.9189
23	8.954	0.1117	79.543	0.0126	8.8832	0.1126	7.1085
24	9.850	0.1015	88.497	0.0113	8.9847	0.1113	7.2881
25	10.835	0.0923	98.347	0.0102	9.0770	0.1102	7.4580
30	17.449	0.0573	164.494	0.0061	9.4269	0.1061	8.1762
35	28.102	0.0356	271.024	0.0037	9.6442	0.1037	8.7086
40	45.259	0.0221	442.593	0.0023	9.7791	0.1023	9.0962
45	72.890	0.0137	718.905	0.0014	9.8628	0.1014	9.3740
50	117.391	0.0085	1163.909	0.0009	9.9148	0.1009	9.5704
55	189.059	0.0053	1880.591	0.0005	9.9471	0.1005	9.7075
60	304.482	0.0033	3034.816	0.0003	9.9672	0.1003	9.8023
65	490.371	0.0020	4893.707	0.0002	9.9796	0.1002	9.8672
70	789.747	0.0013	7887.470	0.0001	9.9873	0.1001	9.9113
75	1271.895	0.0008	12708.954	0.0001	9.9921	0.1001	9.9410
80	2048.400	0.0005	20474.002	0.0000	9.9951	0.1000	9.9609
85	3298.969	0.0003	32979.690	0.0000	9.9970	0.1000	9.9742
90	5313.023	0.0002	53120.226	0.0000	9.9981	0.1000	9.9831
95	8556.676	0.0001	85556.760	0.0000	9.9988	0.1000	9.9889
100	13780.612	0.0001	137796.123	0.0000	9.9993	0.1000	9.9927

12% interest factors for discrete compounding periods

n	F/P, I, n	P/F, I, n	F/A, I, n	A/F, I, n	P/A, I, n	A/P, I, n	A/G, I, n
1	1.120	0.8929	1.000	1.0000	0.8929	1.1200	0.0000
2	1.254	0.7972	2.120	0.4717	1.6901	0.5917	0.4717
3	1.405	0.7118	3.374	0.2963	2.4018	0.4163	0.9246
4	1.574	0.6355	4.779	0.2092	3.0373	0.3292	1.3589
5	1.762	0.5674	6.353	0.1574	3.6048	0.2774	1.7746
6	1.974	0.5066	8.115	0.1232	4.1114	0.2432	2.1720
6	1.974	0.5066	8.115	0.1232	4.1114	0.2432	2.1720
7	2.211	0.4523	10.089	0.0991	4.5638	0.2191	2.5515
8	2.476	0.4039	12.300	0.0813	4.9676	0.2013	2.9131
9	2.773	0.3606	14.776	0.0677	5.3282	0.1877	3.2574
10	3.106	0.3220	17.549	0.0570	5.6502	0.1770	3.5847
11	3.479	0.2875	20.655	0.0484	5.9377	0.1684	3.8953
12	3.896	0.2567	24.133	0.0414	6.1944	0.1614	4.1897
13	4.363	0.2292	28.029	0.0357	6.4235	0.1557	4.4683
14	4.887	0.2046	32.393	0.0309	6.6282	0.1509	4.7317
15	5.474	0.1827	37.280	0.0268	6.8109	0.1468	4.9803
16	6.130	0.1631	42.753	0.0234	6.9740	0.1434	5.2147
17	6.866	0.1456	48.884	0.0205	7.1196	0.1405	5.4353
18	7.690	0.1300	55.750	0.0179	7.2497	0.1379	5.6427
19	8.613	0.1161	63.440	0.0158	7.3658	0.1358	5.8375
20	9.646	0.1037	72.052	0.0139	7.4694	0.1339	6.0202
21	10.804	0.0926	81.699	0.0122	7.5620	0.1322	6.1913
22	12.100	0.0826	92.503	0.0108	7.6446	0.1308	6.3514
23	13.552	0.0738	104.603	0.0096	7.7184	0.1296	6.5010
24	15.179	0.0659	118.155	0.0085	7.7843	0.1285	6.6406
25	17.000	0.0588	133.334	0.0075	7.8431	0.1275	6.7708
30	29.960	0.0334	241.333	0.0041	8.0552	0.1241	7.2974
35	52.800	0.0189	431.663	0.0023	8.1755	0.1223	7.6577
40	93.051	0.0107	767.091	0.0013	8.2438	0.1213	7.8988
45	163.988	0.0061	1358.230	0.0007	8.2825	0.1207	8.0572
50	289.002	0.0035	2400.018	0.0004	8.3045	0.1204	8.1597
55	509.321	0.0020	4236.005	0.0002	8.3170	0.1202	8.2251
60	897.597	0.0011	7471.641	0.0001	8.3240	0.1201	8.2664
65	1581.872	0.0006	13173.937	0.0001	8.3281	0.1201	8.2922
70	2787.800	0.0004	23223.332	0.0000	8.3303	0.1200	8.3082
75	4913.056	0.0002	40933.799	0.0000	8.3316	0.1200	8.3181
80	8658.483	0.0001	72145.693	0.0000	8.3324	0.1200	8.3241
85	15259.206	0.0001	127151.714	0.0000	8.3328	0.1200	8.3278
90	26891.934	0.0000	224091.119	0.0000	8.3330	0.1200	8.3300
95	47392.777	0.0000	394931.472	0.0000	8.3332	0.1200	8.3313
100	83522.266	0.0000	696010.548	0.0000	8.3332	0.1200	8.3321

15% interest factors for discrete compounding periods

n	F/P, I, n	P/F, I, n	F/A, I, n	A/F, I, n	P/A, I, n	A/P, I, n	A/G, I, n
1	1.150	0.8696	1.000	1.0000	0.8696	1.1500	0.0000
2	1.323	0.7561	2.150	0.4651	1.6257	0.6151	0.4651
3	1.521	0.6575	3.473	0.2880	2.2832	0.4380	0.9071
4	1.749	0.5718	4.993	0.2003	2.8550	0.3503	1.3263
5	2.011	0.4972	6.742	0.1483	3.3522	0.2983	1.7228
6	2.313	0.4323	8.754	0.1142	3.7845	0.2642	2.0972
7	2.660	0.3759	11.067	0.0904	4.1604	0.2404	2.4498
8	3.059	0.3269	13.727	0.0729	4.4873	0.2229	2.7813
9	3.518	0.2843	16.786	0.0596	4.7716	0.2096	3.0922
10	4.046	0.2472	20.304	0.0493	5.0188	0.1993	3.3832
11	4.652	0.2149	24.349	0.0411	5.2337	0.1911	3.6549
12	5.350	0.1869	29.002	0.0345	5.4206	0.1845	3.9082
13	6.153	0.1625	34.352	0.0291	5.5831	0.1791	4.1438
14	7.076	0.1413	40.505	0.0247	5.7245	0.1747	4.3624
15	8.137	0.1229	47.580	0.0210	5.8474	0.1710	4.5650
16	9.358	0.1069	55.717	0.0179	5.9542	0.1679	4.7522
17	10.761	0.0929	65.075	0.0154	6.0472	0.1654	4.9251
18	12.375	0.0808	75.836	0.0132	6.1280	0.1632	5.0843
19	14.232	0.0703	88.212	0.0113	6.1982	0.1613	5.2307
20	16.367	0.0611	102.444	0.0098	6.2593	0.1598	5.3651
21	18.822	0.0531	118.810	0.0084	6.3125	0.1584	5.4883
22	21.645	0.0462	137.632	0.0073	6.3587	0.1573	5.6010
23	24.891	0.0402	159.276	0.0063	6.3988	0.1563	5.7040
24	28.625	0.0349	184.168	0.0054	6.4338	0.1554	5.7979
25	32.919	0.0304	212.793	0.0047	6.4641	0.1547	5.8834
30	66.212	0.0151	434.745	0.0023	6.5660	0.1523	6.2066
35	133.176	0.0075	881.170	0.0011	6.6166	0.1511	6.4019
40	267.864	0.0037	1779.090	0.0006	6.6418	0.1506	6.5168
45	538.769	0.0019	3585.128	0.0003	6.6543	0.1503	6.5830
50	1083.657	0.0009	7217.716	0.0001	6.6605	0.1501	6.6205
55	2179.622	0.0005	14524.148	0.0001	6.6636	0.1501	6.6414
60	4383.999	0.0002	29219.992	0.0000	6.6651	0.1500	6.6530
65	8817.787	0.0001	58778.583	0.0000	6.6659	0.1500	6.6593
70	17735.720	0.0001	118231.467	0.0000	6.6663	0.1500	6.6627
75	35672.868	0.0000	237812.453	0.0000	6.6665	0.1500	6.6646
80	71750.879	0.0000	478332.529	0.0000	6.6666	0.1500	6.6656
85	144316.647	0.0000	962104.313	0.0000	6.6666	0.1500	6.6661
90	290272.325	0.0000	1935142.168	0.0000	6.6666	0.1500	6.6664
95	583841.328	0.0000	3892268.851	0.0000	6.6667	0.1500	6.6665
100	1174313.451	0.0000	7828749.671	0.0000	6.6667	0.1500	6.6666

20% interest factors for discrete compounding periods

n	F/P, I, n	P/F, I, n	F/A, I, n	A/F, I, n	P/A, I, n	A/P, I, n	A/G, I, n
1	1.200	0.8333	1.000	1.0000	0.8333	1.2000	0.0000
2	1.440	0.6944	2.200	0.4545	1.5278	0.6545	0.4545
3	1.728	0.5787	3.640	0.2747	2.1065	0.4747	0.8791
4	2.074	0.4823	5.368	0.1863	2.5887	0.3863	1.2742
5	2.488	0.4019	7.442	0.1344	2.9906	0.3344	1.6405
6	2.986	0.3349	9.930	0.1007	3.3255	0.3007	1.9788
7	3.583	0.2791	12.916	0.0774	3.6046	0.2774	2.2902
8	4.300	0.2326	16.499	0.0606	3.8372	0.2606	2.5756
9	5.160	0.1938	20.799	0.0481	4.0310	0.2481	2.8364
10	6.192	0.1615	25.959	0.0385	4.1925	0.2385	3.0739
11	7.430	0.1346	32.150	0.0311	4.3271	0.2311	3.2893
12	8.916	0.1122	39.581	0.0253	4.4392	0.2253	3.4841
13	10.699	0.0935	48.497	0.0206	4.5327	0.2206	3.6597
14	12.839	0.0779	59.196	0.0169	4.6106	0.2169	3.8175
15	15.407	0.0649	72.035	0.0139	4.6755	0.2139	3.9588
16	18.488	0.0541	87.442	0.0114	4.7296	0.2114	4.0851
17	22.186	0.0451	105.931	0.0094	4.7746	0.2094	4.1976
18	26.623	0.0376	128.117	0.0078	4.8122	0.2078	4.2975
19	31.948	0.0313	154.740	0.0065	4.8435	0.2065	4.3861
20	38.338	0.0261	186.688	0.0054	4.8696	0.2054	4.4643
21	46.005	0.0217	225.026	0.0044	4.8913	0.2044	4.5334
22	55.206	0.0181	271.031	0.0037	4.9094	0.2037	4.5941
23	66.247	0.0151	326.237	0.0031	4.9245	0.2031	4.6475
24	79.497	0.0126	392.484	0.0025	4.9371	0.2025	4.6943
25	95.396	0.0105	471.981	0.0021	4.9476	0.2021	4.7352
30	237.376	0.0042	1181.882	0.0008	4.9789	0.2008	4.8731
35	590.668	0.0017	2948.341	0.0003	4.9915	0.2003	4.9406
40	1469.772	0.0007	7343.858	0.0001	4.9966	0.2001	4.9728
45	3657.262	0.0003	18281.310	0.0001	4.9986	0.2001	4.9877
50	9100.438	0.0001	45497.191	0.0000	4.9995	0.2000	4.9945
55	22644.802	0.0000	113219.011	0.0000	4.9998	0.2000	4.9976
60	56347.514	0.0000	281732.572	0.0000	4.9999	0.2000	4.9989
65	140210.647	0.0000	701048.235	0.0000	5.0000	0.2000	4.9995
70	348888.957	0.0000	1744439.785	0.0000	5.0000	0.2000	4.9998
75	868147.369	0.0000	4340731.847	0.0000	5.0000	0.2000	4.9999
80	2160228.462	0.0000	10801137.310	0.0000	5.0000	0.2000	5.0000
85	5375339.687	0.0000	26876693.433	0.0000	5.0000	0.2000	5.0000
90	13375565.249	0.0000	66877821.245	0.0000	5.0000	0.2000	5.0000
95	33282686.520	0.0000	166413427.601	0.0000	5.0000	0.2000	5.0000
100	82817974.522	0.0000	414089867.610	0.0000	5.0000	0.2000	5.0000

25% interest factors for discrete compounding periods

n	F/P, I, n	P/F, I, n	F/A, I, n	A/F, I, n	P/A, I, n	A/P, I, n	A/G, I, n
1	1.250	0.8000	1.000	1.0000	0.8000	1.2500	0.0000
2	1.563	0.6400	2.250	0.4444	1.4400	0.6944	0.4444
3	1.953	0.5120	3.813	0.2623	1.9520	0.5123	0.8525
4	2.441	0.4096	5.766	0.1734	2.3616	0.4234	1.2249
5	3.052	0.3277	8.207	0.1218	2.6893	0.3718	1.5631
6	3.815	0.2621	11.259	0.0888	2.9514	0.3388	1.8683
7	4.768	0.2097	15.073	0.0663	3.1611	0.3163	2.1424
8	5.960	0.1678	19.842	0.0504	3.3289	0.3004	2.3872
9	7.451	0.1342	25.802	0.0388	3.4631	0.2888	2.6048
10	9.313	0.1074	33.253	0.0301	3.5705	0.2801	2.7971
11	11.642	0.0859	42.566	0.0235	3.6564	0.2735	2.9663
12	14.552	0.0687	54.208	0.0184	3.7251	0.2684	3.1145
13	18.190	0.0550	68.760	0.0145	3.7801	0.2645	3.2437
14	22.737	0.0440	86.949	0.0115	3.8241	0.2615	3.3559
15	28.422	0.0352	109.687	0.0091	3.8593	0.2591	3.4530
16	35.527	0.0281	138.109	0.0072	3.8874	0.2572	3.5366
17	44.409	0.0225	173.636	0.0058	3.9099	0.2558	3.6084
18	55.511	0.0180	218.045	0.0046	3.9279	0.2546	3.6698
19	69.389	0.0144	273.556	0.0037	3.9424	0.2537	3.7222
20	86.736	0.0115	342.945	0.0029	3.9539	0.2529	3.7667
21	108.420	0.0092	429.681	0.0023	3.9631	0.2523	3.8045
22	135.525	0.0074	538.101	0.0019	3.9705	0.2519	3.8365
23	169.407	0.0059	673.626	0.0015	3.9764	0.2515	3.8634
24	211.758	0.0047	843.033	0.0012	3.9811	0.2512	3.8861
25	264.698	0.0038	1054.791	0.0009	3.9849	0.2509	3.9052
30	807.794	0.0012	3227.174	0.0003	3.9950	0.2503	3.9628
35	2465.190	0.0004	9856.761	0.0001	3.9984	0.2501	3.9858
40	7523.164	0.0001	30088.655	0.0000	3.9995	0.2500	3.9947
45	22958.874	0.0000	91831.496	0.0000	3.9998	0.2500	3.9980
50	70064.923	0.0000	280255.693	0.0000	3.9999	0.2500	3.9993
55	213821.177	0.0000	855280.707	0.0000	4.0000	0.2500	3.9997
60	652530.447	0.0000	2610117.787	0.0000	4.0000	0.2500	3.9999
65	1991364.889	0.0000	7965455.556	0.0000	4.0000	0.2500	4.0000
70	6077163.357	0.0000	24308649.429	0.0000	4.0000	0.2500	4.0000
75	18546030.753	0.0000	74184119.014	0.0000	4.0000	0.2500	4.0000
80	56597994.243	0.0000	226391972.971	0.0000	4.0000	0.2500	4.0000
85	172723371.102	0.0000	690893480.408	0.0000	4.0000	0.2500	4.0000
90	527109897.162	0.0000	2108439584.646	0.0000	4.0000	0.2500	4.0000
95	1608611746.709	0.0000	6434446982.835	0.0000	4.0000	0.2500	4.0000
100	4909093465.298	0.0000	19636373857.191	0.0000	4.0000	0.2500	4.0000

30% interest factors for discrete compounding periods

n	F/P, I, n	P/F, I, n	F/A, I, n	A/F, I, n	P/A, I, n	A/P, I, n	A/G, I, n
1	1.300	0.7692	1.000	1.0000	0.7692	1.3000	0.0000
2	1.690	0.5917	2.300	0.4348	1.3609	0.7348	0.4348
3	2.197	0.4552	3.990	0.2506	1.8161	0.5506	0.8271
4	2.856	0.3501	6.187	0.1616	2.1662	0.4616	1.1783
5	3.713	0.2693	9.043	0.1106	2.4356	0.4106	1.4903
6	4.827	0.2072	12.756	0.0784	2.6427	0.3784	1.7654
7	6.275	0.1594	17.583	0.0569	2.8021	0.3569	2.0063
8	8.157	0.1226	23.858	0.0419	2.9247	0.3419	2.2156
9	10.604	0.0943	32.015	0.0312	3.0190	0.3312	2.3963
10	13.786	0.0725	42.619	0.0235	3.0915	0.3235	2.5512
11	17.922	0.0558	56.405	0.0177	3.1473	0.3177	2.6833
12	23.298	0.0429	74.327	0.0135	3.1903	0.3135	2.7952
13	30.288	0.0330	97.625	0.0102	3.2233	0.3102	2.8895
14	39.374	0.0254	127.913	0.0078	3.2487	0.3078	2.9685
15	51.186	0.0195	167.286	0.0060	3.2682	0.3060	3.0344
16	66.542	0.0150	218.472	0.0046	3.2832	0.3046	3.0892
17	86.504	0.0116	285.014	0.0035	3.2948	0.3035	3.1345
18	112.455	0.0089	371.518	0.0027	3.3037	0.3027	3.1718
19	146.192	0.0068	483.973	0.0021	3.3105	0.3021	3.2025
20	190.050	0.0053	630.165	0.0016	3.3158	0.3016	3.2275
21	247.065	0.0040	820.215	0.0012	3.3198	0.3012	3.2480
22	321.184	0.0031	1067.280	0.0009	3.3230	0.3009	3.2646
23	417.539	0.0024	1388.464	0.0007	3.3254	0.3007	3.2781
24	542.801	0.0018	1806.003	0.0006	3.3272	0.3006	3.2890
25	705.641	0.0014	2348.803	0.0004	3.3286	0.3004	3.2979
30	2619.996	0.0004	8729.985	0.0001	3.3321	0.3001	3.3219
35	9727.860	0.0001	32422.868	0.0000	3.3330	0.3000	3.3297
40	36118.865	0.0000	120392.883	0.0000	3.3332	0.3000	3.3322
45	134106.817	0.0000	447019.389	0.0000	3.3333	0.3000	3.3330
50	497929.223	0.0000	1659760.743	0.0000	3.3333	0.3000	3.3332
55	1848776.350	0.0000	6162584.500	0.0000	3.3333	0.3000	3.3333
60	6864377.173	0.0000	22881253.909	0.0000	3.3333	0.3000	3.3333
65	25486951.936	0.0000	84956503.120	0.0000	3.3333	0.3000	3.3333
70	94631268.452	0.0000	315437558.172	0.0000	3.3333	0.3000	3.3333
75	351359275.572	0.0000	1171197581.908	0.0000	3.3333	0.3000	3.3333
80	1304572395.051	0.0000	4348574646.838	0.0000	3.3333	0.3000	3.3333
85	4843785982.758	0.0000	16145953272.526	0.0000	3.3333	0.3000	3.3333
90	17984638288.961	0.0000	59948794293.204	0.0000	3.3333	0.3000	3.3333
95	66775703042.233	0.0000	222585676804.110	0.0000	3.3333	0.3000	3.3333
100	247933511096.598	0.0000	826445036985.328	0.0000	3.3333	0.3000	3.3333

35% interest factors for discrete compounding periods

n	F/P, I, n	P/F, I, n	F/A, I, n	A/F, I, n	P/A, I, n	A/P, I, n	A/G, I, n
1	1.350	0.7407	1.000	1.0000	0.7407	1.3500	0.0000
2	1.823	0.5487	2.350	0.4255	1.2894	0.7755	0.4255
3	2.460	0.4064	4.173	0.2397	1.6959	0.5897	0.8029
4	3.322	0.3011	6.633	0.1508	1.9969	0.5008	1.1341
5	4.484	0.2230	9.954	0.1005	2.2200	0.4505	1.4220
6	6.053	0.1652	14.438	0.0693	2.3852	0.4193	1.6698
6	6.053	0.1652	14.438	0.0693	2.3852	0.4193	1.6698
7	8.172	0.1224	20.492	0.0488	2.5075	0.3988	1.8811
8	11.032	0.0906	28.664	0.0349	2.5982	0.3849	2.0597
9	14.894	0.0671	39.696	0.0252	2.6653	0.3752	2.2094
10	20.107	0.0497	54.590	0.0183	2.7150	0.3683	2.3338
11	27.144	0.0368	74.697	0.0134	2.7519	0.3634	2.4364
12	36.644	0.0273	101.841	0.0098	2.7792	0.3598	2.5205
13	49.470	0.0202	138.485	0.0072	2.7994	0.3572	2.5889
14	66.784	0.0150	187.954	0.0053	2.8144	0.3553	2.6443
15	90.158	0.0111	254.738	0.0039	2.8255	0.3539	2.6889
16	121.714	0.0082	344.897	0.0029	2.8337	0.3529	2.7246
17	164.314	0.0061	466.611	0.0021	2.8398	0.3521	2.7530
18	221.824	0.0045	630.925	0.0016	2.8443	0.3516	2.7756
19	299.462	0.0033	852.748	0.0012	2.8476	0.3512	2.7935
20	404.274	0.0025	1152.210	0.0009	2.8501	0.3509	2.8075
21	545.769	0.0018	1556.484	0.0006	2.8519	0.3506	2.8186
22	736.789	0.0014	2102.253	0.0005	2.8533	0.3505	2.8272
23	994.665	0.0010	2839.042	0.0004	2.8543	0.3504	2.8340
24	1342.797	0.0007	3833.706	0.0003	2.8550	0.3503	2.8393
25	1812.776	0.0006	5176.504	0.0002	2.8556	0.3502	2.8433
30	8128.550	0.0001	23221.570	0.0000	2.8568	0.3500	2.8535
35	36448.688	0.0000	104136.251	0.0000	2.8571	0.3500	2.8562
40	163437.135	0.0000	466960.385	0.0000	2.8571	0.3500	2.8569
45	732857.577	0.0000	2093875.934	0.0000	2.8571	0.3500	2.8571
50	3286157.879	0.0000	9389019.656	0.0000	2.8571	0.3500	2.8571
55	14735241.812	0.0000	42100688.035	0.0000	2.8571	0.3500	2.8571
60	66073316.996	0.0000	188780902.847	0.0000	2.8571	0.3500	2.8571
65	296274962.738	0.0000	846499890.681	0.0000	2.8571	0.3500	2.8571
70	1328506839.613	0.0000	3795733824.609	0.0000	2.8571	0.3500	2.8571
75	5957069090.773	0.0000	17020197399.352	0.0000	2.8571	0.3500	2.8571
80	26711696992.525	0.0000	76319134261.500	0.0000	2.8571	0.3500	2.8571
85	119776142486.850	0.0000	342217549959.570	0.0000	2.8571	0.3500	2.8571
90	537080227925.798	0.0000	1534514936927.990	0.0000	2.8571	0.3500	2.8571
95	2408285700639.400	0.0000	6880816287538.290	0.0000	2.8571	0.3500	2.8571
100	10798833608720.200	0.0000	30853810310626.300	0.0000	2.8571	0.3500	2.8571

40% interest factors for discrete compounding periods

n	F/P, I, n	P/F, I, n	F/A, I, n	A/F, I, n	P/A, I, n	A/P, I, n	A/G, I, n
1	1.400	0.7143	1.000	1.0000	0.7143	1.4000	0.0000
2	1.960	0.5102	2.400	0.4167	1.2245	0.8167	0.4167
3	2.744	0.3644	4.360	0.2294	1.5889	0.6294	0.7798
4	3.842	0.2603	7.104	0.1408	1.8492	0.5408	1.0923
5	5.378	0.1859	10.946	0.0914	2.0352	0.4914	1.3580
6	7.530	0.1328	16.324	0.0613	2.1680	0.4613	1.5811
6	7.530	0.1328	16.324	0.0613	2.1680	0.4613	1.5811
7	10.541	0.0949	23.853	0.0419	2.2628	0.4419	1.7664
8	14.758	0.0678	34.395	0.0291	2.3306	0.4291	1.9185
9	20.661	0.0484	49.153	0.0203	2.3790	0.4203	2.0422
10	28.925	0.0346	69.814	0.0143	2.4136	0.4143	2.1419
11	40.496	0.0247	98.739	0.0101	2.4383	0.4101	2.2215
12	56.694	0.0176	139.235	0.0072	2.4559	0.4072	2.2845
13	79.371	0.0126	195.929	0.0051	2.4685	0.4051	2.3341
14	111.120	0.0090	275.300	0.0036	2.4775	0.4036	2.3729
15	155.568	0.0064	386.420	0.0026	2.4839	0.4026	2.4030
16	217.795	0.0046	541.988	0.0018	2.4885	0.4018	2.4262
17	304.913	0.0033	759.784	0.0013	2.4918	0.4013	2.4441
18	426.879	0.0023	1064.697	0.0009	2.4941	0.4009	2.4577
19	597.630	0.0017	1491.576	0.0007	2.4958	0.4007	2.4682
20	836.683	0.0012	2089.206	0.0005	2.4970	0.4005	2.4761
21	1171.356	0.0009	2925.889	0.0003	2.4979	0.4003	2.4821
22	1639.898	0.0006	4097.245	0.0002	2.4985	0.4002	2.4866
23	2295.857	0.0004	5737.142	0.0002	2.4989	0.4002	2.4900
24	3214.200	0.0003	8032.999	0.0001	2.4992	0.4001	2.4925
25	4499.880	0.0002	11247.199	0.0001	2.4994	0.4001	2.4944
30	24201.432	0.0000	60501.081	0.0000	2.4999	0.4000	2.4988
35	130161.112	0.0000	325400.279	0.0000	2.5000	0.4000	2.4997
40	700037.697	0.0000	1750091.741	0.0000	2.5000	0.4000	2.4999
45	3764970.741	0.0000	9412424.353	0.0000	2.5000	0.4000	2.5000
50	20248916.240	0.0000	50622288.099	0.0000	2.5000	0.4000	2.5000
55	108903531.277	0.0000	272258825.693	0.0000	2.5000	0.4000	2.5000
60	585709328.057	0.0000	1464273317.643	0.0000	2.5000	0.4000	2.5000
65	3150085336.530	0.0000	7875213338.824	0.0000	2.5000	0.4000	2.5000
70	16941914960.338	0.0000	42354787398.345	0.0000	2.5000	0.4000	2.5000
75	91117684716.288	0.0000	227794211788.220	0.0000	2.5000	0.4000	2.5000
80	490052776648.528	0.0000	1225131941618.820	0.0000	2.5000	0.4000	2.5000
85	2635621445482.180	0.0000	6589053613702.950	0.0000	2.5000	0.4000	2.5000
90	14175004682950.100	0.0000	35437511707372.700	0.0000	2.5000	0.4000	2.5000
95	76236577186029.300	0.0000	190591442965071.000	0.0000	2.5000	0.4000	2.5000
100	410018608884990.000	0.0000	1025046522212470.000	0.0000	2.5000	0.4000	2.5000

Appendix 2

INDIAN CONTRACT ACT 1872

The important sections of Indian Contract Act 1872 are described briefly.

SECTION 3

Communication, Acceptance and Revocation of Proposals

The communication of proposals, the acceptance of proposals, and the revocation of proposals and acceptances, respectively, are deemed to be made by any act or omission of the party proposing, accepting, or revoking, by which he intends to communicate such proposal, acceptance or revocation, or which has the effect of communicating it.

SECTION 4

When a Communication Is Said to Be Complete

The communication of a proposal is complete when it comes to the knowledge of the person to whom it is made. The communication of an acceptance is complete—

> as against the proposer, when it is put in a course of transmission to him, so as to be out of the power of the acceptor; as against the acceptor, when it comes to the knowledge of the proposer.

The communication of a revocation is complete—

> as against the person who makes it, when it is put into a course of transmission to the person to whom it is made so as to be out of the power of the person who makes it; as against the person to whom it is made, when it comes to his knowledge.

SECTION 5

Revocation of Proposals and Acceptances

A proposal may be revoked at any time before the communication of its acceptance is complete as against the proposer, but not afterwards. An acceptance may be revoked at any time before the communication of the acceptance is complete as against the acceptor, but not afterwards.

SECTION 6

How Revocation Is Made

A proposal is revoked

 i) by the communication of notice of revocation by the proposer to the other party;
 ii) by the lapse of the time prescribed in such proposal for its acceptance, or if no time is so prescribed, by the lapse of a reasonable time, without communication of the acceptance;
 iii) by the failure of the acceptor to fulfil a condition precedent to acceptance; or
 iv) by the death or insanity of the proposer, if the fact of the death or insanity comes to the knowledge of the acceptor before acceptance.

SECTION 7
Acceptance Must Be Absolute
In order to convert a proposal into a promise, the acceptance must
  i) be absolute and unqualified;
  ii) be expressed in some usual and reasonable manner, unless the proposal prescribes manner in which it is to be accepted. If the proposal prescribes a manner in which it is to be accepted, and the acceptance is not made in such a manner, the proposer may, within a reasonable time after the acceptance is communicated to him, insist that his proposal shall be accepted in the prescribed manner, and not otherwise; but, if he fails to do so, he accepts the acceptance.

SECTION 8
Acceptance by Performing Conditions, or Receiving Consideration
Performance of the conditions of proposal, or the acceptance of any consideration for a reciprocal promise which may be offered with a proposal, is an acceptance of the proposal.

SECTION 9
Promises, Expressed or Implied
In so far as the proposal or acceptance of any promise is made in words, the promise is said to be express. In so far as proposal or acceptance is made otherwise than in words, the promise is said to be implied.

SECTION 10
What Agreements Are Contracts
All agreements are contracts if they are made by the free consent of parties competent to contract, for a lawful consideration and with a lawful object, and are not hereby expressly declared to be void.

SECTION 13
Consent
Two or more persons are said to consent when they agree upon the same thing in the same sense.

SECTION 14
Free Consent
Consent is said to be free when it is not caused by coercion, undue influence, fraud, misrepresentation, or mistake. Consent is said to be so caused when it would not have been given but for the existence of such coercion, undue influence, fraud, misrepresentation, or mistake. These terms are defined below:

SECTION 15
Coercion
Coercion is the committing, or threatening to commit, any act forbidden by the Indian Penal Code, or the unlawful detaining, or threatening to detain, any property, to the prejudice of any person whatever, with the intention of causing any person to enter into an agreement.

SECTION 16
Undue Influence
A contract is said to be induced by 'undue influence' where the relations subsisting between the parties are such that one of the parties is in a position to dominate the will of the other and uses that position to obtain an unfair advantage over the other. In particular and without prejudice to the generality of the foregoing principle, a person is deemed to be in a position to dominate the will of another

i) where he holds a real or apparent authority over the other, or where he stands in a fiduciary relation to the other; or

ii) where he makes a contract with a person whose mental capacity is temporarily or permanently affected by reason of age, illness, or mental or bodily distress.

Where a person who is in position to dominate the will of another, enters into a contract with him, and the transaction appears, on the face of it or in the evidence adduced, to be unconscionable, the burden of proving that such contract was not introduced by undue influence shall lie upon the person in a position to dominate the will of other.

SECTION 17
Fraud
Fraud means and includes any of the following acts committed by a party to a contract or with his connivance, or by his agent, with intent to deceive another party thereto or his agent, or to induce him to enter into a contract:-

i) the suggestion, as a fact of that which is not true, by one who does not believe it to be true;

ii) the active concealment of a fact by one having knowledge or belief of the fact;

iii) a promise made without any intention of performing it;

iv) any other act fitted to deceive;

v) any such act or omission as the law specially declares to be fraudulent.

SECTION 18
Misrepresentation
Misrepresentation means and includes-

i) the positive assertion, in a manner not warranted by the information of the person making it, of that which is not true, though he believes it to be true;

ii) any breach of duty which, without an intent to deceive, gains an advantage to the person committing it, or anyone claiming under him; by misleading another to his prejudice, or to the prejudice of anyone claiming under him;
iii) causing, however innocently, a party to an agreement, to make a mistake as to the substance of the thing which is the subject of the agreement.

SECTION 19

Voidability of Agreements without Free Consent

When consent to an agreement is caused by coercion, fraud or misrepresentation, the agreement is a contract voidable at the option of the party whose consent was so caused. A party to a contract, whose consent was caused by fraud or misrepresentation, may, if he thinks fit, insist that the contract shall be performed, and that he shall be put on the position in which he would have been if the representations made would have been true.

SECTION 23

What Considerations and Objects Are Lawful and What Not

The consideration or object of an agreement is lawful unless—

it is forbidden by law; or

is of such a nature that, if permitted, it would defeat the provisions of any law; or

is fraudulent; or

involves or implies injury to the person or property of another; or

the Court regards it as immoral, or opposed to public policy.

In each of these cases, the consideration or object of an agreement is said to be unlawful. Every agreement of which the object or consideration is unlawful is void.

SECTION 24

Agreements Void, if Considerations and Objects Unlawful in Part

If any part of a single consideration for one or more objects, or any one or any part of any one of several considerations for a single object, is unlawful, the agreement is void.

SECTION 37

Obligations of Parties to Contracts

The parties to a contract must either perform, or offer to perform, their respective promises, unless such performance is dispensed with or executed under the provisions of this Act, or any other law. Promises bind the representatives of the promisors in case of the death of such promisors before performance, unless a contrary intention appears from the contract.

SECTION 38

Effect of Refusal to Accept Offer of Performance

Where a promisor has made an offer of performance to the promisee, and the offer has not been accepted, the promisor is not responsible for non-performance, nor does he thereby lose his rights under the contract.

Every such offer must fulfil the following conditions:

i) it must be unconditional;

ii) it must be made at a proper time and place, and under such circumstances that a person to whom it is made is able and willing there and then to do the whole of what he is bound by his promise to do;

iii) if the offer is an offer to deliver anything to the promisee, the promisee must have a reasonable opportunity of seeing that the thing offered is the thing which the promisor is bound by his promise to deliver.

An offer to one of several joint promises has the same legal consequence as an offer to all of them.

SECTION 53

Liability of Party Preventing Event on which the Contract Is to Take Effect

When a contract contains reciprocal promises, and one party to the contract prevents the other from performing his promise, the contract becomes voidable at the option of the party so prevented; and he is entitled to compensation from the other party for any loss that he may sustain in consequence of the non-performance of the contract.

SECTION 55

Effect of Failure to Perform at Fixed Time, in Contract in which Time is Essential

When a party to a contract promises to do a certain thing at or before a specified time, or certain things at or before specified times, and fails to do any such thing at or before the specified time, or the contract, so much of it as has not been performed, becomes voidable at the option of the promisee, if the intention of the parties was that time should be the essence of the contract.

Effect of such failure when time is not essential: If it was not the intention of the parties that time should be the essence of the contract, the contract does not become voidable by the failure to do such thing at or before the specified time; but the promisee is entitled to compensation from the promisor for any loss occasioned to him by such failure.

Effect of acceptance of performance at time other than that agreed upon: If, in case of a contract voidable on account of the promisor's failure to perform his promise at the time agreed, the promisee accepts performance of such promise at any time other than that agreed, the promisee cannot claim compensation of any loss occasioned by the non-performance of the promise at the time agreed, unless, at the time of such acceptance, he gives notice to the promisor of his intention to do so.

SECTION 56

Agreement to Do Impossible Act

An agreement to do an act impossible in itself is void.

Contract to do act after becoming impossible or unlawful: A contract to do an act which, after the contract is made, becomes impossible, or, by reason of some event which the promisor could not prevent becomes unlawful, becomes void when the act becomes impossible or unlawful.

Compensation for loss through non-performance of act known to be impossible or unlawful: Where one person has promised to do something, which he knew, or, with reasonable diligence, might have known, and which the promisee did not know to be impossible or unlawful, such promisor must make compensation to such promisee for any loss the promisee sustains through the non-performance of the promise.

SECTION 62
Effect of Novation, Rescission and Alteration of Contract

If the parties to a contract agree to substitute a new contract for it, or to rescind or alter it, the original contract need not be performed.

SECTION 64
Consequences of Rescission of Voidable Contract

When a person at whose option a contract is voidable rescinds it, the other party thereto need to perform any promise therein contained in which he is promisor. The party rescinding a voidable contract shall, if he has received any benefit thereunder from another party to such contract, restore such benefit, so far as may be, to the person from whom it was received.

SECTION 65
Obligation of Person who Has Received Advantage Under void Agreement or Contract that Becomes Void

When an agreement is discovered to be void, or when a contract becomes void, any person who has received any advantage under such agreement or contract is bound to restore it, or to make compensation for it, to the person from whom he received it.

SECTION 67
Effect of Neglect of Promisee to Afford Promisor Reasonable Facilities for the Performance

If any promisee neglects or refuses to afford the promisor reasonable facilities for the performance of his promise, the promisor is excused by such neglect or refusal as to any non-performance caused thereby.

SECTION 70
Obligation of Person Enjoying Benefit of Non-gratuitous Act

Where a person lawfully does anything for another person, or delivers anything to him, not intending to do so gratuitously, and such other person enjoys the benefit thereof, the latter is bound to make compensation to the former in respect of, or to restore, the thing so done or delivered.

SECTION 73

Compensation for Loss or Damage Caused by Breach of Contract

When a contract has been broken, the party who suffers by such breach is entitled to receive, from the party who has broken the contract, a compensation for any loss or damage caused to him thereby, which naturally arose in the usual course of things from such breach, or which the parties knew, when they made the contract, to be likely to result from the breach of it.

Such compensation is not to be given for any remote and indirect loss or damage sustained by reason of the breach.

Compensation for failure to discharge obligation resembling those created by contract: When an obligation resembling those created by contract has been incurred and has not been discharged, any person injured by the failure to discharge it is entitled to receive the same compensation from the party in default, as if such person had contracted to discharge it and had broken his contract.

SECTION 74

Compensation for Breach of Contract where Penalty Stipulated for

When a contract has been broken, if a sum is named in the contract as the amount to be paid in case of such breach, or if the contract contains any other stipulation by way of penalty, the party complaining of the breach is entitled, whether or not actual damage or loss is proved to have been caused thereby, to receive from the party who has broken the contract reasonable compensation not exceeding the amount so named or, as the case may be, the penalty stipulated for.

Appendix 3

IMPORTANT ACTS APPLICABLE TO ESTABLISHMENTS ENGAGED IN BUILDING AND OTHER CONSTRUCTION WORK

1. WORKMEN COMPENSATION ACT 1923

The Act provides for compensation in case of injury by accident arising out of and during the course of employment.

SECTION 4
Amount of Compensation

Compensation for death — Minimum: ₹ 50,000
Compensation for permanent disablement — Minimum: ₹ 60,000
Temporary disablement — 50% of wages for a maximum period of 5 years

SECTION 10B
Reports of Fatal Accidents and Serious Bodily Injuries

Report to be submitted within 7 days

2. PAYMENT OF GRATUITY ACT 1972

Gratuity is payable to an employee under the Act on satisfaction of certain conditions on separation, if an employee has completed the prescribed minimum years (say, five years) of service or more, or on death the rate of prescribed minimum days' (say, 15 days) wages for every completed year of service. The Act is applicable to all establishments employing the prescribed minimum number (say, 10) or more employees.

3. EMPLOYEES P.F. AND MISCELLANEOUS PROVISION ACT 1952

The Act provides for monthly contributions by the employer plus workers at the rate prescribed (say, 10 per cent or 8.33 per cent). The benefits payable under the Act are:

i) Pension or family pension on retirement or death as the case may be
ii) Deposit-linked insurance on the death in harness of the worker
iii) Payment of P.F. accumulation on retirement/death, etc.

The provisions of the Act

i) apply to factories employing more than 20 persons
ii) can be made applicable by central government to factories employing less than 20 persons or if the majority of employees agree
iii) apply for all persons, employed directly or indirectly, in any kind of work, with pay less than ₹ 3,500 per month
iv) For medical treatment

4. MATERNITY BENEFIT ACT 1951

The Act provides for leave and some other benefits to women employees in case of confinement, miscarriage, etc.

i) Maternity benefits to be provided on completion of 80 days of working

ii) Not required to work during six weeks after day of delivery or miscarriage

iii) No work of arduous nature or long hours of standing likely to interfere with pregnancy/normal development of foetus, or likely to cause miscarriage, or likely to affect health, to be given for a period of one month immediately preceding the period of six weeks before delivery

iv) On medical certificate, advance maternity benefit to be allowed

v) A sum of ₹ 250 as medical bonus to be given in case no prenatal confinement and postnatal care are provided free of charge

5. CONTRACT LABOR (REGULATION AND ABOLITION) ACT 1970

The Act provides for certain welfare measures to be provided by the contractor to contract labour, and in case the contractor fails to provide, the same are required to be provided by the principal employer by law. The principal employer is required to take certificate of registration and the contractor is required to take license from the designated officer. The Act is applicable to the establishments or contractor of principal employer if they employ the prescribed minimum (say, 20) or more contract labour. According to the Act, workers are not to be required to work beyond 9 hours between 6 am and 7 pm, with the exception of midwives and nurses in plantations.

6. MINIMUM WAGES ACT 1948

The employer is to pay not less than the minimum wages fixed by appropriate government as per provisions of the Act, if the employment is a scheduled employment. Construction of buildings, roads and runways is scheduled employment. The Act is applicable to all employees (skilled, unskilled) including out-workers. It also deals with:

i) Fixation of minimum wage of employees

ii) Procedure for fixing and revising minimum wages

iii) Obligation of employees

iv) Rights of workers

7. PAYMENT OF WAGES ACT 1936

It specifies by what date the wages are to be paid, when it will be paid, and what deductions can be made from the wages of the workers. The Act is formulated to ensure regular and prompt payment of wages, and to prevent the exploitation of a wage earner by prohibiting arbitrary fines and deductions from his wages. The Act is not applicable to wages that average ₹ 1,600 per month or more. It recommends equal remuneration for men and women.

8. EQUAL REMUNERATION ACT 1979

The Act provides for payment of equal wages for work of equal nature to male and female workers, and for not making discrimination against female employees in the matters of transfers, training, promotions, etc. There is no discrimination permissible in recruitment and service conditions except where employment of women is prohibited or restricted by or under any law.

9. PAYMENT OF BONUS ACT 1965

The Act is applicable (all factories in India under the Factories Act) to all establishments employing the prescribed minimum (say, 20) or more workmen. The government can extend its coverage to establishments employing between 10 and 20 workers. The Act provides for payments of annual bonus within the prescribed range of percentage of wages to employees drawing up to the prescribed amount of wages, calculated in the prescribed manner. The Act does not apply to certain establishments. The newly set-up establishments are exempted for five years in certain circumstances. States may have different numbers of employment size. The Act covers all workers including supervisors, managers, administrators, and technical and clerical staff employed on salary or wages not exceeding ₹ 2,500 per month.

10. INDUSTRIAL DISPUTES ACT 1947

The Act lays down the machinery and procedure for resolution of industrial disputes and to promote harmonious relation between employers and workers. The Act covers all workers and supervisors drawing salaries up to ₹ 1,600 per month. The Act is not applicable to person employed in managerial and administration capacities. The Act lays down situations in which a strike or lock-out becomes illegal and also specifies the requirements for laying off or retrenching the employees or closing down the establishment.

11. INDUSTRIAL EMPLOYMENT (STANDING ORDERS) ACT 1946

It is applicable to all establishments employing the prescribed minimum (say, 100, or 50). The Act provides for laying down rules governing the conditions of employment by the employer on matters provided in the Act and getting these certified by the designated authority.

12. TRADE UNIONS ACT 1926

The Act lays down the procedure for registration of trade unions of workmen and employers. The trade unions registered under the Act have been given certain immunities from civil and criminal liabilities.

SECTION 15

Objects on which General Funds may be Spent

i) Conduct of trade disputes on behalf of the trade union or any member
ii) Compensation of members for loss arising out of trade disputes
iii) Allowance to members or their dependents on account of death, old age, sickness, accidents, or unemployment of such members
iv) Issuing policies of assurance against sickness, accident, or unemployment

13. CHILD LABOR (PROHIBITION AND REGULATION) ACT 1986

The Act prohibits employment of children below 14 years of age in certain occupations and processes, and provides for regulations of employment of children in all other occupations and processes. Employment of child labour is prohibited in building and construction industry.

14. INTER-STATE MIGRANT WORKMEN'S (REGULATION OF EMPLOYMENT AND CONDITIONS OF SERVICE) ACT 1979

The Act is applicable to an establishment that employs the prescribed minimum (say, five) or more inter-state migrant workmen through an intermediary (who has recruited workmen in one state for employment in the establishment situated in another state). The inter-state migrant workmen, in an establishment to which this Act becomes applicable, are required to be provided certain facilities such as housing, medical aid and travelling expenses from home to the establishment and back. According to the Act, separate toilets and washing facilities to be provided for male and female workers.

15. THE BUILDING AND OTHER CONSTRUCTION WORKERS (REGULATION OF EMPLOYMENT AND CONDITIONS OF SERVICE) ACT 1996 AND THE CESS ACT OF 1996

All the establishments who carry on any building or other construction work and employ the prescribed minimum (say, 10) or more workers are covered under this Act. All such establishments are required to pay cess at the rate not exceeding two per cent of the cost of construction as may be modified by the government. The employer of the establishment is required to provide safety measures at the building or construction work, and other welfare measures such as canteens, first-aid facilities, ambulance and housing accommodations for workers near the workplace. The employer to whom the Act applies has to obtain a registration certificate from the registering officer appointed by the government.

The salient features are given below:

SECTION 28
Fixing Hours of Normal Working Day
(1) Government to fix one day of rest in every period of seven days.

SECTION 30
Maintenance of Registers and Records
(1) Every employers shall maintain records containing employees, work done, hours worked, rest, wages, receipts.
(2) Notices in the prescribed form to be publicly displayed by employer.

SECTION 31
Prohibition of Employment
Persons with deafness, defective vision, or giddiness not allowed to work in any such operation which is likely to involve a risk of any accident.

SECTION 32
Drinking Water
(1) Convenient supply of wholesome drinking water to be made available. All such sources be legibly marked 'Drinking water'
(2) Marked sources of drinking water, not to be within 6 m of source of contamination

SECTION 33
Latrines and Urinals
One for every 50 workers

SECTION 34
Accommodation
(1) Free shelter near work site for all workers
(2) Separate places for cooking, bathing, washing, lavatory

SECTION 35
Crèches
Room for children less than 6 years for more than 50 female workers.

SECTION 36
First Aid
First aid facilities to be provided at work site

SECTION 37
Canteen
Canteen to be provided for workers wherein more than two hundred and fifty workers are employed.

SECTION 38
Safety Committee and Safety Officer
Safety committee with worker representation to be constituted wherein five hundred or more workers are employed.
Appointment of qualified safety officer.

SECTION 39
Notice of Accidents
(1) To give notice in case of death or injury causing disability for a period of forty-eight hours or more immediately following the accident.
(2) Authority to conduct enquiry within one month.

SECTION 40
Power of Government to Make Rules

Rules regarding access, safe means of scaffolding, demolition, explosives, transport equipment, hoists, lighting, dust and fumes, storage, fencing, safe handling of power equipment, fire, lifting loads, transport of workers, electrical safety, ladders and stairs, lifting equipment, concrete work, pile driving, hot asphalt, earthmoving, safety policy, information to be furnished, etc.

16. MANUFACTURE, STORAGE AND IMPORT OF HAZARDOUS CHEMICALS RULES, 1989

SECTION 4
General Responsibility of the Occupier During Industrial Activity

(2) Occupier to provide evidence to show that he has identified major accident hazards; taken adequate steps to prevent; provided to workers with information, training and equipment, including antidotes necessary to ensure their safety.

SECTION 5
Notification of Major Accident

(1) Occupier to notify authority (Schedule S) and furnish report (Schedule 6)
(2) Authority to undertake full analysis and send information to MoEF

SECTION 7
Notification of Sites

Occupier cannot undertake industrial activity without submitting written report (Schedule 7) to authority three months before commencing activity

SECTION 8
Updating of the Site Notification Following Changes in the Threshold Quantity

If occupier makes a change in industrial activity, he shall furnish a further report to the authority

SECTION 10
Safety Reports

Occupier cannot undertake industrial activity without submitting safety report (Schedule 8) to the authority 90 days before commencing activity

SECTION 11
Updating of Reports Under Rule 10

If occupier modifies industrial activity, he shall make a further report to the authority 90 days before making modifications

SECTION 13
Preparation of On-site Emergency Plan By the Occupier

(1) Occupier to prepare and update on-site emergency plan to deal with major accidents with names of responsible persons and those authorized to take action

(2) Every worker to be informed of emergency plan

SECTION 15
Information to Be Given to Persons Liable to Be Affected by a Major Accident

Occupier to inform persons outside the site about the nature of accident hazard and the safety measures to be adopted

SECTION 17
Collection, Development and Dissemination of Information

(2) Occupier to develop information in the form of safety data sheet (Schedule 9)

(4) Container of a hazardous chemical to be clearly labelled about contents, manufacturer, and physical, chemical and toxicological data (Part I of Schedule 1)

17. HAZARDOUS WASTES (MANAGEMENT AND HANDLING) RULES 1989

SECTION 4A
Duties of the Occupier and Operator of a Facility

Take adequate steps to contain contaminants and prevent accidents; and provide workers with safety information, training and equipment

SECTION 7
Packaging, Labelling and Transport of Hazardous Wastes

(1) Occupier to ensure easily visible labelling
(2) Packaging, labelling and transport in accordance with Motor Vehicles Act, 1988
(3) All hazardous waste containers shall be provided with a general label (Form 8)

SECTION 9
Records and Returns

Occupier to maintain records (Form 3)

SECTION 10
Accident Reporting and Follow-up

Occupier to immediately report accident to State Pollution Control Board (Form 5)

18. FACTORIES ACT 1948

The objective of the Act is to ensure adequate safety measures and promote health and welfare of workers. The Act lays down the procedure for approval of plans before setting up a factory, health and safety provisions, welfare provisions, working hours, annual earned leave and rendering information regarding accidents or dangerous occurrences to designated authorities. It is applicable to premises employing the prescribed minimum (say, 10) persons or more with the aid of power, or another prescribed minimum (say, 20) or more persons without the aid of power engaged in manufacturing process. Salient features of the Act are:

SECTION 7A
Duties of Occupier

i) Health, safety and welfare of all workers
ii) (a) Maintenance of safe systems
   (b) Safe use, handling, storage and transport
   (c) Information, instruction, training and supervision for health and safety
   (d) Maintenance of means of safe access and egress
   (e) Safe working environment
   (f) Health and safety policy notified to workers
iii) Prepare written statement of general policy with respect to health and safety of the workers

SECTION 7B
Duties of Manufacturer

i) (a) Ensure articles for use
   (b) Conduct necessary tests and examination
   (c) Ensure adequate information for use of the article in factory
ii) Elimination/minimization of risks

SECTION 11
Cleanliness

i) Every factory shall be kept clean and free from effluvia.
   (a) Daily removal of dirt/refuse
   (b) Weekly washing of floor

(c) Effective drainage of floor

(d) Regular painting of walls, partitions, ceilings

SECTION 12
Disposal of Wastes and Effluents

Effective arrangement to be made for the treatment of wastes and effluents to render them innocuous.

SECTION 13
Ventilation and Temperature

i) (a) Adequate ventilation to be ensured

(b) Comfortable temperature to be maintained

SECTION 14
Dust and Fume

i) Effective arrangement to be made to prevent inhalation and accumulation (exhaust near enclosed point of origin) of dust and fume

ii) Arrangement to be made for the exhaust of internal combustion engine to open air

SECTION 16
Overcrowding

Arrangement to be made for at least 14.2 m² per worker employed

SECTION 18
Drinking Water

i) Sufficient supply of wholesome drinking water to be arranged

ii) Distance of more than 6 m from any source of contamination to be ensured

iii) Arrangement for cool drinking water during summer wherein more than 250 workers are employed

SECTION 19
Latrines and Urinals

i) (a) Conveniently located sufficient latrines and urinals to be provided

(b) Arrangement for separate latrine and urinals for male and female

(c) Latrines and urinals not to be connected directly to work room

(d) Clean and sanitary condition at all times to be ensured

(e) Sweepers to be employed to keep clean

SECTION 21
Fencing of Machinery
Every moving and dangerous part to be securely fenced and regularly examined

SECTION 22
Work On or Near Machinery in Motion
i) Examination/operation only by trained/certified adult worker wearing tight-fitting clothing

ii) All parts in motion to be securely fenced/enclosed to prevent contact

SECTION 24
Striking Gear and Devices for Cutting Off Power
Suitable devices for cutting off power from running machinery in emergencies to be provided

SECTION 29
Lifting Machines, Chains, Ropes and Lifting Tackle
i) (a) All parts of every lifting machines, chains, ropes and lifting tackles to be of good construction, properly maintained and examined every 12 months

(b) The above appliances are not to be loaded beyond marked safe working load

(c) Crane should not approach within 6 m of working place

SECTION 30
Revolving Machinery
Safe working speeds not to be exceeded

SECTION 32
Floors, Stairs and Means of Access
i) Floors, stairs and means of access to be of sound construction, properly maintained, free of obstructions and provided with handrails

ii) Fencing to be ensured for work at a height

SECTION 33
Pits, Sumps Openings in Floors, etc.
Pits, sump openings in floors, etc. to be securely covered/fenced

SECTION 34
Excessive Weights
No lifting/carrying of load likely to cause injury

SECTION 35
Protection of Eyes
i) Risk from fragments to be taken care of
ii) Risk of exposure to excessive light to be taken care of

SECTION 36
Precautions Against Dangerous Fumes, Gases, etc.
i) No entry to be permitted into confined space without manhole
ii) (a) No entry to be permitted unless space certified is free from risk
    (b) Worker to wear suitable breathing apparatus and secure safety belt

SECTION 38
Precautions in Case of Fire
i) (a) Safe means of escape to be provided and maintained
   (b) Equipment for extinguishing fire to be provided and maintained
ii) Adequate training to be provided to all the workers to make them familiar with the means of escape in case of fire

SECTION 40B
Safety Officers
Safety officers to be appointed where in one thousand and or more workers are employed

SECTION 41A
Constitution of Site-appraisal Committees
Site-appraisal committees to be appointed by state government if necessary

SECTION 41B
Compulsory Disclosure of Information by the Occupier
i) Information regarding dangers to be disclosed to workers
ii) Occupier to lay down detailed health and safety policy
iii) Information to include waste-disposal system
iv) Occupier to draw up detailed emergency plan and make it known to workers
v) Occupier to lay down measures for handling, usage, transportation, storage and disposal of hazardous substances, and publicize the same among the workers

SECTION 41C
Specific Responsibility of the Occupier in Relation to Hazardous Processes

i) Occupier to maintain up-to-date health records of exposed workers, and make the same accessible to workers
ii) Appoint qualified persons to supervise handling and provide all the protection facilities
iii) Medical examination of every worker in hazardous jobs before assignment and annual examination of such workers thereafter

SECTION 41F
Permissible Limits of Exposure of Chemical and Toxic Substances

As given in Second Schedule

SECTION 41G
Workers' Participation in Safety Management

Safety committee to have equal number of worker and management representatives

SECTION 41H
Right of Workers to Warn About Imminent Danger

i) Workers can warn about imminent danger to occupier, agent, manager, or inspector
ii) Duty of occupier, agent, manager to take immediate action and/or report to inspector

SECTION 42
Washing Facilities

i) Adequate facilities to be provided and those facilities to be maintained regularly
ii) Such facilities to be separate for male and female workers

SECTION 43
Facilities for Storing and Drying Clothing

Suitable places to be provided for keeping clothing not warm during working hours and for the drying of wet clothing

SECTION 44
Facilities for Sitting

i) For rest facilities for sitting to be provided and maintained for all workers obliged to work in a standing position
ii) For work

SECTION 45
First-aid Appliances
i) One box/cupboard for 150 workers, readily accessible during all working hours
ii) First-aid boxes to be kept in charge of responsible certified (first-aid) person; First-aid appliances to be made readily available during working hours
iii) Ambulance room and medical/nursing staff to be provided wherein more than five hundred workers are employed

SECTION 46
Canteens
Canteens to be provided wherein more than two hundred and fifty workers are employed

SECTION 47
Shelters, Restrooms and Lunch Rooms
Shelters, rest rooms and lunch rooms to be provided wherein more than one hundred and fifty workers are employed

SECTION 48
Crèches
Crèches for the use of children under the age of six years to be provided and maintained wherein more than thirty women workers are employed for the use of children under the age of six years

SECTION 49
Welfare Officers
Welfare officers to be employed wherein five hundred or more workers are employed

SECTION 54
Daily Hours
No adult workers to be permitted to work for more than 9 hours in any day

SECTION 55
Intervals for Rest
Worker to be given rest for at least half an hour every five hours

SECTION 61
Notice of Periods of Work for Adults
Notice of periods of work for adults to be displayed and correctly maintained

SECTION 62
Register of Adult Workers
Register with name, work, group, relay to which the worker is allotted, etc., of worker to be maintained

SECTION 66
Further Restrictions on Employment of Women
Women to be required to work only between 6 a.m. and 7 p.m.

SECTION 67
Prohibition of Employment of Young Children
No child who has not completed his fourteen year to be allowed to work in a factory

SECTION 68
Non-adult Workers to Carry Tokens
Adolescents to have certificate of fitness and to carry token giving a reference to such certificates

SECTION 69
Certificates of Fitness
To be granted by certifying surgeon after examination; valid only for 12 months

SECTION 88
Notice of Certain Accidents
 i) For death and injury
 ii) Authority to make enquiry within one month

SECTION 88A
Notice of Certain Dangerous Occurrences
Notice to be provided to authorities for all dangerous occurrences whether resulting into bodily injury or disability or not.

SECTION 89
Notice of Certain Diseases
 i) As specified in Third Schedule
 ii) Medical practitioner to send report in writing to chief inspector giving name, address and disease of patient, and name and address of factory

iii) If confirmed by certifying surgeon, chief inspector to pay medical practitioner's fees

iv) Medical practitioner who fails to report shall be punishable with fine up to ₹ 1,000

SECTION 90
Power to Direct Enquiry into Cases of Accident or Disease

Power to direct enquiry into cases of accident or disease rest with state Government

SECTION 91A
Safety and Occupational Health Surveys

To be undertaken by Chief Inspector, DGFAS (Director General of Factory Advice Service), DGHS (Director General of Health Service), or authorized officer at their discretion

SECTION 108
Display of Notices

i) Notices containing abstracts of Act and rules and name and address of inspector and certifying surgeon to be displayed

ii) The above mentioned notice to be displayed in English and in language understood by majority of workers near main entrance

SECTION 111A
Right of Workers, etc.

i) Obtain from occupier, information relating to workers' health and safety

ii) Get trained in health and safety

iii) Represent to the inspector of inadequate provision for health or safety

19. MOTOR VEHICLES ACT, 1988

SECTION 56
Certificate of Fitness of Transport Vehicles

To be issued by authorized testing station

SECTION 91
Restriction of Hours of Work of Drivers

As provided in Motor Transport Workers Act, 1961

SECTION 140
Liability to Pay Compensation in Certain Cases on the Principle of No Fault

Compensation for death: ₹ 50,000; for permanent disability: ₹ 25,000

SECTION 215
Road Safety Councils and Committees
To be constituted by government

20. EMPLOYEES STATE INSURANCE ACT, 1948

SECTION 28
Purposes for which the Fund May be Expended
To provide for health cover with contributions both from principal employer and employee

SECTION 46
Benefits
Periodical payments for (as prescribed by central government)
- i) Sickness
- ii) Occupational disease
- iii) Maternity
- iv) Employment injury
- v) Pensions to dependents in case of death or employment injury
- vi) Compensation for death
- vii) Compensation for permanent disablement
- viii) Temporary disablement

21. EMPLOYER'S LIABILITY ACT, 1938
Claim of the employer that some other employee has been entrusted with the responsibility for safety, or that the work has been contracted out, or that the worker knew of the risk (unless so proven), cannot be raised as a defence in suits for damages in respect of injuries sustained by workmen.

22. THE INDUSTRIAL DISPUTES ACT, 1947
To provide machinery for peaceful resolution of disputes and to promote harmonious relation between employers and workers

23. BONDED LABOUR SYSTEM (ABOLITION) ACT, 1976
To provide for the abolition of bonded labour system with a view to preventing the economic and physical exploitation of the weaker sections of the people and for matters connected therewith or incidental thereto

24. THE CHILDREN (PLEDGING OF LABOUR) ACT, 1933
- i) Any agreement to pledge the labour of children is void

25. THE PAYMENT OF GRATUITY ACT, 1972

i) Employees drawing wages not exceeding ₹ 3,500 per month.
ii) Fifteen days' wages for every completed year of service or part thereof in excess of six months, subject to a maximum of ₹ 50,000

26. PUBLIC LIABILITY AND INSURANCE ACT, 1991

i) For death or injury to any person other than a workman, or damage to any property, owner shall be liable
ii) Owner to take out one or more policies to insure against liability

Schedule

i) Reimbursement of medical expenses — maximum: ₹ 12,500
ii) Fatal accidents — ₹ 25,000 + medical expenses (maximum ₹ 12,500)
iii) For permanent total or permanent partial disability or other injury or sickness — medical expenses (maximum ₹ 12,500) + cash relief for percent disability (maximum ₹ 25,000)
iv) For temporary partial disability — monthly relief <₹ 1,000 for maximum 3 months
v) For damage to private property — maximum ₹ 6,000

Appendix 4

BRIEF DESCRIPTION OF CPWD CONTRACT CLAUSES

S. No.	Brief description of contract clauses	Clause number
1	Performance guarantee and recovery of security deposit	Clause 1
2	Compensation for delay and incentive for early completion	Clause 2
3	Clauses indicating 'when contract can be determined'	Clause 3
4	Contractor liable to pay compensation even if action not taken under previous clause, i.e. Clause 3	Clause 4
5	Time and extension for delay	Clause 5
6	Measurements of work done	Clause 6
7	Payment on intermediate certificate to be regarded as advances	Clause 7
8	Completion certificate and completion plans	Clause 8
9	Contractor to keep the site clean	Clause 8A
10	Completion plans to be submitted by the contractor	Clause 8B
11	Payment of final bill and payment of contractor's bills to banks	Clause 9 and 9A
12	Materials supplied by government	Clause 10
13	Materials to be provided by the contractor	Clause 10A
14	Clauses on secured advance i. On non-perishable material ii. Mobilization advance iii. Plant machinery and shuttering material advance iv. Interest and recovery	Clause 10B
15	Escalation clauses: Payment on account of i. Increase in prices/wages due to statutory order(s) ii. Increase/decrease in prices of cement and steel reinforcement bars after receipt of tender iii. Increase/decrease in prices/wages after receipt of tender for works	Clause 10C, 10CA and 10CC
16	Dismantled material treated as government property	Clause 10D
17	Works to be executed in accordance with specifications, drawing orders, etc.	Clause 11
18	Deviation/variations i. Deviation, extra items and pricing ii. Deviation, substituted items, pricing iii. Deviation, deviated quantities, pricing	Clause 12
19	Foreclosure of contract due to abandonment or reduction in scope of work	Clause 13
20	Cancellation of contract in full or part	Clause 14

21	Suspension of work	Clause 15
22	Action in case work not done as per specification	Clause 16
23	Contractor is liable for damages and defects during maintenance period	Clause 17
24	Contractor to supply tools, plants, etc.	Clause 18
25	Recovery of compensation paid to workmen and ensuring payment and amenities to workers if contractor fails	Clause 18A and 18B
26	Labour laws are to be complied with by the contractor and payment of wages to labourers and Minimum Wages Act to be complied with	Clause 19 Clause 20
27	Work not to be sublet and action in case of insolvency	Clause 21
28	Intimation of the change in firm's constitution	Clause 23
29	All works to be executed under the contract shall be executed under the direction, and subject to the approval in all respects, of the engineer-in-charge	Clause 24
30	Settlement of disputes and arbitration	Clause 25
31	Contractor to indemnify government against patent rights	Clause 26
32	Lump-sum provisions in tender	Clause 27
33	Action where no tender specifications are specified, local/district specification to be followed/engineer-in-charge's decision shall be final	Clause 28
34	Owner's lien on withheld amount	Clause 29
35	Employment of coal mining or controlled area labour not permissible	Clause 30
36	Unfiltered water supply and departmental water supply, if available	Clause 31 and 31A
37	Alternate water arrangements	Clause 32
38	Return of surplus materials	Clause 33
39	Hire of plant and machinery	Clause 34
40	Condition relating to use of asphaltic materials	Clause 35
41	Employment of technical staff and employees	Clause 36
42	Levy/tax payable by contractor	Clause 37
43	Conditions for reimbursement of levy/taxes if levied after receipt of tender	Clause 38
44	Termination of contract on death of contractor	Clause 39
45	If relative working in CPWD, then the contractor not allowed to tender	Clause 40
46	No gazetted engineer to work as contractor within two years of retirement	Clause 41
47	Return of material and recovery for excess material issued	Clause 42
48	Compensation during war-like situation	Clause 43
49	Apprentices Act provisions to be complied with	Clause 44
50	Release of security deposit after labour clearance	Clause 45

Appendix 5

STANDARD NORMAL DISTRIBUTION

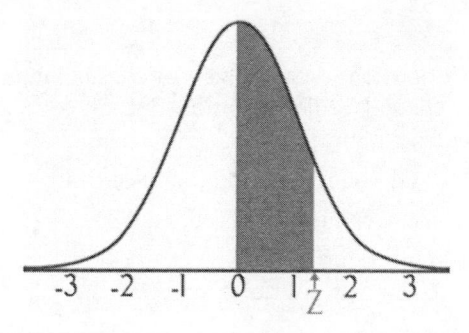

The following table can be used to find the area under the curve from the central line to any Z-value up to 3.
To determine the area under the curve between 0 and 1.35, start at the row for 1.3, and read along until 1.35. The value corresponding to Z = 1.35 is 0.4115.

	0.00	0.01	0.02	0.03	0.04	0.05	0.06	0.07	0.08	0.09
0.0	0.0000	0.0040	0.0080	0.0120	0.0160	0.0199	0.0239	0.0279	0.0319	0.0359
0.1	0.0398	0.0438	0.0478	0.0517	0.0557	0.0596	0.0636	0.0675	0.0714	0.0753
0.2	0.0793	0.0832	0.0871	0.0910	0.0948	0.0987	0.1026	0.1064	0.1103	0.1141
0.3	0.1179	0.1217	0.1255	0.1293	0.1331	0.1368	0.1406	0.1443	0.1480	0.1517
0.4	0.1554	0.1591	0.1628	0.1664	0.1700	0.1736	0.1772	0.1808	0.1844	0.1879
0.5	0.1915	0.1950	0.1985	0.2019	0.2054	0.2088	0.2123	0.2157	0.2190	0.2224
0.6	0.2257	0.2291	0.2324	0.2357	0.2389	0.2422	0.2454	0.2486	0.2517	0.2549
0.7	0.2580	0.2611	0.2642	0.2673	0.2704	0.2734	0.2764	0.2794	0.2823	0.2852
0.8	0.2881	0.2910	0.2939	0.2967	0.2995	0.3023	0.3051	0.3078	0.3106	0.3133
0.9	0.3159	0.3186	0.3212	0.3238	0.3264	0.3289	0.3315	0.3340	0.3365	0.3389
1.0	0.3413	0.3438	0.3461	0.3485	0.3508	0.3531	0.3554	0.3577	0.3599	0.3621
1.1	0.3643	0.3665	0.3686	0.3708	0.3729	0.3749	0.3770	0.3790	0.3810	0.3830
1.2	0.3849	0.3869	0.3888	0.3907	0.3925	0.3944	0.3962	0.3980	0.3997	0.4015
1.3	0.4032	0.4049	0.4066	0.4082	0.4099	0.4115	0.4131	0.4147	0.4162	0.4177
1.4	0.4192	0.4207	0.4222	0.4236	0.4251	0.4265	0.4279	0.4292	0.4306	0.4319
1.5	0.4332	0.4345	0.4357	0.4370	0.4382	0.4394	0.4406	0.4418	0.4429	0.4441
1.6	0.4452	0.4463	0.4474	0.4484	0.4495	0.4505	0.4515	0.4525	0.4535	0.4545
1.7	0.4554	0.4564	0.4573	0.4582	0.4591	0.4599	0.4608	0.4616	0.4625	0.4633
1.8	0.4641	0.4649	0.4656	0.4664	0.4671	0.4678	0.4686	0.4693	0.4699	0.4706
1.9	0.4713	0.4719	0.4726	0.4732	0.4738	0.4744	0.4750	0.4756	0.4761	0.4767
2.0	0.4772	0.4778	0.4783	0.4788	0.4793	0.4798	0.4803	0.4808	0.4812	0.4817
2.1	0.4821	0.4826	0.4830	0.4834	0.4838	0.4842	0.4846	0.4850	0.4854	0.4857
2.2	0.4861	0.4864	0.4868	0.4871	0.4875	0.4878	0.4881	0.4884	0.4887	0.4890
2.3	0.4893	0.4896	0.4898	0.4901	0.4904	0.4906	0.4909	0.4911	0.4913	0.4916
2.4	0.4918	0.4920	0.4922	0.4925	0.4927	0.4929	0.4931	0.4932	0.4934	0.4936
2.5	0.4938	0.4940	0.4941	0.4943	0.4945	0.4946	0.4948	0.4949	0.4951	0.4952
2.6	0.4953	0.4955	0.4956	0.4957	0.4959	0.4960	0.4961	0.4962	0.4963	0.4964
2.7	0.4965	0.4966	0.4967	0.4968	0.4969	0.4970	0.4971	0.4972	0.4973	0.4974
2.8	0.4974	0.4975	0.4976	0.4977	0.4977	0.4978	0.4979	0.4979	0.4980	0.4981
2.9	0.4981	0.4982	0.4982	0.4983	0.4984	0.4984	0.4985	0.4985	0.4986	0.4986
3.0	0.4987	0.4987	0.4987	0.4988	0.4988	0.4989	0.4989	0.4989	0.4990	0.4990

Appendix 6

SALIENT FEATURES OF TENDER AT A GLANCE

1. **General findings**

   i) Information on name and location of work, client detail, details of consultant and architects for the project, commencement date of project mentioned if any, contract duration and milestones if any, estimated cost of project

   ii) Type of work such as civil works, sanitary and plumbing, piling, or electrical

   iii) Type of structure such as RCC framed structure, multi-storeyed, etc.

   iv) Type of contract such as item rate, lump sum, or turnkey, and type of contract conditions such as CPWD, MES, FIDIC, or client's own set of conditions

   v) Information on defects liability period, if any

   vi) Bid submission details such as whether single envelope or two envelopes (comprising technical and financial bid), the place, date and time of submission, various enclosures such as EMD, power of attorney, any other clarification, etc., to be submitted along with tender document. Validity period of the tender prices and opening date and venue are also found out from the set of tender document

2. **Commercial terms/deposits**

   i) Currencies of contract such as INR, dollar and yen, EMD details such as its value, and mode of submission such as demand draft or bank guarantee. In case bank guarantee is needed, the availability of client's own format of bank guarantee should be checked. The validity period and the name of the party in whose favour the bank guarantee is to be made should also be noted. Further, the release details of this bank guarantee should be noted. In some cases, the amount of bank guarantee is partly substituted for the security deposit to be provided by the successful bidder (the one to whom the project has been awarded)

   ii) Collect details on the value, mode of performance security (DD/BG), validity period, and release particulars of performance security

   iii) Collect details on retention money such as whether it is in lump sum or in percent, the limit of retention amount, milestone for its release, and whether retention money can be released against submission of bank guarantee for an equivalent amount. A typical retention clause is produced below:

   > Retention money will be 5% of interim bill subject to a maximum of ₹ 68 lakh (includes EMD of ₹ 50,000). Fifty per cent of the retention money shall be released after issue of completion certificate from architect and balance after rectification of defects pointed out during defects liability period. Fifty per cent of the retention amount can be accepted in the form of BG up to the issue of completion certificate from architect, and balance 50 per cent in the form of cash till the rectification of defects pointed out during defects liability period, which can be put in FD and interest benefits can be given to contractor.

3. **Advance**

   The provision on mobilization advance as well as advance for plant and machinery should be noted from the tender document. The details on advance such as whether lump-sum advance is there or whether any percent is indicated should also be noted. Details like interest charges on advance, their mode of recovery, whether advance is secured against bank guarantee, etc., are noted as well. Sometimes, advances for purchasing materials (called secured advance) are provided by the owners, and the same shall be noted from the document.

4. **Taxes and duties**

   Collect information on whether the price to be quoted has to include taxes such as applicable tax on works contract, value-added tax (VAT), corporate income tax, and any other applicable taxes. Also, read the provision for the eventuality of tax increase/decrease during the contract execution period. In some cases, if fresh taxes are levied after the award of contract, this is reimbursed by the owner if the proof of depositing these additional taxes is furnished.

5. **Terms of payment**

   Collect information on the minimum amount of certification, and the minimum period between two invoices, the time period needed to certify the invoice, and the time required by the owner to pay the contractor after submission of invoice or bill. In normal cases, contractor raises the invoice every month and between 75 per cent and 85 per cent of payments is released within a week of submission of bill, and the balance amount after accounting for retention money is paid within three to four weeks from the date of submission of the bill.

   i) Also, note down the interest clause, if existing, for the delay in payment if any on part of owner

   ii) Collect information on period and payment of final bill

6. **Escalation**

   In some contracts, escalation is payable to the contractors. In case this clause exists in the contract, collect information on the mode of calculation of the escalation. There are different formulas for calculating escalation and these have been discussed elsewhere.

7. **Liquidated damages and bonus**

   The terms of liquidated damages and bonus and their maximum limit are also noted. Normally, liquidity damages are charged on either daily or weekly basis subject to a maximum limit. Bonus is also provided to the contractor on a similar basis. A typical example would be liquidated damage at a rate of ₹ 50,000 per day or ½ per cent of contract value per week, subject to a maximum of 10 per cent of the contract value.

8. **Arbitration and dispute resolution**

   The details of terms of arbitration and clauses pertaining to the settlement of extra claims are also noted. In usual cases, the owner would suggest a sole arbitrator from a panel of arbitrators.

9. **Insurance**

   Information related to different insurances — such as contractor all risk (CAR), third-party insurance (a typical example would be ₹ 2 lakh per person and ₹ 5 lakh per damage of property), workmen compensation, design insurance/guarantee, and employees' state insurance (ESI) — are noted from the contract documents.

10. **Facilities at site**

    The responsibility towards providing water and power, land for site office, staff accommodation, labour camp, etc., should also be noted. Normally, these are provided by the owners on some charges. These charges should also be noted from the tender documents.

11. **Client-issued material and its terms**

    In case some materials such as cement and steel are provided by the owner, the same shall be noted and the terms of issue of these materials — such as charges per ton of cement or per ton of steel, and the allowable wastages — should be noted.

Appendix 7

SITE INVESTIGATION

Site investigation is a process carried out to provide basic inputs and a comprehensive overall picture about the project to the estimator so that a reasonable estimate is prepared. For preparing the estimate, it is desired to find out in detail various parameters required for preparation of bid in a comprehensive manner. It is normally carried out just after collecting and going through the tender document sufficiently in advance and before preparation of bid. A delay in this process may delay the estimate preparation. Further site investigation is carried out by an experienced engineer who is fully conversant with tender scope, conditions and nature of job, and who is capable of being able to go about the entire process in a detailed, systematic and meticulous manner.

Proper and detailed site investigation is a prerequisite to a better estimate, and it offers numerous advantages such as — helping in working out a fair and competitive estimate, giving confidence to the estimator and to the top management about the accuracy of estimate, and better preparation for the site management in case the project is awarded to the contractor, as all relevant information is available since beginning of the project. It also enables the site management to take preventive steps in various matters to avoid occurrence of unwanted and undesirable situations.

The person entrusted for conducting site investigation should possess attributes such as leadership qualities, resourcefulness, communication skill and interpersonal skill. He should be an extrovert, and inquisitive and imaginative, in addition to being an experienced and qualified engineer.

The process of site investigation starts not by reaching at site but while on way to site or approaching the site, as well as by visiting other places in the vicinity of site. The nature of information collected usually varies depending on the type, nature and size of the project as well as a host of other factors; nevertheless, there are some information that need to be collected and are a must for preparing the estimate. In the following section, the information required is categorized based on the place where it is collected, namely on the way to site, at site, and at other places.

Before venturing out for site investigation, it is advisable to note down information such as name of the project, location, name and site address of the client or owner, name and address of the consultant(s), and architect for the project. The investigator also goes through details such as — type of tender, special and general conditions of contract, duration of the project, responsibilities of client and contractor, taxes and duties to be considered in pricing, scope of work and bill of quantities, payment terms, and technical specifications applicable for the contract. In fact, it is advisable to carry a summary of these points in the form of printed notes, and if possible, carry a camera and other electronic devices for taking photographs, making videos, and recording voice description.

INFORMATION COLLECTED ON THE WAY TO SITE OR AT SITE
General Information

i) Collect information on country map and site location map, official language in that area, information on concerned ministries and government departments related to the project, and business hours generally observed.

ii) Collect general information on electricity system (in particular, voltage), public holidays, laws and regulations concerning construction work and workers, prevailing labour laws, banking systems and practices, currency, political and social condition, prevailing economic policy, and general local market conditions.

Information on Taxes, Duties, Tariffs

i) Collect information on import rules, customs duties, import regulation, prevailing tax rates, turnover tax, value-added tax, octroi, entry tax, royalties, stamp duties, income tax on corporate incomes, payroll tax, road taxes on different vehicles, and other municipal taxes.

Information on Laws/Regulation

i) Labour laws and regulations, welfare laws
ii) Traffic regulations
iii) Regulation for storage and handling of explosives
iv) Work permits on public health, water supply and sewage disposal
v) Environmental restrictions on construction work
vi) Is it an existing operating premise? If yes, what is the applicability of specific laws to the project operations?

Meteorological Information

i) Temperature and humidity in different seasons
ii) Rainfall — season and maximum intensity per hour
iii) Wind (maximum speed)
iv) Snow and fog in different seasons
v) Precautions, if any, due to weather

Information Related to Access to Project Site

i) Access by road
Information on nearest bus terminus, nearest container terminal, transport frequency, condition of approach road to site, traffic density, and presence or absence of bridges on the route and their load-carrying capacity

ii) Access by rail
Information on nearest railway station, nearest railway siding, frequency of rail traffic, restrictions on dimensions and weights, if any, and tariffs and fares for materials and men

iii) Access by air
Collect information on nearest airport, frequency of flights, restrictions and hold-ups, and clearances required

iv) Access by water routes
Collect information on nearest port of entry and its capacity, handling capacities of port, frequency of traffic, port hold-ups/operational restrictions, customs clearances

and restrictions, and in case inland waterways are to be used, the maximum capacity and dimensional clearances

Information on Public Utilities and Services

i) Collect information on distance of the nearest police station, post office, bank, petrol pumps, hospital, telephone exchange, water and electricity supply office, and local administration offices

ii) Collect information on material-testing laboratories, explosives licensing, etc.

Information on Material Availability and Their Rates

i) Collect information on quarries, borrow pits, brick kilns, timber depots, hardware shops, workshops and sources for other construction materials. The information on their location and distance from the project site, transport facilities to procure these materials, ownership and rights of use of these materials is also required.

ii) Information on existence of crushers, their capacity, other cost details, quality of these materials, and availability of the required quantity of these materials is to be collected during site investigation.

After reaching the site, the investigator collects the following information.

Information on Site Topography

i) The condition of site — whether it is free or occupied, and whether it is plain, undulating, or levelled. The presence of trees, vegetation, etc., is also noted. The nature of terrain, levels, slopes around site, etc., are recorded as well.

ii) The presence of obstructions such as *nullah*, pond and other structures, of crossings, of power and telephone lines, of open wells and bore wells, etc., is noted.

iii) Information is noted with regard to sub-soil conditions and geotechnical details such as type and nature of soil, levels and condition of groundwater table, suitability of excavated soil for back filling, potential of crushing and producing aggregates from the available rock at site, and need of temporary works/diversions works required to start project operations.

Information on Basic Inputs for Estimating Material Rates

i) For bulk materials such as aggregates, the basic cost, the distance and the cost of transportation from the quarries are found out.

ii) For manufactured items such as cement and steel, the freight on board (FOB)/cost, insurance, and freight (CIF) prices are ascertained. The factors affecting the price variation and the seasons in which lower prices can be obtained are also enquired about.

Information on Site Facilities

i) Type of structure with costs for office, stores, workshop — Whether it will be economical to build these using local building materials or using prefabricated structures

ii) Explosive magazine — This is required in case the project involves excavation using blasting

iii) Fuel and maintenance station

iv) Housing — Location of nearby residential areas, locality and prevailing rents for accommodating staff and labour. The enquiry about housing colony of clients can also be made as sometimes the same can be utilized for accommodating the contractor's staff

v) Space for other site facilities like casting yard, fabrication yard and location proposed by clients

Information on Labour

i) Enquire about works being done by other contractors, and their operating rates with clients

ii) Availability of skilled labour and details on working hours, productivity levels, prevailing labour wages, minimum wages, payments regarding labour benefits, special local customs, practices and holidays, and activities of labour union, if any

iii) In case overseas projects are being investigated, information is required on availability of types of tradesman locally, their normal working time and wages, overtime payment, payment in case of working on holidays, bonus and social benefits

iv) In case the available tradesman does not meet the requirement, the contractor may have to take his workforce from the home country. In such a situation, information on the travel aspect of the workforce, the formalities involved in immigration, their clothing and food requirement, accommodation aspect, etc., need to be carefully studied

Information on Subcontractor

i) Details of subcontractors

ii) Prevailing subcontract rates for various items

iii) Financial resources possessed

iv) Capacity to mobilize

v) For finishing works — collect subcontractor's quotations, giving full details of site, specifications, time of completion and terms and conditions

vi) Evaluate the quotations

Information on Plant and Machinery

Collect information on the type of plant and equipment locally available and their hire charges, the prevailing hiring condition, the condition of equipment, whether the rental charges include operator charges, the availability of plant dealers, the repair facilities, and the availability of tyres, spare parts, fuel and lubricants for the plant and machinery.

Some Good Practices About Site-investigation Process

i) It is always good to meet local government agencies such as electricity board, water board and telephone exchange in order to ascertain the procedures and to find out the prevailing rates for getting temporary connections, installation charges, as well as their consumption charges.

ii) It is preferable to obtain quotations from the registered dealers of sales tax. It should be ensured that the collected rates of the material are as per the desired specification and the quoted rates follow the same unit of measurement as specified in the bill of quantities. The rates should be collected from different agencies in order to have a fair idea of prevailing rates of materials.

iii) All information desired by the estimators should be collected religiously, without ignoring any information as irrelevant since every piece of information is vital for preparing a reasonable estimate.

Appendix 8

PRE-BID MEETING

In order to clarify doubts, clear ambiguities in scope of work and drawings, and resolve anomalies in different tender documents such as drawings, general conditions of contract, special conditions of contract, specifications and bill of quantities, owners usually arrange a meeting prior to the deadline of tender submission. The concerned agencies such as contractors, consultants, architects, structural designers and owner or his representatives are the usual participants at the pre-bid meeting. The date of pre-bid meeting is usually given at the time of issue of tender documents. The participating contractors are requested to communicate their queries well in advance to the owner so that he is ready with the answers/clarification at the pre-bid meeting. It is always advisable to prepare the set of queries in a structured way, mentioning the documents in question and the relevant clauses and page numbers if applicable. Although the nature of queries would depend on a number of factors, some typical queries pertaining to a live project are given below:

1. Can we get a copy of the soil investigation report along with recommendation for foundation for all locations?
2. Can we get the contour levels of the proposed sites along with existing structures and services, if any?
3. Can we submit alternative designs for the roof with either pre-cast concrete members or a combination of RCC and shop fabricated structural steel works?
4. Please indicate the design load to be considered for different floor areas.
5. As the location of expansion joints is not available in the drawings, can we decide on the expansion joint on our own as per IS requirement? Please confirm.
6. Please furnish the requirement/location of mezzanine floor in storeroom.
7. The location of false ceiling is not shown in the drawings.
8. It is mentioned in the bid document that structural steel design should be done as per AISC/DIN code. Kindly confirm whether we can follow IS code.
9. The structural steel design requirement can be met with Fe 250 grade steel manufactured in the country, but grade Fe 345 has to be imported. Given the short duration of this project, can we use steel manufactured in India?
10. Can we get the soft copy of layout drawings?
11. The floor details at 5 m level of equipment are missing. Please furnish the same.
12. The drawings show the spacing of portal as 7.5 m, while the bill of quantity shows this to be 10 m. Please clarify.
13. The nature of bid being EPC, which involves design of civil, mechanical, refrigeration and electrical systems, can we get extension of another 15 days for bid submission?
14. Elaborate on the scope/requirement for external development works.

Appendix 9

A TYPICAL CONSTRUCTION METHODOLOGY

Construction methodology is also referred to as method statement. This normally includes the following:

(1) **Complete description of item/activity** for which the construction methodology is prepared

(2) **The scope** of methodology is to be clearly delineated.

(3) **The references** used for preparation of methodology, for example contract document, bill of quantities, various Indian and International standards are to be clearly mentioned.

(4) **The responsibilities** of persons involved with the preparation, implementation, Supervision, quality control and quality assurance are to be clearly mentioned.

(5) **The work methodology** is described in detail with following sections:

   (a) **Preparation:** Details of all preparatory work associated with the given item/activity are presented in this section.

   (b) **Implementation/Execution:** The details of each step involved with the implementation execution are narrated in this section.

   (c) **Testing:** The method of testing and references for the same are provided. The testing agency including the details of third party if any, are to be provided.

   (d) **Tolerances:** The tolerances level associated with each measurable parameter are to be mentioned.

   (e) **Protection of completed works:** The various steps to be taken to avoid damage of completed works are mentioned in detail.

   (f) **Handing over:** The details of the processes to be used for handing over the completed works after removal of the protection are to be mentioned.

   (g) **Any other detail relevant to the work item:** These are also to be mentioned. For example for tile work, curing method of granting of tiles to be mentioned. Similarly in case of stone work a section on grinding and polishing would be required.

(6) **Supervision:** The details to be observed during supervision are to be described.

(7) **Tools and plant:** The methodology should also include list of tools and plants proposed to be used for the given activity/Items.

(8) **Storage:** The storage method of raw materials proposed to be used for the activity/item is also clearly mentioned.

(9) **Safety measures:** This section would contain all the safety measures required to be taken for the execution of the given activity/item.

(10) **Quality assurance and quality control:** The list of documents (e.g. method statement, shop drawings, material test certificates, non compliance report, etc.) relevant for the activity/item is mentioned. The details of inspection test plan, nature of test, frequency of test etc. are also part of this section. Finally a check list containing the criteria for inspection and their compliance/non compliance is also part of construction methodology/method statement.

The various sections described above are only typical in nature and depending on the item/activity for which the method statement is being prepared, they may vary.

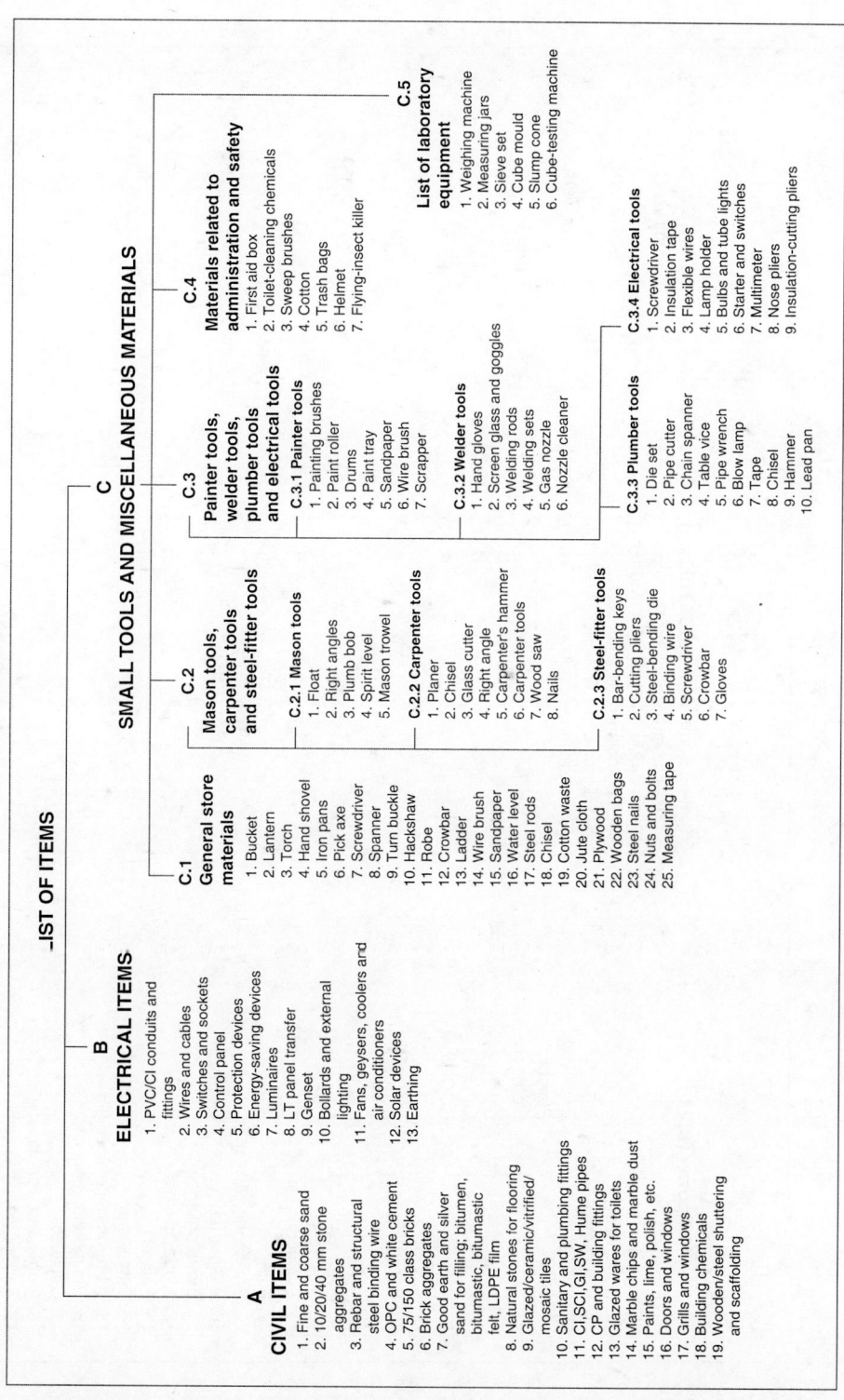

Appendix 10

LIST OF ITEMS

A. CIVIL ITEMS

1. Fine and coarse sand
2. 10/20/40 mm stone aggregates
3. Rebar and structural steel binding wire
4. OPC and white cement
5. 75/150 class bricks
6. Brick aggregates
7. Good earth and silver sand for filling; bitumen, bitumastic, bitumastic felt, LDPE film
8. Natural stones for flooring
9. Glazed/ceramic/vitrified/mosaic tiles
10. Sanitary and plumbing fittings
11. CI, SCI, GI, SW, Hume pipes
12. CP and building fittings
13. Glazed wares for toilets
14. Marble chips and marble dust
15. Paints, lime, polish, etc.
16. Doors and windows
17. Grills and windows
18. Building chemicals
19. Wooden/steel shuttering and scaffolding

B. ELECTRICAL ITEMS

1. PVC/CI conduits and fittings
2. Wires and cables
3. Switches and sockets
4. Control panel
5. Protection devices
6. Energy-saving devices
7. Luminaires
8. LT panel transfer
9. Genset
10. Bollards and external lighting
11. Fans, geysers, coolers and air conditioners
12. Solar devices
13. Earthing

C. SMALL TOOLS AND MISCELLANEOUS MATERIALS

C.1 General store materials

1. Bucket
2. Lantern
3. Torch
4. Hand shovel
5. Iron pans
6. Pick axe
7. Screwdriver
8. Spanner
9. Turn buckle
10. Hacksaw
11. Robe
12. Crowbar
13. Ladder
14. Wire brush
15. Sandpaper
16. Water level
17. Steel rods
18. Chisel
19. Cotton waste
20. Jute cloth
21. Plywood
22. Wooden bags
23. Steel nails
24. Nuts and bolts
25. Measuring tape

C.2 Mason tools, carpenter tools and steel-fitter tools

C.2.1 Mason tools
1. Float
2. Right angles
3. Plumb bob
4. Spirit level
5. Mason trowel

C.2.2 Carpenter tools
1. Planer
2. Chisel
3. Glass cutter
4. Right angle
5. Carpenter's hammer
6. Carpenter tools
7. Wood saw
8. Nails

C.2.3 Steel-fitter tools
1. Bar-bending keys
2. Cutting pliers
3. Steel-bending die
4. Binding wire
5. Screwdriver
6. Crowbar
7. Gloves

C.3 Painter tools, welder tools, plumber tools and electrical tools

C.3.1 Painter tools
1. Painting brushes
2. Paint roller
3. Drums
4. Paint tray
5. Sandpaper
6. Wire brush
7. Scrapper

C.3.2 Welder tools
1. Hand gloves
2. Screen glass and goggles
3. Welding rods
4. Welding sets
5. Gas nozzle
6. Nozzle cleaner

C.3.3 Plumber tools
1. Die set
2. Pipe cutter
3. Chain spanner
4. Table vice
5. Pipe wrench
6. Blow lamp
7. Tape
8. Chisel
9. Hammer
10. Lead pan

C.3.4 Electrical tools
1. Screwdriver
2. Insulation tape
3. Flexible wires
4. Lamp holder
5. Bulbs and tube lights
6. Starter and switches
7. Multimeter
8. Nose pliers
9. Insulation-cutting pliers

C.4 Materials related to administration and safety

1. First aid box
2. Toilet-cleaning chemicals
3. Sweep brushes
4. Cotton
5. Trash bags
6. Helmet
7. Flying-insect killer

C.5 List of laboratory equipment

1. Weighing machine
2. Measuring jars
3. Sieve set
4. Cube mould
5. Slump cone
6. Cube-testing machine

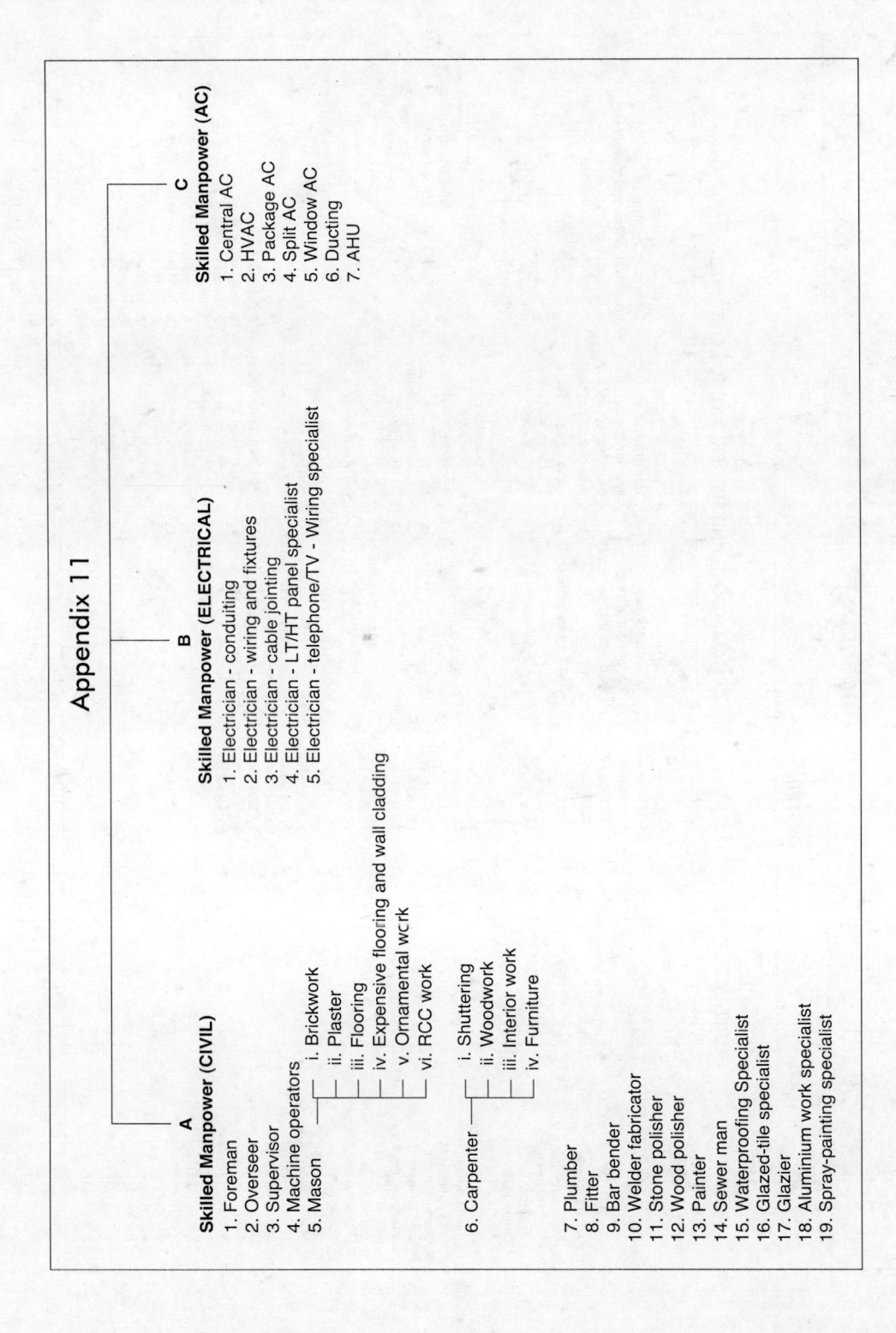

Appendix 11

A

Skilled Manpower (CIVIL)

1. Foreman
2. Overseer
3. Supervisor
4. Machine operators
5. Mason
   - i. Brickwork
   - ii. Plaster
   - iii. Flooring
   - iv. Expensive flooring and wall cladding
   - v. Ornamental work
   - vi. RCC work
6. Carpenter
   - i. Shuttering
   - ii. Woodwork
   - iii. Interior work
   - iv. Furniture
7. Plumber
8. Fitter
9. Bar bender
10. Welder fabricator
11. Stone polisher
12. Wood polisher
13. Painter
14. Sewer man
15. Waterproofing Specialist
16. Glazed-tile specialist
17. Glazier
18. Aluminium work specialist
19. Spray-painting specialist

B

Skilled Manpower (ELECTRICAL)

1. Electrician - conduiting
2. Electrician - wiring and fixtures
3. Electrician - cable jointing
4. Electrician - LT/HT panel specialist
5. Electrician - telephone/TV - Wiring specialist

C

Skilled Manpower (AC)

1. Central AC
2. HVAC
3. Package AC
4. Split AC
5. Window AC
6. Ducting
7. AHU

Appendix 12

TYPICAL PRODUCTIVITY NORMS FOR CIVIL WORKS

Description	Unit	Man hours/unit	Remarks
Excavation			
1. Soft soil (manual)	m³	2.00	Excavation, dressing, ramming, loading, disposal
2. Soft soil using equipment such as (Poclain)	m³	0.07	Same as above
3. Hard soil (manual)	m³	3.5	Same as above
4. Hard soil using equipment such as (Poclain)	m³	0.1	Same as above
5. Hard rock using equipment such as Poclain and bulldozers	m³	20.0	Same as above
RCC			
1. Reinforced concrete for foundations and elevations up to +7 m	m³	13.0	Intercarting, mixing, transporting, laying, vibrating
2. Reinforced concrete for elevations above +7m	m³	15.0	Same as above
3. Reinforced concrete for prefabricated items (columns, beams, etc.)	m³	13.0	Same as above
Brickwork			
1. Foundation level	m³	10.0	Providing, lifting, laying, curing
2. Superstructure	m³	12.0	Same as above
Formwork (in situ)			
1. Foundations	m²	2.5	Making, centering, fixing, removing
2. Superstructure	m²	4.0	Making, staging, centering, fixing, removing
Formwork (pre-cast)			
1. Making	m²	6.0	Making (one time)
2. Fixing and removing	m²	2.0	Centering, fixing, removing
Pre-cast items			
1. Erection of prefabricated reinforced concrete elements	t	1.0	Lifting, placing
Reinforcement			
1. Steel (in situ)	t	44.0	Providing, cutting, bending, straightening, tying

| 2. Steel (pre-cast works) | t | 50.0 | Same as above |

Structural steel

| 1. Steel elements | t | 65.0 | Providing, welding, bending, fixing |

Plastering

| 1. Any level | m² | 1.0 | Making, applying |

INDEX

A

ABC analysis, 433, 440
   item type A, 440
   item type B, 440
   item type C, 440
accelerated depreciation, 345
accident frequency rate, 560
accident root-cause tracing model (ARCTM), 534
accident, 532–535, 547–548, 557–560, 564
   causation theories, 529–531
   causes of, 547
   cost of, 544–547
   investigation and analysis, 557
   statistics and indices, 560
accounting, 15, 19, 22, 268, 324, 391–392, 397–398
   balance sheet, 397–399
   contract status report, 395–397
   defined, 11, 18, 31, 34, 49, 81, 104
   funds flow statement, 421–423
   limitations of, 397
   principles of, 391
   process of,
   profit and loss account, 402
   ratio analysis, 412
   revenue recognition, 394
   working capital, 404, 406
acid test ratio, 414
Activity-on-Arrow (AOA), 168, 187
Activity-on-Node (AON), 170
actual cost of work performed (ACWP), 589
ADR. *See* alternative dispute resolution (ADR)
AGF. *See* Arithmetic gradient factor (AGF)
air-cooled blast furnace slag (AcFBS), 475
allocation, resources, 202–207
all-terrain cranes, 336
alphanumeric codification, 434
alternative dispute resolution (ADR), 609
alternatives by equivalence, evaluation, 75
annual cost and worth comparison, 78
AOA. *See* Activity-on-Arrow (AOA)
AON. *See* Activity-on-Node (AON)
appraisal cost, 493–494
approximate quantity method estimate, arbitration, 127–128, 139, 146–147
   causes leading to, 546
architect, 13, 493, 615, 854

arithmetic gradient factor (AGF), 73
assets, 401–402, 407
   current assets, 407
   fixed assets, 401
   intangible fixed assets, 401
   tangible fixed assets, 401
audit, 502–503, 562
   ISO 9001–2000 requirements for, 502
   need for, 502
average annual rate of return, 51

B

backhoe, 327–328
backward pass, 175
balance quantity, 584
balance sheet, 397–399
balanced matrix, 31, 48
bank finance, 410
bar chart, 180
Basic Guide to Accident Investigation and Loss Control, 533, 566
BCA. *See* Building and Construction Company (BCA)
budgeted cost for work performed (BCWP), 589
BCWS. *See* budgeted cost for work scheduled (BCWS)
benefit–cost ratio, 82
bid price, 262–263, 265, 274
bidding process, 136–138
   agreement, 145
   analysis of submitted tenders, 144
   contractor's estimation and, 256
   evaluation and acceptance, 145
   Indian construction industry, 9, 18
   letter of intent, 145, 237, 582
   notice inviting tender, 142
   submission of bids, 143
   work order, 145
Big-M method, 701, 708, 710, 755
bill of quantities (BOQ), 109, 126, 128
Bonded Labour System (Abolition) Act, 1976, 849
Border Roads Organisation, 2
bottom–slewing tower crane, 2
breakeven analysis, 362–364
breakeven point, 363, 385
budgeted cost for work scheduled (BCWS), 588

buffer stock, 436, 447
Build Operate and Transfer (BOT) Contract, 134
Building and Construction Company (BCA), 501
business and project risk, 522–523
business organization, forms of, 19, 48

C

capital lock-up, 64–66
capital structure ratio, 420
capitalized equivalent method, 77
captim, 66
CAR policy, 519–520
case project, 469–470
carpentry department, 678, 681
cash flow, 53–59
    effect of inflation on, 81
    cash inflow, 58, 237, 245
cash method of revenue recognition, 394
cash outflow, 58, 60, 86, 103, 237, 246
cash-flow diagrams, 53–54, 56, 66, 76, 90, 350
central materials department, 42
    responsibilities of, 429
Central Public Works Department (CPWD), 2
    contract conditions, 146, 150, 161
centralized functional structure, 24
Child Labor (Prohibition & Regulation) Act (1986), 836
Children (Pledging of Labour) Act, The 1976, 849
claims, 7, 14, 16, 19, 59, 153, 235, 414–415, 535, 562, 603
    guidelines to prepare, 605
    management, 604
    clamshell, 327
classical organizations, 26, 29
client's and consultant's requirements, 266
client's definitive estimate, 110
client's detailed estimate, 109
client's indicative cost estimate, 108
client's preliminary cost estimate,
coal fly-ash stabilized bases, 475
commercial paper, 410
common multiple method, 76
company cash-flow diagram, 65
completed accurate method of revenue recognition, 395
conciliation, 608, 616
concrete piers, 470, 472
conducting scenario analysis, 358
conformance cost, 494
construction closure, 612–613

construction company, 18–19
    aim, 18
    defined, 18
    functions, 19
    management levels, 18, 31, 48
    organizing for project management, 26
construction contract, 125–129, 131, 135
Construction Equipment Handbook, 323
construction equipment, 9, 15, 322–323
    factors for selection of, 385
construction industry, 9, 16, 18
    importance, 9
    insurance in, 516
    principles of, 390
    risk in, 15, 291, 511
construction management contract, 133
construction organization, 15, 18–19
    centralized functional structure, 24
    decentralized multidivisional structure, 25
    factors for the success of, 46, 292
    material procurement process, 427, 429
    success variables for, 43
construction planning, 162–163
    bar chart, 180
    critical path method (CPM), 189
    ladder network, 190–193
    network diagram preparation of, 180, 209
    networks, 168, 183
    programme evaluation and review technique (PERT), 184 project plans, 164, 660
    techniques, 166, 209
    work-breakdown structure, 165–166
construction procurement, 136
construction productivity, 503
construction project, 2–4, 6, 11–13
    post-project phase, 8, 17
    pre-project phase, 4–5, 8
    phase, 3–8, 14
    stakeholders/participants, 13
Construction Quality Assessment System (CONQUAS), 501, 506
construction, 1, 15, 12, 489
constructor, 14, 56
contract closure, 612
contract document, 126
    bill of quantities (BOQ), 126
    contract drawings, 126
    general conditions of contract (GCC), 126
    special conditions of contract (SCC), 126
    specifications, 126

contract drawings, 126, 483
Contract Labor (Regulations & Abolition) act (1970), 835
contract revenue, 394
    data for illustration, 394
contract status report, 395–396.
See also work-in-progress schedule
contract-implementation skills, 39, 48
contractor, 6–7, 13–14, 40, 50, 56
    analysis of rates, 267
    bid price determination, 262
    bidding models, 277
    collection of information, 257
    consultation, queries and meetings, 259
    decision to be taken, 278
    game theory models, 278
    mark up, 272
    schedule preparation, 236
    site visit and investigation, 259
    statistical bidding strategy, models, 278
    tendering process, 6, 156, 256
contractor's all-risk insurance (CAR), 518–519
conveyance cost, 265–266
correspondence, 611
Corsby, Philip, 493
cost budgeting, 455
cost codes, 461, 582
cost control, 455, 479
cost control system, 594
    illustrations of, 594
cost of goods sold ratio, 417
cost of labour, 263
cost of material, 263
cost of plants and equipment, 270
cost of quality, 493
cost planning, 454–455
cost plus percentage, 132
cost statement, 461, 463
cost-related information, collection of, 456
    labour cost, 457
    material cost, 458
    overhead cost, 460
    plant and equipment cost, 460
    cost-time trade-off, 246
cost-volume-profit analysis. *See* breakeven analysis coverage ratio, 362
CPI. *See* cost performance index (CPI) CPM. *See* critical path method (CPM)
CPWD. *See* central public works department (CPWD)
cranes, 330–333

    bottom-slewing tower crane, 330
    mobile crane, 333
    top-slewing tower cranes, 331
    tower crane, 330
crawler cranes, 334
crew cost, 262, 264
critical path method (CPM), 189
critical path, 186, 189
cube rate estimate, 106
current assets, 399–400
current liabilities, 399–400
current ratio, 414, 420
custody, 431, 453
cycle stocks, 437

D

debentures, 411
debt asset ratio, 415
debt equity ratio, 414–415
decentralized multidivisional structure, 25
declining balance method, 342–344
deep vibratory compaction, 470
Deming cycle, 489
demobilization, 614
departmental organization, 23
depreciation, 337–338
    accelerated depreciation, 338
    calculations for equipment at site, 340
    calculation methods, 340
        declining balance method, 342
        sinking fund method, 343
        straight line methods, 341
        sum of years digit method, 341–342
    effect on selection of alternative, 342
design, management and construction contract, 133
design–build contract, 133
direct cost, 247, 262, 310
    cost of labour, 263
    cost of material, 263
    cost of plants and equipment, 264
    crew cost, 264
    estimation, 264
    subcontractor cost, 460
discretionary contract, 135
dispute, 154, 605
    courses of, 605
        do's and dont's, 529–530 dispute avoidance, 607–608
dispute resolution, 154, 608

mechanism of, 608–609
distractions theory, 535
dividend coverage ratio, 415
domino theory, 532–533
double declining balance method, 342
dozers, 327
dragline, 326–327
dummy activity, 168, 172
dumper, 269, 300, 327
duration of an activity, 174
duties, 840–841, 858

E

EAC. *See* estimate at completion (EAC)
earned value method, 586, 588, 591
earthwork equipment, 326–328
economic analysis, 15, 75, 355
    sensitivity analysis, 355–362
    breakeven analysis, 362–363
economic decision making, 50
economic order quantity model (EOQ) model, 445–446
education, 541, 554, 752, 800
Employees P.F. and Miscellaneous Provision Act (1952), 834
Employees State Insurance Act (1948), 849
Employer's liability Act (1938), 849
EMR. *See* experience modification rate engineer (consultant), 14
    ethical conduct for, 14
engineering contracts, 128
    discretionary contract, 135
    integrated contract, 133
    management contract, 132–133
    separated contract, 129
Engineering New Record (ENR), 522
EPCAF. *See* Uniform series compound amount factor (EPSCAF or USCAF)
EPSCRF. *See* Capital recovery factor (EPSCRF)
EPSPWF. *See* Uniform series present worth factor (EPSPWF or USPWF) Equal Remuneration Act (1972), 835
equipment, 323–325
    classification, 323
    earthquake, 520–521
    factors for selection, 323
    hosting, 143
    plant and equipment acquisition, 336
equipment-owning cost, 269, 295
    equivalence of alternatives, 67

estimate at completion (EAC), 590
estimates, 108–111, 126, 135, 182, 184
    illustrative cases in preparation of, 104
    types of, 108
estimate to date, 583
estimation, 104, 256
    methods of, 104
    practices, 104
evaluation of alternatives, benefit/cost criteria, 82
expected monetary value model, 279
expenses ratio, 416
experience modification rate (EMR), 562
external failure cost, 495

F

factoring, 410–411
family of curves, 356
feasibility, 4–5, 108
feasible solution, 682, 684, 691, 694, 701
Federation Internationale des Ingenieurs Conseils, 149
    (FIDIC),
    features of, 152
    need and principles of, 151
Ferrel's theory, accident, 534
FIDIC. *See* Federation Internationale des Ingenieurs Conseils (FIDIC)
finance plan, 165
financial closure, 614, 616
financial instruments, 41
financing cost, 265, 318
finish-to-finish (FF), 194, 637
finish-to-start (FS), 191, 193
fire policy, 521
first-party audit, 502
fixed assets, 401, 406
fixed order interval scheduling policy, 437
float time, 182
flooring system, 474
forfaiting, 411
forward pass, 175–176
foundation design, 470
fraud, 829 free float, 179
Friedman's model, 279
FSN analysis, 433, 443, 453
    desirable, 443
    essential, 443
    vital, 441
functional matrix, 31, 48
functional organizations, 26–27

funds flow statement, 419
future worth comparison, 77

G

game theory models, 278
Gate's model, 280
GCC. *See* general conditions of contract (GCC)
geometric gradient factor (GGF), 73
George Dantzig, 677
government enterprises, 21
granular sub-base (GSB), 474
gross domestic product (GDP), 2
gross profit ratio, 416
GSB. *See* granular sub-base (GSB)
guesstimating approach, 182–183
Guide to the Project Management Body of Knowledge(PMBOK), 2

H

Hazardous Wastes (Management and Handling) Rules (1989), 840
health management system, 592
hoists, 330, 554
holding cost, 436
hosting equipment, 330
    cranes, 330
    hoists, 330

I

incentive programmes, 556
incidence rate, 561–562
income and expenditure statement.
    See profit and loss account
incremental rate of return (IRoR), 80
independent float, 179
Indian construction industry, 9, 18, 46, 287
    bidding and estimation practices in, 287
Indian construction industry, 9, 18
Indian Contract Act (1872), 160
indirect cost, 208, 247, 431, 544
    conveyance cost, 265
    estimation, 237
    financing cost, 265
    insurance cost, 265
    progress photographs and video, 265
    travel and transfer cost, 318
induction programmes, 555
Industrial Accident Prevention, 531–532
Industrial Disputes Act (1947), 836
Industrial Employment (Standing Orders) Act (1946), 836

inflation, effect on cash flow, 81
injury, 535–536
inspection, 485
insurance cost, 265
insurer, 515, 517
intangible fixed assets, 401
integrated contract, 133
interdependence, 168, 181, 236
interest computation, 69
    formulation for, 69
    interest coverage ratio, 415
internal failure cost, 495
internal rate of return method (IRR), 78
international construction, 15, 522
    risk in, 522
International Labour Organization (ILO), 16, 531
International Organization for Standardization (ISO), 496
inventory, 418, 435
    functions of, 436
    selective inventory control, 439
inventory management, 435, 440
    holding cost, 436
    inventory-related cost, 436
    set-up cost, 436
    shortage cost, 436
    inventory models, 444
    economic order quantity model (EOQ) model, 445
inventory monitoring and control, 433
inventory policies, 437, 441, 447
    fixed order interval scheduling policy, 437
    lot size reorder point policy, 437–438
    optional replenishment policy, 437–438
    two-bin system, 437
inventory-related cost, 436
investment per annual tonne capacity,
invoice, 237, 239
invoice schedule, 237, 239
IROR. *See* incremental rate of return (IROR) IRR.
    See internal rate of return method (IRR)
    ISO 9000, 485, 496
    benefits of, 496
ISO 9001–2000, 498, 502
ISO standards, 496–497
isoquants, 356, 358
item rate contract, 130
item type A, 441
item type B, 442
item type C, 442

J

Japan Industrial Safety and Health Association (JISHA), 545
joint ventures, 21
Juran, Joseph, 491, 495
just-in-time (JIT) application, 428

L

labour cost, 289–290
labour cost estimation, 289
labour productivity, 261
labour requirement, 242
    schedule of, 242
ladder network, 190–191
Least Cost Method, 760, 763, 765, 769
leverage ratio, 414
liabilities, 400
    current liabilities, 401
    long-term liabilities, 400
Limited Liability Company (LLC), 20
line and staff organization, 22
linear programming, 678, 681, 687, 689
linear programming problem formulation, 16, 680
linear responsibility chart (LRC), 485
line-of-balance (LOB), 203
line-type organization, 22
liquidity, 158, 398
liquidity damages insurance, 521
liquidity ratios, 413
loans, 399, 412
long-term finance, 411, 424
long-term liabilities, 400
lot size reorder point policy, 437
low labour productivity, 504
lump-sum contract, 130–131

M

MAAR. *See* minimum attractive rate of return (MAAR)
MACRS. *See* modified accelerated cost recovery system (MACRS)
management contract, 132–133
management information system (MIS), 597
management levels, 31
    construction management level, 32
    director level, 31
    functional management level, 33
    president level, 32

man-hour estimate, 203
    manpower plan, 203
Manufacture, Storage and Import of Hazardous Chemicals Rules (1989), 839
marine insurance, 520
marine-cum-erection insurance, 519
mark up, 272
    average range of, 289
    break up, 237
    concept of, 273
    determination of, 257
    distribution of, 274
mark up range, 290
material cost, 148, 260
material plan, 164
material procurement process, 429
materials accounting, 432
materials codification, 433
    illustration of, 433
materials management, 431
materials management functions, 431
    computerization, 434
    custody, 431
    disposal, 435
    inventory monitoring and control, 433
    materials accounting, 432
    materials codification, 433
    materials planning, 431
    procurement, 429
    source development, 435
    transportation, 432
    warehousing, 431
materials planning, 431
materials productivity, 260
Maternity Benefit Act (1951), 835
matrix organizations, 29
maximization objective, 684, 693, 698, 701
measurement contract, 130
mediation, 608
merchant bank, 410
milestones, 5, 239
milestone events, 240, 260
Milestone Professional, 620
Military Engineering Services, 2
military organization, 2
minimum attractive rate of return (MARR), 78
Minimum Wages Act (1948), 835
miscellaneous expenses, 266
misrepresentation, 829
mobile crane, 333
    all-terrain cranes, 336

crawler cranes, 334
rough-terrain cranes, 336
truck cranes, 334
modified accelerated cost recovery system (MACRS), 345
money, time value of, 390
monthly progress report, 577
Motor Vehicles Act (1988), 848
MS Project, 620, 624
MSWI bottom ash, 474

N

National Highways Authority of India, 365
National Hydroelectric Power Corporation (NHPC),
negative cash flow, 64
negotiation, 608
net profit ratio, 416
network, 168–169
    ladder, 190–191, 539
    precedence, 171–172
network crashing, 246
network diagram, 180, 205
network logic, 171
network planning, 209
neutral critical activities, 201
nodes, 168, 175
non-conformance cost, 495
normal critical activities, 201
Nuclear Power Corporation of India Limited (NPCIL), 10

O

occupational accident, 532
operating ratio, 416, 420
operational estimating method, 267
optimum crew, 203
optimum feasible solution, 684, 741
optional replenishment policy, 437
original estimate, 584, 587
Out of Crisis, 265
Out-of-Pocket Commitment, 50
output rate, 270
overheads, 262, 264, 272
overhead cost, 456, 460

P

painting department,.678, 680, 682
Park's model, 285

partnership, 19–20
passenger hoist, 330
payback period, 50–51
Payment of Bonus Act (1965), 836
Payment of Gratuity Act (1972), 834, 850
Payment of Wages Act (1936), 835
percentage of completion method of revenue recognition, 394–395
percentage rate contract, 131–132
period of credit, 419
perverse critical activities, 201
petroleum, oil and lubricant (POL), 148
plan, do, check, act (PDCA), 489
plan, do, study, act (PDSA), 489
planning,. *See also* construction planning techniques
plant and equipment cost estimation, 289
plant and equipment cost, 289, 293
plant and machinery insurance, 521
poor scheduling, 235
Pradhan Mantri Gram Sadak Yojana (PMGSY),10
precedence network, 191, 194
    finish-to-finish (FF), 194
    finish-to-start (FS), 193
    start-to-finish (SF), 195
    start-to-start (SS), 194
precedence, 168
precedence network, 191
preliminary estimate, 105, 107
    for buildings, 105
    for industrial structures, 107
present worth comparison, 75
price escalation clause, 148, 160
Primavera, 201, 618–619
private limited company, 20
procurement, 22, 24, 27
product management, 27
productivity ratios, 419
professional indemnity policy, 522
profit and loss account, 390, 392
profit ratios, 416
profitability ratios, 415–417
programme evaluation and review technique (PERT), 184
progress at site, 577
progress photographs and video, 265
project, 454, 575, 621, 657
    defined, 34
    determining the cash requirement for, 67
project cash-flow diagram, 56–57
    factors affecting, 59
    use of, 56

project closure, 111, 603, 605
    construction closure, 612
    contract closure, 614
    financial closure, 612–614
    project manager closure, 111
project closure cost, 111
project control,
    elements, 5, 11
    using Bar chart, 571
    using PERT/CPM, 572
    project coordinator, 38, 40
    traits of, 34, 38, 48
project cost, 104–105, 107
    methods of structuring, 111
project implementation phase, 6
Project Kickstart, 620, 651
project management, 619
    aims, 4, 11, 428
    defined, 34
    elements of, 12
    need, 112
    organizing for, 26
    role, 11
project management cost, 673
Project Management Institute (PMI), 2
project management software, 618–619
    functions of, 619
project manager, 24, 36
    responsibilities, 133, 144
    strategies for enhancing the performance, 37
    traits of, 34
project manager closure, 615
project matrix, 31, 48
project performance, 8, 34
    attributes, 8, 21, 38
    effect of other elements on, 661
    evaluation, 657
    failure attributes, 659
    success factors, 655, 659
project phase, 4, 6, 8, 17
project plans, 164
    construction equipment plan, 165
    finance plan, 165
    manpower plan, 164
    material plan, 164
    time plan, 164
project scheduling, 224–225
    cash inflow and cash outflow schedule, 237
    estimation of direct and indirect costs, 237
    importance, 237
    invoice schedule, 237
    labour schedule, 237
    material schedule, 237
    plant and equipment schedule, 237
    preparation steps in general, 237
    staff schedule, 237
    working capital schedule, 237
project staff, 242, 265, 461
project-organization skills, 40, 48
Public Liability and Insurance Act (1991), 850
public limited company, 20, 48
public projects, 82, 184
    evaluation of, 82
public–private partnership,
pure project, 27–28

Q

QA. *See* quality assurance (QA)
QC. *See* quality control (QC)
quality, 1, 3, 5, 7–8, 482
    concept, 482
    cost of quality, 493
    Crosby on, 493
    defined, 482
    Deming on, 489
    evolution of, 483
    Juran's quality triology, 491
quality assurance (QA), 487
quality control (QC), 487
quality cost breakdown, 493
quality management systems, 497–498
    principles of, 497
quality standards, 485
quarry waste, 475
quick ratio, 414, 420

R

Radosavljevic, 677
rammed aggregate piers (RAP), 470–474
rate of return method, 78
rate, analysis, 120, 274
    operational estimating method, 271
    unit rate estimating method, 270
ratio analysis, 412–413
    activity ratios, 418
    capital structure ratio, 413, 420
    liquidity ratio, 413, 420
    profitability ratio, 413, 420
    supplementary ratios, 413, 420
ratio method, 107

recycled concrete, 474
regional materials department, 429
relative importance index (RII), 43
replacement alternatives, 353
resource allocation, 229, 235
resource leveling, 225
    steps, 225
response management process, 514
retained earnings, 411
return on assets (ROA), 417
return on investment, 408
revenue recognition, 394–395
    cash method, 394
    completed accurate method, 394
    percentage of completion method, 395
    straight accural method, 395
reverse critical activity, 201
    advantages, 201
    challenges, 205
    steps for preparation, 117, 180
revised estimate, 111
rights issue, 411
risk, 274, 509
    factors of, 274
    in international construction, 522–524 risk analysis, 510, 513
risk category summary sheet, 512
risk classification, 512, 527
risk identification process, 511–513
risk management, 47, 509
risk map, 512
ROA. See return on assets (ROA)
roller compactor, 327
rough order of magnitude estimates, 108, 157
rough-terrain cranes, 336

S

safety, 41, 141
    evolution of, 529
    principles of, 529
    research results in, 563–564 safety audit, 562
safety budget, 553
safety committee, 555
safety inspection, 562
safety manual, 555
safety organization, 553
safety personnel, 545
safety plan, 555
safety policy, 552
SCC. See special conditions of contract (SCC)

scenario analysis, 356, 360
schedule performance index, 589
scraper, 327
S-curve, 579–580
second-party audit, 502
Section 40-B of Factories Act (1948), 554
Section 88 of Factories Act (1948), 556
selective inventory control, 439
    ABC analysis, 440
    FSN analysis, 443
    VED analysis, 443–444
sensitivity analysis, 355–356
    different forms of, 356
    problems data for, 356
separated contract. See also engineering contracts
    set-up cost, 436
severity rate, 561
shares, 411, 415
shortage cost, 436
short-term finance, 409
shovel, 327, 865
Simplex method, 16, 677, 689, 692, 712
Single payment compound amount factor (SPCAF), 70, 74
sinking fund method, 343
site investigation, 259, 321, 857, 859–860
six-tenth factor, 107–108
slack time, 178
sole proprietorship, 19, 47
SPCAF. See Single payment compound amount factor (SPCAF)
specialized agencies, 244
standard costing, 579
start-to-start (SS),
start-up phase, 8
statistical bidding strategy, models, 278
statistical quality control (SQC), 484
status report, 395–397
straight accural method of revenue recognition, 394
straight line depreciation, 340, 345
strap footings, 472
study period method, 77
subcontracting, 154–155
subcontractor, 14, 158
    classification of, 155
    guidelines for management, 158
    selection of, 156
subcontractor cost, 264, 318
success variables for, 43
sum of years digit method, 341

supplementary estimate, 111
supplementary ratios, 419

T

tangible fixed assets, 401
task hierarchy, 625
task, defined, 225, 625
tax. *See also* depreciation
    effect on selection of alternative,
tax holiday, 10
taxation, 80–81, 338
    depreciation and, 337
    effect on comparison of alternatives, 80
taxes and duties, 266
team-building skills, 39–40, 48
third-party audit, 502
time plan, 164
time study approach, 182
time-constrained leveling, 228, 640
Times of India, 546
toolbox talks, 555
top-slewing tower cranes, 331
total cost, 110, 262, 269, 275
total float of an activity, 178
total project cost, 249, 253
total quality control (TQC), 484, 506
total quality management (TQM), 488, 507
    elements of, 489
    problems with, 232
tower crane, 330–331, 386
TQC. *See* total quality control (TQC)
TQM. *See* total quality management (TQM)
trade credit, 410
Trade Union Act (1926), 836
training, 489
transactions, 391
transportation, 432
transportation algorithm, 760
transportation problem, 758, 760, 768, 773, 780
travel and transfer cost, 318
treatment strategies, 514–515
truck cranes, 334
turnkey contract, 134, 160
turnover phase, 8
turnover ratio and capital ratio,
two-bin system, 437, 439

U

undue influence, 829
Uniform series compound amount factor, 71, 74 (EPSCAF or USCAF),
Uniform series present worth factor (EPSPWF or USPWF), 71
unit costing, 582
unit rate estimating method, 270
unsafe acts, 538–539
unsafe condition, 538
updating, 570–571
    defined, 571
using precedence network, 202, 574
USCAF. *See* Uniform series compound amount factor (EPSCAF or USCAF)
USPWF. *See* Uniform series present worth factor (EPSPWF or USPWF)
utilities, 266, 644

V

VAC. *See* variance at completion (VAC)
value engineering (VA), 464–465, 469
    application in case project, 469
value index, 464
value management (VM), 461–462
variance at completion (VAC), 590
VED analysis, 443, 453
vendor development, 435
venture capital, 412
vibrated stone columns, 470, 472
Vogel approximation method, 760, 765, 767, 775

W

warehousing, 431–432
work breakdown structure (WBS), 165–166
work order, 145, 157, 456
worker's productivity, 261
working capital, 217, 308, 404
    components of, 406
    determination of, 407
    financing sources of, 408
    need for, 405
    operating cycle, 405
Workmen Compensation Act (1923), 834
World Health organization (WHO), 531, 566